Springer Series in Surface Sciences

Volume 63

This series covers the whole spectrum of surface sciences, including structure and dynamics of clean and adsorbate-covered surfaces, thin films, basic surface effects, analytical methods and also the physics and chemistry of interfaces. Written by leading researchers in the field, the books are intended primarily for researchers in academia and industry and for graduate students.

More information about this series at http://www.springer.com/series/409

Giorgio Benedek · Jan Peter Toennies

Atomic Scale Dynamics at Surfaces

Theory and Experimental Studies with Helium Atom Scattering

Springer

Giorgio Benedek
Università di Milano-Bicocca
Milan, Italy

and

Donostia International Physics Center
Donostia/San Sebastian, Spain

Jan Peter Toennies
Max Planck Institute for Dynamics
and Self-Organization
Göttingen, Germany

ISSN 0931-5195 ISSN 2198-4743 (electronic)
Springer Series in Surface Sciences
ISBN 978-3-662-56441-7 ISBN 978-3-662-56443-1 (eBook)
https://doi.org/10.1007/978-3-662-56443-1

Library of Congress Control Number: 2018932184

This Springer imprint is published by the registered company Springer-Verlag GmbH, DE part of Springer Nature
The registered company address is: Heidelberger Platz 3, 14197 Berlin, Germany

We dedicate this book to our wives Carmen and Monika. We thank them for their selfless patience, their constant support and enduring encouragement.

Preface

So eine Arbeit wird eigentlich nie fertig,
man muss sie für fertig erklären,
wenn man nach Zeit und Umständen
das Möglichste getan hat.

Johann Wolfgang von Goethe

Helium atom diffraction from surfaces was first reported by Otto Stern and colleagues in two short communications (Z. Phys. 53,779; Naturwissenschaften 21,391) in 1929. Two years earlier, Davisson and Germer had observed the diffraction of electrons from a crystal of nickel. These two events ushered in the present era of atomic scale determinations of the structure of surfaces. Many spectacular atom-surface scattering phenomena, including *selective adsorption*, the temporary trapping of atoms in the surface bound states, followed from Stern's laboratory until he was forced to leave Germany in 1933. After World War II, there was a slow renaissance of interest in surface physics and chemistry. In 1970, a visionary quantum mechanical scattering calculation by Cabrera et al. (Surface Sci. 19, 67) predicted that helium atoms in analogy to neutrons, which had then been used to measure bulk phonon dispersion curves, would be the ideal probe to determine the dispersion curves of the phonons at the surfaces of solids. The first successful experiment was then reported in the laboratory of one of the authors in 1981 (Brusdeylins et al., Phys. Rev. Lett. 44, 1417) and the first successful experiment using electrons followed in 1983 (Szeftel et al., Phys. Rev. Lett. 50, 518).

Since then the phonon dispersion curves of more than 200 surfaces of insulators, semiconductors, metals, semimetals, layered crystals, and ultrathin films, and in addition, the vibrations and surface diffusion of a multitude of isolated adsorbed atoms and the phonons of monolayer films of inorganic and organic molecules have been investigated largely using helium atom scattering (HAS) and by the complementary method of electron energy loss spectroscopy (EELS). These experiments were preceded by a decade of extensive theoretical studies of the vibrations at surfaces based on the idealization of the surface as a termination of the bulk. For many substances, these simple model predictions were often found to be in conflict with the experimental results. Moreover in all the types of materials, many unanticipated phenomena were observed. This led to the realization that the interactions between atoms in the surface region frequently differ significantly from those in the bulk. Even in the case of closed-shell insulators such as the solid rare gases and alkali halide crystal surfaces, substantial deviations from the ideal surface behavior are found. Unexpected phenomena also occur in a number of semiconductors, such as Si and GaAs, and charge density wave materials such as $TaSe_2$ and TaS_2. In the case of metals, the HAS experiments invariably reveal an additional unexpected anomalous soft surface phonon branch lying below the expected longitudinal acoustic band, but above the ubiquitous Rayleigh mode with an intensity which can even be larger than that of the Rayleigh mode. The large intensity of this mode, and the fact that it was often not, or only weakly, seen with EELS led ultimately to the understanding that He atoms excite the phonons at the metal surface by coupling predominantly to the spill-out electrons several Ångstroms above the surface atoms. For metal surfaces which are known to reconstruct, such as W(001), the anomalous modes exhibit sharp minima in the phonon frequencies (soft modes) at special wavevectors at temperatures above the structural transition. These and other findings stimulated new theoretical investigations based on microscopic models and first-principle methods where the role of surface electrons is explicitly taken into account. Striking confirmation came in 2009 from helium atom scattering experiments off of thin films of lead in which a large fraction of all of the modes of the entire film were observed and their intensities found to be proportional to the electron–phonon coupling constant. Now it is clear that in the case of metals He atoms interact primarily with the Fermi level electron density at the surface leading to a non-local excitation of electrons over many layers and thereby can excite deep subsurface phonons (*quantum sonar effect*) and even surface electron excitations in the THz energy domain. Thus, it is now firmly established that helium atom scattering is truly unique among all the atomic scale surface-sensitive scattering probes. The purpose of this monograph is to summarize this development and the current state of the field of atomic scale surface dynamics which has come largely from helium atom scattering combined with theory.

This book consists of altogether 14 chapters. They are designed to introduce the reader to the basic and the current microscopic theory. Subsequently, the experimental methods are described in preparation for a comprehensive overview of the experimental results and their current interpretation. The text is designed to provide an in-depth modern account of the atomic scale dynamics at surfaces. It is aimed at advanced graduate students and scientists working in the area of surface physics and chemistry. Each chapter starts with a general discussion so that less initiated readers interested in surfaces will be able to gain an impression of the many remarkable dynamical phenomena occurring in and on surfaces.

As explained in Chap. 1, surface phonons are not only intensively studied for fundamental reasons, but also since they have several perspective technological applications in micro- and nano-devices. Chapter 1 also discusses the advantage of helium atom scattering over other surface-sensitive scattering probes, such as STM and AFM, due to its unique interaction mechanism. Moreover, because of the low energy and the inert nature of helium atoms they are unmatched in being completely nondestructive and insensitive to electric and magnetic fields at the surface. In Chap. 2 the historical evolution of the theory of surface phonons, which goes back to the fundamental research of Rayleigh in 1885, is described. The early developments in atom-surface scattering including the important experiments in Otto

Stern's group in Hamburg in the early 1930s are summarized, and the early surface phonon experiments based on HAS and EELS are described. Then in Chap. 3, the fundamental concepts concerning the origin and nature of surface phonons are surveyed. After presenting the Rayleigh theory of surface dynamics based on the analytical solution of the wave equation, Chap. 3 introduces the reader to the two basic theoretical techniques of surface lattice dynamics, the slab and Green's function method, and to the notation used to label the different types of surface phonons. Chapter 4 outlines how the surface affects the force constants of atoms in the solid and how the perturbed force constants affect the surface phonon dispersion curves. Chapter 5 is devoted to the theoretical methods currently in use for the calculation of surface phonons and how they are derived from methods currently in use in bulk dynamics. Special emphasis is given to the methods which go beyond the Born–von Kármán force-constant model paradigm to account explicitly for the role of electrons in the surface lattice dynamics of semiconductors and metals. Chapter 6 introduces the reader to the frequently used phenomenological potentials which govern the overall interaction between the probe atom and the surface. This chapter also introduces the reader to the important differences between the soft metal surfaces and the relatively hard inert insulator surfaces, a theme which will be dealt with in more detail in Chap. 8. Whereas the Chaps. 4 and 5 are devoted to the interactions within the surface, in Chap. 7 the methods used to describe the inelastic coupling of the He atoms to surface vibrations are discussed at different levels of approximation, with special emphasis to theories used for insulator surfaces. Here also the different mechanisms for multiphonon scattering are briefly reviewed, and the effects of surface temperature on the scattering and the conditions favoring one-phonon inelastic processes are discussed. Chapter 8 emphasizes one of the main themes of the book, namely the very special nature of the interaction of the He atoms with metal and semiconductor surfaces, which is not at all the same as for electrons and neutrons, as assumed in the early history of surface phonon research.

To prepare the reader for the experimental results presented in Chaps. 11 and 12, Chap. 9 describes the apparatus and the experimental methods. The most widespread method of time-of-flight (TOF-HAS) spectroscopy has relatively recently been augmented by spin echo spectroscopy (SE-HAS) with a much higher energy resolution. Chapter 9 also contains a quantitative comparison with electron energy loss spectroscopy and provides a detailed analysis of the factors which affect the resolution in the TOF-HAS experiments. The rich phenomenology accompanying the scattering and interaction of atoms with surfaces is discussed in Chap. 10. Special emphasis is devoted to resonant phenomena involving surface-atom bound states which can affect the elastic and inelastic intensities, and may be used to enhance the response of the scattering atoms from particular surface phonons.

The previous chapters lead up to the important Chap. 11, which contains a critical discussion in the light of the theory of the benchmark experimental results on clean solid surfaces, collected during the last decades with HAS, with frequent comparisons to EELS studies. The chapter first addresses results for the insulator rare gas solids and alkali halide crystals. Next, the experiments on the semiconductor surfaces, silicon and gallium arsenide, are surveyed. The major part of the chapter is devoted to the simple metals and the transition metals. Finally, the results for graphene and graphite surfaces and for other layered materials, including topological insulators, are discussed. Chapter 12 contains a similar description of HAS results on the low-energy vibrations of isolated atoms and molecules. The results on the coupled vibrations of monolayer and thicker films of atoms and molecules are also reviewed. Here again the extreme sensitivity of helium atom scattering has provided unexpected information on the vibrations *normal* to the surface and even on modes at the substrate interface of thin films with up to 20 layers. Chapter 13 is devoted to new applications of helium atom scattering to areas not directly related to phonons, for example, to the study of surface magnetism and plasmarons. Here also the many results on the microscopic dynamics of diffusion, which have come largely from TOF-HAS and SE-HAS studies, are surveyed. Finally in the last chapter, Chap. 14, the vast improvements in both resolution and sensitivity now available or conceivable and possible for the next generation HAS apparatus are briefly outlined. Even with the present technology, there are many areas awaiting exploration with helium scattering such as liquid surfaces, organic films of interest in the new area organic electronics, biological films, and clusters on surfaces forming supraclusters.

The book closes with several appendices for quick reference. In Appendix A, the different theoretical methods and the surfaces investigated by the different methods are listed. Appendix B contains a complete collection of all the results of all the experimental studies with HAS and the calculations of surface phonon dispersion curves, with references to publications up to early 2018. Appendix C contains a tabulation of experimental studies of monolayers and thicker films on single crystal substrates, except for hydrogen-covered surfaces, which are listed together with the clean surfaces in Appendix B. Finally, Appendix D provides a list of surface Debye temperatures compared with the bulk Debye and melting temperatures. We hope that these appendices will help those readers who are mostly interested in the surface phonons of specific surfaces to quickly find the sought after information and the references to the literature.

This book would not have been possible without the contribution of a large number of diploma, master, and Ph.D. students who were at one time working in the groups of the authors. All the important experiments were carried out by students, and much of the insight came from intensive discussions with the students in the countless efforts to clarify the concepts. Among the many students, Bruce Doak deserves special mention since he constructed the first successful TOF-HAS apparatus and succeeded in measuring the first surface phonon dispersion curve. Students also contributed to the development of the theory which went hand in hand with the new phenomena found in the time-of-flight spectra.

Many postdoctoral associates contributed directly or indirectly to the results reported here. In alphabetic order they include Bruce Doak, John Ellis, Anna Franchini, Joost Frenken, Georg Gensterblum, Andrew Graham, Jane Hinch, Angela Lahee, Ninsheng Luo, Alexander Menzel, Salvador Miret-Artés, Vania Panella, Robert Rechsteiner, Paolo Ruggerone, Giorgio Santoro, Patrick Senet, Jochen Vogt, and Helmuth Weiss. Both authors are indebted to many of their more senior colleagues especially to the theory pioneers Vittorio Celli and Joseph (Dick) Manson who accompanied us with their advice and ideas over the entire time that we were working on the book. We are very grateful to Dick Manson for his careful reading of the manuscript.

Other colleagues who provided valuable guidance are in alphabetic order of their institutions: Luciano Colombo (Cagliari), Evgueni Chulkov, Pedro Echenique, Ricardo Diez-Muiño and Irina Sklyadneva (DIPC San Sebastian), Davide Campi (EPFL Lausanne), Andrea Levi and Ugo Valbusa (Genoa), Wolfgang Ernst and Toni Tamtögl (Graz), Chakram Jayanthi (Louisville), Gloria Platero and Victor Velasco (Madrid), Giovanni Onida (Milano), Marco Bernasconi, Gian Paolo Brivio and Leo Miglio (Milano-Bicocca), Virginio (Bibi), Bortolani (Modena), Amand Lucas (Namur), Paolo Santini (Parma), Ulrich Schroeder (Regensburg), Jim Skofronick (Tallahassee), Geza Seriani (Trieste), and Branko Gumhalter (Zagreb). Finally, we thank Gil Alexandrowicz and John Ellis for reading and correcting the discussion of the spin echo experiments.

Without the staff of the Max Planck Institut für Strömungsforschung under the leadership of Wolfgang Sattler, the apparatus, the experiments, and the experimental results reported here would not have been possible. The Göttingen staff scientist Ekkehard Hulpke, who started the surface activities at the very beginning of the establishment of the experimental group of JPT in 1970 and Guido Brusdeylins, were invaluable in providing guidance to the many students.

We thank the Alexander von Humboldt Society for the grant of a Humboldt Senior Award Prize to Giorgio Benedek. During his one-year stay in Göttingen in 1990 and 1991, the collaborative effort leading to the present monograph was initiated. The continuation of this collaboration was greatly strengthened by the award of the Max Planck Prize in 1992 and a short reinvitation awards in 2004 and 2009.

Finally, we would like to acknowledge the suggestion by Charlie Duke after critically reading our joint review for "Surface Science, the First Thirty Years" to write a more complete review with more detail. This got us started in 1994, but in the following intervening years we were regularly distracted by other joint projects which at the time were deemed to be more pressing. Thus as the field developed we were forced to update many sections and passages. Each time when nearly finished new more exciting developments not only in the areas covered by the book but also in other fields led to a renewed postponement. We hope that the constant updating and reviewing of the previous versions have added to the clarity without burdening the text too much.

Last but not least, we wish to thank Claus Ascheron and Angela Lahee of the Springer Verlag in Heidelberg for not having lost hope in the completion of the book and for forgiving us for not meeting several of our self-imposed deadlines.

Göttingen, Germany

Giorgio Benedek
Jan Peter Toennies

Contents

Acronyms

−MP	Negative modified Pandey configuration, buckling
+MP	Positive modified Pandey configuration, buckling
1D	One-dimensional
1DEG	One-dimensional electron gas
2D	Two-dimensional
2DEG	Two-dimensional electron gas
3D	Three-dimensional (also referred to film growth regime)
4NNFC	Fourth nearest neighbor force constant (model)
ABINIT	Ab initio programs implementing DFT on a plane-wave basis
AF	Antiferromagnetic
AFM	Atomic force microscopy
AL	Anomalous longitudinal
AP	Acoustic phason
ARPES	Angle-resolved photoemission spectroscopy
ASP	Acoustic surface plasmon
BCM	Bond charge model
BE	Bose-Einstein
BISSCO	Bismuth strontium calcium cuprate
BMB	Brownian molecular dynamics
BSGF	Breathing-shell-model Green's function calculation
BSM	Breathing shell model
BvK	Born–von Kármán
BZ	Brillouin zone
CASTEP	Cambridge Serial Total Energy Package using DFT on a plane-wave basis
CC	Close-coupling
CDO	Charge density oscillation

CDW	Charge density wave
CE	Continuum elastic theory
CFM	Continued fraction method
C-M	Clausius–Mossotti
CM	Cluster model
CP	Car–Parrinello
CPM	Car–Parrinello method
CPU	Central processing unit
DAS	Dimer-adatom stacking-fault model
DE	Diffuse elastic
DFPT	Density functional perturbation theory
DFT	Density functional theory
DLC	Dispersive linear chain
DOI	Digital object identifier
DOS	Density of states
DR	Density (dielectric) response
DRRS	Double resonance Raman spectroscopy
DSM	Double-shell model
DT	Decanethiol
DW	Debye–Waller
DWBA	Distorted-wave Born approximation
EA	Embedded atom
EAM	Effective mass approximation
EAM	Embedded atom method
EAMD	Embedded atom molecular dynamics
EBA	Exponentiated Born approximation
EBOA	Extended band orbital approximation
EC	Equivalent crystal
EELS	Electron energy loss spectroscopy
EM	Effective medium (theory)
ESI	Electrospray ionization
FC	Force constant (model)
fcc	Face-centered cubic
FLASH	Free-Electron Laser Hamburg
FIR	Focused inelastic resonance
FK	Fuchs and Kliever
FLAPW	Full-potential linearized plane-wave method
FM	Ferromagnetic
FP	First principle
FP-DR	First principle–density response
FPM	Frozen phonon method
FPTB	First-principle tight binding
FR-PAW	Fully relativistic projector-augmented wave method
F-S	Finnis–Sinclair
FS	Focused surfing

FT	Frustrated translation
F-vdM	Frank–van der Merwe
FWHM	Full width at half maximum
GF	Green's function
GGA	Generalized gradient approximation
GM	Glue model
GM-MD	Glue model molecular dynamics
GR	Reciprocal direct space (method)
HAS	Helium atom scattering (also IHAS: inelastic HAS) spectroscopy
HCS	Hard corrugated surface
HCP	Hexagonal close Packing
HD	Hydrodynamic equations
He3SE	3He spin echo scattering spectroscopy
HEIS	High energy ion scattering
HF	Hellmann–Feynmann (forces, theorem)
HFD	Hartree–Fock dispersion (potential, known in different types: B, B2, etc)
HR	Hindered rotation
HREELS	High-resolution electron energy loss spectroscopy
HT	Heptanethiol
HUGO II	Name of a He atom scattering time-of-flight apparatus used in Göttingen
HyAS	Hydrogen atom scattering
IETS	Inelastic electron tunneling spectroscopy
IF	Inelastic focusing
ILPES	Inelastic laser photoemission spectroscopy
IRAS	Infrared absorption spectroscopy
IXS	Inelastic X-ray scattering
KF	Kinematical focusing
KM	Keating model
KS	Kohn–Sham
L	Longitudinal
Ln, L-n	Longitudinally polarized in the n-th plane
LA	Longitudinal acoustic
LB	Langmuir–Blodgett
LCT	Local coupling transfer method
LDA	Local-density approximation
LEED	Low-energy electron diffraction
LEEM	Low-energy electron microscope
LJ	Lennard-Jones
LM	Lucas mode
LMTO	Linear muffin-tin orbital
LO	Longitudinal optic
LOD	Linear optical diffraction
MBE	Molecular beam epitaxy

MD	Molecular dynamics
MDS	Metastable de-excitation spectroscopy
ME	Multipole expansion method
MEIS	Medium energy ion scattering
MEEM	Metastable electron emission microscopy
ML	Monolayer
MP	Modified Pandey (configuration)
NeAS	Neon atom scattering
NEST	National Enterprise for Nanoscience and Nanotechnology
NEXAFS	Near-edge X-ray adsorption fine structure
NIS	Neutron inelastic scattering
NTCDA	Naphthalenetetracarboxylic dianhydride
OFET	Organic field-effect transistor
OI	Ordinary inelastic process
OLED	Organic light-emitting diode
OP	Optical phason
PAA	Phonon-assisted adsorption
PAA-SD	Phonon-assisted adsorption followed by selective desorption
PAD	Phonon-assisted desorption
PAW	Projector-augmented wave
PC	Pseudo charge model
PDF	Partial density formalism
PM	Paramagnetic
PMMA	Polymethylmethacrylate
PPP	Pseudo potential perturbation
PSW	Pseudo surface waves
PTCDA	Perylenetetracarboxylic dianhydride
QHAS	Quasi-elastic helium atom scattering
QI	Quasi-ion (approximation, model)
QSM	Quadrupole shell model
QW	Quantum well
RAIRS	Reflection absorption infrared spectroscopy
RBS	Rutherford backscattering
RE	Resonance enhancement
RHEED	Reflection high energy electron diffraction
RIM	Rigid ion model
RI	Rotational invariance
RL	Rocking libration
RMSD	Root mean square displacement
RPA	Random phase approximation
RW	Rayleigh wave
SA	Selective adsorption
SAM	Self-assembled monolayer
SA-PAD	Selective adsorption followed by phonon-assisted desorption
SAW	Surface acoustic wave

SBZ	Surface Brillouin zone
SC	Scan curve
SCF	Self-consistent field theory
SCMBP	Sutton-Chen many-body potential
SCS	Soft corrugated surface
SD	Selective desorption
SE-HAS	Spin echo ^{3}He atom scattering
SE-QHAS	Spin echo quasi-elastic ^{3}He atom scattering
SETE	Semiempirical total energy approach
SH	Shear horizontal
SHeM	Scanning helium microscope
SHn, SH-n	Shear-horizontally polarized in the n-th plane
S-K	Stranski–Krastanov
SL	Slab (method, diagonalization)
SM	Shell model
S-matrix	Scattering matrix
SMM	Scattering matrix method
SO	Spin–orbit (interaction)
SOC	Spin–orbit coupling
SP	Surface plasmon
$S_{1\text{-}9}$	Surface mode labels (see Table 3.1)
SPLEED	Spin-polarized low energy electron diffraction
SPMDS	Spin-polarized metastable de-excitation spectroscopy
STEM	Scanning transmission electron microscope
STM	Scanning tunnel microscopy
STS	Scanning tunnel spectroscopy
StW	Stoneley wave
SV	Shear vertical
SVn, SV-n	Shear-vertically polarized in the n-th plane
SW	Sezawa wave
SWn	Sezawa wave of *n*-th order
TA	Transverse acoustic
TB	Tight-binding
TBSM	Tight-binding second-moment (semiempirical total energy method)
TF	Trial function
TI	Translational invariance
T-matrix	Transition matrix
TO	Transverse optic
TOF	Time-of-flight
TPPE	Two photon photoemission
TOF HAS	Time-of-flight He atom scattering
UDA	Uniform density approximation
UHV	Ultra-high vacuum
UPS	Ultraviolet photoemission spectroscopy
US	United States

UT	Ultrathin (film)
UV	Ultraviolet
VAS	Vibrations at surface
VASP	Vienna *Ab Initio* Simulation Package for *ab initio* molecular dynamics
vdW	Van der Waals (interaction, forces, potential)
VFF	Valence force field
WKB	Wentzel–Kramers–Brillouin
WM	Wallis mode
XPS	X-ray photoemission spectroscopy
ZA	Z-polarized acoustic (phonon, mode, branch)
ZO	Z-polarized optical (phonon, mode, branch)
ZJM	Zener-Jackson-Mott
ZK	Zaremba-Kohn

Chapter 1
Surface Dynamics, An Introduction

The surfaces of bodies are the field of very powerful forces of whose action we know but little.

Lord Rayleigh, 1842–1919
(from A. Zangwill, Physics at Surfaces)

Abstract Surface elastic waves, discovered by Lord Rayleigh in the late 19th century in the realm of seismology, besides occurring in nature over a wide range of length scales, have found a broad spectrum of technological applications (Sect. 1.1). Surface acoustic waves (SAWs) on the nanometric scale and their quanta, surface phonons, while allowing for signal processing devices operating in the THz domain, hold information on the structure and the interatomic forces at solid surfaces, which generally differ from those of the solid bulk (Sect. 1.2). Inelastic He atom scattering, as compared to other surface phonon probes, offers convenient kinematical conditions for a high-resolution surface phonon spectroscopy over the complete momentum and energy spectral range (Sect. 1.3).

1.1 Surface Acoustic Waves in Nature and Technology

1.1.1 Frequency Range of Surface Acoustic Waves

More than a century ago in 1885 Lord Rayleigh demonstrated mathematically [1, 2] that the long-wave component of earthquakes is due to elastic waves travelling along the surface of the Earth (Fig. 1.1). His theory describes the displacement amplitudes of waves, which are localized in the surface region of a semi-infinite elastic continuum. They are polarized in a plane normal to the surface along the propagation direction called the sagittal plane as shown in Fig. 1.2 [3]. Rayleigh's discovery is commonly taken as the starting point of surface dynamics.

Throughout nature surface waves can exist in all sorts of condensed matter systems, from nuclei to stars, over an extremely wide range of wavelengths covering about 23 orders of magnitude (Fig. 1.3) [4–11]. Since also the speed of

G. Benedek and J. P. Toennies, *Atomic Scale Dynamics at Surfaces*, Springer Series in Surface Sciences 63, https://doi.org/10.1007/978-3-662-56443-1_1

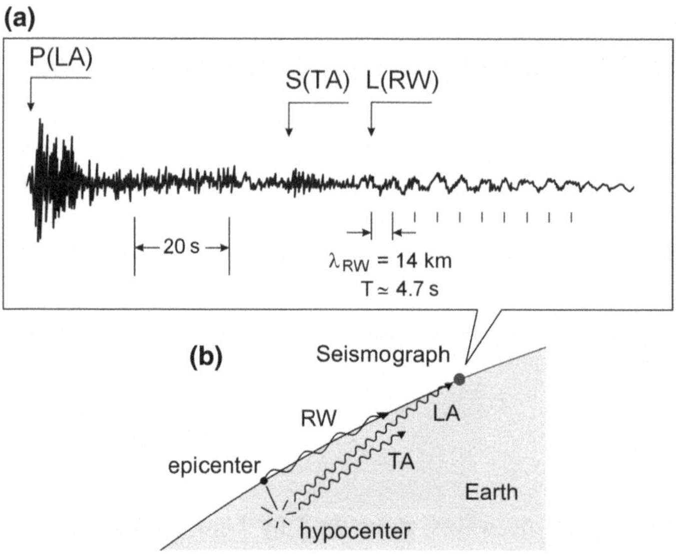

Fig. 1.1 Seismographic record of the earthquake which hit southern Italy in 1980 as detected 420 km away from the epicenter at the Polo Sismico "Alberto Gabriele" of the Ettore Majorana Centre in Erice, Sicily, Italy. **a** The arrival times of the primary (P), secondary (S) and long-wave (L) seismic wave components are indicated by arrows and correspond to longitudinal (LA), transverse (TA) and Rayleigh waves (RW), respectively, in the notation of solid state physics. **b** Since the LA and TA waves travel directly through the earth, practically all frequencies of the spectrum reach the seismograph. The Rayleigh waves, however, travel along the earth's surface. Since the shorter wavelength components are more strongly scattered by the earth surface asperities, only the long waves (about 14 km in the figure) reach the seismograph [Courtesy of Dr. Pino Aceto]

surface waves varies from a few tens of meters per second for sea surface waves [7] to kilometers per second on ordinary solid surfaces [3–5], up to a substantial fraction of the speed of light in nuclei [6]. The range of frequencies is even wider, extending over 26 orders of magnitude.

With regard to waves traveling at the surface of solid elastic bodies, the naturally occurring seismic waves in ordinary earthquakes are at the low frequency end with wavelengths extending up to many kilometers (Fig. 1.1a). The detection of seismic waves produced by nuclear explosions plays an important role in monitoring test bans [12].

The other extreme of acoustic vibrations at short wavelengths has emerged only rather recently. Starting from the sixties the development of interdigital transducers in piezoelectric materials extended the frequencies from the audible region into the ultra- and hypersound regimes, with wavelengths progressively decreasing towards the micrometric and nanometric scales. Interdigital transducers have found many applications and have opened up the technical development of surface acoustic wave (SAW) based devices in delay lines, optoacoustic systems and signal processing devices [13–18], with a wide spectrum of applications, briefly presented in the next subsections.

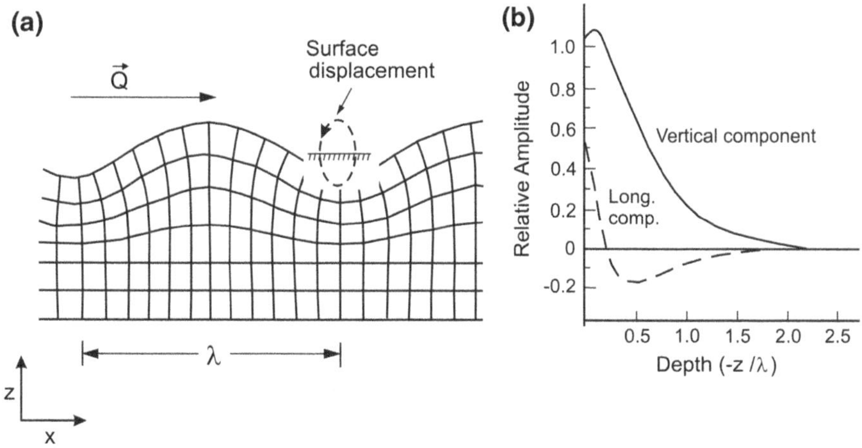

Fig. 1.2 Schematic diagram showing the deformations at the surface and penetration associated with a Rayleigh wave (RW) of a semi-infinite homogeneous elastic solid in the sagittal plane (from Farnell [3]). **a** The displacement field has a dominant component normal to the surface. The smaller longitudinal component has a $\pi/2$ phase shift with respect to the normal one. Together they lead to a retrograde elliptical polarization in the sagittal plane. **b** Both components decay exponentially inside the solid with a decay length proportional to the RW wavelength λ, which is a characteristic of a *macroscopic* surface mode

1.1.2 Surface Acoustic Wave Visualization and Surface Characterization

About 15 years ago it became possible to visualize micrometric surface phonons directly by means of stroboscopic X-ray diffraction: Fig. 1.4 shows an example of the direct observation of a SAW generated in the piezoelectric material $LiNbO_3$ with a micrometric wavelength and a frequency in the sub-GHz domain [19, 20]. Besides offering a view of a SAW, the experiment also images the alteration of the SAW displacement pattern produced by surface defects of size and depth comparable to the SAW wavelength and penetration depth. Thus defects of any extension can be probed by SAWs of appropriate wavelength, and visualized with the technique demonstrated by Shilo et al. (Fig. 1.4) [19, 20]. Moreover, the analysis of SAW pulse propagation has also been shown to be a sensitive tool for studying the mechanical strength and elastic moduli of thin films [5, 21], and for testing the porosity and the microstructure of surface coatings [22–24].

With ultra-short laser pulses in the *ps* to the *fs* time scales and the related techniques of ultra-fast spectroscopy it has become possible to not only excite, but also to monitor SAW's in the THz frequency domain by means of laser pump-probe experiments [5]. A typical pump-probe experimental configuration for SAW studies, as reported in the review article by Hess [5], is depicted in Fig. 1.5. In one configuration two simultaneous *ps* laser pulses interfere at a crystal surface thereby generating two SAW's propagating in opposite directions, which are detected

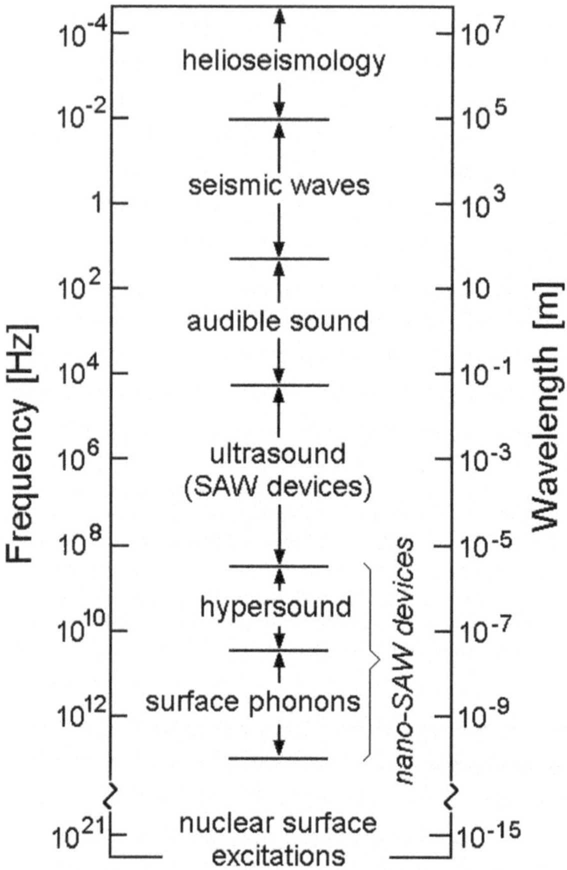

Fig. 1.3 Surface acoustic waves cover a wide range of frequencies and wavelengths extending from 10^7 m long waves involved in the helioseismology on the surface of the sun [9, 10] to 10^{-15} m for the vibrations on the surface of large nuclei [6]. The wavelength of ordinary Rayleigh waves vary over about 13 orders of magnitude, from the long-wave (L) component of earthquakes (or nuclear bomb explosions) to the interatomic distances in crystal lattices. Except for nuclei, where collective modes travel at an appreciable fraction of the speed of light, the speed of surface waves in all the other systems is in the same range from a few meters to kilometers per second so that the magnitude of frequencies is inversely proportional to that of wavelengths over a wide range (adapted from Hess [5])

through the diffraction of the probe laser beam. A similar technique has been used by Wright and Kawashima [25] to determine the surface wave frequency spectrum from the Fourier analysis of a time resolved pump-probe experiment in thin films of Mo and Cr on a silica substrate.

In another experiment a *ps* pulse of the pump laser generates a coherent SAW, which propagates along the surface and is monitored by a continuous probe laser beam through Michelson interferometry. With this method ultrafast generation, imaging and detection of coherent surface phonons emanating from a point source

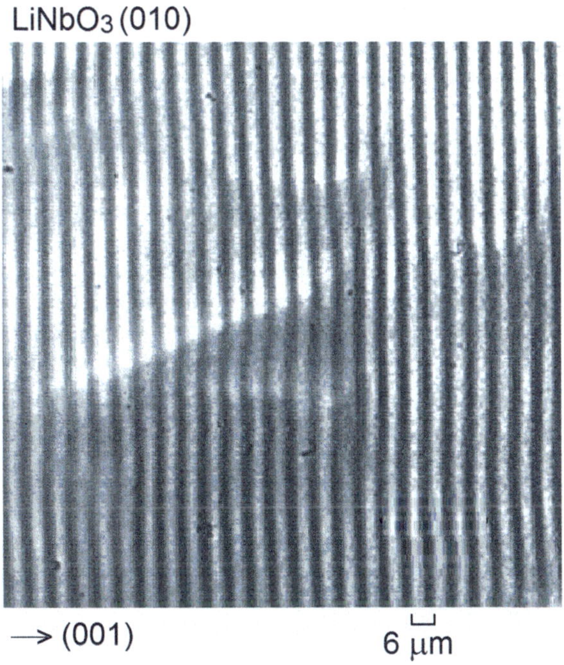

Fig. 1.4 A surface acoustic wave (SAW) of wavelength $\lambda = 6$ μm and frequency $\nu = 0.58$ GHz propagating along the $LiNbO_3(010)$ surface (y-cut) as imaged by stroboscopic X-ray topography. The alterations of the wavefront profile and intensity reveal the presence of extended surface defects, such as dislocations (from Shilo and Zolotoyabko [19, 20])

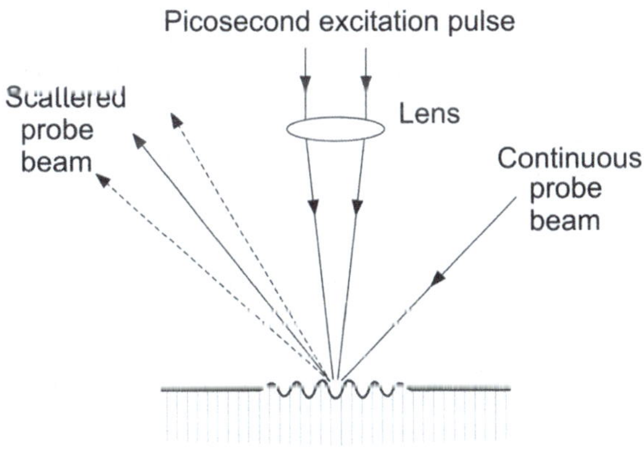

Fig. 1.5 Schematic diagram of a pump-and-probe laser experiment for the generation and detection of surface acoustic waves. Two simultaneous *ps* laser pulses interfere at a crystal surface and the two counterpropagating SAW's which are generated from the interference pattern, are detected through the diffraction of a probe laser beam (from Hess [5]). In other experiments the change in the index of refraction accompanying the SAW is observed

have been demonstrated by Sugawara et al. [26] with a micrometric space resolution and frequencies up to 1 GHz. This pump-probe imaging technique of SAW ripples has also allowed for the dynamic visualization of SAWs on a two-dimensional phononic crystals and to directly measure their acoustic band structure and Bloch harmonics [27, 28].

The advent of scanning probe microscopies [29] has paved the way to nanometric visualization and control of top-down fabrication processes, as well as to the direct manipulation of atoms and molecules in bottom-up nanotechnologies. As demonstrated by Gawronski et al. [30], scanning tunneling spectroscopy (STS) even allows for a direct imaging on the atomic scale of surface phonons generated by inelastic electron tunnel spectroscopy in Au(111) and Cu(111) surfaces (Fig. 1.6). As shown in [31] the phonon wavelength and the energy of about 21 meV observed in Cu(111) permit to associate the major feature to the quasi-longitudinal surface acoustic resonance at the zone boundary, the phonon which actually produces the strongest modulation of the tunneling barrier (Chap. 11). A similar assignment should hold also for Au(111).

1.1.3 SAW Acousto-Electronics

Under the pressure of the rapidly growing industry of mobile telecommunications, the last two decades have witnessed the development of a new generation of SAW devices, such as filters and duplexers for multiband telephones [32–36], low-noise microwave oscillators [37], and multiplexers in frequency modulated continuous-wave radars [38, 39]. In many applications SAW devices combine a small size and integrability with operational frequencies in the GHz range, thus offering capabilities unattainable by other technologies. SAW devices have gained momentum in the broad field of sensors [40–43], with applications to wireless sensors for pressure and temperature [44, 45], bio-sensors for artificial nose and tongue micro-systems [46, 47], for monitoring pathogens [48], proteins [49, 50], and other organic adsorbates of biomedical interest [51, 52], for the quantification of cell growth [53], for environmental purposes like the detection of ozone ultra-low (ppb) concentrations [54], etc. A scalable SAW-based cantilever device for portable biochemical sensors, recently realized by De Simoni et al. [55], was shown to outperform in sensitivity current quartz crystal microbalances.

The extension of surface acoustic waves into the higher GHz and THz spectral regions is opening new perspectives in *acousto-optics* and *phononics*. Modulators of light signals based on Mach-Zehnder interferometry have been implemented in a SAW-driven semiconductor device on a size scale several hundred times smaller than ordinary optical modulators [56]. In a perfect analogy with the photonic crystals [57, 58], the band structure and spectral gaps for SAWs have been demonstrated in piezoelectric phononic crystals [59, 60]. The experimental study and applications of SAWs in the THz region, where wavelenghts reach the

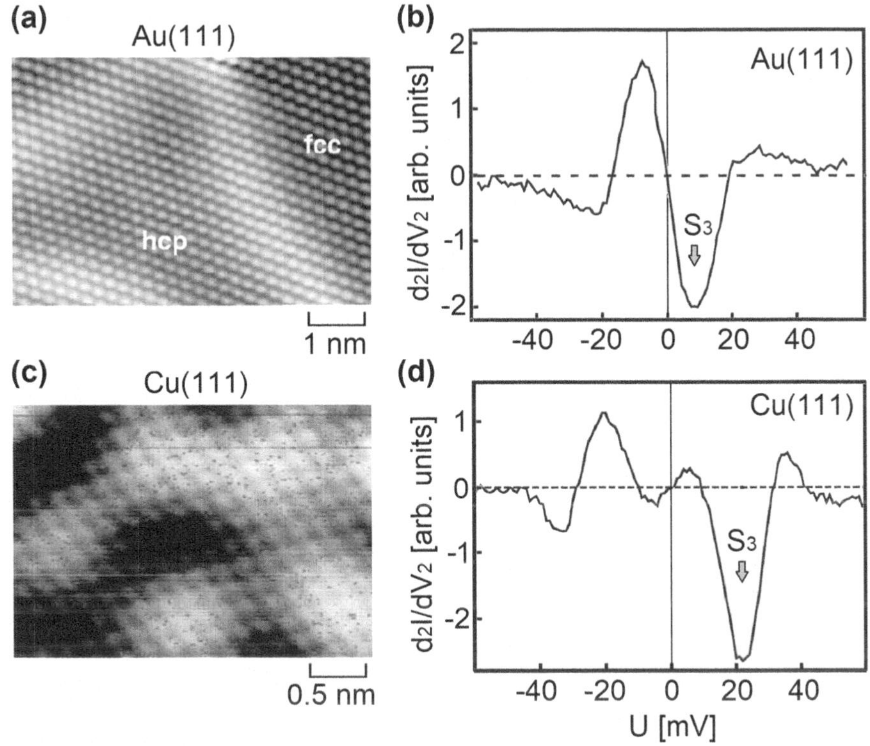

Fig. 1.6 STM images and Inelastic Electron Tunneling Spectroscopy (IETS) spectra of the reconstructed clean Au(111) and Cu(111) measured at 5 K. **a**, **c** STM images in the constant current mode of Au(111) showing the herringbone reconstruction, with boundaries separating fcc from hcp surface domains, and of Cu(111) showing a standard wave pattern produced by adsorbed CO (outside the picture). Both are based on mapping of the inelastic electron spectra in the energy range of the phonons. **c** IETS spectrum of Au(111) taken on the fcc domain. **d** IETS spectrum of the Cu(111) surface. The grey shaded areas mark the regions of calculated and measured phonon energies integrated over the entire Brillouin zone. The major peaks correspond to the energies of surface phonons S_3 (see Chap. 11) producing the largest modulation of the tunneling barrier [31], which confirms the interpretation in terms of surface phonons (adapted from Gawronski et al. [30]

nanometric scale, have been fostered by the availability of nearly ideal defect free surfaces. The band structure of phononic crystals can be designed so as to exhibit topological features similar to the band structure of topological insulators, in particular the presence of phononic helical edge states [61, 62]. As recently shown in a seminal work by Wang et al. [62], one-way edge phononic waveguides bear the promise of a new class of SAW devices endowed with topological robustness against external disturbances, to be used in electronics, telecommunications and acoustic images.

On piezoelectric semiconductor surfaces a SAW produces a periodic acoustic deformation potential which can trap charge carriers and transport them along the surface [25–27, 63, 64]. The charge transport by means of surface acoustic waves is currently known as *acoustoelectric effect*; the influence that SAW's have on the electric conductance through the opening of a ballistic quantum channel goes under the general term of *acousto-conductance* [65]. A single electron transport in a one-dimensional channel by means of high-frequency surface acoustic waves has been achieved in a GaAs/AlGaAs heterostructure in the late nineties at the Cavendish Laboratory [63, 66–68]. In these experiments the number of electrons trapped in the moving potential well of the SAW is controlled by the electron-electron repulsive potential and is well defined. According to the authors it may provide a tool for defining and producing a standard of electric current.

The SAW deformation potential can also separate electrons from holes and prevent recombination of excess carrier pairs [69–71]. An intriguing demonstration of this effect has been given by Rocke et al. [69]. The excess electron-hole pairs are created by a laser pulse at the surface of a ultrathin InGaAs layer in a region which is crossed by a SAW. The SAW, which is produced by a radio-frequency generator, travels along the surface carrying a certain number of excess electrons trapped at its minima and of excess holes trapped at its maxima. The charge carriers are transported along the surface with the SAW velocity until the SAW energy is absorbed by a transducer. At this point the hole-electron pairs suddenly recombine and the absorbed photons are re-emitted. The distance between adsorption and emission is 1 mm, corresponding to a delay of 350 ns. Many applications in composite devices, combining opto-acoustics with electronic functions, can be envisaged for recombination processes controlled by surface phonons. A variety of SAW-driven optoelectronic systems, notably electroluminescent planar light emitting devices, has been developed in recent years [72–78].

The interaction of SAW's with charge carriers in nanostructures under conditions of quantized transport has since been investigated in a number of state-of-the-art experiments concerning resonant tunneling [79], SAW single electron interferometry [80], single electron pumping through quantum dots [81, 82], SAW absorption by metal quantum wires [83], and other effects which the reader may find in the specialized literature. An intriguing example is the experiment by Willett et al. [84], where the presence of a Fermi surface predicted for a half-filled Landau level in a GaAs/AlGaAs heterostructure was demonstrated and probed by a SAW propagating normal to the magnetic field, in resonance with the cyclotron orbits of the charge carriers. Hybrid systems obtained from interfacing GaAlAs thin films with a piezoelectric substrate like $LiNbO_3$ allow for much larger electromechanical coupling than in pure GaAlAs [85–88] with intriguing non-linear effects in the interaction of SAWs with the electron-hole plasma [69].

Surface acoustic waves have also been demonstrated to provide a possible route to quantum computation [89–92]. In an experiment by Beltram and coworkers at NEST in Pisa [93], schematically illustrated in Fig. 1.7, a GHz surface acoustic wave travels at the surface of a GaAs/AlGaAs heterostructure along two parallel rails separated by an impenetrable barrier. The SAW can drive a charge carrier

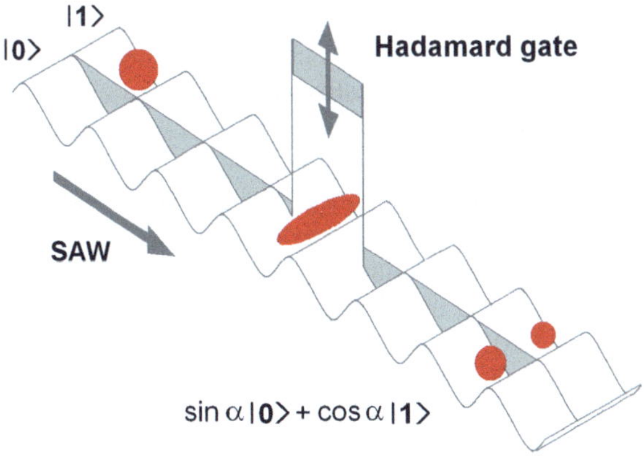

Fig. 1.7 A surface acoustic wave (SAW), travelling along two parallel rails separated by an impenetrable barrier (gray wall), carries an electron (red sphere) trapped at a minimum of the moving SAW periodic potential in either of the two rails. The SAW-driven electron realizes a *flying q-bit*, one trail defining the quantum state $|0\rangle$, the other the quantum state $|1\rangle$. At some point of the transport line a segment of the barrier is lifted, thus allowing a mixing of the binary eigenvectors into a new state $\sin\alpha|0\rangle + \cos\alpha|1\rangle$, with the mixing angle α determined by the structure and operating conditions of the gate. This process is a realization of a *Hadamard gate* for quantum computation [94] (adapted from Piazza et al. [93])

provided by the heterostructure two-dimensional electron gas (2DEG) by trapping it at a minimum of the travelling periodic potential in either of the two rails. Thus the SAW-driven carrier realizes a *flying q-bit*, one trail defining the quantum state $|0\rangle$, the other the quantum state $|1\rangle$. At some point of the transport line a segment of the barrier (*Hadamard gate* [94]) is lifted, thus allowing a mixing of the binary eigenvectors into a new state $\sin\alpha|0\rangle + \cos\alpha|1\rangle$, with the mixing angle α determined by the structure and operating conditions of the gate. Clearly all the basic logic functions of quantum computation [94] can be implemented in this way with a suitable design of the SAW rails. Electrons surfing on a SAW have been shown to provide a platform for quantum optics, with perspectives of applications to quantum computing and single electron spin teleportation [95].

Effects similar to those involved in SAW acousto-electronics can be realized in micro-fluidics. In this case fluid currents on the micro- or nanoscale can be modified by the mechanical surface deformations produced by a SAW, with interesting perspectives in the new field of *acousto-fluidics* [96, 97]. A SAW routing of a fluid flow has been demonstrated together with possible implementation of micro fluidic logic gates [98]. The effects of the SAW deformation potential on surface mobile adsorbates can also be exploited for nano-fabrication purposes. With large-scale molecular dynamic simulations Taillan et al. [99] have demonstrated that SAWs can induce nucleation and a controlled nanoscale self-organization of adsorbed species.

Surface electrons can be trapped by a surface phonon associated with the lattice deformation that the electron itself produces with its coulomb field (*self-trapping*). This special electron-phonon bound state is known as a *small polaron* [100] and results in a comparatively large electron effective mass. In this case the surface phonons are not injected from outside but are generated in situ by the excess electrons. The evidence for such a localization of surface excess electrons due to surface phonons has been obtained by Ge et al. [101] in a state-of-the-art experiment based on the combination of angle-resolved two-photon photoemission (TPPE) and femtosecond laser techniques. Future experiments of this kind will provide important information on charge carrier dynamics in many low dimensional systems including organic light emitting diodes [102]. From these few examples one expects that the dynamical interaction of quantized surface waves, i.e., surface phonons, with electrons and, via the electrons, with photon fields will lead to a wide spectrum of new phenomena and a large variety of new applications and devices.

At phonon frequencies in the THz domain, as occur in the surface phonons discussed in this monograph, the corresponding wavelengths for a typical sound speed of 10^3 m/s become of the order of 1 nm, i.e., comparable to the interatomic distances. With respect to wave propagation the solid does not behave any more as an elastic continuum, but reveals the lattice structure with its well-ordered array of atoms. In this limit the speed of Rayleigh waves, i.e., the group velocity of acoustic surface modes, is no longer constant and the frequencies are no longer inversely proportional to the vibration wavelengths. This phenomenon is known as *dispersion* in analogy with the dispersion of white light crossing a refractive prism as an effect of the different velocities of the rainbow components. Moreover in polyatomic crystals the internal degrees of freedom of the lattice unit cell allow for additional high-frequency surface phonons whose frequencies do not vanish at infinite wavelengths. These microscopic vibrations are termed *optical* surface phonons, since their finite energy at small wavevectors allows them to directly couple with the photon field.

The extension of the accessible surface phonon spectrum into the high-frequency dispersive region of the acoustic branches and possibly into the region of microscopic surface phonons of optical character is one of the prerequisites for the reduction of devices down to nanometric sizes. These technical perspectives have been one of the driving forces behind the progress made in the last three decades in the surface phonon spectroscopy and theory which will be discussed in the following chapters.

1.2 Fundamental Interactions at Surfaces from Surface Phonon Spectroscopy

1.2.1 Physical Properties of Surfaces Compared to the Bulk

As in all physical studies of the microscopic behavior of matter, in the case of solid surfaces, the knowledge of the microscopic atomic forces at solid surfaces is fundamental for understanding not only the structure, but also the dynamical processes occurring at the surface. These include not only the phonons but also diffusion at surfaces, crystal growth, adsorption and epitaxy, heterogeneous catalysis and even macroscopic effects such as rheology and friction. Moreover most nanotechnologies deal with systems where surfaces and interfaces, and the microscopic forces therein, play a major role. The surface structure and changes can now be directly observed on an atomic scale with an arsenal of scanning probe microscopies [103, 104]. It is, however, the spectroscopy of dynamical processes which provides most of the relevant experimental information on the microscopic forces. The experimental data on the dispersion curves of surface phonons, as obtained by high resolution helium atom scattering (HAS) and electron energy loss spectroscopy (EELS), allow for the most direct insight into the interactions between individual atoms in the surface region of a solid.

At first one might expect that the cleavage of a crystal to expose a new surface would have no effect on the structure of the newly formed surface (Figs. 1.8a, b and 1.9a, b). In a first approximation (*ideal surface* approximation) this is the case for the low index metal surfaces and for many insulators. About thirty five years ago, previous to the advent of the spectroscopy of surface vibrations by He atom and electron scattering, a wide variety of experiments (e.g. LEED and ion backscattering) established that the geometrical structure at the surface of a crystal is on closer examination usually somewhat different than in the equivalent planes inside the bulk of the crystal. These changes can be classified according to whether the two-dimensional periodicity of the surface differs from that of the corresponding bulk planes, in which case they are called *reconstruction* (Fig. 1.8e), or the periodicity of the surface plane remains unchanged with respect to the bulk planes but the interlayer spacings near the surface are affected, which is called *relaxation* (Fig. 1.8d). In metals both surface relaxation and reconstruction are accompanied

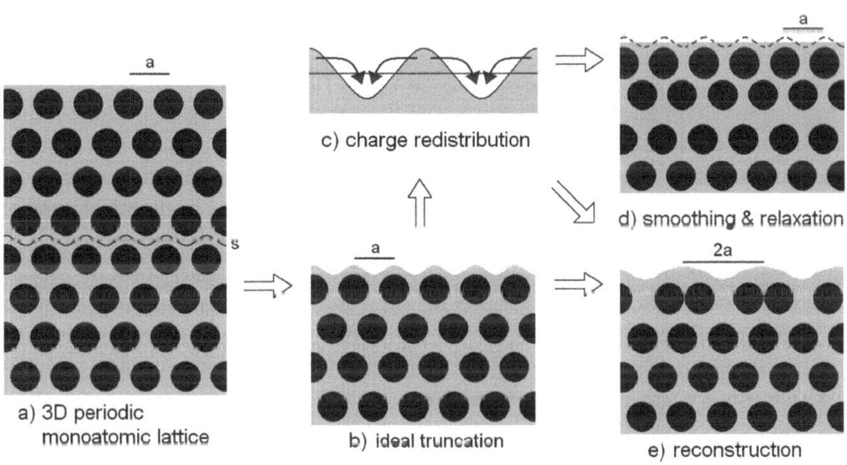

Fig. 1.8 Schematic diagram showing the different structural changes which may occur when a metal crystal is cleaved to expose a free surface

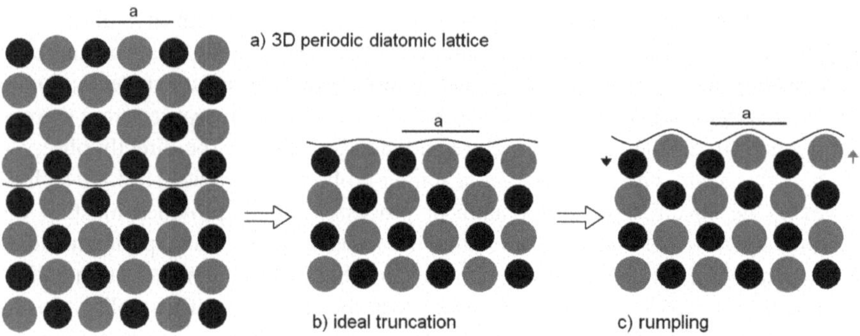

Fig. 1.9 Schematic diagram illustrating the surface structural change of *rumpling* which occurs on an alkali halide crystal surface after cleavage

by, and generally due to, a redistribution of the surface free electrons and a smoothing of the surface charge density (Fig. 1.8c) [103]. Similarly in covalent semiconductors the unpaired electrons of the severed bonds (*dangling bonds*) redistribute, leading to various forms of reconstruction of the clean surface. Reconstruction may, however, not occur if the dangling bonds are saturated by adsorbed atoms, e.g., hydrogen on Si(111).

In closed-shell insulators there is generally no important charge re-distribution, besides some atom polarization, and no reconstruction occurs. However, unreconstructed di- or poly-atomic crystals may nevertheless have different surface relaxations for different atoms, which leads to surface *rumpling* (Fig. 1.9c). Other possible changes of the surface geometry with respect to the order of the bulk planes are order-disorder transitions and related processes such as *roughening*. A roughened surface may be viewed as the 2D analog of a three-dimensional amorphous material [105, 106]. Also melting of the surface may occur at temperatures significantly below the bulk melting point [107].

Another effect of reduced coordination is the change and anisotropy of the thermal root mean square displacements (RMSD) of the surface vibrations of the surface atoms compared to bulk atoms. All the mentioned changes in the equilibrium atomic positions and electron charge distribution occurring at the surface with respect to the bulk can naturally be associated with the reduction in atomic coordination and the force unbalance which would be experienced in the ideal surface (see Chap. 4).

Table 1.1 lists the changes in the first interlayer distance Δd_{12} due to surface relaxation for the (001) surface of NaCl, for the low index surfaces of three face-centered cubic (fcc) metals, which will be extensively discussed in this monograph, and for the H-covered (111) surface of silicon. None of these surfaces are known to reconstruct as compared to some of the heavier metals and clean semiconductor surfaces which are very prone to reconstruction. When compared to the bulk layer spacing the measured first layer spacings listed in Table 1.1 are contracted by about 10% at most. In general the lower coordination of surface atoms is responsible for a larger and anisotropic root mean square displacement

Table 1.1 Surface relaxation, expressed by the relative change $\Delta d_{12}/d_{12}$ of the first interlayer distance, and root mean square displacements (RMSD) of the atomic vibrations for the components parallel ($u_{//}$) and normal ($u_{\perp}$) to the surface at 300 K[§]

Crystal (surface)	Atom	r_0 (Å)	$\frac{\Delta d_{12}}{d_{12}}$ (%)	$\sqrt{<u_{//}^2>}$ (Å)	$\sqrt{<u_{\perp}^2>}$ (Å)	$\sqrt{<u_{\alpha}^2>_{bulk}}$ (Å)
NaCl(001)[a, b]	Na	2.82	−2.05	0.158	0.180	0.146
	Cl		1.12	0.144	0.172	0.131
Ag(001)[c, d]	Ag	2.89	0	0.117	0.123	0.089
Ag(110)[e, k]			−7.8	0.120/0.140	0.136	
Ag(111)[f, h, l]			−2.5	0.106	0.126	
Cu(001)[d, m]	Cu	2.55	−1.2	0.114	0.113	0.084
Cu(110)[d, g]			−8.5	0.106/0.140	0.114	
Cu(111)[e, h, o]			−0.7	0.95	0.127	
Ni(001)[i, p]	Ni	2.49	−3.2	0.089	0.108	0.063
Ni(110)[j, q]			−7.0	0.089/0.106	0.098	
Ni(111)[h, r]			−1.2	0.076	0.095	
Si(111):H 1 × 1[s]	H	–	–	0.219	0.131	–
	Si	2.35	−3.8	0.089	0.098	0.075

[§]For the (110) surfaces the two values for $\sqrt{<u_{//}^2>}$ refer to displacements in the surface plane either parallel or normal to the $[1\bar{1}0]$ atomic rows, respectively. For comparison the bulk nearest neighbour distances r_0 and the bulk RMSD are also reported. For each surface listed in the first column the first (second) reference refers to relaxation (RMSD) data. All the relaxation data, except Si(111):H, are from experiments, whereas the RMSD data are from theory. Acronyms: SM: shell model; LEED: low energy electron diffraction; EA-MD: embedded-atom molecular dynamics; HEIS (MEIS): high (medium) energy ion scattering; SPLEED: spin-polarized LEED; RBS: Rutherford back scattering; DFPT: density functional perturbation theory

[a]F. W. de Wette, W. Kress, and U. Schröder, Phys. Rev. B 32, 4143 (1985) (SM)
[b]T. S. Chen, G. P. Alldredge, and F. W. de Wette, Surf. Sci. 57 (1976) 25 (SM)
[c]H. Li, J. Quimm, Y. S. Li, D. Tian, F. Jona, and P. M. Marcus, Phys. Rev. B 43, 7305 (1991) (LEED)
[d]L. Yang, T.S. Rahman and M.S. Daw, Phys. Rev. B. 44 (1991) 13725 (EA-MD)
[e]T. S. Rahman and Z. J. Tian, J. Electr. Spectr. Rel. Phenom. 64/65 (1993) 651 (EA-MD)
[f]A. N. Al-Rawi, A. Kara and T. S. Rahman, Surf. Sci. 446 (2000) 17 (EA-MD)
[g]L. Yang and T.S. Rahman, Phys. Rev. Lett. 67 (1991) 2327(EA-MD)
[h]A. N. Al-Rawi, A. Kara, and T. S. Rahman, Phys. Rev. B 66, 165439 (2012) (EA-MD)
[i]Y. Beaudet, L. Lewis, and M. Persson, Phys. Rev. B 47, 4127 (1993) (EA-MD)
[j]Y. Beaudet, L. J. Lewis, and M. Persson, Phys. Rev. B 50, 12084 (1994) (EA-MD)
[k]Y. Kuk and L. C. Feldman, Phys. Rev. B 30, 5811 (1984) (HEIS)
[l]P. Statiris, H. L. Lu, and T. Gustafsson, Phys. Rev. Lett. 72, 3547 (1994) (MEIS)
[m]D. M. Lind, F. B. Dunning, G. K. Walter, and H. L. Davis, Phys. Rev. B 35, 9037 (1997) (SPLEED)
[n]D. L. Adams, H. B. Nielsen, J. N. Andersen, I. Stensgaard, R. Feidenhans'l, and J. E. Sorensen, Phys. Rev. Lett. 49, 669 (1982) (LEED, HEIS)
[o]S. A. Lindgren, L. Wallden, J. Rundgren, and P. Westrin, Phys. Rev. B 29, 576 (1984) (LEED)
[p]J. W. M. Frenken, J. F. van der Veen, and G. Allan, Phys. Rev. Lett. 51, 1876 (1983) (RBS)
[q]Y. Gauthier, R. Baudoing, C. Gaubert and L. J. Clarke, J. Phys. C: Solid State Phys. 15, 3223 (1982) (LEED)
[r]J. E. Demuth, P. M. Marcus and D. W. Jepsen, Phys. Rev. 11, 1460 (1975) (LEED)
[s]R. Honke, P.Pavone, and U. Schröder, Surf. Sci. 367 (1996) 75 (DFPT)

(RMSD) of the surface atoms with respect to that of bulk atoms, as illustrated by the room temperature data listed in Table 1.1. The RMSD surface enhancement can however be tempered by the surface relaxation.

For insulators, as exemplified by NaCl in Table 1.1, the calculated relaxation is small and the RMSD at the surface are only slightly larger than in the bulk. The calculated relaxations of the two ions are, however, quite different yielding an appreciable rumpling. For metals the relaxation and the RMSD are larger by up to 50% than those of the corresponding bulk atoms, and the normal components are generally larger than the parallel ones. In face-centered cubic (fcc) metals this is well verified for the close packed (111) surface, whereas for the less densely packed (001) surface the parallel components are comparable to the normal ones suggesting that the in-plane forces between surface atoms are significantly softened compared to the bulk. In the fcc (110) metal surfaces the RMSD $[1\bar{1}0]$ component parallel to the rows is comparable to the parallel component of the other surfaces, whereas the [001] component perpendicular to the rows is 20–30% larger. These differences can in turn be associated with the surface electronic charge redistribution originating from surface-induced hybridization and the appearance of surface electronic states.

In metals, therefore, the interionic forces essentially depend on the interposed free electron charge, and charge redistribution can make these forces quite different in the surface region. As we shall see in this monograph the frequencies and dispersion relations of surface phonons are equally affected by charge redistribution and surface relaxation, and therefore carry important information on the fundamental interactions at crystal surfaces.

1.2.2 Scanning Tunnel Microscopy Probes of Surface Vibrations

Any method which probes the atomic structure of a solid surface provides, at least indirectly, information on the dynamical processes therein. For example, scanning tunneling microscopies such as STM and AFM can also provide information on the frequencies of phonons or adsorbate vibrations. The invention of STM has revived the technique of inelastic electron tunneling spectroscopy (IETS) [108, 109], first applied to vibrational spectroscopy in metal-oxide-metal tunnel junctions by Jaklevich and Lambe in 1966 [110] and then in Josephson junctions [111–113]. The use of a scanning tunnel microscope for IETS experiments with an atomic spatial resolution, also known as scanning tunnel spectroscopy (STM-IETS or simply STS), was first attempted by Smith, Binnig and Quate on the graphite surface [114]. In 1997 the theoretical work by Bo Persson and Alexis Baratoff [115] fostered STS studies of localized surface vibrations. This was achieved experimentally about ten years later in 1998 by Wilson Ho and collaborators. They studied the C–H stretch mode in single C_2H_2 molecules [116] and the vibrations of CO and oxygen on Cu and Ag single crystal surfaces [117, 118].

The atomic scale space resolution achieved in these experiments demonstrated that STS, in combination with STM, allows for the spectroscopy of the vibrations of adsorbed molecules as a function of their geometrical configurations. Real space probes like STS cannot, however, provide information on momentum transfers and are therefore unable to measure the phonon dispersion curves of periodic surfaces. A notable achievement is however the STM direct-space imaging of the squared amplitude of a zone boundary surface phonon excited by the tunnel current [30]. The STS amplitudes, due to their dependence on the phonon-induced modulation of the surface charge density, were shown in a subsequent analysis [31] to provide information on the electron-phonon interaction for that specific phonon.

Information on the electron-phonon interaction for specific short wave subsurface phonons of metal surfaces can now be obtained with inelastic laser photoemission spectroscopy (ILPES), thanks to the great improvement in resolution achieved in recent years in low-energy photoemission studies [119–121].

1.2.3 Scattering Probes of Surface Vibrations

The same scattering probes (electrons, atoms, neutrons, X-rays) currently used for diffraction experiments to determine the atomic structure of periodic surfaces can also be used for energy and momentum-resolved measurements of the dispersion curves of surface phonons. The physical argument is that a surface phonon can be regarded as a moving diffraction grating, and the associated Doppler shift and change in the scattering deflection, respectively, correspond to the frequency and parallel momentum of the phonon, either created or annihilated in the scattering process. This process of *inelastic diffraction* is today still the only method which provides the full information on the dispersion curves of surface phonons.

Figure 1.10 compares the energy-wavevector relationship for various scattering probes which can be used to investigate surface phonons by inelastic scattering experiments and the energy-wavevector domain of surface phonons. Monochromatic beams of light atoms (He isotopes, Ne), molecules (H_2), electrons and, more recently, grazing incidence X rays, have energies and wavevectors in the range of the complete phonon dispersion curves over the entire Brillouin zone. They also have the required high resolution down to the lowest energy losses which are usually of interest. Light atoms and molecules have kinematic conditions close to those of neutrons (Fig. 1.10), which have proven to be ideal for investigating bulk phonons. The kinematic relationships of an incident ^{4}He monochromatic beam linking the de Broglie wavelength λ_i (expressed in Å) and the corresponding wavevector k_i to its kinetic energy E_i (expressed in meV) are given by

$$\lambda_i[\text{Å}] = \frac{2\pi}{k_i\left[\text{Å}^{-1}\right]} = \frac{4.542}{\sqrt{E_i[\text{meV}]}}, \qquad (1.1a)$$

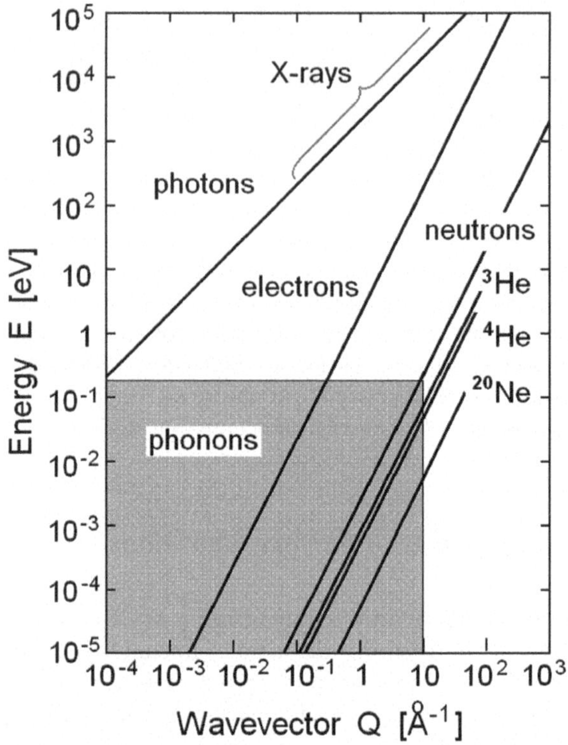

Fig. 1.10 Energy–wavevector relationship of various scattering probe particles used for studying phonons. Neutrons are ideal for measuring the dispersion curves of bulk phonons but their large penetration length makes them virtually insensitive to surface phonons. All the other probes, except neutrons, can be used to probe surface phonons in the dispersive region ($Q > 0.1$ Å^{-1}). The method of grazing incidence X-ray spectroscopy can detect surface phonons, requiring however a very large energy resolution. Low-energy electrons, which penetrate into the solid only a few atomic layers, and helium atoms, which do not penetrate at all, are the best and most used probes of surface phonons in inelastic reflection experiments. ^{3}He spin echo scattering spectroscopy, with a spectacular resolution energy, is ideally suited for the study of slow dynamical processes like ultra-soft surface phonons and diffusion. More massive probes like neon atoms are rarely used due to their intrinsic limitations in momentum and energy resolution and the greater probability of exciting more than a single phonon

$$k_i\left[\text{Å}^{-1}\right] = 1.383\sqrt{E_i[\text{meV}]} \tag{1.1b}$$

At the typical incident energies of He atoms (≈20 meV) the wavelength of a He atom is 1 Å, only slightly smaller than the interatomic distances a at the crystal surface or, more precisely, than the shortest phonon wavelengths $\lambda = 2\pi/a$.

Compared to electrons and X-rays, atoms and molecules have the big advantage that they do not penetrate the surface at all since at incident thermal energies they are reflected at distances of 3–4 Å from the top layer of atoms. Moreover, because of their low thermal collision energies (≤ 60 meV) they are entirely non-disturbing and non-destructive. The light particles He, H_2 and D_2 have the additional advantage that at thermal energies they are repelled by strongly repulsive Pauli forces and thereby exciting mostly single phonons at the surface.

Of the two helium isotopes, the boson ^{4}He allows for highly-monochromatic beams, a necessary condition for high resolution helium atom scattering (HAS) spectroscopy (see Chap. 9). The nuclear magnetic moment of the fermionic ^{3}He is instead exploited in spin echo scattering spectroscopy with a spectacular resolution well in the nano-eV range, which makes it suited for the study of slow surface dynamical processes (see Chap. 13). At energies of 20 meV the phonon energy loss of the He atoms is a substantial fraction of the incident translational energy making the energy loss relatively easy to resolve. Furthermore HAS can be applied to metals as well as semiconductors and insulators because it is not encumbered by surface charging, and can cover the entire range of phonon energies typical of most clean surfaces extending from less than $\hbar\omega \cong 1$ meV up to 30–40 meV. Thanks to its greater energy resolution compared to the other light atoms and molecules, HAS has been largely used to study surface phonons and absorbate vibrations, particularly in the acoustic region, and to detect more subtle effects like Kohn anomalies and soft modes. There is however no basic limitation in performing inelastic HAS with higher incident energies, the main problem being the loss of resolution due to the rise of the multiphonon background.

Other atomic scattering probes with masses close to the neutron mass are hydrogen and deuterium atoms (not shown in Fig. 1.10), which however are much too reactive for surface phonon spectroscopy. The corresponding molecules H_2 and D_2 do provide the intriguing possibilities of undergoing inelastic scattering processes involving a combination of translational and rotational degrees of freedom (see Chap. 13). Heavier rare gas atoms like Ne have been occasionally used in the early days of surface phonon spectroscopy with limited results, due to their intrinsic limitations in momentum and energy resolution, and to the increased role of many-phonon processes.

Electron energy loss spectroscopy (EELS), on the other hand, is able to measure frequencies well above the highest phonon frequencies and is therefore more suitable for optical phonons and adsorbate vibrations. Electrons penetrate no more than a few atomic layers, so that they are in all respects a surface probe. Since electrons are charged particles, EELS is also suitable for detecting surface electromagnetic Fuchs and Kliever (FK) modes in ionic crystals. In order to transfer sufficient momentum to phonons so as to explore the entire Brillouin zone, the energy of the incident electrons needs to be about 100 eV, or three to five orders of magnitude larger than the phonon energy, which explains the limited resolution of EELS in the lower part of the phonon spectrum. Further improvements in EELS have reduced the energy resolution gap [122–124] and EELS can now achieve a resolution better than about 0.65 meV in the impact regime. The competition between ^{4}He atom spectroscopy and EELS, which were both applied to metal surfaces for the first time in 1983, stimulated their further technical development. At the same time the complementarity of the information gained from the two techniques has greatly contributed to our present understanding of surface dynamics. Recently the energetic resolution of scanning transmission electron microscopes (STEM) has been improved to about 10 meV and spatially resolved results have been reported on a number of amorphous systems [125].

Inelastic X-ray scattering (IXS) spectroscopy at grazing incidence can also be surface sensitive and can cover the whole surface phonon wavevector domain in the X-ray spectral region, requiring however, as appears from Fig. 1.10, a very high energy resolution. Such challenging conditions have only recently been met with synchrotron radiation and inelastic X-ray scattering and they are now entering the club of surface phonon spectroscopies [126–131]. An interesting feature of IXS spectroscopy is that at grazing incidence the total reflection conditions may be met for incident energies near the absorption edges, and the penetration of the probing beam can be tuned by changing the kinematic conditions.

Neutron inelastic scattering (NIS), especially in its more recent spin echo version, has become a powerful tool for studying vibrations and dynamical processes in general at the molecular level in low-dimensional soft bulk matter [132]. Incoherent NIS has been used very early to measure the surface-excess phonon density in powders of nanocrystalline MgO [133] and TiN [134] and even of adsorbed H on Raney-Ni catalyst [135]. *Inelastic diffraction* has only been once directly demonstrated in a neutron scattering experiment from coherent SAWs [136–139]. However to our knowledge no NIS measurement of a surface phonon dispersion curve has been reported yet. The reason is, of course, related to the weak interaction of neutrons with the electronic shell of atoms making them ideal for exploring bulk dynamical processes, but totally insensitive to the surface.

Raman and Brillouin scattering, based on UV or visible photon probes, have a far better resolution in the meV scale, but can only probe phonons at very small wavevectors Q (Fig. 1.10). However with these methods portions of the phonon dispersion curves can be measured on surfaces decorated with a periodic grating which provides the reciprocal lattice vectors necessary to probe phonons of finite wavevectors [140–143].

1.2.4 Helium Atom Scattering (HAS) Versus Electron Energy Loss Spectroscopy (EELS)

Although at their best operating conditions the different scattering probes are expected to give the same information on the dispersion curves, the respective inelastic scattering intensities reflect the strength and the rather different mechanisms by which the probing particles interact with the phonon atomic displacements. Due to the strong Pauli repulsion between the electrons of the incident atoms and molecules and the electrons of the solid at these relatively low energies, the atoms can approach the surface at no less than a distance of about 3 Å from the first atomic layer [144–146] corresponding to the classical turning point. In the case of insulator surfaces incident atoms essentially collide only with the outermost tail of the valence electrons of the individual atoms and therefore the inelastic scattering occurs through a direct collision with the surface atoms. Thus in insulators He atoms can only excite surface phonons which have an appreciable amplitude at the surface plane.

For metal surfaces the interaction mechanism is rather different. The tail of the surface free electron charge density probed by He atoms 3 Å from the first atomic layer (Fig. 1.11a) is smeared out over several surface atoms (*Smoluchowski smearing* [147]) and therefore cannot be associated with an individual surface atom. Moreover the electron density deformation induced by a surface phonon cannot be simply associated with the motion of a nearby ion core. Actually the dynamical deformation of the surface charge density depends on the strength and the range of the electron-phonon interaction, so that even the motion of ions which are fairly distant from the surface plane can be perceived by the scattering He atom (*quantum sonar* effect), especially at surfaces with a large electron-phonon interaction such as, e.g. 2D superconductors. In this respect He atoms are, for metals, not a strictly

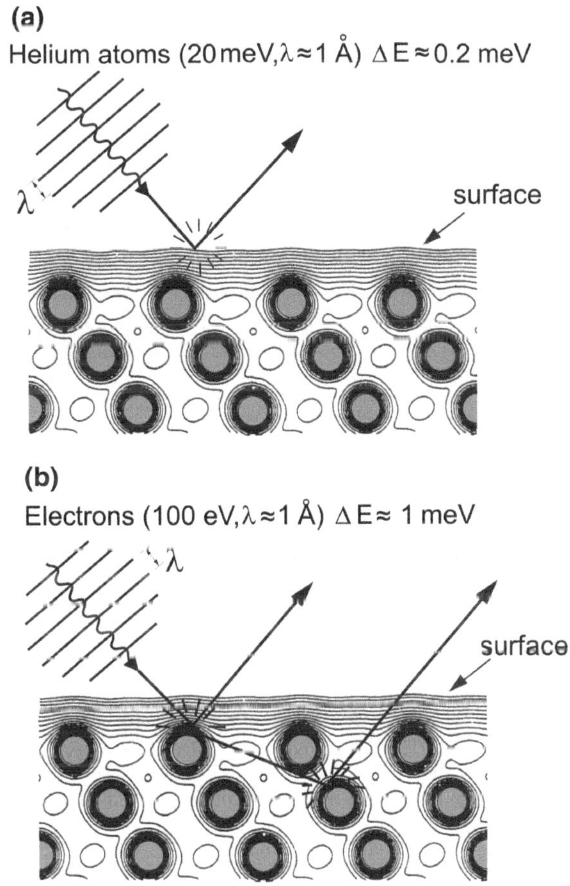

Fig. 1.11 Comparison of the interaction and penetration of helium atoms and electrons scattered from a crystal surface at energies corresponding to a de Broglie wavelength of 1 Å. **a** Helium atoms bounce off the steep repulsive potential which they experience about 3 Å above the surface layer of atoms. **b** Electrons penetrate a few atomic layers into the solid. The contours show lines of equal electron density for the (111) surface of an fcc metal cut along a (110) plane

surface-sensitive probe but can convey information on the dynamics of sub-surface ions through the global surface charge density response to their motions.

On the contrary EELS electrons in the impact regime, as well as neutrons and X-rays, are scattered mostly by the ion cores (Fig. 1.11b), and only by penetration can they interact with sub-surface ions. It is instructive to compare the amplitudes of the accompanying charge-density oscillations seen on metal surfaces by HAS with the vibrational amplitudes of the ion cores seen by EELS. From the comparison valuable information on mode-selected electron-phonon interactions has been obtained and much more can be expected in the future. Whereas the theory for EELS in the impact regime goes back to a long intense development related to understanding low energy electron diffraction (LEED), only recently a full quantum mechanical theory of inelastic HAS interactions has been developed in the framework of density functional perturbation theory (Chaps. 7 and 8).

In addition to the complicated interaction, the theory of inelastic scattering of atoms from crystal surfaces meets with further complications with respect to the theory of inelastic neutron scattering from bulk phonons. First of all the surface phonon spectrum includes, besides a discrete set of surface localized and resonant modes, also the continuum of the bulk phonon bands projected onto the surface. The second complication arises from the surface attractive potential which generally hosts a number of bound states (image states for the electrons). Because of this, specular and diffractive elastic scattering, as well as inelastic scattering from a corrugated surface may occur through two competing channels: the direct scattering and the scattering through surface bound states (see Chap. 10). The interference between the two channels may lead to a resonant enhancement (or depression) of the scattered beam intensity under special kinematic conditions. A third theoretical complication is the coupling strength of atoms, which is substantially larger than for neutrons and allows for multiphonon processes. Thus a complete analysis of inelastic intensities requires an accurate theoretical treatment of surface dynamics (Chaps. 3–5) combined with an accurate description of the surface potential (Chap. 6) and an appropriate scattering theory (Chaps. 7 and 8).

1.2.5 Single Versus Multiphonon Regimes

An important prerequisite for surface phonon spectroscopy using atom scattering is the prevalence of single phonon interactions as opposed to multiphonon scattering. In the latter case kinematics becomes rather complicated and the information on the energy loss and the associated wavevector is largely lost. The scattering theory of single phonon scattering requires a rigorous quantum treatment, whereas under conditions of multiphonon scattering a semi-classical approach may be more expedient. Depending on the nature of the interaction potential, of the kinematic conditions, and the surface temperature scattering events involving a large number of phonons may be better understood in the classical limit.

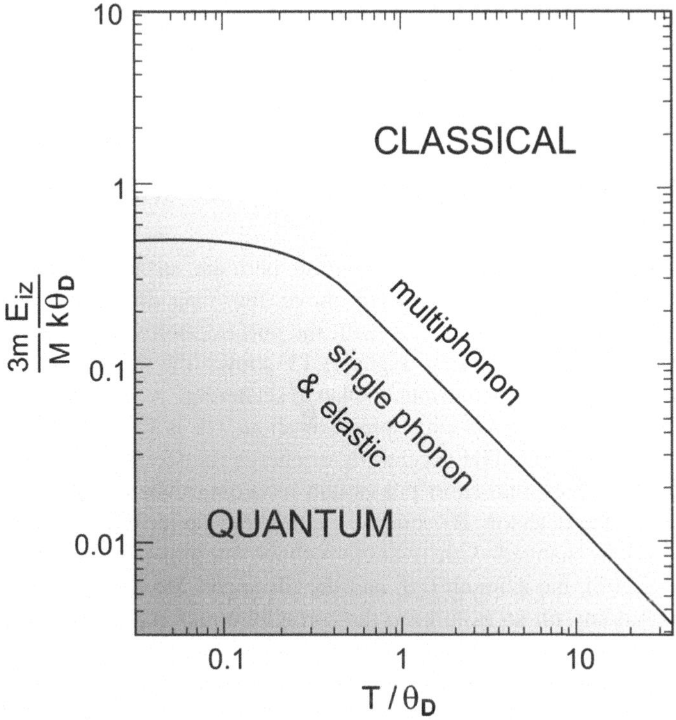

Fig. 1.12 The domains of quantum and classical regimes for scattering of particles of mass m and energy E_{iz} (normal to the surface) from a surface consisting of particles of mass M at temperature T. Both T and the mass weighted energy E_{iz} are expressed in units of the average surface phonon energy $k\theta_D$, where θ_D is the crystal Debye temperature. The solid line curve which shows approximately the border between the two regions corresponds to a Debye-Waller exponent of $W = ½$ (see Sect. 7.6). Only elastic or single phonon processes occur below the curve, whereas above multiphonon processes dominate

The domains of the quantum and classical scattering regimes delineated in Fig. 1.12 depend on the surface temperature and on a mass weighted incident particle energy, $(3\ m/M)E_{iz,}$ where E_{iz} is the component of the incident energy normal to the surface, m is the incident particle mass and M is the mass of a surface atom. By expressing both quantities in units of an average surface phonon energy $k\theta_D$, where θ_D is the crystal Debye temperature, the diagram Fig. 1.12 is universally applicable. The solid line curve separating the two regions corresponds to a Debye-Waller exponent $W = ½$ (see Chap. 7), i.e., to a probability $e^{-W} W \cong 0.3$ of creating one single phonon. The weakly inelastic quantum regime of small E_{iz} and T where only elastic or single phonon processes occur lies below the solid line curve. In the inelastic classical regime of large E_{iz} and T above the solid line curve a large number of phonons is involved. While for normal incidence E_{iz} coincides with the total energy of the incident particle, for grazing incidence E_{iz} may become sufficiently small to fulfill the conditions for quantum single phonon scattering at higher temperatures. In the neighborhood of the boundary line single- and multi-phonon processes may be analyzed in the framework of *semiclassical* theories.

1.2.6 Kinematics of Single Phonon Surface Scattering

The kinematics of scattering from a crystalline periodic surface is conceptually the same for the various probes considered above, the main differences being in the nature and mechanism of interaction with the surface atoms, and therefore in the choice of kinematic parameters. Figure 1.13 shows the basic idea of a HAS experiment in the simplest version of planar scattering. A nearly monoenergetic beam of He atoms with a velocity spread of about 1% is scattered from a crystal surface. Its time-of-flight (TOF) spectrum (or energy transfer spectrum) is measured by chopping the beam into short pulses and recording their time of arrival at the mass spectrometer detector. By simply rotating the target the orientation of the incident beam is changed with respect to the scattering plane as given by the incident angle (θ_i), the azimuth (φ), and the tilt angle. Most experiments are performed without any tilt so as to keep the surface normal in the scattering plane of the incident and outgoing beams (*planar scattering*). Moreover in many HAS apparatus the final angle is complementary to the incident angle, $\theta_i + \theta_f = 90°$ (*90° scattering geometry*). The surface phonon dynamics of both clean and adsorbate-covered surfaces can be studied in this way. The same apparatus also allows for detailed information on the periodic geometry of the surface by measuring the diffraction patterns with a high resolution. Moreover useful information on the surface defects is extracted from the weak diffuse incoherent elastic scattering which appears between the diffraction peaks. In addition the formation of regular layers during the deposition of adsorbates can be monitored through the oscillations of the specular and/or diffraction peak intensities. Thus a rather complete and consistent characterization of both the structural and dynamical properties of a clean or adsorbate covered solid can be achieved with HAS experiments.

In the case of single phonon scattering the energy transfer and the scattering angles are related to the phonon energy $\hbar\omega(\mathbf{Q})$ and the wavevector $\mathbf{Q}$ by the conservation of total energy and parallel momentum in the plane of the surface

$$\Delta E + \hbar\omega(\mathbf{Q}) = 0 \quad \text{or} \quad \hbar\omega(\mathbf{Q}) = \frac{\hbar^2}{2m}\left(k_i^2 - k_f^2\right), \tag{1.2}$$

$$\Delta\mathbf{K} + \mathbf{Q} = -\,\mathbf{G} \quad \text{or} \quad \mathbf{Q} = \mathbf{K}_i - \mathbf{K}_f - \mathbf{G}, \tag{1.3}$$

where $\Delta E = E_f - E_i$ is the energy transferred to the atom and $\mathbf{K_i}$ and $\mathbf{K_f}$ are projections of the initial $\mathbf{k_i}$ and final $\mathbf{k_f}$ wavevectors on the surface and $\mathbf{G}$ is a reciprocal lattice vector of the surface lattice. Here and elsewhere capital letters will be used to denote vectors in the surface plane.

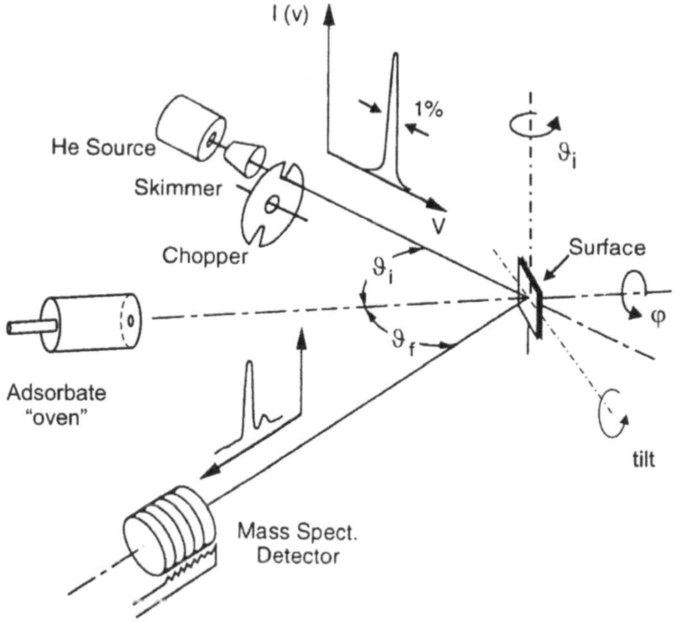

Fig. 1.13 Schematic view of an apparatus for surface phonon spectroscopy by means of helium atom scattering (HAS) time-of-flight (TOF) measurements. A chopper pulses a nearly monoenergetic beam of He atoms (velocity spread $\sim 1\%$) before the atoms are scattered from a crystal surface and their TOF spectrum is measured by a mass spectrometer detector. The incident (θ_i), azimuth (φ), and tilt angles can be accessed by appropriate rotations of the target. In many HAS apparatus the final angle is complementary to the incident angle, $\theta_i + \theta_f = 90°$

Figure 1.14 illustrates the above kinematical conditions as well as the time-of-flight spectra for the three different cases encountered in a typical TOF experiment: creation of a single phonon, specular elastic scattering and annihilation of a single phonon. The diagrams are for the frequently used experimental arrangement, in which the angle between incident and final beams is fixed at 90°. The annihilation and creation of single phonons are identified by additional peaks on the energy gain and energy loss side of the elastic (diffuse) peak of the TOF spectrum. Inelastic scattering from flat smooth surfaces as, e.g., in metals, only occurs for $\mathbf{G} = 0$, whereas scattering processes with both $\mathbf{G} = 0$ and $\mathbf{G} \neq \mathbf{0}$ (*umklapp* processes) are possible with comparable intensities for corrugated surfaces as in insulators and semiconductors.

Four different inelastic processes are encompassed by (1.2) and (1.3):

(a) $\hbar\omega(\mathbf{Q}) > 0$ with $\mathbf{Q}$ in the direction of $\mathbf{K}_i$ corresponds to the *creation* of a phonon travelling in the *forward* direction;
(b) $\hbar\omega(\mathbf{Q}) > 0$ with $\mathbf{Q}$ in the direction of $-\mathbf{K}_i$ corresponds to *creation of* a phonon travelling in the *backward* direction;

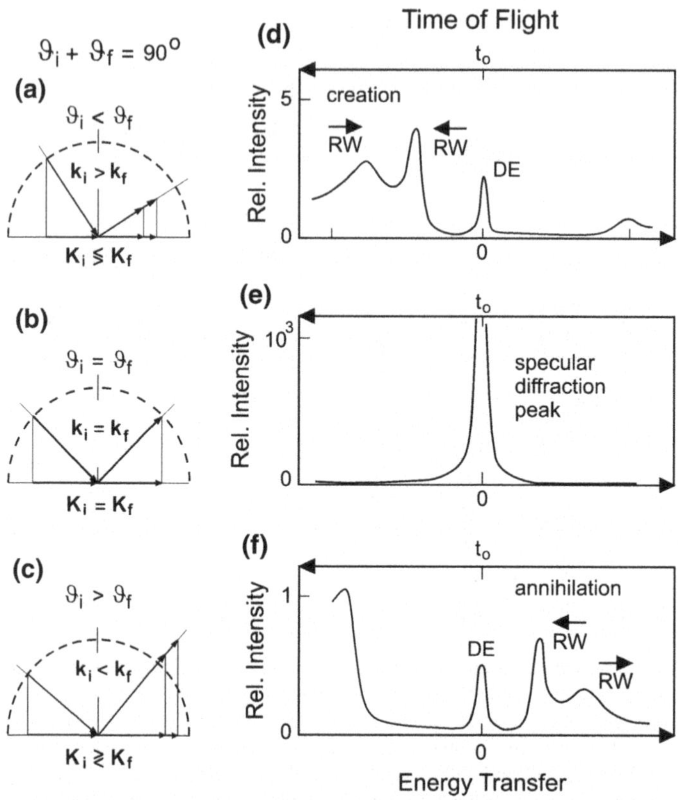

Fig. 1.14 Schematic diagram showing the different kinematic conditions (on the left) for in-plane scattering and the corresponding typical time-of-flight spectra (on the right) for incident angles smaller (**a**, **d**), equal (**b**, **e**) or larger (**c**, **f**) than 45°. The most probable inelastic processes occurring for $\vartheta_i < 45°$ (**a**) or $\vartheta_i > 45°$ (**c**), at small energy transfers and not too large incident energy, involve the creation or annihilation of single Rayleigh phonons (RW), respectively. These processes yield inelastic peaks on the creation (**d**) or annihilation (f) side of the spectrum. At the specular condition $\vartheta_i = \vartheta_f = 45°$ (**b**) only a very intense elastic peak corresponding to pure reflection, about three orders of magnitude larger than the inelastic peaks, is observed (**e**). The parallel wavevector change $\mathbf{K}_f - \mathbf{K}_i$ is opposite to the phonon wavevector $\mathbf{Q}$ for $\mathbf{G} = 0$ (no umklapp) so that both the creation (**a, d**) and annihilation (**c, f**) processes shown in this figure involve two RW phonons moving in opposite directions (The arrows above RW denote forward → or backward ← directed phonons). The small elastic peaks (DE) are due to diffuse incoherent inelastic scattering

(c) $\hbar\omega(\mathbf{Q}) < 0$ with $\mathbf{Q}$ in the direction to $\mathbf{K}_i$ corresponds to the *annihilation* of a phonon that was travelling in the *backward* direction;

(d) $\hbar\omega(\mathbf{Q}) < 0$ with $\mathbf{Q}$ in a direction of $-\mathbf{K}_i$ corresponds to the *annihilation* of a phonon that was travelling in the *forward* direction.

It should be noted that ΔE is always opposite to $\omega(\mathbf{Q})$ and has therefore the meaning of *energy gain* or *energy transfer to the atom*; similarly $\hbar\Delta\mathbf{K}$ has the meaning of *parallel momentum gain* or *parallel momentum transfer to the atom.* The surface reciprocal lattice vector **G** is generally chosen so as to keep the phonon wavevector **Q** within the first surface Brillouin zone, and $\hbar\mathbf{G}$ is the momentum (if $\neq 0$) that the atom transfers to the crystal center of mass.

Figure 1.15 illustrates the kinematical conditions, (1.2) and (1.3), for two examples of inelastic scattering in which a phonon of energy $\hbar\omega(\mathbf{Q})$ is created, either in the backward direction (a) with $\mathbf{G} = 0$, or in the forward direction (b) in an *umklapp* process ($\mathbf{G} \neq \mathbf{0}$).

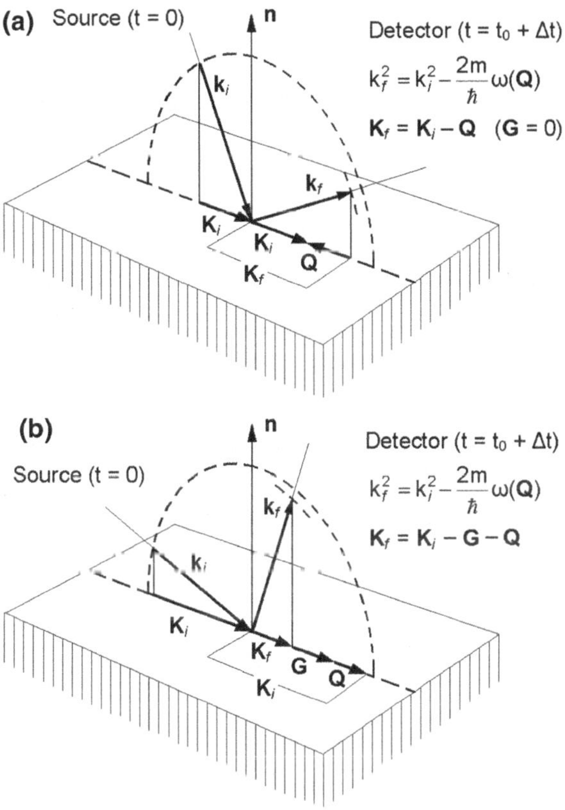

Fig. 1.15 Vector diagram illustrating the kinematics, (1.2) and (1.3), involved in a time-of-flight experiment in the planar scattering configuration (incident wavevector $\mathbf{k}_i$, final wavevector $\mathbf{k}_f$ and normal to the surface **n** in the same plane). With respect to an elastic process (time of flight t_0), inelastic processes are characterized by a longer ($\Delta t > 0$) or shorter ($\Delta t < 0$) time-of-flight corresponding to a phonon creation or annihilation process, respectively. In the example (**a**) a phonon is created with a backward directed momentum **Q** and energy $\hbar\omega(\mathbf{Q})$ in an ordinary ($\mathbf{G} = 0$) scattering process, so that the final wavevector is shorter than the incident wavevector, the difference of their squares being proportional to $\hbar\omega(\mathbf{Q})$. The example (**b**) shows a similar process where a phonon is created with a forward directed wavevector in an umklapp ($\mathbf{G} \neq \mathbf{0}$) process

From this brief introduction it appears that the measurement of the dispersion curves of surface phonons by means of inelastic He atom scattering is a conceptually simple experiment. However the achievement of the present experimental conditions, as far as concerns spectral intensity and resolution, and of the present theoretical understanding of the quantum physics governing the interaction between He atoms and surface atomic vibrations has been a long process, which started in the early days of quantum mechanics. A historical overview is presented in the next chapter.

References

1. J.W. Strutt (Baron Rayleigh), Proc. Lond. Math. Soc. **17**, 4 (1885)
2. J.W.S. Rayleigh, *The Theory of Sound* (MacMillan, London 1896; Unabridged 2nd revised edition: Dover Publications, New York, 1976)
3. G.W. Farnell, in *Physical Acoustics,* vol. 6, ed. by W.P. Mason, R.N. Thurston (Academic Press, New York, 1970), p. 109
4. B.E. Sernelius, *Surface Modes in Physics* (Wiley, Berlin, 2001)
5. P. Hess, Surface acoustic waves in materials science. Phys. Today 42 (2002)
6. R.A. Broglia, Surf. Sci. 500 (2002) 759; G.F. Bertsch, R.A. Broglia, *Oscillations in Finite Quantum Systems* (Cambridge University Press, Cambridge, 1994)
7. H. Huang, *Dynamics of Surface Waves in Coastal Waters* (Springer, Berlin Heidelberg, 2010)
8. P.M. Shearer, *Introduction to Seismology* (Cambridge University Press, Cambridge, 2009)
9. J. Harvey, Sounding out the sun. Phys. Today 32 (1995)
10. L. Gizon, P. Cally, J. Leibhacher (eds.), *Helioseismology, Asteroseismology, and MHD Connections* (Springer, Berlin Heidelberg, 2009)
11. A.L. Piro, L. Bildsten, Surface modes on bursting neutron stars. Astrophys. J. **629**, 438 (2005)
12. P. G. Richards and W-Y. Kim, Nature Physics 3, 4 (2007)
13. A.A. Oliner (ed.), *Acoustic Surface Waves* (Springer, Berlin, Heidelberg, 1978)
14. G.W. Farnell, in *Acoustic Surface Waves*, ed. by A.A. Oliner (Springer, Berlin Heidelberg, 1978), p. 13
15. H.M. Gerald, in *Acoustic Surface Waves*, ed. by A.A. Oliner (Springer, Berlin, Heidelberg, 1978), p. 61
16. E.A. Ash, in *Acoustic Surface Waves*, ed. by A.A. Oliner (Springer, Berlin, Heidelberg, 1978), p. 97
17. A.A. Maradudin, in *Nonequilibrium Phonon Dynamics*, ed. by W.E. Bron (Plenum, New York, 1985), p. 395
18. A.A. Maradudin, G.I. Stegeman, in *Surface Phonons*, vol. 27, Springer Series in Surface Sciences, ed. by F.W. deWette, W. Kress (Springer, Berlin, Heidelberg, 1991), p. 5
19. D. Shilo, E. Zolotoyabko, Phys. Rev. Lett. **91**, 115506 (2003)
20. D. Shilo, E. Zolotoyabko, J. Phys. D Appl. Phys. **36**, A 122 (2003)
21. F. Yang, D.J. Dorantes-Gonzalez, K. Chen, Z. Lu, B. Jin, Y. Li, Z. Chen, X. Hu, Sensors **12**, 122 (2012)
22. A.J. Ricco, G.C. Frye, S.J. Martin, Langmuir **5**, 273 (1989)
23. D. Schneider, E. Hensel, A. Leson, *Testing Thin Films Surfaces* (Wiley, Weinheim, 2007)

24. L.-M. Berger, D. Schneider, T. Großer, in *Proceedings of the 2007 International Thermal Spray Conference: Global Coating Solutions, Beijing, China*, ed. by B.R. Marple, M.M. Hyland, Y.-C. Lau, C.-J. Li, R.S. Lima, G. Montavon (ASM International, Materials Park, Ohio, USA, 2007)
25. O.B. Wright, K. Kawashima, Phys. Rev. Lett. **69**, 1668 (1992)
26. Y. Sugawara, O.B. Wright, O. Matsuda, M. Takigahira, Y. Tanaka, S. Tamura, V.E. Gusev, Phys. Rev. Lett. **88**, 185504 (2002)
27. D.M. Profunser, O.B. Wright, O. Matsuda, Phys. Rev. Lett. **97**, 055502 (2006)
28. D.M. Profunser, E. Muramoto, O. Matsuda, O.B. Wright, U. Land, Phys. Rev. B **80**, 014301 (2009)
29. E. Meyer, H.J. Hug, R. Bennewitz, *Scanning Probe Microscopy* (Springer, Berlin Heidelberg, 2004)
30. H. Gawronski, M. Mehlhorn, K. Morgenstern, Science **319**, 930 (2008)
31. G. Benedek, M. Bernasconi, V. Chis, E. Chulkov, P.M. Echenique, B. Hellsing, J. P. Toennies, J. Phys.: Cond. Matter **22**, 084020 (2010)
32. C.K. Campbell, *Surface Acoustic Wave Devices for Mobile and Wireless Communications* (Academic, New York, 1998)
33. K.-Y. Hashimoto, *Surface Acoustic Wave Devices in Telecommunications Modeling and Simulation* (Springer, Berlin, 2000)
34. C.C.W. Ruppel, T.A. Fjeldly (eds.), *Advances in Surface Acoustic Wave Technology, Systems and Applications,* vol. 1 (World Scientific, Singapore, 2000); vol. 2 (World Scientific, Singapore, 2001)
35. D. Morgan, *Surface Acoustic Wave Filters*, 2nd edn. (Academic Press, London, 2007)
36. X. Lu, J. Galipeau, K. Mouthaan, E. Henry Briot, B. Abbott, in *Power Amplifiers for Wireless and Radio Applications (PAWR)* (IEEE, 2013), p. 82
37. L. Eichinger, S. Sischka, G. Olbrich, R. Weigel, in *Proceedings IEEE Ultrasonics Symposium* (IEEE, Piscataway, NJ, 2000) p. 29
38. L. Reindl, C.C.W. Ruppel, S. Berek, U. Knauer, M. Vossiek, P. Heide, L. Oreans, IEEE Trans. Microw. Theory Tech. **49**, 787 (2001)
39. S. Scheiblhofer, S. Schuster, A. Stelzer, IEEE Trans. Microw. Theory Tech. **54**, 1477 (2006)
40. See, e.g., the papers in the Special issue on acoustic wave sensors and applications. IEEE Transact. Ultrason. Ferroelectr. Freq. Control **51**, 1365–1469 (2004)
41. K. Länge, B.E. Rapp, M. Rapp, Anal. & Bioanal. Chem. **391**, 15099 (2008)
42. J. Filipiak, L. Solarz, G. Steczko, Sensors **11**, 11809 (2011)
43. T. Venkatesan, H.M. Pandya, J. Environ. Nanotechnol. **2**, 81 (2013)
44. A. Pohl, I.E.E.E. Transact, Ultrason. Ferroelectr. Freq. Control **47**, 317 (2000)
45. L. Reindl, A. Pohl, G. Scholl, R. Weigel, IEEE Sens. J. **1**, 69 (2001)
46. D.D. Stubbs, S. Lee, W.D. Hunt, IEEE Sens. J. **2**, 394 (2002)
47. S. Jacesko, J.K. Abraham, T. Ji, V.K. Varadan, M. Cole, J.W. Gardner, Smart Mater. Struct. **14**, 1010 (2005)
48. M.-I. Rocha-Gaso, C. March-Iborra, Á. Montoya-Baides, A. Arnau-Vives, Sensors **9**, 5740 (2009)
49. K. Länge, F.J. Gruhl, M. Rapp, Meth. Molec. Biol. **949**, 491–505 (2012)
50. S. Hohmann, S. Kögel, Y. Brunner, B. Schmieg, C. Ewald, F. Kirschhöfer, G. Brenner-Weiß, K. Länge, Sensors **15**, 11873–11888 (2015)
51. M.I. Rocha Gaso, Y. Jiménez, L. A. Francis, A. Arnau, in *State of the Art in Biosensors—General Aspects*, ed. by T. Rinken (InTech, Rijeka, 2013), Chap. 11
52. T. Reusch, F.J.R. Schuelein, J.D. Nicolas, M. Osterhoff, A. Beerlink, H.J. Krenner, M. Müller, A. Wixforth, T. Salditt, Phys. Rev. Lett. **13**, 118102 (2014)
53. T. Wang, R. Green, R.R. Nair, M. Howell, S. Mohapatra, R. Guldiken, S.S. Mohapatra, Sensors **15**, 32045 (2015)
54. R.S. Westafer, G. Levitin, D.W. Hess, M.H. Bergin, W.D. Hunt, Sensors and Actuators B: Chemical **192**, 406 (2014)

55. G. De Simoni, G. Signore, M. Agostini, F. Beltram, V. Piazza, Biosensors & Bioelectr. **68**, 570 (2015)
56. M.M. de Lima Jr., M. Beck, R. Hey, P.V. Santos, Appl. Phys. Lett. **89**, 121104 (2006)
57. J.D. Joannopoulos, R.D. Meade, J.N. Winn, *Photonic Crystals* (Princeton University Press, Princeton, 1995)
58. K. Inoue, K. Ohtaka, *Photonic Crystals: Physics, Fabrication, and Applications* (Springer, Berlin, 2004)
59. V. Laude, M. Wilm, S. Benchabane, A. Khelif, Phys. Rev. E **71**, 036607 (2005)
60. S. Benchabane, A. Khelif, J.-Y. Rauch, L. Robert, V. Laude, Phys. Rev. E **73**, 065601 (2006)
61. R. Susstrunk, S.D. Huber, Science **349**, 47 (2015)
62. P. Wang, P.L. Lu, K. Bertoldi, Phys. Rev. Lett. **115**, 104302 (2015)
63. V.I. Talyanskii, J.M. Shilton, M. Pepper, C.G. Smith, C.J.B. Ford, E.H. Linfield, D.A. Ritchie, G.A.C. Jones, Phys. Rev. B **56**, 15180 (1997)
64. R.B. Dunford, M.R. Gates, C.J. Mellor, V.W. Rampton, J.S. Chauhan, J.R. Middelton, M. Henini, Physica E **12**, 462 (2002); Physica B **219**, 316–317 (2002)
65. H. Totland, Ø.L. Bø, Y.M. Galperin, Phys. Rev. B **56**, 15299 (1997)
66. J.M. Shilton, V.I. Talyanskii, M. Pepper, D.A. Ritchie, J.E.F. Frost, C.J.B. Ford, C.G. Smith, G.A.C. Jones, J. Phys.: Cond. Matter **8**, L531 (1996)
67. J. Cunningham, V.I. Talyanskii, J.M. Shilton, M. Pepper, A. Kristensen, P.E. Lindelof, Phys. Rev. B **62**, 1564 (2000)
68. M. Kataoka, C.H.W. Barnes, H.E. Beere, D.A. Ritchie, M. Pepper, Phys. Rev. B **74**, 085302 (2006)
69. C. Rocke, S. Zimmermann, A. Wixworth, J.P. Kotthaus, G. Bohm, G. Weinman, Phys. Rev. Lett. **78**, 4009 (1997)
70. F. Alsina, P.V. Santos, H.-P. Schönherr, R. Nötzel, K.H. Ploog, Phys. E **21**, 430 (2004)
71. P. Kleinert, A. Garcìa-Cristòbal, P.V. Santos, Solid State Commun. **134**, 535 (2005)
72. M. Cecchini, G. De Simoni, V. Piazza, F. Beltram, H.E. Beere, D.A. Ritchie, Appl. Phys. Lett. **85**, 3020 (2004)
73. M. Cecchini, V. Piazza, F. Beltram, D.G. Gevaux, M.B. Ward, A.J. Shields, H.E. Beere, D. A. Ritchie, Appl. Phys. Lett. **86**, 241107 (2005)
74. M. Cecchini, V. Piazza, F. Beltram, M. Ward, A. Shields, H. Beere, D. Ritchie, Mater. Res. Soc. Sump. Proc. **869**, D4.3.1 (2005)
75. M. Cecchini, G. De Simoni, V. Piazza, F. Beltram, H.E. Beere, D.A. Ritchie, Appl. Phys. Lett. **88**, 212101 (2006)
76. J.R. Gell, M.B. Ward, P. Atkinson, S.P. Bremmer, D. Anderson, C.E. Norman, M. Kataoka, C.H.W. Barnes, G.A.C. Jones, A.J. Shields, D.A. Ritchie, Phys. E **40**, 1775 (2008)
77. G. De Simoni, V. Piazza, L. Sorba, G. Biasiol, F. Beltram, Appl. Phys. Lett. **94**, 121103 (2009)
78. S. Roddaro, E. Strambini, L. Romeo, V. Piazza, K. Nilsson, L. Samuelson, F. Beltram, Semicond. Sci. Technol. **25**, 024013 (2010)
79. A.B. Hutchinson, V.I. Talyanskii, M. Pepper, G. Gumbs, G.R. Aizin, D.A. Ritchie, E.H. Linfield, Phys. Rev. B **62**, 6948 (2000)
80. R. Rodriquez, D.K.L. Oi, M. Kataoka, C.H.W. Barnes, T. Ohshima, A.K. Ekert, Phys. Rev. B **72**, 085329 (2005)
81. J. Ebbecke, N.E. Fletcher, T.B.M. Janssen, F.J. Ahlers, M. Pepper, H.E. Beere, D.A. Ritchie, Appl. Phys. Lett. **84**, 4319 (2004)
82. M. Kataoka, R.J. Schneble, A.L. Thorn, C.H.W. Barnes, C.J.B. Ford, D. Anderson, G.A.C. Jones, I. Farrer, D.A. Ritchie, M. Pepper, Phys. Rev. Lett. **98**, 046801 (2007)
83. M. Rokni, Y. Levinson, Phys. B: Cond. Mat. **263**, 180 (1999)
84. R.L. Willet, R.R. Ruel, K.W. West, L.N. Pfeiffer, Phys. Rev. Lett. **71**, 3846 (1993)
85. M. Rotter, A. Wixforth, W. Ruile, D. Bernklau, H. Riechert, Appl. Phys. Lett. **73**, 2128 (1998)

86. M. Rotter, A.V. Kalameitsev, A.O. Govorov, W. Ruile, A. Wixforth, Phys. Rev. Lett. **82**, 2171 (1999)
87. M. Rotter, A. Wixforth, A.O. Govorov, W. Ruile, D. Bernklau, H. Riechert, Appl. Phys. Lett. **75**, 965 (1999)
88. A.V. Kalameitsev, A.O. Govorov, H.-J. Kutschera, A. Wixforth, JETP Lett. **72**, 190 (2000)
89. A. Bertoni, P. Bordone, R. Brunetti, C. Jacoboni, S. Reggiani, Phys. Rev. Lett. **84**, 5912 (2000)
90. C.H.W. Barnes, J.M. Shilton, A.M. Robinson, Phys. Rev. B **62**, 8410 (2000)
91. C.H.W. Barnes, Philos. T. Roy. Soc. A **361**, 1487–1492 (2003)
92. S. Furuta, C.H.W. Barnes, C.J.L. Doran, Phys. Rev. B **70**, 205320 (2004)
93. V. Piazza, G. De Simoni, E. Strambini, M. Cecchini, F. Beltram, *SAW-Driven Opto-Electronic Devices* (NEST Scientific Report, Pisa, 2005–2006), p. 41
94. G. Benenti, G. Casati, G. Strini, *Principles of Quantum Computing and Information, vol. 1: Basic Concepts, vol. 2: Basic Tools and Special Topics* (World Scientific, Singapore, 2004/2007)
95. S. Hermelin, S. Takada, M. Yamamoto, S. Tarucha, A.D. Wieck, L. Saminadayar, C. Bauerle, T. Meunier, Nature **477**, 435 (2011)
96. T.-D. Luong, N.-T. Nguyen, Micro Nanosyst. **2**, 1–9 (2010)
97. J. Friend, L.Y. Yeo, Rev. Mod. Phys. **83**, 647 (2011)
98. M. Travagliati, G. De Simoni, C.M. Lazzarini, V. Piazza, F. Beltram, M. Cecchini, Lab Chip **12**, 2621 (2012)
99. C. Taillan, N. Combe, J. Morillo, Phys. Rev. Lett. **106**, 076102 (2011)
100. G.D. Mahan, *Many-Particle Physics* (Plenum Press, New York, 1981), Sect. 6.2
101. N.-H. Ge, C.M. Wong, R.L. Lingle Jr., J.D. McNeill, K.J. Gaffney, C.B. Harris, Science **279**, 202 (1998)
102. U. Höfer, Science **279**, 190 (1998)
103. A. Zangwill, *Physics at Surfaces* (Cambridge University Press, Cambridge, 1996)
104. H. Ibach, *Physics of Surfaces and Interfaces* (Springer, Berlin Heidelberg, 2007)
105. J.F. van der Veen, B. Pluis, A.W. Denier van der Gon, in *Chemistry and Physics of Solid Surfaces VII.*, vol. 10, Springer Series in Surface Sciences, ed. by R. Vanselow, R.F. Howe (Springer, Heidelberg, 1988), p. 455
106. J. Lapujoulade, B. Salanon, The roughening transition on surfaces, in *Phase Transitions in Surface Films 2*, ed. by H. Taub, G. Torzo, H.J. Lauter, S.C. Fain, Jr. (Plenum Press, New York 1991), p. 217
107. M. Bernasconi, E. Tosatti, Surf. Sci. Rep. **17**, 363 (1993)
108. R.M. Feenstra, J.A. Stroscio, Methods of tunneling spectroscopy, in *Scanning Tunneling Microscopy*, ed. by W.J. Kaiser, J.A. Stroscio (Academic Press Inc., San Diego, 1993)
109. H.J.W. Zandvliet, A. van Houselt, Scanning Tunneling Spectroscopy. Ann. Rev. Anal. Chem. **2**, 37–55 (2009)
110. R.C. Jaklevich, J. Lambe, Phys. Rev. Lett. **17**, 1139 (1966)
111. P.K. Hansma, Inelastic electron tunneling. Phys. Rep. **30**, 145–206 (1977)
112. T. Wolfram (ed.), *Inelastic Electron Tunneling Spectroscopy* (Springer, New York, 1978)
113. K.W. Hipps, U. Mazur, *Inelastic Electron Tunneling Spectroscopy*. Handbook of Vibrational Spectroscopy, vol. 1: Theory and Instrumentation (Wiley, Hoboken 2001)
114. D.P.E. Smith, G. Binnig, C.F. Quate, Appl. Phys. Lett. **49**, 1641 (1986)
115. B.N.J. Persson, A. Baratoff, Phys. Rev. Lett. **59**, 339 (1987)
116. B.C. Stipe, M.A. Rezhaei, W. Ho, Science **280**, 1732 (1998)
117. L.J. Lauhon, W. Ho, Phys. Rev. B **60**, R8525 (1999-II)
118. J.R. Hahn, H.J. Lee, W. Ho, Phys. Rev. Lett. **85**, 1914 (2000)
119. R. Arafune, K. Hayashi, S. Ueda, Y. Uehara, S. Ushioda, Phys. Rev. Lett. **95**, 07601 (2005)
120. R. Arafune, M.Q. Yamamoto, N. Takagi, M. Kawai, Phys. Rev. B **80**, 073407 (2009)
121. E. Minamitani, R. Arafune, M.Q. Yamamoto, N. Takagi, M. Kawai, Y. Kim, Phys. Rev. B **88**, 224301 (2013)

122. H. Ibach, D.L. Mills, *Electron Energy Loss Spectroscopy and Surface Vibrations* (Academic Press, Inc., New York, 1982)
123. H. Ibach, *Electron Energy Loss Spectrometers: The Technology of High Performance*, vol. 63, Springer Series in Optical Sciences (Springer, Berlin Heidelberg, 1991)
124. H. Ibach, *Physics of Surfaces and Interfaces* (Springer, Berlin Heidelberg, 2007), Chap. 7
125. O.L. Krivanek, T.C. Lovejoy, N. Dellby, T. Aoki, R.W. Carpenter, P. Rez, E. Soignard, J. Zhu, P.E. Batson, M.L. Lagos, R.F. Egerton, P.A. Crozier, Nature **514**, 209 (2014)
126. B.M. Murphy, H. Requardt, J. Stettner, M. Müller, M. Krisch, W. Press, Phys. Rev. Lett. **95**, 256104 (2005)
127. B.M. Murphy, M. Müller, J. Stettner, H. Requardt, J. Serrano, M. Krisch, W. Press, J. Phys.: Cond. Matter 20, 224001 (2008)
128. T. Ślęzak, J. Lazewski, S. Stankov, K. Parlinski, R. Reitinger, M. Rennhofer, R. Rüffer, B. Sepiol, M. Slezak, N. Spiridis, M. Zając, A.I. Chumakov, J. Korecki, Phys. Rev. Lett. **99**, 066103 (2007)
129. S. Stankov, R. Rohlsberger, T. Ślęzak, M. Sladecek, B. Sepiol, G. Vogl, A.I. Chumakov, R. Ruffer, N. Spiridis, J. Łażewski, K. Parlinski, J. Korecki, Phys. Rev. Lett. **99**, 185501 (2007)
130. F. Weber, S. Rosenkranz, J.-P. Castellan, R. Osborn, R. Hott, R. Heid, K.-P. Bohnen, T. Egami, A.H. Said, D. Reznik, Phys. Rev. Lett. **107**, 107403 (2011)
131. F. Weber, R. Hott, R. Heid, K.-P. Bohnen, S. Rosenkranz, J.-P. Castellan, R. Osborn, A.H. Said, B.M. Leu, D. Reznik, Phys. Rev. B **87**, 245111 (2013)
132. H. Frielinghaus, O. Holderer, F. Lipfert, M. Monkenbusch, N. Arend, D. Richter, Nucl. Instr. Meth. Phys. Res. A **686**, 71–74 (2012)
133. K.H. Rieder, E.M. Hörl, Phys. Rev. Lett. **20**, 209 (1968)
134. K.H. Rieder, W. Drexel, Phys. Rev. Lett. **34**, 148 (1975)
135. R.R. Cavanagh, R.D. Kelley, J. J. Rush, J. Chem. Phys. 77, 1540 (1982)
136. W.A. Hamilton, A.G. Klein, G.I. Opat, P.A. Timmins, Phys. Rev. Lett. **58**, 2770 (1987)
137. W.A. Hamilton, M. Yethiraj, Phys. B **241**, 1080 (1997)
138. W.A. Hamilton, M. Yethiraj, Phys. Rev. B **59**, 3388 (1999)
139. W.A. Hamilton, A. Cimmino, A. Nelson, M. Lesha, in *ACNS 2014*, Abstract B4.02
140. R. Dutcher, S. Lee, B. Hillebrands, G.J. McLaughlin, B.G. Nickel, G.I. Stegeman, Phys. Rev. Lett. **68**, 2464 (1992)
141. L. Giovannini, F. Nizzoli, A.M. Marvin, Phys. Rev. Lett. **69**, 1572 (1992)
142. S. Lee, L. Giovannini, J.R. Dutcher, F. Nizzoli, G.I. Stegeman, A.M. Marvin, Z. Wang, J.D. Ross, A. Amoddeo, L.S. Caputi, Phys. Rev. B **49**, 2273 (1994)
143. G. Benedek, J.P. Toennies, Probing the dispersion of surface phonons by light scattering, in *Epioptics 9*, ed. by A. Cricenti (World Scientific, Singapore, 2008), pp. 162–179, and references therein
144. G.G. Kleinman, U. Landman, Phys. Rev. B **8**, 5484 (1973)
145. R.B. Laughlin, Phys. Rev. B **25**, 2222 (1982)
146. E. Zaremba, W. Kohn, Phys. Rev. B **15**, 1769 (1977)
147. R. Smoluchowski, Phys. Rev. **60**, 661 (1941)

Chapter 2
History of Surface Phonons and Helium Atom Scattering

Science grows slowly and gently; reaching the truth by a variety of errors. One must prepare the introduction of a new idea through long and diligent labor; then at a given moment, it emerges as if compelled by a divine necessity.

Carl Jacobi, 1832

Abstract The chapter starts with an overview of the history of the theory of surface phonons. Next the history of helium atom and electron inelastic scattering experiments are briefly reviewed. Present day helium atom scattering experiments are described next and some important related concepts are introduced. The chapter closes with a short history of the classical and quantum theory of inelastic surface scattering.

2.1 History of Surface Phonon Theory

2.1.1 Continuum Versus Atomistic Models

In the historical evolution of our present understanding of surface phonons the theory was developed very early, much before the experiments became available. In this Subsection only an overview of the highlights in the development of various aspects of the theory of surface phonons will be touched upon. No attempt is made here to provide the details of the various approaches. The in-depth discussion of the theory of surface phonons is presented in Chaps. 3–5.

Present day theoretical research was initiated by Rayleigh's fundamental work of 1885 [1] mentioned in Sect. 1.1. Soon afterwards Rayleigh's theory was extended along two different paths. In the long wavelength limit the theory was applied to different geometries, as well as to anisotropic and piezoelectric materials [2, 3]. In the short wavelength limit the effect of the discrete lattice structure of solids was considered. In 1912 Born and von Kármán (BvK) provided the first atomic scale theory of lattice vibrations in a three-dimensional crystal and introduced harmonic force constant models [4, 5].

G. Benedek and J. P. Toennies, *Atomic Scale Dynamics at Surfaces*, Springer Series in Surface Sciences 63, https://doi.org/10.1007/978-3-662-56443-1_2

Once the atomistic point of view was adopted in studying the dynamics of solids, it was realized that at the surface there are two basically different types of modes. The modes of the first type are called *microscopic* modes because their penetration inside the crystal remains comparable to the interatomic distances even at infinite wavelengths. In contrast, the penetration depth of modes like the Rayleigh waves is proportional to the wavelength in the long wave (continuum) limit and, in this case, the waves are referred to as *macroscopic modes*. The scaling law of the macroscopic surface modes, by which the penetration length is proportional to the wavelength λ along the surface, is an obvious property of the semi-infinite homogeneous continuum, which is invariant to any change of length scale. This is illustrated in Fig. 1.2b of the previous chapter, where the RW amplitudes are plotted as functions of z/λ. Thus the displacement field shown in Fig. 1.2a holds for the 14 km RW component of the earthquake of Fig. 1.1a as well as for the 6 μm SAW of Fig. 1.4, provided the length scale is reduced by a factor 2.3×10^{10} and assuming that the elastic moduli of the two systems are of the same order of magnitude.

The theory of microscopic surface vibrational modes was first reported in 1948 by Lifshitz and Rosenzweig [6–8] who used the Green's function (GF) method [9]. They considered the free surface as a perturbation of a lattice with cyclic boundary conditions. This made it possible to derive the frequencies of surface lattice vibrations from the spectrum of bulk vibrations. In the 1960s this method was the subject of various elegant formulations and extensions by the following authors: Maradudin et al. [10, 11], Dobrzynski, Leman and Masri (phase shift method [12, 13]), Garcia-Moliner (GF matching method [14]), Armand (GF generating coefficient method, [15, 16]) and by Allen (continued fraction method [17]). During this period an appreciable number of applications to systems with short-range interactions or short range surface perturbations were reported [18–23]. The GF method has the advantage that the size of the dynamical problem is reduced from that of the whole crystal to that of the perturbation space corresponding to the surface and is, therefore, particularly suitable for materials with short range forces. In 1973 one of the authors devised an extension of the GF method for ionic crystals with long range Coulomb forces, such as alkali halides [24–26], which later proved to be efficient also for crystals with complex many-body interactions such as the refractory superconductors [27].

Another theoretical method for treating macroscopic surface lattice dynamics, conceptually similar to Lord Rayleigh's approach for continuous media, was based on the matching of trial surface-localized waves [*trial function* (TF) method] to the bulk vibrations with a number of adjustable parameters which increases with the number of interacting neighbors. An extension to long range forces was first given in 1967 by Feuchtwang [28, 29], and about ten years later applied to metal surfaces by Bortolani et al. [30]. Since then Khater, Szeftel and coworkers in 1987 combined this real space matching method with a group theoretical analysis with successful applications to superstructures, defects and disorder on metal and alloy surfaces [31–33]. A special and convenient choice of trial functions is that of Gottlieb polynomials, proposed earlier by Trullinger [34] and subsequently applied by Fasolino et al. [35] and by Black [36] to the dynamics of metal surfaces.

An important test for the early models of surface lattice dynamics was the long wave limit where the exact results of Rayleigh continuum theory should be recovered. Such a comparison was possible with the analytical solutions obtained for simple solvable monoatomic lattices by Gazis and Wallis [37, 38] in the late 1950s. By means of perturbative and GF techniques [39] Wallis obtained the exact analytical solutions for a semi-infinite diatomic linear chain and extended them to the diatomic square and simple cubic lattices. This led to the prediction, in diatomic crystals, of optical surface modes with polarizations normal to the surface, also called shear vertical (SV) optical surface modes or Wallis modes (WM) [39, 40]. The existence of optical surface modes polarized in the surface plane was first theoretically demonstrated by Lucas in 1968 [41]. In cubic crystals these modes form a degenerate pair with finite frequencies at infinite wavelength, which have been termed Lucas modes (LM).

Wallis and Lucas modes, which also exist in non-polar crystals like silicon, are examples of microscopic modes. Also optical modes with a macroscopic character can exist in polar crystals in the form of mixed phonon-photon modes known as surface- phonon-polaritons. Surface-phonon-polaritons were discovered theoretically, in the late 1960s, by Fuchs and Kliever (FK modes) [42–44]. Subsequently Jones and Fuchs discussed the non-retarded FK modes in comparison with the microscopic optical modes [45]. A theory of surface dynamics including the effects of retardation was first developed by Bryksin and Firsov in 1969 [46]. An exhaustive account of surface electromagnetic modes encompassing the wide family of surface polaritons, associated not only with phonons but also with other elementary excitations like plasmons, excitons and spin waves, is found in a book edited by Boardman [47].

2.1.2 Evolution of Present Day Surface Phonon Theories

Following the pioneering work of Clark et al. [48], in the late 1960s and early 1970s de Wette et al. [49–51] extended the Born-von Kármán theory to surface vibrations. Their method, which requires the direct diagonalization of the dynamical matrix for a sufficiently thick slab consisting of many infinitely extended two-dimensional layers of interacting atoms, has become one of the basic modern computational techniques. The advent of fast computers has made this approach, frequently called the slab (SL) method, more expedient than the earlier GF method. Indeed the advantages of the Green's function method in using small diagonalization spaces are more than compensated for by the computational difficulties connected with the complex and singular nature of Green's functions, which make the numerical extraction of surface phonon frequencies and eigenvectors a fairly involved numerical task. On the contrary, the SL method permits a straightforward determination of the dispersion curves and eigenvectors of surface phonons.

In 1969 de Wette and coworkers used this method to perform the first realistic calculations of a complete set of surface phonon dispersion curves. Their systematic studies on the low index surfaces of Lennard-Jones fcc crystals [49–54] and of the alkali halides [55–59] have served as a basic reference point for later calculations. With the SL method Alldredge discovered additional surface acoustic modes with a polarization in the surface plane, but normal to the propagation direction, which are called shear horizontal (SH) acoustic surface modes [60, 61]. The subsequent study of surface dynamics in layered structures, where the soft interlayer vibrations are weakly coupled to the stiff intralayer vibrations [62], led to an expedient version of the direct slab diagonalization known as the dispersive linear chain (DLC) method. This method is based on a theorem showing that a three-dimensional dynamical problem can be reduced to that of a linear chain normal to the surface. In this case the force constants, corresponding to the interplanar force constants of the original solid, depend on a two-dimensional wavevector **Q** normal to the chain direction [63]. The DLC method appears to be convenient whenever the **Q**-dependence of the linear-chain force constants can be approximated by an analytical form.

At the present time the SL method, in combination with the Born-von Kármán force constant models, constitutes the basic tool for the phenomenological analysis of surface phonon dispersion curves. Moreover the slab method for surface dynamics is currently used with *ab initio* quantum theoretical methods for the determination of the electronic structure and the dielectric response, although being computationally more demanding.

The implementation of first principle surface dynamics has been made possible by the progress in the direct *ab initio* calculation of the interatomic force constants based on the density functional theory (DFT), which has evolved along three main lines. The first line is based on the frozen phonon method (FPM), consisting in the direct calculation of the change in the total energy produced by suitably selected atomic displacements [64–67]. The second approach is based on the density functional perturbation theory (DFPT). Through an iterative procedure DFPT facilitates the calculation of the electron screening potential originating from the atomic displacements to the first and, if needed, higher orders of the perturbation [68–72].

With the presently available vector and massively parallel computing machines, a third method based on a unified approach involving large scale molecular dynamics (MD) and DFT simulations is currently also used for surface dynamics. In parallel with the extension of MD methods based on classical [73–79] or quantum semi-empirical forces [80–90] to systems with large numbers of atoms ($N \sim 10^2$–10^7), quantum MD has eFxperienced an impressive development in the last three decades, starting from the 1985 breakthrough by Car and Parrinello (CP) (Sect. (5.4.3) [91]. Since then Car-Parrinello molecular dynamics (CP-MD) has been largely applied, though on a smaller scale ($N \sim 10^2$) than classical MD, to many problems of surface dynamics, the first relevant steps being made in the early nineties [92–97]. Besides a series of improvements and extensions of CP-MD codes, other powerful *ab initio* MD methods similar conceptually to CP-MD have been developed in the nineties with the aim of improving the efficiency and predictability of the original CP method [98, 99]. Either classical or quantum MD methods are extending the domain of surface dynamical theory to the study of temperature-dependent phenomena driven by surface phonons, like anharmonic effects, surface reconstruction, roughening and surface melting, as well as of the dynamical phenomena occurring at complex surfaces like those of organic crystals, self-assembled overlayers, etc. The present day atomistic theory of surface phonons is the subject of Chap. 5 where an in-depth discussion of some of the topics discussed here can be found.

2.2 History of Surface Phonon Spectroscopy by Helium Atom Scattering

2.2.1 Early Developments

In the early 1970s, as a result of these theoretical developments there was considerable information available on the nature and dispersion of long wavelength surface phonons for ideal surfaces in different classes of crystals [100–106], but not a single experiment providing evidence on the dispersive wavevector dependence of surface phonons. Their elusive nature challenged experimentalists for fifteen years from about 1965 to 1980. As discussed in Sect. 1.2.3 neutron scattering, successfully employed since 1955 as the most efficient tool for the study of bulk phonons [62, 107–109], is insensitive to surfaces owing to the very small cross sections of neutrons. Attempts to measure surface phonons with neutron scattering from powders with a large specific surface area have met, however, with some success in providing information on the surface excess phonon density [110–112].

Surface sensitive probes like low-energy electrons were considered for surface phonon spectroscopy already as early as 1965 [113–115]. The suggestion to use helium atom scattering (HAS) for measuring surface phonons goes back to two theoretical studies by Cabrera et al. (1969) [116] and Manson and Celli (1971) [117]. A personal history of the developments which led to this visionary theory, which is at the root of much of the research in this monograph can be found in the 2010 autobiographical narrative of Manson [118]. Cabrera and his colleagues realized the great potential of helium atoms as a uniquely surface sensitive inelastic probe of dynamical processes occurring at surfaces. Their distorted-wave Born approximation theory provided the first convincing demonstration that the inelastic diffraction of helium atoms from single phonons can be expected to be the dominant inelastic process at thermal collision energies. As discussed in Sect. 1.2.5, for the detection of single phonons the incident energy of the probe particles and surface temperature must be properly chosen. The scattering regime fulfilling these conditions is customarily designated as quantum as opposed to the classical regime.

2.2.2 Early Helium Atom Scattering Experiments

The history of molecular beams in surface science goes back to the 1915 experiments of Wood [119, 120] and Knudsen [121, 122] following closely on the first demonstration of molecular beams in 1911 by Dunoyer [123]. Starting from 1929 Stern et al. [124–126] in Hamburg and Johnson [127, 128] in Philadelphia were the first to detect the diffraction of matter waves as predicted by the wave particle duality proposed by de Broglie in his 1924 Sorbonne dissertation [129, 130]. The book by Otto Frisch describes his personal experiences in the exciting 1930s in Stern's lab in Hamburg [131]. The early history of surface scattering and of diffraction experiments has been reviewed in the 1931 book by Fraser [132], the 1935 monograph by Fues [133] and the books by von Laue [134], Kenneth Smith [135], and in the review by Comsa [136].

Early experimental studies of energy transfer in scattering from surfaces are described in the reviews by Beder [137] and Stickney [138–140]. Then, as mentioned above, in 1969 Cabrera, Celli and Manson pointed out the possibility to observe single phonon excitations in the angular distribution of a scattered He atom beam [116, 117, 141]. This immediately stimulated several groups to carry out the corresponding experiments.

Fisher and coworkers [142, 143] in 1971 were the first to use time-of-flight (TOF) scattering spectroscopy. Although they succeeded in detecting energy losses in the TOF spectra of He atoms scattered from LiF(001), the resolution was not sufficient to obtain information on the dispersion of surface phonons. At about the same time Miller and coworkers at La Jolla also succeeded in resolving inelastic structures associated with single surface phonons near the zone origin [144–146]. Although these early experiments provided clear evidence for an interaction of He and Ne atoms with surface phonons, they were, however, severely hampered by inadequate velocity resolution ($\Delta v/v \geq 5\%$) and detector sensitivity [147].

The first experiments to provide information on the Rayleigh wave dispersion curve were reported by Williams at Ottawa in 1971 [148]. Through an ingenious analysis of angular distributions Williams together with Mason obtained the first information on the long wavelength part of the dispersion curves of LiF(001) [148, 149] and NaF(001) [150, 151]. Because of special kinematic conditions at certain incident and/or final directions, the latter being either in the incident plane (*in-plane* scattering) or outside (*out-of-plane* scattering), the inelastic scattering from surface phonons leads to additional small peaks in the angular distributions. In the experiments of Mason and Williams the out-of-plane angular distributions were analyzed, whereas the effects of inelasticity on the in-plane scattering were reported soon after in 1974 by Boato and Cantini [152] in high resolution angular distributions of helium and neon scattered from LiF(001). The neon angular distributions observed by Boato and Cantini exhibited complex inelastic features which were attributed to van Hove singularities [153] associated with the Rayleigh and Lucas modes near the symmetry points of the surface Brillouin zone [154, 155]. This mechanism was later designated as *kinematical focusing*. These developments were one of the hot subjects at the 1973 session of the LVIII "Enrico Fermi" International School at Varenna on "Dynamic Aspects of Surface Physics" [156].

In 1981 Avila and Lagos [157] showed that the out-of-plane kinematical focusing effect was able to explain also the earlier data of Mason and Williams [148–151]. Shortly afterwards Cantini et al. [158] provided an alternative explanation of most inelastic features in terms of scattering resonances involving Rayleigh waves out to the zone boundary. For a detailed discussion the reader is referred to Sect. 10.2.

In 1980 Brusdeylins, Doak and Toennies at Göttingen reported the first successful helium atom TOF measurements of surface phonon dispersion curves out to the zone boundary on LiF(001) [159–161]. The following year Feuerbacher and Willis were able to measure the Ni(111) Rayleigh wave dispersion curve out to the zone boundary with a pulsed supersonic Ne beam, though with a lower resolution [162]. Their measurements attracted however much attention because of the intriguing observation of additional inelastic processes at very small momentum transfers, which they attributed to the large size of the Ne atom. Although an *ab initio* study by Gunnarsson and Schönhammer [163] soon excluded that such processes could be attributed to surface electron-hole excitations, the electronic mechanism was later recognized as the most efficient one for the energy transfer to the elementary excitations of metal surfaces. This is discussed in Chaps. 8 and 13.

The first high resolution dispersion curves of a metal surface were reported in 1983 by the Göttingen group of Doak, Harten and Toennies for the scattering of helium atoms from the Ag(111) surface [164]. In addition to observing the full dispersion curve of the Rayleigh mode they also discovered what has now become an important anomalous phonon resonance. In these HAS measurements the incident energies of the beam were between typically 60 meV ($k_i \cong 11$ Å^{-1}) and 20 meV ($k_i \cong 6$ Å^{-1}). Energy transfers from less than 1 meV up to 30 meV were observed with a resolution of 0.3–1.0 meV, which is similar to the typical resolution in neutron studies of bulk phonons.

These advances by the Göttingen group were made possible by the introduction of new technology. The He nozzle beam source was operated at much higher pressures of several hundred bar, compared to pressures of several hundred mbar in earlier studies [165]. To cope with the flux the orifice diameter was reduced to 5 μm compared to 0.5 mm used previously. This provided a relative velocity resolution of 1% [166, 167] which is a factor 5–10 better than in the previous experiments described above. The second improvement consisted of a carefully designed system of differential pumping vacuum stages to greatly reduce the helium background in the detector, thereby significantly enhancing the signal-to-noise ratio of the mass spectrometer detector [168, 169]. The experimental aspects of HAS spectroscopy are discussed in Chaps. 9 and 10.

2.2.3 Early Electron Scattering Experiments

As mentioned previously the first attempts with low-energy electron scattering go back even earlier. In 1965 the weak thermal diffuse scattering observed between diffraction peaks in LEED experiments was initially attributed to inelastic scattering from Rayleigh modes [113, 114, 170]. Then in 1967 Probst and Piper [171] demonstrated for the first time the ability of electrons to resolve adsorbate vibrations in the energy distribution of the scattered electrons and thereby launched the field of electron energy loss spectroscopy (EELS). In these early experiments the energies were relatively low (E = 2–10 eV) and because of the predominant dipole scattering mechanism the electrons appeared near the specular direction. EELS experiments performed by Ibach on polar crystals in 1970 were the first to show the expected FK modes [172, 173], in agreement with earlier observations by Andreas Otto on the spectra of attenuated total reflection of light [174]. Related EELS experiments showed also for non-polar crystal surfaces, such as Si(111)2x1, a strong, almost isotropic, optical surface mode at 55 meV [175]. The latter experiments were regarded as providing the first evidence for genuine microscopic *optical* surface modes [176].

In the *dipole regime*, where the electrons interact via their long range electrostatic potential with the oscillating dipoles at the surface, the parallel momentum transferred by the electrons is practically zero and no information on the dispersion curves can be obtained. The advantages of using electrons at higher incident electron energies (100–200 eV), in the so-called *impact regime*, for investigating surface phonon dispersion curves were first pointed out in 1980 by Li et al. [177]. The first such measurements were reported in 1983 by Ibach and coworkers [178, 179]. They succeeded in measuring a full dispersion curve for one direction in Ni (100) using a 180 eV electron beam with an energy halfwidth of only 7 meV (4×10^{-5} resolution), which at that time was much less than previously possible. For more details on the history of EELS the reader is referred to a 1994 survey by Ibach, which includes experimental and theoretical developments related to adsorbate vibrational spectroscopy [180]. Further information may be found in other reports and books by Ibach and Mills [181–184].

2.3 History of Inelastic Atom Scattering Theory

2.3.1 Classical Scattering Theories

The first treatments of energy transfer in surface collisions date back to before quantum mechanics was discovered and for this reason the first theories were classical. As seen in Fig. 1.12 the quantum scattering regime suitable for single phonon spectroscopy is restricted to atoms of small mass and low incident energy interacting with a sufficiently cold surface. When any of these three conditions are not fulfilled, gas surface collisions are still today best treated with classical mechanics and possible quantum effects can be accounted for within a semi-classical approximation. Maxwell already in 1879 was probably the first to approach the problem of the interaction of a gas with a solid wall by assuming that thermal equilibrium between the gas and a surface is reached through a temporary residence of most of the gas atoms on the surface [185]. The theoretical studies of inelastic collisions of gas atoms on solid surfaces started with the simple classical model of Baule in 1914 [186]. This and subsequent classical approaches addressing the problems of energy transfer and sticking early in the 20th century are found in the 1967 volume edited by Saltsburg [187] and are thoroughly discussed in the 1973 review by Logan [188]. Other important reviews were published by Stickney et al. [138, 140, 189, 190]. The simplest classical approach was to consider the surface atoms as an assembly of independent oscillators (a type of Einstein model) behaving in the collision with the impinging atoms as hard cubes [191]. The hard-cube model evolved then into the soft-cube model [192], in order to account for the finite interaction time, and into the hard-sphere models [193], to account for the corrugation effects within a pure impulsive collision model.

Parallel to the above developments a classical theory suitable for surface collisions of fast and/or heavy atoms was proposed by Landau in 1935 [194]. After World War II the Russian school [195] continued the development of theories in connection with aerospace research. The unexpected launching of the Sputnik satellite in 1957 stimulated extensive activities in the area of rarefied gas dynamics especially in the United States. An important issue was the frictional heating of supersonic missiles and space craft as they enter the stratosphere. The Sputnik shock also lead to a sudden strong interest in molecular beam surface scattering experiments in the US. These activities in turn stimulated new theoretical work based not only on classical mechanics [194–200], but also on the generalized Langevin equation [201–210], the response function formalism [211, 212] and molecular dynamics [213]. A summary of the classical scattering theories can be found in the reviews by Barker and Auerbach [214], Gerber [210] and Brenig [215]. Only recently, Maxwell's 1879 conjecture [185] that a fraction of scattered gas particles would be directly scattered with little change of state and a fraction trapped and subsequently evaporated has been tested and the conditions for its validity established by Fan and Manson, who used a classical scattering theory and an iterative algorithm to account for multiple collisions [216, 217]. This approach allows for an accurate description of the energy accommodation coefficient of heavy rare gas atoms on metal surfaces at temperatures above θ_D, so that no detailed knowledge of the surface vibrational spectrum is required [218]. Classical scattering theories, as well as the semi-classical approximations discussed in the next Subsection, continue to be the choice methods for this class of collisional processes involving heavy atoms [219, 220].

2.3.2 *Semiclassical Scattering Theories*

Under conditions intermediate between the quantum and classical regimes (see Fig. 1.12), quantum effects have been handled in the semi-classical approximation in which the quantum transition probabilities are calculated for the classical trajectories of scattered atoms [221, 222]. In 1981 Meyer [223] derived an elegant semi-classical theory of inelastic atom-surface scattering within the scattering matrix (S-matrix) formalism developed somewhat earlier for gas phase collisions by Miller [224–226] and Marcus [227, 278]. Meyer's theory provided new insight into the theory of the Debye-Waller factor and an analytical solution of the inelastic scattering problem for a flat hard surface model [223, 229]. The validity conditions and approximations involved in the semi-classical approach have been thoroughly discussed in reviews by Celli and Evans [230] and by Manson et al. [231–233]. Further theoretical developments are found in the publications by Gumhalter and coworkers [234–238] and Billing and coworkers [239–243].

Multiphonon processes, which can occur in inelastic atom scattering spectra are often involved in integral properties such as the sticking and energy accommodation coefficients at $T < \theta_D$, require a quantum mechanical treatment of the surface vibrational field. The motion of the atomic probes can however still be described semi-classically in terms of trajectories. This approach has been adopted by Brenig and coworkers [244–247], and applied to heavy atom scattering by Brako and Newns [248–251] and by Persson and Harris [252]. Many authors [231, 232, 244, 253–262] have approached the theory of multiphonon scattering from the quantum side with the powerful scattering-matrix (S-matrix) formalism, from which the semiclassical trajectory approximation can be conveniently derived. For lighter probe atom masses, such as He atoms, Šiber and Gumhalter [263, 264] showed that the semi-classical approximations deviate substantially from the results of a full quantum theory even for integral properties such as the Debye-Waller factor. *A fortiori* the full quantum treatment, discussed next, is more reliable for the analysis of the inelastic He atom scattering spectra.

2.3.3 Quantum Scattering Theories

In the early 1930s soon after the advent of quantum mechanics the theory of inelastic scattering of atoms involving single vibrational quanta of the solid surface was formulated by Zener [265, 266], and Jackson et al. [267–269]. These early theories were stimulated by the experiments of Knudsen [121, 122] for understanding energy accommodation. They used a one-dimensional (1D) model in the distorted-wave Born approximation (DWBA) with either a power-law (Zener) or an exponential repulsive potential (Jackson et al.). Since, however, important phenomena, like sticking, selective adsorption and energy accommodation depend on the attractive part of the atom-surface potential, this part was later included, in the form of a Morse potential [270], by Lennard-Jones, Strachan, Devoushire and Goodwin [271–278, 289, 290]. These authors in altogether twelve seminal theoretical articles, which appeared in the late thirties under the common title "The Interaction of Atoms and Molecules with soild Surfaces" with Roman numerals I-XII covered a wide range of dynamical phenomena involving light atoms and molecules interacting with solid surfaces. They also provided the explanation for the strange dips in the intensity observed by Stern and coworkers in their angular distributions of scattered He atoms and H_2 molecules [279, 280] in terms of a temporary resonant trapping on the surface, a phenomenon now called *selective adsorption*. These authors were also the first to use the DWBA and to develop approximate methods for calculating realistic atom-surface potentials. These publications have become classical reference points for understanding a wide range of scattering resonance phenomena and provided the basis for all subsequent quantum treatments. World War II put a temporary end to these developments.

As mentioned earlier renewed theoretical activities were stimulated by the launching of the Sputnik in the late 1950s. Most of the theoretical work in the 1960s was carried out within quantum theory [281–288] with the aim to overcome the approximations made in the original work of Lennard-Jones et al. [271–278, 289, 290]. The DWBA quantum theory of inelastic scattering by Cabrera, Celli and Manson in 1969 and 1971, referred to in the previous Section [116, 117], was an important breakthrough. Within a fully 3D approach the surface potential was represented by the sum of an average uniform static but realistic potential and a periodic part describing the surface corrugation, the latter acting as a perturbation and being dynamically modulated by surface phonons. The authors used the two-potential formalism of Gell-Mann and Goldberger [291, 292], generalizing previous treatments of particle-surface inelastic scattering from single phonons. This new theory incorporated the correct surface dynamics and was therefore appropriate for predicting and later analyzing the experimental one-phonon time-of-flight (TOF) HAS spectra, as well as the phonon-induced effects in diffraction (e.g., the Debye-Waller factor and line shapes) [141, 293, 294]. With this method Goodman in 1971 [295] amended the original quantum theory of Lennard-Jones's and coworkers and analyzed the then available inelastic HAS data for silver [296] obtained by Subbarao and Miller [144, 145], and for LiF(001) [297, 298], reported by Williams [148, 149]. Benedek and Seriani with a DWBA calculation of inelastic HAS spectra for LiF(001) based on BSM-GF surface dynamics, Lennard-Jones interionic potentials and WKB distorted wavefunctions [299] confirmed that a substantial fraction (>60%) of the relatively small (<1%) transferred energy must be carried away by surface phonons [281, 300]. Similar conclusions were reached by Lagos and Birstein [301, 302] who reproduced the inelastic data of Williams for LiF(001) [148, 149] by using only Rayleigh waves and Tsuchida's Morse potential parameters [303]. This early phase of theoretical studies of inelastic HAS is documented in several reviews [304–307] and the important monograph by Goodman and Wachman [308].

2.3.4 Quantum Scattering Theory: Metal Surfaces and Debye-Waller Theory

Since the short range repulsive atom-atom potential arises from the Pauli exchange repulsion between the outer electrons at distances of about 3–4 Å, atoms impinging onto a surface are repelled by the surface electron density above the surface. For this reason the potential was frequently approximated by a continuous hard (or soft) wall, with or without corrugation. In the case of metals the electrons at the surface are strongly delocalized as first pointed by Smoluchowski already in 1941 [309]. For this reason a continuous smooth potential is a better representation of the surface potential than obtained by assuming a sum of phenomenological two-body interatomic potentials. The basic contributions in this framework were made by Beeby who solved the Lippmann-Schwinger equation for scattering from a flat hard wall for one-phonon [310] and two-phonon processes [311]. The theory agreed qualitatively with Subbarao and Miller's data [144, 145] on phonon inelasticity induced angular distributions of He atoms scattered from Ag(111).

In the early 1970s there was much discussion as to why diffraction peaks had so far not been seen on the low index surfaces of metals [312, 313] and only on insulator surfaces. Beeby [314] also addressed the question of why in 1971 atom scattering experiments from metals with the exception of the highly corrugated (112) tungsten surface [315] did not reveal diffraction. A semi-classical extension to a hard surface with a small corrugation was tackled by Levi and coworkers in the early 1970s [316–318]. They developed an elegant T-matrix formalism and obtained important results within the eikonal approximation [319]. On this basis Levi and Suhl [318] formulated a detailed theory of the Debye-Waller factor, while the complete theory of the differential reflection coefficient was reported by Levi [320] and thoroughly discussed in Bortolani and Levi's 1986 review [253]. A similar approach, using the assumption of impulsive collisions (sudden approximation) was formulated by Gerber et al. [321, 322] and by Miller et al. [323, 324].

A crucial question for the above models is the manner in which single surface atoms concur to give a hard- or a soft-wall potential, and the extent to which the potential is modulated by the atomic displacements. For long wave RW's the potential modulation can be explicitly derived from the continuum elastic theory, as shown by Van Trong [325], and similarly by Conn in an earlier theory based on GF diagrammatic techniques [326]. For surface phonons of any wavelength the problem was solved for insulator surfaces in the eikonal approximation by directly deriving the phase shifts from the individual atom-ion forces [327, 328]. In metals, however, the mediating effect of the conduction electrons [316, 329–332] was shown to spread the interaction over many atoms of the surface (Armand effect [333–335]). Armand and Manson have also shown that the addition of an attractive well has a dramatic effect on inelastic HAS from a flat metal surface [336, 337]. With a virtually exact theory of multiphonon scattering processes Armand and Manson obtained the temperature dependence of the Debye-Waller factor and the specular intensities for Cu(001) in excellent agreement with the 1980 HAS measurements by Lapujoulade et al. [338] and provided a precise determination of the attractive potential. An alternative quantum approach based on an exactly soluble surface potential for the probe atom embedded in the continuum of the free atom states was suggested by Doyen and Grimley [339].

The assumption of a hard corrugated potential surface (HCS) greatly facilitates the calculation for surfaces with large corrugations as on the alkali halides or, in general, for many-phonon processes [283]. A HCS theory for elastic and one-phonon scattering was developed by Armand and Manson [340], with a discussion of the Debye-Waller factor. Practical applications, however, could first be achieved by Garcia et al. [341–343] through the new GR method, which is a powerful numerical technique providing exact solutions to the scattering problem from a HCS. This method was extended to the theory of inelastic scattering by Benedek and Garcia [344, 345].

2.3.5 Quantum Scattering Theory: Developments Since 1980

In 1980 Benedek and Garcia [346, 347] applied their theory to the analysis of the first high resolution TOF-HAS spectra of the surface phonons on LiF(001) [159–161]. A subsequent systematic analysis [348–350] of the same data provided close agreement between calculated and experimental TOF spectra and confirmed: (i) the dominance of one-phonon processes for scattering of low energy beams from surfaces at or below room temperature, in accord with the theoretical criteria established by Weare [351] and Lin et al. [352] (see Sect. 7.5); (ii) the dominance of the RW with respect to other surface and bulk phonons [353, 354]. By exploiting a procedure to account for inelastic corrections due to Celli et al. [355, 356], Evans et al. [357] could reproduce the exact shape of bound state resonances in inelastic spectra and could demonstrate the striking effect of (selective adsorption) resonance enhancement on the one-phonon scattering. By means of extensive HCS calculations with an attractive well Nichols and Weare [358] were able to reproduce the inelastic HAS angular distributions from LiF(001) including all the features arising from bound state inelastic resonances, kinematical focussing and even the huge anomalous peaks, called *supernovae*, discovered by Lilienkamp and Toennies [359, 360] (see Chap. 10). In a subsequent study such anomalous features have been assigned to a new resonant process called focused inelastic resonance (FIR) [361, 362] (see Chap. 10). The theoretical understanding of the inelastic bound state resonances paved the way to a novel method for obtaining information on the dispersion curves of optical surface modes by exploiting the resonance enhancement in TOF spectra [363, 364].

One-phonon inelastic processes can be calculated more rigorously quantum-mechanically to first order in the inelasticity, but with all orders of the diffraction channels using close-coupling (CC) calculations. The feasibility of CC calculations for inelastic scattering was first demonstrated in 1974 by Wolken who used a discretization of the momentum space [365, 366]. The first state-of-the-art CC calculations of inelastic HAS-TOF spectra and angular distributions from LiF (001) was accomplished 10 years later by Eichenauer and Toennies [367, 368] using the best available additive He atom-surface ion potentials and including bound states. Although the CC-calculations of Eichenauer and Toennies were restricted to one-phonon processes, the CC approach is suitable, in principle, to a full quantum treatment of deep inelastic processes involving many phonons [366]. A detailed derivation of the general theory of elastic and inelastic processes combining a Green's function scattering theory with the close-coupling approach, suitable to treat also bound state resonances associated with a general potential, can be found in two papers by Choi and Poe [369, 370]. Another apparently powerful method, which incorporates a detailed phonon dynamics, realistic two-body potentials, the bound states and possibly multiphonon processes, is the quantum statistical theory of Horia Metiu [371]. It is regrettable that numerical applications of these two powerful approaches to existing experiments have not been reported so far and still appear to be rather challenging.

The inelastic reflection coefficient obtained with DWBA turns out to be an exponential form of the displacement-displacement correlation function. Its power expansion provides multiphonon contributions to all orders, even in the ordinary case of linear atom-phonon coupling (see Sect. 7.5). The exponentiated Born approximation (EBA), developed by Brenig et al. [234, 236, 244, 246, 256, 257, 372], is an elegant realization of this concept and a powerful tool in the analysis of high resolution multiphonon HAS spectra. Šiber and Gumhalter have shown, in the case of He scattering from a Xe overlayer, that the non-linear coupling yields only a marginal contribution to zero-, one- and two-phonon processes, the major part being accounted for by EBA and a linear coupling [238, 373]. EBA is not sufficient, however, for highly corrugated surfaces due to multiple scattering and the accessibility of bound states. In this case virtually exact coupling-channel calculations are needed. Within the last 15 years a further step in the theory of inelastic HAS which elucidates the role of phonon-mediated bound state resonances in multiphonon spectra has been reported by Brenig [374] and by Šiber and Gumhalter [375].

2.3.6 Summary and Outlook

The above review briefly summarizes the evolution of the theory for inelastic atom-surface scattering which anticipated and accompanied the early phases of surface phonon HAS spectroscopy. A lively account on this period was given by Robinson in a 1982 Science paper [376]. A new challenge for inelastic HAS theory, and consequently for the theory of surface dynamics, started in 1983 with the discovery of surface phonon anomalies in metals and their anomalous scattering intensities [164, 377]. This led to a controversy concerning the interpretation in terms of the force constants at the surface compared to EELS measurements. This was dubbed the Bortolani-Mills paradox [378], which is discussed in more detail in Chap. 8. The observation by HAS of an anomalous surface longitudinal resonance [164], which in the course of time was observed in practically all metal surfaces raised a long discussion among theoreticians about its origin and nature [379, 380]. The association of the anomalous resonance with a subsurface mode predicted by different phenomenological models was at odds with the common wisdom that HAS only probes surface atom displacements.

Only recently has it been possible to convincingly demonstrate by means of state-of-the-art first principle calculations, that in metals even deep sub-surface modes can induce sufficiently large electronic charge density oscillations at the surface so as to scatter He atoms inelastically [381]. This recent discovery, besides solving a long-standing issue and shedding new light on the range and strength of the electron-phonon interaction at metal surfaces, has revealed another aspect (or "facet") of HAS as a unique probe of the dispersion of sub-surface metal phonon branches and of mode-selected electron-phonon coupling strengths [380–384].

The problem of the atom-surface potential as mediated by conduction electrons is a crucial issue in the quantum theory of HAS, much in the same way as for the lattice dynamics theory of metal surfaces. The new theory resulting from the recent understanding of how He atoms couple to sub-surface modes is discussed at lengths in Chap. 8. This is one of the main issues of this book and various aspects will be dealt with extensively in the following chapters.

2.4 Further Reading

Although the historical developments in surface phonon theory on the one hand, and of experimental and theoretical studies of gas-surface collisions on the other hand have been dealt with separately for practical purposes, they were strictly and necessarily entangled, as normally happens in any field of physics. Present understanding of surface phonon spectroscopy by helium atom scattering has benefited greatly from the continuous feedback between experiment and theory over the last 35 years.

A number of reviews and books have kept pace with the developments on both fronts of this progress. Following the proceedings of two historical schools on gas-surface interaction (1974, 1982) [307, 385] and the founding book by Goodman and Wachman (1976) [308], we mention the review articles by Celli (1984, 1992) [386, 387], Barker and Auerbach (1994) [229], Garcia-Moliner (1990) [388], Wöll (1991) [389], Doak [390], Toennies (1989, 1990, 1993) [168, 377, 391, 392], Hoffmann and Toennies (1996) [393], Safron (1996) [394], Gumhalter (2000) [236], Rusina and Chulkov (2013) [395] and Manson (2006) [233]. The recent related developments using the He-3 spin echo technique have recently been reviewed by Jardine et al. (2009) [396]. The book by Billing [243], the volumes edited by de Wette and Kress (1991) [59], by Hulpke (1992) [397], and in the Landolt-Börstein collections compiled by Celli (1994/1995) [398] Wallis and Tong (1994/1995) [399], Rocca (1994) [400] and by Benedek, Campi and Toennies [401] also contain extensive discussions of surface phonons and HAS experiments. Research on surface phonons have also been reported at a number of conference series especially at the "Workshop on Surface Dynamics" also called "Surphon".

The proceedings of several conferences have been published in Journal of Physics: Condensed Matter e.g. Vol. 16, Nr. 29, July 2002 (11th); Vol. 16, Nr. 29, July 2004; Vol. 14, Nr. 24, August 2007 (12th); Vol. 21, Nr. 26, July 2009 (13th). The proceedings of the biannual international conference series "Vibrations at Surfaces (VAS)" also are regularly published either as conference proceedings or in various journals such as Surface Science.

The initial motivation for this monograph comes from the 1989 survey by Toennies [377] and the 1994 review by the present authors in the volume dedicated to the first "Thirty Years of Surface Science" [402]. Additional information and insight into the various aspects of surface phonons and surface phonon spectroscopy by atom scattering can be found in other more specialized review articles and edited volumes which are quoted in the next chapters.

Detailed dispersion curves of surface phonons for different crystals can be found in some of the quoted volumes. In Appendix A a complete list of crystals is given for which the surface phonon dispersion curves have been measured with all available techniques and/or calculated with all possible theoretical methods, with the respective references.

References

1. J.W. Strutt (Baron Rayleigh): Proc. London Math. Soc., **17**, 4 (1885); see also: *The Theory of Sound* (MacMillan, London 1896; unabridged 2nd revised edition: Dover Publ., New York 1945)
2. G.W. Farnell, in *Physical Acoustics*, vol. 6, ed. by W.P. Mason, R.N. Thurston (Academic Press, New York, 1970), p. 109
3. A.A. Maradudin, in *Nonequilibrium Phonon Dynamics*, ed. by W.E. Bron (Plenum, New York, 1985), p. 395
4. M. Born, Th von Kármán, Phys. Zeit. **13**, 297 (1912)
5. M. Born, K. Huang, *Dynamical Theory of Crystal Lattices* (Oxford University Press, London, 1954)
6. I.M. Lifshitz, L.M. Rozenzweig, Zh. Eksp. Teor. Fiz. **18**, 1012 (1948)
7. I.M. Lifshitz, Nuovo Cimento Suppl. **3**, 732 (1956)
8. I.M. Lifshitz, A.M. Kosevich, Rep. Progr. Phys. **29**(I), 217 (1966)
9. E.N. Economou, *Green's Functions in Quantum Physics* (Springer, Berlin, 1979)
10. A.A. Maradudin, E.W. Montroll, G.H. Weiss and I.P. Ipatova: *Theory of Lattice Dynamics in the Harmonic Approximation*, Solid State Physics, suppl. 3 (Academic Press, New York, 1971)
11. A.A. Maradudin, R.F. Wallis, L. Dobrzynski, *Handbook of Surfaces and Interfaces*, vol. 3 (Garland, New York, 1980)
12. L. Dobrzynski, G. Leman, J. Phys. (Paris) **30**, 116 (1970)
13. P. Masri, L. Dobrzynski, J. Phys. (Paris) **32**, 295 (1971)
14. F. Garcia-Moliner, Ann. Phys. (Paris) **2**, 179 (1977)
15. G. Armand, Phys. Rev. B **14**, 2218 (1976)
16. G. Armand, P. Masri, Surf. Sci. **130**, 89 (1983)
17. R.A. Allen, Surf. Sci. **76**, 91 (1978)
18. A.A. Maradudin, J. Melngailis, Phys. Rev. **133**, A1188 (1964)
19. T.P. Martin, Phys. Rev: B **1**, 3480 (1970)

20. S.W. Musser, K.H. Rieder, Phys. Rev. B **2**, 3034 (1970)
21. M. Croitoru, D. Grecu, Surf. Sci. **38**, 60 (1973); Rev. Roum. Phys. **18**, 805 (1973)
22. J.E. Black, B. Lacks, D.L. Mills, Phys. Rev. B **22**, 1818 (1980)
23. W. Goldhammer, W. Ludwig, W. Zierau, C. Falter, Surf. Sci. **141**, 139 (1984)
24. G. Benedek, Phys. Status Sol. **B58**, 661 (1973)
25. G. Benedek, Surf. Sci. **61**, 603 (1976)
26. G. Benedek, L. Miglio, in *Surface Phonons*, ed. by F.W. de Wette, W. Kress, Springer Ser. Surf. Sci., vol. 27 (Springer, Berlin, 1991), p. 37
27. G. Benedek, M. Miura, W. Kress, H. Bilz, Phys. Rev. Lett. **52**, 1907 (1984)
28. T.E. Feuchtwang, Phys. Rev. **155**, 731 (1967)
29. A simplified approach based on one single exponential function, used by B.N.N. Achar, G. R. Barsch, Phys. Rev. **188**, 1356, 1361 (1969) within the scattering-matrix method (SMM), has been argued to be incorrect by R.E. Allen, Phys. Rev. B **3**, 3580 (1971)
30. V. Bortolani, F. Nizzoli, G. Santoro, in *Lattice Dynamics*, ed. by M. Balkanski (Flammarion, Paris, 1978), p. 302
31. J. Szeftel, A. Khater, J. Phys. C **20**, 4725 (1987)
32. J. Szeftel, A. Khater, F. Mila, S. d'Addato, N. Auby, J. Phys. C **21**, 2133 (1988)
33. A. Khater, N. Auby, R.F. Wallis, Phys. B **167**, 273 (1991)
34. S. Trullinger, J. Math. Phys. **17**, 1884 (1976)
35. A. Fasolino, G. Santoro, E. Tosatti, J. Phys. **42**, C6–846 (1981)
36. J.E. Black, in *Dynamical Properties of Solids* (Chap. 5), ed. by G.K. Horton, A.A. Maradudin (Elsevier, Amsterdam, 1990)
37. D.C. Gazis, R. Herman, R.F. Wallis, Phys. Rev. **119**, 533 (1960)
38. D.C. Gazis, R.F. Wallis, Surf. Sci. **5**, 482 (1966)
39. R.F. Wallis, Phys. Rev. **105**, 540 (1957)
40. R.F. Wallis, Phys. Rev. **116**, 302 (1959)
41. A.A. Lucas, J. Chem. Phys. **48**, 3156 (1968)
42. R. Fuchs, K.L. Kliever, Phys. Rev. **140**, A2076 (1965)
43. K.L. Kliever, R. Fuchs, Phys. Rev. **144**, 495 (1966)
44. K.L. Kliever, R. Fuchs, Phys. Rev. **150**, 573 (1966)
45. W.E. Jones, R. Fuchs, Phys. Rev. B **4**, 3581 (1971)
46. V.V. Bryksin, Y.A. Firsov, Fiz. Tverd. Tela **11**, 2167 (1969) [Sov. Phys. Solid State **11**, 1751 (1970)]
47. A.D. Boardman (ed.), *Electromagnetic Surface Modes* (Wiley, New York, 1982)
48. B.C. Clark, R. Herman, R.F. Wallis, Phys. Rev. **139**, A860 (1965)
49. R.E. Allen, F.W. de Wette, Phys. Rev. **179**, 873 (1969)
50. R.E. Allen, F.W. de Wette, A. Rahman, Phys. Rev. **179**, 887 (1969)
51. G.P. Alldredge, R.E. Allen, F.W. de Wette, J. Acoust. Soc. Am. **49**, 1453 (1971)
52. R.E. Allen, G.P. Alldredge, F.W. de Wette, Phys. Rev. Lett. **23**, 1285 (1969)
53. R.E. Allen, G.P. Alldredge, F.W. de Wette, Phys. Rev. B **4**, 1648 (1971)
54. R.E. Allen, G.P. Alldredge, F.W. de Wette, Phys. Rev. B **4**, 1661 (1971)
55. T.S. Chen, G.P. Alldredge, F.W. de Wette, R.E. Allen, Phys. Rev. Lett. **26**, 1543 (1971)
56. T.S. Chen, G.P. Alldredge, F.W. de Wette, Solid State Commun. **10**, 941 (1972)
57. T.S. Chen, G.P. Alldredge, F.W. de Wette, Phys. Lett. **40** A, 401 (1972)
58. T.S. Chen, F.W. de Wette, G.P. Alldredge, Phys. Rev. B **15**, 1167 (1977)
59. F.W. de Wette, W. Kress (eds.), in *Surface Phonons*, Springer Ser. Surf. Sci., vol. 21 (Springer, Berlin, 1991)
60. G.P. Alldredge, Phys. Lett. **41A**, 291 (1972)
61. V.L. Zoth, G.P. Alldredge, F.W. de Wette, Phys. Lett. **47A**, 247 (1974)
62. H. G. Smith, N. Wakabayashi, in *Dynamics of Solids and Liquids by Neutron Scattering*, ed. by S. W. Lovesey, T. Springer (Springer, Berlin, 1977), p. 78
63. L. Miglio, L. Colombo, Phys. Rev. B **37**, 3025 (1988)
64. K.M. Ho, K.P. Bohnen, Phys. Rev. Lett. **56**, 934 (1986)
65. K.P. Bohnen, K.M. Ho, Surf. Sci. **207**, 105 (1988)

66. K.P. Bohnen, K.M. Ho, Surf. Sci. Rep. **19**, 99 (1993)
67. K.P. Bohnen, K.M. Ho, in *Tutorials on Selected Topics in Modern Surface Science*, ed. by T. Barr, D. Saldin (Elsevier, Dordrecht, 1993), pp. 100–117
68. C. Bungaro, S. de Gironcoli, S. Baroni, Phys. Rev. Lett. **77**, 2491 (1996)
69. S. Baroni, S. de Gironcoli, A. dal Corso, P. Giannozzi, Rev. Mod. Phys. **73**, 515 (2001)
70. M. Lazzeri, S. de Gironcoli, Phys. Rev. Lett. **81**, 2096 (1998)
71. J. Fritsch, C. Eckl, P. Pavone, U. Schröder, Festkörperprobleme **36**, 135 (1997)
72. J. Fritsch, U. Schröder, Phys. Rep. **309**, 209 (1999)
73. R. Ravelo, M. El-Batanouny, Phys. Rev. B **40**, 9574 (1989)
74. R. Ravelo, M. El-Batanouny, J. Electr. Spectr. Rel. Phenom. **54**(55), 255 (1990)
75. T.E. Karakasidis, D.G. Papageorgiou, G.A. Evangelakis, Appl. Surf. Sci. **162**, 233 (2000)
76. P.C. Weakliem, E.A. Carter, J. Chem. Phys. **96**, 3240 (1992)
77. M.R. Baldan, E. Granato, S.C. Ying, Phys. Rev. B **62**, 2146 (2000)
78. D.D. Koleske, S.J. Sibener, Surf. Sci. **268**, 406 (1992)
79. D.D. Koleske, S.J. Sibener, Surf. Sci. **268**, 418 (1992)
80. X.Q. Wang, G.L. Chiarotti, F. Ercolessi, E. Tosatti, Phys. Rev. B **38**, 8131 (1988)
81. C.Z. Wang, E. Tosatti, A. Fasolino, Phys. Rev. Lett. **60**, 2661 (1988)
82. L. Yang, T.S. Rahman, Phys. Rev. Lett. **67**, 2327 (1991)
83. L. Yang, T.S. Rahman, M.S. Daw, Phys. Rev. B. **44**, 13725 (1991)
84. T.S. Rahman, Z.J. Tian, J. Electr. Spectr. Rel. Phenom. **64**(65), 651 (1993)
85. D.D. Koleske, S.J. Sibener, Surf. Sci. **290**, 179 (1993)
86. Y. Beaudet, L.J. Lewis, M. Persson, Phys. Rev. B **50**, 12084 (1994)
87. R. Zivieri, G. Santoro, V. Bortolani, Phys. Rev. B **59**, 15959 (1999)
88. K.S. Smirnov, G. Raseev, Surf. Sci. **459**, 124 (2000)
89. A.N. Al-Rawi, A. Kara, T.S. Rahman, Surf. Sci. **446**, 17 (2000)
90. A. Kara, T.S. Rahman, Surf. Sci. Rep. **56**, 159 (2005)
91. R. Car, M. Parrinello, Phys. Rev. Lett. **55**, 2471 (1985)
92. F. Ancillotto, W. Andreoni, A. Selloni, R. Car, M. Parrinello, Phys. Rev. Lett. **65**, 3148 (1990)
93. F. Ancillotto, A. Selloni, W. Andreoni, S. Baroni, R. Car, M. Parrinello, Phys. Rev. B **43**, 8930 (1991)
94. C.M. Bertoni, A.I. Shkrebtii, R. Di Felice, F. Finocchi, Prog. Surf. Sci. **42**, 319 (1993)
95. D.R. Alfonso, D.A. Drabold, S.E. Ulloa, Phys. Rev. B **51**, 1989 (1995)
96. D.R. Alfonso, D.A. Drabold, S.E. Ulloa, Phys. Rev. B **51**, 14669 (1995)
97. G. Kresse, J. Furthmüller, J. Hafner, Europhys. Lett., 729 (1995)
98. D. Marx, J. Hutter, in *Modern Methods and Algorithms of Quantum Chemistry*, ed. by J. Grotendorst, John von Neumann Institute for Computing, Jülich, NIC Series, vol. 1, ISBN 3-00-005618-1, pp. 301–449, 2000; and references therein
99. D. Marx, J. Hutter, *Ab-Initio Molecular Dynamics* (Cambridge University Press, Cambridge, 2009)
100. R.F. Wallis, in *Atomic Structure and Properties of Solids*, ed. by E. Burstein (Academic, New York, 1972)
101. R.F. Wallis, in *Progress in Surface Science*, ed. by S.G. Davison, vol. 4, (Pergamon, New York, 1974), p. 234
102. R.F. Wallis, in *Dynamical Properties of Solids*, ed. by G.K. Horton, A.A. Maradudin, vol. 2, (North-Holland, Amsterdam, 1975), p. 443
103. M.G. Lagally, in *Surface Physics of Materials*, ed. by J.M. Blakeley, vol. II, (Academic, New York, 1975), p. 419
104. G.I. Stegeman, F. Nizzoli, in *Surface Excitations* (Chap. 2), ed. by V.M. Agranovich, R. Loudon (North-Holland, Amsterdam 1984), p. 195
105. J.E. Black, in *Structure and Dynamics of Surfaces I*, ed. by W. Schommers, P. von Blanckenhagen (Springer, Berlin, 1986), p. 153
106. F.W. de Wette, in *Dynamical Properties of Solids* (Chap. 5), ed. by G.K. Horton, A.A. Maradudin (Elsevier, Amsterdam, 1990)

107. B.N. Brockhouse, A.T. Stewart, Phys. Rev. **100**, 756 (1955)
108. R.R. Cavanagh, R.D. Kelley, J.J. Rush, J. Chem. Phys. **77**, 1540 (1982)
109. B.N. Brockhouse, S. Hanteder, H. Stiller, in *Interaction of Radiation with Solids*, ed. by Strumane et al. (North-Holland, Amsterdam, 1963)
110. B. Dorner, *Coherent Inelastic Neutron Scattering in Lattice Dynamics* (Springer, Berlin, 1982)
111. K.H. Rieder, E.M. Hörl, Phys. Rev. Lett. **20**, 209 (1968)
112. K.H. Rieder, W. Drexel, Phys. Rev. Lett. **34**, 148 (1975)
113. E.R. Jones, J.T. McKinney, M.B. Webb, Bull. Am. Phys. Soc. **10**, 324 (1965)
114. J. Aldag, R.M. Stern, Phys. Rev. Lett. **14**, 857 (1965)
115. D.L. Huber, Phys. Rev. **153**, 772 (1967)
116. N. Cabrera, V. Celli, J.R. Manson, Phys. Rev. Lett. **22**, 346 (1969)
117. J.R. Manson, V. Celli, Surf. Sci. **24**, 495 (1971)
118. J.R. Manson, J. Phys.: Condens. Matter **22**, 300401 (2010)
119. R.W. Wood, Phil. Mag., London, **30**, 300–304 (1915)
120. R.W. Wood, Phil. Mag., London, **32**, 364 (1916)
121. M. Knudsen, Ann. Phys. (Leipzig) **47**, 697–708 (1915)
122. M. Knudsen, Ann. Phys. (Leipzig) **48**, 1113–1121 (1915)
123. L. Dunoyer, Radium **8**, 142 (1911)
124. O. Stern, Naturwiss. **17**, 391 (1929)
125. I. Estermann, O. Stern, Z. Phys. **61**, 95 (1930)
126. I. Estermann, R. Frisch, O. Stern, Z. Phys. **73**, 348 (1931)
127. T.H. Johnson, J. Franklin Inst. **210**, 135 (1930)
128. T.H. Johnson, Phys. Rev. **37**, 847–861 (1931)
129. L. De Broglie, *Recherches sur la Théorie des Quanta* (Thesis (Sorbonne), Paris, 1924)
130. De Broglie, L., Ann. de Phys. **10**(3), 22 (1925)
131. O.R. Frisch, *What Little I Remember* (Cambridge University Press, Cambridge, 1979)
132. R.G.J. Fraser, *Molecular Rays* (Cambridge University Press, Cambridge, 1931)
133. E. Fues, *Beugungsversuche mit Materiewellen: Einführung in die Quantenmechanik* (Akad. Verlangsges, Leipzig, 1935)
134. M. Von Laue, in *Materiewellen und ihre Interferenzen* (Akadem. Verl.-Ges. Becker & Erler, 1944; Geest und Portig, 1948)
135. K.F. Smith, *Molecular Beams* (Wiley, New York, 1955)
136. G. Comsa, Surf. Sci. **299**, 77 (1994)
137. E.C. Beder, Advan. At. Mol. Phys. **3**, 205–290 (1967)
138. R.E. Stickney, R.M. Logan, S. Yamamoto, J.C. Keck, in *Fundamentals of Gas-Surface Interactions,* ed. by H. Saltsburg (Academic Press Inc., New York 1967), p. 422. *Quantenmechanik* (Akad. Verlangsges., Leipzig 1935)
139. R.E. Stickney, J. Vac. Sci. Technol. **6**, 920 (1969)
140. R.E. Stickney, J. Vac. Sci. Technol. **7**, 90 (1970)
141. N. Cabrera, V. Celli, F.O. Goodman, R. Manson, Surf. Sci. **19**, 67 (1970)
142. S.S. Fisher, J.R. Bledsoe, J. Vac. Sci. Technol. **9**, 814 (1971)
143. S.S. Fisher, J.R. Bledsoe, Surf. Sci. **46**, 129 (1974)
144. R.E. Subbarao, D.R. Miller, J. Chem. Phys. **51**, 4679 (1969)
145. R.E. Subbarao, D.R. Miller, J. Vac. Sci. Technol. **9**, 808 (1972)
146. J.M. Horne, D.R. Miller, Phys. Rev. Lett. **41**, 511 (1978)
147. G. Benedek, G. Boato, Europhys. News **8**, 5 (1977)
148. B.R. Williams, J. Chem. Phys. **55**, 3220 (1971)
149. B.R. Williams, J. Chem. Phys. **56**, 1895 (1972)
150. B.F. Mason, B.R. Williams, J. Chem. Phys. **61**, 2765 (1974)
151. B.F. Mason, B.R. Williams, Jpn. J. Appl. Phys., suppl., Pt. 2, 557 (1974)
152. G. Boato, P. Cantini, in *Dynamics Aspects of Surface Physics*, ed. by F.O. Goodman (Compositori, Bologna, 1974), p. 707
153. L. Van Hove, Phys. Rev. **89**, 1189 (1953)

154. G. Benedek, Phys. Rev. Lett. **35**, 234 (1975)
155. G. Benedek, L. Miglio, G. Seriani, in *Helium Atom Scattering from Surfaces*, ed. by E. Hulpke (Springer, Berlin, 1992), p. 207
156. F.O. Goodman (ed.), in *Dynamic Aspects of Surface Physics*, vol. 19 (Compositori, Bologna, 1974)
157. R. Avila, M. Lagos, Surf. Sci. **103**, L104 (1981)
158. P. Cantini, G.P. Felcher, R. Tatarek, Phys. Rev. Lett. **37**, 606 (1976)
159. G. Brusdeylins, R.B. Doak, J.P. Toennies, Phys. Rev. Lett. **44**, 1417 (1980)
160. G. Brusdeylins, R.B. Doak, J.P. Toennies, Phys. Rev. Lett. **46**, 437 (1981)
161. G. Brusdeylins, R.B. Doak, J.P. Toennies, Phys. Rev. B **27**, 3662 (1983)
162. B. Feuerbacher, R.F. Willis, Phys. Rev. Lett. **47**, 526 (1981)
163. O. Gunnarsson, K. Schönhammer, Phys. Rev. B **25**, 2514 (1982)
164. R.B. Doak, U. Harten, J.P. Toennies, Phys. Rev. Lett. **51**, 578 (1983)
165. J.B. Anderson, J.B. Fenn, Phys. Fluids **8**, 780 (1965)
166. J.P. Toennies, K. Winkelmann, J. Chem. Phys. **66**, 3965 (1977)
167. G. Brusdeylins, H.-D. Meyer, J.P. Toennies, K. Winkelmann, in *Progress in Astronautics and Aeronautics*, vol. 51, ed. by J.L. Poetter (AIAA, New York, 1977), p. 1047
168. J.P. Toennies, in *Surface Phonons*, ed. by F.W. de Wette, W. Kress, Springer Ser. Surf. Sci., vol. 21 (Springer, Berlin, 1991)
169. J.P. Toennies, Europhys. News **23**, 63 (1992); A. Lahee, J. P. Toennies, Physics World, April 1993, p. 61
170. R.F. Wallis, A.A. Maradudin, Phys. Rev. **148**, 962 (1966)
171. F.M. Probst, T.C. Piper, J. Vac. Sci. Technol. **4**, 53 (1967)
172. H. Ibach, Phys. Rev. Lett. **24**, 1416 (1970)
173. H. Ibach, J. Vac. Sci. Technol. **9**, 713 (1971)
174. A. Otto, Z. Phys. **216**, 398 (1968)
175. H. Ibach, Phys. Rev. Lett. **27**, 253 (1971)
176. W. Ludwig, Jpn. J. Appl. Phys., suppl. **2**, Pt. 2 (1974), p. 879
177. C.H. Li, S.Y. Tong, D.L. Mills, Phys. Rev. B **21**, 3057 (1980)
178. J.M. Szeftel, S. Lehwald, H. Ibach, T.S. Rahman, J.E. Black, D.L. Mills, Phys. Rev. Lett. **51**, 268 (1983)
179. S. Lehwald, J.W. Szeftel, H. Ibach, T.S. Rahman, D.L. Mills, Phys. Rev. Lett. **50**, 518 (1983)
180. H. Ibach, Surf. Sci. **299–300**, 116 (1994)
181. H. Ibach, D.L. Mills, *Electron Energy Loss Spectroscopy and Surface Vibrations* (Academic Press, Inc., New York, 1982)
182. H. Ibach, in *Electron Energy Loss Spectrometers: The Technology of High Performance,* Springer Series in Optical Sciences, vol 63 (Springer, Berlin, 1991)
183. H. Ibach, in *Physics of Surfaces and Interfaces* (Chap. 7) (Springer, Berlin, 2007)
184. H. Ibach, J. Chem Soc.-Faraday Trans. **92**, 4771–4774 (1996)
185. J.C. Maxwell, Phil. Trans. R. Soc. Lond. **170**, 231 (1879)
186. B. Baule, Ann. der Phys. **44**, 145 (1914)
187. H. Saltsburg (ed.), *Fundamentals of Gas-Surface Interactions* (Academic Press, New York, 1967)
188. R.M. Logan, in *Theory of Gas-Surface Scattering and Accommodation*, Solid State Surface Science, ed. by M. Green, vol. 3 (Dekker, New York, 1973), p. 1
189. R.E. Stickney, Adv. At. Mol. Phys. **3**, 143–204 (1968)
190. S. Yamamoto, R.E. Stickney, J. Chem Phys. **53**, 1594 (1970)
191. R.M. Logan, R.E. Stickney, J. Chem. Phys. **44**, 195 (1966)
192. R.M. Logan, J.C. Keck, J. Chem. Phys. **49**, 860 (1968)
193. J.A. Barker, D.J. Auerbach, Chem. Phys. Lett. **67**, 393 (1979)
194. L.D. Landau, Phys. Z. Sowjet. **8**, 489 (1935)
195. See, e.g., B.L. Bonch-Bruevich, Usp. Fiz. Nauka **40**, 369 (1950) [English translation: National Research Council of Canada TT.509]

196. R.W. Zwanzig, J. Chem. Phys. **32**, 1173 (1960)
197. B. McCarroll, J. Chem. Phys. **39**, 1317 (1963)
198. F.O. Goodman, in *Rarefied Gas Dynamics*, ed. by J.H. De Leeuw, vol. 2 (Academic Press, New York, 1966), p. 336
199. C. Caroli, B. Roulet, D. Saint-James, Phys. Rev. B **18**, 545 (1978)
200. D.R. Dion, J.A. Barker, R.P. Merril, Chem. Phys. Lett. **57**, 218 (1978)
201. S.A. Adelman, J.D. Doll, J. Chem. Phys. **64**, 2375 (1976)
202. S.A. Adelman, J. Chem. Phys. **71**, 4471 (1979)
203. S.A. Adelman, Adv. Chem. Phys. **44**, 143 (1980)
204. M. Shugard, J.C. Tully, A. Nitzan, J. Chem. Phys. **66**, 2534 (1977)
205. M. Shugard, J.C. Tully, A. Nitzan, J. Chem. Phys. **69**, 336 (1978)
206. M. Shugard, J.C. Tully, A. Nitzan, J. Chem. Phys. **69**, 2525 (1978)
207. J.C. Tully, J. Chem. Phys. **73**, 1975 (1980)
208. J.C. Tully, Surf. Sci. **111** (1981)
209. C.W. Muhlhausen, J.A. Serri, J.C. Tully, G.E. Becker, M.J. Cardillo, Israel J. Chem. **22**, 315 (1982)
210. R.B. Gerber, Chem. Rev. **87**, 29–79 (1987)
211. F.O. Goodman, Surf. Sci. **3**, 386 (1965)
212. J.A. Barker, W.A. Steele, Surf. Sci. **74**, 596 (1978)
213. J.A. Barker, D.R. Dion, R.P. Merrill, Surf. Sci. **95**, 15 (1980)
214. J.A. Barker, D.J. Auerbach, Surf. Sci. Rep. **4**, 1 (1984)
215. W. Brenig, Z. Phys. B **48** (1982), p. 127
216. G. Fan, J.R. Manson, Phys. Rev. Lett. **101**, 063202 (2008)
217. G. Fan, J.R. Manson, Phys. Rev. B **79**, 045424 (2009)
218. G. Fan, J.R. Manson, Chem. Phys. **370**, 175 (2010)
219. W.W. Hayes, J.R. Manson, Phys. Rev. Lett. **109**, 063203 (2012)
220. W.W. Hayes, J.R. Manson, Phys. Rev. B **89**, 045406 (2014)
221. E. Drauglis, Phys. Rev. **178**, 1527 (1969)
222. J.I. Kaplan, E. Drauglis, Surf. Sci. **36**, 1 (1973)
223. H.-D. Meyer, Surf. Sci. **104**, 117 (1981)
224. W.H. Miller, J. Chem. Phys. **53**, 1949, 3578 (1970)
225. W.H. Miller, in *Classical-Limit Quantum Mechanics and the Theory of Molecular Collisions, Advances in Chemical Physics*, vol. 25, ed. by S.A. Rice (Wiley, New York, 1974), p. 69
226. W.H. Miller, *The Classical S-Matrix in Molecular Collisions, Advances in Chemical Physics*, vol. 30, ed. by K.P. Lawley (Wiley, New York, 1975), p. 77
227. R.A. Marcus, Chem. Phys. Letters **7**, 525 (1970)
228. R.A. Marcus, J. Chem. Phys. **54**, 3965 (1971)
229. H.D. Meyer, R.D. Levine, Chem. Phys. **85**, 189 (1984)
230. V. Celli, D. Evans, in *Dynamics of Gas-Surface Interaction*, ed. by G. Benedek, U. Valbusa (Springer, Berlin, 1982), p. 2
231. J.R. Manson, Phys. Rev. B **43**, 6924 (1991)
232. J.R. Manson, V. Celli, D. Himes, Phys. Rev. B **49**, 2782 (1994)
233. J. R. Manson, *Energy Transfer to Phonons in Atom and Molecule Collisions with Surfaces*, in Handbook of Surface Science, vol. 3, Surface Dynamics, ed. by W.N. Unertl (North Holland, Amsterdam, 2006)
234. A. Bilić, B. Gumhalter, Phys. Rev. B **52**, 12307 (1995)
235. B. Gumhalter, A. Šiber, J.P. Toennies, Phys. Rev. Lett. **83**, 1375 (1999)
236. B. Gumhalter, Phys. Rep. **351**, 1 (2000)
237. A. Šiber, B. Gumhalter, Ch. Wöll, J. Phys. Cond. Matter **14**, 5193 (2002)
238. A. Šiber, B. Gumhalter, Phys. Rev. Lett. **90**, 126103 (2003)
239. G.D. Billing, J. Phys. Chem. **99**, 15378 (1995)
240. G.D. Billing, Appl. Surf. Sci. **142**, 7 (1999)
241. G.D. Billing, J. Chem. Phys. **112**, 335 (2000)

242. S. Adhikari, G.D. Billing, J. Chem. Phys. **112**, 3884 (2000)
243. G.D. Billing, *Dynamics of Molecule Surface Interactions* (Wiley, New York, 2000)
244. W. Brenig, Z. Phys. B **36**, 81 (1976)
245. R. Sedlmeir, W. Brenig, Z. Phys. B **36**, 235 (1976)
246. J. Böheim, W. Brenig, Z. Phys. B **41**, 243 (1981)
247. J. Stutzki, W. Brenig, Z. Phys. B **45**, 49 (1981)
248. D.M. Newns, Surf. Sci. **154**, 658 (1985)
249. R. Brako, D.M. Newns, Phys. Rev. Lett. **48**, 1859 (1982)
250. R. Brako, D.M. Newns, Surf. Sci. **117**, 42 (1982)
251. R. Brako, Surf. Sci. **123**, 439 (1982)
252. M. Persson, J. Harris, Surf. Sci. **187**, 67 (1987)
253. V. Bortolani, A.C. Levi, Riv. Nuovo Cim. **9**, 1 (1986)
254. V. Celli, D. Himes, P. Tran, J.P. Toennies, Ch. Wöll, G. Zhang, Phys. Rev. Lett. **66**, 3160 (1991)
255. D. Himes, V. Celli, Surf. Sci. **272**, 139 (1992)
256. K. Burke, B. Gumhalter, D.C. Langreth, Phys. Rev. B **47**, 12852 (1993)
257. B. Gumhalter, K. Burke, D.C. Langreth, Surf. Rev. Lett. **1**, 133 (1994)
258. J.R. Manson, Comp. Phys. Comm. **80**, 145 (1994)
259. W. Brenig, B. Gumhalter, J. Phys. Chem. B **108**, 14549 (2004)
260. A. Muis, J.R. Manson, Phys. Rev. B **54**, 2205 (1996)
261. A. Muis, J.R. Manson, J. Chem. Phys. **107**, 1655 (1996)
262. F. Hofmann, J.P. Toennies, J.R. Manson, Surf. Sci. **349**, L184 (1996)
263. A. Šiber, B. Gumhalter, Surf. Sci. **385**, 270 (1997)
264. A. Šiber, B. Gumhalter, Phys. Rev. Lett. **81**, 1742 (1998)
265. C. Zener, Phys. Rev. **37**, 557 (1931)
266. C. Zener, Proc. Roy. Soc. (London) **40**, 178, 335(1932)
267. J.M. Jackson, Proc. Cambridge Phil. Soc. **28**, 136 (1932)
268. J.M. Jackson, N.F. Mott, Proc. Roy. Soc. (London) A **137**, 703 (1932)
269. J.M. Jackson, A. Howarth, Proc. Roy. Soc. (London) A **142**, 447 (1933)
270. P. Morse, Phys. Rev. **34**, 57 (1929)
271. J.E. Lennard-Jones, C. Strachan, Proc. Roy. Soc. (London) A **150**, 442 (1935) (I)
272. C. Strachan, Proc. Roy. Soc. (London) A **150**, 456 (1935) (II)
273. J.E. Lennard-Jones, A.F. Devonshire, Proc. Roy. Soc. (London) A **156**, 6 (III), 29 (IV) (1936)
274. J.E. Lennard-Jones, A.F. Devonshire, Nature **137**, 1069 (1936)
275. A.F. Devonshire, Proc. Roy. Soc. (London) A **156**, 37 (1936) (V)
276. J.E. Lennard-Jones, A.F. Devonshire, Proc. Roy. Soc. (London) A **158**, 242 (VI), 253 (VII) (1937)
277. A. F. Devonshire, Proc. Roy. Soc. (London) A **158**, 269 (1937) (VIII)
278. C. Strachan, Proc. Roy. Soc. (London) A **158**, 591 (1937) (IX)
279. R. Frisch, O. Stern, Z. Phys. **84**, 430 (1933)
280. R. Frisch, Z. Phys. **84**, 443 (1933)
281. D.M. Gilbey, J. Phys. Chem. Solids **23**, 1453 (1962)
282. R.T. Allen, P. Feuer, J. Chem. Phys. **43**, 4500 (1965)
283. A.J. Howsmon, in *Proc. IV. Int. Symp. on Rarefied Gas Dynamics*, vol. 1, Rarefied Gas Dynamics, suppl. 4 (1967), p. 67
284. A.E. Sisson, A.J. Howsmon, J. Vac. Sci. Technol. **6**, 921 (1969)
285. J.W. Gadzuk, Phys. Rev. **153**, 759 (1967)
286. D. Secrest, B.R. Johnson, J. Chem. Phys. **45**, 4556 (1966)
287. D.W. Roach, R.E. Harris, J. Chem. Phys. **51**, 3404 (1969)
288. C.B. Duke, G.E. Laramore, Phys. Rev. B **2**, 4765 (1970)
289. J.E. Lennars-Jones, E.T. Goodwin, Proc. Roy. Soc. (London) A **163**, 101 (1937) (X)
290. J.E. Lennard-Jones, Proc. Roy. Soc. (London) **163**, 127 (XI), 132 (XII) (1937)
291. M. Gell-Mann, M.L. Goldberger, Phys. Rev. **91**, 398 (1953)
292. M.L. Goldberger, K.M. Watson, *Collision Theory* (Wiley, New York, 1964)

293. F.O. Goodman, Surf. Sci. **19**, 93 (1970)
294. F.O. Goodman, Surf. Sci. **26**, 327 (1971)
295. F.O. Goodman, Surf. Sci. **24**, 667 (1971), **26**, 364 (1971)
296. F.O. Goodman, Surf. Sci. **30**, 1 (1972)
297. F.O. Goodman, W.-K. Tan, J. Chem. Phys. **59**, 1805 (1973)
298. F.O. Goodman, Surf. Sci. **46**, 118 (1974)
299. G. Benedek, G. Seriani, Jpn. J. Appl. Phys. **Suppl 2**(Pt. 2), 545 (1974)
300. G.F. Miller, H. Pursey, Proc. Roy. Soc. (London) A **223**, 521 (1975)
301. M. Lagos, L. Birstein, Surf. Sci. **51**, 469 (1975)
302. M. Lagos, L. Birstein, Surf. Sci. **52**, 391 (1975)
303. A. Tsuchida, Surf. Sci. **14**, 375 (1969)
304. J.N. Smith, Surf. Sci. **34**, 613 (1973)
305. J.P. Toennies, Appl. Phys. **3**, 114 (1974)
306. W.H. Weinberg, Adv. Colloid Surf. Sci. **4**, 301 (1975)
307. F.O. Goodman, H. Wachman, in *Dynamics Aspects of Surface Physics*, ed. by F.O. Goodman (Compositori, Bologna, 1974)
308. F.O. Goodman, H. Wachman, *Dynamics of Gas-Surface Scattering* (Academic Press, Boston, 1976)
309. R. Smouchowski, Phys. Rev. **60**, 661 (1941)
310. J.L. Beeby, J. Phys. C: Solid State Phys. **5**, 3438 (1972)
311. J.L. Beeby, J. Phys. C: Solid State Phys. **5**, 3457 (1972)
312. D.R. Okeefe, J.N. Smith, R.L. Palmer, H. Saltsburg Surf. Sci. **20**, 27 (1970)
313. R.L. Palmer, D.R. O'Keefe, H. Saltsburg, J.N. Smith, J. Vac. Sci. Technol. **7**, 91 (1970)
314. J.L. Beeby, J. Phys. C: Solid State Phys. **4**, L359 (1971)
315. D.V. Tendulkar, R.E. Stickney, Surf. Sci. **27**, 516 (1971)
316. U. Garibaldi, A.C. Levi, R. Spadacini, G.E. Tommei, Surf. Sci. **38**, 269 (1973)
317. U. Garibaldi, A.C. Levi, R. Spadacini, G.E. Tommei, Surf. Sci. **48**, 649 (1975)
318. A.C. Levi, H.G. Suhl, Surf. Sci. **88**, 221 (1979)
319. J.D. McClure, J. Chem. Phys. **52**, 2712 (1970). See also Rayleigh's theory of sound wave scattering from an undulated surface in [1], p. 272
320. A. C. Levi, Nuovo Cim. **54** B, 357 (1979)
321. R.B. Gerber, A.T. Yinnon, J.N. Murrel, Chem. Phys. **31**, 1 (1978)
322. R. Schinke, R.B. Gerber, J. Chem. Phys. **82**, 1567 (1985)
323. J.E. Adams, W.H. Miller, Surf. Sci. **85**, 77 (1979)
324. R.I. Masel, R.P. Merrill, W.H. Miller, J. Chem. Phys. **65**, 2690 (1976)
325. N. Van Trong, Phys. Stat. Solidi (b) **93**, 45 (1979)
326. R. Conn, Phys. Rev. A **6**, 439 (1972)
327. G. Platero, V.R. Velasco, F. Garcia-Moliner, G. Benedek, L. Miglio, Surf. Sci. **143**, 244 (1984)
328. A.C. Levi, G. Benedek, L. Miglio, G. Platero, V.R. Velasco, F. Garcia-Moliner, Surf. Sci. **143**, 253 (1984)
329. H. Hoinkes, H. Nahr, H. Wilsch, Surf. Sci. **33**, 516 (1972)
330. H. Hoinkes, H. Nahr, H. Wilsch, Surf. Sci. **40**, 457 (1973)
331. A.A. Marvin, F. Toigo, Nuovo Cim. B **53**, 25 (1979)
332. M. Lagos, Surf. Sci. **71**, 414 (1978); see, however, Levi's comment in [279]
333. G. Armand, J. Lapujoulade, Y. Lejay, J. Phys. (Paris) **37**, L187 (1976)
334. G. Armand, J. Lapujoulade, Y. Lejay, Surf. Sci. **63**, 143 (1977)
335. J. Lapujoulade, Surf. Sci. **134**, 529 (1983)
336. G. Armand, J.R. Manson, Phys. Rev. Lett. **53**, 1112 (1984)
337. G. Armand, J.R. Manson, in *Dynamics on Surfaces*, ed. by B. Pullman, J. Jortner, A. Nitzan, B. Gerber (Reidel, Dordrecht, 1984), p. 59
338. J. Lapujoulade, Y. Lejay, G. Armand, Surf. Sci. **95**, 107 (1980)
339. G. Doyen, T.B. Grimley, Surf. Sci. **91**, 51 (1980)
340. G. Armand, J.R. Manson, Surf. Sci. **80**, 532 (1979)

341. N. Garcia, J. Ibanez, J. Solana, N. Cabrera, Surf. Sci. **60**, 385 (1976)
342. N. Garcia, Phys. Rev. Lett. **37**, 912 (1976)
343. N. Garcia, J. Chem. Phys. **67**, 897 (1977)
344. G. Benedek, N. Garcia, Ned. Tijdsch. Vacuumtechniek **16**, 344 (1978)
345. G. Benedek, N. Garcia, Surf. Sci. **80**, 543 (1979)
346. G. Benedek, N. Garcia, Le Vide **Suppl. 201**, 818 (1980)
347. G. Benedek, N. Garcia, Surf. Sci. **103**, L143 (1981)
348. G. Benedek, in *Dynamics of Gas-Surface Interaction*, ed. by G. Benedek, U. Valbusa (Springer, Berlin, 1982), p. 227
349. G. Benedek, G. Brusdeylins, R.B. Doak, J.P. Toennies, J. Phys. (Paris) **42**, C6–793 (1981)
350. G. Benedek, R.B. Doak, J.P. Toennies, Phys. Rev. B **28**, 7277 (1983)
351. J.H. Weare, J. Chem. Phys. **61**, 2900 (1974)
352. Y.W. Lin, S.A. Adelman, G. Wolken, Surf. Sci. **66**, 376 (1977)
353. E. Semerad, E.M. Hörl, Surf. Sci. **126**, 661 (1983), observed an enhanced role of bulk phonons for scattering of Ne atoms in the one-phonon regime. However, under kinematic focussing conditions RW's and also the optical Lucas modes (LM) seem to recover their major weight even for Ne scattering, as seen in [122, 125–126]
354. It was discovered by Doak, Harten and Toennies, [139], that in metals the surface longitudinal acoustic (LA) resonance (see Chapter 3) can yield large HAS intensities, sometimes even larger than RW's (anomalous longitudinal resonance). G. Benedek, L. Miglio, Z. Phys. B **50**, 93 (1983), have shown that in those symmetry directions where the RW is weakly localized the LA resonance may be dominant even in insulators
355. J. Hutchison, V. Celli, N. R. Hill, M. Haller, in *Rarified Gas Dynamics*, ed. by S.S. Fisher, Progr. Astron. Aeron. **74**, 129 (1981)
356. V. Celli, N. Garcia, J. Hutchison, Surf. Sci. **87**, 112 (1979)
357. D. Evans, V. Celli, G. Benedek, J.P. Toennies, R.B. Doak, Phys. Rev. Lett. **50**, 1854 (1983)
358. W.L. Nichols, J.H. Weare, Phys. Rev. Lett. **56**, 753 (1986)
359. G. Lilienkamp, J.P. Toennies, Phys. Rev. B **26**, 4752 (1982)
360. G. Lilienkamp, J.P. Toennies, J. Chem. Phys. **78**, 5210 (1983)
361. G. Benedek, S. Miret-Artes, Surf. Sci. **339**, L935 (1995)
362. G. Benedek, R. Gerlach, A. Glebov, G. Lange, S. Miret-Artes, J.G. Skofronick, J. P. Toennies, Phys. Rev. B **53**, 11211 (1996)
363. G. Brusdeylins, R. Rechsteiner, J.G. Skofronick, J.P. Toennies, G. Benedek, L. Miglio, Phys. Rev. Lett. **54**, 466 (1985)
364. G. Bracco, M. D'Avanzo, C. Salvo, R. Tatarek, F. Tommasini, Surf. Sci. **189**(190), 684 (1987)
365. G. Wolken, J. Chem. Phys. **60**, 2210 (1974)
366. Y.W. Lin, G. Wolken Jr., J. Chem. Phys. **65**, 2634 (1976)
367. D. Eichenauer, J.P. Toennies, in *Dynamics on Surfaces*, ed. by B. Pullman, J. Jortner, A. Nitzan, B. Gerber (Reidel, Dordrecht, 1984), p. 1
368. D. Eichenauer, J.P. Toennies, J. Chem. Phys. **85**, 532 (1986)
369. B.H. Choi, R.T. Poe, J. Chem. Phys. **83**, 1330 (1985)
370. B.H. Choi, R.T. Poe, J. Chem. Phys. **83**, 1344 (1985)
371. H. Metiu, J. Chem. Phys. **67**, 5456 (1977)
372. W. Brenig, B. Gumhalter, J. Phys. Chem. B **108**, 14549 (2004)
373. B. Gumhalter, D.C. Langreth, Phys. Rev. B **60**, 2789 (1999)
374. W. Brenig, Phys. Rev. Lett. **92**, 056102 (2004)
375. A. Šiber, B. Gumhalter, J. Phys.: Condens. Matter **20**, 224002 (2008)
376. A.L. Robinson, Science **218**, 40 (1982)
377. J.P. Toennies, in *Solvay Conference on Surface Science*, ed. by F.W. de Wette, Springer Ser. Surf. Sci., vol. **14** (Springer, Berlin, 1989)
378. C. Kaden, P. Ruggerone, J.P. Toennies, G. Zhang, G. Benedek, Phys. Rev. B **46**, 13509 (1992)
379. R. Heid, K.P. Bohnen, Phys. Rep. **387**, 151 (2003)

380. G. Benedek, M. Bernasconi, V. Chis, E. Chulkov, P.M. Echenique, B. Hellsing, J. P. Toennies, J. Phys. Cond. Matter **21**, 084020 (2010)
381. V. Chis, B. Hellsing, G. Benedek, M. Bernasconi, E.V. Chulkov, J.P. Toennies, Phys. Rev. Lett., **101**, 206102 (2008); **103**, 069902 (2009)
382. J. Braun, P. Ruggerone, G. Zhang, J.P. Toennies, G. Benedek, Phys. Rev. B **79**, 205423 (2009)
383. I.Y. Sklyadneva, G. Benedek, E.V. Chulkov, P.M. Echenique, R. Heid, K.-P. Bohnen, J. P. Toennies, Phys. Rev. Lett. **107**, 095502 (2011)
384. G. Benedek, M. Bernasconi, K.-P. Bohnen, D. Campi, E.V. Chulkov, P.M. Echenique, R. Heid, IYu. Sklyadneva, J.P. Toennies, Phys. Chem. Chem. Phys. **16**, 7159 (2014)
385. G. Benedek, U. Valbusa (eds.), *Dynamics of Gas-Surface Interaction,* Springer Series in Chemical Physics, vol. 21. (Springer, Berlin, 1981)
386. V. Celli, in *Many-Body Phenomena at Surfaces*, ed. by D. Langreth, H. Suhl (Academic, Orlando, 1984), p. 315
387. V. Celli, in *Surface Phonons* (Chap. 6), ed. by F.W. de Wette, W. Kress (Springer, Berlin, 1992)
388. F. Garcia-Moliner, in *Interaction of Atoms and Molecules with Solid Surfaces*, ed. by V. Bortolani, N.H. March, M.P. Tosi (Plenum, New York, 1990), pp. 71–116
389. Ch. Wöll, Appl. Phys. A **53**, 377 (1991)
390. R.B. Doak, in *Atomic and Molecular Beam Methods*, vol. 2, ed. by G. Scoles, D. Lané, U. Valbusa (Oxford University Press, New York, 1992), p. 384
391. J.P. Toennies, Superlattices Microstr. **7**(3), 193 (1990)
392. J.P. Toennies, J. Phys.: Condens. Matter **5**, A25–A40 (1993)
393. F. Hofmann, J.P. Toennies, Chem. Rev. **96**, 1307 (1996)
394. S.A. Safron, in *Advances in Chemical Physics*, vol. XCV, ed. by I. Prigogine and S.A. Rice (Wiley, New York, 1996)
395. G.G. Rusina, E.V. Chulkov, Russ. Chem. Rev. **82**(6), 483–510 (2013)
396. A.P. Jardine, H. Hedgeland, G. Alexandrowicz, W. Allison, J. Ellis, Prog. Surf. Sci. **84**, 323 (2009)
397. E. Hulpke (ed.), *Helium Atom Scattering from Surfaces* (Springer, Berlin, 1992)
398. V. Celli, *Interaction of Atoms with Surfaces*, in Landolt-Börnstein, New Series III, vol. 24: Physics of Solid Surfaces, ed. by G. Chiarotti (Springer, Berlin, 1994/1995), Part c, p. 278
399. R.F. Wallis, S.Y. Tong, *Surface Phonons*, in Landolt-Börnstein, New Series III, vol. 24: Physics of Solid Surfaces, ed. by G. Chiarotti (Springer, Berlin, 1994/1995) Part b, p. 434
400. M. Rocca, *Surface Phonon Dispersion,* in Landolt-Börstein, New Series III/42,A2/4,5 (Springer, Berlin, 2002)
401. G. Benedek, D. Campi, J.P. Toennies, In Physics of Solid Surfaces. Subvol. A, Landolf-Börnstein, New Series 45A ed. by P. Chiaradia and G.F. Ciarotti, Chap. 10, p. 572–646 (2015)
402. G. Benedek, J.P. Toennies, in *The First Thirty Years of Surface Science*, ed. by C.B. Duke (Elsevier, Amsterdam, 1994), and Surf. Sci. **299/300**, 587 (1994)

Chapter 3
The Physics of Surface Phonons

For those well ordered motions, and regular paces,
though they give no sound unto the ear,
yet to the understanding
they strike a note most full of harmony.

Sir Thomas Browne
(from Ziman, Principles of the Theory of Solids, 1972)

Abstract The effect of the surface termination on the phonon dispersion curves is illustrated by comparison with the bulk dispersion curves. After a review of Rayleigh's theory of surface waves in elastic media the atomistic Born-von Karman theory for bulk phonons is presented. How the surface affects the surface vibrations is illustrated by the Green's function theory and slab calculations. The chapter closes with some illustrations of surface phonon dispersion curves and their classification.

3.1 The Dispersion Curves of Surface Versus Bulk Phonons

A basic goal of surface phonon studies is to obtain detailed microscopic insight into the forces between atoms at the surface of a crystal. These forces are of fundamental importance for understanding the static and dynamic properties at surfaces. Since at present much is known about these forces in the bulk it is expected that this goal can be reached by understanding the origin of the similarities and differences in not only the phonons, but also the structures occurring at the surface with respect to the bulk.

As discussed in Chap. 1 surface phonons are vibrational modes not present in the bulk of a crystal which appear at the surface as a result of the surface boundary conditions. Of course the bulk modes which propagate inside the solid also contribute to the vibrational amplitudes of the surface atoms and are slightly affected by the surface boundary conditions. Some examples of important similarities and differences between surface and bulk phonon dispersion curves are illustrated for

G. Benedek and J. P. Toennies, *Atomic Scale Dynamics at Surfaces*, Springer Series in Surface Sciences 63, https://doi.org/10.1007/978-3-662-56443-1_3

four systems representing different types of materials in Fig. 3.1. For each system the dispersion curve of the lowest frequency bulk mode, which is one of the two orthogonally polarized transverse acoustic (TA) modes measured using neutrons, is compared with the dispersion curve of the corresponding surface phonon obtained by HAS measurements along an equivalent crystallographic direction. The correspondence between the directions of the bulk phonon wavevector in the 3D Brillouin zone (BZ) and those of the surface phonon wavevector in the surface Brillouin zone is shown in Fig. 3.2 for the different surfaces considered in Fig. 3.1 [1–3].

As illustrated in Fig. 3.1a for NaF(001) [4, 5] the phonon dispersion curves and correspondingly the interatomic forces at the surface planes of crystals made of

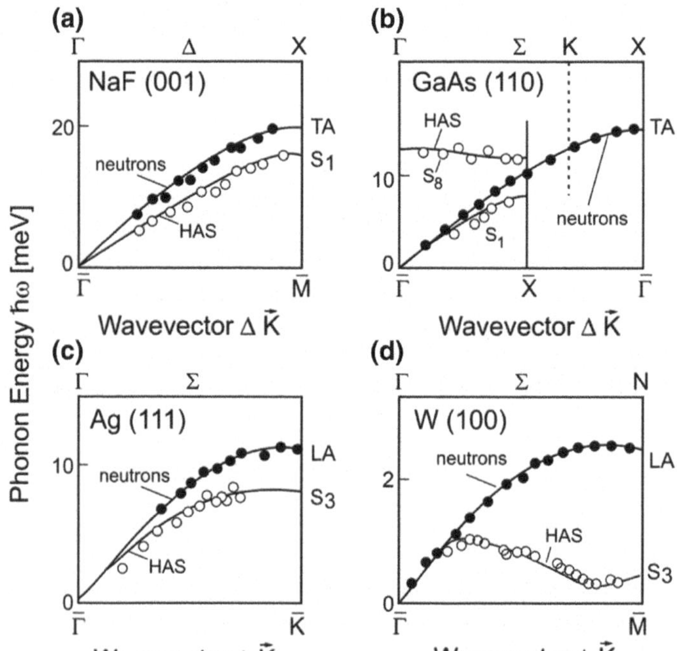

Fig. 3.1 Acoustic phonon dispersion curves measured by helium atom scattering (open circles) at the surface of various types of solids are compared to those obtained with neutron scattering inside the bulk (solid circles) along the same crystallographic direction. (The labelling of the dispersion curves is summarized in Table 3.1 at the end of the chapter.) **a** For the closed-shell ionic NaF(001) [4, 5] the surface Rayleigh wave (RW, S_1) dispersion curve is similar to the transverse acoustic (TA) bulk mode dispersion curves. **b** For GaAs(110) [6, 7] reconstruction leads to a back folding of the Rayleigh wave into an upper branch S_8 with a splitting at $\overline{X}$ and the surface Brillouin zone is only half that in the bulk. **c** For the fcc noble metal Ag(111) [8, 9] and **d** the bcc transition metal W (001) at 500 K [10, 11] the anomalous longitudinal (S_3) surface branches are compared with the corresponding longitudinal acoustic (LA) bulk modes. For W(001) the S_3 branch eventually softens to zero energy at a lower temperature leading to a 2×1 surface reconstruction and a halving of the BZ similar to that of panel (**b**) (see Chap. 11)

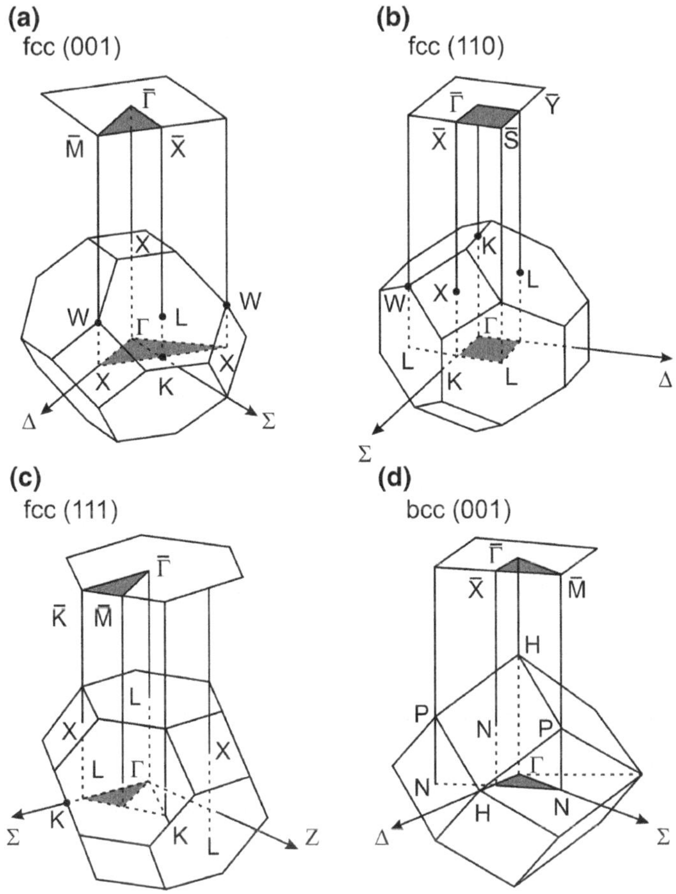

Fig. 3.2 Correspondence between the bulk and the surface Brillouin zones for the four surfaces considered in Fig. 3.1: **a** fcc(001), **b** fcc(110), **c** fcc(111) and **d** bcc(001) (adapted from Plummer and Eberhardt [1], see also [2])

closed shell ions or atoms, such as the alkali halides, or the rare gas solids, differ only slightly with respect to the corresponding atomic planes inside the bulk. Apart from some hybridization induced by the lower point-group symmetry, the atoms or ions at the surface of a closed shell solid have about the same electronic structure as in the bulk. The observed reduction in the frequency of the surface modes thus results largely from the reduced number of nearest neighbors at the surface.

The surfaces of crystals which are not composed of closed shell atoms or ions show a much more complex behaviour as illustrated in Fig. 3.1b for GaAs. In covalent homopolar semiconductors, like diamond, Si and Ge, the equal share of the bond charge between the two newly created surfaces of the cleaved crystal initially leaves half filled bands consisting of dangling bonds at the surfaces. This situation is unstable against any reconstruction which reduces the local symmetry and causes a splitting of the half-filled band, thereby lowering the energy of the filled (bonding) surface states and raising the energy of the empty (antibonding) ones. If the surface temperature is much less than the splitting energy between the bonding and antibonding surface states the surface can reconstruct and form a variety of new surface structures. A similar reconstruction occurs in heteropolar tetrahedrally-coordinated semiconductors, such as GaAs(110), where in the bulk the bonding charge is not equally shared by neighboring atoms. In this case the charge transfer between surface ions suffices to stabilize the surface into a buckled, unreconstructed phase [12] with a surface BZ which is twice as small as the corresponding section of the bulk BZ. Consequently, as shown in Fig. 3.1b, the surface phonon dispersion curve of GaAs(110) along the $[1\bar{1}0]$ direction is folded at the surface BZ edge [6], whereas the corresponding bulk mode extends up to the bulk BZ edge [7].

Despite the complexity of the associated structures the bond lengths show only little variations as compared to the larger changes in bond angles resulting from the reduced coordination. The resulting buckling at the surface enlarges the gap between the folded surface phonon branches [6, 12, 13]. Thus the buckling affects the force constants mostly through the rotational invariance condition (Sect. 4.1) referred to the new equilibrium configuration [13]. This example illustrates how it has been possible to discriminate between subtle structural differences from an analysis of the frequencies of zone boundary surface phonons. For example, the buckling angles of π-bonded chains in Si(111)2×1 or GaAs(110) have been inferred in this way [6, 13]. For a more extended discussion see Sect. 11.2.

The notation used for crystallographic directions and planes throughout this monograph are those of the International Union of Crystallography [14]. Except for hexagonal lattices, three index notations are used also for specific crystallographic directions [*lmn*] on a given surface (*ijk*) so as to fulfill the orthogonality condition $li + mj + nk = 0$. Occasionally families of equivalent directions on a surface may also be needed and indicated by <*lmn*>, being intended in this case that only equivalent directions which fulfill the orthogonality conditions are included in the family. For hexagonal lattices four index notations, [*lmnt*] with $l + m + n = 0$, are used with the same conventions for families.

The effects found in metals are illustrated in Fig. 3.1c, where the anomalous longitudinal (AL) surface phonon branch S_3 observed in Ag(111) along the $[1\bar{1}0]$ direction at room temperature [8] is compared with the bulk longitudinal acoustic (LA) branch [9]. In metals the partitioning of the free electrons between the two newly created surfaces and the resulting rearrangement of electrons in real space and on the Fermi surface leads to a lattice instability. This intimate connection between structure and electronic distortion is not surprising since the interactions between ion cores in metals are mediated by the interposed electrons. Even for unreconstructed metal surfaces the changes in surface force constants required to explain the experimental surface phonon frequencies are so significant that they can only be described in terms of many-body effects. Moreover the redistribution of the electronic charge at the surface also invariably leads to an inward structural relaxation of the outermost layer [15], which also affects the surface force constants (see Table 1.1). In the case of Ag(111) these effects explain why the surface phonon branch in Fig. 3.1c shows a sizeable softening towards the zone edge as will be discussed in Sect. 8.3. Another related quantity is the surface stress (see Chap. 4), which, in turn, can also explain certain anomalies in surface phonon dispersion curves [16].

For some metal surfaces the wavevectors of electronic transitions between surface states near the Fermi level may be concentrated around certain wavevector values (*nesting*). Since the energies, although negligible on the electronic energy scale, can be comparable to the phonon energies a singularity appears in the electron susceptibility. At these nesting wavevectors a strong electron-phonon coupling occurs, leading to phonon softening, i.e., to depressions in the surface phonon dispersion curves, referred to as *Kohn anomalies* [17]. A spectacular case is that of W(001), where the surface branch along the [100] direction at 500 K [10, 18, 19] shows a large depression whereas the corresponding bulk LA phonons show a regular behaviour [11]. The anomalous surface branch eventually softens to zero energy at a lower temperature, thus leading to a 2×1 surface reconstruction and a halving of the BZ similar to that illustrated in Fig. 3.1b. A complete discussion is presented in Sect. 11.3.5.

In a few extreme cases, such as W(110): H(1×1), a soft mode appears at some intermediate wavevector, which leads to a vanishing of the surface phonon frequency. The associated *surface soft mode* can drive a second-order phase transition leading to a reconstructed phase, whose point symmetry group is a subgroup of the unreconstructed phase. A surface soft mode can appear at some wavevector under certain external conditions, e.g., at a given critical temperature or stress. Indeed the remarkable softening of the surface phonon branch of W(001) (Fig. 3.1d) eventually drives a $\sqrt{2} \times \sqrt{2}$ surface reconstruction at temperatures below a critical temperature of about 280 K (see Sect. 11.3) [10, 18, 19].

3.2 Rayleigh Theory of Surface Waves

3.2.1 Isotropic Elastic Continuum

This section is devoted to the theory developed by Lord Rayleigh in 1885 (see Chap. 1) for the semi-finite isotropic elastic continuum. It provides an instructive introduction to the lattice dynamics of surfaces presented in the next sections. Rayleigh was the first to predict the propagation velocity and the displacement field of surface vibrations in the extreme acoustic limit of very large wavelengths, $Q = 2\pi/\lambda \to 0$ [20–22]. His theory considers a semi-infinite homogeneous isotropic cubic crystal filling the space at $z \leq 0$ below a free planar surface at $z = 0$. The elastic constants are assumed to satisfy the following isotropy condition

$$\eta \equiv 2c_{44}/(c_{11} - c_{12}) = 1, \tag{3.1}$$

where η is the isotropy factor, also denoted as the anisotropy ratio. Thus in the bulk of an isotropic crystal only two elastic constants are independent. These are usually taken to be c_{11} and c_{44} since they determine the phase velocities of the longitudinal (L) and transverse (T) polarized sound waves in the bulk

$$v_L = (c_{11}/\rho)^{1/2}, \quad v_T = (c_{44}/\rho)^{1/2}, \tag{3.2}$$

respectively, where ρ is the crystal density. Due to isotropy both v_L and v_T are independent of the direction of propagation in the bulk.

The special sound wave predicted by Rayleigh, now called the Rayleigh wave (RW), propagates along the surface with a phase velocity v_R smaller than both v_L and v_T, which, in the isotropic bulk, is also independent of the propagation direction. The displacement field of the RW is largely transverse polarized and localized at the surface and decays exponentially inside the crystal as illustrated in Fig. 1.2. The reduced squared RW velocity, defined by $x_T \equiv v_R^2/v_T^2$, is a solution of Rayleigh's cubic equation [20–23]

$$\frac{1}{16}x_T^3 - \frac{1}{2}x_T^2 + \left(\frac{3}{2} - \frac{v_T^2}{v_L^2}\right)x_T + \frac{v_T^2}{v_L^2} - 1 = 0. \tag{3.3}$$

This equation has only one meaningful solution, which can be approximated by [24]

$$\frac{v_R}{v_T} \cong \frac{0.72 - (v_T/v_L)^2}{0.75 - (v_T/v_L)^2}. \tag{3.4}$$

In an isotropic solid the ratio $(v_T/v_L)^2 = c_{44}/c_{11}$ is (for positive c_{12}) less than 0.5, and therefore the reduced RW velocity varies within the narrow range $0.88 < v_R/v_T < 0.9$. In real anisotropic solids these limits are relaxed and v_R/v_T may be smaller than 0.88 or between 0.90 and unity, depending on the anisotropy. In some directions, v_R can be even larger than the phase velocity of the bulk transverse modes propagating in the same direction. In this case the RW is referred to as a pseudosurface wave (see also Sect. 3.6).

The RW field is a plane wave propagating with a surface (two-dimensional) wavevector $\mathbf{Q} = (Q_x, Q_y)$ in the surface plane $\mathbf{R} = (x, y)$ and an amplitude given by

$$\mathbf{e}(\mathbf{R}, z, t) = \frac{1}{2\pi}\mathbf{e}(\mathbf{Q}, z)\exp[i(\mathbf{Q}\cdot\mathbf{R} - \omega t)], \tag{3.5}$$

where the dependence on the depth z is expressed by the displacement field amplitude $\mathbf{e}(\mathbf{Q}, z)$ (Fig. 1.2b). The wavevector modulus Q is related to the RW wavelength λ_R by $Q = 2\pi/\lambda_R$. Since the solid is assumed to be homogeneous and continuous, the RW field has a linear dispersion, in which the frequency is linearly related to the wavevector by $\omega = v_R Q$. In this case the phase and group velocities are equal. For propagation along the x-axis, $\mathbf{Q} = (\pm Q, 0)$, the displacement field amplitude has the following Cartesian components [22]:

$$e_x(\mathbf{Q}, z) = \mp(A\omega)^{1/2}(\beta_T^{1/2}\mathrm{e}^{Q\beta_T z} - \beta_L^{-1/2}\mathrm{e}^{Q\beta_L z}) \tag{3.6}$$

$$e_y(\mathbf{Q}, z) = 0 \tag{3.7}$$

$$e_z(\mathbf{Q}, z) = i(A\omega)^{1/2}(\beta_T^{-1/2}\mathrm{e}^{Q\beta_T z} - \beta_L^{1/2}\mathrm{e}^{Q\beta_L z}), \tag{3.8}$$

where A is a normalization constant and

$$\beta_T \equiv \left(1 - \frac{v_R^2}{v_T^2}\right)^{1/2}, \quad \beta_L \equiv \left(1 - \frac{v_R^2}{v_L^2}\right)^{1/2}. \tag{3.9}$$

The normalization of the squared RW displacement field amplitude to unity requires

$$A\omega = \frac{Q(\beta_T\beta_L)^2}{(\beta_T\beta_L)^2 - 2(\beta_T\beta_L)^{3/2} + \frac{1}{2}(\beta_T^2 + \beta_L^2)}. \tag{3.10}$$

In the case of a discrete lattice $z = -al_3$, with $l_3 = 0, 1, 2, \ldots$, and a the interlayer distance, the normalization expressed by (3.10) remains valid in the long wave limit $Qa \ll 1$, provided the factor Q is replaced by Qa. In this case the polarization vector $\mathbf{e}(\mathbf{Q}, l_3)$ is dimensionless since the integration over z is replaced by a sum over l_3.

Equations (3.6)–(3.8) show that the RW is polarized in the sagittal (xz) plane with the vertical z-component oscillating out of phase by $\pi/2$ with respect to the longitudinal x-component, as indicated by the factor $i \equiv \mathrm{e}^{i\pi/2}$ in the expression for $e_z(\mathbf{Q}, z)$ (3.8). This dephasing is responsible for the *retrograde elliptical polarization* of the RW illustrated in Fig. 1.2a of Chap. 1. The ellipticity of the polarization at the surface ($z = 0$) for the RW propagating in the $\pm x$ direction is expressed by the amplitude ratio

$$e_x(\mathbf{Q}, 0)/e_z(\mathbf{Q}, 0) = \pm i(\beta_T/\beta_L)^{1/2} \tag{3.11}$$

and is independent of Q. From (3.6) and (3.8) (see also Fig. 1.2b) the RW field decays exponentially inside the crystal for $z \to -\infty$ and each Cartesian component is the sum of two terms with different decay lengths Λ_T and Λ_L,

$$\Lambda_T = \frac{1}{Q\beta_T} = \frac{\lambda_R}{2\pi\beta_T}, \quad \Lambda_L = \frac{1}{Q\beta_L} = \frac{\lambda_R}{2\pi\beta_L}, \tag{3.12}$$

where $\lambda_R = 2\pi/Q$ is the RW wavelength. Since in general $v_L \gg v_T$, from (3.9) it follows that β_L is appreciably larger than β_T and therefore the term with Λ_T has a slower decay and is, asymptotically, the dominant one. Moreover the term proportional to $\beta_T^{-1/2}$ in (3.8) dominates over the other three terms in (3.6) and (3.8) because v_R/v_T is only slightly less than unity, and therefore the vertical component $e_z(\mathbf{Q}, z)$ is at all depths larger in absolute value than the longitudinal component $e_x(\mathbf{Q}, z)$. This gives the RW a quasi-shear vertical character with a vertically elongated polarization ellipse throughout the surface region (see Fig. 1.2).

An important feature of a RW, which derives from its *macroscopic* character, is the proportionality of the decay lengths to the RW wavelength, (3.12). This implies that the RW amplitudes at the surface ($z = 0$)

$$e_x(\mathbf{Q}, 0) = \mp(A\omega)^{1/2}(\beta_T^{1/2} - \beta_L^{-1/2}), \tag{3.13a}$$

$$e_z(\mathbf{Q}, 0) = i(A\omega)^{1/2}(\beta_T^{-1/2} - \beta_L^{1/2}), \tag{3.13b}$$

vanish for $\omega \to 0$, due to the diverging decay lengths, and increase with the frequency like $\omega^{1/2}$. This behaviour has an interesting consequence for the effect of the RW field, on the actual quantum zero point displacements of the surface atoms. For a monatomic crystal with atoms of mass M the quantum zero point displacements are given by $\mathbf{e}(\mathbf{Q}, 0)$ multiplied by the zero point displacement of the harmonic oscillator of mass M and angular frequency ω [25],

$$\mathbf{u}_{RW}(\mathbf{Q}, 0) = \left(\frac{\hbar}{2M\omega}\right)^{1/2} \mathbf{e}(\mathbf{Q}, 0). \tag{3.14}$$

Thus $\mathbf{u}_{RW}(\mathbf{Q}, 0)$ is *independent of the frequency* since the factor $\omega^{1/2}$ in $\mathbf{e}(\mathbf{Q}, 0)$ (3.13a, b) cancels out in (3.14).

3.2.2 Dispersive Effects

Although these properties of Rayleigh waves are only strictly valid for an isotropic continuum, they can be extended in a first approximation to the dispersive region of a discrete lattice by assuming a simple sinusoidal dispersion relation for the RW frequency

$$\omega(Q) \cong \omega_o \sin\left(\frac{1}{2}Qa\right), \tag{3.15}$$

where a is the interplanar distance and ω_o is the maximum frequency corresponding to the minimum wavelength $\lambda_{R,\min} = 2\pi/Q_{\max} = 2a$. At $\omega(\pi/a) = \omega_0$ the group velocity vanishes and the amplitudes, (3.6–3.8), at $\mathbf{Q} = (\pi/a, 0)$ and $-\mathbf{Q}$ must be equal. Thus the dispersion introduced by the lattice requires $e_x(\pm\pi/a, 0) = 0$, and therefore the elliptical polarization evolves into a shear-vertical linear polarization at $\mathbf{Q} = (\pm\pi/a, 0)$. This dispersive effect may be accounted for by multiplying the r.h. members of (3.6), (3.11) and (3.13a, b) by the correcting factor $\cos^{1/2}\left(\frac{1}{2}Qa\right)$. By replacing (3.15) into the corrected (3.13a, b) a useful approximation for the RW amplitudes in the dispersive case is obtained:

$$e_x(\mathbf{Q},0) \cong \mp(A\omega_o)^{1/2}(\beta_T^{1/2} - \beta_L^{-1/2})\left[\frac{1}{2}\sin(Qa)\right]^{1/2}, \tag{3.16a}$$

$$e_z(\mathbf{Q},0) \cong i(A\omega_o)^{1/2}(\beta_T^{-1/2} - \beta_L^{1/2})\left[\sin\left(\frac{1}{2}Qa\right)\right]^{1/2}. \tag{3.16b}$$

Also with an assumed dispersion it follows from (3.14) that the RW quantum zero point displacements of the surface atoms are approximately independent of Q over the entire surface BZ.

The atomistic theory of surface lattice dynamics, which considers the discrete atomic structure of the crystal, predicts in general as many surface modes for $Q \to 0$ as there are degrees of freedom in the crystal unit cell (see next section). As a consequence of the invariance of the total energy for infinitesimal rigid translations in the three Cartesian directions three of these modes, the so-called acoustic modes, must have a vanishing frequency for $Q \to 0$. One of these modes with a quasi-shear vertical polarization corresponds to the Rayleigh mode described above, whereas the other two modes, which have quasi-longitudinal and shear-horizontal polarizations, are not predicted by the continuum theory. As a result for $Q \to 0$ they degenerate into a longitudinal and a transverse *bulk* mode, respectively. As shown by Alldredge [26] their limiting propagation velocities coincide with the respective bulk velocities and their penetration lengths, unlike (3.12), diverge at least as fast as Q^{-2}. Moreover their polarization vectors at the surface layer behave approximately like $\sin^n(Qa/2)$ with $n \geq 1$. For a detailed discussion of the nature of the different polarizations of surface modes the reader is referred to Figs. 3.9, 3.10 and Sect. 3.7.

3.2.3 *Anisotropic Crystals*

Single crystals of cubic symmetry are in general anisotropic ($\eta \neq 1$) with respect to the elastic properties since the three independent elastic constants do not fulfill, if not fortuitously, the isotropy condition, (3.1). Whether the anisotropy is positive ($\eta > 1$) or negative ($\eta < 1$) depends on the microscopic constitution of the solid, i.e., on the nature of interatomic forces. For anisotropic cubic crystals, however, many of the general properties discussed for the isotropic (semi-infinite) continuum above remain approximately valid, though no simple analytical expression can be obtained for the RW velocity and amplitudes. The bulk anisotropy implies an anisotropy also for the Rayleigh waves with respect to the surface orientation and to the propagation direction on a given surface [22, 23]. For example, for the (001) surface of a cubic crystal v_R increases with rotation of the azimuth from the [100] to the [110] direction for $\eta < 1$ (Fig. 3.3a) and decreases for $\eta > 1$ (Fig. 3.3b). Also the bulk transverse velocities, $v_{T\perp}$ and $v_{T\|}$, for polarizations normal (shear vertical, SV) and parallel (shear horizontal, SH) to the surface, depend on the propagation direction and are generally different (see Sect. 3.7). Since in the [100] direction they are degenerate, $v_{T\perp} = v_{T\|} \equiv v_T$, this velocity is taken as a reference value. With rotation of the azimuth from the [100] to the [110] direction, $v_{T\perp}$ remains constant, whereas $v_{T\|}$ changes to $v_{T\perp}/\eta^{1/2}$. Thus for $\eta < 1$ (Fig. 3.3a) the SH bulk velocity $v_{T\|}$ is always greater than v_T and therefore does not interfere with the RW velocity which is always below v_T. For $\eta > 1$ the SH bulk velocity, however, falls below v_T and may interfere with the RW velocity. In this case (Fig. 3.3b) a strong hybridization may occur between the RW and the bulk modes whose three-dimensional wavevectors have the same parallel component **Q** as the RW. This leads to an avoided crossing of the RW branch with the bulk modes and the appearance of a pseudo-surface wave (PSW) branch as a prolongation of the RW branch into the continuum of bulk modes. The PSW is interpreted as a resonance between the RW and a bulk mode of equal frequency and parallel wavevector, as will be discussed in more detail in Sect. 3.7.1.

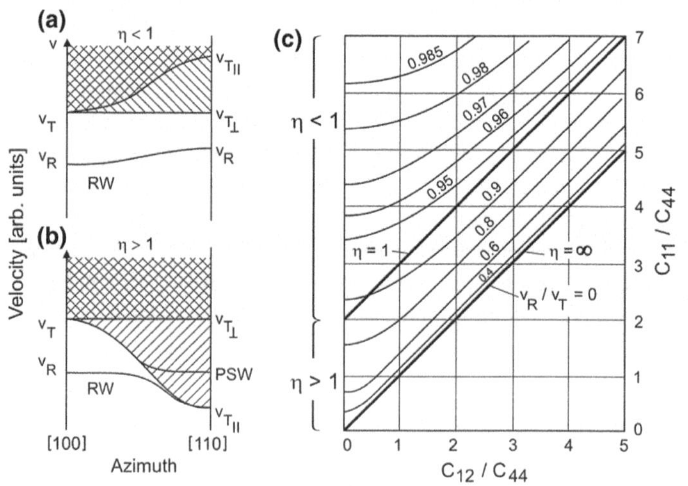

Fig. 3.3 Rayleigh wave and bulk transverse velocities at the (001) surface of a cubic crystal as a function of the propagation azimuth for an anisotropy constant $\eta < 1$ (**a**) and for $\eta > 1$ (**b**), where η is defined in (3.1). In **b** PSW stands for *pseudo-surface wave*. In **c** the reduced RW velocity v_R/v_T, is plotted as a function of two ratios of bulk elastic constants. v_R/v_T increases from the straight line for $v_R/v_T = 0$ and approaches unity in the upper left-hand corner. The two heavy straight lines in **c** indicate the isotropy condition ($\eta = 1$), according to (3.1), and the extreme case of infinite anisotropy $\eta = \infty$, which corresponds to $v_R/v_T = 0$ (adapted from Farnell [22, 23])

The relative RW velocity v_R/v_T along the [100] azimuth for the (001) surface of a cubic crystal can be obtained as a function of the reduced bulk elastic constants from the curves plotted in Fig. 3.3c [22, 23]. The isotropy condition $\eta = 1$ is represented by a heavy line separating an upper region for $\eta < 1$ from a lower region for $\eta > 1$. Unlike the isotropic case, where v_R/v_T was confined to a narrow range of values, now v_R/v_T can take any value between 0 (for $\eta \to \infty$) and 1 (for $\eta \to 0$). Figure 3.3 is very useful for checking the continuum limit of the RW frequencies obtained from HAS or EELS spectroscopy for the (001) surfaces of cubic crystals in the [100] direction. For other surfaces and directions similar numerical results are available [22, 23] owing to the great importance of surface acoustic waves (SAW) for many technical applications [24], some of which are discussed in Sect. 1.2.

3.3 Born-von Kármán Theory: Bulk Phonons

In this section the atomic treatment of surface lattice dynamics is discussed. This theory emerged out of the first atomic theory of bulk lattice dynamics introduced by Born-von Kármán [27, 28]. Then the modifications needed for the description of surface phonons will be presented.

For an array of interacting atoms the equations of motion for the harmonic vibrations are derived from a phenomenological expression for the total potential energy

$$\phi = \frac{1}{2}\sum\nolimits_{lj} \mathrm{v}(r_{lj}) + \frac{1}{2}\sum\nolimits_{ljk} \mathrm{w}(\mathbf{r}_{jl}, \mathbf{r}_{kl}), \tag{3.17}$$

where $\mathrm{v}(r_{lj})$ is the two-body radial potential between the pair of atoms at the equilibrium positions $\mathbf{r}_j$ and $\mathbf{r}_l$, $r_{jl} = |\mathbf{r}_j - \mathbf{r}_l|$ being their separation (Fig. 3.4a). The three-body potential $\mathrm{w}(\mathbf{r}_{kl}, \mathbf{r}_{jl})$ describes the interaction between the atoms j and k via the atom l (Fig. 3.4a) and depends on the vectors $\mathbf{r}_{kl} = \mathbf{r}_k - \mathbf{r}_l$ and $\mathbf{r}_{jl} = \mathbf{r}_j - \mathbf{r}_l$ connecting atoms k and j to atom l [29, 30]. It should be noted that the prefactor of

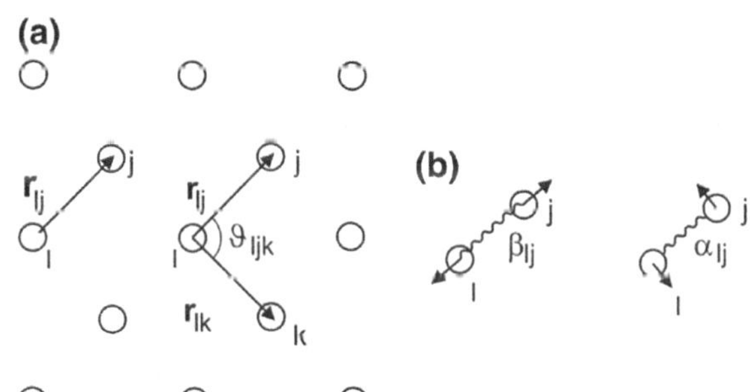

Fig. 3.4 Schematic diagram illustrating force constants. **a** Two-body and three-body angle-bending interactions between atoms in a lattice; **b** radial (β) and tangential (α) components of the interatomic two-body force constants

the 3-body potential is ½ and not 1/6 because it describes the interaction between one single atom pair j, k via a third atom l. The three-body potentials provide the simplest model for describing many-body forces between atoms. Other more realistic theories, which account for the internal electronic degrees of freedom and the effective electron-mediated interactions between atoms in metals, will be introduced in Chap. 5. For the basic foundations of lattice dynamics in the harmonic approximation and the connection to elastic continuum theory the reader is refereed to several classical textbooks, e.g., by Born and Huang [31], Maradudin et al. [28], Cochran [32], Brüesch [33], Böttger [34], Askar [35], and Dove [36].

In order to describe the vibrational dynamics, the potential energy, (3.17), is expanded around the respective equilibrium positions $\mathbf{r}_l$, defined by the conditions $\partial\phi/\partial\mathbf{r} = 0$, in a power series of the time-dependent atomic displacements $\mathbf{u}(l) = \mathbf{u}(l, t)$.

$$\phi = \phi_0 + \frac{1}{2}\sum\nolimits_{lj}\frac{\partial^2\phi}{\partial\mathbf{r}_l\partial\mathbf{r}_j}\mathbf{u}(l)\mathbf{u}(j) + \frac{1}{6}\sum\nolimits_{lj}\frac{\partial^3\phi}{\partial\mathbf{r}_l\partial\mathbf{r}_j\partial\mathbf{r}_k}\mathbf{u}(l)\mathbf{u}(j)\mathbf{u}(k) + \ldots, \quad (3.18)$$

where ϕ_0 is the static potential energy. The two-body derivatives,

$$\frac{\partial^2\phi}{\partial r_{l\alpha}\partial r_{j\beta}} \equiv R_{\alpha\beta}(l,j), \quad (3.19)$$

define the harmonic atom-atom *force constant matrix*. In (3.19) and in the following, whenever α, β are used as *indices*, they denote any one of the three Cartesian coordinates. The cubic and further higher order terms in the expansion of the potential energy determine a variety of important anharmonic effects in lattice dynamics, such as thermal expansion, thermal resistance and phonon-phonon interaction. They can however usually be neglected (*harmonic approximation*) in the theories of surface dynamics and inelastic atom scattering necessary for the analysis of current HAS experiments. A brief discussion of anharmonic effects will be given in Sect. 13.3.

According to the Born-von Kármán theory, the equations of motion in the harmonic approximation for the atomic vibrations of a crystal lattice are given by

$$-M_l\ddot{u}_\alpha(l,t) = \Sigma_{j\beta}R_{\alpha\beta}(l,j)u_\beta(j,t), \quad (3.20)$$

where M_l is the mass of the l-th atom.

For a *two-body central interatomic potential* $\mathrm{v}(r_{lj})$, which depends only on the distance r_{lj} between two atoms and not on the orientation of the distance vector $\mathbf{r}_{lj}$, the force constant matrix can be expressed in terms of derivatives of the potential as

$$R_{\alpha\beta}(l,j) = \frac{r_\alpha r_\beta}{r^2}\beta_{lj} + \left(\delta_{\alpha\beta} - \frac{r_\alpha r_\beta}{r^2}\right)\alpha_{lj}, \quad (3.21)$$

where the r_α are the Cartesian components of $\mathbf{r} \equiv \mathbf{r}_{lj}$ and the β_{lj} and α_{lj} are the two-body force constants between atoms l and j defined by

$$\beta_{lj} \equiv \frac{\partial^2 \mathrm{v}}{\partial r_{lj}^2}, \quad \alpha_{lj} \equiv \frac{1}{r_{lj}} \frac{\partial \mathrm{v}}{\partial r_{lj}}. \tag{3.22}$$

They are referred to as radial and tangential force constants, respectively, and are illustrated in Fig. 3.4b.

Similarly the derivatives of the three-body potential provide a set of angle-bending force constants. There are several different ways for defining the three-body angular force constants, which have been thoroughly discussed in DeLaunay's review [29]. Three-body force constants are often introduced in lattice dynamical models for semiconductors and metals in order to account for the indirect interactions between ions which are mediated by valence and/or free electrons. The simplest case of three-body force constants of the angle-bending type may be found in the Keating model [37, 38] for tetrahedrally coordinated semiconductors where the three-body part of the potential is written as (see Fig. 3.4a)

$$\mathrm{w}(r_{kl}, r_{jl}, \vartheta_{klj}) = w_0 \left(\cos \vartheta_{klj} + \frac{1}{3} \right)^2 \tag{3.23}$$

and w_0 is a constant. Rather lengthy expressions for the three-body force constants are obtained from (3.23) by deriving $\mathrm{w}(r_{kl}, r_{jl}, \vartheta_{klj})$ with respect to the atomic positions after setting $\cos \vartheta_{klj} = \mathbf{r}_{lj} \times \mathbf{r}_{lk} / r_{lj} r_{lk}$ in (3.23). The force constants of a Born-von Kármán model for fcc transition metals including three-body terms have been derived by Bortolani et al. and applied to the dynamics of Ni(001) surface [39].

For dealing with the vibrations of an infinite three-dimensional (3D) periodic crystal lattice it is useful to define a set of translations

$$\mathbf{r}_l \equiv \mathbf{r}(l_1 l_2 l_3) = l_1 \mathbf{a}_1 + l_2 \mathbf{a}_2 + l_3 \mathbf{a}_3 \tag{3.24}$$

under which the lattice transforms into itself. Here $\mathbf{a}_m$ (m = 1, 2, 3) are the basis vectors of the lattice and index $l \equiv (l_1 l_2 l_3)$ is now a set of any three positive, null or negative integer numbers. The translation vectors $\mathbf{r}_l$ define the conventional positions of the 3D unit cells which constitute the crystal (Fig. 3.5a). For a monoatomic crystal they define also the positions of the atoms. For a polyatomic crystal with s atoms in each unit cell the actual positions of the atoms are given by

$$\mathbf{r}_{lk} \equiv \mathbf{r}(l_1 l_2 l_3 \kappa) = \mathbf{r}(l_1 l_2 l_3) + \mathbf{d}(\kappa), \tag{3.25}$$

where the vectors $\mathbf{d}(\kappa)$ (κ = 1, 2, … s) define the relative positions of the s atoms inside each unit cell.

As a consequence of the Bloch theorem the solutions of (3.20) have the form

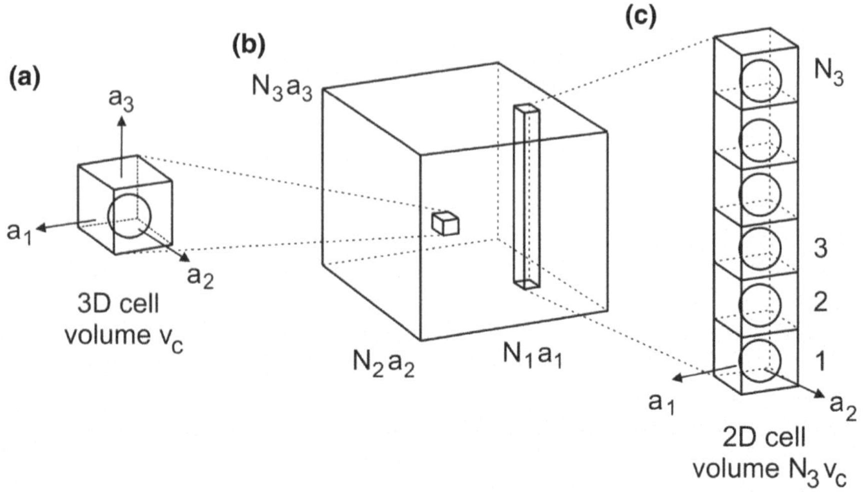

Fig. 3.5 Schematic diagram showing different representations of a crystalline solid of parallelepiped shape shown in (**b**). **a** Representation by three-dimensional translations (basis vectors $\mathbf{a}_1$, $\mathbf{a}_2$ and $\mathbf{a}_3$) of a small unit cell (volume $v_c = \mathbf{a}_3 \cdot (\mathbf{a}_2 \times \mathbf{a}_2)$). **c** Representation by two-dimensional translations (basis vectors $\mathbf{a}_1$ and $\mathbf{a}_2$) of an N_3-times larger unit cell (volume N_3v_c). The latter is known as the slab representation. For simplicity the unit cell shown in this figure contains only one atom

$$u_a(l\kappa, t) \propto u_a(\kappa, \mathbf{q}\nu)\, exp\big(i\mathbf{q}\cdot\mathbf{r}_l - i\omega_{\mathbf{q}\nu} t\big), \tag{3.26}$$

where $\mathbf{q} = (q_x, q_y, q_z)$ is a 3D wavevector. In general for a lattice with s atoms per unit cell there are $3s$ independent eigenfrequencies $\omega_{\mathbf{q}\nu}$ for each value of $\mathbf{q}$ labeled by the *branch index* ν. In either monoatomic ($s = 1$) or polyatomic lattices ($s > 1$) three values of ν correspond to the three acoustic phonons (where $\omega_{\mathbf{q}\nu} \rightarrow 0$ for $\mathbf{q} \rightarrow 0$), one of which is longitudinal ($\nu = LA$) and two transverse ($\nu = TA_1, TA_2$). The additional $3(s - 1)$ phonons which exist for polyatomic lattices are the optical phonons, also of longitudinal ($\nu = LO_{s-1}$) or transverse ($\nu = TO_{1,s-1}, TO_{2,s-1}$) character. For $s > 2$, an index $s - 1$ has to be added to label the different sets of optical phonons.

The three-dimensional Born-von Kármán theory has provided the basis for analyzing a large number of the bulk phonon dispersion curves measured by neutron scattering. A minimal set of effective force constants is obtained by fitting the phonon dispersion curves over the entire Brillouin zone [40]. In general the elements of the atom-atom force constant matrix $R_{\alpha\beta}(l, j)$ (3.19) are linear combinations of the β's and the α's defined by (3.22). Only for the special case when the interatomic distance vectors $\mathbf{r}_{lj}$ are parallel to a Cartesian axis are β_{lj} and α_{lj} identical to one of the elements of the force constant matrix $R_{\alpha\beta}$ (l, j). Otherwise their linear combinations provide a convenient set of effective force constants for central potentials and are currently used in the analysis of neutron dispersion curves as well as in surface dynamics [41, 42]. For *cubic monoatomic crystals* the set consists of only two effective force constants for each order of nearest neighbors. They are usually denoted by a_k (k = 1, 2) for the first nearest neighbors, b_k for the 2nd nearest neighbors and c_k for the 3rd nearest neighbors (when needed) and are defined by the following equations:

simple cubic (sc):

$$\begin{aligned} &a_1 \equiv \beta_{nn}, \quad a_2 \equiv \alpha_{nn}, \\ &b_1 \equiv 1/2(\beta_{2n} + \alpha_{2n}), \quad b_2 \equiv 1/2(\beta_{2n} - \alpha_{2n}); \end{aligned} \tag{3.27}$$

face-centered cubic (fcc):

$$\begin{aligned} &a_1 \equiv 1/2(\beta_{nn} + \alpha_{nn}), \quad a_2 \equiv 1/2(\beta_{nn} - \alpha_{nn}), \\ &b_1 \equiv \beta_{2n}, \quad b_2 \equiv \alpha_{2n}; \end{aligned} \tag{3.28}$$

body-centered cubic (bcc):

$$\begin{aligned} &a_1 \equiv 1/3(\beta_{nn} + 2\alpha_{nn}), \quad a_2 \equiv 1/3(\beta_{nn} - \alpha_{nn}), \\ &b_1 \equiv \beta_{2n}, \quad b_2 \equiv \alpha_{2n}; \\ &c_1 \equiv 1/2(\beta_{3n} + \alpha_{3n}), \quad c_2 \equiv 1/2(\beta_{3n} - \alpha_{3n}), \end{aligned} \tag{3.29}$$

where the α_{lj} and β_{lj} are defined by (3.22), but with the indices lj = nn, $2n$ and $3n$ referring to first, second and third nearest neighbors, respectively.

3.4 Surface-Projected Bulk Phonons

In the application of Born-von Kármán theory to the calculation of surface phonons it is convenient to split the theory into two parts: first the phonons in the bulk are *projected* on to new coordinates appropriate to 2D boundary conditions. By maintaining 3D cyclic boundary conditions the reduction in the coordination of atoms exposed by the creation of a physical surface is not accounted for. In the following Sect. 3.5 the influence of the perturbation introduced by the termination of the bulk at the surface is introduced. This step leads to new surface localized phonons, e.g., the Rayleigh waves. The projection of bulk phonons on to a 2D plane within the bulk is discussed first in this Section. Then the influence of the surface perturbation is dealt with in subsequent sections.

To account for 2D boundary conditions the crystal is approximated by a slab consisting of N_3 parallel atomic layers normal to the direction of $\mathbf{a}_3$ and of N_1 and N_2 unit cells along the other two directions $\mathbf{a}_1$, $\mathbf{a}_2$ parallel to the layers, respectively, so that the crystal has a finite number $N_1 \times N_2 \times N_3$ of unit cells. As illustrated in Fig. 3.5, in a finite 3D crystal the periodicity in three dimensions is established through the *cyclic* boundary conditions which require that the translations $N_1\mathbf{a}_1$,

$N_2\mathbf{a}_2$, $N_3\mathbf{a}_3$ (or any combination thereof) brings any atom into itself. A three-dimensional lattice of basis vectors ($\mathbf{a}_1$, $\mathbf{a}_2$, $\mathbf{a}_3$), with a unit cell which is assumed to contain s atoms, (Fig. 3.5a) can also be viewed as generated by two-dimensional (2D) translations in the ($\mathbf{a}_1$, $\mathbf{a}_2$) plane of a large unit cell containing sN_3 atoms (Fig. 3.5c). Such a 2D representation of a crystal is called the slab representation of a crystal and provides an optimal basis for adapting the Born von Kármán theory to surface phonons.

The solutions of the atomic equations of motion, (3.20), appropriate to the 2D boundary conditions, are Bloch waves with 2D wavevectors $\mathbf{Q}$ in the $(\mathbf{a}_1, \mathbf{a}_2)$ plane and angular frequencies $\omega_{\mathbf{Q}\nu}$:

$$u_\alpha(l\kappa, t) \propto u_\alpha(l_3\kappa, \mathbf{Q}\nu) \exp\left[i\mathbf{Q}\cdot(l_1\mathbf{a}_1 + l_2\mathbf{a}_2) - i\omega_{\mathbf{Q}\nu}t\right]. \quad (3.30)$$

Here the sets (l_1, l_2) and (l_3, κ) are used to label the slab cells in two dimensions and the atoms inside each slab cell, respectively, whereas the branch index ν labels the $3sN_3$ phonon solutions for each wavevector $\mathbf{Q}$.

Inserting (3.30) into (3.20) leads to a coupled set of equations for the slab which can be written as the following eigenvalue problem [43, 44],

$$M_\kappa\omega_{\mathbf{Q}\nu}^2 u_\alpha(l_3\kappa, \mathbf{Q}\nu) = \sum\nolimits_{\beta l_3'\kappa'} R_{\alpha\beta}(l_3\kappa, l_3'\kappa', \mathbf{Q}) u_\beta(l_3'\kappa', \mathbf{Q}\nu), \quad (3.31)$$

where

$$R_{\alpha\beta}(l_3\kappa, l_3'\kappa', \mathbf{Q}) = \sum\nolimits_{l_1 l_2} R_{\alpha\beta}(l\kappa, l'\kappa') \exp[i\mathbf{Q}\cdot(l_1\mathbf{a}_1 + l_2\mathbf{a}_2)] \quad (3.32)$$

is the 2D Fourier transform of the force constant given by (3.19) between the (l, κ)-th and the (l', κ')-th atoms, with the indices $l \equiv (l_1 l_2 l_3)$ and $l' \equiv (l_1' \; l_2' \; l_3')$. Equation (3.32) defines the set of $\mathbf{Q}$-*dependent interplanar force constants*. As in the 3D case the set of modes of given ν for all values of $\mathbf{Q}$ forms the ν-th phonon branch and the function $\omega_{\mathbf{Q}\nu}$ is its dispersion relation (Fig. 3.6b). From the solution of (3.31) the eigenfrequencies $\omega_{\mathbf{Q}\nu}$ and the displacement vectors $u_\alpha(l_3\kappa, \mathbf{Q}\nu)$, known as the $\mathbf{Q}$-*dependent planar displacements*, are obtained. Since the $u_\alpha(l_3\kappa, \mathbf{Q}\nu)$ have the dimensions of a length, they are conveniently expressed in analogy to (3.14) as

$$u_\alpha(l_3k, \mathbf{Q}\nu) = \left(\frac{\hbar}{2M_\kappa\omega_{\mathbf{Q}\nu}}\right)^{\frac{1}{2}} e_\alpha(l_3k, \mathbf{Q}\nu), \quad (3.33)$$

where the dimensionless eigenvectors $e_\alpha(l_3\kappa, \mathbf{Q}\nu)$ form an orthogonal complete set in their $3sN_3$ space and are normalized to unity:

$$\sum\nolimits_{l_3\kappa\alpha} e_\alpha^*(l_3\kappa, \mathbf{Q}\nu)\, e_\alpha(l_3\kappa, \mathbf{Q}\nu') = \delta_{\nu\nu'} \quad \text{(orthonormality)} \quad (3.34)$$

$$\sum\nolimits_\nu e_\alpha^*(l_3\kappa, \mathbf{Q}\nu)\, e_\beta(l_3'\kappa', \mathbf{Q}\nu) = \delta_{l_3 l_3'}\delta_{\kappa\kappa'}\delta_{\alpha\beta} \quad \text{(completeness)}. \quad (3.35)$$

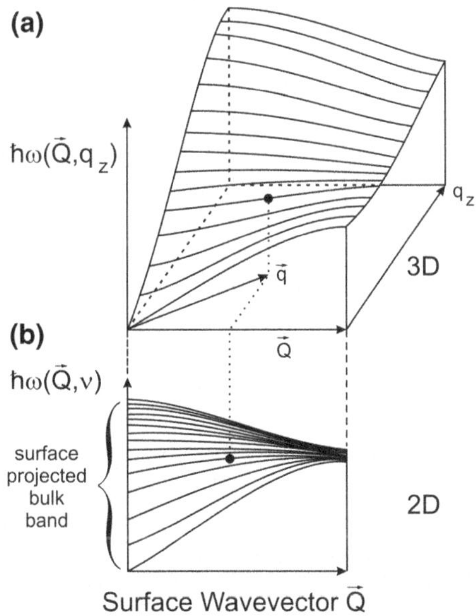

Fig. 3.6 Schematic diagram showing the projection of bulk phonons on the surface. **a** The energy surface $\hbar\omega(\mathbf{Q}, q_z)$ generated by the dispersion curves of an acoustic bulk phonon within a cyclic slab with N_3 layers is plotted as a function of parallel (**Q**) and normal (q_z) wavevector components. **b** The same dispersion curves as a function of the surface wavevector **Q** form a two-dimensional band $\hbar\omega_{\mathbf{Q}\nu}$. The normal component q_z is replaced by an index ν labeling all the N_3 modes of the band having the same parallel wavevector **Q**. In this example there are 15 dispersion curves corresponding to the $N_3 = 15$ atomic layers of the slab. If the unit cell of the original periodic crystal contains s atoms there will be $3s$ such bands and index ν will label altogether $3s\ N_3$ modes for each value of the parallel wavevector **Q**

The eigenvalues determine the polarization of the displacement field amplitude and are therefore termed *polarization vectors*. The quantized physical displacement of the (l, κ)-th atom, with $l = (l_1 l_2 l_3)$, for the $(\mathbf{Q},\nu)$-th phonon mode is thus given by

$$u_\alpha(l\kappa, \mathbf{Q}\nu, t) = \left(\frac{\hbar}{2N_1 N_2 M_\kappa \omega_{\mathbf{Q}\nu}}\right)^{1/2} e_\alpha(l_3\kappa, \mathbf{Q}\nu)\ \exp[i\mathbf{Q}\cdot(l_1\mathbf{a}_1 + l_2\mathbf{a}_2) - i\omega_{\mathbf{Q}\nu}t]. \tag{3.36}$$

The factor $(N_1N_2)^{-1/2}$ ensures the normalization to unity of the exponential plane waves, so that the eigenwaves $(N_1N_2)^{-1/2}e_\alpha(l_3\kappa, \mathbf{Q}\nu)\ \exp[i\mathbf{Q}\cdot(l_1\mathbf{a}_1 + l_2\mathbf{a}_2)]$ form an orthonormal, complete set in the full crystal space (*normal modes*).

Since the cyclic boundary condition is still maintained also for the third translation $N_3\mathbf{a}_3$ the above solutions are just those of the 3D lattice which have been relabeled to correspond to a 2D representation. The eigenfrequencies $\omega_{\mathbf{Q}\nu}$, for each **Q** and for ν varying from 1 to $3sN_3$, now form $3s$ bands, each one containing N_3 modes. For an infinitely thick slab ($N_3 \to \infty$) the individual modes can no longer be resolved and the bands become continuous. They are referred to as *bulk phonon bands* projected onto the ($\mathbf{a}_1$, $\mathbf{a}_2$) plane.

The correspondence between the two- and three-dimensional representations of the phonon branches inside a solid with 3D cyclic boundary conditions is illustrated in Fig. 3.6, where the surface formed by the dispersion curves of a bulk acoustic phonon mode in 3D is shown in Fig. 3.6a. The 3D dispersion surface in Fig. 3.6a is generated by the phonon energies $\hbar\omega$ as a function of the 3D wavevector $\mathbf{q}$. In order to obtain the 2D representation $\mathbf{q}$ is split into a 2D component $\mathbf{Q}$, which is parallel to the atomic planes of the slab, and a normal component q_z, namely $\mathbf{q} = (\mathbf{Q}, q_z)$. Each curve in the 2D representation (Fig. 3.6b) represents the phonon energy as a function of $\mathbf{Q}$ for a given value of q_z. Thus each bulk phonon branch generates just as many curves as the number of possible values of q_z, that is N_3. The resulting set of curves is referred to as the *surface-projected* bulk phonon band.

If the unit cell contains s atoms then it has $3s$ degrees of freedom and the total phonon spectrum is made up of $3s$ different, partially superimposed surface-projected bands. Thus for each value of $\mathbf{Q}$ in the 2D representation there are altogether $3sN_3$ modes, which are conveniently labeled by a single index ν. The corresponding frequency distribution within all the bands is given by the *total* **Q**-*selected phonon density of states* (DOS)

$$D_0(\mathbf{Q},\omega) = \frac{1}{N_3}\sum\nolimits_{\nu}\delta\left(\omega - \omega_{\mathbf{Q},\nu}\right). \tag{3.37}$$

Similarly the **Q**-selected DOSs projected onto either a specific plane l_3 ($l_3 = 1$ for the surface layer), or a specific plane l_3 and sublattice κ, the latter for any polarization or a given polarization α, are defined as

$$D_0(l_3\mathbf{Q},\omega) = \frac{1}{N_3}\sum\nolimits_{\kappa\alpha,\nu}|e_\alpha(l_3\kappa,\mathbf{Q}\nu)|^2\delta(\omega - \omega_{\mathbf{Q},\nu}), \tag{3.38}$$

$$D_0(\kappa l_3\mathbf{Q},\omega) = \frac{1}{N_3}\sum\nolimits_{\alpha,\nu}|e_\alpha(l_3\kappa,\mathbf{Q}\nu)|^2\delta(\omega - \omega_{\mathbf{Q},\nu}), \tag{3.39}$$

$$D_{0\alpha}(\kappa l_3\mathbf{Q},\omega) = \frac{1}{N_3}\sum\nolimits_{\nu}|e_\alpha(l_3\kappa,\mathbf{Q}\nu)|^2\delta(\omega - \omega_{\mathbf{Q},\nu}), \tag{3.40}$$

respectively. Clearly

$$\sum\nolimits_{l_3\kappa\alpha}D_{0\alpha}(l_3\kappa\mathbf{Q},\omega) = D_0(\mathbf{Q},\omega) \tag{3.41}$$

due to eigenvector normalization.

3.5 Effects of the Surface

3.5.1 The Rayleigh Theorem

So far only a macroscopic sized crystal slab with three-dimensional cyclic boundary conditions has been dealt with, where the N_3-th layer ($l_3 = N_3$) is connected to the first layer ($l_3 = 1$) by the same bulk interatomic forces which act between any other pair of adjacent layers (Fig. 3.7a). If long range forces act in the bulk between non-adjacent layers the same is true for any pair of layers which are connected across the boundary Σ separating the N_3-th from the first layer of the slab. By the requirements of cyclic boundary conditions along the $\mathbf{a}_3$ (z) direction the reduction in the coordination of the atoms at a physical surface and the changes in the force constants at the surface were not accounted for in the above discussion.

A surface is created at a 2D boundary Σ, if all the interatomic forces connecting atoms across Σ are broken apart. In this way the cyclic boundary conditions along the z-axis are lifted and two free surfaces are created at $l_3 = 1$ and N_3 (Fig. 3.7). The original cyclic lattice has now been transformed into a finite slab with two free surfaces. Only the atoms in the two surface regions A and B are affected by the cut of forces across Σ. This operation leads to a perturbation of the force constant matrix and a consequent change of the eigenfrequencies and eigenvectors of (3.31). The perturbation matrix has non-zero elements only within a subspace σ which encompasses the regions A and B. When the two surfaces are identical the problem can be reduced to that of a single surface of a semi-infinite lattice by using a new set

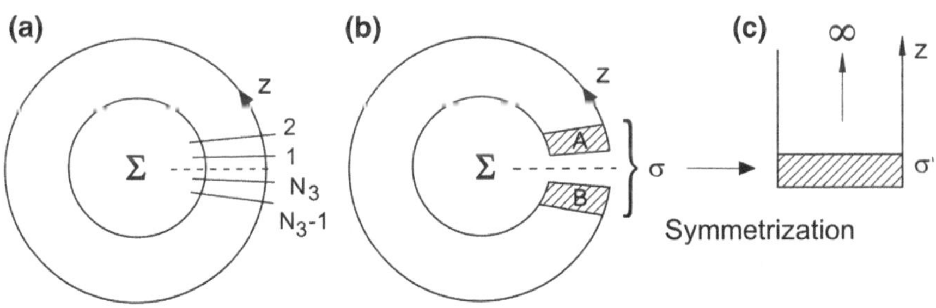

Fig. 3.7 Representation of a lattice with cyclic boundary conditions in the z direction. **a** The broken line indicates the boundary plane Σ located between the 1st and the N_3 th lattice planes. **b** When all the interactions connecting atoms across the boundary plane Σ are set to zero the cyclic boundary conditions in the z direction normal to boundary plane are lifted and the original periodic lattice is transformed into a slab with two parallel free surfaces. The periodicity is preserved in two independent directions parallel to the surfaces. The atoms affected by the perturbation are included in the surface regions A and B. **c** For identical surfaces a symmetry transformation allows for the reduction of the perturbation subspace σ to one half of it (σ'), and for $N_3 \to \infty$ the slab problem transforms into that for a semi-infinite lattice with one single surface

of symmetrized coordinates and letting $N_3 \to \infty$ (Fig. 3.7c) [45, 46, 47]. The size of the new perturbation subspace σ' is one half of that for a finite slab.

According to a theorem by Lord Rayleigh in 1896 [21, 48] a surface localized perturbation affecting a finite number n of the total number of degrees of freedom produces a separation of n eigenfrequencies from the quasi-continuum of the surface-projected bulk band. In this way the appearance of n localized modes is explained. The implications of the Rayleigh theorem can be nicely illustrated by considering, for example, the highly idealized case of the (001) surface of a *monoatomic simple cubic* lattice with one radial nn force constant β (> 0). Assume that the symmetrization depicted in Fig. 3.7c has been made so that a single surface is involved.

Let $R_{\alpha\beta}(l_3 l_3', \mathbf{Q})$ be the dynamical matrix of the unperturbed lattice and $\Lambda_{\alpha\beta}(l_3 l_3', \mathbf{Q})$ the perturbation matrix associated with the free surface. For this simple model the perturbation matrix involves only one degree of freedom at the surface plane ($l_3 = 1$) along the z axis and is therefore written as $\Lambda\delta_{l_3 1}\delta_{l_3' 1}\delta_{\alpha z}\delta_{\beta z}$ where the perturbation constant is just $\Lambda = -\beta$. This means that one force constant β has been subtracted from those acting on each surface atom as a consequence of the cut across the boundary plane Σ. In this model no relaxation of the surface atoms from their original positions in the unperturbed cyclic lattice is considered, so that the only effect of the surface is to reduce the coordination of surface atoms and this leads to a negative Λ. It is also conceivable, however, that the freshly created surface may undergo a strong inward relaxation—so strong that the local force constants are stiffened and the diagonal perturbation element Λ becomes positive. In the derivation leading to the Rayleigh theorem in either case the corresponding eigenvalue equation including the perturbation,

$$\sum\nolimits_{l_3'\beta}\left[R_{\alpha\beta}(l_3, l_3', \mathbf{Q}) + \Lambda\delta_{l_3 1}\delta_{l_3' 1}\delta_{\alpha z}\delta_{\beta z}\right]u_{l_3'\beta} = M\omega^2 u_{l_3\alpha}, \qquad (3.42)$$

is transformed into

$$u_{l_3\alpha} = -\left[\mathrm{R}(\mathbf{Q}) - \mathrm{M}\omega^2\right]^{-1}_{l_3 1, \alpha z}\Lambda\, u_{1z}, \qquad (3.43)$$

where the quantities inside the square brackets denote the matrix $R_{\alpha\beta}(l_3, l_3', \mathbf{Q}) - \delta_{l_3 l_3'}\delta_{\alpha\beta}M\omega^2$, hereafter written in matrix notation. Next it is convenient to introduce the spectral representation of the resolvent matrix $-[\mathrm{R}(\mathbf{Q}) - \mathrm{M}\omega^2]^{-1}$ [28]

$$-\left[\mathrm{R}(\mathbf{Q}) - \mathrm{M}\omega^2\right]^{-1}_{l_3 l_3', \alpha\beta} = \frac{1}{M}\sum\nolimits_{\nu}\frac{e_\alpha^+(l_3, \mathbf{Q}\nu)e_\beta(l_3', \mathbf{Q}\nu)}{\omega^2 - \omega_\nu^2}, \qquad (3.44)$$

where $\omega_\nu^2 \equiv \omega_\nu^2(\mathbf{Q})$ and $e_\alpha(l_3, \mathbf{Q}\nu)$ are the squared eigenfrequencies and the normalized eigenvectors (3.33) of $R(\mathbf{Q})$, respectively. The new perturbed squared eigenfrequencies can then be obtained as the solutions of the determinantal equation in ω^2

$$\det\left|\delta_{l_3 l_3'}\,\delta_{\alpha\beta} - \frac{\Lambda}{M}\sum\nolimits_{\nu}\frac{e_{\alpha}^{+}(l_3,\mathbf{Q}\nu)\,e_z(0,\mathbf{Q}\nu)}{\omega^2-\omega_{\nu}^2}\,\delta_{1 l_3'}\,\delta_{z\beta}\right| = 1 - \frac{\Lambda}{M}\sum\nolimits_{\nu}\frac{\left|e_z(0,\mathbf{Q}\nu)\right|^2}{\omega^2-\omega_{\nu}^2} = 0. \tag{3.45}$$

A graphical solution of (3.45) is shown in Fig. 3.8 for the simplest case of a one-dimensional perturbation affecting one single degree of freedom. In this case the perturbation matrix in (3.42) is 1×1 and the determinant in (3.45) is simple. The vertical asymptotes are located at the N_3 unperturbed eigenvalues ω_{ν}^2. The new eigenvalues are obtained from the intersections of the curve representing the function $\sum_{\nu}\left|e_z(0,\mathbf{Q}\nu)\right|^2/\left(\omega^2-\omega_{\nu}^2\right)$ with a horizontal line located at M/Λ. Of the altogether N_3 perturbed eigenvalues, $N_3 - 1$ fall between two adjacent unperturbed ones and are therefore shifted by a small amount of order $o(1/N_3)$. The remaining eigenvalue is shifted by a finite amount away from the edge of the unperturbed band. If $\Lambda < 0$, as is expected if only the degree of coordination is reduced at the surface, the localized eigenvalue is shifted below the bulk band. If, on the other hand, $\Lambda > 0$, as is the case for a strong inward relaxation producing an overall

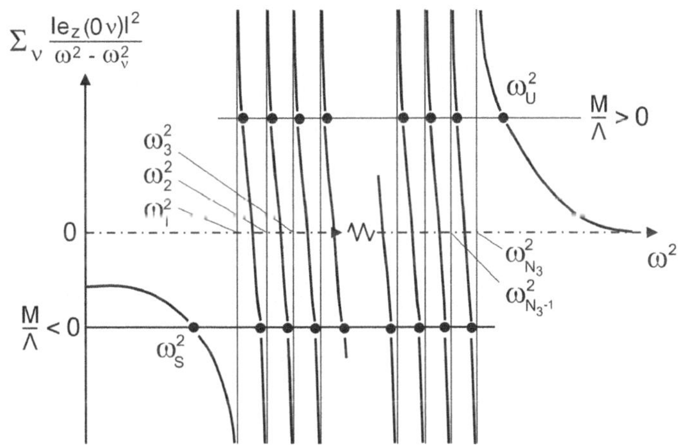

Fig. 3.8 Graphical solution of the determinantal equation (3.45) giving the perturbed eigenvalues of a lattice dynamical system with N_3 degrees of freedom for a surface perturbation Λ affecting one single degree of freedom. The vertical asymptotes are located at the unperturbed eigenvalues ω_ν^2 ($\nu = 1, 2, \ldots, N_3$). The perturbation, besides shifting $N_3 - 1$ of the N_3 eigenvalues by a small amount proportional to $1/N_3$, yields one localized eigenvalue, denoted ω_S^2 or ω_U^2. They are shifted by a finite amount from either the lower or upper band edge, according to whether the perturbation M/Λ is negative or positive, respectively. The former case corresponds to the Rayleigh wave. Such a diagrammatic solution of (3.45) holds for each value of $\mathbf{Q}$, the quantities $u_z(0,\nu)$, ω_ν^2, ω_S^2, ω_U^2 and Λ all being functions of $\mathbf{Q}$.

stiffening of the force constants acting on the surface atoms, the localized eigenvalue is shifted above the bulk band. Of course, with increasing magnitude of Λ/M, the separation of the localized mode away from the band edge will also increase.

A more extended perturbation produces as many localized modes as the number of degrees of freedom that are affected. For example, if the perturbation would affect both the first layer, due to the creation of a free surface, and the second layer, due to the inward relaxation of the first one, then two localized modes would be expected, one below and one above the band edge.

3.5.2 The Green's Function Method

The transformation of (3.42) into the matrix (3.43) has the advantage that it reduces the perturbed problem from the full space of the given system, where one has to diagonalize an $N_3 \times N_3$ matrix, to the perturbation subspace, which, in the present example, is just 1×1. This is, in fact, the central idea of the Green's function (GF) method [28, 45,46, 49–52], which is discussed next. The Green's function method also provides the suitable framework to describe another important effect of the surface perturbation, the emergence of resonances in the spectrum of bulk mode frequencies. For $N_3 \to \infty$ the set of eigenfrequencies $\{\omega_\nu\}$ constituting the spectrum of the unperturbed matrix $\mathrm{R}(\mathbf{Q})$ tends to a continuum. Solutions of a perturbed linear equation like (3.42), now written as

$$\mathrm{R}(\mathbf{Q})\mathbf{u} - \mathrm{M}\omega^2\mathbf{u} = -\Lambda\mathbf{u}, \tag{3.46}$$

are now sought outside the unperturbed spectrum $\{\omega_\nu\}$ as well as inside $\{\omega_\nu\}$. The former solutions are just the localized surface modes discussed above, whereas the latter are called *resonances*. For a frequency ω falling inside $\{\omega_\nu\}$ (3.46) can be transformed into

$$\mathbf{u} = \mathbf{u}_0 + \mathrm{G}_0\left(\mathbf{Q}, \omega^2\right)\Lambda\mathbf{u}, \quad \omega \in \{\omega_\nu\} \tag{3.47}$$

where $\mathbf{u}_0$ is any eigenvector of $\mathrm{R}(\mathbf{Q})$ for an eigenfrequency equal to ω, and

$$\mathrm{G}_0\left(\mathbf{Q}, \omega^2\right) = \left[\mathrm{M}\left(\omega^2 + i0^+\right) - \mathrm{R}(\mathbf{Q})\right]^{-1}, \tag{3.48}$$

is the unperturbed GF matrix, defined as the analytical continuation of the resolvent matrix, (3.44), into the complex plane. The shift of the frequency ω away from the real axis by an infinitesimal quantity $i0^+$ into the upper complex half-plane allows for the inversion in (3.48) for ω^2 inside the spectrum of $\mathrm{R}(\mathbf{Q})$. In scattering theory (3.47) is known as the Lippmann-Schwinger equation [53]. Its formal solution is given by

$$\begin{aligned}\mathbf{u} &= \left[\mathrm{I} - \mathrm{G}_0\left(\mathbf{Q}, \omega^2\right)\Lambda\right]^{-1}\mathbf{u}_0 \\ &= \mathbf{u}_0 + \mathrm{G}_0(\mathbf{Q}, \omega^2)\mathrm{T}(\mathbf{Q}, \omega^2)\mathbf{u}_0,\end{aligned} \tag{3.49}$$

where

$$\mathrm{T}(\mathbf{Q}, \omega^2) \equiv \Lambda\left[I - \mathrm{G}_0(\mathbf{Q}, \omega^2)\Lambda\right]^{-1}, \tag{3.50}$$

is the transition matrix (or simply T-matrix) and has non-zero elements only in the subspace of the perturbation Λ. Equation (3.47) represents indeed the solution of a scattering problem with the eigenvector $\mathbf{u}$ given by the sum of the incident wave $\mathbf{u}_0$ and the scattered wave $\mathrm{G}_0\mathrm{T}\mathbf{u}_0$ produced by the perturbation Λ, here acting as a scattering potential. In general the perturbed wave $\mathbf{u}$ in (3.49) is not normalized in the same way as $\mathbf{u}_0$ and may need to be multiplied by a normalization constant.

Owing to the factor $[\mathrm{I} - \mathrm{G}_0\Lambda]^{-1}$, called the *resonant denominator*, the elements of the T-matrix are all proportional to

$$\frac{1}{\det\left(\mathrm{I} - \mathrm{G}_0\Lambda\right)} = \frac{\mathrm{e}^{i\theta}}{\left|\mathrm{Re}\det\left(\mathrm{I} - \mathrm{G}_0\Lambda\right) + i\,\mathrm{Im}\det\left(\mathrm{I} - G_0\Lambda\right)\right|}, \tag{3.51}$$

where the argument

$$\theta = -\arctan\frac{\mathrm{Im}\det\left(\mathrm{I} - \mathrm{G}_0\Lambda\right)}{\mathrm{Re}\det\left(\mathrm{I} - \mathrm{G}_0\Lambda\right)} \tag{3.52}$$

gives the phase shift between the incident and the scattered wave [49, 50]. According to (3.50) a resonance enhancement of the T-matrix, and therefore of the amplitude of the perturbed wave with respect to the incident wave, occurs at frequencies ω where

$$\mathrm{Re}\det\left[\mathrm{I} - \mathrm{G}_0(\mathbf{Q}, \omega^2)\Lambda\right] = 0. \tag{3.53}$$

This equation is a generalization of (3.45). It gives the frequencies of both surface resonances, for ω inside $\{\omega_\nu\}$, and localized modes, for ω outside $\{\omega_\nu\}$. In the latter case the determinant of (3.53) is a real function. According to (3.52) a resonance occurs when the phase shift θ between the incident and the scattered wave is equal to $\pi/2$. This corresponds exactly to the case of a forced mechanical oscillator, where the external oscillating force corresponds to the incident wave and the oscillator response to the scattered wave. The scattered part of the wave becomes very large with respect to the incident wave.

Surface localized modes and resonances yield important features in the perturbed density of phonon states, as appears from the following definitions. The replacement $\omega^2 \to \omega^2 + i0^+$ in (3.44) provides the spectral representation of the unperturbed Green's function, (3.48). Then, by means of the identities

$$\frac{1}{\omega^2 - \omega_\nu^2 + i0^+} = \frac{1}{\omega^2 - \omega_\nu^2} - i\pi\delta(\omega^2 - \omega_\nu^2), \tag{3.54}$$

$$\delta(\omega^2 - \omega_\nu^2) = \frac{1}{2\omega}[\delta(\omega - \omega_\nu) + \delta(\omega + \omega_\nu)], \tag{3.55}$$

the *total unperturbed* **Q**-selected DOS, see (3.37), can be written as

$$D_0(\mathbf{Q}, \omega) = -\frac{2\omega}{\pi} Tr\, \mathrm{Im}\left[\mathrm{MG}_0\left(\mathbf{Q}, \omega^2\right)\right]. \tag{3.56}$$

The *total perturbed* **Q**-selected DOS, $D(\mathbf{Q}, \omega)$, is defined in a similar way

$$D(\mathbf{Q}, \omega) = -\frac{2\omega}{\pi} Tr\, \mathrm{Im}\left[\mathrm{MG}\left(\mathbf{Q}, \omega^2\right)\right], \tag{3.57}$$

where

$$\begin{aligned} G(\mathbf{Q}, \omega^2) &= \left[\mathrm{M}(\omega^2 + i0^+) - \mathrm{R}(\mathbf{Q}) - \Lambda\right]^{-1} \\ &= \left[\mathrm{I} - \mathrm{G}_0\left(\mathbf{Q}, \omega^2\right)\Lambda\right]^{-1} \mathrm{G}_0\left(\mathbf{Q}, \omega^2\right), \end{aligned} \tag{3.58}$$

is the *perturbed GF matrix*. The change in the DOS can be expressed, after some algebra, as

$$\begin{aligned} \Delta D\left(\mathbf{Q}, \omega^2\right) &\equiv D\left(\mathbf{Q}, \omega^2\right) - D_0\left(\mathbf{Q}, \omega^2\right) \\ &= -\frac{1}{\pi}\frac{\partial}{\partial\omega} \mathrm{Im}\left\{\ln \det\left[\mathrm{I} - G_0\left(\mathbf{Q}, \omega^2\right)\Lambda\right]\right\} \\ &= \frac{1}{\pi}\frac{\partial\theta}{\partial\omega}. \end{aligned} \tag{3.59}$$

The diagonal elements:

$$D_{0,l_3\kappa\alpha}(\mathbf{Q}, \omega) = -\frac{2\omega}{\pi} \mathrm{Im}\left[\mathrm{MG}_0\left(\mathbf{Q}, \omega^2\right)\right]_{l_3\kappa\alpha, l_3\kappa\alpha}, \tag{3.60}$$

$$D_{l_3\kappa\alpha}(\mathbf{Q}, \omega) = -\frac{2\omega}{\pi} \mathrm{Im}\left[\mathrm{MG}\left(\mathbf{Q}, \omega^2\right)\right]_{l_3\kappa\alpha, l_3\kappa\alpha} \tag{3.61}$$

are the unperturbed **Q**-selected DOS *projected* onto the l_3-th plane, κ-th sublattice and polarization α, see (3.40), and the corresponding *perturbed projected* **Q**-selected DOS, respectively.

In the limit $N_3 \to \infty$ the densities of phonon states become continuous functions for ω inside $\{\omega_\nu\}$. Moreover the perturbed phonon densities display δ-functions at certain frequencies ω_S outside $\{\omega_\nu\}$, which correspond to the localized surface modes, and peaks at the resonance frequencies ω_R falling in the continuum $\{\omega_\nu\}$ of bulk modes. By expanding the resonant denominator around ω_R, where its real part, (3.53), vanishes, a Lorentzian peak is obtained in the phonon density change [45, 46]

$$\Delta D(\mathbf{Q},\omega) \cong \frac{1}{\pi}\frac{\Gamma/2}{(\omega-\omega_R)^2+(\Gamma/2)^2}, \quad \omega\sim\omega_R \tag{3.62}$$

with a FWHM given by

$$\Gamma = \frac{2\mathrm{Im}\,\det\left[\mathrm{I}-\mathrm{G}_0\left(\mathbf{Q},\omega_R^2\right)\Lambda\right]}{\frac{\partial}{\partial\omega}\,\mathrm{Re}\,\det[\mathrm{I}-\mathrm{G}_0(\mathbf{Q},\omega^2)\Lambda]\big|_{\omega=\omega_R}}. \tag{3.63}$$

From (3.61)–(3.62) it follows that the perturbed GF has a pole in the complex plane at $z = \omega_R - i\Gamma/2$ corresponding to a resonance. In order to fulfill causality [49, 50], by which the response (the scattered wave) follows the stimulus (the incident wave), Γ has to be positive. The same holds also for localized modes, for which, however, $\Gamma \to 0^+$ and the corresponding peaks in $D(\mathbf{Q}, \omega)$, (3.61), become δ-functions. Note that a solution ω_R of (3.53) which yields a negative Γ does not contribute a real resonance but an *anti-resonance*, i.e., a depletion region in $D(\mathbf{Q}, \omega)$ with respect to $D_0(\mathbf{Q}, \omega)$. Such depletion regions compensate for the density of states, which has been transferred to localized and real resonant modes.

The eigenvectors of the perturbed dynamics can be obtained with the GF method by means of (3.47). For the solutions outside the unperturbed spectrum, i.e., for $\omega \notin \{\omega_\nu\}$ (*surface localized modes*), $\mathbf{u}_0$ is set to zero, whereas for the solutions inside the unperturbed spectrum, $\omega \in \{\omega_\nu\}$ (*surface resonance modes*), $\mathbf{u}_0$ is an unperturbed eigenwave of frequency ω.

From the computational point of view the GF method has the great advantage that all the diagonalizations are performed in the perturbation subspace rather than in the whole slab space. This advantage, however, is lost to some extent due to the comparatively difficult task of calculating the matrix elements of the complex-valued Green's functions, which require integrations over the bulk Brillouin zone with special care for frequencies around van Hove singularities [34, 54, 55]. For this reason the GF method is less popular than the direct diagonalization of the whole slab dynamical matrix (the so-called *slab method* discussed next) [44, 56], the latter method is more expedient thanks to the availability of fast computers.

On the other hand the GF method provides a convenient framework for the calculation of the various vibrational response functions of the surface. An example is the inelastic atom scattering cross section (see Chap. 7). Such response functions can in general be expressed in terms of the time-dependent displacement-displacement correlation function $<u_\alpha(l\kappa,t)u_\beta(l'\kappa',0)>_T$, where

$$\widehat{u}_\alpha(l\kappa, t) = \sum\nolimits_{\mathbf{Q}\nu} u_\alpha(l_3\kappa, \mathbf{Q}\nu)\, e^{i\mathbf{Q}\cdot\mathbf{r}_l} \left(b^+_{\mathbf{Q}\nu} e^{i\omega_{\mathbf{Q}\nu}t} + b_{\mathbf{Q}\nu} e^{-i\omega_{\mathbf{Q}\nu}t}\right) \tag{3.64}$$

is the displacement field operator, $b^+_{\mathbf{Q}\nu}$ and $b_{\mathbf{Q}\nu}$ are phonon creation and annihilation operators, respectively, and $<...>_T$ means the thermal average at temperature T. With some little algebra it is found that

$$\int_{-\infty}^{\infty} dt\, e^{i\omega t} \left\langle \widehat{u}_\alpha(l\kappa, t)\widehat{u}_\beta(l'\kappa', 0)\right\rangle_T = -\frac{2\hbar}{1 - e^{-\hbar\omega/k_BT}} \operatorname{Im} G_{\alpha\beta}(l\kappa, l'\kappa', \omega^2), \tag{3.65}$$

where $G_{\alpha\beta}(l\kappa, l'\kappa', \omega^2)$ are the GF matrix elements in the direct lattice space and the 2D Fourier transforms of the elements of $\mathrm{G}(\mathbf{Q}, \omega^2)$ [28, 34].

3.5.3 The Slab Method

In the slab method the vibrations of a finite crystal slab consisting of N_3 parallel layers of atoms are calculated by solving the coupled equations (3.31) numerically [56]. The number of layers N_3 has to be chosen large enough so that the spectral features of the surface layers are clearly distinguished from those of the bulk. In practice, however, N_3 is limited by the size of the force constant matrix which can be diagonalized with the available computational methods. Typically N_3 lies somewhere between 10^1 and 10^2 [56–58]. However, by using algebraic codes to set up the force constant matrix it is possible to go to larger numbers of layers. This has been especially useful for calculating stepped surfaces [59], where, because of the substantially larger unit cells, N_3 must be increased correspondingly.

The slab method and the physical nature of surface phonons can be nicely illustrated by first considering the simple case of an ideal two-dimensional slab of a *single layer* of densely packed atoms. Figure 3.9 shows the calculated dispersion curves for a single layer of atoms with a mass corresponding to that of nickel, arranged according to the close packed (111) surface of an fcc lattice [60, 61]. The two-dimensional Brillouin zone and the labeling of the symmetry directions are illustrated in Fig. 3.9a. The dispersion curves in Fig. 3.9b are for a nearest neighbor (*nn*) pairwise interaction characterized by a single radial force constant β_1. For simplicity the *nn* tangential force constant was set to zero, $\alpha_1 = 0$. In this case only two modes with in-plane longitudinal (L) polarization and an in-plane transverse (T) polarization, commonly referred to as a shear horizontal (SH) mode, have non-vanishing frequencies.

With a finite value of α_1 the third mode with a transverse out-of-plane polarization [shear vertical (SV)] now has a finite frequency, while the other modes are only weakly affected, as shown in Fig. 3.9c. The displacement patterns for all three modes at the zone boundary point $\overline{\mathrm{M}}$ are also shown in Fig. 3.9d. This simple model shows the importance of the tangential force constant for out-of-plane

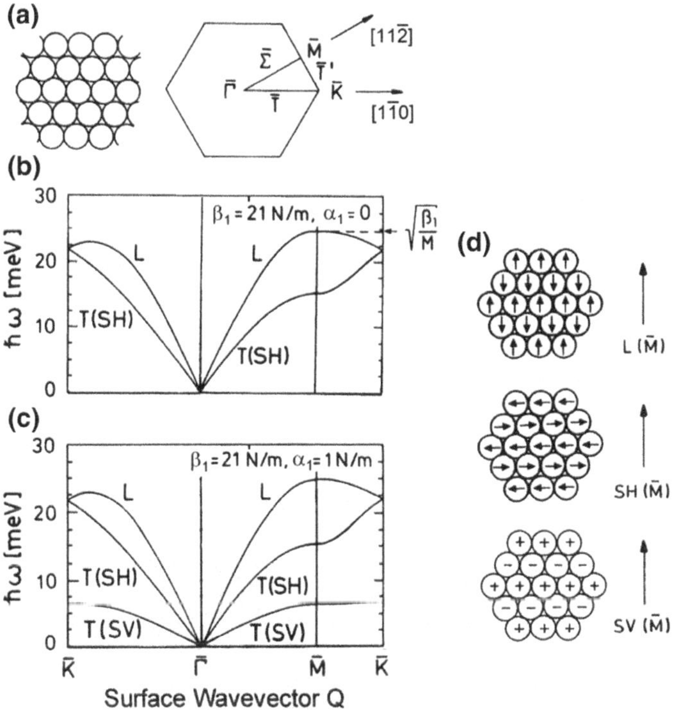

Fig. 3.9 Calculated dispersion curves for a single (111) atomic layer equivalent to the surface layer of an fcc crystal. **a** The surface geometry of the closely packed hexagonal structure of atoms and the first Brillouin zone. The triangle $\overline{\Gamma MK}$ shows the irreducible part of the zone boundary. **b** The dispersion curves for nearest neighbors interactions with only radial forces β_1. **c** The effect of adding the nearest neighbor tangential force constant α_1. L denotes longitudinal, T (SV) transverse shear vertical (out-of-plane) and T(SH) transverse shear horizontal (in-plane) polarizations. **d** Illustration of the displacement patterns as seen from above for the three modes at the $\overline{M}$ symmetry point. Small arrows indicate displacements parallel to the surface while ± indicate displacements perpendicular to the surface. The long arrows on the right of each set indicate the direction of the wavevector

vibrations. It also finds some use in estimating the dispersion curves of monolayers of adsorbates which are only weakly coupled to the substrate. Analytical solutions for monolayer simple cubic [62] and hexagonal close packed [63] free standing films have been reported in the literature.

The slab method has the further advantage that it provides direct insight into how some of the normal modes of a thin film gradually evolve into surface-projected bulk bands and others into surface localized modes as the thickness of the film increases layer by layer to macroscopic dimensions. This is illustrated in Fig. 3.10a for the case in which the same two force constants β_1 and α_1 are used throughout, not only within the layers but also between the layers. With a second layer ($N_3 = 2$) the expected $3N_3 = 6$ modes appear. The additional dispersion curves are qualitatively similar to those of the single layer except that they now have finite frequencies at the

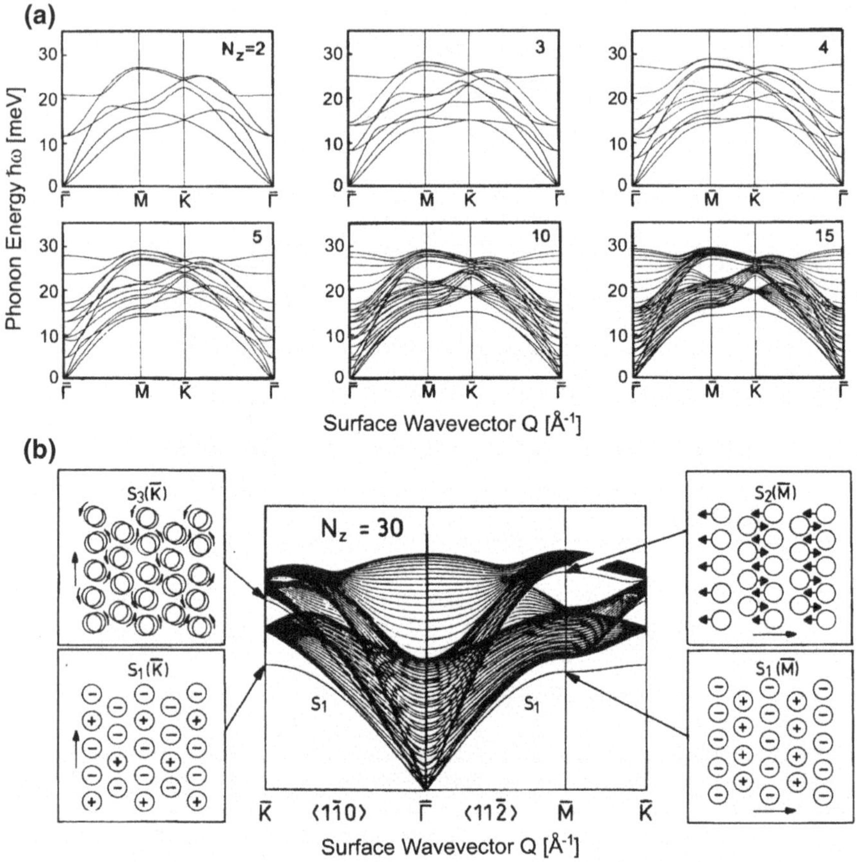

Fig. 3.10 Surface phonon dispersion curves of a fcc(111) slab with increasing numbers of layers and displacement patterns of surface phonons. **a** The evolution of surface phonon dispersion curves of a fcc(111) slab as additional layers are added to the single layer shown in Fig. 3.9a. **b** The dispersion curves for $N_3 = 30$ approach those of an infinitely thick slab and the distinction between the isolated surface modes and the closely spaced surface-projected bulk bands becomes apparent. The insets to the left and right display the displacement patterns of the surface atoms for the surface modes at the zone-boundary symmetry points. The geometry of the surface layer and the first Brillouin zone in reciprocal lattice space are the same as in Fig. 3.9a. The mode $S_3(\overline{K})$ is circularly polarized. The long arrows indicate the direction of the wavevector

zone origin $\overline{\Gamma}$. The finite frequencies can be attributed to vibrations of the two planes with respect to each other. With additional layers the number of modes increases according to $3N_3$. The finite-frequency zone-origin modes evolve into standing waves along a direction normal to the planes [64]. Beyond about 10 layers a new structure in the distribution of the dispersion curves becomes apparent with some modes well separated from dense bands. In the limit of $N_3 = \infty$ the former are the surface modes, while the bands are attributed to the surface projected bulk modes.

3.6 Localization of Surface Modes

As mentioned in Sect. 3.5 the appearance of the surface localized modes can be understood as due to the negative perturbation of the force constant matrix in the first few atomic layers of the surface region resulting from the reduced coordination due to the removal of all the atomic layers above the surface. According to the Rayleigh theorem [48] discussed in Sect. 3.5 no more than one surface localized phonon branch is expected for each band per surface, provided the perturbation, as in an ideal case, is restricted to the first layer with no significant relaxation or reconstruction effects. Since in the slab model of a surface there are, in fact, usually two identical surfaces, the localized frequencies occur as pairs of nearly degenerate branches with a splitting which decreases exponentially for $N_3 \to \infty$ and thus disappears with increasing thickness of the slab.

In this context the term *localization* has two different meanings. From the point of view of the frequency spectrum, isolated δ-functions are *spectrally localized* outside the continuous DOS of the bulk bands. In the spatial sense, surface phonons have a displacement field which decays exponentially with increasing distance from the surface, and therefore are *spatially localized* at the surface. This is the case for both the *macroscopic* modes like the Rayleigh waves (Fig. 2.1) or the *microscopic* modes like the optical surface phonons [65, 66]. The spectral localization of the eigenfrequencies is related to the spatial localization of the eigenvectors in the surface region. The larger the separation of the surface phonon frequency away from the bulk band edge, the greater the localization of the corresponding displacement field at the surface of the crystal. This is clearly seen for Rayleigh waves in the continuum limit, Sect. 3.2, where the parameter β_T expressing the separation of the RW velocity from the bulk velocity v_T, (3.9), also describes the inverse of the penetration length, (3.12). It can also be understand on the basis of the following general qualitative argument. The squared frequency of any bulk phonon branch can be represented as an expansion in the 3D wavevector $\mathbf{q} = (\mathbf{Q}, q_z)$

$$\omega^2(\mathbf{q}) = \omega^2(0) + cq^2 + \cdots = \omega^2(0) + c(Q^2 + q_z^2) + \cdots, \tag{3.66}$$

where c is a constant. The surface-projected bulk phonon band is obtained for any Q by letting q_z have all possible real values within the bulk Brillouin zone, as illustrated in Fig. 3.6. The surface mode associated with this branch occurs for a negative $q_z^2 = -\kappa^2$. Consequently with increasing depth inside the crystal the surface mode amplitude decays for $z \to -\infty$ like $\exp(iq_z z) = \exp(\kappa z)$ and its squared frequency is separated from the bulk edge by an amount equal to $-c\kappa^2$. For the Rayleigh wave, where $\omega^2(0) = 0$, (3.12) and (3.9) give $\kappa = Q\beta_T$ and $c = v_T^2$.

When the penetration length of the surface wave is much less than the slab thickness the pair of modes arising at the two surfaces of the slab are practically degenerate. In a slab of infinite thickness the two surfaces are independent of each other and the surface mode frequencies are equal, which is equivalent to replacing the thick slab with a semi-infinite crystal with a single surface. Only in the case of ionic crystals, in which the long-wave longitudinal optical modes carry a macroscopic electric field (Fuchs and Kliever modes or *surface phonon polaritons* [67]), does the transformation of a thick slab into a semi-infinite crystal require some care. This case has been discussed in the review by Benedek and Miglio [46].

Figure 3.11a illustrates the decrease of the surface phonon amplitudes with increasing penetration into the bulk for the modes of a 21 layer thick fcc(111) slab. The attenuation of the Rayleigh wave (S_1) at the $\overline{\mathrm{M}}$ point appears to be purely exponential and faster than that of a RW with a wavevector at an intermediate point along the $\overline{\Gamma\mathrm{M}}$ direction. This is consistent with the fact that the RW frequency has a larger separation from the band edge at the $\overline{\mathrm{M}}$ point than at smaller wavevectors. The S_2 mode at $\overline{\mathrm{M}}$ and the S_3 mode at an intermediate point in the $\overline{\Gamma\mathrm{K}}$ direction also show an average exponential decay, but with an oscillatory behavior. The presence of such oscillations depends on the value of the wavevector component q_z perpendicular to the surface of the bulk mode at the edge of the band from which the localized surface mode originates. The RW originates from the bulk mode with the lowest frequency which, as illustrated in Fig. 3.6, has $q_z = 0$. In this case the decay does not exhibit oscillations. On the other hand the longitudinal modes S_2 and S_3 are associated with $q_z = \pi/d$ and $q_z = 2\pi/3d$ modes at the corresponding bulk edges, d being the interlayer distance. This is because the bulk longitudinal mode for $\mathbf{Q}$ near the $\overline{M}$ or the $\overline{K}$ point has a downward dispersion for increasing q_z. Therefore

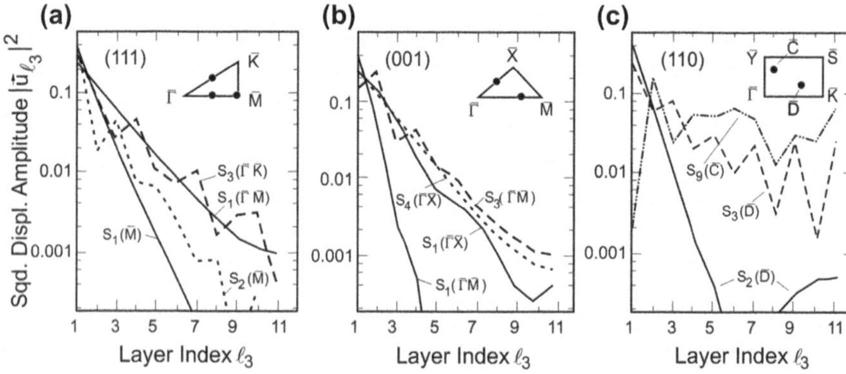

Fig. 3.11 Displacement amplitudes as a function of layer thickness index. **a** The squared displacement amplitudes of the Rayleigh wave (S_1) at the $\overline{\mathrm{M}}$ point and in the middle of the $\overline{\Gamma\mathrm{K}}$ direction, the longitudinal S_2 mode at $\overline{\mathrm{M}}$ and the quasi longitudinal S_3 mode in the middle of the $\overline{\Gamma\mathrm{K}}$ direction (see Fig. 3.10) are plotted as a function of the layer index l_3 for a 21-layer thick fcc(111) slab. **b** The squared amplitudes for some surface modes of a 21-layer thick fcc(001) slab in the middle of the $\overline{\Gamma\mathrm{X}}$ and at 2/3 of the $\overline{\Gamma\mathrm{M}}$ direction. **c** The squared displacement amplitudes as a function of the layer index l_3 for some resonant surface modes of a thick fcc(110) slab out of the symmetry directions. The S_9 resonance is taken at the point $\overline{\mathrm{C}}$ of the surface BZ indicated in the inset, while the other resonances S_2 and S_3 are taken at the point $\overline{\mathrm{D}}$ (adapted from Allen et al. [44]). Note that the labels of the modes S_1 through S_4 correspond to the convention adopted in Table 3.1 and may be different from those used by Allen et al. [44]

the decay of the modes S_2 and S_3 oscillates with wavelengths $2d$ and $3d$, respectively.

Figure 3.11c also shows the peculiar behavior of surface resonances inside a slab with a fcc(110) surface, which have a large amplitude in the first surface layer and then go over into a periodic bulk wave. In these calculations the resonant modes have been chosen at two low symmetry points of the Brillouin zone $\overline{D}$ (for the S_2 and S_3 resonances) and $\overline{C}$ (for the S_9 resonance), as indicated in the inset of Fig. 3.11c, in order to ensure a mixing of SH and sagittal components and hence a strong coupling to bulk modes. The S_9 resonance is shown as an extreme example of weak surface localization, where the maximum amplitude occurs on the second layer and is less than an order of magnitude larger than the wave-like amplitude in the limit $l_3 \rightarrow \infty$.

The decay length of Rayleigh waves in the long-wave limit ($\mathbf{Q} \rightarrow 0$) diverges, according to (3.12), like $1/Q$, as expected for a macroscopic mode (see Chap. 2), whereas all the optical surface modes maintain a very short penetration length also for $\mathbf{Q} \rightarrow 0$, consistent with their microscopic character. Away from the long wavelength non-dispersive region where the solid behaves as a continuum, also the Rayleigh wave becomes microscopic with a decay length rapidly decreasing to values of the order of the interatomic distances. Note that the FK surface phonon polaritons, i.e., the surface electromagnetic modes associated with the optical phonons of polar crystals, are also macroscopic [67–69].

These considerations have implications on the sensitivity and resolution of the HAS and EELS experiments. For example, a surface sensitive spectroscopy like HAS becomes less sensitive to Rayleigh waves which penetrate more than just a few tens of atomic layers, because of the reduction of their amplitudes at the surface. HAS is most sensitive to the microscopic surface excitations, whereas it practically does not couple to the macroscopic Fuchs and Kliever modes, even though they can have some interesting mixing with the microscopic surface optical modes [46]. EELS, due to the greater penetration of electrons, is in principle more sensitive to surface phonons with a longer decay length or with a larger amplitude in the second layer. The observation of deeply penetrating RW modes with small excitation energies may, however, be hampered by the inability to resolve them. EELS is also a method of choice for detecting macroscopic Fuchs and Kliever modes, which couple directly with the charge of the electron. Microscopic optical modes can be probed equally well by HAS and EELS, although HAS intensities rapidly decrease for phonons with energies comparable to that of the incident He atoms. It should be noted, however, that these arguments hold within the classical picture of HAS as an atom-atom collision process, which only applies to closed shell crystal surfaces. For metal surfaces, where the He phonon coupling is mediated by the surface free-electron density, HAS can also reveal subsurface phonons with their maximum amplitude several layers beneath the surface plane.

3.7 Classification of Surface Modes

3.7.1 *Polarizations and Displacement Patterns*

The displacement patterns shown in Fig. 3.10b suggest that surface modes can be classified according to their polarization, very much in the same way as bulk modes. Since the point symmetries with respect to surface atoms are necessarily reduced with respect to the symmetries *at* the equivalent atoms in the bulk of the infinite crystal, the polarizations of surface modes are in general more complicated than for bulk modes. For example, Rayleigh waves (S_1 in Fig. 3.10b) are characterized by an elliptical polarization of quasi-shear vertical ($\sim$SV) character (see Eqs. 3.6–3.8), while the mode S_2 has also an elliptical though quasi-longitudinal ($\sim$L) polarization. Only at the $\overline{M}$ symmetry point do both modes acquire a linear shear-vertical (SV) or longitudinal (L) polarization, respectively, as depicted in Fig. 3.10b, whereas the mode S_3, also at the symmetry point $\overline{K}$, shows an unusual circular polarization.

From the example of Fig. 3.10b it appears that the polarization of each surface branch reflects approximately that of the parent bulk band from which it originates, although it may be strongly affected by the symmetry with respect to the direction of propagation on the surface. In this connection the plane defined by the wavevector direction and the normal to the surface, which is called the *sagittal plane* (see Sect. 9.1), plays an important role. As illustrated in Fig. 3.12a, in both the high symmetry directions of the fcc(100) surface the sagittal plane is a mirror symmetry plane of the crystal, whereas on the fcc(111) surface (Fig. 3.12b) this is true for the six surface directions equivalent to $[11\bar{2}]$, but not for the six directions equivalent to $[1\bar{1}0]$. On the fcc(110) surface the sagittal plane is a mirror symmetry plane for the symmetry directions [001] and $[1\bar{1}0]$, whereas along $[1\bar{1}1]$ it is not.

First the case is considered in which the sagittal plane coincides with (or is parallel to) a mirror symmetry plane of the crystal. For directions containing a mirror symmetry plane the displacement fields of the surface modes have either even or odd symmetry with respect to reflection through to the sagittal plane. The former are fully contained in the sagittal plane and are called *sagittal modes* ($\perp$), while the latter are perpendicular to the sagittal plane. They are called *shear horizontal modes* ($\|$ or SH) because they are perfectly transverse and parallel to the surface plane.

Since the surface itself is not a mirror symmetry plane the shear vertical and longitudinal components of the displacements in the sagittal plane are coupled together and out-of-phase with respect to each other, leading to an *elliptical* polarization. This geometrical property is common to all kinds of surface waves having a SV component, e.g., surface plasmons [70], surface polaritons [71], ocean waves [72], etc. Depending on the dominant component in the sagittal plane the modes with sagittal polarization are designated as either quasi-transverse ($\sim$SV) or quasi-longitudinal ($\sim$L).

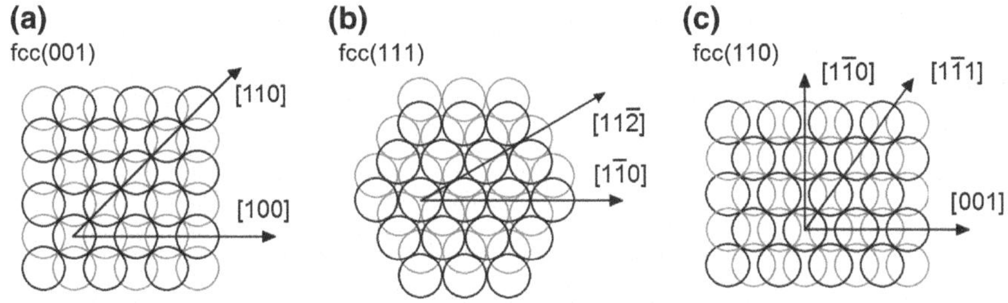

Fig. 3.12 The low-index surfaces (001), (111) and (110) of a face-centered cubic (fcc) crystal showing the atoms in the first (black circles) and second layer (light grey circles) and the inequivalent symmetry directions. **a** For the (001) surface the sagittal plane, defined by the surface wavevector and the surface normal, is a mirror symmetry plane along both the [100] and [110] directions. **b** For the (111) surface only along [11$\bar{2}$] is the sagittal plane a mirror plane, whereas along [1$\bar{1}$0] it is not. **c** For the (110) surface along both the [001] and [1$\bar{1}$0] directions the sagittal plane is a mirror symmetry plane, whereas along [1$\bar{1}$1] it is not. The same considerations as for fcc (111) also hold for the (0001) surface of a hexagonal close packed (hcp) crystal. Surface phonons propagating in a mirror symmetry plane have either pure sagittal or shear horizontal polarization

Figure 3.13a illustrates the displacement fields for the three surface acoustic modes along a mirror symmetry plane, each for a specified surface wavevector. Their locations in the dispersion curve are shown in Fig. 3.13b. As illustrated in Fig. 1.2 and discussed in some detail in Sect. 3.2 the RW (which is ~SV) has a retrograde elliptical polarization with a larger vertical than longitudinal component in the sagittal plane. Because of the surface induced coupling between SV and L components the ~L modes also exhibit a retrograde elliptical polarization in the sagittal plane but with a longitudinal component which is greater than the vertical component. As illustrated in Fig. 3.13a, the SH mode is linearly polarized within the surface plane. It is important to note, however, that at high symmetry points in the reciprocal space, either at the origin or the boundaries of the Brillouin zone, where the sagittal modes have a vanishing group velocity, an exactly linear polarization is recovered, which is either SV or L. This is however not true for the RW near $\bar{\Gamma}$, where the polarization remains elliptical, consistently with the finite value of the group velocity which tends to v_R.

The polarization of surface modes is more complicated if the sagittal plane is not parallel to a mirror symmetry plane of the crystal. A classical example is that of the fcc(111) shown in Fig. 3.12b. While the sagittal plane in the [11$\bar{2}$] direction is clearly a mirror symmetry plane of the crystal, that in the [1$\bar{1}$0] direction is not. In this case the broken symmetry couples the sagittal and SH components. In general in this case the polarization of each mode is elliptical in a plane which is tilted with respect to the sagittal plane. However for crystal slabs having an axial-inversion symmetry [73] one axis of the ellipse is normal to the surface while the other axis is parallel to the surface and rotated with respect to the longitudinal direction. An additional complication arises when, as is usually the case, the bulk bands of different polarization partially overlap each other. Then a surface mode originating

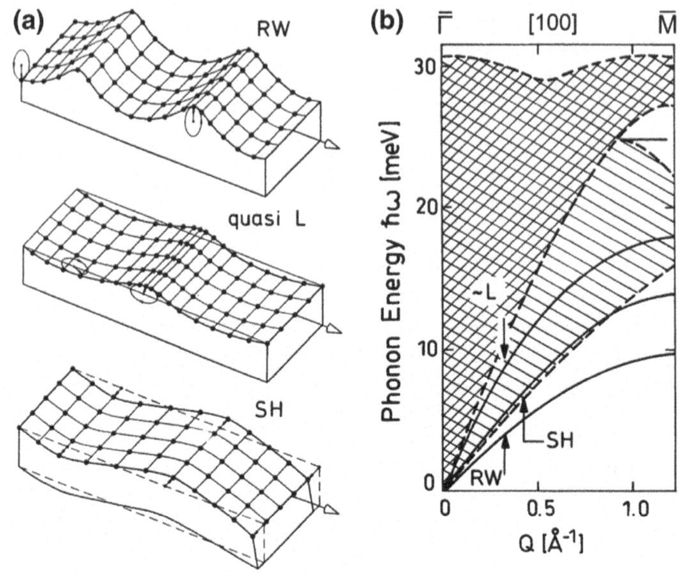

Fig. 3.13 Schematic diagram illustrating the displacement patterns of the atoms of the topmost layer for three polarizations of the surface modes of the fcc(001) surface of a monoatomic crystal along the high symmetry [100] direction. **a** Both the quasi-shear vertical ($\sim$SV) Rayleigh waves (RW) and the quasi-longitudinal ($\sim$L) modes have an elliptical polarization in the sagittal plane with the major axis either normal or parallel to the surface, respectively. The two modes are independent, with mutually orthogonal eigenvectors and their phases shown in the figure are arbitrary. The shear-horizontal (SH) modes have a linear, purely transverse polarization within the surface plane. **b** The arrows show the wavelength of the sagittal modes RW, L and the shear SH mode as shown in **a**. The RW and L modes have wavelengths equal to eight times the interatomic distance and are therefore located at one fourth of the surface Brillouin zone (SBZ). The SH mode shown in **a** has a wavelength six times the interatomic distance and is located at one third of the SBZ. At the zone boundary the three modes in **a** would acquire, from top to bottom, a pure SV, L and SH linear polarization, respectively. A ball model simulation of the three main polarizations of surface phonons on several flat and stepped surfaces is available as a video at https://youtu.be/LTOLkhVEIGk

from a given band may overlap with another bulk band and as a result of hybridization may no longer be entirely localized.

In summary the following three different situations are encountered [46, 44]:

(1) *Localized surface modes* (solutions of (3.49) with $\mathbf{u}_0 = 0$) have frequencies for a given wavevector $\mathbf{Q}$ which lie outside the spectrum of all the bulk modes having the same parallel wavevector $\mathbf{Q}$. As a result no mixing (hybridization) can occur and their displacement fields must decay on average exponentially inside the crystal, whether they are macroscopic or microscopic. Examples of such exponential decay are shown in Fig. 3.11a, b. Localized surface modes contribute δ-functions in the DOS.

(2) *Surface resonances* or mixed modes (solutions of (3.49) with $\mathbf{u}_0 \neq 0$) have frequencies for a given wavevector which lie inside a band of bulk modes and

have a polarization which is not orthogonal to that of the bulk modes. As a result of mixing their displacement fields are large in the surface region but they do not decay to zero inside the crystal, tending rather in an oscillatory fashion to some bulk mode of equal frequency and polarization. Examples of the attenuation of surface resonance displacement fields within the bulk are shown in Fig. 3.11c. The resonance peaks in the phonon density of states have a Lorentzian profile given by (3.62), with a finite width Γ given by (3.63).

(3) *Pseudo-surface modes* (PSW) have for a given wavevector frequencies which lie inside a band of bulk modes of orthogonal polarization. Due to orthogonality no mixing occurs and therefore these modes are still localized and decay exponentially inside the crystal. Such orthogonality, however, may only occur along a symmetry direction, and any infinitesimal rotation of the propagation direction away from that direction leads to a mixing with the bulk bands and transforms these modes into surface resonances described in (2) above.

3.7.2 An Illustrative Example: NaF(001)

These different types of modes are best illustrated by the full set of phonon dispersion curves calculated for the surface of a diatomic crystal, NaF(001) [46, 74] shown in Fig. 3.15. For reference the arrangement of Na and F ions in the (001) surface is illustrated in Fig. 3.14a together with the irreducible part of the corresponding surface Brillouin zone (BZ) b and one eighth of the fcc Brillouin zone for the three-dimensional NaF crystal c. In Fig. 3.14b the shaded triangle $\overline{\Gamma\text{MX}}$ depicting the irreducible part of the surface BZ in the basal xy plane also corresponds to one eighth of the entire surface zone. The surface phonon dispersion curves shown in Fig. 3.15a were calculated by the Green's function (GF) method [16, 75, 76] for the sagittal polarization (⊥) and in Fig. 3.16a for the shear-horizontal polarization (||) along the symmetry directions $\overline{\Gamma\text{M}}$ and $\overline{\Gamma\text{X}}$. Figure 3.16b provides the frequencies at the boundary of the surface Brillouin zone connecting $\overline{\text{M}}$ with $\overline{\text{X}}$. Since the sagittal plane for any wavevector along the zone boundary, $\overline{\text{M}}$ and $\overline{\text{X}}$ points excluded, is not a mirror plane the phonon polarizations are neither sagittal nor shear-horizontal, and therefore the entire set of surface phonons is displayed. Since there are two atoms per unit cell, altogether six different bulk bands for the acoustical and optical modes exist for each symmetry direction. The heavy lines in Figs. 3.15 and 3.16 are the surface modes, whereas thin lines mark the upper and lower boundaries of the bulk bands of the shaded areas. The complex pattern of the avoided crossings occurring for sagittal surface modes along [100] is shown in more detail in Fig. 3.15b. It results from the hybridization between the SV and L polarizations and between modes of acoustic and optical characters. The experimental points from HAS measurements [4, 75, 77], are plotted in Fig. 3.15a for the [100] and the [110] directions and appear to be in good agreement with the calculated dispersion curves calculated for sagittal

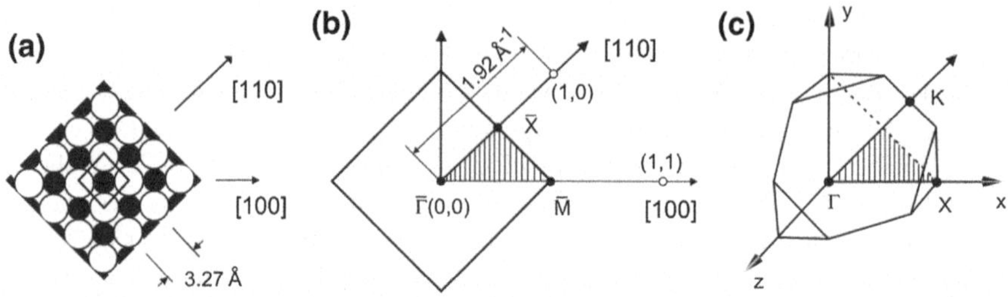

Fig. 3.14 Structure and Brillouin zone of the fcc NaF(001) surface **a** The arrangement of sodium (full circles) and fluorine (open circles) ions with the symmetry directions and the length of the unit cell edge indicated. **b** The corresponding surface Brillouin zone showing the irreducible part (shaded area) and the symmetry points $\overline{\Gamma}$, $\overline{\mathrm{M}}$ and $\overline{\mathrm{X}}$, and the **G**-vectors (1,0) and (1,1) in 2D notation and the crystallographic directions [100] and [110]. **c** The location of the surface Brillouin zone inside the 3D bulk Brillouin zone for the positive octant. Γ, X and K are the 3D symmetry points in the (x, y) plane

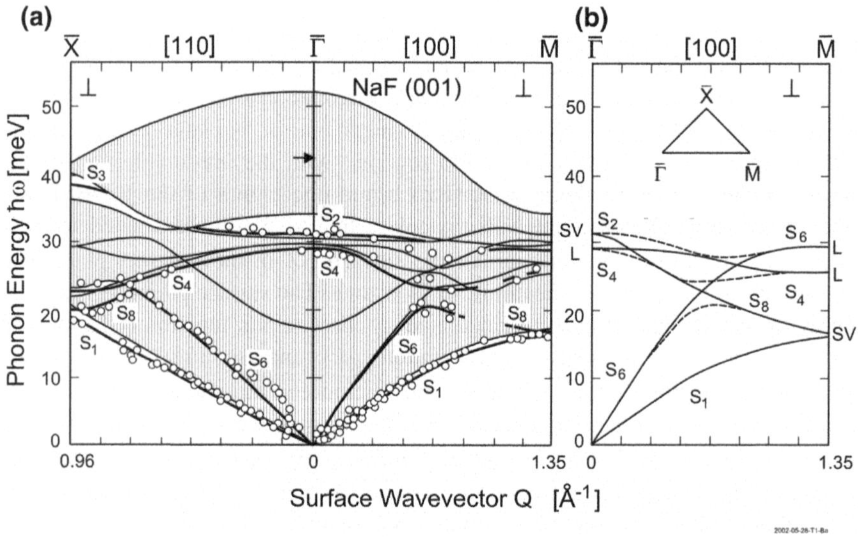

Fig. 3.15 Experimental and theoretical surface phonon dispersion curves with sagittal polarization (⊥) for fcc NaF(001) along the symmetry directions $\overline{\Gamma\mathrm{M}} \equiv [100]$ and $\overline{\Gamma\mathrm{X}} \equiv [110]$. **a** Heavy lines are the surface modes and the shaded areas the bulk bands projected onto the surface calculated with the GF method [74]. Open circles show the HAS measurements for the RW's [4, 77] and the higher frequency modes [75]. The small horizontal arrow at the $\overline{\Gamma}$ point indicates the frequency of the macroscopic FK electromagnetic mode [67, 68]. **b** The complex hybridization scheme of the sagittal surface modes along [100]. Full lines show the ideal dispersion curves in the absence of hybridization; broken lines show the avoided crossing of the actual dispersion curves due to hybridization. The inset shows the symmetry points of the irreducible surface Brillouin zone. The corresponding dispersion curves for *shear horizontal polarization* are shown in Fig. 3.16a

polarization. The avoided crossings depicted in Fig. 3.15b are also rather well reproduced by the experimental data.

As seen in Fig. 3.16a there are far fewer dispersion curves for SH surface phonons than for sagittal polarized modes shown in Fig. 3.15. The optical surface mode S_5 (Fig. 3.16b) arises from the bulk band of the TO phonons polarized in the surface plane and exists over the entire surface Brillouin zone including the $\overline{MX}$ direction, where for surface wavevectors between 1.35 and 1.05 $Å^{-1}$ it becomes a resonance. The acoustic surface mode S_7 in Fig. 3.16a arises from the TA bulk phonons polarized in the surface plane and only exists as a localized surface mode in the [110] direction and only for a short distance starting at $\overline{X}$ along the $\overline{MX}$ direction (Fig. 3.16b) before disappearing into the bulk band. Along the $\overline{\Gamma X}$ direction, as shown by Alldredge [26], the S_7 mode is an example of a microscopic acoustic surface mode since its phase velocity tends to the bulk transverse velocity v_T for $Q \rightarrow 0$ and therefore is not a surface mode in the continuum limit. The lack of an S_7 mode in the [100] direction can be explained by the fact that the bulk TA phonon at the low frequency band edge fulfills the stress-free surface boundary conditions as required for a surface mode [78]. For this reason it coincides with the S_7 mode, which in this direction completely loses its surface localized character and degenerates into an ordinary bulk mode. On the other hand the acoustic SH mode exists as a surface mode in the continuum limit, acquiring a macroscopic character, in other directions away from [100] (Fig. 3.16a) [78].

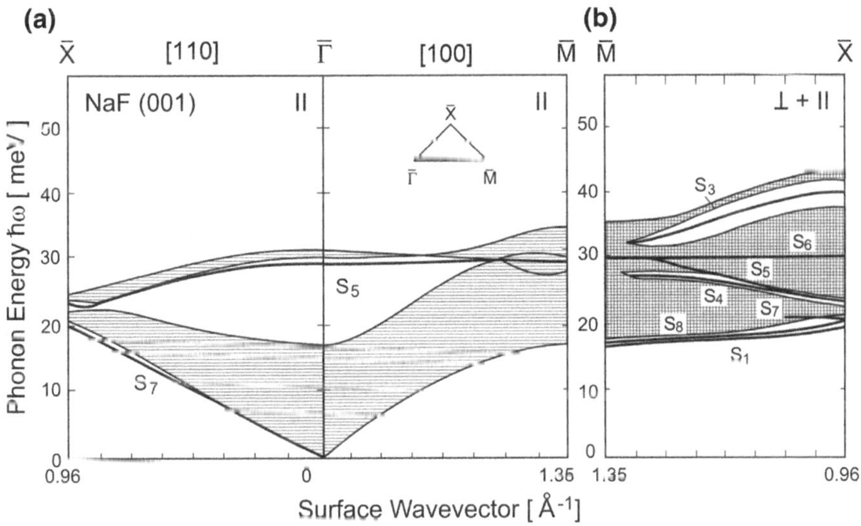

Fig. 3.16 Surface phonon dispersion curves calculated with the GF method [77] for NaF(001). **a** Along the symmetry directions $\overline{\Gamma M} \equiv [100]$ and $\overline{\Gamma X} \equiv [110]$ for SH polarization (||) **b** Along the boundary $\overline{MX}$ of the surface Brillouin zone for all polarizations. Along $\overline{MX}$ the polarization of surface modes is neither sagittal nor shear-horizontal

For symmetry reasons at the symmetry points $\overline{\mathrm{M}}$ and $\overline{\mathrm{X}}$ each SH mode becomes degenerate with a sagittal mode (see Fig. 3.15b), which at these points acquires a purely L polarization. Thus at the $\overline{\mathrm{M}}$ point S_5 is degenerate with S_6. At the $\overline{\mathrm{X}}$ point S_5 is degenerate with S_4, and S_7 is degenerate with S_8. Also at the $\overline{\Gamma}$ point each optical SH mode is degenerate with a mode of L polarization and together they form a *Lucas mode pair*. On the NaF(001) surface there is one pair of optical Lucas modes formed by S_4 and S_5 at the $\overline{\Gamma}$ point.

The optical surface modes occurring in diatomic crystals are characterized by an anti-phase motion of the two ions of the unit cell. This is schematically illustrated in Fig. 3.17 for **Q** at the midpoint of the $\overline{\Gamma\mathrm{M}}$ direction and for the polarizations ~SV, corresponding to the Wallis mode S_2 [79], and for ~L and SH polarizations, corresponding to the two Lucas modes S_4 and S_5, respectively [66, 80]. For simplicity the first two diagrams (a) and (b) only show the dominant SV and L

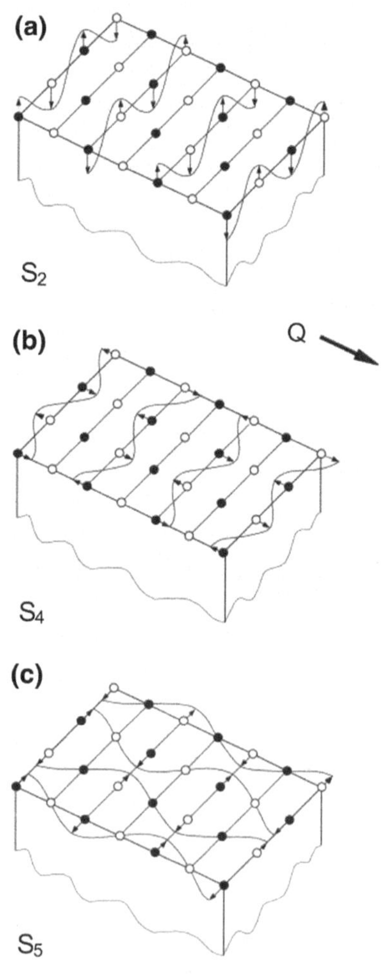

Fig. 3.17 Schematic diagram illustrating the displacement patterns of the cations (•) and the anions (○) of the topmost layer for the optical surface modes in a (001) surface of a NaCl-type crystal, for **Q** at the midpoint of the $\overline{\Gamma\mathrm{M}}$ direction. **a** The ~SV polarized Wallis mode S_2 [79]. **b, c** The ~L and SH polarized Lucas modes S_4 and S_5, respectively [66, 80]. For simplicity the elliptical polarization of the S_2 and S_4 modes, resulting from a slight mixing between the SV and L displacements, is not shown

components, respectively. The sagittal modes S_2 and S_4 have in actual fact an elliptical polarization, which means that their respective displacement patterns illustrated in Fig. 3.17a, b are slightly mixed together and out of phase.

Finally, the surface-projected phonon densities of phonon states (DOS) for the sagittally polarized modes of the NaF(001) surface calculated with the GF method [74] are displayed in Fig. 3.18 for the $\overline{\Gamma}$- and $\overline{M}$-points. The total unperturbed DOS projected on the surface plane and the total perturbed surface-projected DOS, as defined by (3.56) and (3.57), respectively, are compared in the upper and lower parts of the Fig. 3.18, respectively. For $\hbar\omega \to 0$ the projected unperturbed DOS at $\overline{\Gamma}$ vanishes linearly (unlike the total bulk DOS which is quadratic in ω) while the projected perturbed DOS is constant and finite due to the presence of Rayleigh waves. The localized surface modes S_2 at $\overline{\Gamma}$, S_1 and S_8 at $\overline{M}$, as well as the pseudo-surface mode S_4 at $\overline{\Gamma}$ and the surface resonances S_4 and S_6 at $\overline{M}$ are clearly seen outside and inside the perturbed surface-projected phonon bands. The labels S_j ($j = 1, 2, \ldots$) relate the DOS spectral features to the dispersion curves shown in Figs. 3.15 and 3.16.

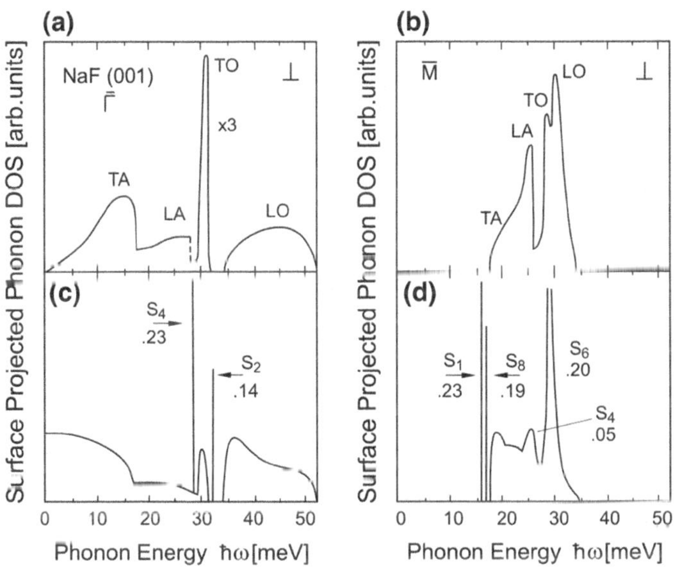

Fig. 3.18 Comparison of the unperturbed and perturbed surface-projected phonon density of states (DOS) with sagittal polarization (⊥) for the NaF(001) surface at the $\overline{\Gamma}$- and $\overline{M}$-points along the [100] direction as calculated by the GF method [74]. **a**, **b** are for the unperturbed and **d**, **e** are for the perturbed surface. The contributions of the bulk bands TA, LA, TO and LO to the total unperturbed densities in **a**, **b** are indicated. The vertical bars in the perturbed densities **d**, **e** represent the δ-function peaks of localized surface modes. Their intensities are indicated as a fraction of the integrated density. The S_4 and S_6 modes at the $\overline{M}$-point are surface resonances

3.7.3 Surface Phonon Labels

For future reference Table 3.1 summarizes the conventions used in this monograph for labeling the surface modes. These designations follow the general principles, laid out by Miglio and Benedek [45], that the labels should correspond to well defined spectroscopic entities, with special regard to the polarization and the optical or acoustic character. Often the labels do not follow continuously the respective branches across the entire Brillouin zone since they may be interchanged due to the avoided crossing behavior or hybridization, as, for example, in the case of the quasi-SH and quasi-sagittal branches in non-symmetry directions. Also along the symmetry directions the various branches may have a complicated non-crossing behavior, as shown in the hybridization scheme of Fig. 3.15b for the NaF(001) sagittal surface phonons in the $\overline{\Gamma M}$ direction.

According to the conventions in Table 3.1 the sagittally polarized S_1 mode corresponds to a Rayleigh wave in the continuum limit ($\mathbf{Q} \rightarrow 0$). In monatomic crystals a quasi-longitudinal surface mode (S_2) is normally found in the LA-TA gap occurring at the edge at the Brillouin zone (Fig. 3.10b) [44]. There is also a microscopic longitudinal acoustic mode peeled off from the lower edge of the LA band, called S_3 (Fig. 3.10b) (S_6 in diatomic crystals). It is a resonance due to the mixing with the TA bulk band modes. For metals the S_3 mode exhibits often a comparatively large softening and appears in the HAS time-of-flight spectra as a strong resonance. This mode is referred to as an *anomalous longitudinal (AL) resonance* [8] because for the ideal surface no such mode is predicted to be split off by such a large amount from the lower edge of the longitudinal acoustic band [81]. Much of the discussion about surface dynamics and the role of surface electrons in metals is related to this resonance, as will be explained in the following chapters.

The SH surface mode S_4 for a monatomic surface (or S_7 for diatomic crystals, Fig. 3.16a), is an acoustic surface mode which in the (001) surface of cubic crystals occurs along the $\overline{\Gamma X}$ direction. In the $\overline{\Gamma M}$ direction it degenerates into a bulk TA mode, whereas in the other non-symmetry directions it acquires a mixed character. As explained in Sect. 3.1, for $\mathbf{Q} \rightarrow 0$ this mode tends to an ordinary TA bulk mode, which for an anisotropy $\eta > 1$ lies below the RW branch. Thus in some cubic crystals the SH mode is below the RW in the $\overline{\Gamma X}$ direction and above it in the $\overline{\Gamma M}$ direction, with an avoided crossing at some intermediate direction. For this reason some authors, at variance with the present convention, have labeled the SH mode as S_1 when it lies below the RW mode. Additional surface modes are induced by the surface relaxation (see the discussion of the Rayleigh theorem, Sect. 3.5) such as the mode S_5 appearing in monatomic crystals below the TA_1 band at the zone boundary [44, 56]. Monoatomic (especially metal) surfaces with an appreciable contraction of the first and a dilation of the second interplanar spacing may support bilayer surface modes similar to those of a diatomic lattice, e.g., sagittal Lucas (S_4) and Wallis (S_2) resonances with a larger amplitude on the second layer (subsurface modes). Surface perturbation extending for a few layers below the surface can induce several

Table 3.1 Classification and designation of surface modes for low-index surfaces of monatomic and diatomic crystals according to their bulk band origin. For surface modes the polarization of the dominant component in the first plane is assigned following the convention proposed by Miglio and Benedek [45, 76]

Bulk band[a]	Surface Mode		Polarization	Conventional name
	Monatomic	Diatomic[b]		
TA_1	S_1	S_1	~SV	Rayleigh wave [20]
	–	S_8	~SV	Folded Rayleigh mode [77]
	S_5	–	~SV	Relaxation induced [44]
TA_2	S_4	S_7	SH	Alldredge mode [26]
	S_7	–	Mixed[c]	Gap mode (zone edge) [44]
LA	S_3	S_6	~L	Longitudinal mode[d] [43, 44]
	S_2	–	~L	Gap mode (zone edge) [44]
	S_6, S_8	–	Mixed[c]	Gap modes (zone edge) [44]
	S_9	–	Mixed[c]	Second-layer resonance [44]
TO_1	–	S_4^b	~L	Sagittal Lucas mode [66]
TO_2	–	S_5	SH	Shear horizontal Lucas mode [66]
LO		S_2^b	~SV	Wallis mode [79]
	–	S_3	~SV	Gap mode (zone edge) [43]

[a]TA_1 and TO_1 label the transverse acoustic and optical bulk modes polarized in the sagittal plane, respectively, whereas TA_2 and TO_2 label the transverse bulk modes with polarization parallel to the surface. LA and LO label the longitudinal acoustic and optical bulk modes, respectively
[b]Monoatomic (especially metal) surfaces with an appreciable contraction of the first interplanar spacing may support bilayer surface modes similar to those of a diatomic lattice, e.g., Lucas and Wallis modes (resonances) with a larger amplitude on the second layer (subsurface modes)
[c]Modes occurring out of symmetry directions with polarizations neither parallel nor normal to the sagittal plane
[d]In general the longitudinal surface mode is a resonance. In most metals this mode is considerably softened with respect to the bulk band edge and will be referred to as *anomalous longitudinal resonance* (AL)

additional branches (see, e.g., in Si(111)2×1, Sect. 11.2) with the largest amplitude on some subsurface layer, as for example the S_9 resonance plotted in Fig. 3.11c.

Other surface modes are found in model calculations for monatomic crystals within the gaps occurring at the edge of the Brillouin zone edge out of the symmetry directions [44]. Example of such modes are S_6 and S_8 in the LA-TA_2 gap and S_7 in the TA_2-TA_1 gap [44]. These modes have polarizations which are neither parallel nor perpendicular to the sagittal plane but are mixtures of the L, SV and SH components.

In diatomic crystals the sagittal resonance S_8, crossing the LA bulk band in both directions, may appear in some crystals as a folded prolongation of the Rayleigh wave. Its intensity is appreciable in crystals with nearly equal ionic masses such as NaF (Fig. 3.15a), KCl, RbBr [77] and MgO [82]. Other mechanisms connected with the hybridization of the bulk phonon branches may give similar surface resonances, which cross the acoustic bulk bands with a negative group velocity, also in crystals with unequal masses such as NaCl and KBr [83]. As noted above, in diatomic crystals the optical surface modes which are associated with the transverse optical phonon bands form the pair of Lucas modes (LM) S_4 and S_5 [66]. In the (001) and (111) surfaces of cubic crystals the Lucas modes are degenerate at the zone center ($\overline{\Gamma}$ point) (see Figs. 3.14a and 3.15a). For $\mathbf{Q} \rightarrow 0$ along any direction, S_4 and S_5 become linearly polarized and exactly longitudinal or transverse polarized in the surface plane, respectively.

The optical mode S_2 is quasi-SV everywhere and only at $\overline{\Gamma}$ and $\overline{M}$ is it polarized perpendicular to the surface (SV) [78, 79]. While at $\overline{\Gamma}$ the S_2 mode is associated with the LO bulk band, at the zone boundary S_2 (or S_6 after hybridization, as seen in Fig. 3.16 along $\overline{\Gamma M}$) is associated with the bulk TO_1 mode, whereas another mode, S_3, originates from the edge of the bulk LO mode. Actually the modes S_2, S_3 and S_6 have the same symmetry and can hybridize with each other by interchanging their optical/acoustic characters. In the intermediate BZ regions they turn into broad resonances and eventually disappear into the continuum of bulk optical modes. In general these modes are split from the lower edge of their respective bulk bands. However in RbBr(001) and RbI(001) the S_2 mode lies above the LO band as a result of stiffening induced by surface relaxation [83, 84]. Also in diatomic crystals the relaxation and rumpling, as well as reconstruction, may induce further modes involving second or deeper layers below the surface, similar to the S_9 resonance of Fig. 3.11c. An interesting example is offered by the deep reconstruction occurring in Si(111)2×1 which induces an impressive number of optical surface phonons [13]. Some of these modes have the largest amplitude well below the topmost layer at the interface between the reconstructed and the normal lattice. In this case the classification is conveniently based on the displacement pattern of the topmost atoms of the unit cell (see Fig. 11.10 and related discussion in Sect. 11.2).

The analysis and classification of optical surface modes is readily extended to polyatomic crystals with s atoms per unit cell. In general they will have $s - 1$ pairs of Lucas modes and $s - 1$ Wallis modes. Examples of polyatomic crystals whose optical surface modes have been thoroughly investigated by HAS are the layered hexagonal crystals $TaSe_2$(0001) [85], with three atoms per unit cell in each layer, and GaSe(0001) [86, 87], with four atoms per unit cell in each layer (see Sect. 11.4). In these crystals the upper bulk phonon bands are in general well separated by wide gaps. Thus the surface phonon branches are expected to strictly retain the character of the parent bands all along the surface Brillouin zone and are classified accordingly.

References

1. E. W. Plummer, W. Eberhardt, in: *Advances in Chemical Physics*, vol. XLIX, ed. by I. Prigogine, S. A. Rice (Wiley, New York, 1982) p. 1
2. H. Lüth, *Surfaces and Interfaces of Solid Materials*, 3rd edn. (Springer, Heidelberg, 1995)
3. H. Lüth, *Solid Surfaces, Interfaces and Thin Films* (Springer, Heidelberg, 2001)

4. G. Brusdeylins, R.B. Doak, J.P. Toennies, Phys. Rev. B **27**, 3662 (1983)
5. W.J.L. Buyers, Phys. Rev. **153**, 923 (1967)
6. P. Santini, L. Miglio, G. Benedek, U. Harten, P. Ruggerone, J.P. Toennies, Phys. Rev. B **42**, 11942 (1990)
7. J.L.T. Waugh, G. Dolling, Phys. Rev. **132**, 2410 (1963)
8. R.B. Doak, U. Harten, J.P. Toennies, Phys. Rev. Lett. **51**, 578 (1983)
9. W. Drexel, Z. Phys. **255**, 281 (1977)
10. H.-J. Ernst, E. Hulpke, J.P. Toennies, Phys. Rev. Lett. **58**, 1941 (1987)
11. A. Larose, B.N. Brockhouse, Can. J. Phys. **54**, 1819 (1976)
12. Y.R. Wang, C.B. Duke, Surf. Sci. **205**, L755 (1988)
13. L. Miglio, P. Santini, P. Ruggerone, G. Benedek, Phys. Rev. Lett. **62**, 3070 (1989)
14. K. Hermann, *Crystallography and Surface Structure: An Introduction for Surface Scientists andNanoscientists*, 2nd edn. (Wiley, Berlin, 2016)
15. M.W. Finnis, V. Heine, J. Phys. F (Metal Phys). **4**, L37 (1974)
16. S. Lehwald, F. Wolf, H. Ibach, B.M. Hall, D.L. Mills, Surf. Sci. **192**, 131 (1987)
17. W. Kohn, Phys. Rev. Lett. **2**, 393 (1959)
18. J. Ernst, E. Hulpke, J.P. Toennies, Europhys. Lett. **10**, 747 (1989)
19. H.-J. Ernst, E. Hulpke, J.P. Toennies, Phys. Rev. B **46**, 16081 (1992)
20. J.W. Strutt (Baron Rayleigh), Proc. London Math. Soc. **17**, 4 (1885)
21. J.W. Strutt (Baron Rayleigh), *The Theory of Sound* (MacMillan, London 1896; reprinted by Dover Publ., New York, 1945)
22. G.W. Farnell, in *Physical Acoustics*, vol. 6, ed. by W.P. Mason, R.N. Thurston (Academic Press, New York, 1970) p. 109
23. G.W. Farnell, in *Acoustic Surface Waves*, ed. by A.A. Oliner (Springer, Berlin, 1978) p. 13
24. A.A. Oliner (ed.), *Acoustic Surface Waves* (Springer V, Berlin, 1978)
25. B.H. Brandsen, C.J. Joachain, *Introduction to Quantum Mechanics* (Longman, New York, 1989)
26. G.P. Alldredge, Phys. Lett. **41A**, 281 (1972)
27. M. Born, Th von Kármán, Phys. Zeit. **13**, 297 (1912)
28. A.A. Maradudin, E.W. Montroll, G.H. Weiss and I.P. Ipatova, in *Theory of Lattice Dynamics in the Harmonic Approximation*, Solid State Physics, suppl. 3 (Academic Press, New York, 1971)
29. J. DeLaunay, Solid State Phys. **2**, 219 (1956)
30. V. Bortolani, A. Franchini, G. Santoro, in Electronic Structure, Dynamics and Quantum Structural Properties of Condensed Matter, ed. by J.T. Devreese, P. van Camp (Plenum Press, New York, 1984) p. 401
31. M. Born, K. Huang, *Dynamical Theory of Crystal Lattices* (Clarendon Press, 1954 and 1998)
32. W. Cochran, *The Dynamics of Atoms in Crystals* (Arnold, London, 1973)
33. P. Brüesch, *Phonons: Theory and Experiments. Lattice Dynamics and Models of Interatomic Forces* (Springer, Berlin, 1982)
34. H. Böttger, *Principle of the Theory of Lattice Dynamics* (Akademie-Verlag, Berlin, 1983) Sec. 1.4
35. A. Askar, *Lattice Dynamical Foundations of Continuum Theories* (World Sci., Singapore, 1986)
36. M. T. Dove, *Introduction to Lattice Dynamics* (Cambridge Univ. Press, 1993)
37. P.N. Keating, Phys. Rev. **140A**, 369 (1965)
38. P.N. Keating, Phys. Rev. **145**, 637 (1966)
39. V. Bortolani, A. Franchini, F. Nizzoli, A. Franchini, J. Electr. Spectr. Rel. Phenom. **29**, 219 (1983)
40. For a compilation of bulk phonon data see H.R. Schober, P.H. Dedrich in Landolt-Börnstein, Neue Serie Bd. 13a (Springer, Berlin, 1981)
41. R.F. Wallis in *Dynamical Properties of Solids*, vol. 2, ed. by G.K. Horton, A.A. Maradudin (North-Holland, Amsterdam, 1975) p. 443
42. J.E. Black, in *Dynamical Properties of Solids*, vol. 6, ed. by G.K. Horton and A.A. Maradudin (Elsevier, Amsterdam, 1990) p. 179

43. R.E. Allen, G.P. Alldredge, F.W. de Wette, Phys. Rev. Lett. **23**, 1285 (1969)
44. R.E. Allen, G.P. Alldredge, F.W. de Wette, Phys. Rev. B **4**, 1648, 1661(1971)
45. L. Miglio, G. Benedek, in *Structure and Dynamics of Surfaces II*, ed. by W. Schommers, P. von Blanckenhagen (Springer, Heidelberg, 1987) p. 35
46. G. Benedek, L. Miglio, in *Surface Phonons*, vol. 27, ed. by F.W. de Wette, W. Kress, Springer Ser. Surf. Sci., (Springer, Berlin, 1991) p. 37
47. G. Benedek, Phys. Status Sol. **B58**, 661 (1973)
48. A.A. Maradudin, P. Mazur, E.W. Montroll, G.H. Weiss, Rev. Mod. Phys. **30**, 173 (1958), App. A
49. S. Doniach, E.H. Sondheimer, in *Green's Funtion for Solid State Physics* (Benjamin, London, 1974; reprinted by Imperial College Press, London, 1998)
50. E.N. Economou, *Green's Function in Quantum Physics* (Springer, Berlin, 1978)
51. I.M. Lifshitz and L.M. Rozenzweig, Zh. Eksp. Teor. Fiz. 18 (1948) 1012 (in Russian); I.M. Lifshitz: Nuovo Cimento, suppl. 3 (1956) p. 732
52. I.M. Lifshitz, A.M. Kosevich, Rep. Progr. Phys. **29**(I), 217 (1966)
53. B.A. Lippmann, J. Schwinger, Phys. Rev. **79**, 469 (1950)
54. L. van Hove, Phys. Rev. **95**, 249 (1954)
55. G. Benedek, Surf. Sci. **61**, 603 (1976)
56. F.W. de Wette, W. Kress (eds.), *Surface Phonons*, Springer Ser. Surf. Sci., vol. 21 (Springer, Berlin, 1991) p. 67
57. F.W. de Wette, G.P. Alldredge, in *Methods in Computational Physics*, vol. 15, ed. by G. Gilat, B. Alder, S. Fernbach, M. Rotenberg (Academic, New York, 1976) p. 163
58. F.W. de Wette, in *Dynamical Properties of Solids*, ed. by G.K. Horton, A.A. Maradudin (Elsevier, Amsterdam, 1990) Chap. 5
59. A. Lock, J.P. Toennies, G. Witte, J. Electr. Spectr. Rel. Phenom. **54**(55), 309 (1990)
60. Ch. Wöll, Thesis (University of Göttingen, 1987); Max-Planck-Institut für Strömungsforschung, Report 18/1987
61. The calculation of a one-layer two-dimensional lattice is described in O. Madelung, *Festkörpertheorie II* (Springer, Berlin, 1972)
62. O. Madelung, *Festkörpertheorie I* (Springer, Berlin, 1972) p. 127ff
63. Ch. Wöll, Appl. Phys. A **53**, 377 (1991)
64. G. Benedek, J. Ellis, A. Reichmuth, P. Ruggerone, H. Schief, J.P. Toennies, Phys. Rev. Lett. **69**, 2951 (1992)
65. R.F. Wallis, Phys. Rev. **105** (1957) 540; 116 (1959) 302
66. A.A. Lucas, J. Chem. Phys. **48**, 3156 (1968)
67. R. Fuchs, K.L. Kliever, Phys. Rev. **140**, A2076 (1965)
68. K.L. Kliever, R. Fuchs, Phys. Rev. **144**, 495 (1966); **150**, 573 (1966)
69. A discussion of non-retarded FK modes vs. microscopic optical modes is given by W.E. Jones, R. Fuchs, Phys. Rev. **B4**, 3581 (1971). The surface dynamics theory including retardation has been developed first by V.V. Bryksin, Y.A. Firsov, Fiz. Tverd. Tela **11**, 2167 (1969) [Sov. Phys. Solid State **11**, 1751 (1970)]
70. H. Raether, *Surface Plasmons* (Springer, Berlin, 1988)
71. A.D. Boardman (ed.), *Electromagnetic Surface Modes* (Wiley, New York, 1982)
72. J. Pedlosky, *Waves in the Ocean and Atmosphere: Introduction to Wave Dynamics* (Springer, Berlin, 2003)
73. F.W. de Wette, in *Surface Phonons*, ed. by F.W. de Wette, W. Kress, Springer Series on Surf. Sci. Vol. 27 (Springer, Berlin, 1991) p. 67
74. G. Benedek, G.P. Brivio, L. Miglio, V.R. Velasco, Phys. Rev. B **26**, 497 (1982)
75. G. Brusdeylins, R. Rechsteiner, J.G. Skofronick, J.P. Toennies, G. Benedek, L. Miglio, Phys. Rev. Lett. **54**, 466 (1985)
76. G. Benedek, L. Miglio, in *Ab-Initio Calculation of Phonon Spectra*, ed. by J.T. Devreese, V. E. van Doren, P.E. Van Camp (Plenum, New York, 1983)
77. G. Benedek, L. Miglio, G. Brusdeylins, J.G. Skofronick, J.P. Toennies, Phys. Rev. B. **35**, 6593 (1987)

78. T.S. Chen, F.W. de Wette, G.P. Alldredge, Phys. Rev. B **15**, 1167 (1977)
79. R.F. Wallis, Phys. Rev. **105**, 540 (1957); **116**, 302 (1959)
80. G. Benedek, Physicalia Magazine. **18**, 205 (1996)
81. G. Armand, Phys. Rev. B **14**, 2218 (1976)
82. G. Benedek, G. Brusdeylins, V. Senz, J.G. Skofronick, J.P. Toennies, F. Traeger, R. Vollmer, Phys. Rev. B **64**, 125421 (2001)
83. W. Kress, F.W. de Wette, A.D. Kulkarni, U. Schröder, Phys. Rev. B **35**, 5783 (1987)
84. F.W. de Wette, A.D. Kulkarni, U. Schröder, W. Kress, Phys. Rev. B **35**, 2467 (1987)
85. G. Benedek, L. Miglio, G. Brusdeylins, C. Heimlich, J.G. Skofronick, J.P. Toennies, Europhys. Lett. **5**, 253 (1988)
86. G. Brusdeylins, R. Rechsteiner, J.G. Skofronick, J.P. Toennies, G. Benedek, L. Miglio, Phys. Rev. B **34**, 902 (1986)
87. V. Panella, A. L. Glebov, J.P. Toennies, C. Sebenne, C. Eckl, C. Adler, P. Pavone, U. Schöder, Phys. Rev **B59**, 15772 (1999)

Chapter 4
The Effect of the Surface Termination on Force Constants and Dispersion Curves

As regards the Atomicists, it is not only clear what their explanation is;
it is also obvious that it follows with tolerable consistency from the assumptions they employ.

Aristotle 384–332 BC

Abstract The Born-von Kármán model introduced in the previous Chapter, is used to illustrate how the invariance properties of the total energy, the equilibrium conditions and the emergence of the surface stress affect the surface force constants in the surface region and the surface phonon dispersion curves. Then the effects of extended surface and interplanar force constant perturbations, which in the case of metals may involve several layers, are discussed.

4.1 Surface Equilibrium

The important link between force constants and the equilibrium structure can be understood from two important constraints derived from the invariance of the total energy against infinitesimal rigid body translations and rotations [1]. Infinitesimal rigid body translations and rotations are defined by the displacement fields

$$\mathbf{u}(l) = \boldsymbol{\varepsilon} \quad \text{and} \quad \mathbf{u}(l) = \boldsymbol{\varepsilon} \times \mathbf{r}_{lj}, \tag{4.1}$$

respectively, where $\mathbf{r}_{lj} \equiv \mathbf{r}_l - \mathbf{r}_j$, and $\boldsymbol{\varepsilon}$ is a constant infinitesimal vector of arbitrary direction. For rotations $\boldsymbol{\varepsilon}$ defines the direction of the rotation axis passing through $\mathbf{r}_j$. When these rigid body displacements are inserted into (3.18), the invariance of the energy $U = U_0$, equivalent to the condition that no net force nor torque is acting on the solid, yields to the lowest order in $\boldsymbol{\varepsilon}$ two constraints on the harmonic force constants $R_{\alpha\beta}(l,j)$. These constraints, known as the *translational invariance* (TI) and *rotational invariance* (RI) conditions, are expressed by

G. Benedek and J. P. Toennies, *Atomic Scale Dynamics at Surfaces*, Springer Series in Surface Sciences 63, https://doi.org/10.1007/978-3-662-56443-1_4

$$\text{TI:} \quad \sum_l R_{\alpha\beta}(l,j) \equiv \Im_T R = 0, \quad \forall j, \tag{4.2}$$

$$\text{RI:} \quad \sum_l [R_{\alpha\beta}(l,j) r_{l\gamma} - R_{\alpha\gamma}(l,j) r_{l\beta}] \equiv \Im_R R = 0, \quad \forall j, \tag{4.3}$$

where the sums are over all the N atoms in the lattice. Equations (4.2) and (4.3) define two linear projectors $\Im_T$ and $\Im_R$ which reduce the $3N\times3N$ force constant matrix R to two rectangular $3\times3N$ matrices.

For a two-body central interaction the substitution of the atom-atom force constant matrix given by (3.21) into the RI condition, (4.2), yields

$$\sum_l \alpha_{lj}(\mathrm{I} \times \mathbf{r}_{lj}) = \sum_l \frac{\partial \mathrm{v}}{\partial r_{lj}} \left(\mathrm{I} \times \frac{\mathbf{r}_{lj}}{r_{lj}} \right) = 0. \tag{4.4}$$

where I denotes the unit tensor. Since the cross product in (4.4) gives an anti-symmetric tensor of components $\pm r_{lj,\alpha} / r_{lj}$, it follows that the components of the forces exerted by all atoms j on atom l sum to zero. Thus for two-body central forces the rotational invariance is equivalent to the condition of equilibrium (see Sect. 3.3).

In general at equilibrium the potential energy U must be invariant with respect to any infinitesimal lattice strain, e.g., with respect to an isotropic deformation around the j-th atom $\mathbf{u}(l) = \varepsilon\,\mathbf{r}_{lj}$, with ε an infinitesimal constant. In this case the tangential force constants $\alpha_{lj,}$ defined by (3.22), turn out to be linked to the atomic positions through another equilibrium condition [1]

$$\sum_{lj} \alpha_{lj} r_{lj}^2 = 0. \tag{4.5}$$

To illustrate the effects of (4.5) consider first a crystal with only nearest neighbor (nn) forces: in this case equilibrium requires $\alpha_{lj} = 0$ and (4.4) and (4.5) are trivially fulfilled. However in a monoatomic simple cubic (*sc*) model with first (nn) and second nearest neighbor (2nn) force constants (see Sect. 3.3), (4.5) gives $\alpha_{2n} = -\alpha_{nn}/4$ in the bulk and $\alpha_{2n} = -5\alpha_{nn}/16$ at the (001) surface, whereas (4.4) is trivially fulfilled in the bulk and gives $\alpha_{2n} = -\alpha_{nn}/4$ at the surface. This conflicting result indicates that the ideal (001) surface of a *sc* crystal with only nn and 2n interactions cannot be in equilibrium and some relaxation must occur. Indeed (4.4) and (4.5) can be reconciled by allowing a reduction of the interplanar distance between the first two surface planes from a_0 to $a = \sqrt{(2/3)}a_0$ (i.e., −18%). This automatically implies a change in the tangential force constants, e.g., $\alpha_{nn}(a) \cong \alpha_{nn}(a_0)(a/a_0) + \beta_{nn}(a_0)(a/a_0 - 1)$, whereas, in the absence of anharmonicity, the radial force constants remain unchanged. This simple example shows that (4.4) and (4.5) impose two important constraints on the tangential force constants.

It should be noted that in general first principle calculations of the dynamical matrix based on a preliminary equilibration of the structure fulfill automatically, within the numerical errors, all invariance conditions. On the contrary surface dynamics calculations based on empirical force constant models need to ensure that TI and RI conditions are fulfilled in order to give reliable results.

4.2 The Intrinsic Surface Perturbation

As discussed in Chap. 3 the surface may be considered as an intrinsic defect. Within the Green's function approach (see Sect. 3.5.2) the perturbation matrix Λ accounts for the effect of the surface upon the bulk dynamics. Indeed the intrinsic perturbation matrix can be derived from the bulk dynamic structure through the elements of the Green's function matrix projected onto the perturbation subspace [2]. It is assumed, as happens in most cases, that the crystal slab has two identical surfaces and an inversion symmetry point (or a mirror symmetry plane) in the middle. Thus all solutions of the dynamical problem have either an even or an odd symmetry with respect to the inversion point (or the mirror plane), namely a well defined parity $p = \pm$. There is therefore a symmetry transformation S producing a simultaneous diagonalization of the Green's function and perturbation matrices into two blocks of defined parity $\mathrm{G}_{0\,p}$ and Λ_p, respectively. Similarly for the perturbed Green's function, (3.54):

$$\mathrm{G}_p = (\mathrm{I} - \mathrm{G}_{0p}\Lambda_p)^{-1}\mathrm{G}_{0p}. \tag{4.6}$$

This is just the symmetrization illustrated in Fig. 3.7c, leading to a perturbation subspace restricted to a single surface region (reduced perturbation subspace). It is then possible to transform the two surface slab problem into the problem of a single surface semi-infinite crystal by setting the condition

$$\mathrm{G}_+ = \mathrm{G}_- \equiv \tilde{\mathrm{G}} \tag{4.7}$$

which is equivalent to assuming that the slab is so thick that there is no interaction between the two surfaces and the slab solutions are degenerate in pairs. This condition holds over the entire spectrum of surface phonons except, in the case of ionic crystals, for the macroscopic FK modes. As anticipated in Sect. 3.5, they can however be included in the GF theory of the semi-infinite crystal by adding to the surface perturbation Λ the contribution of the macroscopic field [3].

Equation (4.7) implies that

$$\tilde{\mathrm{G}} - (\mathrm{I} - \bar{\mathrm{G}}\bar{\Lambda})^{-1}\bar{\mathrm{G}} \tag{4.8}$$

where

$$\bar{\Lambda} \equiv \frac{1}{2}(\Lambda_+ + \Lambda_-), \tag{4.9}$$

$$\bar{G} \equiv 2(G_{0+}^{-1} + G_{0-}^{-1})^{-1}, \tag{4.10}$$

are the perturbation and surface-projected unperturbed Green's function matrices for the semi-infinite crystal. Note that all matrix inversions are performed in the reduced perturbation subspace. All the above matrices can be expressed in the surface wavevector representation since the two-dimensional periodicity is preserved in the semi-infinite crystal, and both $\bar{G}$ and $\tilde{G}$ are functions of $\mathbf{Q}$ and ω^2. Then the corresponding equation giving the frequencies of localized and resonant surface modes is

$$\text{Re det}\left[I - \bar{G}(\mathbf{Q}, \omega^2)\bar{\Lambda}(\mathbf{Q})\right] = 0. \tag{4.11}$$

In ionic crystals the symmetrization has the advantage of producing a cancellation of the long-range Coulomb terms in the perturbation matrix so that the elements of $\bar{\Lambda}$ decay exponentially with l_3 [4], which allows to restrict the reduced perturbation space to only a few layers, or just to the first layer.

The elements of $\bar{\Lambda}$ can now be deduced by applying the TI and RI vector operators, defined in the previous section, to the equation

$$\Lambda_\pm = \bar{\Lambda} + \frac{1}{2}(G_{0\pm}^{-1} - G_{0\mp}^{-1}), \tag{4.12}$$

which can be derived from (4.7) and (4.9). The symmetrization of the TI and RI operators, $\Im_T S^{-1}$ and $\Im_R S^{-1}$, yields the new TI and RI operators $\Im_{T,p}$ and $\Im_{R,p}$ in the reduced perturbation subspace. By applying them to both members of (4.12) and noting that $\Im_{T,p}\Lambda_p = \Im_{R,p}\Lambda_p = 0$ one obtains

$$\begin{aligned}\Im_{T,\pm}\bar{\Lambda} &= -\Im_{T,\pm}(G_{0\pm}^{-1} - G_{0\mp}^{-1})\\ \Im_{R,\pm}\bar{\Lambda} &= -\Im_{R,\pm}(G_{0\pm}^{-1} - G_{0\mp}^{-1}).\end{aligned} \tag{4.13}$$

This is a set of linear inhomogeneous equations whose solution gives the elements of the intrinsic perturbation $\bar{\Lambda}$. The static nature of the perturbation and the unchanged periodicity of the surface with respect to the equivalent planes in the bulk (i.e., no reconstruction) imply that the inverse projected Green's function elements in (4.13) are calculated at $\omega = 0$ and $\mathbf{Q} = 0$. The extension of this method in order to include in the perturbation the effects of the electronic response and its surface modifications, as required, e.g., in the shell model treatments, is discussed in [5, 6]. A further extension to the multipole expansion method (see next chapter), in order to apply the GF method to the dynamics of metal surfaces, where Friedel oscillations may extend to several layers below the surface, has not been investigated yet.

The above method has been successfully applied to the GF calculation of the surface phonons in alkali halides [2–6]. The equivalence between the RI and equilibrium conditions, which holds for two-body central forces, leads to an interesting simplification in the Green's function calculations of weakly relaxed surfaces [2–6]. Instead of changing the positions of the surface atoms, allowing them to relax to the equilibrium positions, the equilibrium is restored for the ideal (unrelaxed) surface by a suitable change of the tangential force constants within the surface plane, with a procedure which was first devised by Lengeler and Ludwig for *sc* lattices [7]. In practice the additional stabilizing force constants mimics the effect of surface relaxation on the surface dynamics. The stabilizing tangential force constants in the surface plane are obtained from the RI condition (4.4), as simple combinations of the interplanar tangential force constants between the first and the underlying planes. Indeed this way of constructing a rotationally invariant surface perturbation for the surface phonons of alkali halides was shown to guarantee a good agreement with the slab calculations for the relaxed surface [8, 9].

4.3 Surface Stress

There is a simple relationship between the microscopic tangential force constant within the surface plane (denoted by $\alpha_{\parallel}^{s}$) and the *macroscopic* surface stress tensor $\sigma_{\alpha\beta}$. The concept of surface stress goes back to the seminal studies of Shuttelworth [10] and Herring [11, 12] from the early 1950s. At equilibrium there must, of course, be an overall cancellation of all forces acting on each surface atom. However, not all of these forces arise within the surface layer, part of them being exerted by the substrate. The forces acting within the surface alone may not be sufficient to keep the surface atoms in their actual equilibrium positions. If so they do not cancel and give rise to a surface stress. The surface stress is *tensile* (positive) when it favors a contraction of the surface layer, as happens on most metal surfaces, or *compressive* (negative) when it favors an expansion of the surface layer, as occurs in many ionic crystals [10].

Surface stress is defined as the derivative of the surface free energy $F_s = \gamma S$ due to an in-plane strain $\varepsilon_{\alpha\beta}$ per surface area, where S is the surface area and γ the surface tension. The surface tension is a measure of either the force per unit increase of peripheral length or the increase in total free energy with area, and is related to the surface stress tensor $\sigma_{\alpha\beta}$ by the equation [11, 12]

$$\sigma_{\alpha\beta} = \frac{1}{S}\frac{\partial F_s}{\partial \varepsilon_{\alpha\beta}} = \gamma\,\delta_{\alpha\beta} + \frac{\partial \gamma}{\partial \varepsilon_{\alpha\beta}}. \tag{4.14}$$

In a liquid the surface stress is isotropic and coincides with the surface tension, because $\partial\gamma/\partial\varepsilon_{\alpha\beta} = 0$ as a consequence of the high mobility of molecules and the lack of shear stiffness. For a solid, however, both terms in (4.14) make nearly equal contributions [10]. For a rare gas solid at 0 °K Shuttelworth [10] already in 1950 calculated the surface tension assuming central pairwise additive forces. His formula when applied to a simple nearest neighbor interaction model leads to the following simple proportionality

$$\sigma_{\alpha\alpha} = \frac{1}{2}\alpha^{s}_{||}. \tag{4.15}$$

Here it is important to note that $\alpha^{s}_{||}$ describes an interatomic interaction in the surface layer (superscript s) whereas the surface stress is a macroscopic concept.

The effects of surface stress on surface dynamics have been investigated in Ni (110) and in Ni(111) by Lehwald et al. [13] and Menezes et al. [14], respectively and in Cu(001) by Wuttig et al. [15]. In these cases the experimental dispersion curves obtained by EELS have been shown to be indicative of a positive surface stress, as expected for clean metal surfaces. On the other hand fcc Fe films of a few monolayers grown on a Cu(001) substrate were shown to be affected by a negative, compressive stress which accounts for the pronounced softening of the RW dispersion curve at the $\overline{\mathrm{M}}$ point [16]. Daum et al. [16] have identified the compressive surface stress as the driving force for the structural instabilities observed in fcc-Fe films [17, 18]. Subsequent HAS measurements of the RW dispersion curve as a function of temperature and layer thickness suggested a magnetic origin for the compressive stress in these films [19] (see discussion in Chap. 13).

Important effects of compressive surface stress have been observed with HAS in the commensurate p(3×3) phase of Pb ultrathin films (0.5–4 monolayers) on Cu (111) [20]. Due to the larger thermal expansion of lead with respect to copper, the compressive surface stress increases with temperature, leading to the expected softening of the ~SV surface phonon branches, but a remarkable anomalous stiffening of some ~L branches. Surface stress plays an essential role in determining the growth regimes on thin films on metal surfaces under different temperature and exposure conditions. In the case of Pb films on copper, while at low temperature a layer-by-layer Frank-van-der-Merwe growth regime is observed [21, 22], at room temperature the Stranski-Krastanov layer-plus-island growth occurs where islands of different size and height are formed above the wetting layer [23]. In particular the p(3×3) 2ML Pb film on Cu(111) is found to be unstable even at low temperature. Surface dynamics calculations [20] indicate the large compressive surface stress, eventually leading to imaginary surface phonon frequencies, as the cause of instability. Actually by alloying Pb with 20% of the slightly smaller Tl ions, the surface stress is reduced and the 2 ML film is stabilized [24], with all the surface phonon branches in agreement with HAS data [20].

In ionic crystals the dispersion relation of the SH Lucas mode S_5, due to its strong localization at the surface and in-plane polarization (Fig. 3.16), contains useful information on the surface stress [25]. A simple example serves to qualitatively illustrate the concept. Consider the (001) surface of an fcc alkali halide crystal with in-plane radial and shear nearest neighbor force constants $\beta^{s}_{||}$ and $\alpha^{s}_{||}$, respectively, and interplane radial and shear force constants $\beta^{s}_{\perp}$ and $\alpha^{s}_{\perp}$ linking each ion to the nn ion in second layer. Force constants α and β are defined in Sect. 3.3. By assuming for the S_5 mode an ideal localized displacement pattern where only the surface atoms of mass M_1 and M_2 move in the surface plane against the underlying atoms kept at rest, the following expression linking the frequencies of the S_5 mode at the $\bar{\Gamma}$ and $\bar{M}$ points is obtained [25]

$$\omega^2(\bar{\mathrm{M}}) - \omega^2(\bar{\Gamma}) = -\frac{16\alpha_{\parallel}^{s}\beta_{\parallel}^{s}}{\omega^2(\bar{\mathrm{M}})M_1M_2}. \tag{4.16}$$

This equation shows that the dispersion of the shear SH Lucas mode S_5 along the <100> direction depends, in a first approximation, only on the in-plane shear and radial force constants. Since $\beta_{\parallel}^{s}$ is positive, (4.16) implies that the frequency dispersion of the S_5 mode along the <100> directions has a sign opposite to that of $\alpha_{\parallel}^{s}$, namely, according to (4.15), opposite to the sign of the surface stress. The latter can thus be derived from the magnitude of the dispersion by means of (4.15) provided $\beta_{\parallel}^{s}$ is known. The calculated dispersion curves of the SH modes in NaF(001), Fig. 3.16a, show practically no dispersion for S_5, i.e., no surface stress, whereas all the other alkali halide (001) surfaces so far calculated [6, 26], show an upward dispersion indicating a negative (compressive) surface stress. Such anomalous behavior of NaF(001) is consistent with the early analysis of experimental data reported by Shuttelworth [10] which indicates a negative stress for all potassium and sodium halides except for NaF which has a positive (tensile) surface stress.

In other fcc diatomic crystals such as MgO(001) [27, 28] the dispersion of S_5 is also upward and comparatively strong, indicating a large negative surface stress and a tendency to "rumpling", which describes the structure resulting from different relaxations of the anions and cations of the first layer. A similar situation occurs in refractory superconductors such as TiN(001) [29]. Here the increase of nitrogen dipolar and quadrupolar polarizabilites, besides enhancing the superconducting anomalies, yields a larger slope of the S_5 dispersion with a dramatic increase in the (negative) surface stress. However, not all systems having a phase transition related in some way to the electron-phonon interaction have a negative surface stress. For example in cubic perovskites which show a soft mode behaviour, like $KZnF_3$, or a structural phase transition, like $KMnF_3$ and $SrTiO_3$, the S_5 mode exhibits a strong downward dispersion along the <100> directions for both types of (001) surfaces [30, 31]. This indicates a strong, positive (tensile) surface stress, which tends to contract the surface. Actually the rigid rotations of the anion octahedra surrounding transition metal ions, occurring in these materials at the phase transition, yield such a contraction. These last examples illustrate the role played by surface stress as a latent driving force for surface phase transitions.

The microscopic foundations of surface stress have been clarified through several *ab initio* calculations for metal surfaces [32–34] and thin slabs [35–39]. The concept of surface stress has been discussed in connection with the driving forces of clean-surface [40–42] and adsorbate-induced reconstructions [43]. The crucial role that surface stress plays in surface reconstruction, epitaxial growth and stabilization of mesoscopic structures has been reviewed by Ibach [44].

4.4 The Effect of Force Constant Changes on Surface Phonon Dispersion Curves

4.4.1 *Surface Force Constant Perturbations*

A knowledge of the Rayleigh wave velocity in the continuum limit, which is an intrinsic property of the bulk crystal, is very useful in the preliminary analysis of inelastic HAS and EELS data. However little information can be extracted from the RW velocity about the microscopic force constants unless one deals with a simple model where the number of force constants does not exceed that of the elastic constants. The relationship between the bulk force constants and the elastic constants is well established for all crystal symmetries [45]. As seen in Sect. 3.2, the elastic continuum theory provides a relationship, though not explicit, between the Rayleigh wave velocity in the long wave limit and the ratios among elastic constants [46]. The RW velocity divided by the bulk transverse velocity, v_R/v_T, as a function of the ratios c_{11}/c_{44} and c_{12}/c_{44} is plotted in Fig. 3.2 for the <100> propagation directions on the (001) surface of cubic crystals. Similar diagrams can be constructed for other propagation directions and surfaces of cubic crystals. For non-cubic crystals the number of independent elastic constants increases and therefore the RW velocities in different directions and surfaces are needed in order to fit the full set of elastic constants.

In the *ideal surface* model the force constant changes occurring at the surface are neglected in a first approximation. In this case a much richer information on the bulk force constants can be obtained from the dispersion of Rayleigh waves and other surface phonons, notably from the surface phonon frequencies at the high symmetry points of the Brillouin zone. A systematic numerical study has been reported by Black [47] for bcc monoatomic crystals with two-body interatomic potentials up to third nearest neighbors in which the surface mode frequencies at the zone-edge symmetry points were calculated for the three low-index surfaces of tungsten. For different sets of *bulk* force constants, which equally well reproduce the bulk phonon frequencies, large differences in the surface phonon frequencies were obtained. Such ideal surface studies, although of tutorial value, are of limited use for understanding real crystals since surface modes at the zone boundary penetrate little into the bulk and therefore mostly depend on the actual interatomic force constants at the surface, which are generally different from those in the bulk. Moreover in metals three-body forces are normally needed to account for the usually large deviations from the Cauchy relations in the bulk elastic constants. Three-body forces are also severely affected at the surface by the reduced coordination and relaxation.

A tutorial demonstration of the extent to which surface force constants can affect the surface phonon frequencies has been given by Musser and Rieder [48] for the Ni (001) surface. Their calculation is based on the DeLaunay model [49], which includes first and second nearest neighbor central force constants, β_{nn} and β_{2n} (3.22), and a nearest-neighbor angle bending force constant derived from the angle-bending potential, (3.23), which we call here γ. The three force constants were first obtained by fitting the bulk dispersion curves.

The surface phonon dispersion curves were then calculated with the GF method for a surface model in which the force constant $\beta_{\perp} \equiv \beta_{nn}$ between each atom in the topmost layer and all nearest-neighbor atoms in the second layer are changed to a new value denoted $\beta^{s}_{\perp}$, while all the other central force constants remain unchanged. In order to fulfill rotational invariance (4.2), also the angle-bending force constant γ had to be changed in the surface region [50]. The resulting dispersion curves of surface phonons in the $\bar{\Gamma}\bar{M}$ direction are shown in Fig. 4.1a for three assumed values of the surface force constant ratio, $\beta^{s}_{\perp}/\beta_{\perp}$ = 1, 0.68 and 1.3. For unchanged force constants ($\beta^{s}_{\perp}/\beta_{\perp}$ = 1) only the Rayleigh wave $S_1(1)$ is found. For a reduction of the surface force constant ($\beta^{s}_{\perp}/\beta_{\perp}$ = 0.68) the RW branch, $S_1(0.68)$, is considerably softened with respect to $S_1(1)$ with a frequency shift increasing at large wavevectors due to the smaller penetration length. Moreover two new surface localized phonons, $S_3(0.68)$ and $S_5(0.68)$, appear in the spectrum, the first one in the gap being peeled off from the lower edge of the LA bulk band, the second one from the lower edge of the TA band at large wavevector.

An increase of the surface force constant ($\beta^{s}_{\perp}/\beta_{\perp}$ = 1.3) leads to a stiffening of the RW branch, $S_1(1.3)$ and the appearance of the $S_3(1.3)$ mode above the LA band, in agreement with the discussion of the Rayleigh theorem in Sect. 3.5. Musser and Rieder's calculations also demonstrate the important role of rotational invariance for the correct prediction of the RW phase velocity in the continuum limit.

In Fig. 4.1b the phase velocities of Rayleigh waves in units of the bulk TA velocity (full lines) corresponding to the RW dispersion curves in (a) are compared with the calculated values for an unequilibrated surface (broken lines) in which the angle-bending force constant is not changed at the surface and the rotational invariance condition is not fulfilled [48]. It follows that the RW velocity for $Q \rightarrow 0$ turns out to be independent of the surface change in central force constants, since the penetration length diverges, but decreases dramatically for the unequilibrated surface for which the RI condition is violated.

4.4.2 *Merits and Limits of BvK Force Constant Fits*

Largely stimulated by the extensive new data on surface phonon dispersion curves in metals a multitude of additional numerical studies based on phenomenological Born-von Kármán models have been reported by many authors [47, 51, 52]. The Modena group of Bortolani, Santoro and colleagues, which has been especially prominent in this area [53–57], have pioneered the use of a large number of two-body as well as angle-bending three-body force constants including interactions between distant neighbors. In this way the long-range coupling implied by the Friedel oscillations in metals could be accounted for. Although the physical basis of such force constants is not transparent, especially for metals, these calculations

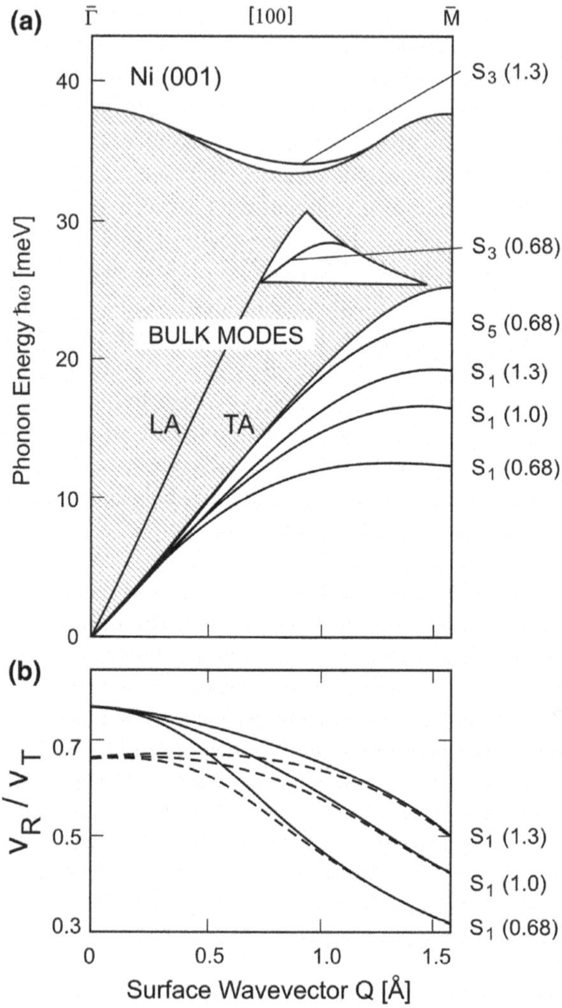

Fig. 4.1 Dispersion curves of surface localized phonons and Rayleigh wave velocities calculated with the GF method and a three force constant model in the direction [100] on Ni(001) for three assumed values of the ratios of the surface-to-bulk force constants, $\beta_{\perp}^{s}/\beta_{\perp}$, = 1.0, 0.68 and 1.3. **a** For unchanged force constants ($\beta_{\perp}^{s}/\beta_{\perp}$, = 1.0) the Rayleigh wave $S_1(1.0)$ is the only surface mode outside the bulk bands. With reduction of the surface force constant ratio ($\beta_{\perp}^{s}/\beta_{\perp}$, = 0.68) the RW branch $S_1(0.68)$ is softened with respect to $S_1(1.0)$, and two new surface localized phonons, $S_3(0.68)$ and $S_5(0.68)$, appear. With increasing ratio, $\beta_{\perp}^{s}/\beta_{\perp}$, = 1.3, the RW stiffens and a new mode appears above the bulk modes. **b** The phase velocities of Rayleigh waves in units of the bulk TA velocity (full lines) corresponding to the RW dispersion curves in (**a**) are compared with the calculated values for an unequilibrated surface (broken lines) in which the angle-bending force constant is not changed at the surface and the rotational invariance condition is not fulfilled [48]

provide quantitative information on the extent to which they are modified at the surface compared to their bulk values, as well as pedagogical insight into the effect of such force constant changes on the surface phonon frequencies.

The surface dispersion curves of the Pt(111) surface, nicely illustrate these model calculations [55, 56]. In Fig. 4.2 the $\overline{\mathrm{M}}$- and $\overline{\mathrm{K}}$-point frequencies are plotted as a function of the changes of the surface top layer radial force constants $\beta_{\perp}^{s}$ and $\beta_{\parallel}^{s}$ for nearest neighbor (nn) inter planar (perpendicular) and intra-planar (in the surface plane) displacements, respectively, relative to the bulk nearest neighbor radial force constant β^{b}. Only the surface mode frequencies are shown in Fig. 4.2 since the bulk bands are not significantly affected by the force constants at the surface. At both $\overline{\mathrm{M}}$ and $\overline{\mathrm{K}}$ the reduction of either $\beta_{\parallel}^{s}$ and $\beta_{\perp}^{s}$ leads to a linear reduction in ω^2 as expected for a simple harmonic oscillator model. The mode AL which is shown in Fig. 4.2a is a new mode which appears only when the first nn interplanar radial force constant $\beta_{\perp}^{s}$ is significantly reduced with respect to the first nn bulk radial force constant. It involves vertically polarized displacements localized in the second plane and longitudinal displacements in the first and third planes, and for this reason it has been called *anomalous longitudinal* (AL) resonance (also referred to as the S_3 mode in monoatomic surfaces, see Table 3.1) [58]. Its appearance can be explained by the increasingly loose coupling of the topmost layer to the substrate which leads to a transfer of the motion of the Rayleigh wave away from the first to the second layer. At the $\overline{\mathrm{K}}$ point (Fig. 4.2c, d) similar new modes, designated by K_2 and K_3, appear. K_2 is similar to AL discussed above while K_3 is a mode in which the transverse vertical vibrational motion has been transferred into the third layer.

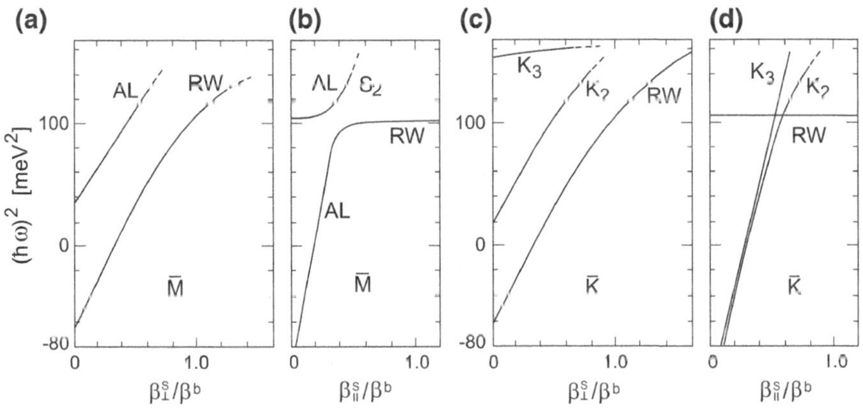

Fig. 4.2 The calculated zone boundary surface mode frequencies for Pt(111) are plotted as a function of the force constant between the first two layers $\beta_{\perp}^{s}$ and the lateral intra-planar surface force constant $\beta_{\parallel}^{s}$ both relative to the bulk nearest neighbor force constant β^{b}. [54]. **a** and **b** for the M-point. **c** and **d** for the $\overline{\mathrm{K}}$-point. The other force constants are held fixed at the values obtained for the bulk dispersion curve with interactions extending up to 6 nearest neighbors

Figure 4.2d confirms that the in-plane force constant $\beta_{\parallel}^{s}$ has absolutely no effect on the Rayleigh mode at the zone boundary, indicating that in this approximation the transverse and longitudinal displacements are not coupled. The strong effect of a reduction of $\beta_{\parallel}^{s}$ at the $\overline{K}$ point on the new modes K_2 and K_3 is attributed to the large longitudinal amplitudes of these modes in the surface layer. At the $\overline{M}$ point there is a gap mode S_2 which originates from the lower edge of the longitudinal bulk band at a frequency of about 20 meV and is strongly depressed by a reduction of $\beta_{\parallel}^{s}$ down into the TA band where it becomes an AL resonance. Since this mode and the Rayleigh mode are both sagittal, there is an avoided crossing of their branches at about $\beta_{\parallel}^{s} / \beta^{b} = 0.4$. For smaller $\beta_{\parallel}^{s}$ the lower mode transforms into an AL localized mode and becomes unstable at $\beta_{\parallel}^{s} / \beta^{b} \cong 0.4$, whereas the upper mode now has a constant energy and the ~SV character of a Rayleigh wave.

The anomalous longitudinal resonance (AL) was observed already in the first HAS experiments on Ag(111) in 1983 [59] and since then it has been found in practically all metal surfaces (see Sect. 11.3), thus qualifying as a new intrinsic property of metal surfaces. As discussed in Chap. 11 the analysis of both the surface phonon dispersion curves and the corresponding inelastic HAS peak intensities using the Born-von Karman (BvK) force constant models yielded rather large surface force constant changes. The changes found at the time in most of these metals now appear to be unphysical [60–64]. This effect, which is referred to as Bortolani-Mills paradox (see Sect.11.3.3), is now understood to be due to the incorrect assumption that the coupling of the He atoms to the phonons could be explained in terms of the two-body repulsive interaction between the He atoms and the surface atoms. As discussed in Chaps. 6 and 8 the interaction of the atoms with a metal surface is, in fact, more complicated. In particular their coupling to the electronic degrees of freedom at the surface needs to be included, since, for example, this affects the intensities of the AL mode TOF peaks to a much greater extent than for the RW TOF peaks. Thus the force constants in the BvK fits had to be adjusted to unphysical values to compensate for the incorrect coupling.

Despite its limitations, the BvK model is frequently used to quantify the results of surface phonon dispersion curve measurements because of the widespread availability of the corresponding force constants for the bulk phonons. In any case the BvK force constant fits are a valid procedure for a wide class of closed-shell solids, ranging form rare-gas crystals to molecular crystals and ionic compounds in the point-ion approximation. For closed-shell solids additive two-body interatomic potentials can also be justified for describing the interaction with an external He atom, so that the surface atom displacements obtained from BvK surface dynamics normally provide correct inelastic HAS intensities for all kinds of surface phonons. For all the other solids where the electronic degrees of freedom play an essential role in lattice dynamics, strong effects are also produced in the inelastic HAS intensities due to the strong coupling of the He atoms with the surface electron density. These two closely entangled aspects of inelastic HAS theory are the subject of the next two chapters.

References

1. A. A. Maradudin, E.W. Montroll, G.H. Weiss, I.P. Ipatova, *Theory of Lattice Dynamics in the Harmonic Approximation*, Solid State Physics, Suppl. 3 (Academic Press, New York, 1971)
2. G. Benedek, Surf. Sci. **61**, 603 (1976)
3. L. Miglio, G. Benedek, in *Structure and Dynamics of Surfaces II*, ed. by W. Schommers, P. von Blackenhagen. Topics in Current Physics, vol 43 (Springer, Heidelberg, 1987)
4. G. Benedek, Phys. Status Sol. **B58**, 661 (1973)
5. G. Benedek, G.P. Brivio, L. Miglio, V.R. Velasco, Phys. Rev. B **26**, 497 (1982)
6. G. Benedek, L. Miglio, in *Surface Phonons*, eds. by F.W. deWette, W. Kress, Springer Series on Surface Science, vol 27 (Springer, Berlin, Heidelberg, 1991), p. 37
7. B. Lengeler, W. Ludwig, Solid State Comm. **2**, 83 (1964)
8. W. Kress, F.W. de Wette, A.D. Kulkarni, U. Schröder, Phys. Rev. B **35**, 5783 (1987)
9. F.W. de Wette, A.D. Kulkarni, U. Schröder, W. Kress, Phys. Rev. B **35**, 2467 (1987)
10. R. Shuttelworth, Proc. Phys. Soc. London **A63**, 444 (1950)
11. C. Herring, in *The Physics of Powder Metallurgy*, ed. by W.E. Kingston (McGraw-Hill, New York, 1951), p. 162
12. For an introductory treatment of surface stress see A. Zangwill, *Physics at Surfaces* (Cambridge Univ. Press, Cambridge, 1989), p. 10
13. S. Lehwald, F. Wolf, H. Ibach, B.M. Hall, D.L. Mills, Surf. Sci. **192**, 131 (1987)
14. W. Menezes, P. Knipp, G. Tisdale, S.J. Siebener, Phys. Rev. B **41**, 5648 (1990)
15. W. Wuttig, R. Franchy, H. Ibach, Z. Physik B **65**, 71 (1986)
16. W. Daum, C. Stuhlmann, H. Ibach, Phys. Rev. Lett. **60**, 2741 (1988)
17. S. Müller, P. Bayer, C. reischl, K. Heinz, B. Feldmann, H. Zillgen, W. Wuttig, Phys. Rev. Lett. **74**, 765 (1995)
18. M. Zharnikov, A. Dittschar, W. Kuch, C.M. Schneider, J. Kirschner, Phys. Rev. Lett. **76**, 4620 (1996)
19. G. Benedek, E. Hulpke, W. Steinhögl, Phys. Rev. Lett. **87**, 027201 (2001)
20. G.G. Rusina, S.D. Borisova, S.V. Eremeev, IYu. Sklyadneva, E.V. Chulkov, G. Benedek, J.P. Toennies, J. Phys. Chem. C **120**, 22304 (2016)
21. B.J. Hinch, C. Koziol, J.P. Toennies, G. Zhang, Europhys. Lett. **10**, 341 (1989)
22. J. Braun, J.P. Toennies, Surf. Sci. **384**, L858 (1997)
23. M. Schackert, T. Märkl, J. Jandke, M. Hölzer, S. Ostanin, E.K.U. Gross, A. Ernst, W. Wulfhekel, Phys. Rev. Lett. **114**, 047002 (2015)
24. J. Braun, J.P. Toennies, Surf. Sci. **368**, 226 (1996)
25. G. Benedek, Physicalia Mag. **18**, 205 (1996)
26. F.W. de Wette, in *Dynamical Properties of Solids*, ed. by G.K. Horton, A.A. Maradudin (Elsevier, Amsterdam, 1990) Chap. 5
27. W. Kress, F.W. de Wette, A.D. Kulkarni, U. Schröder, Phys. Rev. B **35**, 5783 (1987)
28. F.W. de Wette, in *Surface Phonons*, ed. by F.W. de Wette, W. Kress, Springer Series on Surface Science, vol 27 (Springer, Berlin, Heidelberg, 1991), p. 67
29. G. Benedek, M. Miura, W. Kress, H. Bilz, Phys. Rev. Lett. **52**, 1907 (1984)
30. R. Rieger, J. Prade, U. Schröder, F.E. de Wette, A.D. Kulkarni, W. Kress, Phys. Rev. B **39**, 7938 (1989)
31. J. Prade, U. Schröder, W. Kress, A.D. Kulkarni, F.W. de Wette, in *Phonons 1989*, ed. by S. Hunklinger, W. Ludwig, G. Weiss (World Sci., Singapore, 1990), p. 841
32. R.J. Needs, M.J. Godfrey, M. Mansfield, Surf. Sci. **242**, 215 (1991)
33. R.J. Needs, Phys. Rev. Lett. **58**, 53 (1987)
34. M.C. Payne, N. Roberts, R.J. Needs, N. Needels, J.D. Joannopoulos, Surf. Sci. **211**(212), 1 (1989)
35. R.J. Needs, M.J. Godfrey, Phys. Rev. B **42**, 10933 (1990)
36. M. Mansfield, R.J. Needs, Phys. Rev. B **43**, 8829 (1991)
37. P.J. Feibelman, Phys. Rev. B **56**, 2175 (1997)

38. P.M. Marcus, X. Qian, W. Hübner, Phys. Rev. B **60**, 16088 (1999)
39. P.M. Marcus, X. Qian, W. Hübner, J. Phys.: Condens. Matter **12**, 5541 (2000)
40. O.L. Alerhand, D. Vanderbilt, R.D. Meade, J.D. Joannopoulos, Phys. Rev. Lett. **61**, 1973 (1988)
41. M. Mansfield, R.J. Needs, J. Phys.: Condens. Matter **2**, 2361 (1990)
42. D. Wolf, Phys. Rev. Lett. 70 (1993) 627, with comments by R.J. Needs, 71 (1993) 460 and by M.Y. Chou, S. Wei and D. Vanderbilt, 71 (1993) 461 and D. Wolf's reply 71 (1993) 462
43. J.E. Müller, M. Wuttig, H. Ibach, Phys. Rev. Lett. **56**, 1583 (1986)
44. H. Ibach, Surf. Sci. Rep. **29**, 196 (1997)
45. M. Born, K. Huang, *Dynamical Theory of Crystal Lattices* (Oxford University Press, London, New York, 1954)
46. G.W. Farnell, in *Physical Acoustics*, vol. 6, ed. by W.P. Mason, R.N. Thurston (Academic Press, New York, 1970), p. 109
47. J.E. Black, in *Dynamical Properties of Solids*, vol. 4, ed. by G.K. Horton, A.A. Maradudin (Elsevier, Amsterdam, 1990), p. 179
48. S.W. Musser, K.H. Rieder, Phys. Rev. B **2**, 3034 (1970)
49. J. DeLaunay, Solid State Phys. **2**, 219 (1956)
50. J.M. Morabito, Jr. R.F. Steiger, G.A. Somorjai, Phys. Rev. **179**, 638 (1969)
51. M.H. Mohamed, L.L. Kesmodel, B.M. Hall, D.L. Mills, Phys. Rev. B **37**, 2763 (1988)
52. B.M. Hall, D.L. Mills, M.H. Mohamed, L.L. Kesmodel, Phys. Rev. **38**, 5856 (1988)
53. V. Bortolani, A. Franchini, F. Nizzoli, G. Santoro, Phys. Rev. Lett. **52**, 429 (1984)
54. V. Bortolani, A. Franchini, G. Santoro, Surf. Sci. **189**(190), 675 (1987)
55. V. Bortolani, G. Santoro, U. Harten, J.P. Toennies, Surf. Sci. **148**, 82 (1984)
56. G. Santoro, A. Franchini, V. Bortolani, U. Harten, J.P. Toennies, Ch. Wöll, Surf. Sci. **183**, 180 (1987)
57. V. Bortolani, A. Franchini, F. Nizzoli, G. Santoro, in *Dynamics of Gas Surface Interactions*, ed. by G. Benedek, U. Valbusa (Springer, Heidelberg, 1982), p. 196
58. G. Brusdeylins, R.B. Doak, J.P. Toennies, Phys. Rev. B **27**, 3662 (1983)
59. R.B. Doak, U. Harten, J.P. Toennies, Phys. Rev. Lett. **51**, 578 (1983)
60. C.S. Jayanthi, H. Bilz, W. Kress, G. Benedek, Phys. Rev. Lett. **59**, 795 (1987)
61. C. Kaden, P. Ruggerone, J.P. Toennies, G. Zhang, G. Benedek, Phys. Rev. B **46**, 13509 (1992)
62. S.Y. Tong, Y. Chen, K.P. Bohnen, T. Rodach, K.M. Ho, Surf. Rev. Lett. **1**, 97 (1994)
63. G. Santoro, A. Franchini, V. Bortolani, Phys. Rev. Lett. **80**, 2378 (1998)
64. R. Heid, K.-P. Bohnen, Phys. Rep. **387**, 151–213 (2003)

Chapter 5
Theoretical Methods of Surface Dynamics

> Facts without theory is chaos, but theory without facts is fantasy.
>
> Charles Otis Whitman, 1894

Abstract The phenomenological models and the quantum theoretical methods which are currently used to calculate the dispersion curves of surface phonons are reviewed for the various classes of solids.

5.1 Introductory Comments

In Chap. 2 the historical development of surface dynamics was outlined and the evolution of the basic techniques to solve the surface dynamical equations were reviewed. The techniques run from early refined mathematical treatments, such as the use of trial functions and the Green's function (GF) methods, within either the continuum elastic (CE) theory or Born-von Karman (BvK) force constant (FC) models. More recent but more computationally demanding techniques are the direct slab (SL) diagonalization, the frozen phonon methods (FPM) and molecular dynamics (MD). In this Section attention is drawn to the methods of surface dynamics by briefly surveying the phenomenological models and the quantum theoretical methods which are currently used to calculate the dispersion curves of surface phonons for the various classes of solids. Of course, most of the phenomenological and *ab initio* theories were first devised for bulk dynamics. These theories and the extent to which they can be applied to the surface dynamical problem in combination with some of the above techniques are reviewed in this chapter. A compendium may be found in a short review recently published in the Landolt-Börstein handbook on the Physics of Solid Surfaces [1].

G. Benedek and J. P. Toennies, *Atomic Scale Dynamics at Surfaces*, Springer Series in Surface Sciences 63, https://doi.org/10.1007/978-3-662-56443-1_5

Table 5.1 Theoretical methods in surface dynamics

Approach	Mostly conductors	Mostly insulators and semiconductors	
Continuum	Continuum Elastic theory (CE)		
	Hydrodynamic Equations (HD)	Electromagnetic (Fuchs-Kliever) modes (FK)	
Only atomic coordinates	Born-von Kármán (BvK) Force Constants (FC)	Rigid Ion Model (RIM)	Keating Model (KM) Valence Force Field (VFF)
Atomic + electronic coordinates	Double Shell Model (DSM) Cluster Model (CM)	Shell Model (SM) Breathing Shell Model (BSM) Quadrupolar Shell Model (QSM)	Bond Charge Model (BCM)
Quantum semi-empirical	Finnis-Sinclair (FS) Glue Model (GM) Effective Medium (EM) Embedded Atom (EA)	Quasi-Ion Approach (QI) Pseudocharge Model (PC) Multipole Expansion (ME) Method	Tight-Binding (TB) Semi-Empirical Total Energy (SETE) Method
Quantum first-principles (FP)	Pseudopotential Perturbation theory (PPP) Total-energy Frozen Phonon Method (FPM) Density Response (DR) First-Principle Tight-Binding (FPTB) Density Functional Perturbation Theory (DFPT) Density functional molecular dynamics (Car-Parrinello method (CPM))		

Some methods are common to different classes of solids. For example, the Pseudocharge (PC) and the Multipole Expansion (ME) methods, though conceptually related to the shell model, have been used also for noble metals

A comprehensive list of theoretical methods and their acronyms which have been and are currently used in surface dynamics are listed in Table 5.1 according to a logical sequence, from the simplest models at the top to the recent *ab initio* methods at the bottom. Only the general theoretical concepts are discussed, with emphasis to the increasing complexity of the surface dynamical problem resulting from the need to include the response of the valence and conduction electrons. For the detailed theoretical formulation of each method the reader is addressed to the abundant list of references in the following sections.

The computational techniques by which the above theoretical methods can be implemented are listed in Table 5.2, from the most traditional at the top downwards to the most advanced ones. The set of coordinates (i.e., the size of the operational space) to be dealt with is indicated for each technique, together with the acronyms of the theoretical methods given in Table 5.1 to which it applies. The crystal surfaces and the adsorbed layers for which phonons have been measured with HAS up to 2017 are listed in Appendix A with the theoretical methods and computational techniques, and corresponding references.

Table 5.2 Classification of computational techniques in surface dynamics according to their operational space with examples of the corresponding methods

Technique and methods (acronyms from Table 5.1)	Operational space
Trial Functions (TF) CE, FC, HD	(**Q**, parameters)
Effective Mass Approximation (EMA) FC	(**Q**, parameters)
Green's function (GF) method CE, FC, VFF, BSM, CM	(**Q**, z of perturbed layers)
Dispersive Linear Chain (DLC) FC, SM	(**Q**, z of all layers)
Slab diagonalization (SL) All	(**Q**, z of all layers)
Frozen Phonon Method (FPM) FP	Displacement configurations
Molecular Dynamics (MD) FC, GM, EA, TB, FP (FPTB, CPM)	x, y, z

5.2 The Important Role of Electrons in the Dispersion of Surface Phonons

The Lattice Force Constant (FC) models consider atoms as hard massive spheres (Rigid Ion Model (RIM)) disregarding the actual shape of the valence electron distribution and its deformability during the atomic motion. Historically the rigid ion model was the first approach to calculate the bulk dynamics [2, 3] and surface dynamics of ionic crystals [4]. Due to its simplicity it is still in use as a first approach to the surface dynamics especially in complex systems [5].

Electrons play a crucial role in determining the phonon dispersion relations in all three classes of solids, insulators, semiconductors and metals. The simplest picture is provided by a Continuum Elastic (CE) model, where the electrons and atoms form two interacting fluids and the degrees of freedom of the electronic and atomic densities are treated via linearly coupled Hydrodynamical (HD) equations [6–9]. This early approach deserves more attention today for providing a useful starting point in the theory of surface plasmon-phonon excitations.

Since in lattice dynamics atoms are treated as discrete entities, the many-body effects induced by valence electrons on the motion of the atom cores can be modelled either by non-central two-body force constants or by three-body terms, e.g., angle bending force constants. These terms characterize the Valence Force Field (VFF) [10, 11] and the Keating Model (KM) [12, 13], which were found to be useful in the dynamical theory of covalent semiconductors [14]. Though these methods are not currently applied to surface dynamics, they provided the basis for

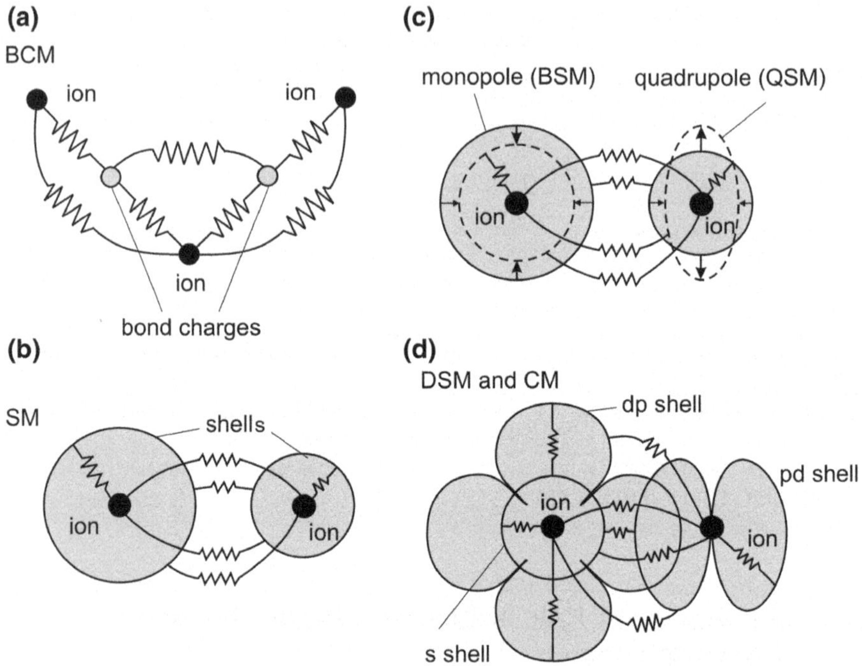

Fig. 5.1 Force constant models involving the electronic coordinates. **a** The bond charge model suitable for covalent semiconductors, treats the bonding charges as massless "atoms" coupled to the ions and each other by two-body and angle bending force constants [19–21]. **b** The shell model considers the dipolar polarization of atomic shells [26–30]. **c** The breathing and quadrupolar shell models [31–33] include also monopolar and quadrupolar deformations of atomic shells. All these models are especially suitable for closed shell insulators. **d** The double shell and cluster shell models apply to open shell systems like the transition metal compounds [34, 35]

the Bond Charge Model (BCM) [15–21], which includes new adiabatic degrees of freedom associated with the valence electrons of the covalent bonds (Fig. 5.1a). Its straight forward extension to treat π-bonding electrons [22] makes Weber's formulation of the BCM [19, 20, 21] a reliable and efficient tool for surface dynamics of covalent semiconductors [22], III–V compounds [23] and graphite [24]. The first principle quantum mechanical basis of BCM was demonstrated by Fleszar and Resta [25]. Some BCM results for semiconductors and graphite are discussed in Chap. 11, and compared to *ab initio* calculations as well as to HAS and EELS experiments.

The BCM belongs to a wide class of models, depicted in Fig. 5.1, where many-body effects are accounted for by a set of electronic coordinates harmonically coupled to the nuclear coordinates in the adiabatic approximation [36–39, 40].

Among these the Shell Model (SM) gives excellent performance for the surface dynamics of ionic crystals within the slab method [26–30, 41]. In the SM the valence electron shells can be rigidly displaced with respect to ion cores to mimic the ionic dipolar polarizabilities [Fig. 5.1b]. The SM was formulated independently by Dick and Overhauser [42] and by Cochran [43, 44] for the analysis of the phonon dispersion curves in insulators as obtained from neutron scattering. The SM originates from the theoretical studies on the ionic polarization and charge transfer in lattice vibrations by Szigeti [45, 46], Tolpygo [47, 48], and Yamashita and Kurosawa [49].

5.3 The Multipole Expansion (ME) Method and the Pseudocharge (PC) Model

Shell models and bond charge models of all types, which describe the electronic response to the atomic motion through suitable dynamical coordinates harmonically coupled to the ion coordinates, are all encompassed by the Multipole Expansion (ME) method. The ME method for lattice dynamics was originally formulated by P.B. Allen [50] as a special representation of the general microscopic theory constructed within a set of local basis functions [51–65]. Such a local representation allows for a general and efficient phenomenological treatment of the electronic degrees of freedom. Originally this method proved very successful in reproducing the anomalous dispersion curves of transition metals such as niobium [50] and quite effective for the analysis of surface phonon anomalies in noble metals [66]. The microscopic basis of the ME method provides a scheme suitable to either a first principle calculation or a phenomenological parametrization known, in metal surface dynamics, as the Pseudocharge (PC) model [66–68]. Here the ME method and its phenomenological version, the PC model, are described in detail since they provide a convenient framework for the calculation of dispersion curves and inelastic HAS amplitudes for complex surfaces, which are still computational demanding for DFPT. Recent examples include the analysis of HAS dispersion curves and Kohn anomalies of topological insulator surfaces [69–72].

More sophisticated versions of the SM including shell deformabilities of either the breathing (Breathing Shell Model, BSM [31, 32]) or the quadrupolar (Quadrupolar Shell Model, QSM [33]) types (Fig. 5.1c) may be needed for a more precise description of the dynamical polarization phenomena in alkali halides as well as in noble and transition metal halides and other classes of ionic compounds. The BSM in combination with the GF method gives excellent results in the calculation of surface phonons for several alkali halide surfaces [73–76], as seen in Fig. 3.15a, b for NaF(001) and in Sect. 11.1 (Fig. 11.3) for LiF(001), and is computationally fast enough to allow for extensive calculations of their inelastic HAS amplitudes [75, 77].

In open shell systems, such as the transition metal compounds, the many-body effects depend more on the electronic susceptibility of the ion-ligand complexes than on the individual ion polarizabilities. In this case more degrees of freedom are needed for the valence electrons (Fig. 5.1d) and the shell model concept evolves into that of the Double Shell Model (DSM), where more than one shell per atom is included [78], and the Cluster Model (CM), where hybridized shells of the transition metal atom with those of the surrounding ligand ions are considered [79]. The latter model proved efficient for the study of bulk and surface phonon anomalies in a superconducting material such as TiN(001) [80, 81].

In the ME method, as in DFT, the total energy of the lattice system is written as the sum of the energy E_{ion} due to the direct interaction of the ion cores plus the energy of the electrons, the latter being expressed by a functional of the electron charge density $n(\mathbf{x})$ at the position $\mathbf{x}$ [82, 83]:

$$E_{tot} = E_{ion} + F[n(\mathbf{x})] + \int n(\mathbf{x})\mathbf{v}_{ion}(\mathbf{x})d^3x. \tag{5.1}$$

The functional $F[n(\mathbf{x})]$ includes the electron-electron potential energies as well as the kinetic and exchange-correlation energies. Finally $V_{ion}(\mathbf{x})$ is the external potential acting on the electron system, which in the present context is the sum of the individual potentials of the ions at their instantaneous positions:

$$V_{ion}(\mathbf{x}) = \sum_l v_l(\mathbf{x} - \mathbf{r}_l - \mathbf{u}(l)). \tag{5.2}$$

The electron density is assumed to be partitioned into cellular contributions, sometimes referred to as Pseudocharges (PC), as illustrated in Fig. 5.2. The pseudocharges are centered at convenient positions within the unit cells given by $\mathbf{r}_{l\lambda} = \mathbf{r}_l + \mathbf{r}_\lambda$, where $\mathbf{r}_l$ are lattice translation vectors and the $\mathbf{r}_\lambda$ are the special points within the unit cell. In the simplest case of a monoatomic crystal, the electron density in each unit cell is expanded in a multipole series about each of these special points:

$$n_l(\mathbf{x}) = \sum_{\Gamma\lambda} C_\Gamma(l\lambda) Y_\Gamma(\mathbf{x} - \mathbf{r}_{l\lambda}). \tag{5.3}$$

In (5.3) each basis function $Y_\Gamma(\mathbf{x})$ is the product of an appropriate radial function and an angular harmonic function which transforms according to the Γ-th irreducible representation of the point group at $\mathbf{r}_{l\lambda}$ [9, 85]. For instance an orthonormal set of Wannier functions may be chosen for Y_Γ. The expansion coefficients $C_\Gamma(l\lambda)$ transform in turn according to the irreducible representations Γ, i.e., as either a scalar or a vector or a tensor component of increasing rank, and are therefore the components of a pseudovector $\mathbf{C}(l\lambda)$.

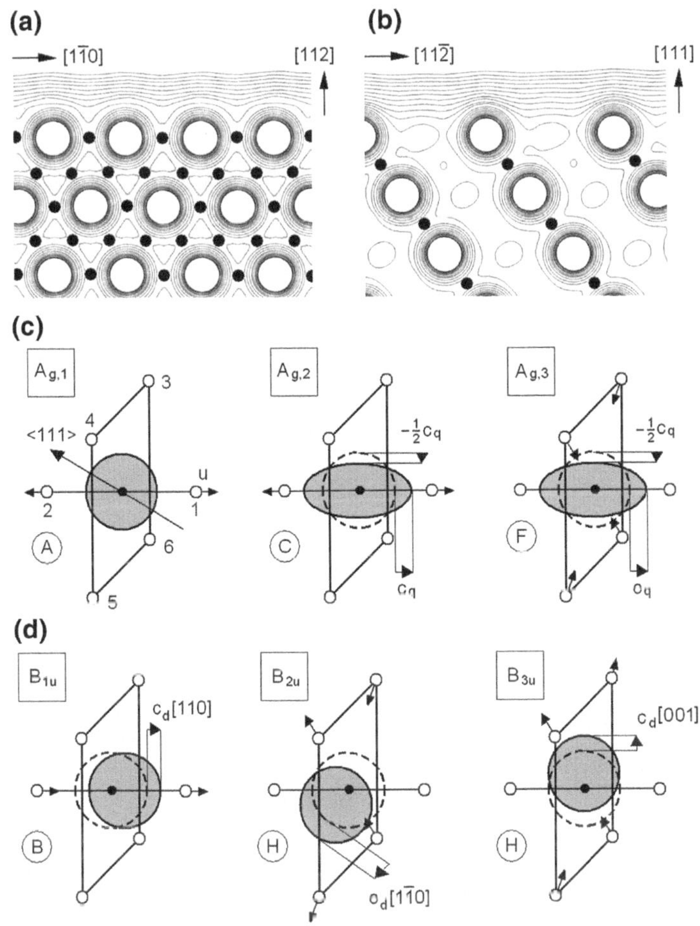

Fig. 5.2 Pseudocharge locations and displacements. **a** Charge density contours in the first few layers below the (111) surface of Cu in the [1$\bar{1}$0] direction for an oblique section in the [112] direction [plane (11$\bar{1}$)]. The charge density contours have been derived from the calculations by Euceda et al. based on the self-consistent linear combination of Gaussian orbitals [84]. The black circles, at the midpoints between nearest neighbor ions, represent a possible choice for the locations of pseudocharges (PC's). **b** Same as **a** for a normal section in the direction [11$\bar{2}$]. **c**, **d** Dynamical effects of the PC's are described by their monopolar ($A_{g,1}$), quadrupolar ($A_{g,2}$, $A_{g,3}$) and dipolar (B_{1u}, B_{2u}, B_{3u}) deformations coupled to the ionic displacements by suitable force constants (encircled capital letters). **c** The symmetric displacements of two nn ions 1 and 2, besides being directly coupled by a force constant A, also interact with a PC quadrupolar deformation c_q with a force constant *C*. Moreover the PC quadrupolar deformation is coupled to the breathing motion of other four neighbor ions (3–6) with a force constant *F*. These displacements belong to the same irreducible representation A_{1g} of the local point group (D_{2h}). **d** The three PC dipolar deformations c_q along the principal axes [110], [1$\bar{1}$0] and [001] couple instead to the ion odd-symmetry displacements of B_{1u}, B_{2u} and B_{3u} symmetry with two independent force constants *B* and *H*

The choice of the ME basis is, of course, somewhat arbitrary. A quantum mechanical procedure has been derived by Falter et al. for bulk silicon and other semiconductors [65, 86–91]. Their procedure, which is based on the acoustic sum rule for partitioning the electronic charge density into localized contributions, is illustrated in Fig. 5.3. Their partitioning leads to the concept of *Quasi-Ions* (QI), which provides a microscopic basis for the electronic charge localization in crystals and for the electron-mediated many-body force constants. The QI concept is in some respects a generalization in the framework of Ziman's DFT of pseudo-atoms [92] and of the dielectric pseudopotential theory of valence charge distributions [93, 94]. It has been shown to give a good description of the bulk phonon dispersion curves in covalent semiconductors [65].

In lattice vibrations the PC multipolar components in (5.3) are expected to follow the ionic motion by performing oscillations of amplitude $c_\Gamma(l\lambda)$ around the equilibrium values $C_{0,\Gamma}(l\lambda)$. Thus

$$C_\Gamma(l\lambda) = C_{0,\Gamma}(l\lambda) + c_\Gamma(l\lambda), \tag{5.4}$$

where the components $C_{0,\Gamma}(l\lambda)$ correspond to the ground state density $n_0(\mathbf{x})$, which minimizes the total crystal energy when the lattice ions are at their equilibrium positions. The variations $c_\Gamma(l\lambda) \equiv \Delta C_\Gamma(l\lambda)$ of the expansion coefficients in (5.3) may be looked upon as the coordinates of the electronic motion and act in the lattice dynamics as time-dependent variables harmonically coupled to the ion coordinates. In the multipole representation the PC electronic ground state can be defined by the conditions

$$\partial E_{tot}/\partial c_\Gamma(l\lambda) = 0 \quad \text{for all } \Gamma, l\lambda. \tag{5.5}$$

The effective forces acting on the ions as well as on the degrees of freedom of the pseudocharges are obtained in an analogous fashion as in the conventional BvK force constant model. The total energy is expanded not only with respect to the ionic displacements $\mathbf{u}(l)$, as in the BvK model, but also with respect to the time-dependent multipole oscillation amplitudes $c_\Gamma(l\lambda)$:

$$\begin{aligned} E_{tot} = E_o &+ \frac{1}{2}\sum\nolimits_{l\alpha,l'\beta} R_{\alpha\beta}(l,l')u_\alpha(l)u_\beta(l') \\ &+ \frac{1}{2}\sum\nolimits_{\Gamma}\sum\nolimits_{l\lambda,l'\alpha}\left[T_{\alpha\Gamma}(l,l'\lambda) + T^{+}_{\alpha\Gamma}(l'\lambda,l)\right]u_\alpha(l)c_\Gamma(l'\lambda) \\ &+ \frac{1}{2}\sum\nolimits_{\Gamma\Gamma'}\sum\nolimits_{l\lambda,l'\lambda j} H_{\Gamma\Gamma'}(l\lambda,l'\lambda')c_\Gamma(l\lambda)c_{\Gamma'}(l'\lambda'). \end{aligned} \tag{5.6}$$

The ion-ion force constant matrix $R_{\alpha\beta}(l,l')$ by virtue of (5.1) and (5.2) is given by a sum of two terms:

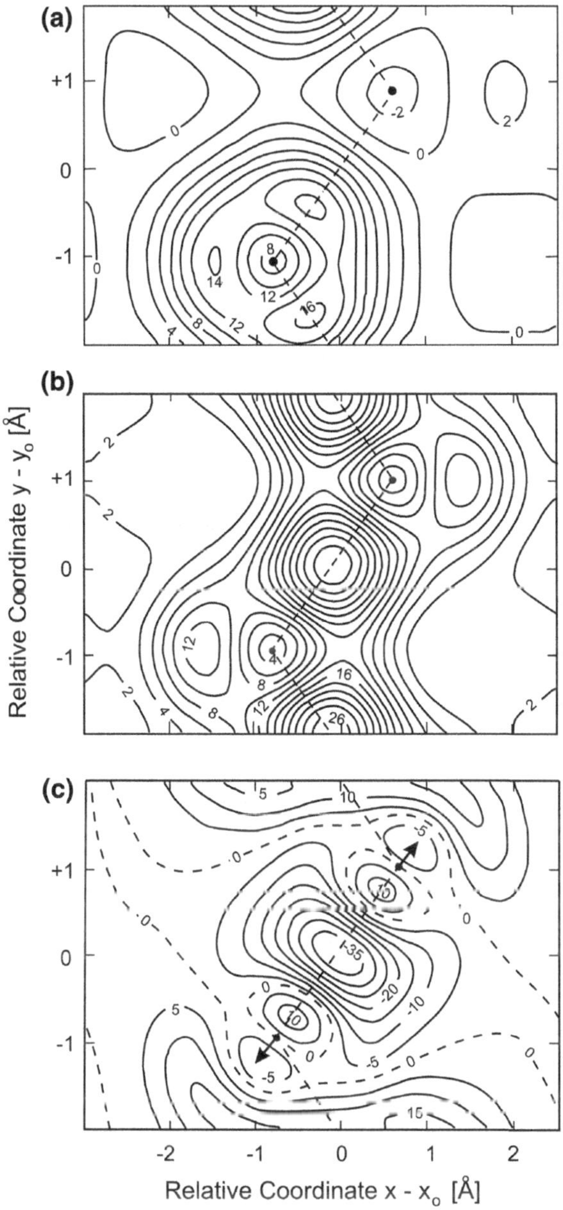

Fig. 5.3 Deformation of bond charges in semiconductors. **a** Calculated partial charge density associated with one sublattice of Si in the (110) plane (in units of electrons per cell) illustrating the concept of a quasi-ion at the atomic sites, according to Falter et al. [89]. **b** Total charge density resulting from the addition of the two sublattices reveals the bond charge at the midpoint of two neighbor atoms. **c** Deformation of the bond charge in **b** induced by the displacement field of a TO phonon has a clear quadrupolar character, whereas the quasi-ion charges around the atoms shows a dipolar deformation [89]

$$R_{\alpha\beta}(l,l') = \frac{\partial^2 E_{tot}}{\partial u_\alpha(l)\partial u_\beta(l')} = \frac{\partial^2 E_{ion}}{\partial u_\alpha(l)\partial u_\beta(l')} + \delta_{ll'}\int d^3x\, n(\mathbf{x})\frac{\partial^2 V_l(\mathbf{x}-\mathbf{r}_l)}{\partial r_\alpha \partial r_\beta} \\ \equiv R^{ion}_{\alpha\beta}(l,l') + R^{el}_{0,\alpha\beta}(l,l'). \tag{5.7}$$

The first term defines the matrix $R^{ion}_{\alpha\beta}$, which accounts for the direct interaction between the ion cores. The second term, $R^{el}_{0,\alpha\beta}$, is a diagonal matrix in the atomic index l expressing the restoring force constant exerted by the electron gas against the displacement of a single ion. The new effective force constant matrices T and H describe the multipole-ion and multipole-multipole interactions. T and H are defined in an analogous way as the harmonic force constants, and can also be expressed as integrals over the functional derivatives of the energy with respect to $n(\mathbf{x})$:

$$\begin{aligned} T_{\alpha\Gamma}(l',l\lambda) &= \partial^2 E_{tot}/\partial u_\alpha(l')\partial c_\Gamma(l\lambda) \\ &= \frac{1}{V}\int d^3x[\partial\delta E_{tot}/\delta n(\mathbf{x})\partial u_\alpha(l')]Y_\Gamma(\mathbf{x}-\mathbf{r}_{l\lambda}) \\ &= \int d^3x[\partial V_l(\mathbf{x})/\partial u_\alpha(l')]Y_\Gamma(\mathbf{x}-\mathbf{r}_{l\lambda}), \end{aligned} \tag{5.8}$$

$$\begin{aligned} H_{\Gamma\Gamma'}(l\lambda,l'\lambda') &= \partial^2 E_{tot}/\partial c_\Gamma(l\lambda)\partial c'_\Gamma(l'\lambda') \\ &= \frac{1}{V^2}\iint d^3x\, d^3x' \left[\delta^2 E_{tot}/\delta n(\mathbf{x})\delta n(\mathbf{x}')\right] Y_\Gamma(\mathbf{x}-\mathbf{r}_{l\lambda})Y_{\Gamma'}(\mathbf{x}'-\mathbf{r}_{l'\lambda'}), \end{aligned} \tag{5.9}$$

where V is the crystal volume. The following two coupled equations now describe the motion of the ionic and electronic lattices, respectively:

$$-M\ddot{u}_\alpha(l) = \sum_{l'\beta} R_{\alpha\beta}(l,l')u_\beta(l') + \sum_{\Gamma l'\lambda'} T_{\alpha\Gamma}(l,l'\lambda)c_\Gamma(l'\lambda) \tag{5.10}$$

$$-\mu_\Gamma \ddot{c}_\Gamma(l\lambda) = \sum_{l'\alpha} T^+_{\alpha\Gamma}(l\lambda,l')u_\alpha(l') + \sum_{\Gamma'l'\lambda'} H_{\Gamma\Gamma'}\left(l\lambda,l'\lambda'\right)c_{\Gamma'}\left(l'\lambda'\right) \tag{5.11}$$

where μ_Γ is a generalized mass which accounts for the inertial properties of the Γ-th multipole. For a finite value of μ_Γ and $u_\alpha(l') = 0$, equation (5.11) describes a harmonic equation of motion for the multipolar oscillations of the electron density in the rigid lattice and is thus able to account in the vibrating lattice for nonadiabatic effects.

After Fourier transforming with respect to space and time, the multipolar oscillations of the electronic density resulting from the ionic displacements are obtained by solving (5.11) for $c_\Gamma(l\lambda)$ with the adiabatic condition (written in vector form):

$$\mathbf{c} = -H^{-1}T^{+}\mathbf{u}. \tag{5.12}$$

The matrix $H^{-1}T^{+}$ will be referred to in the following as the electronic response function. By substituting (5.12) into (5.10) the following eigenvalue equation for the ionic displacement vector is obtained

$$M\omega^{2}_{\mathbf{Q}\nu}\mathbf{u}(\mathbf{Q}\nu) = \Phi\mathbf{u}(\mathbf{Q}\nu) = (R^{ion} + R_{0}^{el} - TH^{-1}T^{+})\mathbf{u}(\mathbf{Q}\nu). \tag{5.13}$$

Thus the effective force constant matrix Φ contains, besides the direct interionic term, the additional terms $R_{0}^{el} - TH^{-1}T^{+}$ accounting for those interactions between ions which are mediated by the interposed electrons and depend on the response of the electrons to the atomic motion. It is important to note that the term $-TH^{-1}T^{+}$ is responsible for effective long range ion-ion interactions, even though the multipole interaction matrices themselves T and H only have a short range character.

The ME method for lattice dynamics described above is in fact no more than a DFT formulation of the standard microscopic theory based on the dielectric response function [56, 65]. In the dielectric response theory of lattice dynamics the electron density oscillations $\delta n(\mathbf{x})$ induced by the ion displacements $\mathbf{u}(l)$ from the equilibrium positions $\mathbf{r}_l$ are given by

$$\delta n(\mathbf{x}) = -\int d^3r'\chi(\mathbf{x},\mathbf{x}')\sum\nolimits_l [\partial V_{ion}(\mathbf{x}')/\partial \mathbf{r}_l]\cdot\mathbf{u}(l), \tag{5.14}$$

where $\chi(\mathbf{x},\mathbf{x}')$ is the nonlocal dielectric response function (susceptibility) of the electrons. When projected onto the space of basis functions, (5.14) becomes

$$c_{\Gamma}(l\lambda) = -\frac{1}{V}\iint d^3xd^3x' Y_{\Gamma}(\mathbf{x}-\mathbf{r}_{l\lambda})\chi(\mathbf{x},\mathbf{x}')\sum\nolimits_l [\partial V_{ion}(\mathbf{x}')/\partial \mathbf{r}_l]\cdot\mathbf{u}(l), \tag{5.15}$$

and, from (5.8) and (5.12),

$$H^{-1}_{\Gamma\Gamma'}(l\lambda,l'\lambda') = \iint d^3x\,d^3x' Y_{\Gamma}(\mathbf{x}-\mathbf{r}_{l\lambda})\chi(\mathbf{x},\mathbf{x}')Y^{+}_{\Gamma'}(\mathbf{x}-\mathbf{r}_{l'\lambda'}). \tag{5.16}$$

Thus the inverse of the H matrix turns out to be the electron nonlocal susceptibility in the representation of the basis functions Y_{Γ}. This implicitly shows that the electron susceptibility is a ground state property of the system. In the adiabatic approximation it is defined as the inverse of the second order functional derivative of the total energy with respect to the electronic density [65].

By virtue of (5.16) the matrix H can then be derived from a microscopic theory of the electron susceptibility. In the Self-Consistent Field (SCF) framework [95] this is related to the band structure through summations over all relevant intra- and interband electronic transitions across the Fermi level. Thus the ME method, like the dielectric response methods described below, includes, through (5.13), all the dynamical effects originating from special features of the Fermi surface. These include Kohn anomalies [96–99] which appear in the phonon dispersion as a consequence of **Q**-space singularities in the susceptibility due to Fermi surface nesting [100, 101], and effective long range force constant perturbations due to Friedel oscillations [102, 103]. Even in crystals where direct long range interactions are apparently screened out, the complex **Q**-dependence of H^{-1} generally leads to effective long range force constants in the direct-space dynamical matrix, thus providing a clear physical meaning to the extended sets of phenomenological force constants which need to be used for metals in BvK models.

Moreover the ME method offers a conceptually simple, though computationally demanding, scheme for the treatment of non-adiabatic effects, e.g., non-adiabatic Kohn anomalies through the replacement in (5.13) of the H^{-1} matrix with the corresponding frequency dependent susceptibility matrix:

$$[H_{\Gamma\Gamma'}(l\lambda, l'\lambda')]^{-1} \rightarrow [H_{\Gamma\Gamma'}(l\lambda, l'\lambda') - \delta_{\Gamma\Gamma'}\delta_{ll'}\delta_{\lambda\lambda'}\mu_{\lambda}\omega^2]^{-1}. \tag{5.17}$$

This requires a definition of the effective PC inertial masses μ_{λ}: they are in some way related to the electron effective masses in the Fermi level states involved in the response function, but may be treated as phenomenological parameters. A tutorial description based on a simplified non-adiabatic PC model of the HAS observed sharp Kohn anomalies in W(110):H(1×1) and Mo(110):H(1×1) [104–106] (see Sect. 11.3.3) is found in [106].

The ME method has so far been used in its phenomenological adiabatic PC version with good results in the analysis of the dispersion curves and HAS spectra of 3d noble-metal surfaces [66, 107] and topological insulator surfaces [69–71], respectively discussed in Sects. 11.3.2 and 11.4.3. The connections between semi-empirical and fully *ab initio* methods are summarized in the next section.

5.4 From Semi-empirical to *Ab initio* Methods

5.4.1 Tight-Binding Methods

The great progress made in the theory and calculations of the electronic band structure in all kinds of solids during the last three decades has made it possible to evaluate the electronic contributions to the total energy of solids as a function of the atomic positions. This development has opened up a new phase in the theory of surface dynamics. In insulators and semiconductors short range repulsive force constants can be derived directly from the overlap of atomic orbitals [78]. In more general terms the interatomic force constants can be obtained from a Tight-Binding (TB) scheme for the electronic structure, provided the dependence on the interatomic distances of the Coulomb and hopping integrals between neighboring atoms are known [108]. Such integrals are in turn determined either from fitting procedures (empirical TB) or from first principle calculations of the electronic structure.

The TB methods give reliable results for the surface phonon frequencies of covalent and III–V semiconductors [109, 110] with a reasonable computational effort. In the Semi-Empirical Total Energy method (SETE) developed by Pollmann and coworkers [111, 112] an empirical TB scheme is used to calculate the band-structure contribution to the total energy, whereas the short-range repulsive contribution is accounted for by phenomenological pair potentials. Some examples are quoted in Chap. 11 in connection with the experiments.

First-principle Tight-Binding (FPTB) methods derived from the Varma-Weber TB theory of transition metal dynamics [113, 114] have been used to approach difficult problems such as the dynamics and instabilities of tungsten and molybdenum (001) surfaces [109, 115]. Complex surface dynamics problems such as those involving extensively reconstructed surfaces have also been dealt with using quantum molecular dynamics based on the tight-binding Hamiltonian [116]. Indeed the TB scheme in combination with molecular dynamics is very promising thanks to the invention of linear scaling simulation codes where the computational time increases linearly with the number N of atoms [O(N) codes] [117–121].

5.4.2 Pseudopotentials and Density Response

For free-electron metals, such as alkali and alkaline earth metals and aluminum, a significant progress in lattice dynamics was achieved with the use of pseudopotentials [122–124]. With pseudopotentials the effects of electronic screening on the electron-ion interactions can be accounted for in a perturbation calculation. Within the Pseudopotential Perturbation (PPP) theory the electron-mediated interaction is shown to be proportional to the Fourier transformed product of the squared pseudopotential times the density (dielectric) response function of the interacting electrons [100, 101]. Along these lines a microscopic theory of lattice dynamics based on the Density Response (DR) method can be formulated in general terms and implemented at various levels of approximation [51–65]. The DR-PPP theory has been applied to the surface dynamics of alkali metals by Calandra et al. [125–127], who derived the screening from an infinite barrier model, and by Eguiluz et al. [128–133] in a state-of-the-art, self-consistent calculation for *sp*-bonded surfaces such as Al(001), (110) and (111) [132]. The latter calculations are discussed and compared with experiments and a BvK analysis [133] in Sect. 11.3. The PPP theory is less successful with noble and transition metals where the localized d electrons contribute a strong pseudopotential which cannot be treated within a perturbation theory [134, 135].

The total energy of an atom in a host electronic system and the many-body interactions with the surrounding atoms mostly depend upon the immersion energy of the atom in the electron density that the neighbouring atoms contribute at its position. From the knowledge of this immersion energy and its dependence on the positions of neighbouring atoms the dynamical matrix can be obtained for the calculation of both bulk and surface phonons. The first principle basis of this approach can be traced back to (i) Kohn-Sham (KS) theory of impurities in jellium [136], (ii) to the subsequent calculations for hydrogen [137–139] and helium [140, 141] impurities, (iii) and to the theory of chemisorption by Nørskov and Lang [142] and (iv) the quasi-atom concept introduced by Stott and Zaremba [143, 144] within the uniform-density approximation (UDA). Various semiempirical methods based on this concept are now available [145]. These include the Embedded Atom (EA) method [146–149], the Effective Medium (EM) theory [150–152], and the methods relying on special forms of the many-body interaction such as the Finnis-Sinclair (FS) [153] or Tight-Binding Second Moment (TBSM) potentials [154–156] and the Glue Model (GM) [157, 158].

These quantum semi-empirical approaches are easily implemented in efficient molecular dynamics schemes and have produced a considerable number of interesting results for a wide spectrum of surface dynamical phenomena, mostly for metals. Many valuable results have been produced with the EA approach for sp metals as well as for transition and noble metals. Examples of EM and EA calculations of surface phonons in noble metals are discussed in Chap. 11 in connection with the problem of the anomalous longitudinal resonance [159]. Semiempirical methods are particularly valuable in the study of complex surfaces with large surface unit cells, too large for present first principle methods. Relevant examples are the calculation of the surface phonon structure for vicinal metal surfaces by means of the EA method [160] and the GM Molecular Dynamics (GM-MD) simulation of the dynamics of the $23 \times \sqrt{3}$ reconstructed Au(111) surface [161]. Its large-period reconstruction allows for new branches of low energy excitations known as phasons, which can have either acoustic or optical character [162, 163].

5.4.3 *Phonons out of DFT*

Conceptually the above methods, on the one hand, provide a microscopic basis for the phenomenological models and, on the other hand, surrogate for the challenging *ab initio* calculations based on the Density Functional Theory (DFT) [82, 164, 165]. The DFT for electrons in solids is founded on the Hohenberg-Kohn theorem [164], which states that the total energy E of an electronic system is related in a unique way to its electron density function $n(\mathbf{r})$ via a (generally unknown) functional $E[n(\mathbf{r})]$. A trial functional is constructed through the Kohn-Sham eigenfunctions of an effective one electron Hamiltonian [136], where the Coulomb interactions are derived from efficient pseudopotentials and the density $n(\mathbf{x})$ itself. The exchange and correlation term, which represent the unknown part of the functional E, can be approximated with suitable local or semi-local functionals of the density [166], commonly within the Local Density Approximation (LDA) [167] or the Generalized Gradient Approximation (GGA) [168–170]. Through a self-consistency loop the functional E is minimized and the electronic density

$n(\mathbf{r})$ determined. This makes DFT the most powerful, albeit the most computationally demanding method for a first principle calculation of the ground state total energy.

There are different ways to calculate the phonon frequencies from DFT. A first route to a First-Principle (FP) DFT lattice dynamics is based on the determination, through molecular dynamics simulations at sufficiently low temperature, of the velocity-velocity correlation function, from which the phonon spectrum is derived by a Fourier transform. A major breakthrough in this direction is the Car-Parrinello Method (CPM) [171], where the Kohn-Sham wavefunctions are expanded in plane waves and the expansion coefficients are treated as classical time-dependent variables, together with the nuclear coordinates. The evolution of this set of coordinates in time, within the adiabatic approximation, fully describes the quantum dynamics of the entire system. The CPM is highly demanding in terms of computer time and memory and is therefore restricted to comparatively small numbers of atoms in the periodic cell. This limitation imposes the need of a clever and cautious use of the CPM, since the periodic cell has to be chosen to simulate as closely as possible the behavior of a real system. When this condition is satisfied, the CPM provides an important reference frame for many surface dynamical problems, including the calculation of surface phonons in semiconductors [172–174]. It is presently a widespread application of density functional theory to time-dependent problems.

In 1989 Sankey and Niklewski developed a less computationally-demanding approach to *ab initio* molecular dynamics based on the already defined FPTB scheme [175]. The tight-binding matrix elements are constructed within a DF-LDA theory where the total energy is represented by a Harris functional [176] and the Kohn-Sham wavefunctions are approximated by a set of local pseudo-atomic orbitals [175]. This theory was successfully applied by D. R. Alfonso et al. to the calculation of the surface phonon densities for the low index surfaces of diamond [177, 178].

A second approach to FP lattice dynamics is the Frozen Phonon Method (FPM) [179–183] where the effective interatomic force constants are obtained from a calculation of the total energy $E[n(\mathbf{r})]$ or of the Hellmann-Feynmann (HF) forces [184, 185] for the atoms slightly shifted from their equilibrium positions. There are two main variants of this method. In the first, one-phonon energies are determined by a quadratic fit of the total energy changes produced by displacing the atoms according to a pattern which corresponds to a normal mode. This procedure can be applied only in those cases in which phonon eigenvectors are completely determined by symmetry and thus known *a priori*. Moreover only phonons at $\mathbf{q} = 0$ can be studied using a single unit cell, whereas to deal with phonons at finite $\mathbf{q}$ a supercell commensurate with $\mathbf{q}$ is needed. In surface dynamics this technique encounters the difficulty that the displacement pattern of any surface phonon, even at symmetry points, is not precisely known in principle. Thus its application is limited to particular cases, e.g., for adsorbates which are largely decoupled from the substrate.

The second variant exploits the possibility of calculating the forces with little computational effort by means of the Hellmann-Feynmann theorem. Once the ground state properties have been determined, the force acting on each atom *l'* as an effect of a small displacement of atom *l*, and thus an entire row of the dynamical matrix, can be computed. The complete dynamical matrix can be constructed by performing all the symmetry independent displacements. This method has the advantage that it does not need an *a-priori* knowledge of the phonon eigenvectors and allows to construct the whole real space force constant matrix. Hence its Fourier transform provides the dynamical matrix for any arbitrary wavevector **q**. In metals, however, effective interatomic force constants have long range terms due to conduction electrons which develop Friedel oscillations around a local disturbance [102, 103]. Consequently computationally demanding large supercells must be used to assure a good convergence. Nevertheless, Bohnen, Ho and coworkers have successfully employed these methods using a pseudopotential density functional constructed on a plane wave basis, for the surfaces of simple metals [186–192]. This method has also been applied to the semiconductors GaP and InA [193]. For noble metals Bohnen, Ho and coworkers obtained good results using a mixed basis of plane waves and localized orbitals [194–198]. Total energy DFT calculations for Rh(111) of the force constants up to the 12th neighbours have been performed by derivation of HF forces, where both norm-conserving and ultrasoft non-norm-conserving pseudopotentials have been used [199]. The calculation based on the norm-conserving pseudopotential yields a good fit of the experimental surface phonon dispersion curves for both clean and hydrogen-covered Rh(111) surfaces [200], thus supporting the argument that hydrogen surface passivation does not alter appreciably the force constants of noble metal surfaces.

The third route to *ab initio* surface lattice dynamics originates from the Density Response (DR) method in which the dynamical matrix is obtained by calculating the second-order derivatives of the total energy directly with respect to the atomic positions. By virtue of the Hellman-Feyman theorem the derivatives of $E[n(\mathbf{x})]$ become integral expressions of the potential and density derivatives and the latter, within the linear response theory, can be expressed in terms of a DR function. In this way a first principles formulation of the DR theory is obtained for lattice dynamics, where the DR function is expressed as the second order functional derivative of E with respect to $n(\mathbf{r})$ [65]. In practice no exact form of E is available for such a derivation and therefore the DR function has to be obtained self-consistently by summing over matrix elements of the perturbing potential between occupied and unoccupied Kohn-Sham states [133, 201–204]. With such a non-perturbative pseudopotential approach Eguiluz and Quong have obtained very good results for the dynamics of the three low index surfaces of aluminum [132, 204]. Since the DR is a ground state property the sum over unoccupied Kohn-Sham states can be avoided via the use of Green's functions, as done in various implementations of the DR method. The Eguiluz-Quong FP-DR approach is in many respects a theory of the dielectric function. In other respects it may be regarded as a different formulation of the density functional perturbation theory (DFPT) developed by Baroni and coworkers [205–208].

The good performance of DFPT in reproducing the experimental HAS and EELS dispersion curves for most solid surfaces relies on the preliminary accurate determination of the equilibrium surface relaxation and its temperature dependence [209]. Dal Corso et al. have successfully calculated the bulk phonon dispersion curves of Cu, Ag and Au using a DFPT based on Vanderbilt's ultra-soft pseudopotentials [210]. The DFPT method has been further applied by Dal Corso to the surface dynamics of the Cu(001) surface [211] in combination with either the Perdew, Burke and Ernzerhof generalized gradient approximation (GGA) [170] or LDA [167]. The GGA yields a substantially better agreement with HAS and EELS experimental data. A thorough analysis of the above developments in *ab initio* phonon dynamics of metal surfaces up to DFPT is found in Heid and Bohnen's review [192] while the complete set of *ab initio* calculations presently available for different classes of crystal surfaces is collected in Appendix A.

The next Section provides a summary of the DFPT formalism not only for the surface phonon dispersion curves but also for the mode-selected electron-phonon coupling strengths. As discussed in Chap. 8, they are necessary ingredients in the calculation of inelastic HAS intensities for conducting surfaces.

5.5 Density Functional Pertubation Theory (DFPT)

DFT allows to efficiently investigate the ground state properties of a system of N interacting electrons in a generic external potential $V(\mathbf{x})$ by considering the electronic density $n(\mathbf{x})$ instead of the wavefunctions $\Psi_i(\mathbf{x})$ as the fundamental quantity to describe the system [165]. The theoretical basis of DFT rests on the Hohenberg and Kohn theorems [164, 212] which, together with the Rayleigh-Ritz variational principle, state that the electronic part $E[n(\mathbf{x})]$ of the total energy, (5.1), is a unique functional of the electron density $n(\mathbf{x})$,

$$E[n(\mathbf{x})] = F[n(\mathbf{x})] + \int n(\mathbf{x}) V_{ion}(\mathbf{x}) d\mathbf{x}, \tag{5.18}$$

which is minimized by the ground state electronic density corresponding to the external (ion) potential $V_{ion}(\mathbf{x})$, and its minimum correspond to the ground state energy. The functional $F[n(\mathbf{x})]$ is a sum of the kinetic energy functional $T[n(\mathbf{x})]$ for the interacting system, of the classical Hartree term $J[n(\mathbf{x})]$ for the electron electron interaction and the non-classical contribution $E_{ncl}[n(\mathbf{x})]$ to the electron-electron interaction. Of these functionals only $J[n(\mathbf{x})]$ is known. The Kohn-Sham theorem [136] suggests a reasonable approximation to the $F[n(\mathbf{x})]$ functional. Their *ansatz* is to calculate the exact kinetic energy of a non-interacting system $T_s[n(\mathbf{x})]$ with the same density as the interacting one and rewrite the functional as $F[n(\mathbf{x})] = T_s[n(\mathbf{x})] + J[n(\mathbf{x})] + E_{xc}[n(\mathbf{x})]$ where $E_{xc}[n(\mathbf{x})]$ incorporates the unknown terms arising from the exchange and correlation energy. Reliable approximations are available for the exchange-correlation functional, such as the Local Density Approximation (LDA) [167] and the Generalized Gradient Approximation (GGA) [170]. They have been found to be particularly good in many cases except for highly correlated systems.

The variation of the energy functional (5.18), with respect to $n(\mathbf{x})$ and with the constraint that the number of electrons is fixed leads to the Kohn-Sham equations

$$\left(-\frac{\hbar^2}{2m}\nabla + V_{SCF}(\mathbf{x})\right)\psi_n(\mathbf{x}) = \varepsilon_n\psi_n(\mathbf{x}), \tag{5.19}$$

where the effective (self-consistent) potential V_{SCF} is given by

$$V_{SCF}(\mathbf{x}) = V_{ion}(\mathbf{x}) + e^2\int\frac{n(\mathbf{x}')}{|\mathbf{x}-\mathbf{x}'|}d\mathbf{x}' + \frac{\delta E_{xc}}{\delta n(\mathbf{x})}, \tag{5.20}$$

and

$$n(\mathbf{x}) = 2\sum_{n=1}^{N/2}|\psi_n(\mathbf{x})|^2. \tag{5.21}$$

The self-consistent solution of (5.19–5.21), normally obtained by iteration from an initial guess of $n(\mathbf{x})$, yields the ground-state charge density. It is straightforward to show that the Hellmann-Feynman theorem holds in the framework of DFT. Thus the forces $\mathbf{F}_l$ and the force constant matrix, $\Phi(l,l')$ introduced in (5.13), given by the first and second derivatives of the total energy with respect to the ionic positions, can be written as functions of the ground state charge density as

$$\mathbf{F}_l = -\frac{\partial E_{tot}}{\partial \mathbf{r}_l} = -\frac{\partial E_{ion}}{\partial \mathbf{r}_l} - \int n(\mathbf{x})\frac{\partial V_{ion}(\mathbf{x})}{\partial \mathbf{r}_l}d\mathbf{x}, \tag{5.22}$$

and

$$\begin{aligned}\Phi(l,l') &= \frac{\partial^2 E_{tot}(\mathbf{r})}{\partial \mathbf{r}_l\partial \mathbf{r}_{l'}} = \frac{\partial^2 E_{ion}(\mathbf{r})}{\partial \mathbf{r}_l\partial \mathbf{r}_{l'}} + \int n(\mathbf{x})\frac{\partial^2 V_{ion}(\mathbf{x})}{\partial \mathbf{r}_l\partial \mathbf{r}_{l'}}d\mathbf{x} + \int\frac{\partial n(\mathbf{x})}{\partial \mathbf{r}_l}\frac{\partial V_{ion}(\mathbf{x})}{\partial \mathbf{r}_{l'}}d\mathbf{x}\\ &= \frac{\partial^2 E_{ion}(\mathbf{r})}{\partial \mathbf{r}_l\partial \mathbf{r}_{l'}} + \int n(\mathbf{x})\frac{\partial^2 V_{ion}(\mathbf{x})}{\partial \mathbf{r}_l\partial \mathbf{r}_{l'}}d\mathbf{x} - \int\int\frac{\partial V_{ion}(\mathbf{x})}{\partial \mathbf{r}_l}\chi(\mathbf{x},\mathbf{x}')\frac{\partial V_{ion}(\mathbf{x}')}{\partial \mathbf{r}_{l'}}d\mathbf{x}\,d\mathbf{x}'.\end{aligned} \tag{5.23}$$

The second and third terms in the force constant matrix (5.23), expressing the electron-phonon coupling, are conveniently represented by the two Feynman diagrams of Fig. 5.4, where the broken and full lines represent the phonon and electron-hole pair propagation, respectively.

In the last term of (5.23) the density derivative has been explicitly written in terms of the electron susceptibility $\chi(\mathbf{x},\mathbf{x}')$ (independent to the lowest order of ion

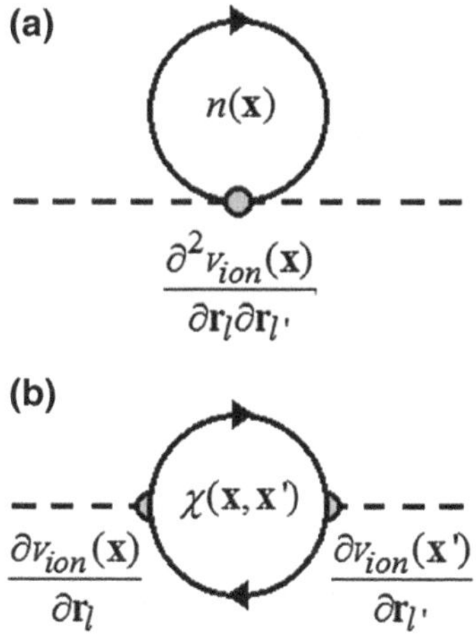

Fig. 5.4 Feynman diagram representation of the two electron-phonon terms contributing to the force constant matrix in equation (5.23). The broken lines represent the phonon propagation, while the full line loops describe virtual electron-hole excitations. **a** One-vertex terms proportional to the second derivative of the ion potential. **b** Two vertex terms proportional to the square of the first ion-potential derivative give contributions of the same order but with opposite sign

positions) in order to show the one-to-one correspondence of (5.23) with the ME force constant matrix, (5.13). The comparison actually provides a transcription rule of the semi-empirical PC method into *ab initio* DFPT, although the ME method works on a set of local basis functions.

In the calculation of the DFPT force-constant matrix it is not necessary to first evaluate the electron susceptibility. To first order the derivatives $\partial V_{SCF}/\partial \mathbf{r}_l$, $\partial\psi/\partial \mathbf{r}_l$, and $\partial n(\mathbf{x})/\partial \mathbf{r}_l$, the latter needed in (5.24), are just as well obtained by solving self-consistently the variation of Kohn-Sham equation (5.19)

$$(H_{SCF} - \varepsilon_n)\frac{\partial\psi_n(\mathbf{x})}{\partial \mathbf{r}_l} = -\left(\frac{\partial V_{SCF}(\mathbf{x})}{\partial \mathbf{r}_l} - \frac{\partial\varepsilon_n}{\partial \mathbf{r}_l}\right)\psi_n(\mathbf{x}), \tag{5.24}$$

where

$$\frac{\partial V_{SCF}(\mathbf{x})}{\partial \mathbf{r}_l} = \frac{\partial v_{ion}(\mathbf{x})}{\partial \mathbf{r}_l} + e^2\int\frac{1}{|\mathbf{x}-\mathbf{x}'|}\frac{\partial n(\mathbf{x}')}{\partial \mathbf{r}_l}d\mathbf{x}' + \int\frac{\delta^2 E_{xc}}{\delta n(\mathbf{x})\delta n(\mathbf{x}')}\frac{\partial n(\mathbf{x}')}{\partial \mathbf{r}_l}d\mathbf{x}', \tag{5.25}$$

$$\frac{\partial\varepsilon_n}{\partial \mathbf{r}_l} = \left\langle\psi_n\left|\frac{\partial V_{SCF}}{\partial \mathbf{r}_l}\right|\psi_n\right\rangle. \tag{5.26}$$

This approach is particularly effective since only occupied states are needed to evaluate the right side of (5.25) [213], whereas the standard pertubation theory approach, based on the expansion in eigenfunctions of the unperturbed Hamiltonian would require a sum over the full spectrum of the zero order Hamiltonian.

An alternative formulation of DFPT can be obtained on the basis of a variational principle as reported by Gonze [214, 215]. It can be shown that the two formulations are equivalent [208] but the latter implies a minimization rather than the self-consistent solution of linear equations. As shown in [208] DFPT has the advantage, with respect to other *ab initio* methods requiring supercells to treat a $\mathbf{q} \neq 0$ perturbation, that in the $\mathbf{q}$-space the Fourier components of the perturbation are, up to a reciprocal lattice vector $\mathbf{G}$, independent. Moreover to first order a $\mathbf{q}$-component of the perturbation can only admix an electronic eigenfunction of wavevector $\mathbf{k}$ with eigenfunctions of wavevector $\mathbf{k} + \mathbf{q}$, which allows to readily obtain the Fourier components of the force constant matrix, (5.23), and the phonon frequencies at an arbitrary $\mathbf{q}$-point.

The above formulation is rigorously applicable to solids where long range Coulomb forces are either absent, as in non-polar insulators and semiconductors, or screened out by free electrons. In polar materials the contribution from the macroscopic electric field is naturally present in the ion-ion part of the force constant matrix. The only effect of the electronic density response is the replacement of bare ion charges with effective Born charges, as usually done in any shell model treatment [95, 216].

In the case of conducting materials there could be the problem that the number of $\mathbf{k}$-points needed to describe correctly the effects related to the presence of a Fermi surface is prohibitively large. An efficient implementation of DFPT for metallic systems, based on the energy level broadening by a smearing function has been proposed by de Gironcoli [217]. Another effective approach, based on a different treatment of the perturbation, is that of Equiluz and Quong mentioned above [204]. Even with use of the latter technique, however, the phonon dispersion curves for metals within DFPT can critically depend on the sampling of $\mathbf{k}$- and $\mathbf{q}$-points, as required, e.g., to properly describe effects such as Kohn anomalies.

In practical calculations the bare electron-nucleus interaction $V_{ion}(\mathbf{x})$ is often replaced by a pseudopotential that takes into account effects of core electrons and only valence electrons are treated explicitly. DFPT was at first implemented within the framework of a plane-wave basis set expansion for Kohn-Sham states [218, 219] and the use of norm-conserving pseudopotentials [220–223] which was extensively used with great success to study bulk and surface phonons of semiconductors and simple metals.

The need to deal also with transition metals and other *hard* refractory materials, using a pseudopotential and a plane-wave approach without requiring a huge basis set to ensure transferability, led to the development of ultrasoft pseudopotentials [224]. Their use within DFPT was implemented by dal Corso et al. [210, 211]. Recently the development of fully relativistic pseudopotentials and their inclusion in the linear-response code [225] made it possible to account for spin-orbit coupling and its effect on phonon modes and electron-phonon interactions. An alternative approach to hard materials is the use of a local or mixed basis set instead of plane waves. Applications to such methods to DFTP have to face the additional problem of the Pulay forces [226], an additional term in (5.22) that arises because of the dependence of the basis set on ionic positions. Significant progress in this field has been made by Savrasov and Andersen [227] who implemented a new full-potential Linear-Muffin-Tin-Orbital (LMTO) method and applied it to the doped high-T_c superconductor $CaCuO_2$. It is hoped that this powerful method for treating "difficult" ions such as copper and other 3d metals, presently applied to a layered crystal, will soon be extended to the dynamics of thin slabs. To date the few available HAS experiments on the surface phonons of layered superconducting cuprates have been analyzed in the framework of the shell model [228, 229]. In recent times the introduction of non-local pseudopotentials [230–232] provide an effective way to properly treat van der Waals interactions within the framework of DFT. Their inclusion into DFPT codes would improve considerably the performance of DFPT in the description of weakly-bound layered crystals.

DFPT is presently one of the most popular methods for the calculations of phonons and is implemented in some of the most widely used simulation codes like Quantum Espresso [233], ABINIT [234], VASP [235] and CASTEP [236]. The excellent performance of DFPT in the analysis of recent HAS data will be illustrated in Chap. 11 with a few paradigmatic examples for metal and semimetal surfaces.

References

1. G. Benedek, D. Campi, J.P. Toennies, in *Physics of Solid Surfaces*, Subvol. A, Landolt-Börnstein, Numerical Data and Functional Relationships in Science and Technology – New Ser., Subvol. 45A, ed. by P. Chiaradia, G.F. Chiarotti, Chap. 10, 572–646 (2015)
2. E.W. Kellermann, Phil. Trans. Roy. Soc. (London) **A238**, 513 (1940)
3. M. Born, K. Huang, *Dynamical Theory of Crystal Lattices* (Clarendon Press, Oxford, 1954)
4. S.Y. Tong, A.A. Maradudin, Phys. Rev. **181**, 1318 (1969)
5. L. Philippe, Ph Lambin, P. Senet, A.A. Lucas, Appl. Supercond. **2**, 135 (1994)
6. J.B. Mc Bride, P. K. Kaw, Phys. Lett. A **33**, 72 (1970)
7. W.L. Schaich, Phys. Rev. B **24**, 686 (1981)
8. C. Schwartz, W.L. Schaich, J. Physique **42**, C6–828 (1981)
9. W.L. Schaich, C. Schwarz, Phys. Rev. B **25**, 7365 (1982)
10. E.B. Wilson, Jr., J.C. Decius, P.C. Cross, *Molecular Vibrations* (McGraw-Hill, New York, 1955)
11. R. Tubino, L. Piseri, G. Zerbi, J. Chem. Phys. **6**, 1022 (1972)
12. P.N. Keating, Phys. Rev. **140A**, 369 (1965)
13. P.N. Keating, Phys. Rev. **145**, 637 (1966)
14. S.K. Sinha, Crit. Rev. Solid State Sci. **3**, 273 (1973)
15. R.M. Martin, Chem. Phys. Lett. **2**, 268 (1968)
16. J.C. Phillips, Phys. Rev. **166**, 832 (1968)
17. R.M. Martin, Phys. Rev. **186**, 871 (1969)
18. F.A. Johnson, Proc. R. Soc. (London) Ser. A **339**, 73 (1974)
19. W. Weber, Phys. Rev. Lett. **33**, 371 (1974)
20. K.C. Rustagi, W. Weber, Solid State Comm. **18**, 1027 (1976)
21. W. Weber, Phys. Rev. B **15**, 4789 (1977)

22. L. Miglio, P. Santini, P. Ruggerone, G. Benedek, Phys. Rev. Lett. **62,** 3070 (1989)
23. P. Santini, L. Miglio, G. Benedek, U. Harten, P. Ruggerone, J.P. Toennies, Phys. Rev. B **42**, 11942 (1990)
24. G. Benedek, G. Onida, Phys. Rev. B **47**, 16471 (1993)
25. A. Fleszar, R. Resta, Phys. Rev. B **34**, 7140 (1986)
26. T.S. Chen, G.P. Alldredge, F.W. de Wette, R.E. Allen, Phys. Rev. Letters **26**, 1543 (1971)
27. T.S. Chen, G.P. Alldredge, F.W. de Wette, Solid State Commun. **10**, 941 (1972)
28. T.S. Chen, G.P. Alldredge, F.W. de Wette, Phys. Lett. **40A**, 401 (1972)
29. V.L. Zoth, G.P. Alldredge, F.W. de Wette, Phys. Lett. **47A**, 247 (1974)
30. T.S. Chen, F.W. de Wette, G.P. Alldredge, Phys. Rev. B **15**, 1167 (1977)
31. U. Schröder, Solid State Commun. **4**, 347 (1966)
32. U. Schröder, V. Nüsslein, Phys. Status Solidi **21**, 309 (1967)
33. K. Fischer, H. Bilz, R. Haberkorn, W. Weber, Phys. Status Solidi (b) **54**, 285 (1972)
34. W. Weber, Phys. Rev. B **8**, 5082 (1973)
35. W. Kress, P. Roedhammer, H. Bilz, W.D. Teuchert, A.N. Christensen, Phys. Rev. B **17**, 111 (1978)
36. W. Cochran, R.A. Cowley, *Encyclopedia of Physics 25/2a* (Springer, Berlin Heidelberg, 1967), p. 83
37. H. Bilz, B. Gliss, W. Hanke, in *Dynamical Properties of Solids*, vol. 1, eds. by G.K. Horton, A.A. Maradudin (North-Holland, Amsterdam, 1974), p. 343
38. J. Hardy, in *Dynamical Properties of Solids*, vol. 1, ed. by G.K. Horton, A.A. Maradudin (North-Holland, Amsterdam, 1974), p. 157
39. H. Bilz, W. Kress, *Phonon Dispersion Relations in Insulators* (Springer, Berlin Heidelberg, 1976)
40. J. Hardy, A.M. Karo, *The Lattice Dynamics and Statics of Alkali Halide Crystals* (Plenum Press, New York, 1979)
41. G.P. Alldredge, Phys. Lett. **41A**, 291 (1972)
42. B.C. Dick, A.W. Overhauser, Phys. Rev. **112**, 90 (1958)
43. W. Cochran, Phys. Rev. Lett. **2**, 495 (1959)
44. W. Cochran, Adv. Phys. **9**, 387 (1960)
45. B. Szigeti, Trans. Faraday Soc. **45**, 155 (1949)
46. B. Szigeti, Proc. Roy. Soc. (London) **A204**, 51 (1950)
47. K.B. Tolpygo, Zh Eksp, Theor. Fiz. **20**, 497 (1950)
48. K.B. Tolpygo, Fiz. Tverd. Tela **3**, 943 (1961)
49. J. Yamashita, T. Kurosawa, J. Phys. Soc. Jpn. **10**, 610 (1950)
50. P.B. Allen, Phys. Rev. B **16**, 5139 (1977)
51. L.J. Sham, Phys. Rev. **188**, 1431 (1969)
52. R.M. Pick, M.H. Cohen, R.M. Martin, Phys. Rev. B **1**, 910 (1970)
53. R. Pick, in *Phonons*, ed. by M. A. Nusimovici (Flammarion, Paris, 1971), p. 20
54. W. Hanke, Phys. Rev. B **8**, 4585 (1973)
55. W. Hanke, Phys. Rev. B **8**, 4591 (1973)
56. L.J. Sham, in *Dynamical Properties of Solids*, vol. 1, ed. by G.K. Horton, A.A. Maradudin (North-Holland, Amsterdam, 1974), p. 301
57. W. Hanke, L.J. Sham, Phys. Rev. B **12**, 4501 (1975)
58. P.E. van Camp, V.E. van Doren, J.T. Devreese, Phys. Rev. Lett. **42**, 1224 (1979)
59. P.E. van Camp, V.E. van Doren, J.T, Devreese, Phys. Rev. B **25**, 4270 (1982)
60. W. Hanke, L.J. Sham, Phys. Rev. B **21**, 4656 (1980)
61. R. Resta, A. Baldereschi, Phys. Rev. B **24**, 4839 (1981)
62. P.E. van Camp, V.E. van Doren, J.T Devreese, Phys. Rev. B **25**, 4270 (1982)
63. R. Resta, Phys. Rev. B **27**, 3620 (1983)
64. P.E. van Camp, V.E. van Doren and J.T, Devreese, Phys. Rev. B **31**, 4089 (1985)
65. C. Falter, Phys. Rep. **164**, 1 (1988)
66. C.S. Jayanthi, H. Bilz, W. Kress, G. Benedek, Phys. Rev. Lett. **59**, 795 (1987)

67. C. Kaden, P. Ruggerone, J.P. Toennies, G. Zhang, G. Benedek, Phys. Rev. B **46**, 13509 (1992)
68. C. Kaden, P. Ruggerone, J.P. Toennies, G. Benedek, Nuovo Cim. **14D**, 627 (1992)
69. X. Zhu, Y. Yao, H.C. Hsu, F.C. Chou, M. El-Batanouny, Surf. Sci. **605**, 376 (2011)
70. X. Zhu, L. Santos, R. Sankar, S. Chikara, C. Howard, F.C. Chou, C. Charmon, M. El-Batanouny, Phys. Rev. Lett. **107**, 186102 (2011)
71. X.T. Zhu, L. Santos, C. Howard, R. Sankar, F.C. Chou, C. Chamon, M. El- Batanouny, Phys. Rev. Lett. **108**, 185501 (2012)
72. C. Howard, M. El-Batanouny, R. Sankar, F.C. Chou, Phys. Rev. B **88**, 035402 (2013)
73. G. Benedek, Phys. Status Sol. **B58**, 661 (1973)
74. G. Benedek, Surf. Sci. **61**, 603 (1976)
75. G. Benedek, R.B. Doak, J.P. Toennies, Phys. Rev. B **28**, 7277 (1983)
76. G. Benedek, L. Miglio, in *Surface Phonons*, ed. by F.W. deWette, W. Kress, Springer Ser. Surf. Sci., vol. 27 (Springer, Berlin, Heidelberg, 1991), p. 37
77. D. Eichenauer, J. P. Toennies, in *Dynamics on Surfaces*, ed. by B. Pullman, J. Jortner, A. Nitzan, B. Gerber (Reidel, Dordrecht, 1984), p. 1
78. R. Zeyher, Phys. Rev. Lett. **35**, 174 (1975)
79. M. Miura, W. Kress, H. Bilz, Z. Phys. B **54**, 103 (1984)
80. G. Benedek, M. Miura, W. Kress, H. Bilz, Phys. Rev. Lett. **52**, 1907 (1984)
81. G. Benedek, M. Miura, W. Kress, H. Bilz, Surface Sci. **148**, 92 (1984)
82. R.G. Parr, W. Yang, *Density Functional Theory of Atoms and Molecules* (Oxford University Press, New York, 1989)
83. R.M. Dreizer, E.K.U. Gross, *Density Functional Theory* (Springer, Berlin Heidelberg, 1990)
84. A. Euceda, D.M. Bylander, L. Kleinman, Phys. Rev. B **28**, 528 (1983)
85. M. Hamermesh, in *Group Theory and Its Applications to Physical Problems* (Dover Publ. Inc., New York, 1989), Chap. 4
86. C. Falter, W. Ludwig, M. Selmke, W. Zierau, Phys. Lett. A **105**, 139 (1984)
87. C. Falter, M. Selmke, W. Ludwig, Solid State Commun. **52**, 949 (1984)
88. C. Falter, W. Ludwig, M. Selmke, Solid State Commun. **54**, 321 (1985)
89. C. Falter, W. Ludwig, M. Selmke, Solid State Commun. **54**, 497 (1985)
90. C. Falter, W. Ludwig, A.A. Maradudin, M. Selmke, W. Zierau, Phys. Rev. B **32**, 6510 (1985)
91. C. Falter, M. Selmke, W. Ludwig, K. Kunc, Phys. Rev. B **32**, 6581 (1985)
92. J.M. Ziman, *Principles of the Theory of Solids* (Cambridge Univ. Press, Cambridge, 1972)
93. E. Tosatti, C. Calandra, V. Bortolani, C.M. Bertoni, J. Phys. C **5**, L299 (1972)
94. C.M. Bertoni, V. Bortolani, C. Calandra, F. Nizzoli, J. Phys. C **6**, 3612 (1973)
95. H. Böttger, *Principle of the Theory of Lattice Dynamics* (Akademie-Verlag, Berlin, 1983)
96. A.B. Migdal, Sov. Phys. JETP **34**, 996 (1958)
97. W. Kohn, Phys. Rev. Lett. **2**, 393 (1959)
98. E.J. Woll Jr., W. Kohn, Phys. Rev. **126**, 1693 (1962)
99. A. M. Afanasev, Yu. M. Kagan, Sov. Phys. JETP **16**, 1030 (1963)
100. E.G. Brovman, Yu. M. Kagan, Zh. Eksp. Teor. Fiz. **57**, 1329 (1970) [Engl. Transl.: Sov. Phys. JETP 30 (1970) 721]
101. E.G. Brovman, Yu. M. Kagan, in *Dynamical Properties of Solids*, vol. 1, ed. by G.K. Horton, A.A. Maradudin (North-Holland, Amsterdam, 1974), p. 191
102. J. Friedel, Nuovo Cim., Suppl. **7**, 287 (1968)
103. W.A. Harrison, in *Solid State Theory* (Dover, New York, 1979), Ch. II
104. E. Hulpke, J. Lüdecke, Phys. Rev. Lett. **68**, 2846 (1992)
105. E. Hulpke, J. Lüdecke, Surf. Sci. **287**(288), 837 (1993)
106. G. Benedek, M. Pardo, J.P. Toennies, in *Highlights on Spectroscopies of Semiconductors and Nanostructures*, ed. by G. Guizzetti, A.C. Andreani, F. Marabelli, M. Patrini, Conf. Proc., vol. 94 (SIF, Bologna 2007), pp. 151–167
107. G. Benedek, M. Bernasconi, V. Chis, E. Chulkov, P.M. Echenique, B. Hellsing, J.P. Toennies, J. Phys.: Cond. Matter **22**, 084020 (2010)

108. W. Harrison, *Electronic Structure and the Properties of Solids* (Dover Publ, New York, 1988)
109. X.W. Wang, C.T. Chan, K.-H. Ho, W. Weber, Phys. Rev. Lett. **60**, 2066 (1988)
110. Y.R. Wang, C.B. Duke, Surf. Sci. **205**, L755 (1988)
111. B. Sandfort, A. Mazur, J. Pollmann, Phys. Rev. B **51**, 7139 (1995)
112. A. Mazur, B. Sandfort, V. Gräschus, J. Pollmann, Festkörperprobleme/Adv. Solid State Phys. **36**, 181 (1997)
113. C.M. Varma, W. Weber, Phys. Rev. B **21**, 6142 (1979)
114. W. Weber, in *Electronic Structure of Complex Systems*, ed. by S. Phariseau, W.M. Temmermann (Plenum, New York, 1984), p. 345
115. X.W. Wang, W. Weber, Phys. Rev. Lett. **58**, 1452 (1987)
116. J. Kim, F.S. Khan, M. Yeh, J.W. Wilkins, Phys. Rev. B **52**, 14709 (1995)
117. F. Mauri, G. Galli, R. Car, Phys. Rev. B **47**, 9973 (1993)
118. P. Ordejòn, D. Drabold, M. Grunbach, R. Martin, Phys. Rev. B **48**, 14646 (1993)
119. S. Goedecker, L. Colombo, Phys. Rev. Lett. **74**, 122 (1994)
120. S. Goedecker, Rev. Mod. Phys. **71**, 1085 (1999)
121. L. Colombo, M. Rosati, Comp. Phys. Comm. **128**, 108 (2000)
122. L.J. Sham, Proc. Royal Soc. (London) **A283**, 33 (1965)
123. A.O.E. Animalu, F. Bonsignori, V. Bortolani, Nuovo Cim. B **44** 159 (1966)
124. W. Harrison, *Pseudopotentials in the Theory of Metals* (Benjamin, New York, 1966)
125. C. Beatrice, C. Calandra, Phys. Rev. B **10**, 6130 (1983)
126. C. Calandra, A. Catellani, C. Beatrice, Surf. Sci. **148**, 90 (1984)
127. C. Calandra, A. Catellani, C. Beatrice, Surf. Sci. **152**(153), 814 (1985)
128. A.G. Eguiluz, Phys. Rev. Lett. **51**, 1097 (1983)
129. A.G. Eguiluz, Phys. Rev. B **31**, 3303 (1985)
130. A.G. Eguiluz, Phys. Scripta **36**, 651 (1987)
131. A.G. Eguiluz, A.A. Maradudin, R.F. Wallis, Phys. Rev. Lett. **60**, 309 (1988)
132. J.A. Gaspar, A.G. Eguiluz, M. Gester, A. Lock, J.P. Toennies, Phys. Rev. Lett. **66**, 337 (1991)
133. R.F. Wallis, A.A. Maradudin, V. Bortolani, A.G. Eguiluz, A.A. Quong, A. Franchini, C. Santoro, Phys. Rev. B **48**, 6043 (1993). In this paper there is an instructive comparison of empirical calculations with perturbative and non-perturbative pseudopotential DR results
134. W.A. Harrison, Phys. Rev. **181**, 1036 (1969)
135. J.A. Moriarty, Phys. Rev. B **26**, 1754 (1982)
136. W. Kohn, L.J. Sham, Phys. Rev. A **140**, 1133 (1965)
137. Z.D. Popovic, M.J. Stott, Phys. Rev. Lett. **33**, 1164 (1974)
138. E. Zaremba, L.M. Sander, H.B. Shore, J.H. Rose, J. Phys. F **7**, 1763 (1977)
139. D.S. Larsen, J.K. Nørskov, J. Phys. F: Met. Phys. **9**, 1975 (1980)
140. M.D. Whitmore, J. Phys. F **6**, 1259 (1976)
141. J.K. Nørskov, Sol. State Commun. **24**, 691 (1977)
142. J.K. Nørskov, N.D. Lang, Phys. Rev. B **21**, 2131 (1980)
143. M.J. Stott, E. Zaremba, Solid State Commun. **32**, 1297 (1979)
144. M.J. Stott, E. Zaremba, Phys. Rev. B **22**, 1564 (1980)
145. A.E. Carlsson, in *Beyond Pair Potentials in Elemental Metals and Semiconductors,* Solid State Physics, vol. 43, ed. by H. Ehrenreich, D. Turnbull (Academic Press, Boston, 1990)
146. M.S. Daw, M.I. Baskes, Phys. Rev. Lett. **50**, 1285 (1983)
147. M.S. Daw, M.I. Baskes, Phys. Rev. B **29**, 6443 (1984)
148. J.S. Nelson, E.C. Sowa, M.S. Daw, Phys. Rev. Lett. **61**, 1977 (1987)
149. P.R. Underhill, Surf. Sci. **200**, L441 (1988)
150. W.K. Jacobsen, J.K. Nørskov, M.J. Puska, Phys. Rev. B **35**, 7423 (1987)
151. P.D. Ditlevsen, J.K. Nørskov, J. El. Spectr. Rel. Phenom. **54**(55), 237 (1990)
152. P.D. Ditlevsen, J.K. Nørskov, Surf. Sci. **254**, 261 (1991)
153. M.W. Finnis, J.E. Sinclair, Phil. Mag. A **50**, 45 (1984)
154. V. Rosato, M. Guillope, B. Legrand, Phil. Mag. A **59**(2), 321 (1989)

155. J. Guevara, A.M. Llois, M. Weismann, Phys. Rev. B **52**, 11509 (1995)
156. F. Raouafi, C. Barreteau, M.C. Desjonquères, D. Spanjaard, Surf. Sci. **507**, 748 (2002)
157. E.F. Ercolessi, E. Tosatti, M. Parrinello, Surf. Sci. **177,** 314 (1986)
158. F. Ercolessi, M. Parrinello, E. Tosatti, Phys. Rev. Lett. **57**, 719 (1986) and Phil. Mag. A **58**, 513 (1988)
159. R.B. Doak, U. Harten, J.P. Toennies, Phys. Rev. Lett. **51**, 578 (1983)
160. I.Y. Sklyadneva, G.G. Rusina, E.V.Chulkov, Surf. Sci. **146**, 17 (1998)
161. X.Q. Wang, Phys. Rev. Lett. **67**, 1294 (1991)
162. R. Ravelo, M. El-Batanouny, Phys. Rev. B **40**, 9574 (1989)
163. E.M. McIntosh, P.R. Kole, M. El-Batanouny, D.M. Chisnall, J. Ellis, W. Allison, Phys. Rev. Lett. **110**, 086103 (2013)
164. P. Hohenberg, W. Kohn, Phys. Rev. **136**, 864 (1964)
165. E.K.U. Gross, R.M. Dreizler (eds.), *Density Functional Theory* (Plenum, New York, 1994)
166. R. M. Martin, in *Electronic Structure: Basic Theory and Practical Methods,* (Cambridge University Press, 2004)
167. J.P. Perdew, A. Zunger, Phys. Rev. B **23**, 5048 (1981)
168. A.D. Becke, Phys. Rev. A **38**, 3098 (1988)
169. J.P. Perdew, in *Electronic Structure of Solids,* ed. by P. Ziesche, H. Eschrig (Berlin Academic, 1991)
170. J.P. Perdew, K. Burke, M. Ernzerhof, Phys. Rev. Lett. **77**, 3865 (1996)
171. R. Car, M. Parrinello, Phys. Rev. Lett. **55**, 2471 (1985)
172. F. Ancillotto, W. Andreoni, A. Selloni, R. Car, M. Parrinello, Phys. Rev. Lett. **65**, 3148 (1990)
173. A. Ancillotto, W. Andreoni, A. Selloni, R. Car, M. Parinello, in *Phonons 89*, ed. by S. Hunklinger, W. Ludwig, G. Weiss (World Scientific, Singapore, 1990), p. 931
174. R. di Felice, A.I. Shkrebtii, F. Finocchi, C.M. Bertoni, G. Onida, J. El. Spect. Rel. Phenom. **64**(65), 697 (1995)
175. O.F. Sankey, D.J. Niklewski, Phys. Rev. B **40**, 3979 (1989)
176. J. Harris, Phys. Rev. B **31**, 1770 (1985)
177. D.R. Alfonso, D.A. Drabold, S.E. Ulloa, Phys. Rev. B **51**, 1989 (1995)
178. D.R. Alfonso, D.A. Drabold, S.E. Ulloa, Phys. Rev. B **51**, 14669 (1995)
179. V. Heine, D. Weaire, in *Solid State Physics*, vol. 24, ed. by H. Ehrenreich, F. Seitz, T. Turnbull (Academic, New York, 1970)
180. H. Wendel, R.M. Martin, Phys. Rev. B **19**, 5251 (1979)
181. K. Kunc, R.M. Martin, Phys. Rev. B **24**, 2311 (1981)
182. K.M. Ho, C.L. Fu, B.N. Harmon, W. Weber, D.R. Hamann, Phys. Rev. Lett. **49**, 673 (1982)
183. M.T. Yin, M.L. Cohen, Phys. Rev. B **26**, 3259 (1982)
184. H. Hellmann, *Einführung in die Quantenchemie* (Deuticke, Leipzig, 1937)
185. R.P. Feynman, Phys. Rev. **56**, 340 (1939)
186. K.M. Ho, K.P. Bohnen, Phys. Rev. Lett. **56**, 934 (1986)
187. K.P. Bohnen, K.M. Ho, Surf. Sci. **207**, 105 (1988)
188. K.M. Ho, K.P. Bohnen, Phys. Rev. B **38**, 12897 (1988)
189. Th Rodach, K.P. Bohnen, K.M. Ho, Surf. Sci. **209**, 481 (1989)
190. K.P. Bohnen, K.M. Ho, Surf. Sci. Rep. **19**, 99 (1993)
191. K.P. Bohnen, K.M. Ho, in *Tutorials on Selected Topics in Modern Surface Science*, ed. by T. Barr, D. Saldin (Elsevier, Dordrecht, 1993), p. 100–117
192. R. Heid, K.-P. Bohnen, Phys. Rep. **387**, 151 (2003)
193. C. Eckl, J. Fritsch, P. Pavone, U. Schroeder, Surf. Sci. **394**, 47 (1997)
194. A.M. Lahee, J.P. Toennies, C. Woll, K.P. Bohnen, K.M. Ho, Europhys. Lett. **10**, 261 (1989)
195. K.M. Ho, K.P. Bohnen, J. Electron Spectrosc. Rel. Phen. **54**(55), 229 (1990)
196. Th Rodach, K.P. Bohnen, K.M. Ho, Vacuum **41**, 751 (1990)
197. Th Rodach, K.P. Bohnen, K.M. Ho, Surf. Sci. **286**, 66 (1993)
198. T. Rodach, K.P. Bohnen, K.M. Ho, Surf. Sci. **296**, 123 (1993)
199. K.-P. Bohnen, A. Eichler, J. Hafner, Surf. Sci. **368**, 222 (1996)

200. G. Witte, J.P. Toennies, Ch. Wöll, Surf. Sci. 323, 228 (1995)
201. A.A. Quong, B.M. Klein, Phys. Rev. B **46**, 10734 (1992)
202. A.A. Quong, A.G. Eguiluz, Phys. Rev. Lett. **70**, 3955 (1993)
203. A.A. Quong, Phys. Rev. B **49**, 3226 (1994)
204. A.E. Eguiluz, A.A. Quong, in *Dynamical Properties of Solids*, vol. 7, ed. by G.K. Horton, A.A. Maradudin (North-Holland, Amsterdam, 1995)
205. M. Buongiorno Nardelli, S. Baroni, P. Giannozzi, Phys. Rev. Lett. **69**, 1069 (1992)
206. P. Giannozzi, S. de Gironcoli, P. Pavone, S. Baroni, Phys. Rev. B **43**, 7231 (1993)
207. C. Bungaro, S. de Gironcoli, S. Baroni, Phys. Rev. Lett. **77**, 2491 (1996)
208. S. Baroni, S. de Gironcoli, A. Dal Corso, P. Giannozzi, Rev. Mod. Phys. **73**, 515 (2001)
209. J. Xie, S. de Gironcoli, S. Baroni, M. Scheffler, Phys. Rev. B **59**, 965 (1999)
210. A. dal Corso, A. Pasquarello, A. Baldereschi, Phys. Rev. B **56**, 11369 (1997)
211. A. dal Corso, Phys. Rev. B **64**, 235118 (2001)
212. E. Runge, E.K.U. Gross, Phys. Rev. Lett. **52**, 997 (1984)
213. S. Baroni, P. Giannozzi, A. Testa, Phys. Rev. Lett. **70**, 1863 (1987)
214. X. Gonze, Phys. Rev. A **52**, 1086 (1995)
215. X. Gonze, Phys. Rev. A **52**, 1096 (1995)
216. A.A. Maradudin, E.W. Montroll, G.H. Weiss, I.P. Ipatova, in *Theory of Lattice Dynamics in the Harmonic Approximation*, Solid State Physics, Suppl. 3 (Academic Press, New York, 1971)
217. S. de Gironcoli, Phys. Rev. B **51**, 6773 (1995)
218. W.E. Pickett, Comput. Phys. Rep. **9**, 115 (1989)
219. M.C Payne, M.P. Teter, D.C. Allen, T.A. Arias, J.D. Joannopoulos. Rev. Mod. Phys. **64**, 1045 (1992)
220. D.R. Hamann, M. Schluter, C. Chiang, Phys. Rev. Lett. **43**, 1494 (1979)
221. L. Kleinman, Phys. Rev. B **26**, 1055 (1982)
222. L. Kleinman, D.M. Bylander, Phys. Rev. Lett. **48**, 1425 (1982)
223. N. Troullier, J.L. Martins, Phys. Rev. B **43**, 1993 (1991)
224. D. Vanderbilt, Phys. Rev. B **41**, 7892 (1990)
225. A. Dal Corso, J. Phys.: Condens. Matter **20**, 445202 (2008)
226. P. Pulay, Mol. Phys. **17**, 197 (1969)
227. S.Y. Savrasov, O.K. Andersen, Phys. Rev. Lett. **77**, 4430 (1996)
228. U. Paltzer, D. Schmicker, F.W. de Wette, U. Schröder, J.P. Toennies, Phys. Rev. B **54**, 11989 (1996)
229. U. Paltzer, F.W. de Wette, U. Schroder, E. Rampf, Physica C **301**, 55 (1998)
230. M. Dion, H. Rydberg, E. Schroeder, D.C. Langreth, B.I. Lundquist, Phys. Rev. Lett. **92**, 246401 (2004)
231. K. Lee, E.D. Murray, L. Kong, B.I. Lundquist, D.C. Langreth, Phys. Rev. B **82**, 081101 (2010)
232. M. Wijzenbroek, D.M. Klein, B. Smits, M.F. Somers, G.-J. Kroes, J. Phys. Chem. A **119**, 12146 (2015)
233. P. Giannozzi et al., J. Phys.: Condens. Matter **21**, 395502 (2009)
234. X. Gonze et al., Comput. Phys. Commun. **180**, 2852 (2009)
235. G. Kresse, J. Furtmüller, Phys. Rev. B **54**, 11169 (1996)
236. S.J. Clark, M.D. Segall, C.J. Pickard, P.J. Probert, K. Refson, M.C. Payne, Z. Kristall. **220** (5–6), 567 (2005)

Chapter 6
The Atom-Surface Potential

> *... All things are made of atoms attracting each other when they are a little distance apart, but repelling upon being squeezed into one another. In that one sentence, there is an enormous amount of information about the world, ...*
>
> R. Feynmann, Lectures in Physics, Sects. 6.1–6.2

Abstract In this chapter the various models and approximations used for describing the static atom-surface potential are reviewed and analyzed. The potentials are of fundamental importance for the calculation of the elastic scattering of atoms from surfaces. These potentials also are the basis for the calculations of the inelastic scattering from surface phonons on insulator surfaces, discussed in detail in Chap. 7. In the case of metal and semiconductor surfaces the semi-empirical parameters used to describe the modifications of the potential resulting from the spill-out of the conduction electrons are discussed. The theory of inelastic surface phonon scattering from metals and semiconductors is the subject of Chap. 8.

6.1 Introduction

6.1.1 Atom-Surface Scattering Coordinates

As will be discussed in Chap. 7, the conventional theory of inelastic scattering is generally based on the Distorted-Wave Born Approximation (DWBA). In this theory the interaction potential between the atom and the surface enters in two ways. A *static potential* determines the predominant effect of elastic reflection from the surface. An additional potential term, which depends on the amplitude of the atomic vibrations at the surface, describes the inelastic interaction with the surface phonons, which is treated as a perturbation. The latter potential is frequently called the *dynamic* (or *inelastic*) *potential*. Both types of potentials will be surveyed in this

G. Benedek and J. P. Toennies, *Atomic Scale Dynamics at Surfaces*, Springer Series in Surface Sciences 63, https://doi.org/10.1007/978-3-662-56443-1_6

Chapter. More recent theories more appropriate for describing the interaction with conducting or highly polarizable surfaces will be discussed in Chap. 8.

Figure 6.1 shows a schematic diagram of the surface scattering process. An atom of mass m arrives at the surface with an incident wavevector $\mathbf{k}_i$ and kinetic energy $E_i = \hbar^2\mathbf{k}_i^2/2m$ and is scattered with a final wavevector $\mathbf{k}_f$ and energy E_f, with a change of wavevector given by $\Delta\mathbf{k} \equiv \mathbf{k}_f - \mathbf{k}_i$. As shown in Fig. 6.1 the momentary position of the atom $\mathbf{r} = (\mathbf{R}, z)$ is conveniently expressed in terms of coordinates in the surface plane (capital letters) and normal to it along the z-axis, likewise $\mathbf{k} = (\mathbf{K}, k_z)$. The atom is assumed to interact with all the atoms of the semi-infinite crystal via a total potential $V(\mathbf{r})$.

In general the structure of the crystal surface is not an orthogonal Bravais lattice like that of Fig. 6.1, but may be accounted for by a 2-dimensional glide vector $\mathbf{b}$, by which the lattice atom positions are $\mathbf{r}_l = (\mathbf{R}_L - l_3\mathbf{b}, z_{l_3})$, where $l_3 = 0$ is the surface plane and $l_3 = 1, 2\ldots$ denote the planes below the surface (plane index). If surface relaxation is neglected then $z_{l_3} = l_3 d$ with d the interlayer distance. For crystals with s atoms per unit cell, $\mathbf{r}_l$ is the position of the cell and a set of vectors $\mathbf{d}_\kappa$ ($\kappa = 1, 2\ldots s$) define the positions of the atoms in each cell with respect to its position. Thus the atom positions are $\mathbf{r}_{l\kappa} = \mathbf{r}_l + \mathbf{d}_\kappa$, consistent with the definition given in (3.25). The cell position is often chosen as that of one of its atoms, e.g., $\kappa = 1$, so that $\mathbf{d}_1 = 0$.

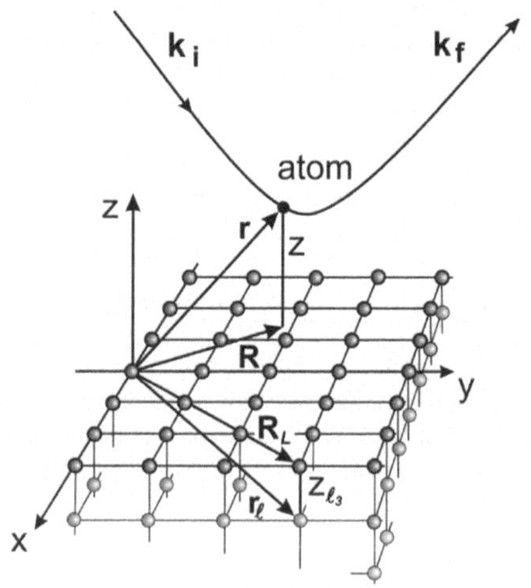

Fig. 6.1 Schematic diagram showing the coordinates used in surface scattering from a simple cubic lattice. $\mathbf{r}_l \equiv (\mathbf{R}_L, z_{l_3})$ denotes the equilibrium location of the l-th lattice atom, where l is the set of integer indices (l_1, l_2, l_3), $\mathbf{R}_L$ is its projection onto the surface layer, with $L = (l_1, l_2)$, and z_{l_3} is its component normal to the surface with l_3 the layer index. Here $\mathbf{r} \equiv (\mathbf{R}, z)$ is the instantaneous location of the scattering atom. The atom incident wavevector is $\mathbf{k}_i$ while the final wavevector is $\mathbf{k}_f$. The interaction of the scattering atom with the atoms of the surface is described by the total potential $V(\mathbf{r})$

The various possible elementary scattering processes are briefly introduced in Chap. 1. In the following discussion some concepts in describing the overall corrugation of the atom-surface potentials of insulators are presented. Next the models used to describe the attractive and repulsive potentials in atom-atom interactions are reviewed. Section 6.2 describes the models for the long range attractive atom-surface potential. Sections 6.3, 6.4, and 6.5 are devoted to models for the total atom-surface potential in increasing order of sophistication. The final Sect. 6.4 discusses how metal surfaces differ from the insulator surfaces. The corresponding experimental apparatus and scattering phenomena are discussed in Chaps. 9 and 10 respectively.

6.1.2 The Corrugation of the Atom-Surface Potential

By far the most probable scattering channel is elastic and is determined by the time average of the total potential $\langle V(\mathbf{r}, t)\rangle$ over the relatively small lattice vibrations. The time averaged potential invariably has the periodicity of the static surface and is only slightly smeared out with respect to the static surface potential $V(\mathbf{r})$, because of the small vibrational amplitudes (Table 1.1). Thus the potential is largely determined by the lattice atoms fixed at their equilibrium positions at a given temperature. The effect of the temperature induced motional smearing on the elastic scattering amplitudes is customarily accounted for by the Debye-Waller factor, which is introduced in Chap. 7 and discussed in detail in Sects. 7.6.3 and 8.4.3.

The interaction of a closed-shell atom like He or any other rare-gas atom, with the surface consists of a short-range repulsive term $V_{rep}(\mathbf{r})$ and of a long-range attractive term $V_{att}(\mathbf{r})$. The former arises from the Pauli repulsion of the electrons of the probe atom with the electrons of the surface; the latter is due to dispersion forces and to induction forces resulting from the electrostatic polarization of the atom induced by the surface charge distribution:

$$V(\mathbf{r}) = V_{rep}(\mathbf{r}) + V_{att}(\mathbf{r}). \tag{6.1}$$

The equipotential surface at which the potential $V(\mathbf{r}) \equiv V(\mathbf{R}, z)$ equals the He atom incident energy for the vertical component of motion $E_{iz} = \hbar^2 k_{iz}^2/2m$ is obtained by solving

$$V(\mathbf{R}, z) = E_{iz}. \tag{6.2}$$

The resulting surface $z = z(\mathbf{R}, E_{iz})$ defines the closest approach of the atom to the surface, i.e., the locus of the classical turning points for the given surface potential and vertical incident energy. Figure 6.2a shows a schematic diagram with the profile of a few equipotential surfaces mostly in the well region, and Fig. 6.2b the typical z-dependence of $V(\mathbf{R}, z)$ for two different lateral positions $\mathbf{R}$.

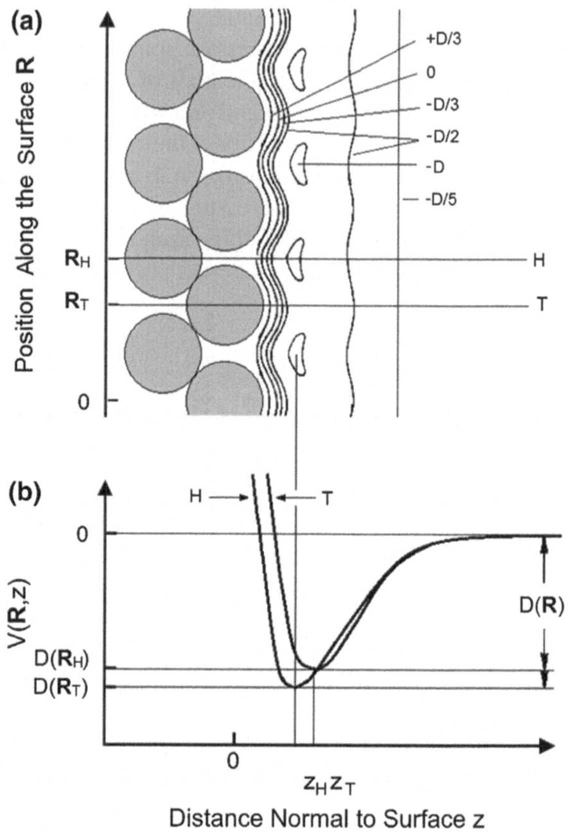

Fig. 6.2 Schematic diagrams showing a typical interaction potential of a neutral atom with a solid surface. **a** Equipotential contour lines for positive, zero, and negative energies given in fractions of the potential depth D. **b** Potential as a function of z for two different lateral positions H (hollow position) and T (top position) [1]

The two-dimensional periodicity of the surface, makes it possible to express the static potential in terms of a Fourier representation

$$V(\mathbf{r}) \equiv V(\mathbf{R}, z) = \sum_{\mathbf{G}} V_{\mathbf{G}}(z) e^{i\mathbf{G}\cdot\mathbf{R}}, \tag{6.3}$$

where the sum runs over the surface reciprocal lattice vectors **G**. The 2-dimensional Fourier transforms $V_{\mathbf{G}}(z)$ of the potential appearing in (6.3) is an important quantity which will be encountered often in the following discussion. In general the elastic diffraction amplitudes from the static surface potential [2] are proportional to the matrix elements of the Fourier components $V_{\mathbf{G}}(z)$ between the final- and initial-translational states of the atom in the z-direction (see Chap. 7). The Fourier components are related to the potential by

$$V_{\mathbf{G}}(z) = \frac{1}{a_c} \int_{cell} d^2 R e^{-i\mathbf{G}\cdot\mathbf{R}} V(\mathbf{r}), \tag{6.4}$$

where the integral extends over the surface unit cell of area a_c.

The $\mathbf{G} = 0$ component in (6.4), denoted $V_{00}(z)$, describes the flat *laterally-averaged* surface potential. The $\mathbf{G} \neq 0$ components describe the periodic undulation of the potential along the surface and are used to account for the static surface corrugation. This is defined by the function

$$\zeta(\mathbf{R}, E_{iz}) = z(\mathbf{R}, E_{iz}) - \bar{z}(E_{iz}), \tag{6.5}$$

where $z(\mathbf{R}, E_{iz})$ is obtained from (6.2) and $\bar{z}(E_{iz})$ is the laterally-averaged turning-point distance, which is given by

$$\bar{z}(E_{iz}) = \frac{1}{a_c} \int_{cell} d^2R\, z(\mathbf{R}, E_{iz}). \tag{6.6}$$

In the semi-classical scattering theory, known as the *eikonal approximation* [3], the diffraction amplitudes are proportional to the $\mathbf{G} \neq 0$ Fourier components of the corrugation function

$$\zeta_{\mathbf{G}}(E_{iz}) = \frac{1}{a_c} \int_{cell} d^2R^{-i\mathbf{G}\cdot\mathbf{R}} \zeta(\mathbf{R}, E_{iz}), \tag{6.7}$$

where the integral is over the surface unit cell. At thermal incident energies the atoms are largely reflected by the short-range repulsive wall, which determines the surface corrugation, whereas the attractive part is practically flat. The main effect of the long range attractive part is to accelerate the incoming particle and decelerate it on its way out from the surface (*Beeby correction* [4]). As will become clear from the discussion of the inelastic scattering from surface phonons in Chap. 7, the repulsive potential is also largely responsible for the phonon excitation processes.

The shape and the amplitude of the corrugation function depend directly on the nature of the interaction of the atom with the solid. The explicit dependence of $\zeta_{\mathbf{G}}(E_{iz})$ on the incident energy indicates that the corrugation is however not entirely an intrinsic property of the surface.

The corrugation amplitude, defined as

$$\zeta_0 \equiv \frac{1}{2}[\max\zeta(\mathbf{R}, E_{iz}) - \min\zeta(\mathbf{R}, E_{iz})] \tag{6.8}$$

depends on the vertical component of the incident energy, though weakly, due to the steepness of the repulsive potential. It also depends on the size and nature of the scattering atom or molecule, and on the surface temperature. In general at higher surface temperatures, the vibrations will tend to smooth the surface and reduce the corrugation amplitude. Values of the surface corrugation amplitude as measured by He or hydrogen molecule diffraction amplitudes for various insulator and metal surfaces are collected in Table 6.1. Insulator surfaces generally have large corrugation amplitudes as compared to the metal surfaces. Oxides are however less corrugated than alkali halides since the additional electrons of the O^{2-} ion are less

Table 6.1 Experimental corrugation amplitudes[a] in Å from He atom and hydrogen molecule diffraction experiments. The data are mostly from the analysis of diffraction peak intensities[b]. For some surfaces values for two different directions are given in the respective order:

Surface and Azimuth	He	H_2/D_2
LiF(001)	0.325[c]	0.17[d]
NaF(001)	0.289[e]	–
NaCl(001)	0.34[f]	–
NiO(001)	0.14[g]	0.25[h]
MgO(001)	0.18[i]	–
AgBr(001)	0.16[q]	–
Graphite	0.21[j]	–
Ag(111)	0.011[k]	0.044[k]
Ag(110) [001]	0.13[l]	–
Cu(001)	< 0.005[m]	0.024[m]
Cu(110) [001], [1 $\bar{1}$ 0]	0.10[n], -	0.108[o], 0.019[p]
Ni(110)	0.08[r]	–
Ni(111)	0.02[s]	0.16[s]
Pd(110) [001], [1 $\bar{1}$ 0]	0.21[t], 0.02[u]	–, –
Rh(110) [001], [110]	0.125[v], 0.066[v]	0.166[v], 0.131[v]
Bi(111) [E_i = 17 meV]	0.726[w]; 0.554 [123 K], 0.581 [300 K][x]; 0.75[y]	–

[a]The corrugation amplitude ξ_0 is defined by (6.8)
[b]Most of the data are for $Ei \cong 60$ meV. For Bi(111) (E_i = 17 meV) data from [24] are given for two different surface temperatures.
[c][5], [d][6], [e][7], [f][8], [g][9], [h][10], [i][11], [j][12], [k][13], [l][14], [m][15], [n][16], [o][13], [p][15], [q][17], [r][18], [s][19], [t][20], [u][21], [v][22], [w][23], [x][24], [y][25]

localized at the surface (the free O^{2-} is unstable), thus contributing some smoothing of the corrugation. On the contrary semimetals like bismuth exhibit a larger corrugation due to the concentration of Fermi level electrons (and/or holes) within pockets in the reciprocal space around zone-boundary symmetry points.

6.1.3 Additive Potentials

The static potential $V(\mathbf{r})$ can, to a good approximation, be reduced to a sum of direct two-body interatomic potentials when the solid is also a closed-shell system, like the rare-gas or molecular solids or ionic crystals. In all the other cases (semiconductors, metals, etc.) the interaction of the atom with the surface is to a large extent mediated by the bonding or conduction electrons and acquires a many-body character (see Sect. 6.6). Nevertheless, for HAS surface crystallography and inelastic scattering it is essential to know the dependence of $V(\mathbf{r})$ on the surface atomic positions $\{\mathbf{R}_i\}$, even in the case of metals as a first approximation. For this purpose it is customary to approximate the total atom-surface potential $V(\mathbf{r})$ by a sum over *effective* two-body potentials $v(\mathbf{r}-\mathbf{r}_l)$ between the scattering atom and the atoms of the solid at their equilibrium positions $\mathbf{r}_l$:

$$V(\mathbf{r}) = \sum\nolimits_l v(\mathbf{r} - \mathbf{r}_l). \tag{6.9}$$

In general the largest contribution to (6.9) comes from the atoms of the top surface layer ($l_3 = 0$), so that the contributions from deeper layers are frequently neglected. In this case, and if the first layer positions $\mathbf{R}_L$ form a Bravais lattice, then it can be shown that the total potential $V_{\mathbf{G}}(z)$ equals the two-dimensional Fourier transform $v_{\mathbf{G}}(z)$ of the two-body potential $v(\mathbf{r})$:

$$\begin{aligned} V_{\mathbf{G}}(z) &= \frac{1}{A}\int d^2R\, \mathrm{e}^{-i\mathbf{G}\cdot\mathbf{R}} V(\mathbf{r}) \\ &= \frac{1}{a_c}\int d^2R\, \mathrm{e}^{-i\mathbf{G}\cdot\mathbf{R}} v(\mathbf{r}) \equiv v_{\mathbf{G}}(z), \end{aligned} \tag{6.10}$$

where both integrals extend over the surface area A.

The equivalence in (6.10), which may seem surprising at first glance, holds for all the discrete wavevectors corresponding to surface reciprocal vectors $\mathbf{G}$. For any wavevector $\mathbf{K}$ different from a $\mathbf{G}$-vector the Fourier transform $V_{\mathbf{K}}(z)$ vanishes as a result of the averaging over the two-dimensional periodicity of $V(\mathbf{r})$. For the potential between the scattering atom and a single surface atom $v(\mathbf{r})$, on the other hand, its two-dimensional Fourier component $v_{\mathbf{K}}(z)$ is in general different from zero for any given $\mathbf{K}$. Therefore there is an infinite number of possible forms of the effective two-body potential $v(\mathbf{r})$ which can sum up to a given total surface potential $V(\mathbf{r})$, because only the Fourier components $v_{\mathbf{K}}(z)$ for $\mathbf{K}$ equal to a surface reciprocal lattice vector $\mathbf{G}$ concur to give $V(\mathbf{r})$. This result can be likened to diffraction where the sum over a discrete periodic set of lattice sites leads to the cancellation of the Fourier components for all $\mathbf{K}$ which are not a $\mathbf{G}$-vector. In other words the diffraction amplitudes do, in fact, not provide enough information for reconstructing a unique two-body potential between the probe atom and the surface atoms. There is however a positive practical aspect in the above arbitrariness: even an atom-surface potential $V(\mathbf{r},\{\mathbf{r}_l\})$ with a strong many-body character, as is the case, e.g., for a free-electron metal surface, can be expressed, thanks to its periodicity, in the form of (6.9) as a sum of well-chosen *effective* two-body potentials in many different possible ways. It should be noted that for metals restricting the lattice sum of (6.9) to the first layer may not be justified since the interatomic forces mediated by conduction electrons can have a long range. The inclusion of many layers in the lattice sum leads however to the following simple extension of (6.10):

$$V_{\mathbf{G}}(z) = \sum\nolimits_{l_3} v_{\mathbf{G}}(z - z_{l_3})\, e^{il_3\mathbf{b}\cdot\mathbf{G}}, \tag{6.11}$$

where $\mathbf{b}$ is a glide vector.

6.1.4 Two-Body Atom-Atom Potentials

In the case of an insulator surface or a surface consisting of closed shell atoms and molecules the interaction with the surface can be accurately described by a sum of two-body potentials $v(\mathbf{r}-\mathbf{r}_l)$ between the He atom and the closed shell atoms of surface. The appropriate potentials are assumed to be the same as the corresponding gas phase two-body potentials, for which an extensive literature exists [26, 27]. Since at the present time even the gas phase interatomic potentials for most systems are not known with sufficient accuracy, the effective two-body potentials must be estimated by some type of approximate model [28–33].

The frequently used two-body potential models assume simple analytical forms for the attractive and repulsive parts, which are partly based on quantum mechanical arguments. The most important surface-atom attractive forces originate from interatomic London dispersion (van der Waals) potentials [34]. They are expressed (neglecting retardation) for non-overlapping atoms by even powers $\propto r^{-2(\ell+\ell'+1)}$ with $\ell, \ell' = 1, 2, 3\ldots$ labeling the orders of the interacting *dynamic* multipoles [34, 35]. For two ground state atoms they are given by

$$V_{disp}(r) = -\sum_{n=3}^{\infty} \frac{C_{2n}}{r^{2n}}, \quad n = l + l' + 1. \tag{6.12}$$

The dispersion coefficients C_{2n} can be accurately calculated from the optical properties of the separate atoms and are known for a large number of systems [37, 38, 67]. The leading dipole-dipole coefficient C_6 depends on the dielectric response of the interacting atoms through the Casimir-Polder equation [32, 33, 35, 39]. For the scattering of a He atom from a surface atom,

$$C_6 = \frac{3\hbar}{\pi} \int_0^{\infty} d\omega \, \alpha_{He}(i\omega)\alpha(i\omega), \tag{6.13}$$

where $\alpha_{He}(i\omega)$ and $\alpha(i\omega)$ are the frequency-dependent dipolar polarizabilities of He and one of the surface atoms, respectively, evaluated along the imaginary axis. Similar expressions hold, up to a numerical factor, for the higher-order coefficients C_{2n}, with $\alpha_{He}(i\omega)$ or/and $\alpha(i\omega)$ replaced by multipolar polarizabilities [32, 33, 35, 39]. Recently it has been demonstrated that reasonably accurate dispersion coefficients can also be obtained by the density functional method [40–43, 44]. The theoretical dispersion coefficients are now considered to be very accurate for a large number of ground state atoms, some ions and excited atoms, and have been recently compiled [45].

In ionic crystals the surface atoms, besides the ionic charge, may have *static* multipoles due to the local gradients of the Madelung electric field, which also act on the probe atom via similar attractive power-law induction potentials [46, 47]. Also the two-body electrostatic potentials between the crystal ion charges and induced dipole of the atom are proportional to an inverse power of the distance, $\propto r^{-4}$. However, due to global neutrality, 2D periodicity and the Laplace equation, the sum over all positive and negative ions yields a total surface potential which decays exponentialy with distance and can normally be neglected.

For the Pauli short-range repulsion between the probe and the surface atom the simple exponential form has been found to be a very good approximation, consistent with quantum-mechanical calculations. A pure exponential

$$V_{BM}(r) = B\mathrm{e}^{-\beta r}, \tag{6.14}$$

referred to here as a Born-Mayer repulsion, or the screened Coulomb potential generated by a charge of

$$V_{SC}(r) = q\mathrm{e}^{-\beta r}/r, \tag{6.15}$$

B, β (*hardness parameter*) and q are phenomenological parameters, (6.15) is also known as Yukawa potential, which provides the interaction between two particles via the exchange of a massive boson (β is proportional to the boson mass and is zero for the bare Coulomb potential, which is due to the exchange of massless photons). For the Coulomb potential of a charge screened by a free electron gas of density n and chemical potential μ the constant β coincides where the Thomas-Fermi screening wavevector $k_0 = (4\pi\mathrm{e}^2\partial n/\partial\mu)^{1/2}$.

Born and Mayer in 1932 obtained a good description of the cohesive energy and the dynamical properties of ionic crystals [48–51] by combining the exponential repulsion, (6.14), with the electrostatic and dispersion attractive terms expressed by power-laws. Actually the purely repulsive atom-atom potential was introduced in the same year in the first quantum theories of atom scattering by Zener [52] and Jackson and Mott [53]. For the interaction with surfaces the phenomenological constants B and β can be calculated using Hartree-Fock self-consistent field (SCF) theory [32] or estimated from the calculated charge density via the Esbjerg-Nørskov approximation [36] (see Sect. 6.6)

In dealing with two-body phenomenological potentials a substantial algebraic simplification is obtained by combining a power-law r^{-m} for the van der Waals attraction with a steeper power-law r^{-n} for the repulsion. The first potential of this kind, using integer exponents $n > m$, was introduced by Gustav Mie in 1903 [54]. The most popular choice, with $m = 6$ representing the lowest order (dipole-dipole) London dispersion forces and $n = 12$ chosen for algebraic convenience, is the so called *12–6 Lennard-Jones* potential [55, 56], discussed in connection with atom-surface potential in Sect. 6.3. One of the advantages of this choice is that the bound states for the Lennard-Jones atom-surface potential can be calculated analytically within the Wentzel-Kramers-Brillouin (WKB) approximation (see Sect. 10.1).

A combination of two exponential forms for both the repulsive and attractive two body potentials, known as the *Morse potential* [57, 58], is also frequently used in the literature due to its algebraic simplicity. The Morse potential, however, does not treat correctly the long range attraction (see Sect. 6.4.2) and is not so appropriate to the study of inelastic HAS resonance and focusing effects determined by the He-surface bound states (see Sect. 10.2).

A particularly successful model of atom-atom interactions which has been applied to calculating the atom-surface potentials is the so-called Tang-Toennies atom-atom potential. This model combines a simple exponential repulsion with the long range attractive dispersion given by the usual London power-law expansion, corrected, however, by damping functions $f_{2n}(\beta r)$. These account for the overlap effects at the intermediate distances. This model and its application to describe the atom-surface potential is discussed in the Sect. 6.5.

6.2 The Long Range Attractive Atom-Surface Potential

6.2.1 *Continuum Model*

At distances sufficiently larger than the lattice interatomic spacing, the corrugation of the surface potential has negligible effects and only the *laterally averaged* potential, equal to the **G** = 0 Fourier component of the total potential needs to be considered. Moreover at such distances only the long range attractive part of the potential survives and only its **G** = 0 component is important. In the case of conducting surfaces, where the major contribution to the dispersive forces comes from free electrons, also the short range repulsive potential is generally flat, even at distances comparable to those between surface atoms. In both cases the semi-infinite crystal lattice can be viewed as a continuum (Fig. 6.3), and the lattice sums over the interatomic two-body potentials (whether real, in case of insulators, or effective, in case of metals) can be approximated by 3D integrals. The comparison with the exact lattice sums provides information about the limits of validity of the continuum model.

In general, when z is much larger than the interlayer distance d and when only the laterally-averaged surface potential is relevant, the sum over the atomic layers

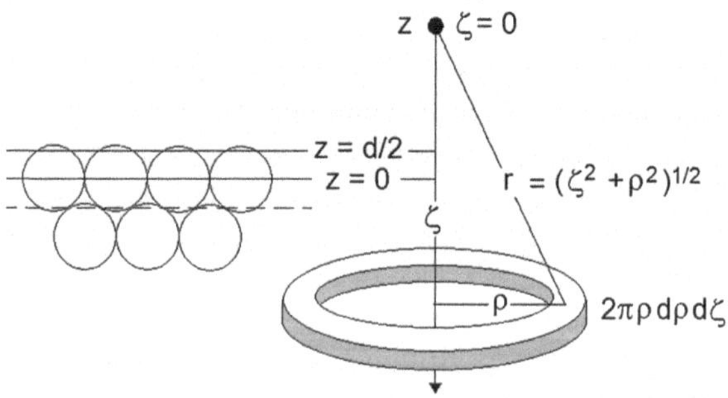

Fig. 6.3 Schematic diagram showing the continuum model. The interaction potentials of the He atom at position z with all the atoms (circles) of the semi-infinite lattice is modeled by an integral over the volume of a semi-infinite continuum. The infinitesimal ring-shaped integration element contains $2\pi n_0 \rho d\rho d\zeta$ atoms, with n_0 the atom density. In this model the continuum extends from $z = d/2$ ($\zeta = z - d/2$) to $z = -\infty$ ($\zeta = \infty$), while in the Celli model (Sect. 6.2.3) the first atomic layer (z = 0) is treated separately and the continuum extends from $\zeta = z + d/2$ (broken line) to $\zeta = \infty$

(see (6.11) for $\mathbf{G} = 0$) can be replaced by an integral over $l_3 d \equiv \zeta - z$, with l_3 a continuous variable. In any case the continuum approximation provides the correct asymptotic behavior of the long range attractive atom-surface potential. For this reason it is largely used for a straightforward analysis of the scattering data, and is therefore discussed next.

As shown by London already in 1930 [34, 62, 63], the leading term of the attractive dispersion potential between the probe atom and all the atoms of the semi-infinite crystal can be easily calculated in the continuum approximation by integrating the leading term of the atom-atom dispersion potential $-C_6/r^6$, (6.12). By assuming a homogeneous distribution of atoms with a number density n_0, it is found (cfr. Fig. 6.3 for the integration variables)

$$V^{(3)}_{disp,\,L}(z) = -2\pi n_0 C_6 \int\limits_{z-d/2}^{\infty} d\zeta \int\limits_{0}^{\infty} \frac{\rho d\rho}{\left(\rho^2 + \zeta^2\right)^3}, \tag{6.16}$$

where n_0 is the particle number density of the solid. The lower integration limit $z - d/2$ in (6.16) was introduced by London to account for the finite size of the surface atoms (Fig. 6.3). The integration yields

$$V^{(3)}_{disp,\,L}(z) = -\frac{c_3}{\left(z - d/2\right)^3}, \tag{6.17}$$

where

$$c_3 = \frac{\pi}{6} n_0 C_6. \tag{6.18}$$

As in the Casimir-Polder equation, (6.13), in which C_6 is expressed in terms of the complex dielectric functions of the two interacting atoms, c_3 can be expressed as well in terms of the dielectric response of the surface. For helium atom scattering c_3 is given by

$$c_3 = \frac{\hbar}{4\pi} \int\limits_{0}^{\infty} d\omega \frac{\varepsilon\left(i\omega\right) - 1}{\varepsilon\left(i\omega\right) + 1} \alpha_{He}(i\omega), \tag{6.19}$$

where $\varepsilon\left(i\omega\right)$ and $\alpha_{He}(i\omega)$ are the complex dielectric function of the semi-infinite solid and the frequency-dependent dipole polarizability of the He atom $\alpha_{He}(i\omega)$, respectively, both at imaginary frequencies [64]. This result, derived by Lifshitz in 1956 [65] and known as the Lifshitz formula, is universally applicable to insulators as well as to metals, as long as their dielectric function is known analytically on the complex plane [66]. In this derivation and throughout this Section the effects of retardation are neglected. The higher order multipolar terms in the atom-atom dispersion potential, (6.12), yield a corresponding extension of the atom-surface potential for a continuous solid [31]

$$V_{disp}^{(n)}(z) = -\frac{c_{2n-3}}{(z-d/2)^{2n-3}}, \quad n \geq 3, \tag{6.20}$$

with the coefficients

$$c_{2n-3} = \frac{\pi n_0}{(n-1)(2n-3)} C_{2n}. \tag{6.21}$$

In this way the coefficients c_{2n-3} can be obtained from the tabulated values of C_{2n} [37, 38, 45, 61]. Values for c_5 have also been reported for the insulators LiF, MgO and graphite using an extension of the Lifshitz formula to the dipole-quadrupole terms [68, 69].

In recent years much progress has been reported for the inclusion of the leading and higher-order dispersion terms in *ab initio* DFT calculations of atom-surface interaction [44, 70–77]. A thorough account of these recent achievements is found in a review by Berland et al. [78], as well as in recent papers by Tao and Rappe [79, 80] and references therein.

6.2.2 Layer Model

The continuum approximation fails to correctly represent the atom-surface potential when the distance from the surface becomes comparable to the interlayer spacing *d*. While averaging over the atomic distribution on each atomic layer (plane) is implicit in the **G** = 0 Fourier component of the potential, the sum over the discrete set of atomic layers each separated by a constant distance *d* has to be performed exactly in order to correctly describe the dispersion forces at short distances which are relevant, e.g., in inelastic atom scattering and bound state resonances. Luckily the corresponding series can be summed and expressed in terms of the Hurwitz (generalized Riemann) zeta functions $\zeta(n, z)$ [81]. The total atom-surface dispersion potential summed over all layers derived from (6.12) is [82]:

$$V_{disp}(z) = \sum_{n \geq 3} V_{disp}^{(2n-3)}(z), \tag{6.22a}$$

$$V_{disp}^{(2n-3)}(z) = -\frac{\pi n_0 C_{2n}}{d^{2n-3}(n-1)} \zeta\left(2n-2, \frac{z}{d}\right) \tag{6.22b}$$

$$= -\frac{c_{2n-3}}{z^{2n-3}}\left[1 + (2n-3)\frac{d}{2z} + o(z^{-2})\right], \quad z \to \infty \tag{6.22c}$$

$$= -(2n-3)\frac{c_{2n-3}d}{z^{2n-2}} + o(1), \quad z \to 0 \tag{6.22d}$$

The second term in the asymptotic expression for $z \to \infty$, (6.22c), contributes an even-power term z^{-2n+2} in the expansion, followed however by only odd-power terms. The same power z^{-2n+2} gives the divergence for $z \to 0$ (6.22d). Thus at short distance the potential diverges with a power which is increased by one unit, indicating, as expected, that the effect of the first plane overwhelms that of all the other subsurface planes. Note that for $d/2$ and any value of n the second term in the squared parenthesis of (6.22c) is exactly the second term in the expansion with respect to $d/2z$ of the London potentials, (6.17) and (6.20). Thus London's Ansatz, consisting in shifting the integration limit from z = 0 to z = $d/2$ in order to account for the finite size of atoms, describes correctly the first two terms of the asymptotic expansion at large distances of the dispersion forces. It is clear however that this is in general not enough to reasonably represent the potential at comparatively short distances like those of the bound states or the scattering turning point, where the inelastic interaction is strongest. At these short distances the deviations from the exact potential based on the Hurwitz zeta functions may amount to more than 50%.

6.2.3 *Celli Potential*

Celli [30, 83] devised a brilliant but simple remedy to this apparent limitation of the London potentials without dealing with special functions. In order to remove the unphysical divergence of the London potential at $z = d/2$, Celli suggested a dispersion potential where the contribution from the surface atomic layer is treated separately and taken equal to the first term of the asymptotic expansion for $z \to 0$, (6.22d), while the rest of the crystal is treated in the continuum approximation by integrating (6.16) from the second layer, i.e., from $z + d/2$ to ∞. The dipole-dipole Celli potential is then written as:

$$V^{(3)}_{disp,C}(z) = -\frac{c_3}{(z+d/2)^3} - \frac{3c_3 d}{z^4} \tag{6.23a}$$

$$= -\frac{c_3}{z^3} - \frac{3c_3 d}{2z^4} - \frac{3c_3 d^2}{2z^5} - o(z^{-6}), \quad z \to \infty \tag{6.23b}$$

In this way the Celli potential interpolates with a simple analytical form the exact short distance behaviour with that at long distance, so as to deviate almost imperceptibly from the exact potential, (6.22b) down to $z = 0$. Note that the residue $o(z^{-6})$ from the expansion of the London term brings in all powers of $1/z$, whereas the exact potential residue has only odd powers. The Celli potentials for higher multipoles can be defined in the same way:

$$V_{disp,C}^{(2n-3)}(z) = -\frac{c_{2n-3}}{(z+d/2)^{2n-3}} - (2n-3)\frac{c_{3n-3}d}{z^{2n-2}}. \tag{6.24}$$

Thanks to the excellent performance of the long range attractive Celli potentials at short and intermediate distances, they offer practical advantages in the calculation of scattering matrix elements, especially for processes involving bound state resonances. It should be considered however that at short distances the dispersion forces are perturbed by overlap, many-body and non-locality effects in the electronic response, especially in metals but also in insulators with highly-polarizable atoms (ions) [84]. In principle the Lifshitz formula is applicable to any type of solid, provided its dielectric response function is known.

6.2.4 *Zaremba-Kohn Potential*

In a seminal 1976 article Zaremba and Kohn (ZK) [84], in analyzing the effect of the spatial non-locality of the surface electronic response function along the lines of the previous jellium theory by Lang and Kohn [85], introduced the concept of a reference plane. In this way also the spill-out of the electronic charge density characteristic of metals is accounted for and, at the same time, the canonical z^{-3} dipole-dipole and the next term are condensed into an effective dipole-dipole form

$$V_{disp,ZK}^{(3)} = -\frac{c_3}{(z-z_0)^3} \cong -\frac{c_3}{z^3} - \frac{3c_3z_0}{z^4} - \ldots, \tag{6.25}$$

where the origin of the z-axis (called the *reference plane*) is shifted by z_0 above the top layer of atoms or, in the case of a jellium model, above the sharp edge of the positive charge density. The coefficient of the z^{-4} term from the asymptotic expansion of (6.25) $c_4 = 3z_0c_3$, derived by ZK from a quantum-mechanical calculation to second order in the perturbation theory, corrects the potential down to a sufficiently short distance so as to consistently reproduce the adsorption energies of Xe on the (111) surface of noble metals.

Table 6.2 compares values of $z_0 - d/2$ from the most recent theory for a number of metal surfaces with one half the layer spacing $d/2$. There it is seen that for the noble metals z_0 is only slightly larger than $d/2$ and that London's idea leading to (6.17) was indeed very farsighted. In the case of aluminum and the alkali metals z_0 is larger by between 20% and 50% than $d/2$ (depending on the surface) reflecting the comparatively greater electron spill-out in these free-electron metals. Another interesting consequence of the ZK theory is that for insulating crystals z_0 is essentially equal to $d/2$, in complete agreement with London's conjecture, (6.17).

It is important to note that the first two asymptotic terms of the Celli potential, (6.23b), coincide with those of the ZK potential after replacing $d/2$ with z_0. Thus the

Table 6.2 Location of the positive background edge *d*/2, where d is the layer spacing, for the jellium model and the relative location of the reference plane z_0 for the interaction of He and H_2 with metal surfaces

Surface	*d*/2 [Å]	He: $z_0 - d/2$ [Å]	H_2: $z_0 - d/2$ [Å]
Cu(001)	0.904	0.171[a]	0.284[c]
Cu(110)	0.635	0.171[a]	0.284[c]
Cu(111)	1.044	0.171[a]	0.284[c]
Ag(001)	1.022	0.160[a]	–
Ag(110)	0.723	0.160[a]	–
Ag(111)	1.180	0.160[a]	–
Au(001)	1.020	0.121[a]	–
Au(110)	0.721	0.121[a]	–
Au(111)	1.177	0.121[a]	–
Al(110)	0.715	0.38[b]	–
Na(110), (001)	1.519, 1.074[d]	0.31[b]	–
K(110), (001)	1.849, 1.307[d]	0.30[b]	–

[a][86] These values are identical to those reported in [87]
[b]Taken from Table 2.2 of [31]
[c][76]
[d]Unspecified surface for Na and K: d/2 is given for the two most compact surfaces

Celli potential provides also for metals a more physical description of the dispersion potential close to the surface, besides eliminating the $z = z_0$ divergence. In practice the atom–metal surface dispersion potential is best modelled by an effective Celli potential, where overlap, all many-body effects and non-locality effects are encompassed by the two constants c_3 and z_0. This is equivalent to assuming an effective pairwise additive potential between the probe and the semi-infinite crystal atoms, as already discussed in connection with (6.9). Several studies have addressed the effects of three-body interactions on the value of c_3 estimated for a number of metal surfaces using the pairwise additive approximation (see, e.g., [88, 89]).

Only a few calculations have been reported for the dipole-quadrupole dispersion coefficient c_5 for metal surfaces [68]. It is important to note that the z^{-5} term in the total dispersion potential includes the lowest-order term of the dipole-quadrupole interaction and the third order term of the dipole-dipole asymptotic expansion. Thus the effective coefficient for the z^{-5} term, is given by: $c_5^* = c_5 + 6c_3z_0^2$. Similarly the effective quadrupole-quadrupole and dipole-octupole terms will be expressed by a term $-c_7^*/z^7$ including the higher-order z^{-7} terms from the asymptotic expansions of dipole-dipole and dipole-quadrupole potentials, etc. Such small additions to the total dispersion potential, which only contribute at shorter distances, may well be overwhelmed by other effects.

6.2.5 Experimental Data on the Long Range Attractive Potential

On the experimental side most of the current information on the coefficient c_3 for insulators stems from the analysis of the bound state resonance energies derived from selective adsorption measurements [90]. Moreover several groups have succeeded in more directly measuring c_3 in elegant grating diffraction experiments. The first results were obtained for the rare gases and for H_2 and D_2 interacting with silicon nitride [59]. This work, together with more recent results for Li and Na atoms also interacting with gold surfaces, have been reviewed in [91]. The effect of the surface and retardation effects on c_3 coefficients is discussed in [92]. Tables 6.3 and 6.4 present a list of the c_3 coefficients for the interactions of He atoms and H_2 molecules for a number of representative insulator surfaces in addition to some semiconductor and metal surfaces. The data for the well depth D and minimum position z_m referring to the He atom (Table 6.3) and hydrogen molecule (Table 6.4) in the Lennard-Jones potential are presented in the next Section.

6.3 Lennard-Jones Atom-Surface Potential

The addition of a positive r^{-12} repulsive term to the atom-atom dipole-dipole dispersion potential discussed in the previous section, (6.12) for $n = 3$, leads to the popular Lennard-Jones (LJ) 12-6 potential [82, 109]

$$V_{LJ}(r) = \varepsilon\left[\left(\frac{r_m}{r}\right)^{12} - 2\left(\frac{r_m}{r}\right)^{6}\right]. \tag{6.26}$$

The choice of the exponent 12 for the repulsive part is dictated by algebraic simplicity. The potential well minimum is at $r = r_m$, the well depth is ε, and $C_6 = 2\varepsilon\, r_m^6$. The LJ laterally averaged ($\mathbf{G} = 0$) atom-surface potential can then be calculated from (6.22b) using the continuum method

$$V_{LJ,00}(z) = \frac{\pi}{5}\frac{r_m^2}{a_c}\varepsilon\left[\left(\frac{r_m}{d}\right)^{10}\zeta\left(10,\frac{z}{d}\right) - 5\left(\frac{r_m}{d}\right)^{4}\zeta\left(4,\frac{z}{d}\right)\right], \tag{6.27}$$

where $a_c = (n_0 d)^{-1}$ is the area of the surface unit cell. Retaining only the first term in the asymptotic expansion with respect to $1/z$ of both the repulsive and attractive potentials leads to

$$V_{LJ}(z) = \frac{\pi}{45} n_0 r_m^3 \varepsilon\left[\left(\frac{r_m}{z}\right)^{9} - 15\left(\frac{r_m}{z}\right)^{3}\right]. \tag{6.28}$$

Table 6.3 Dispersion coefficients c_3, the potential well depth D and location z_m for the static Lennard-Jones potentials of He atoms with insulator, semiconductor and metal single crystal surfaces

Surface	c_3 (meV Å^3)	D (meV)	z_m (Å)
Insulators			
LiF(001)	103[a]	8.38[b]	2.98–3.05[a, d]
MgO(001)	151[e]	12.5[f]	3.5–3.7[g]
NaCl(001)	106[e]	6.1[e]	3.49[e]
NaF(001)	73	7.58[e]	3.00[e]
NiO(001)	220 ± 110	10.1	–
Silicon nitride	100[h]	–	–
SiO_2(0001)	–	8.00[i]	2.98[i]
Xe(111)	–	5.625[j]	3.514[j]
Semiconductors and semimetals			
GaAs(110)	173 ± 8[e]	4.2[e], 4.4–8.7[k]	4.0–5.0[k]
Ge	178	≈6.2[l]	–
Graphite	180 ± 15[e]	16.6 ± 0.4[e]	2.65[e]
Si(111)–(IXI)H	–	7.5[m]	–
Metals			
Ag(110)	249[e]	6.0[e], 5.94[n]	3.44[n]
Ag(111)	249[e]	6–7, 6.97[n] 7.5[q]	3.57[n]
Al(110)	209 ± 9[e]	5.2	4.47
Al(001)	–	7.0	–
Au(111)	250[p]	7.67	
Cu(001)	–	6.5[o]	3.68[o]
Cu(110)	235[e]	6.27 ± 0.8[e], 5.79[n] 6.05[q]	3.65[e], 3.28[n]
Cu(111)	234[e]	6.63[n], 5.75[o]	3.41[n], 4.02
K(111)	70	0.6	5.09
Mo(110)	–	3.7	3.86
Na(111)	92	0.9	–
Ni(100)	–	6.5[o]	3.68[o]
Ni(110)	218	4.2	–
Ni(111)	–	5.75[o]	4.01[o]
Pd(110)	211	8.05	–
Pt(111)	251	5.9 ± 3.0	–
W(110)	265	3.5	

Values without references are from [90]
[a][93], [b][94], [c]Average of values in [90], [d][95], [e]"Best Estimate" from [90], [f][96], [g][74], [h][97], [i][98], [j][99], [k][100], [l][101], [m][102], [n][103], [o][104], [p][105], [q][106]

Table 6.4 Same as Table 6.3 but for the static Lennard-Jones potentials of H_2 molecules[a]

Surface	c_3 (meV ·Å^3)	D (meV)	z_m (Å)
LiF(001)	‖ 349[a] ⊥275[a]	32 (Li^+) 12(F^-)[a] 10 (Li^+) 32(F^-)[a]	–
MgO(001)	372	48	–
NaCl(001)	310	37.4	–
NiO	800 ± 190[b]	63.8[b]	–
Graphite (0001)	520 ± 30	51.7 ± 0.5	2.87
Silicon nitride	320 ± 80[c]	–	–
Ag(110)	714	31.7	3.26
Au(111)	770 ± 112	40.7[d]	2.24[d]
Al(110)	–	40.4[e]	3.46
Cu(001)	–	31.3[e]	3.26[e]
Cu(110)	689[f]	32.1[e]	2.97[e]
Cu(111)	673[f]	29.0[e]	3.52[e]

[a]‖ and ⊥denote the orientation of the H_2 bond with respect to the surface, Li^+ and F^- denote the ion on the surface [107]
[b]Average of values listed in [10]
[c][97], [d][86Har], [e][108], [f][16, 76]

This potential is frequently referred to as the Lennard-Jones (9-3) potential [110]. It is useful to note that the powers for an atomic potential (12-6) are reduced by as many units as the number of dimensions on which lattice sums are performed. The LJ potential of a 1D string of atoms would be (11-5), that of a single atomic surface (10-4), whereas the integration over a semi-infinite array of atomic planes leads to the above (9-3) potential. Dimensional scaling laws like this one are actually independent of the lattice order, and also hold for quasicrystalline and amorphous atomic arrays of any dimension.

For the atom-surface potential of (6.28) the position of the potential minimum z_m and the surface well depth D are, respectively, related to the two-body parameters by

$$z_m = 5^{-1/6} r_m, \qquad D = \frac{2}{9}\sqrt{5}\,\pi n_0 r_m^3 \varepsilon, \tag{6.29}$$

and allow to re-write the L-J (9-3) potential in the more useful form

$$V_{LJ}(z) = \frac{D}{2}\left[\left(\frac{z_m}{z}\right)^9 - 3\left(\frac{z_m}{z}\right)^3\right]. \tag{6.30}$$

When the detailed lattice structure of the semi-infinite solid is considered, the 2D Fourier components of the total L-J atom-surface potential can be calculated explicitly for any given lattice without resorting to integrations over the semi-infinite continuum [82, 109]. They are given by

$$V_{LJ,\mathrm{G}}(z) = \frac{\pi}{5}\frac{r_m^2}{a_c}\varepsilon \sum\nolimits_j S_{j\mathrm{G}} \left[\left(\frac{G r_m^2}{2z_j}\right)^5 K_5(G z_j) - 5 \left(\frac{G r_m^2}{2z_j}\right)^2 K_2(G z_j) \right], \tag{6.31}$$

where $z_j \equiv z + j\,d$ and $K_n(x)$ is the n-th order modified Bessel function of the second kind. $S_{j\mathbf{G}}$ is a *structure factor* defined by [111]

$$S_{j\mathrm{G}} = \sum\nolimits_{\kappa} \exp[-i\mathbf{G}\cdot(j\mathbf{a}_3 + \mathbf{d}_\kappa)]. \tag{6.32}$$

For the j-th layer the structure factor contains information on how the $\mathbf{G} \neq 0$ components depend on the crystal structure, i.e., on the actual positions $\mathbf{r}_{l\kappa} = l_1\mathbf{a}_1 + l_2\mathbf{a}_2 + j\mathbf{a}_3 + \mathbf{d}_\kappa$ of the atoms in the lattice.

At large distances from the surface only the second term, the attractive part, in the square brackets of (6.30) and (6.31), is important and can be expressed through the respective asymptotic forms [112]

$$\zeta(n+1, x) \approx x^{-n}/n,\ K_\nu(x) \approx (\pi/2x)^{1/2}\mathrm{e}^{-x},\ x \to \infty. \tag{6.33}$$

In this limit the $\mathbf{G} = 0$ term becomes, as expected,

$$V_{LJ,00}(z) \sim -c_3/z^3. \tag{6.34}$$

The $\mathbf{G} \neq 0$ terms of the attractive dispersion potential at large distances from the surface ($z \gg d$) decay exponentially and are therefore negligible. This reflects the fact that at long range the potential samples many surface atoms so that the resulting corrugation is averaged out. At short ranges the corrugation increases and therefore is largely determined by the repulsive part, which is the first term in the bracketed expression on (6.31).

6.4 Other Atom-Surface Potentials

The steep rise of the repulsive part of the potentials shown in Fig. 6.2 is characteristic of atom-solid surface interactions, especially for closed-shell atom-surface. In all cases it is due to the Pauli repulsion originating from the overlap of the outer He electrons with the electrons of the surface. Since the inelastic scattering processes depend essentially on the repulsive part of the surface potential and its corrugation, special care is needed in modeling the short-range interaction. Only in recent years could many important inelastic effects such as the excitation of the anomalous longitudinal resonances in metals or the quantum sonar effect (see Chap. 8) be unraveled through a full quantum-mechanical calculation of the short-range atom-surface interaction. However much understanding of inelastic processes and intensities can be gained on the basis of simple analytical models for the short-range repulsive potential. The simplest of these is the r^{-12} repulsive term of the LJ potential discussed in the previous Section; others are presented below.

6.4.1 Hard Corrugated Surface (HCS) Potential

Often there is need of some simple approximation for this part of the atom-surface potential. The simplest one is the *hard corrugated surface* (HCS) potential defined as [113]:

$$V^{HCS}(z) = \begin{cases} \infty, & z \leq \xi(\mathbf{R}) \\ -D, & \xi(\mathbf{R}) < z < z_1 \\ V_{disp}(z), & z \geq z_1 \end{cases} \tag{6.35}$$

where, as illustrated in Fig. 6.4, the long range attractive potential is connected to the repulsive barrier by a flat bottom located at the assumed well depth $D = -V_{disp}(z_1)$. The value of z_1 and the position of the undulated vertical barrier, determining the extension of the flat bottom, is chosen to provide the right number and best fit of the bound states as known from selective adsorption measurements. For metal surfaces where the ZK potential is used for $V_{disp}(z)$ it is implicit that $z_0 < \zeta(\mathbf{R})$. Despite its simplicity this potential model has been found to give good results in many calculations of elastic diffraction from insulators and other surfaces [114], as well as in the calculation of the inelastic scattering intensities (see next chapter).

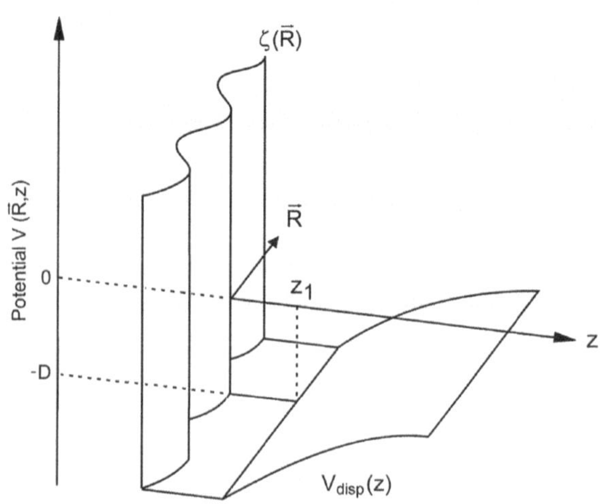

Fig. 6.4 Hard corrugated surface (HCS) model potential with an attractive well of depth D

From the point of view of quantum mechanics the exponential forms provide however a more realistic representation of the surface repulsive potential. Some examples of such *soft-wall potentials* are discussed in the next section.

6.4.2 Exponential Potential

The main justification for a power-law repulsion in the L-J potential is the algebraic simplification which allows to use as working parameters the equilibrium distance and the potential depth. However, as argued above the Born-Mayer potential, but also for the more realistic Tang-Toennies potential discussed below, quantum mechanical arguments suggest an exponential form for the short range repulsive potentials. It is worth noting that an algebraic simplification can also be obtained for an exponential repulsion by assuming also for the attractive part an exponential form (Morse potential [57, 58]).

Here it is convenient to discuss first the *screened Coulomb* potential, (6.15), as it encompasses the pure exponential as well as the bare Coulomb potentials as limiting cases. The Fourier components of the screened Coulomb potential for all values of **G** are given by [115]

$$\begin{aligned} V_{SC,\mathbf{G}}(z) &= \frac{2\pi q}{a_c}\sum\nolimits_j S_{j\mathbf{G}}\,\beta_{\mathbf{G}}^{-1}\,\mathrm{e}^{-jd\beta_{\mathbf{G}}}\,\mathrm{e}^{-\beta_{\mathbf{G}}z} \\ &= \frac{2\pi q}{a_c}\frac{\beta_{\mathbf{G}}^{-1}\mathrm{e}^{-\beta_{\mathbf{G}}z}}{1-\exp(-d\beta_{\mathbf{G}}+i\mathbf{G}\cdot\mathbf{a}_3)}\sum_{\kappa}\mathrm{e}^{i\mathbf{G}\cdot\mathbf{d}_{\kappa}} \end{aligned} \tag{6.36}$$

with $\beta_{\mathbf{G}} \equiv (\beta^2+G^2)^{1/2}$. With respect to the power-law potentials the exponential form has the advantage that the sum over the layer index j is straightforward. Note that the κ-th atom sublattice displaced by $\mathbf{d}_{\kappa}$ with respect to the conventional origin of the unit cell simply contributes a phase factor $\exp[i\mathbf{G}\cdot\mathbf{d}_{\kappa}]$. The $\mathbf{G} = 0$ component in the continuum limit ($d \to 0$ with $q/da_c \to n_p$, a constant with the meaning of screened charge density) is given by a pure exponential:

$$V_{sc,00}(z) = \frac{2\pi n_q}{\beta^2}\,\mathrm{e}^{-\beta z}. \tag{6.37}$$

Consider now a *bare Coulomb* potential produced by a semi-infinite ionic lattice with the rocksalt structure terminating at the (001) surface, made of *rigid* ions with effective ionic charges $q = \pm e^*$. By using the simple cubic 3D cell with four molecules (lattice constant $2r_0$, with r_0 the interionic distance), then summing the screened Coulomb potentials $V_{SC,\mathbf{G}}(z)$ for the eight sub-lattices, the G component of the bare Coulomb potential is readily obtained:

$$V_{coul,\mathbf{G}}(z) = \frac{2\pi e*}{r_0^2 G} \frac{e^{-Gz}}{1+e^{-Gr_0}} \sin\frac{r_0 G_x}{2} \sin\frac{r_0 G_y}{2}. \tag{6.38}$$

Note that the $\mathbf{G} = 0$ component vanishes because of charge neutrality, whereas the $\mathbf{G} \neq 0$ components are non-vanishing and decay exponentially in vacuum as exp($-Gz$) for increasing distance from the surface z. This result, which could be more directly obtained from the Laplace equation, $\nabla^2 V_{coul}(\mathbf{r}) = 0$, implies that a probe atom scattered from an ionic crystal surface is also subject to an attractive potential $-\alpha_{atom} E^2_{coul}(\mathbf{r})$, where α_{atom} is the probe atom polarizability and $\mathbf{E}_{coul}(\mathbf{r}) = -\nabla V_{coul}(\mathbf{r})$ is the static electric field. Due to the fast exponential decay of $\mathbf{E}_{coul}$, the static Coulomb potential becomes negligible at sufficiently large distances as compared to the attractive power-law dispersion terms, but it may be appreciable, especially in the case of selective adsorption, where the periodic lattice structure of the surface is distinctly perceived by the probe atom.

The *Born-Mayer* repulsion, (6.14), can also be related to the screened Coulomb potential through the derivative $V_{BM}(r) = -(B/q)\, \partial V_{SC}(r)/\partial\beta$. Since the $V_{SC,\mathbf{G}}(z)$ are linear combinations of the two-body potentials, the components $V_{BM,\mathbf{G}}(z)$ are readily obtained by taking the derivative of (6.36):

$$\begin{aligned} V_{BM,\mathbf{G}}(z) &= \frac{2\pi B}{a_c \beta^2_{\mathbf{G}}} \sum\nolimits_j S_{j\mathbf{G}} \beta(\beta^{-1}_{\mathbf{G}} + jd + z)\, e^{-jd\beta_{\mathbf{G}}}\, e^{-\beta_{\mathbf{G}} z} \\ &= \frac{2\pi B}{a_c \beta^2_{\mathbf{G}}} \beta \left(\frac{\beta^{-1}_{\mathbf{G}} + z}{1 - e^{i\,\mathbf{G}\cdot\mathbf{a}_3} e^{-d\beta_{\mathbf{G}}}} + \frac{d}{4} \mathrm{csch}^2 \frac{i\mathbf{G}\cdot\mathbf{a}_3 - d\beta_{\mathbf{G}}}{2} \right) e^{-\beta_{\mathbf{G}} z} \sum\nolimits_\kappa e^{i\,\mathbf{G}\cdot\mathbf{b}_\kappa}. \end{aligned} \tag{6.39}$$

The Born-Mayer potential for a flat surface in the continuum approximation can be obtained by replacing B with the potential energy density $b \equiv B/da_c$ and letting $d \to 0$ with $\mathbf{G} = 0$. In this limit (6.39) reduces to

$$V_{BM,00}(z) = \frac{2\pi b}{\beta^3} (1 + \beta z)\; e^{-\beta z}. \tag{6.40}$$

The same result can be directly obtained by integrating $V_{BM}(|\mathbf{r} - \mathbf{r}'|)$ with respect to **r'** over the semi-infinite continuum. Note that even in this limit the Born-Mayer surface potential is not a simple exponential, as frequently assumed, due to the factor $1 + \beta z$.

The **G**-components $V_{Morse,\mathbf{G}}(z)$ of the two-body *Morse* potential [57, 58]

$$V_{Morse}(r) = \varepsilon\, [e^{-\beta(r-r_m)} - 2e^{-\beta(r-r_m)/2}], \tag{6.41}$$

with ε the potential depth and r_m the two-body equilibrium distance, can be readily expressed in a closed form by means of (6.39) or, in the continuum limit, by means of (6.40). In the latter case the $\mathbf{G} = 0$ component is given by

$$V_{Morse,00}(z) = \frac{2\pi b_M}{\beta^3}[(1+\beta z)e^{-\beta(z-r_m)} - 8\,(2+\beta z)e^{-\beta(z-r_m)/2}], \tag{6.42}$$

with $b_M \equiv \varepsilon/da_c$ the potential energy density. This potential, as the pure Born-Mayer atom-surface potential, are often written without the terms proportional to βz [110]. Note however that the minimum of $V_{Morse,00}(z)$ is at $z_e = r_m - 2\beta^{-1}\ln 4$, thus a positive z_e requires $\beta r_m > 2\ln 4 = 2.77\ldots$. This means that already at a distance from the surface as short as the two-body equilibrium distance r_m the product βz is much larger than 1. Thus neglecting the βz terms should be avoided in dealing with atom-surface scattering, while it may be acceptable in the study of adsorption states if $z_e \ll r_m$. It is a general fact that $z_e < r_m$ because the short range repulsion jointly exerted by all the surface atoms on the probe atom is only slightly larger than the two-body repulsion between the probe atom and its surface neighbor, while the joint long range attraction of the surface atoms largely exceeds that exerted on the probe atom by a single surface neighbor. If the Born-Mayer repulsive parameter for closed shells β^{-1} = 0.34 Å [116] is used for the Morse potential, there is a substantial reduction of almost 1 Å of the equilibrium distance from the two-body to the atom-surface value. This may be even larger for the softer potentials of metal surfaces, where $\beta^{-1} \sim$ 0.5 Å, as discussed below in Sect. 6.6.

Despite of these limitations of the pure exponential form for the atom-surface potential, many simple and useful results, discussed in Sect. 6.6, can be obtained for metal surfaces. It may be useful to remark that a pure exponential atom-surface potential would be exactly derived from a Yukawa potential for the corresponding two-body potential.

6.5 Tang-Toennies Potential

All the atom-atom potential models discussed so far are based on the assumption that the interaction can be described by the *addition* of the repulsive and attractive potential terms. Since, however, the terms in the London power-law dispersion expansion, (6.12), are only valid for negligible overlap, they fail to properly describe the potential in the important well region where the repulsive potential becomes appreciable. An atom-atom potential function, which provides an accurate description in the intermediate overlap region and therefore covers the full range of interatomic distances, has been developed by Tang and Toennies [162, 117]. In this potential model, which provides simple, yet physically correct two-body van der Waals potentials between atoms, the repulsion is given by a simple exponential and the long range attractive dispersion terms are those of the usual London power-law expansion, corrected, however, by damping functions $f_{2n}(\beta r)$ which account for the overlap effects at intermediate distances. The atom-atom potential is then given by

$$\begin{aligned} V(r) &= V_{rep}(r) + V_{att}(r) \\ &= Be^{-\beta r} - \sum_{n\geq 3} f_{2n}(\beta r)\frac{C_{2n}}{r^{2n}}, \end{aligned} \tag{6.43}$$

where the C_{2n} are the theoretical coefficients in the expansion of the attractive dispersion potential, (6.12). The damping functions f_{2n} (βr) are defined as

$$f_{2n}(x) = 1 - \mathrm{e}^{-x}\sum_{k=0}^{2n}\frac{x^k}{k!}. \tag{6.44}$$

With this model it is not straightforward to calculate the pairwise additive interaction with a discrete lattice since the products $\mathrm{e}^{-\beta r}r^{-2n+k}$ appearing in (6.43) yield lattice sums which cannot be expressed by easy-to-handle functions. The $\mathbf{G} = 0$ Fourier component of the atom-surface potential can, however, be obtained to a good approximation in a closed form if the semi-infinite lattice is replaced by a continuous medium so that the lattice sums transform into tractable integrals [62]. More precisely, as done for the Celli attractive potential, the topmost atomic layer is treated separately as a two-dimensional continuum floating at a distance $d/2$ above a three-dimensional semi-infinite continuous medium representing the rest of the lattice. The following result is obtained:

$$\begin{aligned} V_{00}(z) = &\frac{2\pi B}{a_c\beta^2}(1+\beta z)\,\mathrm{e}^{-\beta z} + \frac{2\pi B}{a_c\beta^3 d}(2+\beta z_d)\mathrm{e}^{-\beta z_d} \\ &- \sum_{n\geq 3}\left[\frac{c_{2n-2}}{z^{2n-2}}F_{2n-2}(\beta z) + \frac{c_{2n-3}}{z_d^{2n-3}}F_{2n-3}(\beta z_d)\right], \end{aligned} \tag{6.45}$$

where $z_d \equiv z + d/2$. It is interesting to point out that when d becomes small, so that the entire lattice tends to a continuous medium, the Born-Mayer potential (6.39) for reduces, for $\mathbf{G} = 0$, to the first two terms of (6.45).

The new even-index atom-surface coefficients c_{2n-2} are defined, cfr. (6.24), by

$$c_{2n-2} = (2n-3)d\,c_{2n-3}. \tag{6.46}$$

The damping functions for the $\mathbf{G} = 0$ component of the surface potential, (6.45), are related to those for the two-body potential, (6.43), by

$$F_{2n-2}(x) = f_{2n-3}(x) - \frac{2n-2}{(2n)!}(2n+1+x)x^{2n-2}\mathrm{e}^{-x} \tag{6.47}$$

and

$$F_{2n-3}(x) = f_{2n-4}(x) - \frac{(2n-2)(2n-3)}{(2n)!}(2n+2+x)x^{2n-3}e^{-x}. \quad (6.48)$$

The Tang-Toennies potential model has also been used to calculate numerically the full atom-surface potential including the corrugation. Figure 6.5 illustrates the results for the extensively studied benchmark He-LiF surface [95], using a

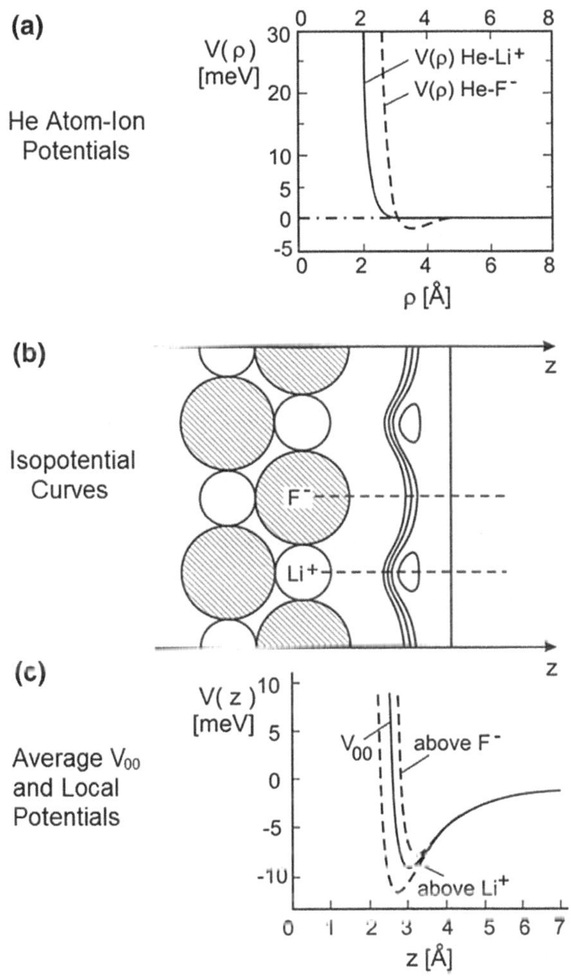

Fig. 6.5 The He-LiF (001) surface potential based on the Tang-Toennies model. **a** Two-body potentials between He–Li^{+} and He–F^{-} as a function of the distance between the nuclei *R*. **b** The geometry and some iso-potential curves for the interaction of the He atom with the surface from a summation of the atom-ion potentials in (**a**). **c** The solid line curve shows the $V_{00}(z)$ Fourier component of the potential between an He atom and the LiF surface on the same distance scale as in (**b**). The two dashed line curves show the local potentials above the F^{-} and Li^{+} ions. Adapted from [113]

procedure similar to that described above. Figure 6.5a shows the two-body He–Li^+ and He-F^- potentials calculated with (6.43) and (6.44). When summed over all the ions of the lattice following (6.9), the potential $V(\mathbf{R}, z)$ is found to be highly corrugated. This is clearly apparent from the equipotential lines depicted in Fig. 6.5b. Thus the Fourier expansion of $V(\mathbf{R}, z)$, as given by (6.3), contains a substantial number of components. The predicted lowest-order Fourier component $V_{00}(z)$ of the static potential, as given by (6.45), is shown by the solid line curve in Fig. 6.5c. Also shown in Fig. 6.5c are the local potentials above the F^- and Li^+ surface ions. The repulsive potential from the larger F^- ion is more extended outwards than for He–Li^+ (Fig. 6.5a) and thus makes the dominant contribution to both the elastic and the inelastic scattering. Using this potential model Celli, Eichenauer, Kaufhold and Toennies [95] calculated both the bound states and the diffraction intensities with inclusion of the effects of the bound states resonances.

A similar but more detailed calculation of the He-LiF(001) potential was subsequently reported by Fowler and Hutson [118]. These authors included many-body effects by calculating the repulsive potential using the self-consistent-field method with the individual ions perturbed by their surroundings. Many-body effects and higher order dispersion terms were also accounted for in the long range attractive potential. Subsequently Eichenauer and Toennies simplified the model of Fowler and Hutson to calculate the interaction of He atoms with the (001) surfaces of LiF, NaF, NaCl and LiCl [119].

Very precise measurements of selective adsorption resonances (Sect. 10.1) on LiF using the spin echo technique (see Sect. 9.4.3) have been reported for a wide range of incident beam energies [120]. The results were found to be in quite good agreement with the predictions based on the above model potentials. A detailed fit of these very extensive data led to a precise potential well depth of $D = 8.38$ meV, which is less than 5% smaller than the well depths predicted by the above calculations [94]. At the present time this is the most precise determination of any atom/molecule surface potential. Similar model potentials have also been reported for H_2 molecules interacting with LiF (001) [107].

This trend to full scattering simulations based on realistic two-body potentials model potentials continues due to the persistent difficulty in deriving reliable atom-surface potentials from first principle calculations. In 2001 Siber et al. reported a simulation of high resolution diffraction and surface phonon data of He atoms from the (111) surface of a thick slab of Xe atoms based on the latest Xe–Xe and He–Xe two- body potentials [99]. A general very good agreement was found with high resolution experimental data on diffraction intensities. However for inelastic scattering intensities even the latest two-body potential models appear to be unsuitable for surfaces with highly polarizable atoms like Xe due to important non-local many-body effects. In this case *ab initio* methods like the density functional perturbation theory are needed for a correct interpretation of experimental data [121]. This holds *a fortiori* for metal surfaces, due to the predominant many-body effects. Nevertheless many valuable results concerning the short-range repulsive potential, reviewed in the next Section, have been obtained by models and approximations inspired by *ab initio* quantum mechanical theories.

6.6 Metal Surfaces

6.6.1 *Smoothing of Metal Surfaces by Conduction Electrons*

The low-index metal surfaces, such as the (111) surfaces, especially in simple metals with *sp* conduction electrons, appear to be nearly perfectly flat mirrors for He atoms. Thus the diffraction intensities, typically are less than 10^{-3} of the specular peak intensities. As a result diffraction was only first observed in 1976 [122, 123] and previous to their first observation there was considerably speculation why the diffraction intensities were so weak. The smoothness is due to the substantial redistribution of the conduction electrons in the surface region, also called *Smoluchowski* smoothening [124], so that at the surface the distributions of the more tightly bound core electrons of the underlying ions are completely concealed [125]. Thus the effective interaction between the He atom and the surface ion cores is mostly mediated by the interposed electrons and acquires a substantial many-body character.

The delocalization of the interatomic forces is expected to be especially large for the low-index surfaces of free-electron metal surfaces such as aluminium. Figure 6.6 compares the results of an EM calculation of the equipotential lines for the repulsive part of the interaction of He with the low-index (111) surface of Cu with the rough surface of Al(110) along the [001] direction perpendicular to the close-packed rows [126]. Since for Cu(111) the well position is at 3.52 Å (Table 6.4) it is clear from Fig. 6.6a that the repulsive potential, which is proportional to the charge density, this far from the outer layer is very smooth. Surprisingly even along the [001] direction perpendicular to the rows of the (110) Al surface, which is expected to exhibit the largest corrugation among all the

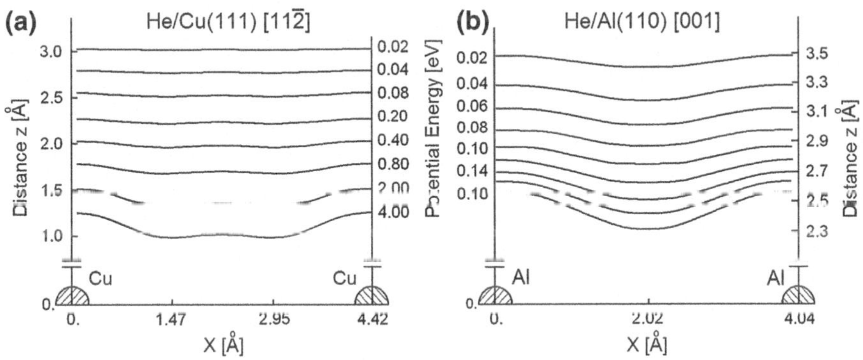

Fig. 6.6 Theoretically calculated equipotential lines for the repulsive part of the interaction of He atoms with two metal surfaces. **a** For the Cu(111) surface in the [11$\bar{2}$] direction. **b** For the Al(110) surface in the [001] direction perpendicular to the close packed rows. In the [1$\bar{1}$0] direction along the close-packed rows of the Al(110) surface the equipotential lines would appear to be completely flat on this scale, as for He/Cu(111) [126]

low-index surfaces, the effective corrugation at 20 meV, at about the well position at 3.46 Å (Table 6.4), is less than 0.1 Å.

The weaker localization of conduction electrons with respect to the core electrons also explains why the metal surfaces have considerably softer potentials for He atoms than closed-shell surfaces. This is an important difference since, as mentioned earlier, the rebound of the probe particle from the repulsive potential is largely responsible for the phonon excitation. This also holds for the effective two-body potentials used in the inelastic DWBA theory. Figure 6.7a compares a log-plot of the He-Cu effective repulsive two-body potential with that for He-F^- as derived from scattering data for the Cu(111) and LiF(001) surfaces [128]. Both potentials have the same typical Born-Mayer $e^{-\beta z}$ behaviour. The repulsive parameter for LiF(001) is $1/\beta = 0.250$ Å, whereas for He-Cu(111) it is about a factor 2 softer, with $1/\beta = 0.470$ Å. The Fourier components $V_K(z)$ of the interatomic potentials also decrease exponentially as also their slopes, expressed by the derivatives $dV_K(z)/dz$ (Fig. 6.7b). As a consequence of the greater softness of Cu, its slope at larger wavevectors K is also smaller than for F^- [128, 129]. The smoothness of the metal surfaces and the small diffraction intensities compared to the insulator surfaces have the additional effect that bound state resonances play hardly any role in the scattering from the low index metal surfaces.

The pairwise additive potential model has also been used to predict the interactions with metal surfaces. Some typical potential well depths for the He atom and H_2 molecules on metal surfaces are listed in Tables 6.3 and 6.4. For example, Eichenauer et al. used the same Tang-Toennies potential model to predict the potentials between an He atom and the Cu(110), Cu(111), Ag(110) as well as the

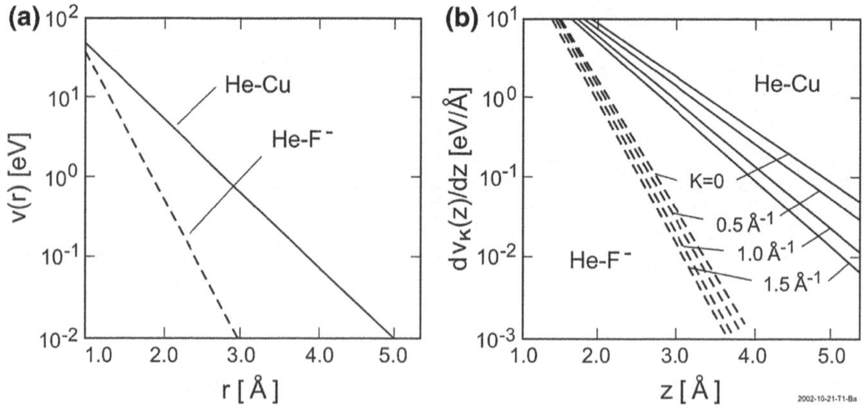

Fig. 6.7 Comparison of the He-F^- repulsive potentials of the LiF(001) surface with the He-Cu atom repulsive potentials of the Cu(111) surface. **a** Two-body Born-Mayer exponential potentials as a function of the interatomic distance. **b** The gradients of their Fourier components for different values of the wavevector K as functions of the distance from the surface. Both figures indicate that the interaction potential of the He atom with a closed shell ion (F^-) is more repulsive (stiffer) than with an open shell noble metal atom (Cu). Adapted from Bortolani et al. [128]

Ag(111) surfaces [103]. Without modifications, however, it was not possible to fit both the bound states and the diffraction intensities which were too large in comparison to the experiments. Using an Ansatz suggested by Takeda and Kohn [130] to account for the quasi-free electrons at the surface they could dampen the corrugation to make it consistent with the very small diffraction intensities. This illustrates once more the important difference between insulator and metal surfaces.

6.6.2 Atom-Metal Surface Potential in the Local Density Approximation

Different *ab initio* approaches have been used to calculate the potentials of atom-metal surface potentials, ranging from Hartree-Fock [131] to density functional [132, 133] methods up to the DFT treatments of dispersion forces [134, 135]. For non-insulators, for which the pairwise-additive model is inadequate, the repulsive as well as the long range attractive interactions with the surface can be derived from the electronic response function of the surface. For the short range repulsive interaction an effective two-body He-metal atom potential mediated by the conduction electrons derived in the framework of the DFT [136] can be written as

$$V_{eff}(\mathbf{r},\mathbf{r}_l) = \int d^3x \; V_{ps}(\mathbf{x},\mathbf{r}) \int d^3x' \chi(\mathbf{x},\mathbf{x}') \, V_l(\mathbf{x}'-\mathbf{r}_l) \tag{6.49}$$

where V_{ps} is the pseudopotential of the He atom at position $\mathbf{r}$ acting on a conduction electron at position $\mathbf{x}$, $V_l(\mathbf{x}'-\mathbf{r}_l)$ is the potential between a surface ion at $\mathbf{r}_l$ and an electron at $\mathbf{x}'$, and χ $(\mathbf{x}, \mathbf{x}')$ is the non-local electron susceptibility. In general $V_{eff}(\mathbf{r},\mathbf{r}_l)$ is not purely a function of the distance $\mathbf{r} - \mathbf{r}_l$ because χ $(\mathbf{x}, \mathbf{x}')$ is not just a function of $\mathbf{x}-\mathbf{x}'$ especially in the direction normal to the surface. However, setting $\mathbf{x} \equiv (\mathbf{X}, \zeta)$, for an electron gas homogeneous in two dimensions the translational symmetry along the surface implies the dependencies $\chi(\mathbf{X}-\mathbf{X}';\zeta,\zeta')$ and $V_{ps}(\mathbf{X}-\mathbf{R};\zeta,z)$ and therefore an effective two-body potential $V_{eff}(\mathbf{R}-\mathbf{R}_L;z,l_3d) = V_{eff}(\mathbf{r}-\mathbf{r}_l;l_3)$ with the specification of the layer which the lattice atom belongs to. Thus an effective two-body potential may be used, provided the (eventually small) difference between lattice atoms on different layers is taken into account.

Due to the short range nature of the pseudopotential $V_{ps}(\mathbf{x},\mathbf{r})$ it is often approximated by a delta function, $V_{ps}(\mathbf{x},\mathbf{r}) \cong A_n \delta\,(\mathbf{x}-\mathbf{r})$ (local density approximation (LDA)) where A_n is a semiempirical constant known as the Esbjerg and Nørskov constant [126]. By inserting this into (6.49), a simple expression for the atom metal surface interaction potential is obtained [126]:

$$V(\mathbf{r}) = \sum\nolimits_l V_{eff}(\mathbf{r}, \mathbf{r}_l) = A_n n(\mathbf{r}), \tag{6.50}$$

where

$$n(\mathbf{r}) = \sum\nolimits_l \int d^3x' \chi(\mathbf{r}, \mathbf{x}') \, V_l(\mathbf{x}' - \mathbf{r}_l). \tag{6.51}$$

Thus within LDA the atom-surface interaction potential $V(\mathbf{r})$ is proportional to the total electron density at the position $\mathbf{r}$. Since the original Esbjerg and Nørskov work suggesting A_n = 750 eV/a.u. [126], the correct value of A_n has been extensively discussed in the literature [30, 90, 137–139]. Actually a much smaller value, A_n = 364 eV/a.u., is considered to be more appropriate for most metal surfaces [90]. According to (6.50), He atoms at thermal energy (E_i = 60 meV) are already repelled by the comparatively weak electron density of about 10^{-4} $(\text{a.u.})^{-3}$ which occurs at distances of about 3 Å above the top plane of atoms [84, 140, 141]. The linearity between the charge density and the ion potentials $V_l(\mathbf{x}' - \mathbf{r}_l)$, (6.51), relies on the assumption of a potential-independent susceptibility. Any implicit dependence of $\chi(\mathbf{x}, \mathbf{x}')$ on the ion distribution, as expected with some degree of electron localization, would yield nonlinear corrections, for example three-body corrections to the effective two-body potential. Note that in the extreme localization limit, where the response is purely local, $\chi(\mathbf{x}, \mathbf{x}') \propto \delta(\mathbf{x} - \mathbf{x}')$, $V_l(\mathbf{x}' - \mathbf{r}_l)$ recovers a central character, which may be removed, however, by three-body terms due to non-locality effects (see below).

The important role of the Esbjerg-Nørskov potential defined by (6.50) in the formulation of the electron-phonon theory of inelastic scattering of atoms from conducting surfaces is discussed in Chap. 8.

6.6.3 Soft Corrugated Surface (SCS) Potential and Cut Off Effects

Already the pioneering group of Hoinkes, Nahr and Wilsch pointed out in 1972 that because of the similar size of an impinging He atom and that of the atoms on the surface the interaction is not restricted to a single surface atom but involves several atoms of the surface [142], with corresponding implications for the interaction potential (see Fig. 6.8) and the Debye-Waller factor. The same effect was elaborated on by Armand et al. [143] and sometimes is referred to as the *Armand effect* (see Sect. 2.3). Levi and Suhl [127] also discussed how this delocalisation effect would influence the inelastic scattering. Although the effect occurs on all surfaces including insulators it is especially important for metal surfaces because of the smoothness and hardness of the potential.

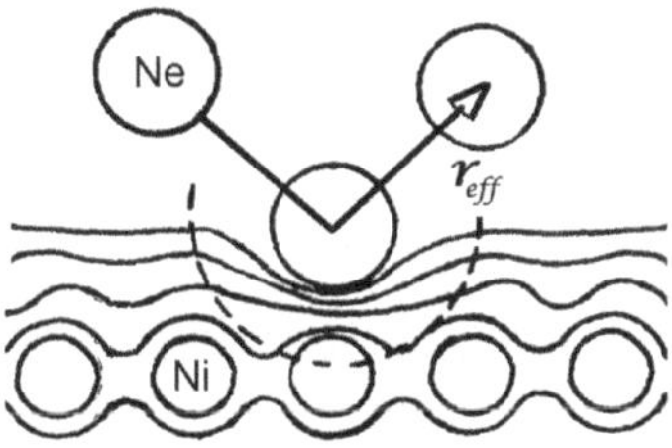

Fig. 6.8 Schematic diagram illustrating the cut-off effect in scattering from metal surfaces. In an inelastic collision the interposed conduction electrons transmit the energy to surface atoms over an extended area, which suppresses excitations at large wavevectors. Adapted from Feuerbacher and Willis [144]

A simple way of modeling the atom-metal surface potential, in analogy to the HCS potential for insulators (Fig. 6.4), is provided by a purely repulsive *soft corrugated surface* (SCS) potential. In view of the small probability of bound state resonances and the predominant importance of the repulsive potential in this model the attractive part is neglected altogether. In the SCS potential the interatomic two-body potential between the impinging atom and a surface atom at position $\mathbf{R}_l$ is factorized into a surface part and a z-dependent part

$$V(\mathbf{r}-\mathbf{r}_l) = A(\mathbf{R}-\mathbf{R}_l)\, e^{-\beta z}, \tag{6.52}$$

where $\mathbf{r}_l = (\mathbf{R}_l, 0)$ (see Fig. 6.1). The factor $A(\mathbf{R}-\mathbf{R}_l)$ in (6.52) defines the extent of delocalization of the interaction (*Armand effect*). Thus $A(\mathbf{R}-\mathbf{R}_l)$ extends from a surface atom at $\mathbf{R}_l$ a distance which is comparable to or larger than the classical scattering turning point from the surface atom layer ($\sim$3 Å at thermal incident energies). An especially convenient functional for $A(\mathbf{R}-\mathbf{R}_l)$ is a Gaussian:

$$A(\mathbf{R}) = A_0 Q_c^2 \exp(-Q_c^2 R^2/2), \tag{6.53}$$

where the adjustable parameter Q_c is called the cut-off wavevector. Equation (6.53) has the advantage that its two-dimensional Fourier transform $A_\mathbf{Q}$ is also a Gaussian:

$$A_Q = \frac{1}{2\pi}\int d^2R\, A(\mathbf{R})\, e^{-i\mathbf{Q}\cdot\mathbf{R}} = A_0 \exp(-Q^2/2Q_c^2). \tag{6.54}$$

The same separable form as (6.52), also leading to (6.54), can be derived from a pure exponential (Born-Mayer) two-body potential

$$\begin{aligned} V_{BM}(r) = A_0 Q_c^2 e^{-\beta r} &= A_0 Q_c^2 exp[-\beta\sqrt{R^2+z^2}] \\ &\simeq A_0 Q_c^2 e^{-\beta z} \exp[-(\beta/2z) R^2], \end{aligned} \tag{6.55}$$

where the second expression is obtained by expanding the exponent with respect to R/z. Since z is equal to or larger than the turning point distance $\bar{z}(\mathbf{R})$, the expansion is justified as long as R is smaller than the average turning point distance $\bar{z}$. Since the interaction is largely determined by the classical turning point, by setting

$$Q_c^2 = \beta/\bar{z}, \tag{6.56}$$

Equation (6.55) can be made approximately equal to (6.52) and (6.53).

Another way to incorporate the many-body nature of the potential is to account for the fact that the interaction between the probe atom (labelled by an index a) and some specific surface atom l can depend on the actual position of a neighboring atom l'. In this way the influence of atom l' on the electronic charge distributed between atom l and the probe atom is considered. Such three-body effects have been investigated for the case of Ne scattering from Ni(111) with the two-body interatomic potential model [145]

$$V(\mathbf{r}-\mathbf{r}_l) = A\sum_{l'} exp\left[-\frac{1}{2}\beta(r_{al}+r_{al'})-\alpha r_{ll'}^2\right], \tag{6.57}$$

where β is, as before, the inverse of the range of the two-body potential, α is the inverse squared three-body potential range, $r_{al} = |\mathbf{r}-\mathbf{r}_l|$, $r_{al'} = |(\mathbf{r}-\mathbf{r}_l)+(\mathbf{r}_l-\mathbf{r}_{l'})|$ and $r_{ll'} = |\mathbf{r}_l-\mathbf{r}_{l'}|$. The resulting effective three-body potential $V(\mathbf{r}-\mathbf{r}_l)$ has an apparent two-body central character because it depends only on the distance between the He atom at $\mathbf{r}$ and a surface atom at $\mathbf{r}_l$. However the coefficient multiplying the two-body central term $\exp(-\beta r_{al}/2)$ implicitly contains three-body effects through the dependence on the distances $\mathbf{r}_l - \mathbf{r}_{l'}$ between l and neighboring l' atoms. The term for $l = l'$ is independent of α and is simply an ordinary two-body exponential potential, which will be denoted V_2. The total effect of the three-body terms, given by the sum of all the other terms with $l \neq l'$ will be denoted V_3. The Gaussian factor $\exp(-\alpha r_{ll'}^2)$ in (6.57) expresses the rapid decay of the three-body interaction for increasing distance of the third atom l' from atom l. Since the range of the three-body terms is equal to α^{-1} they are dominant for $\alpha \to 0$. They become comparable to the two-body term when α is of the order of $\beta/r_{ll'}$ and vanish for $\alpha \to \infty$. In (6.57) the prefactor A can be chosen to be independent of $\mathbf{R}$ because the delocalization of $A(\mathbf{R}-\mathbf{R}')$ can now be accounted for by the three-body part, i.e., by the summation over the neighbour atoms l'.

The potential of (6.57) leads to the following more complicated expression for the cut off wavevector in terms of the potential constants and the average turning point distance $\bar{z}$ [145]:

$$Q_c^2 = \frac{\beta}{\bar{z}}\frac{8\,\alpha\bar{z}+\beta}{8\,\alpha\bar{z}+2\,\beta}. \tag{6.58}$$

The parameter α appearing in (6.58) can be used to estimate the effect of three-body forces on the classical turning point. In the extreme case $\alpha \to 0$ the turning point distance $\bar{z}$ decreases from the value for a pure two-body potential $((\alpha = \infty))$, to

$\bar{z} = \beta/2Q_c^2$. Thus the additional three-body forces allow the probe atom to come closer to the surface.

Inelastic phonon scattering of He atoms, when combined with a reliable theory, provides a tool for measuring the cut-off factor. Usually, however, the data are analyzed by assuming two-body forces only, and in this limit ($\alpha = \infty$) Q_c^2, given by (6.56), only depends on the potential parameters β and $\bar{z}$. These parameters have been determined empirically for many metal surfaces as well as for several insulators, such as LiF(001) and KCN(001) from an analysis of inelastic HAS intensities. They are reported in Table 6.5 for phonons measured for different surfaces along different directions after a compilation by Hofmann et al. [146]. Appreciable differences occur in the values of β and Q_c for different surfaces and directions of the same crystal. This apparent inconsistency suggests that the simple additive two-body potential model is not sufficiently realistic and a more refined analysis, including three body-forces, should be used on the basis of (6.57). In insulators the unscreened closed shell interaction yields, as expected and as shown by Bortolani et al. [147] for LiF(001), a significantly larger value of Q_c than in most metals (Table 6.5, last row). Indeed insulator surfaces are characterized by a strong corrugation which reinforces the scattering at large wavevectors beyond the first Brillouin zone, thus implying a larger cut-off wavevector Q_c, i.e., a more localized scattering center.

The effect of three-body terms on the surface corrugation of the model potential (6.57) can be seen by examining the second derivative of the potential $\partial^2 V(\mathbf{r}-\mathbf{R}_l)/\partial R^2$ calculated for $z = \bar{z}$ above one of the surface atoms at $\mathbf{R} = \mathbf{R}_l$. With only four nearest neighbors in a square surface lattice contributing to the three-body term, the second derivative is given by

Table 6.5 Survey of the experimentally determined potential softness parameters β, cut-off factors Q_c and the corresponding average turning point distances $\bar{z}$ [a] derived from an analysis of inelastic surface phonon scattering of He atoms from different metal and insulator surfaces in alphabetical order. Adapted from [146]

	Direction	β(Å^{-1})	Q_c(Å^{-1})	$\bar{z}$(Å)
Al(111)	⟨112⟩,⟨110⟩	4.0[c]	0.92[c]	4.73
Ag(001)	⟨100⟩	2.77[d]	0.87[d]	3.67
Ag(001)	⟨110⟩	4.12[d]	1.06[d]	3.67
Ag(111)	⟨112⟩,⟨110⟩	2.21[e]	0.74[e]	3.99[e]
Au(111)	⟨112⟩,⟨110⟩	2.12[e]	0.74[e]	4.15[e]
Cu(001)	⟨100⟩	3.00[b]	1.02[b]	2.86
Cu(001)	⟨110⟩	3.35[b]	1.08[b]	2.86
Cu(111)	⟨112⟩,⟨110⟩	2.08[e]	0.74[e]	3.84[e]
Ni(110)	⟨100⟩	2.92[f]	0.84[f]	4.14
Pt(111)	⟨112⟩,⟨110⟩	1.83[g]	0.57[g]	5.67
Rh(111)	⟨112⟩,⟨110⟩	3.15[j]	0.82[j]	4.68
LiF(001)	⟨100⟩	3.21[e]	1.22[e]	2.29[e]

[a]Unless explicit reference is given the average turning point distance $\bar{z}$ is derived from (6.38)
[b][148], [c][149], [d][150], [e][147], [f][151], [g][152], [h][153]

$$\frac{\partial^2 V(\mathbf{r} - \mathbf{R}_l)}{\partial R^2} \cong -\frac{\beta}{\bar{z}} V_2 + \frac{\beta}{4\bar{z}} (\beta \bar{z} - 9) V_3, \tag{6.59}$$

where V_2 and V_3 (both positive) are the two-body and three-body parts of $V(\mathbf{r} - \mathbf{R}_l)$, respectively. A negative $\partial^2 V/\partial R^2$ indicates a maximum of the potential at the top position, i.e., a normal corrugation, whereas a positive value indicates what is sometimes called *anti-corrugation* [154]. When $\beta \bar{z} > 9$, as happens in most of the cases shown in Table 6.5, a sufficiently large three-body potential can reverse the sign of $\partial^2 V/\partial R^2$ and lead to an anti-corrugation. Evidence for anti-corrugation has been obtained from an analysis of He atom diffraction intensities measured for certain transition metal surfaces. For the same surfaces in the case of Ne scattering a normal corrugation has been found [154]. This implies that the He atoms (unlike the more polarizable Ne) experience a stronger repulsion at the surface hollow or bridge sites than on the top sizes. An explanation comes from first principle [155] and Green's function embedded-atom model (GF-EAM) calculations [131, 156, 157]. DFT calculations by Jean et al. [158] have confirmed the anti-corrugation effect for the He-Cu(111) potential. Santoro et al. [159–161] have conjectured that the anti-corrugation effect is also related to the large inelastic HAS response from the longitudinal resonance of metal surfaces.

Some of the concepts discussed in this subsection are now superseded by present *ab initio* methods in the theory and calculation of the inelastic atom scattering intensities, specifically in the electron-phonon theory of inelastic scattering from conducting or highly polarizable surfaces which will be introduced in Chap. 8. Nevertheless those preliminary phenomenological studies of inelastic atom scattering from metal surfaces have provided the basic concepts and the appropriate lexicon for the next steps.

References

1. J.P. Toennies, Appl. Phys. **3**, 91 (1974)
2. V. Celli, D. Evans, in *Dynamics of Gas-Surface Interaction*, ed. by G. Benedek, U. Valbusa (Springer, Berlin Heidelberg, 1982), p. 2
3. V. Bortolani, A.C. Levi, Riv. Nuovo Cim. **9**, 1 (1986)
4. J.L. Beeby, J. Phys. Part C Solid State Phys. **4**(18), L359–000 (1971)
5. H. Legge, J.R. Manson, J.P. Toennies, J. Chem. Phy. **110**(17), 8767–8785 (1999)
6. U. Garibaldi, A.C. Levi, R. Spadacini, G.E. Tommei, Surf. Sci. **55**(1), 40–60 (1976)
7. G. Boato, P. Cantini, L. Matera, Surf. Sci. **55**(1), 141–178 (1976)
8. G. Benedek, G. Brusdeylins, R.B. Doak, J.G. Skofronik, J.P. Toennies, Phys. Rev. B **28**(4), 2104–2113 (1983)
9. P. Cantini, R. Tatarek, G.P. Felcher, Phys. Rev. B **19**(2), 1161–1171 (1979)
10. H. Hoinkes, Rev. Mod. Phys. **52**(4), 933–970 (1980)
11. R. Martinez-Casado, B. Meyer, S. Miret-Artes, F. Traeger, C. Woll, J. Phys.-Condensed Matter **19**(30), (2007)
12. W.E. Carlos, M.W. Cole, Surf. Sci. **77**(1), L173–L176 (1978)
13. L. Mattera, R. Musenich, C. Salvo, S. Terreni, Faraday Discuss. **80**, 115–126 (1985)

14. A. Luntz, L. Mattera, M. Rocca, S. Terreni, F. Tommasini, U. Valbusa, Surf. Sci. **126**(1–3), 695–701 (1983)
15. J. Lapujoulade, J. Perreau, Phys. Scr. **T4**, 138–140 (1983)
16. K.H. Rieder, W. Stocker, Phys. Rev. B **31**(6), 3392–3497 (1985)
17. A.L. Glebov, J.P. Toennies, F. Träger, Phys. Rev. Lett. **82**, 4492 (1999)
18. K.H. Rieder, Surf. Sci. **117**(1–3), 13–22 (1982)
19. D.O. Hayward, A.O. Taylor, J. Phy. C-Solid State Phys. **19**(15), L309–L314 (1986)
20. G. Parschau, E. Kirsten, A. Bischof, K.H. Rieder, Physi. Rev. B **40**(9), 6012–6017 (1989)
21. K.H. Rieder, W. Stocker, J. Phys. C-Solid State Phys. **16**(22), L783–L788 (1983)
22. D. Cvetko, A. Morgante, A. Santaniello, F. Tommasini, J. Chem. Phy. **104**(19), 7778–7783 (1996)
23. A. Tamtögl, M. Mayrhofer-Reinhartshuber, N. Balak, W.E. Ernst, K.H. Rieder, J. Phys.: Condens. Matter **22**, 304019 (2010)
24. M. Mayrhofer-Reinhartshuber, A. Tamtögl, P. Kraus, K.H. Rieder, W.E. Ernst, J. Phys.: Condens. Matter **24**, 104008 (2012)
25. A. Tamtögl, P. Kraus, M. Mayrhofer-Reinhartshuber, D. Campi, M. Bernasconi, G. Benedek, W.E. Ernst, Phys. Rev. B **87**, 035410 (2013)
26. U. Buck, Adv. Chem. Phys. **36**, 313 (1975)
27. K.T. Tang, J.P. Toennies, J. Chem. Phys. **118**, 4976 (2003)
28. W.A. Steele, *The Interaction of Gases with Solid Surfaces* (Pergamon Press, Oxford, 1974), Chap. 2
29. V. Celli, in *Surface Phonons,* ed. by W. Kress, F.W. de Wette (Springer, Berlin, 1991), p. 167
30. V. Celli, in *Helium Atom Scattering from Surfaces*, ed. by E. Hulpke (Springer, Berlin, 1992), p. 25
31. L.W. Bruch, M.W. Cole, E. Zaremba, *Int. Series of Monographs on Chemistry*, vol. 33 (Clarendon, New York, 1997), Chap. 2
32. A.J. Stone, *The Theory of Intermolecular Forces* (Clarendon, Oxford, 1996)
33. I.G. Kaplan, *Intermolecular Interactions* (Wiley, Hoboken, NJ, 2006); *Theory of Molecular Interactions* (Elsevier, Amsterdam 1986).
34. F. London, Z. Physik, 63, 245 (1930) and Z. Physik. Chemie B 11, 222 (1930); English translations in H. Hettema, *Quantum Chemistry, Classic Scientific Papers,* World Scientific, Singapore (2000)
35. H.B.G. Casimir, D. Polder, Phys. Rev. **73**, 360 (1948)
36. N. Eshjerg, J.K. Nørskov, Phys. Rev. Lett. **45**, 807 (1980)
37. K.T. Tang, J.M. Norbeck, P.R. Certain, J. Chem. Phys. **64**, 3063 (1976)
38. A.A. Radzig, B.M. Smirnov, *Reference Data on Atoms, Molecules, and Ions* (Springer, Berlin, Heidelberg, New York, Tokyo, 1980)
39. S.H. Patil, K.T. Tang. *Asymptotic Methods in Quantum Mechanics* (Springer, 2008)
40. K. Rosciszewski, B. Paulus, P. Fulde, H. Stoll, Phys. Rev. B **60**, 7905 (1999)
41. P.L. Silvestrelli, Phys. Rev. Lett. **100**, 053002 (2008)
42. A. Tkatchenko, M. Scheffler, Phys. Rev. Lett. **102**, 073005 (2009)
43. T. Sato, H. Nakai, J. Chem Phys. **131**, 224104 (2009)
44. N.A. de Lima, J. Chem. Phys. **132**(1) (2010)
45. J. Jiang, J. Mitroy, Y. Cheng, M.W.J. Bromley, At. Data Nucl. Data Tables **101**, 158 (2015)
46. L. Miglio, F. Quasso, G. Benedek, Surf. Sci. **136**, L9 (1984)
47. L. Miglio, F. Quasso, G. Benedek, J. Chem. Phys. **83**, 913 (1985)
48. M. Born, J.E. Mayer, Z. Physik (1932) 1
49. E.S. Rittner, J. Chem. Phys. **19**, 1030 (1951)
50. M.P. Tosi, F.G. Fumi, J. Phys. Chem. Solids **25**, 45 (1964)
51. M.P. Tosi, Solid State Physics **16**, 1–120 (1964)
52. C. Zener, Phys. Rev. **40**, 178 and 335 (1932)
53. J.M. Jackson, N.F. Mott, Proc. Roy. Soc. (London) A **137**, 703 (1932)
54. G. Mie, Ann. Phys. **11**, 657 (1903)

55. The 12–6 Lennard-Jones potential is a special case of the general *n-m* potential first introduced by Mie [Ann. der Physik 11 (1903) 657] and thoroughly discussed in E.A. Moelwyn-Hughes, *Physical Chemistry* (Pergamon, Oxford, 1978), p. 311
56. J.E. Lennard-Jones, Proc. Royal Soc. Lond. A **106**, 463 (1924)
57. P.M. Morse, Phys. Rev. **34**, 57–64 (1929)
58. I.G. Kaplan, in *Handbook of Molecular Physics and Quantum Chemistry* (Wiley, 2003), p. 207
59. R.E. Grisenti, W. Schollkopf, J.P. Toennies, G.C. Hegerfeldt, T. Köhler, Phys. Rev. Lett. **83** (9), 1755–1758 (1999)
60. R. Bruhl, P. Fouquet, R.E. Grisenti, J.P. Toennies, G.C. Hegerfeldt, T. Köhler, M. Stoll, C. Walter, Europhys. Lett. **59**(3), 357–363 (2002)
61. K.T. Tang, J.P. Toennies, J. Chem. Phys. **118**(11), 4976 (2003)
62. H. Patil, K.T. Tang, J.P. Toennies, J. Chem. Phys. **116**, 8118 (2002)
63. R. Eisenschitz, F. London, Zeitschrift für Physik, vol. 60, p. 491 (1930). English translation in H. Hettema, Quantum Chemistry, Classic Scientific Papers, World Scientific, Singapore (2000)
64. Note that the classical Clausius-Mossotti (CM) relation $4\pi\,\alpha(i\omega)/3v_c = [\varepsilon(i\omega) - 1]/[\varepsilon(i\omega) + 2]$ is obtained by integrating over a spherical distribution of dipoles around a given probe dipole. This holds for gases and homogeneous bulk media, whereas for the surface geometry the integration gives the CM relation in the form $2\pi\,\alpha(i\omega)/v_c = [\varepsilon(i\omega) - 1]/[\varepsilon(i\omega) + 1]$. Substitution of this equation into (6.19) yields the Casimir-Polder formula for $C_6 = (6v_c/\pi)c_3$, (6.13)
65. E.M. Lifshitz, Sov. Phys. **2**, 73 (1956)
66. G. Vidali, M.W. Cole, Surf. Sci. **110**, 10 (1981)
67. J. Jiang, J. Mitroy, Y. Cheng, M.W.J. Bromley, At. Data Nucl. Data Tables **101**, 158 (2015)
68. X.-P. Jiang, F. Toigo, M.W. Cole, Surf. Sci. **145**, 281 (1984)
69. J.M. Hutson, P.W. Fowler, E. Zaremba, Surf. Sci. **175**(2), L775–L781 (1986)
70. M. Dion, H. Rydberg, E. Schröder, D.C. Langreth, B.I. Lundqvist, Phys. Rev. Lett. **92**, 246401 (2004)
71. D.C. Langreth, M. Dion, H. Rydberg, E. Schröder, P. Hyldgaard, B.I. Lundqvist, Int. J. Quantum Chem. **101**, 599 (2005)
72. S. Grimme, J. Comput. Chem. **27**, 1787 (2006)
73. A. Tkatchenko, M. Scheffler, Phys. Rev. Lett. **102**, 073005 (2009)
74. R. Martinez-Casado, G. Mallia, D. Usvyat, L. Maschio, S. Cassasa, M. Schutz, N.M. Harrison, J. Chem. Phys. **134**(1), (2011)
75. J. Klimes, A. Michaelides, J. Chem. Phys. **137**(12) (2012)
76. K. Lee, K. Berland, M. Yoon, S. Andersson, E. Schroder, P. Hyldgaard, B.I. Lundqvist, J. Phys.-Condensed Matter **24**(42), 424213 (2012)
77. P.L. Silvestrelli, A. Ambrosetti, S. Grubisic, F. Ancilotto, Phys. Rev. B **85**(16) (2012)
78. K. Berland, V.R. Cooper, K. Lee, E. Schröder, T. Thonhauser, P. Hyldgaard, B.I. Lundqvist, Rep. Prog. Phys. **78**, 066501 (2015)
79. J. Tao, A.M. Rappe, Phys. Rev. Lett. **111**, 106101 (2014)
80. J. Tao, A.M. Rappe, J. Chem. Phys. **144**, 031102 (2016)
81. A. Erdélyi (ed.), *Higher Trascendental Functions* (McGraw Hill, New York, 1953)
82. A. Tsuchida, Sci. Papers Inst. Phys. Chem. Res. (Tokyo) 64, 75 (1970); J. Cryst. Soc. Japan, 14, 17 (1972); Surf. Sci. 46, 611 (1974).
83. V. Celli, Interaction of Atoms with Surfaces. in *Landolt-Börnstein* ed. by G. Chiarotti, vol. 24 (Springer, Berlin, Heidelberg, 1994/1995), pp. 278–328,
84. E. Zaremba, W. Kohn, Phys. Rev. B **13**, 2270 (1976)
85. N.D. Lang, W. Kohn, Phys. Rev. B **7**, 3541 (1973)
86. A. Chizmeshya, E. Zaremba, Surf. Sci. **220**(2–3), 443–470 (1989)
87. A. Liebsch, Phys. Rev. B **33**(10), 7249–7251 (1986)
88. O.A. von Lilienfeld, A. Tkatchenko, J. Chem. Phys. **132**, 234109 (2010)
89. V.V. Gobre, A. Tkatchenko, Nat. Commun. **4**, 2341 (2013)

90. G. Vidali, G. Ihm, H.-Y. Kim, M.W. Cole, Surf. Sci. Rep. **12**, 133 (1991)
91. S. Lepoutre, V.P.A. Lomij, H. Jelassi, J. Trenec, M. Buchner, A.D. Cronin, J. Vigue, Eur. Phys. J. D **62**, 309 (2011)
92. V.P.A. Lomij, E.K. Catherine, W.F. Holmgren, A.D. CroninJ, Phys. Chem A **115**, 71234 (2011)
93. P.W. Fowler, J.M. Hutson, Phys. Rev. B **33**, 3724 (1986) (for He-LiF)
94. D.J. Riley, A.P. Jardine, S. Dworski, G. Alexandrowicz, P. Fouquet, J. Ellis, W. Allison, J. Chem. Phys. **126**, 104702 (2007)
95. V. Celli, D. Eichenauer, A. Kaufhold, J.P. Toennies, J. Chem. Phys. **86**, 2504 (1985)
96. G. Benedek, G. Brusdeylins, V. Senzet, J.G. Skofronik, J.P. Tonnies, F. TRaeger, F. Vollmer, Phys. Rev. B **64**(12) (2001)
97. R.E. Grisenti, W. Schöllkopf, J.P. Toennies, G.C. Hegerfeldt, T. Köhler, Phys. Rev. Lett. **83**, 1755 (1999)
98. J.A. Kunc, D.E. Shemansky, Surf. Sci. **163**(1), 237–248 (1985)
99. A. Siber, B. Gumhalter, A.P. Graham, J.P. Toennies, Phys. Rev. B **63**(11) (2001)
100. R.B. Laughlin, Phys. Rev. B **25**(4), 2222–2247 (1982)
101. W.R. Lambert, P.L. Trevor, M.J. Cardillo, A. Sakai, D.R. Hamann, Phys. Rev. B **35**(15), 8055–8064 (1987)
102. J.R. Buckland, W. Allison, J. Chem. Phy. **112**(2), 970–978 (2000)
103. D. Eichenauer, U. Harten, J.P. Toennies, V. Celli, J. Chem. Phys. **86**, 3693 (1987)
104. K. Lenarcicpoljanec, M. Hodoscek, D. Lovrik, B. Gumhalter, Surf. Sci. **251**, 706–711 (1991)
105. G. Lach, M. Dekieviet, U.D. Jentschura, Int. J. Mod. Phys. A **25**(11), 2337–2344 (2010)
106. T. Andersson, P. Linde, M. Hassel, S. Andersson, J. Chem. Phys. **124**(11) (2006)
107. G.-K. Kroes, R.C. Mowrey, J. Chem. Phys. **103**, 2186 (1995)
108. S. Andersson, M. Persson, J. Harris, Surf. Sci. **360**, L499–L504 (1996)
109. W.A. Steele, Surf. Sci. **36**, 317 (1973)
110. M.W. Cole, T.T. Tsong, Surf. Sci. **69**(1), 325–335 (1977)
111. C. Kittel, *Introduction to Solid State Physics* (John Wiley and Sons, Inc. 2005), Chap. 2
112. A. Erdélyi, *Asymptotic expansions* (Dover Publications Inc., New York, 1956)
113. V. Celli, N. Garcia, J. Hutchison, Surf. Sci. **87**, 112 (1979)
114. D. Farias, K.-H. Rieder, Rep. Prog. Phys. **61**, 1575 (1998)
115. H. Jónnson, J.H. Weare, Surf. Sci. **181**, 495 (1987)
116. F.G. Fumi, M.P. Tosi, J. Phys. Chem. Solids **25**, 31 (1964)
117. K.T. Tang, J.P. Toennies, J. Chem. Phys. **80**, 3726 (1984)
118. J.M. Hutson, P.W. Fowler, Surf. Sci. **173**, 337 (1986) (for He-NaCl)
119. D. Eichenauer, J.P. Toennies, Surf. Sci. **197**, 267 (1988)
120. A.P. Jardine, S. Dworski, P. Fouquet, G. Alexandrowicz, D.J. Riley, G.Y.H. Lee, J. Ellis, W. Allison, Science **304**, 1790 (2004)
121. D. Campi, M. Bernasconi, G. Benedek, J.P. Toennies, J. Phys. Chem. C **119**, 14579 (2015)
122. G. Boato, P. Cantini, R. Tatarek, J. Phys. F: Metal Phys. **6**, L237 (1976)
123. J.M. Horne, D.R. Miller, Surf. Sci. **66**, 365 (1977)
124. R. Smoluchowski, Phys. Rev. **60**, 661 (1941)
125. M.W. Finnis, V. Heine, J. Phys. F (Metal Phys). **4**, L37 (1974)
126. N. Esbjerg, J.K. Nørskov, Phys. Rev. Lett. **45**, 807 (1980)
127. A.C. Levi, H. Suhl, Surf. Sci. **88**, 221–254 (1979)
128. J.F. Annett, R. Haydock, Phys. Rev. Lett. **53**, 838 (1984)
129. N. Garcia, J.M. Soler, Surface Sci. **126**, 689 (1983)
130. Y. Takeda, W. Kohn, Phys. Rev. Lett. **54**, 470 (1985)
131. I.P. Batra, P.S. Bagus, J.A. Barker, Phys. Rev. B **31**, 1737 (1985)
132. G.P. Brivio, M.I. Trioni, Rev. Mod. Phys. **71**, 231 (1999)
133. P. Nordlander, J. Harris, J. Phys. C-Solid State Phys. **17**(6), 1141–1152 (1984)
134. J.L.F. Da Silva, C. Stampfl, M. Scheffler, Phys. Rev. Lett. **90**, 066104 (2003)
135. P.L. Silvestrelli, Phys. Rev. Lett. **100**, 053002 (2008); J. Phys. Chem. A **113**, 5224 (2009)

136. J.P. Senet, G. Benedek, J.P. Toennies, Europhys. Lett. **57**, 430 (2002)
137. J. Harris, A. Liebsch, Phys. Rev. Lett. **49**, 341 (1982)
138. A. Liebsch, J. Harris, Surface Sci. **111**, 2721 (1981)
139. A. Liebsch, Phys. Rev. B **33**, 7249 (1986)
140. N.D. Lang, Phys. Rev. Lett. **46**, 842 (1981)
141. E. Zaremba, W. Kohn, Phys. Rev. B **15**, 12769 (1977)
142. H. Hoinkes, H. Nahr, H. Wilsch, Surf. Sci. **33**, 516–524 (1972)
143. G. Armand, J. Lapujoulade, Y. Lejay, Surf. Sci. **63**, 143–152 (1977)
144. B. Feuerbacher, R.F. Willis, Phys. Rev. Lett. **47**, 526 (1981)
145. V. Bortolani, A. Franchini, F. Nizzoli, G. Santoro, G. Benedek, V. Celli, Surf. Sci. **128**, 249 (1983)
146. F. Hofmann, J.P. Toennies, J.R. Manson, J. Chem. Phys. **101**, 20155 (1994)
147. V. Bortolani, A. Franchini, N. Garcia, F. Nizzoli, G. Santoro, Phys. Rev. B **28**, 7358 (1983)
148. A. Lock, Diplom Thesis, University of Göttingen, (1987)
149. N. Bunjes, N.S. Luo, P. Ruggerone, J.P. Toennies, G. Witte, Phys. Rev. B **50**, 8897 (1994-II)
150. V. Celli, G. Benedek, U. Harten, J.P. Toennies, R.B. Doak, V. Bortolani, Surf. Sci. **143**, L376 (1984)
151. D.R. Hamann, Phys. Rev. Lett. **46**, 1227 (1981)
152. K. Kern, Ph.D. Thesis, University of Bonn, (1986)
153. J.P. Toennies, G. Witte, Ch. Wöll, Surf. Sci. **323**, 1228 (1995)
154. K.H. Rieder, G. Parschau, B. Burg, Phys. Rev. Lett. **71**, 1059 (1993)
155. M. Petersen, S. Wilke, P. Ruggerone, B. Kohler, M. Scheffler, Phys. Rev. Lett. **76**, 995 (1996)
156. F. Montalenti, M.I. Trioni, G.P. Brivio, S. Crampin, Surf. Sci. Lett. **364**, L595 (1996); M.I. Trioni, F. Montalenti and G.P. Brivio, *ibidem* **401**, L383 (1998)
157. M.I. Trioni, S. Marcotulio, G. Santoro, V. Bortolani, G. Palumbo, G.P. Brivio, Phys. Rev. B **58**, 11043 (1998)
158. N. Jean, M.I. Trioni, G.P. Brivio, V. Bortolani, Phys. Rev. Lett. **92**, 013201 (2004)
159. G. Santoro, A. Franchini, V. Bortolani, Phys. Rev. Lett. **80**, 2378 (1998)
160. G. Santoro, A. Franchini, V. Bortolani, D.L. Mills, R.F. Wallis, Surf. Sci. **478**, 99 (2001)
161. A. Franchini, G. Santoro, J. Phys.-Cond. Matter **14**, 5881 (2002)
162. K.T. Tang, J.P. Toennies, Surf. Sci. Lett. **279**, L203 (1992)

Chapter 7
Theory of Atom Scattering from Surface Phonons: Basic Concepts and Temperature Effects

... in the likeness of the waves which in May the course of the wind makes in the cornfields, when one sees the waves running over the fields without the ears of the corn changing their place.

Leonardo da Vinci (1452–1519).
Del moto e misura dell'acqua (Ms F, 87 v).
[cited in Leonardo da Vinci,
by L. H. Heydenreich, Vol 1, p 145 (1954)]

Abstract The reflection coefficient determining the inelastic scattered intensity is defined and analysed theoretically. Various approximations including the Born approximation, the Distorted-Wave Born approximation (DWBA), the GR method and the eikonal approximation are discussed and illustrated with examples. From an expression for the DWBA inelastic reflection coefficient for n-phonon processes, factors determining the relative role of temperature for multiphonon processes and one-phonon scattering are analysed. Several approximate expressions for the Debye-Waller factor are derived and used to define experimental criteria favouring one-phonon scattering. Most of the examples dealt with are for insulator surfaces. Chapter 8 then describes the scattering theory for metal and semiconductor surfaces.

7.1 Inelastic Scattering from Phonons

The dynamics of atom scattering from the surface of a crystal lattice is governed by the total Hamiltonian

$$H = H_A + H_L + V(\mathbf{r}, t), \tag{7.1}$$

where H_A and H_L are the unperturbed Hamiltonian operators of the free atom and of the vibrating semi-infinite crystal lattice, respectively, in the absence of any

---

The original version of this chapter was revised: Equations (7.33a) and (7.82) were incorrect, these have been corrected. The correction to this chapter is available at https://doi.org/10.1007/978-3-662-56443-1_15

G. Benedek and J. P. Toennies, *Atomic Scale Dynamics at Surfaces*, Springer Series in Surface Sciences 63, https://doi.org/10.1007/978-3-662-56443-1_7

atom-crystal interaction. $V(\mathbf{r}, t)$ is the time-dependent potential coupling the probe atom at position $\mathbf{r}$ with the vibrating surface atoms of the lattice. This is decomposed as

$$V(\mathbf{r}, t) = V(\mathbf{r}) + \delta V(\mathbf{r}, t), \tag{7.2}$$

where $V(\mathrm{r})$ is the surface static potential discussed in the previous Chapter and $\delta V(\mathbf{r}, t)$ is the dynamic, time-dependent perturbation of the potential arising from the vibrations of the lattice atoms. The static potential is responsible for the elastic scattering, whereas the time-dependent potential is responsible for the inelastic scattering processes in which the probe atom exchanges a certain amount of energy with the crystal. Only processes producing a change in the vibrational energy of the crystal are considered.

Since the overall vibrational motion of a periodic lattice can be decomposed, by virtue of the Fourier theorem, into harmonically oscillating plane waves, the inelastic scattering can be understood as a diffraction process from a moving grating, in a perfect analogy with the elastic diffraction from a static grating. The change in energy of the probe atom is analogous to the change of frequency of a diffracted wave from a moving grating due to the Doppler effect.

Since the amplitudes of the atomic vibrations are much smaller than the interatomic distances, the dynamic perturbation is conveniently expanded in a Taylor series with respect to the time-dependent displacements $\mathbf{u}_l(\mathrm{t})$ of the atoms around their equilibrium positions $\mathbf{r}_l$

$$\delta V(\mathbf{r}, t) = \sum\nolimits_l \frac{\partial V(\mathrm{r})}{\partial \mathbf{r}_l} \mathbf{u}_l(t) + \frac{1}{2} \sum_{ll'} \frac{\partial^2 V(\mathbf{r})}{\partial \mathbf{r}_l \, \partial \mathbf{r}_{l'}} \mathbf{u}_l(t) \mathbf{u}_{l'}(t) + \cdots \tag{7.3}$$

In perturbation theory each term of the above expansion contributes to the inelastic scattering probability. In itself each term can, in addition, contribute to a first order, second order, and higher order effects. Thus one-phonon processes involve only the linear term to the first order, whereas two-phonon processes can originate either from the linear term to second order or from the quadratic term to first order, etc. In general many-phonon processes are treated by keeping only the first term on the right hand side of (7.3) and the higher order terms are neglected. Higher order terms may also include many-phonon processes induced by anharmonicity.

The modelling of the effective He-surface inelastic potential should be conveniently consistent with the model used to describe the surface phonons. Because of this requirement the successful calculation of both the dispersion curves and the inelastic intensities in the HAS spectra provides a severe test of the theoretical methods and approximations adopted in surface dynamics. For closed shell ionic surfaces, such as the alkali halides, rare gas solids, etc., the standard pairwise additive interatomic potentials (Born-Mayer, Lennard-Jones, etc.), used to simulate the interactions between the probe atom and the surface atoms, as discussed in the previous chapter and illustrated in Fig. 6.3, are generally considered to be appropriate also for describing the lattice dynamics. For example, the steepness of the closed shell repulsion potential, which forms the basis for approximating the atom-surface potential by a hard corrugated surface (HCS) model (see Fig. 6.4) is consistent with the simple rigid ion model used to describe the lattice dynamics of the alkali halide crystals. In the next level of lattice dynamical approximations such as the shell models, dipolar and quadrupolar ion deformabilities [1, 2], must also be accounted for in treating the scattering. These require additional terms in the He-surface potential. The long range attraction used in the static atom-surface potential model and the pairwise additive potential models such as (6.9) do not, in fact, take full account of the multipolar deformations of surface ions. The inclusion of such effects could be of importance since they account for another mechanism by which the probe particle is coupled to the surface lattice vibrations.

At metal and semiconductor surfaces similar pairwise additive models were routinely used up to about 1990 to describe the coupling of atoms to the phonons [3]. Although this can work to a certain extent for diffraction [4] it is now generally appreciated that this approach cannot be correct for inelastic scattering. In addition to the direct pairwise additive interactions with the surface ions, the coupling of the scattering atom via the electrons at the surface has to be accounted for. As discussed in Chap. 8 this coupling can be conveniently described by the multipolar expansion (ME) method and the electron-phonon theory of inelastic HAS.

Quantum semi-empirical methods such as effective medium (EM) and embedded atom (EA) approximations, still currently used in surface dynamics [5–11], also have been employed to approximate the interaction energy of an external atom with a free electron metal surface [12–15]. Still the difficulty of these methods in correctly accounting for Fermi-surface effects and phonon anomalies also affects the calculation of the associated HAS inelastic intensities, because the same mechanism which produces Kohn anomalies, also leads to large oscillations of the surface electron densities and therefore to large inelastic scattering amplitudes.

In the last decades the DF-LDA theory has been frequently applied to gas-surface interactions [16–23]. In the so-called *evanescent wave approximation* [24] the repulsive He-metal surface potential is described in terms of the scattering of the surface electrons near the Fermi edge from the He atom. These electrons with energy between E_F and E_F-$\hbar\omega$ are just those which participate in the modulation of the charge density induced by a phonon of frequency ω. Total energy calculations based on **Q**-dependent interplanar Hellmann-Feynman forces, while correctly reproducing the phonon anomalies of noble metals [25], provided no convincing conclusion as regards the HAS intensities [26, 27]. It is with the density response methods such as the DFPT formalism that it became possible to explicitly calculate the phonon-induced modulation of the atom-surface potential, thus fully accounting for the nonlocality effects [28] and for the observed inelastic HAS intensities from metal surface [29–31] The DFPT approach provides the basis to the modern electron-phonon theory of inelastic HAS from conducting surfaces, presented in the next Chap. 8.

7.2 Inelastic Reflection Coefficient

The *reflection coefficient* $\Re$ for a given inelastic transition $i \rightarrow f$ is defined as the transition rate w_{fi} (number of inelastic events per unit time) divided by the incident flux J_i (number of incident particles in the state i per unit time)

$$\Re(\mathbf{k}_f, \mathbf{k}_i) = w_{fi}/J_i. \tag{7.4}$$

The particle flux incident on the surface is given by [32]

$$J_i = \frac{\hbar|k_{iz}|}{m}\rho A, \tag{7.5}$$

where k_{iz} is the component of $\mathbf{k}_i$ normal to the surface, A is the surface area exposed to the incident beam and ρ is the density of particles just above that area (see Fig. 7.1).

The transition rate w_{fi} from an initial to a final state of total energies E_i^{tot} and E_f^{tot}, respectively, has the general form [33]

$$w_{fi} = \frac{2\pi}{\hbar}|T_{fi}|^2\delta\left(E_f^{tot} - E_i^{tot}\right), \tag{7.6}$$

where T_{fi} is the matrix element of the transition operator between the final and initial states. The total energy E_i^{tot} of the particle-crystal system in the initial state is the sum of the total energies of the particle, $E_i = \hbar^2 k_i^2/2m$, and of the phonon gas

$$E_i^{tot} = E_i + \sum_{Q\nu}\hbar\omega_{Q\nu}\left(n_{Q\nu,i} + \frac{1}{2}\right). \tag{7.7}$$

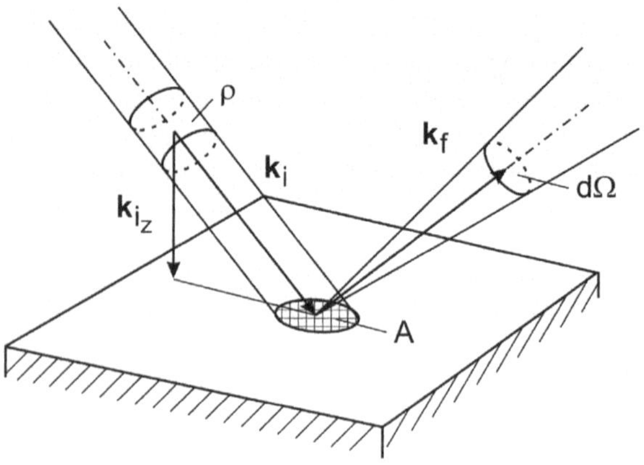

Fig. 7.1 Schematic diagram showing the important parameters used in defining the reflection coefficient of an atom with incident wavevector $\boldsymbol{k}_i$ and final wavevector $\boldsymbol{k}_f$

The same holds for the energy of the final state, E_f^{tot}, and of any other intermediate state s, e.g., a bound state of total energy E_b^{tot} ($s = b$). In an inelastic scattering process involving phonons, the initial set $\{n_i\}$ of phonon occupation numbers $n_{\mathbf{Q}\nu,i}$ changes into a final set $\{n_f\}$ where the new occupation numbers are $n_{\mathbf{Q}\nu,f}$. For example, in a one-phonon processes only one of the occupation numbers changes by ± 1. With (7.4) to (7.6) the reflection coefficient then becomes

$$\Re\left(\mathbf{k}_f, \mathbf{k}_i\right) = \frac{2\pi m}{\hbar^2 |k_{iz}| \rho A} \sum_{\{n_i\}\{n_f\}} P_{\{n_i\}} \left|T_{fi}\right|^2 \delta(E_f^{tot} - E_i^{tot}), \tag{7.8}$$

where

$$P_{\{n_i\}} = \prod\nolimits_{\mathbf{Q}\nu} P(n_{\mathbf{Q}\nu,i}) \tag{7.9}$$

is the probability of finding in the initial state a given distribution of phonon occupation numbers $n_{\mathbf{Q}\nu,i}$ on the different phonon states $\mathbf{Q}\nu$. $P(n_{\mathbf{Q}\nu,i})$ is the Bose-Einstein probability of finding, in the initial state and at the surface temperature T, the phonon oscillator of frequency $\omega_{\mathbf{Q}\nu}$ in its $n_{\mathbf{Q}\nu,i}$-th state:

$$\begin{aligned} P(n_{\mathbf{Q}\nu,i}) &= \frac{\exp[-(n_{\mathbf{Q}\nu,i} + \frac{1}{2})(\hbar\omega_{\mathbf{Q}\nu}/kT)]}{\sum\limits_{m=0}^{\infty} \exp[-(m + \frac{1}{2})(\hbar\omega_{\mathbf{Q}\nu}/kT)]} \\ &= \exp[-n_{\mathbf{Q}\nu,i}(\hbar\omega_{\mathbf{Q}\nu}/kT)][1 - \exp(-\hbar\omega_{\mathbf{Q}\nu}/kT)] \,. \end{aligned} \tag{7.10}$$

The occupation number for a phonon of energy $\hbar\omega_{\mathbf{Q}\nu}$, thermally averaged over the Bose-Einstein distribution appropriate for phonons, is then given by

$$n(\hbar\omega_{\mathbf{Q}\nu}) = \langle n_{\mathbf{Q}\nu}\rangle - \sum\nolimits_{\{n\}} P_{\{n\}} n_{\mathbf{Q}\nu}, \tag{7.11}$$

where the brackets <···> define the thermal average at the equilibrium temperature T. By means of (7.9) and (7.10) one obtains

$$n(\hbar\omega_{\mathbf{Q}\nu}) = \sum\nolimits_{n_{\mathbf{Q}\nu}} P(n_{\mathbf{Q}\nu}) n_{\mathbf{Q}\nu} = \frac{1}{exp(\hbar\omega_{\mathbf{Q}\nu}/kT) - 1} \,. \tag{7.12}$$

This function has the interesting property that

$$n(-h\omega) = -[1 + n(\hbar\omega)]. \tag{7.13}$$

The functions $n(\hbar\omega)$ and $1 + n(\hbar\omega)$, plotted in Fig. 7.2 as functions of the reduced inverse temperature $\hbar\omega/kT$, are the Bose-Einstein factors which weigh the probability that either a phonon annihilation or creation process occurs, respectively. By virtue of (7.13) an annihilation process can be viewed as a creation of a phonon of negative frequency, and viceversa.

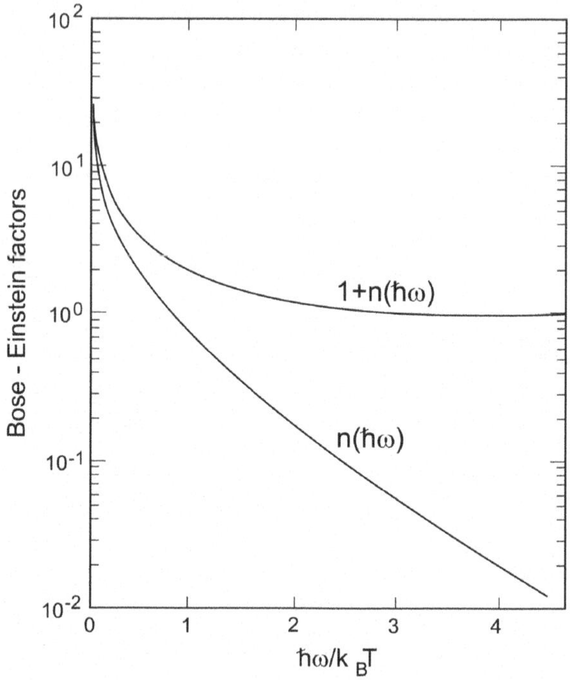

Fig. 7.2 The Bose-Einstein factors $n(\hbar\omega)$ and $1+n(\hbar\omega)$, weighing phonon annihilation and creation processes, respectively, are plotted versus the reduced inverse temperature $\hbar\omega/k_BT$

The signal measured in inelastic scattering experiments [34–36] is proportional to the product of the reflection coefficient $\Re$ times the differential phase space volume occupied by the final states of the scattered atom with momentum between $\mathbf{k}_f$ and $\mathbf{k}_f + d\mathbf{k}_f$ within the solid angle $d\Omega_f$ spanned by the detector (Fig. 7.1)

$$d^2\Re = \Re\frac{k_f^2}{8\pi^3\rho}dk_f d\Omega_f = \Re\frac{mk_f}{8\pi^3\hbar^2\rho}dE_f d\Omega_f. \tag{7.14}$$

The inelastic differential reflection coefficient is then given by

$$\frac{d^2\Re}{dE_f d\Omega_f} = \frac{k_f}{|k_{iz}|}\frac{m^2}{4\pi^2\hbar^4\rho^2 A}\sum_{\{n_i\}\{n_f\}} P_{\{n_i\}}|T_{fi}|^2\delta(E_f^{tot} - E_i^{tot})\,. \tag{7.15}$$

This equation can be cast in a more convenient form through the van Hove transformation [37]. By using the Fourier representation of the δ-function

$$\delta(E) = \lim_{\eta\to 0^+}\int_{-\infty}^{+\infty}\frac{dt}{2\pi\hbar}e^{itE/\hbar-\eta|t|} \tag{7.16}$$

and considering that the time evolution of the transition operator T is governed by the lattice dynamics,

$$T_{fi}(t) = \langle f|\exp(iH_L t/\hbar)T(0)\exp(-iH_L t/\hbar)|i\rangle = T_{fi}(0)\exp(i\Delta E_L t/\hbar), \quad (7.17)$$

where $\Delta E_L = E_f^{tot} - E_i^{tot} - (E_f - E_i)$ is the change of the lattice energy due to the creation or annihilation of phonons, (7.15) becomes

$$\frac{d^2\Re}{dE_f d\Omega_f} = \frac{k_f}{|k_{iz}|}\frac{m^2}{8\pi^3\hbar^5\rho^2 A}\int_{-\infty}^{+\infty} dt\, e^{-i(E_f-E_i)t/\hbar}\left\langle T_{fi}(0)\, T_{fi}^+(t)\right\rangle. \quad (7.18)$$

The thermal average originates from the sum over the phonon distributions, according to (7.11), and defines the time-dependent *transition matrix correlation function*. It is important to remark that (7.18) is completely general and holds for the collisional coupling to the phonon system to any order of approximation.

The matrix elements of the transition operator T_{fi} are obtained by solving the dynamical problem of the interaction of the scattering particle with the vibrating surface, expressed by the Hamiltonian of (6.1). The elements T_{fi} are the solutions of the matrix equation

$$T_{fi} = V_{fi} + \sum\nolimits_s V_{fs}(E_i^{tot} - E_s^{tot} + i\delta)^{-1} T_{si}. \quad (7.19)$$

They can formally be expressed by the iteration

$$\begin{aligned} T_{fi} &= V_{fi} + \sum\nolimits_s V_{fs}(E_i^{tot} - E_s^{tot} + i\delta)^{-1} V_{si} \\ &\quad + \sum\nolimits_{ss'} V_{fs}(E_i^{tot} - E_s^{tot} + i\delta)^{-1} V_{ss'}(E_i^{tot} - E_{s'}^{tot} + i\delta)^{-1} V_{s'i} + \cdots \end{aligned} \quad (7.20)$$

The positive constant δ appearing in (7.19) and (7.20) is $\hbar$ times the inverse lifetime of the intermediate state s. For the bound states δ determines the resonance half-width, as appears, e.g., in selective adsorption experiments [38]. From (7.19) it appears that the elements T_{fi} owe their time dependence to the lattice dynamics through the time-dependence of the scattering potential $V(\mathbf{r}, t)$ which is modulated, according to (7.3), by the vibrations of the ion cores in the surface region. The matrix elements of the time-dependent potential $V(\mathbf{r}, t)$ in (7.19) can be split, according to (7.2) and (7.3), into a static (elastic) part $V_{0,fj}$ and a time-dependent (inelastic) part δV_{fj} as given by

$$\begin{aligned} V_{fj} &= \langle \Psi_f |\, V(\mathbf{r}, t)\, | \Psi_j \rangle = V_{0,fj} + \delta V_{fj} \\ &= \int d^3r \psi_f^* V_0(\mathbf{r})\, \psi_j + \sum\nolimits_l \int d^3r \psi_f^* \frac{\partial V(\mathbf{r})}{\partial \mathbf{r}_l}\psi_j \langle \Phi_f |\, \mathbf{u}_l(t) | \Phi_j \rangle + \cdots \end{aligned} \quad (j = i, s), \quad (7.21)$$

where the integration space for the matrix elements in (7.21) is defined by the volume per atom in the incident beam $\int d^3 r = \rho^{-1}$ and the exposed surface area $\int d^2R = A$ (see Fig. 7.1).

In the above expansion with respect to the atomic displacements $\mathbf{u}_l(t)$ the unperturbed total wavefunctions Ψ_m, with m referring either to the final (f), initial

(i) or intermediate (s) states, have been factorized into the unperturbed atom and phonon wavefunctions as

$$\Psi_m(k, r) = \psi_m(\mathbf{k}, \mathbf{r})\Phi_m \quad (m = f, i, s)\,, \tag{7.22}$$

where Φ_m is the total phonon wavefunction for the set $\{n_m\}$ of occupation numbers in the state m and

$$\psi_m(\mathbf{k}, \mathbf{r}) = \frac{1}{\sqrt{A}}\zeta_{m,\mathbf{K}}(k_z; \mathbf{R}, z)\mathrm{e}^{i\mathbf{K}\cdot\mathbf{R}} \tag{7.23}$$

is the unperturbed *atom* wavefunction. The form of $\psi_m(k, r)$ is that of a two-dimensional Bloch function describing the motion of the atom along the periodic potential of the surface. Thus, according to the Bloch theorem, $\zeta_{m,\mathbf{K}}(k_z; \mathbf{R}, z)$ is periodic in $\mathbf{R}$ with the periodicity of the potential and may be expressed in terms of its Fourier components as

$$\zeta_{m,\mathbf{K}}(k_z; \mathbf{R}, z) = \sum\nolimits_{\mathbf{G}} \zeta_{m,\mathbf{K}+\mathbf{G}}(k_z, z)\mathrm{e}^{i\mathbf{G}\cdot\mathbf{R}}, \tag{7.24}$$

where $\mathbf{G}$ are surface reciprocal lattice vectors. For not too large surface corrugations the motion of the atom in the directions parallel to the surface may be decoupled from the motion along the surface normal, which allows for the factorization

$$\zeta_{m,\mathbf{K}}(k_z; \mathbf{R}, z) = \chi_m(k_z, z)\,\varphi_{\mathbf{K}}(\mathbf{R}), \tag{7.25}$$

where $\chi_m(k_z, z)$ is a solution of a one-dimensional Schrödinger equation for the motion of the atom along z. $\varphi_{\mathbf{K}}(\mathbf{R})\mathrm{e}^{i\mathbf{K}\cdot\mathbf{R}}$ is a purely two-dimensional Bloch wave, with $\varphi_{\mathbf{K}}(\mathbf{R})$ a periodic function and $\varphi_{\mathbf{K}+\mathbf{G}}$ its Fourier components. If the surface potential is flat $\varphi_{\mathbf{K}}(\mathbf{R})\mathrm{e}^{i\mathbf{K}\cdot\mathbf{R}}$ is a pure plane wave [$\varphi_{\mathbf{K}}(\mathbf{R})$ = const.].

7.3 The Distorted-Wave Born Approximation (DWBA)

7.3.1 General Formulation

The choice of the unperturbed atom wave function, (7.23), is to some extent arbitrary and depends on what is chosen as the perturbation potential $\delta V(\mathbf{r}, t)$ and the scattering approximation. In a brute force approach one might adopt as unperturbed states those of the free atom and the entire (static and dynamic) crystal potential as the perturbing potential. This is inconvenient, however, as it would lead in general to a slow convergence of the iteration (7.20) for T_{fi}. It is preferable to choose as the unperturbed problem one which approximates the desired solution as much as possible in order to ensure a rapidly converging iteration. Whenever a rapid convergence occurs, the simplest approximation is to take only the first term, $T_{fi} \cong V_{fi,}$ which is also called the *(first order) Born approximation.*

In general the dynamic part of the potential is weak enough to be dealt with as a perturbation. In some inelastic scattering problems, however, also the static potential of the solid is sufficiently weak so that its particle eigenfunctions can still be approximated by plane waves and the Born approximation works quite well. This is the case for neutron scattering, electron-phonon scattering in the effective mass approximation, etc., but unfortunately not for atom scattering from a compact solid surface. Atoms interact strongly with the surface atoms of a solid to the extent that they are fully reflected ($\Delta k_z \sim 2k_{zi}$) by the surface. In such a situation only the dynamical part of the potential is adopted as a perturbation and its matrix elements are taken between eigenfunctions distorted by the static surface potential. The distortion accounts for the reflection. Now the iteration for T_{fi} truncated at the first term, $T_{fi} \cong V_{fi,}$ defines the *(first order) distorted-wave Born approximation* (DWBA).

Since the general form of the pure Born approximation, remains valid in further approximations, the differential reflection coefficient is first derived in the pure Born approximation for pedagogical purposes. In the first order Born approximation $T_{fi}(t)$ is replaced by $V_{fi}(t)$, which means that in (7.19) all channels involving intermediate states, namely multiple scattering and resonances with bound states, are neglected. Then $V(\mathbf{r}, t)$ is described as a sum of two-body potentials $v(\mathbf{r} - \mathbf{r}_l - \mathbf{u}_l(t))$ between the atom at position $\mathbf{r}$ and the l-th atom in the crystal at the instantaneous position $\mathbf{r}_l + \mathbf{u}_l(\mathrm{t})$, where $\mathbf{r}_l$ is the equilibrium position and $\mathbf{u}_l(\mathrm{t})$ the displacement at time t. As usual for a lattice with more than one atom per surface unit cell index l has to be replaced by (l,κ) with κ labelling the atoms within the unit cell. In the Born approximation the initial and final free atom states are given by $\rho^{1/2}\exp(i\mathbf{k}_i\cdot\mathbf{r})$ and $\rho^{1/2}\exp(i\mathbf{k}_f\cdot\mathbf{r})$, respectively, and the matrix element of the perturbing potential is given by

$$V_{fi}(t) = \rho \int d^3r\ e^{i(\mathbf{k}_f - \mathbf{k}_i)\cdot\mathbf{r}} \sum\nolimits_l V(\mathbf{r} - \mathbf{r}_l - \mathbf{u}_l(t))\ . \tag{7.26}$$

Next the three-dimensional Fourier transform of the individual two-body potential calculated at the equilibrium position is introduced

$$v(\Delta\mathbf{k}) = \rho \int d^3r'\ e^{i\Delta\mathbf{k}\cdot\mathbf{r}'} V(\mathbf{r}'), \tag{7.27}$$

where $\mathbf{r}' = \mathbf{r} - \mathbf{r}_l - \mathbf{u}_l(t)$ and $\Delta\mathbf{k} \equiv \mathbf{k}_f - \mathbf{k}_i$. Substituting into (7.26) leads to

$$V_{fi}(t) = \sum_l e^{i\Delta\mathbf{k}\cdot(\mathbf{r}_l + \mathbf{u}_l(t))} V(\Delta\mathbf{k})\, , \tag{7.28}$$

where the phonon displacements $\mathbf{u}_l(t)$ now appear in the argument of the exponential function. Thus, with the replacement $T \to V$, the transition matrix correlation function appearing in (7.18) reduces to a thermal average over a time-dependent exponential form

$$\begin{aligned}&\left\langle T_{fi}(0)\,T_{fi}^{+}(t)\right\rangle \rightarrow \left\langle V_{fi}(0)\,V_{fi}^{+}(t)\right\rangle\\&=|\,V(\Delta\mathbf{k})\,|^{2}\sum\nolimits_{ll'}\mathrm{e}^{i\Delta\mathbf{k}\cdot(\mathbf{r}_{l}-\mathbf{r}_{l'})}\left\langle \mathrm{e}^{i\Delta\mathbf{k}\cdot\mathbf{u}_{l}(0)}\mathrm{e}^{-i\Delta\mathbf{k}\cdot\mathbf{u}_{l'}(t)}\right\rangle.\end{aligned}\tag{7.29}$$

The correlation function in the thermal average on the right hand side contains all the information on the lattice dynamics. It can be calculated by expanding the displacement vectors $\mathbf{u}_l(t)$ in terms of the normal modes of the crystal [39]

$$\mathbf{u}_{l}(t)=\sum\nolimits_{\mathbf{Q}\nu}\mathbf{u}_{\mathbf{Q}\nu}\,\mathrm{e}^{i\mathbf{Q}\cdot\mathbf{r}_{l}}\left(\,b_{\mathbf{Q}\nu}^{+}\mathrm{e}^{i\omega_{\mathbf{Q}\nu}t}+\,b_{\mathbf{Q}\nu}\mathrm{e}^{-i\omega_{\mathbf{Q}\nu}t}\right),\tag{7.30}$$

where $\mathbf{u}_{\mathbf{Q}\nu}$ is an abbreviation for the planar displacements $\mathbf{u}(l_3\kappa,\mathbf{Q}\nu)$, (3.33), for the surface layer ($l_3 = 1$), which is assumed to be the only one interacting with the He atoms. In (7.30) $\omega_{\mathbf{Q}\nu}$ are the phonon eigenfrequencies and the $b_{\mathbf{Q}\nu}^{+}$ and $b_{\mathbf{Q}\nu}$ are the creation and annihilation operators, respectively. By expanding the exponentials in the correlation function of (7.29) and exploiting the commutation rules for $b_{\mathbf{Q}\nu}^{+}$ and $b_{\mathbf{Q}\nu}$ the resulting series can be rearranged and resummed [40] to give

$$\left\langle \mathrm{e}^{i\Delta\mathbf{k}\cdot\mathbf{u}_{l}(0)}\mathrm{e}^{-i\Delta\mathbf{k}\cdot\mathbf{u}_{l'}(t)}\right\rangle=\mathrm{e}^{-W_{l}(\Delta\mathbf{k})-W_{l'}(\Delta\mathbf{k})}\mathrm{e}^{\Delta\mathbf{k}\cdot\langle\mathbf{u}_{l}(0)\mathbf{u}_{l'}(t)\rangle\cdot\Delta\mathbf{k}},\tag{7.31}$$

where the terms $W_l(\Delta\mathbf{k})$ in the first exponent are given by

$$W_{l}(\Delta\mathbf{k})\equiv\frac{1}{2}\left\langle |\Delta\mathbf{k}\cdot\mathbf{u}_{l}(t)|^{2}\right\rangle.\tag{7.32}$$

As a result of the thermal averaging the $W_l(\Delta\mathbf{k})$ are independent of time and depend only on the position of the atom and temperature. For this reason the first exponential can be treated separately. It corresponds to the Debye-Waller factor which is discussed in more detail below. Here it is only noted that for a periodic monoatomic surface $W_l(\Delta\mathbf{k}) = W(\Delta\mathbf{k})$; for a polyatomic one $W_l(\Delta\mathbf{k}) = W_\kappa(\Delta\mathbf{k})$. Thus all the dynamical effects are now expressed by the displacement-displacement correlation functions $\langle\mathbf{u}_l(0)\mathbf{u}_{l'}(t)\rangle$ in the second exponential on the right side in (7.31). Its expansion in powers of $\langle\mathbf{u}_l(0)\mathbf{u}_{l'}(t)\rangle$ leads then to a series of multiphonon processes. The first order linear term in this expansion corresponds to one-phonon processes (see Sect. 7.6).

The differential reflection coefficient for one-phonon processes in the Born approximation is then obtained by substituting (7.28)–(7.31) into (7.18)

$$\frac{d^{2}\mathfrak{R}^{(1)}}{dE_{f}d\Omega_{f}}=\frac{m^{2}}{(2\pi\hbar)^{3}\rho^{2}\mathrm{A}}\frac{k_{f}}{|k_{iz}|}\,w(\mathbf{k}_{f},\mathbf{k}_{i}),\tag{7.33a}$$

with

$$w(\mathbf{k}_f, \mathbf{k}_i) = \frac{1}{\hbar} \sum_{Qv} \left| \mathbf{F}_{fi} \cdot \mathbf{u}_{Qv} \right|^2 \{ n(\hbar\omega_{Qv})\, \delta(E_f - E_i - \hbar\omega_{Qv}) + (1 + n(\hbar\omega_{Qv}))\, \delta(E_f - E_i + \hbar\omega_{Qv}) \}, \tag{7.33b}$$

where the Bose-Einstein occupation number $n(\hbar\omega_{Qv})$ is defined by (7.12), and

$$\mathbf{F}_{fi} \equiv -ie^{-W(\Delta\mathbf{k})} V(\Delta\mathbf{k})\Delta\mathbf{k} \tag{7.34}$$

is the atom-phonon linear coupling force in the Born approximation. Equation (7.33b) introduces the frequently used transition rate $w(\mathbf{k}_f, \mathbf{k}_i)$, defined by (7.6) and now expressed as a function of final and initial atom wavevectors. The terms proportional to $1 + n(\hbar\omega_{Qv})$ and $n(\hbar\omega_{\mathbf{Q}v})$ describe one-phonon creation and annihilation processes, respectively. The temperature dependence of these terms are shown in Fig. 7.2.

For a polyatomic crystal the scalar product in (7.33b) extends over all degrees of freedom of the surface unit cell and, with the addition of index κ labelling the atoms in each unit cell, it becomes

$$\mathbf{F}_{fi} \cdot \mathbf{u}_{Qv} \rightarrow \sum_{\kappa} \mathbf{F}_{fi,\kappa} \cdot \mathbf{u}_{\mathbf{Q}v,\kappa} = -i \sum_{\kappa} \mathrm{e}^{-W_\kappa(\Delta\mathbf{k})} V_\kappa(\Delta\mathbf{k})\Delta\mathbf{k} \cdot \mathbf{u}_{\mathbf{Q}v,\kappa}, \tag{7.35}$$

where $V_\kappa(\Delta\mathbf{k})$ is the three-dimensional Fourier transform of the potential $V_\kappa(\mathbf{r})$ between the atom and the κ-th atom of the surface unit cell. Equation (7.35) can be interpreted as the work done by the impinging atom on the vibrating surface. The convention that the phonon wavevector $\mathbf{Q}$ is restricted to lie within the first Brillouin zone requires $\mathbf{Q} + \mathbf{G} = \mathbf{K}_i - \mathbf{K}_f$, with $\mathbf{G}$ a surface reciprocal lattice vector, while the change in the normal wavevector $\Delta k_z = k_{fz} - k_{iz} = k_{fz} + |\, k_{iz}|$ is then determined by energy conservation for the given scattering geometry.

As pointed out at the beginning of this section the Born approximation derived here is too crude to be of direct use for calculating the inelastic differential reflection coefficient. It does, however, have the important advantage that it illustrates the important connections between the reflection coefficient and the physical processes. Equation (7.33a, b) indicates that the differential reflection coefficient is proportional to the surface phonon spectrum weighted by the square of the coupling potential energy $\mathbf{F}_{fi} {\cdot} \mathbf{u}_{\mathbf{Q}v}$. Thus if the phonon spectrum is known from lattice dynamical calculations, it is clear that the inelastic scattering intensities provide information about the atom-surface interaction potential. In practice the relationship between scattering intensities and potentials may not, however, be as simple and transparent as described by the Born approximation. Nevertheless the formal structure of (7.33a, b) remains valid for further levels of approximation, provided the atom-phonon linear coupling force $\mathbf{F}$ is given the appropriate form.

In the distorted-waves Born approximation (DWBA) the inelastic scattering of the atoms by the surface is accounted for by the matrix element of the time-dependent perturbing potential. This, however, is no longer taken between the plane waves $\rho^{1/2}\exp(i\mathbf{k}_{i,f}\mathbf{r})$, as in the pure Born approximation, but between initial and final waves $\psi_m(\mathbf{k},\mathbf{r})$ ($m = i,f$) defined by (7.23), which are severely distorted by the reflection from the static surface potential. According to (7.23) and (7.24), the matrix element $V_{fi}(t)$ is now given by a more complicated equation than (7.28), that is

$$V_{fi}(t) = \sum\nolimits_{l\kappa,\mathbf{GG}'} \exp\left[\, i(\mathbf{K}_f - \mathbf{K}_i + \Delta\mathbf{G})\cdot \mathbf{r}_{l\kappa}\right] \int dq_z V_\kappa(\mathbf{q}) s_{fi}^{\mathbf{GG}'}(q_z) e^{i\mathbf{q}\cdot\mathbf{u}_{l\kappa}(t)}, \tag{7.36}$$

where $\mathbf{q} \equiv (\mathbf{Q}, q_z)$, $\Delta\mathbf{G} = \mathbf{G}' - \mathbf{G}$, and

$$s_{fi}^{\mathbf{GG}'}(q_z) \equiv \int dz\ e^{-iq_z z} \zeta^*_{\mathbf{K}_f+\mathbf{G}'}(k_{fz}, z)\zeta_{\mathbf{K}_i+\mathbf{G}}(k_{iz}, z). \tag{7.37}$$

Note that q_z in (7.36) is an integration variable and therefore $\mathbf{q} \neq \Delta\mathbf{k}$. For a highly corrugated static potential a rather complicated close-coupling calculation may be required in order to obtain the Fourier component waves $\zeta_{m,\mathbf{K}+\mathbf{G}}(k_z, z)$ for a suitable number of $\mathbf{G}$ vectors [41].

A significant simplification is possible when also the corrugated part of the static potential is assumed to be small and is therefore incorporated into the perturbation. This is the first order DWBA as originally derived by Cabrera, Celli and Manson [34, 35]. In this case the distorted-waves are used to solve the Schrödinger equation for the flat surface and therefore all components $\zeta_{m,\mathbf{K}+\mathbf{G}}(k_z, z)$ for $\mathbf{G} \neq 0$ are neglected. The factorization (7.24) holds, with a complete decoupling between the normal component of the atom motion and the free propagation along the surface [$\varphi_{\mathbf{K}}(\mathbf{R}) = 1$]. The matrix elements of (7.37) then reduce to

$$s_{fi}^{00}(q_z) \equiv s_{fi}(q_z) = \int dz\ e^{-iq_z z} \chi^*_{\mathbf{K}_f}(k_{fz}, z)\chi_{\mathbf{K}_i}(k_{iz}, z) \tag{7.38}$$

and the DWBA coupling force is found to be

$$\mathbf{F}_{fi,\kappa} = -i \int dq_z\, s_{fi}(q_z) e^{-W_\kappa(\mathbf{q})} \mathbf{q} V_\kappa(\mathbf{q})\,. \tag{7.39}$$

It is important to remark that the Debye-Waller exponent $W_\kappa(\mathbf{q})$ depends now on the integration variable q_z. In this case no simple separation of the Debye-Waller factor from the potential components can be made as in (7.34).

It is an easy matter to prove that if the initial and final waves are not distorted, then $\chi(k_z, z) = exp\,(ik_z z)$ and the matrix element in (7.38) becomes $s_{fi}(q_z) = \delta(q_z - k_{fz} + k_{iz})$, which simply accounts for conservation of momentum. Thus with $q_z = \Delta k_z$ (7.39) yields the simple Born approximation result, (7.34). Note that for incident and final plane waves distorted so as to decay exponentially inside the crystal (with the substitution $k_z \to k_z + i\,\Theta(-z)\kappa$ for $z<0$, $\kappa > 0$) the function $s_{fi}(q_z)$ remains sharply peaked around Δk_z, so that the above Born-approximation expression for the force $\mathbf{F}_{fi}$ may be an acceptable approximation in a first analysis of inelastic intensities.

7.3.2 Jackson-Mott Analytical Solution

To illustrate the merits of the DWBA it is useful to consider the example of atom scattering from a monoatomic, low index metal surface, where the surface corrugation is very small, and to adopt the separable, purely repulsive exponential potential $V(z) \propto \exp(-\beta z)$ described in Chap. 6. As first shown by Zener [42] and Jackson and Mott [43] already in 1932 and 1933, this approximation allows for an analytical solution. In fact the distorted incident and final waves, $\chi(k_{iz}, z)$ and $\chi(k_{fz}, z)$, respectively, are simplified since they describe the motion of the atom normal to the surface in the average repulsive potential. Under these circumstances the coupling matrix elements for the exponential potential can be calculated in closed form [43].

For a simplified Debye-Waller factor which is assumed independent of q_z and for $\mathbf{G} = 0$, which is valid for a flat surface, the coupling force is found to be

$$\mathbf{F}_{fi} = a_c^{-1} S(p_i, p_f) A_Q \, \mathrm{e}^{-W}(-i\mathbf{Q},\beta), \tag{7.40}$$

where $(-i\mathbf{Q},\beta)$ is an abbreviation for a three-dimensional vector of components $(-iQ_x, -iQ_y, \beta)$, A_Q is the cut off factor given by (6.54), and

$$S(p_i, p_f) = \frac{p_f^2 - p_i^2}{\sinh^2 p_f - \sinh^2 p_i} \left(\frac{\sinh 2p_f \sinh 2p_i}{4 p_f p_i} \right)^{1/2} \tag{7.41}$$

is called the Jackson-Mott *factor* [43]. It depends on the potential parameter β via the dimensionless parameters $p_i \equiv \pi |k_{iz}|/\beta$ and $p_f \equiv \pi |k_{fz}|/\beta$ and its dependence on the sum and difference of p_i and p_f is shown in Fig. 7.3. For elastic scattering

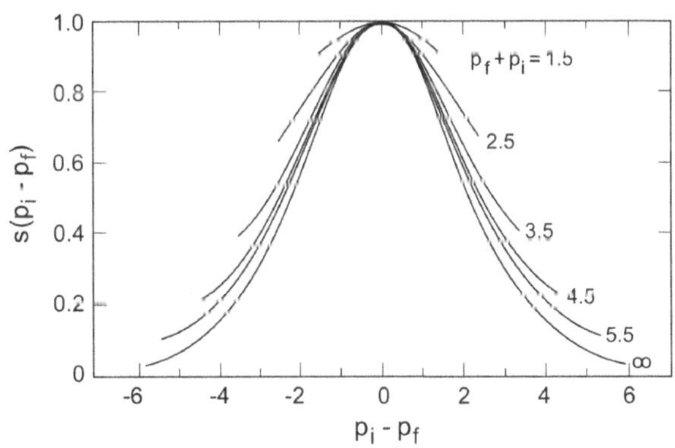

Fig. 7.3 The function appearing in (7.41), which contains the deviations from the hard wall limit of the Jackson-Mott matrix elements, is plotted as a function of $p_i - p_f \equiv \pi(|k_{iz}| - |k_{fz}|)/\beta$ for different values of $p_i + p_f \equiv \pi(|k_{iz}| + |k_{fz}|)/\beta$

($p_f = p_i$), $S = 1$. Otherwise $1 \geq S \geq 0$, with the limits $S = 1$ for the hard-wall case ($\beta \to \infty$, i.e., $p_i \to 0$ and $p_f \to 0$) and $S = 0$ for an infinitely soft potential ($\beta \to 0$, i.e., $p_i \to \infty$ and $p_f \to \infty$ with $p_i \neq p_f$). In the limit of very small corrugations the coupling force is normal to the surface and only surface modes with a normal displacement component can be excited.

7.3.3 Cabrera-Celli-Manson DWBA for Corrugated Surfaces

The force matrix elements for ionic crystal surfaces, (7.39), cannot be calculated analytically as for the Jackson-Mott case due to the complications arising from the significant surface corrugation. There are however numerical results which have been obtained within the DWBA in the Cabrera-Celli-Manson form [34, 35] for He scattering from the LiF(001) surface [44]. In the example shown in Fig. 7.4 the calculated matrix elements of the forces exerted by the He atom on the Li^+ and F^- ions in a one-phonon inelastic scattering process are plotted as functions of the transferred energy. The calculations were based on Lennard-Jones interatomic potentials, Sect. 6.3, with the distorted wavefunctions expressed in the semiclassical Wentzel-Kramers-Brillouin (WKB) approximation. The incident energy is $E_i = 112$ meV, the incident angle $\theta_i = 45°$ and the final angle chosen so as to correspond to a fixed phonon wavevector $\mathbf{Q} = (^3/_8,0)(2\pi/a)$ (a = lattice constant) for two different **G**-vectors.

In Fig. 7.4 full and dashed lines represent the z- and the x-components of the forces, to be read on the left- and right-hand ordinate scales, respectively. It appears that the coupling forces decrease rapidly for increasing energy transfer at $\mathbf{G} = 0$, whereas a more complex behavior is found for $\mathbf{G} \neq 0$. As expected the forces are larger for F^- ions than for Li^+ ions due to the much larger size of the former. Moreover at $\mathbf{G} = 0$ the z-components of $\mathbf{F}_{fi,\kappa}$ are much larger than the x-components, but they become comparable at larger **G**-vectors as an effect of the surface corrugation. Thus the coupling to quasi-SV modes is favored at small parallel momentum transfers $\Delta\mathbf{K}$, whereas at larger $\Delta\mathbf{K}$, involving one or more **G**-vectors, quasi-longitudinal modes should be more easily observed. Since the components of $\mathbf{F}_{fi,\kappa}$ for either $\kappa = Li^+$ or F^- have the same sign, the sum over κ in (7.35) is larger for acoustic surface modes, where the ions move almost in phase, than for optical modes, where the ions move with almost opposite phases, and their contributions partially cancel. Such a partial cancellation in the interaction with optical modes and the weakening of the coupling forces for increasing energy transfer yield a rapid decrease in the phonon response at high energies. This effect is clearly seen in Fig. 7.5 where the calculated surface phonon DOS for acoustical and optical surface modes on LiF(001) at the $\overline{M}$ point are compared with the DWBA inelastic scattering intensities calculated for a constant-**Q** scattering configuration ($\theta_f = 0$, cfr. Fig. 9.2b) [45]. For an incident energy of 65 meV a parallel momentum transfer corresponding

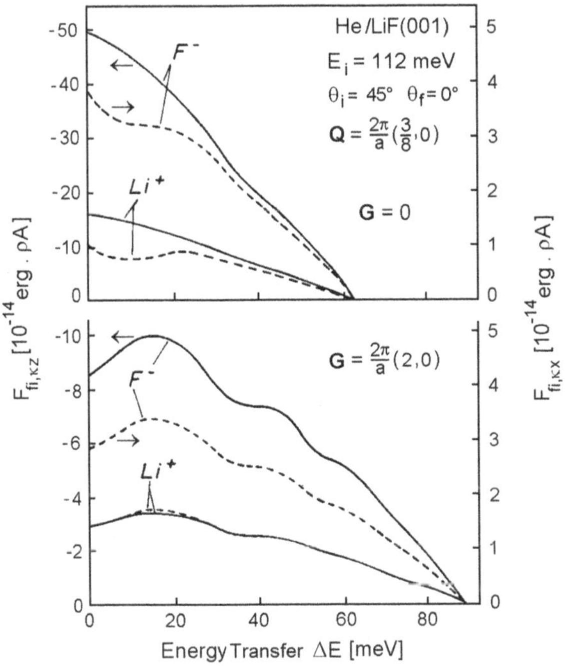

Fig. 7.4 Matrix elements of the forces, (7.39), exerted by the He atom on the Li^+ and F^- ions of the LiF(001) surface in a one-phonon inelastic scattering process as functions of the transferred energy. The calculation, based on DWBA and Lennard-Jones interatomic potentials [44], is made for an incident energy E_i = 112 meV, incident angle $\theta_i = 45°$ and final angle θ_f chosen so as to correspond to a fixed phonon wavevector $\mathbf{Q} = (^3/_8,0)(2\pi/a)$ (a = lattice constant) for two different **G**-vectors, $G = 0$ and $G = (2\pi/a)(2,0)$. Full and dashed lines represent the z-component (left scale) and the x-component (right scale) of the forces, respectively

to the $\overline{M}$ point is obtained with an incident angle of 6.14°. While the RW (S_1) contributes about 15% to the DOS its contribution to the scattering intensity rises to 50% at the expense of the optical region which is severely depressed.

By integrating the inelastic HAS intensity over the transferred energy one obtains the angular reflection coefficient

$$\frac{d\Re^{(1)}}{d\Omega_f} = \int dE_f \frac{d^2\Re^{(1)}}{dE_f d\Omega_f}. \tag{7.42}$$

Calculations performed by Wang et al. [46] for the Cu(111) surface at different surface temperatures on the basis of a DBWA formalism [47] show that the angular reflection coefficient decays very rapidly, at a more than exponential rate for increasing parallel wavevector. Though this behavior is expected for a flat surface like that of Cu(110), a similar result was also found for a corrugated surface such as LiF(001).

This is illustrated in Fig. 7.6 which shows a DWBA calculation of $d\Re^{(1)}/d\Omega_f$ for one-phonon inelastic helium atom scattering from LiF(001) as a function of the

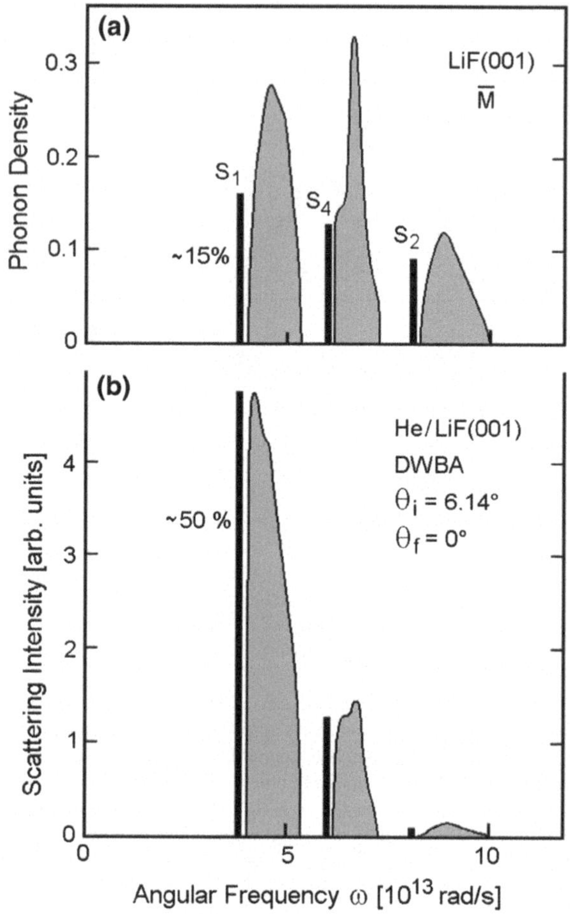

Fig. 7.5 Comparison of the density of states of surface-localized modes (vertical bars) and bulk bands projected on the surface (grey areas) with the HAS intensities calculated with DWBA for LiF(001). **a** The density of states of the S_1 (RW) and of the two optical surface modes S_2 and S_4 and the surface-projected phonons at the $\overline{M}$-point normalized to unity. **b** The calculated intensities of the same phonons for an arrangement in which the incident angle $\theta_i = 6.14°$ and the final angle $\theta_f = 0°$. In this arrangement (Fig. 9.2a $\mathbf{K}_i = \mathbf{Q}$. Since in the optical modes the adjacent ions move oppositely their contributions to the inelatic intensity in (**b**) are greatly suppressed compared to the RW

parallel momentum transfer [44]. The latter increases as the final angle decreases away from the incident angle of $\theta_i = 45°$. At the same time the angular reflection coefficient decreases dramatically, by over 8 orders of magnitude. The total reflection coefficient $\Re^{(1)}$, resulting from an integration with respect to the final solid angle over 2π, is about 0.18. Despite the comparatively high incident energy of $E_i = 112$ meV and low surface temperature of 0 K. only 18% of the total flux goes into inelastic one-phonon channels. This is consistent with the fact that LiF is a

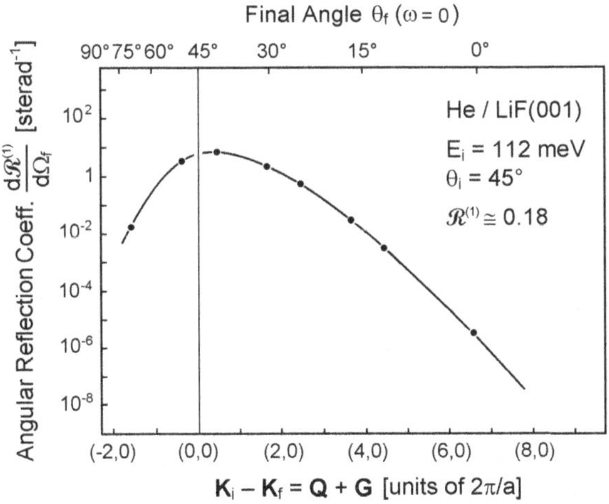

Fig. 7.6 The angular reflection coefficient for inelastic one-phonon HAS from LiF(001) calculated in the DWBA as a function of the parallel momentum transfer (lower abscissa) and of the corresponding final angle for quasi-elastic processes (upper abscissa). The calculation has been performed at distinct values of **Q** + **G** (black points, interpolated by the solid curve) for an incident energy $E_i = 112$ meV, incident angle $\theta_i = 45°$, the scattering plane in the [100] direction, and a surface temperature of 0 K [44]. The total one-phonon reflection coefficient, obtained by integration over the final solid angle, is $\mathfrak{R}^{(1)} \cong 0.18$

rather stiff crystal with a maximum phonon frequency at 65 meV. The contribution of inelastic processes in the total scattering intensity will be discussed as a function of surface temperature and crystal stiffness in connection with the Debye-Waller factor in the next Sect. 7.6.

Although in good qualitative agreement with the subsequent experimental findings, the DWBA calculations of the inelastic HAS intensities were questioned in all cases where the static surface corrugation could not be treated as a weak perturbation, e.g., in those ionic crystals where the difference between anion and cation radii is large (as in LiF) [45]. The next Section illustrates alternative approaches in which the static surface corrugation can be treated exactly by numerical methods or within the eikonal approximation and the DWBA approximation is restricted to the small dynamical corrugation induced by surface vibrations.

7.4 The GR Method and the Eikonal Approximation

For corrugated surfaces an exact solution of the scattering problem can be obtained with numerical techniques such as the GR method introduced by Garcia et al. [49–51]. This approach is a valid alternative to the DWBA, though at the cost of a simplified picture for the atom-surface potential. The results are particularly simple for the hard corrugated surface (HCS) potential which is shown in Fig. 6.4. In the following a time-dependent HCS potential without an attractive well will be considered first:

$$V(\mathbf{r},t) = \begin{cases} \infty & \text{for} \quad z \le D(\mathbf{R},t), \\ 0 & \text{for} \quad z > D(\mathbf{R},t), \end{cases} \tag{7.43}$$

where

$$D(\mathbf{R},t) = D_0(\mathbf{R}) + \delta D(\mathbf{R},t) \tag{7.44}$$

is the surface corrugation function. It is the sum of the static surface corrugation $D_0(\mathbf{R})$ and the dynamic surface corrugation $\delta D(\mathbf{R}, t)$ produced by the vibrations of the surface atoms. The dynamic positions

$$\mathbf{d} = (\mathbf{R}, D(\mathbf{R},t)) \tag{7.45}$$

define the locus where the potential becomes infinite or, equivalently, the locus of the classical turning points. The decomposition (7.44) of the HCS potential with a static part has the advantage that the corresponding T-matrix can be evaluated exactly, though numerically, for any corrugation and the DWBA is only needed to treat the small dynamic perturbation. This approach, first developed for inelastic scattering by Benedek and Garcia [52–55] and by Armand and Manson [56], is here only briefly formulated as given in [57] with a slight change in notation.

The atom wave function is represented as a mixture of the incident and all the final state plane waves

$$\psi(\mathbf{r},t) = e^{i\mathbf{k}_i\cdot\mathbf{r} - iE_i t/\hbar} + \int d^3k_f \, A_{fi}(t) \, e^{i\mathbf{k}_f\cdot\mathbf{r} - iE_f t/\hbar}, \tag{7.46}$$

where the time-dependent amplitudes $A_{fi}(t)$ are related to the T—matrix elements, (7.19), by

$$A_{fi}(t) = \frac{i}{4\pi^2\hbar\rho} \int_{-\infty}^{t} dt' \, e^{i\omega t'} \, T_{fi}(t'). \tag{7.47}$$

Thus $A_{fi}(+\infty) = 0$, while $A_{fi}(+\infty)$ is proportional to the Fourier transform of $T_{fi}(t)$. It is convenient to express $T_{fi}(t)$ as a generalized Fourier transform in the two-dimensional surface space:

$$T_{fi}(t) = \int d^2R \, e^{i\Delta\mathbf{k}\cdot\mathbf{d}} f(\mathbf{R},t), \tag{7.48}$$

where $f(\mathbf{R}, t)$ is a surface energy density carried by the incident atoms and redistributed among all scattered waves. In analogy to the Fresnel picture in classical optics it is called the *source function.*

The scattering amplitudes are determined by the surface boundary conditions and their derivation is complicated by the fact that the surface is oscillating in space and time. The HCS model, however, imposes the simple boundary condition $\psi(\mathbf{d}, t) = 0$ at all times and surface positions, which leads, in combination with (7.46–7.48), to an integral equation for the source function

$$\int d^2R'dt' \; M(\mathbf{R},\mathbf{R}',t,t')f(\mathbf{R}',t') = -1, \qquad \forall \mathbf{R}, t \tag{7.49}$$

where

$$M(\mathbf{R},\mathbf{R}',t,t') = \frac{i}{4\pi^2\hbar\rho}\,\Theta(t-t')\int d^3k_f \; e^{i\Delta\mathbf{k}\cdot(\mathbf{d}-\mathbf{d}')}e^{-i\omega(t-t')} \tag{7.50}$$

and $\Theta(t)$ is the Heaviside unit step function. At this stage of the calculation the DWBA is used by expanding M and $f(\mathbf{R}, t)$ to first order in $\delta D(\mathbf{R}, t)$ as

$$M = M_0 + \mu, \qquad f = f_0 + \varphi, \tag{7.51}$$

where the kernel M_0 for the static potential is defined by (7.50) with $\mathbf{d} = \mathbf{d}_0 \equiv (\mathbf{R}, D_0(\mathbf{R}))$, and the static source function f_0 solves the integral equation

$$\int d^2R' \; M_0(\mathbf{R},\mathbf{R}')f_0(\mathbf{R}') = -1, \qquad \forall \mathbf{R}. \tag{7.52}$$

Thus φ can be formally expressed as

$$\varphi = -M_0^{-1}\mu f_0 \tag{7.53}$$

The inverse of the matrix M_0 as well as the static source function f_0 can be obtained numerically from (7.52) by means of the GR method [49–51]. From this the time-dependent source function and the T-matrix elements, (7.48), are obtained. The coupling forces, readily derived from the dynamic one-phonon part of the T-matrix, are then given by [54, 55, 57–59]

$$\mathbf{F}_{fi,\kappa} = \frac{\Delta k_z}{2\pi}\int \mathrm{d}^2R \; e^{i\Delta\mathbf{k}\cdot\mathbf{d}(\mathbf{R})}e^{-W(\mathbf{R})}f_o(\mathbf{R})\,\frac{\partial D_o(\mathbf{R})}{\partial \mathbf{r}_\kappa}, \tag{7.54}$$

where $\partial D_o(\mathbf{R})/\partial \mathbf{r}_\kappa$ represents the variation of surface corrugation induced by the displacement of the κ-th atom in the surface unit cell and

$$W(\mathbf{R}) = \frac{1}{2}(\Delta k_z)^2\left\langle(\delta D(\mathbf{R},t))^2\right\rangle \tag{7.55}$$

is the Debye Waller exponent for the HCS model.

A systematic analysis of HAS spectra from LiF(001) in the HCS scheme [57] has proven to be quite satisfactory with regard to the general features, particularly in the acoustic region and away from bound state resonances which arise from the potential well, which was not included. This is illustrated in Fig. 7.7 which shows a direct comparison between the HAS inelastic scattering intensities with the calculated differential reflection coefficient. The good agreement between the positions and relative intensities of the observed peaks with the calculations enables an assignment of the main experimental features to the Rayleigh (S_1) phonons.

For soft corrugated repulsive potentials (7.53) can still be used provided the surface profile $D_0(\mathbf{R})$ is interpreted as the locus of classical turning points at a given incident energy of the atoms. However, the implicit dependence of the surface profile on the incident energy introduces a substantial complication in the calculation of $f_0(\mathbf{R})$ with the GR method. Thus the use of more realistic soft potentials may cost some approximation for $f_0(\mathbf{R})$. Levi has shown that the *eikonal approximation*, according to which

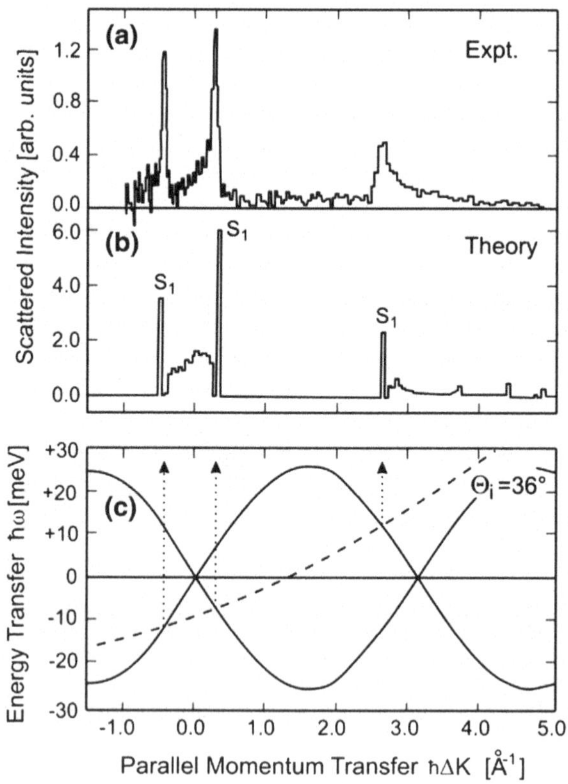

Fig. 7.7 Comparison of an experimental inelastic HAS spectrum with a calculation based on the HCS potential model [55]. **a** The time-of-flight spectrum projected on the parallel momentum transfer ΔK for the LiF(001) room temperature surface along the [60] direction at an incident angle $\theta_i = 36°$, and incident wavevector $k_i = 6.10$ Å^{-1} [61]. **b** The HCS calculation based on the GF method and the breathing shell model for the surface lattice dynamics [57]. **c** The Intersections of the scan curve (broken line) with the theoretical RW (S_1) dispersion curves (full lines) plotted on both creation and annihilation sides over two surface Brillouin zones, indicate that the three main peaks are due to RWs

$$f_0(\mathbf{R}) \sim \frac{2\pi \hbar^2 \rho}{m} (ik_{iz}), \tag{7.56}$$

works quite well if the surface corrugation is not too large [36, 62]. Levi et al. [63] performed systematic calculations of the inelastic HAS spectra for the NaF(001) surface based on the eikonal scattering approximation and two-body Morse potentials for the He–Na^+ and the He–F^- interactions. They treated the surface lattice dynamics with the surface GF matching technique and the BSM [64]. NaF (001) was the first surface where optical surface modes were observed by HAS [65], and provides an ideal benchmark for the inelastic scattering theories. As compared to HCS a soft corrugated potential does not overweigh the high frequency part of the phonon spectrum. Indeed calculations by Levi et al. were found to be in a very

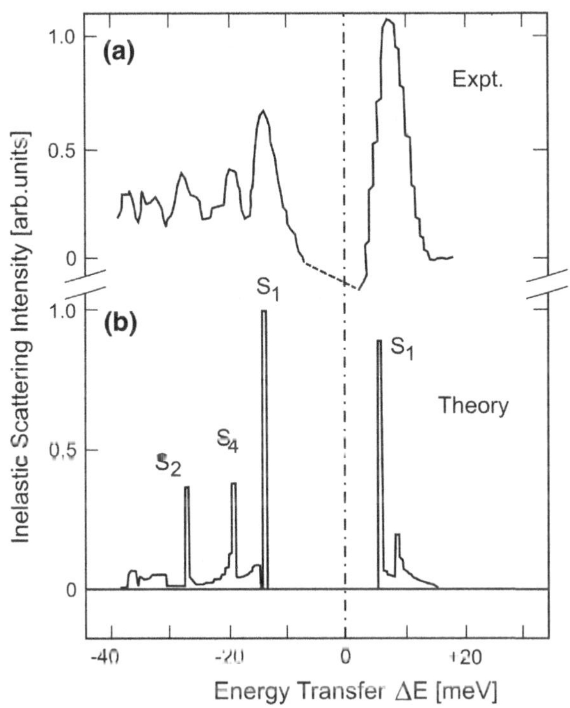

Fig. 7.8 Comparison of an experimental and theoretical inelastic HAS energy loss spectrum. **a** The experimental spectrum from the room temperature (001) surface of NaF along the (100) azimuth for an incident wavevector k_i = 13.36 Å^{-1} (E_i = 93.0 meV) [65]. The spectrum was corrected by subtraction of the multiphonon background and of the incoherent elastic peak. The peaks labelled S_1 correspond to Rayleigh waves and S_2 and S_4 to surface optical modes (see Table 3.1). **b** Calculation based on a soft corrugated surface potential and the eikonal approximation for the scattering theory, and the GF matching method for the surface lattice dynamics [63]

good agreement with the experimental TOF spectra at least in all cases where inelastic bound state resonances were absent [63]. This is illustrated in the example shown in Fig. 7.8, which provides convincing evidence that under suitable kinematical conditions surface optical modes like S_4 and S_2 can indeed be detected with HAS [65].

For the observation of the optical surface modes the incident energy of the helium beam was increased from typically 60 meV or less to 93 meV with the disadvantage of an increased multiphonon background. This together with the diffuse elastic peak has been empirically subtracted from the experimental HAS spectrum of Fig. 7.8a for a closer comparison with the calculated one-phonon spectrum. The multiphonon background involves, with the appropriate weights, all possible combinations of two or more surface phonon wavevectors and energies which sum up to the actual transferred wavevector and energy, and is therefore an unstructured broad function of ΔE. A more physical procedure requires a theory of the multiphonon scattering, as, for example, outlined in Sect. 7.6.

7.5 Selection Rules and Propensities

7.5.1 Insulator Surfaces and Pairwise Additive Potentials

An important question in connection with the interpretation of one-phonon time-of-flight spectra is the dependence of the inelastic intensities on the polarization of the phonon, e.g. transverse sagittal (SV), longitudinal sagittal (L) or shear horizontal (SH), and the character (acoustical or optical) (see Sect. 3.7.1). As is well known from neutron studies of bulk modes the proper assignment of the observed phonons is unique only if the polarization of the modes is properly accounted for [66–69]. Since the polarization cannot be measured directly it may nevertheless be inferred from the dependence of the inelastic intensities of the different types of modes on the kinematical conditions and on the wavevector transfer.

These selection rules and propensities can be most clearly formulated for insulator or surfaces which are well characterized by pairwise additive potentials. In this case the excitation probabilities can be understood by examining the DWBA expression for the work exerted by the scattering atom. This can be written with the aid of (7.35) and (7.39), as

$$\sum_{\kappa} \mathbf{F}_{fi,\kappa} \cdot \mathbf{u}_{Q\nu,\kappa} = \int dq_z s_{fi}(q_z) \sum_{\kappa} V_{\kappa}(\mathbf{q})\, e^{-W_{\kappa}(\mathbf{q})} \left(\Delta\mathbf{K} \cdot \mathbf{u}_{Q\nu,\kappa} + q_z u_{Q\nu,\kappa z} \right), \tag{7.57}$$

where $\mathbf{q} \equiv (\Delta\mathbf{K}, q_z)$. In order for one of the terms in the brackets on the right hand side of (7.57) to remain finite the phonon displacement vector $\mathbf{u}_{Q\nu,\kappa}$ must have a component in the scattering plane determined by $\Delta\mathbf{K}$ and q_z. If $\Delta\mathbf{K} \equiv \mathbf{Q} + \mathbf{G}$ is parallel to one of the crystallographic symmetry directions, as in the case of inelastic HAS experiments in the planar scattering geometry, then the He atom couples only to the sagittal modes. Under this condition the excitation of the shear horizontal mode is forbidden since $\Delta\mathbf{K}.\ \mathbf{u}_{\mathbf{Q}\nu,\kappa} = 0$.

This argument is strictly valid only when the phonon wavevector, and hence the scattering plane, are in a symmetry direction of the surface of the crystal surface and the sagittal plane is a mirror symmetry plane of the crystal, so that surface phonons are either sagittal or SH. If the sagittal plane is not a mirror plane, as, e.g., in the $\overline{\Gamma\mathrm{K}}$ direction of the (110) surface of a monoatomic fcc crystal, then the surface modes are only approximately sagittal or SH, the latter having a small sagittal component (see Fig. 3.12). In this case a weak scattering intensity is expected also from nominal SH modes even in a planar scattering configuration along a symmetry direction. Away from the symmetry directions, mirror symmetry is always lost, and the surface phonon polarizations deviate substantially from the pure sagittal or SH orientations, and therefore can all become active in inelastic HAS. These arguments are obviously independent of the probe and therefore apply equally well to EELS in the impact scattering regime [70–72].

For a scattering plane orientation away from a symmetry direction, or nonplanar scattering ($\mathbf{K}_f$ not parallel to $\mathbf{K}_i$) a pure SH mode with $\mathbf{G} \neq 0$ can be excited by choosing a $\mathbf{G}$ vector perpendicular to $\mathbf{Q}$ [73]. This method and other experimental aspects of SH-mode spectroscopy by HAS and EELS are discussed in the Chap. 9.

Most experiments are, however, carried out with the scattering plane along symmetry directions and are sensitive only to sagittal modes. As explained in Sect. 3.7 and Table 3.1, sagittal modes are either quasi-SV ($\sim$SV, e.g., the RW) or quasi-longitudinal ($\sim$L, e.g. the S_3 mode). Both have components in the surface and normal to it, which are out of phase so as to give an elliptical polarization in the sagittal plane. The $\sim$SV modes have their major axis normal to the surface whereas the $\sim$L modes have their major axis in the surface along the direction of $\mathbf{Q}$ (see Fig. 3.13).

To understand their contributions it is useful to transform (7.57) into a more explicit form. By inserting the expression of $s_{fi}(q_z)$ given by (7.38) for $\mathbf{G} = \mathbf{G}' = 0$, neglecting the q_z-dependence of $W_\kappa(\mathbf{q})$ and integrating by parts, (7.57) becomes

$$\sum_\kappa \mathbf{F}_{fi,\kappa} \cdot \mathbf{u}_{\mathbf{Q}\nu,\kappa} = \sum_\kappa e^{-W_\kappa} v_{\kappa,fi}(\Delta\mathbf{K})[\Delta\mathbf{K} \cdot \mathbf{u}_{\mathbf{Q}\nu,\kappa} - i\beta_\kappa(\Delta\mathbf{K})u_{\mathbf{Q}\nu,\kappa z}]. \qquad (7.58)$$

The coefficients

$$\beta_\kappa(\Delta\mathbf{K}) = -V'_{\kappa,fi}(\Delta\mathbf{K})/V_{\kappa,fi}(\Delta\mathbf{K}), \qquad (7.59)$$

give the slope of the $\Delta\mathbf{K}$-th Fourier component of the potentials $V_\kappa(\mathbf{r})$ between the atom and the κ-th ion of the surface unit cell, $V'_{\kappa,fi}(\Delta\mathbf{K})$ and $V_{\kappa,fi}(\Delta\mathbf{K})$ being the matrix elements of the 2D Fourier transforms of $\partial V_\kappa(\mathbf{r})/\partial z$ and $V_\kappa(\mathbf{r})$, respectively.

For small momentum transfers $\Delta\mathbf{K} = \mathbf{Q}$ in the first Brillouin zone and for a simple exponential repulsive potential with $\beta_\kappa(\mathbf{Q}) = \beta$ of the order of 2 to 4 Å^{-1} (see Table 6.5), the second term on the right hand side of (7.58) is the larger and the vertical component of a sagittal phonon $u_{\mathbf{Q}\nu,\kappa z}$ is dominant. For larger momentum transfers $\Delta\mathbf{K} = \mathbf{Q} + \mathbf{G}$ the first term which includes longitudinal components is more important. In principle the comparison of scattering amplitudes for the same sagittal phonon measured in two different Brillouin zones should make it possible to determine the two Cartesian components of the polarization vector. Also the comparison with measurements made outside the direction of symmetry which provides only the vertical component $u_{\mathbf{Q}\nu,\kappa z}$ (see above) could help.

When there are more atoms in the surface unit cell (κ = 1, 2,…, s) and optical surface phonons occur, the scattering amplitude will contain both SV and L contributions arising from the different atoms, which are now weighted by the potential matrix elements $V'_{\kappa,fi}(\Delta\mathbf{K})$ and $V_{\kappa,fi}(\Delta\mathbf{K})$, respectively. For atoms of the same size the elements $V'_{\kappa,fi}(\Delta\mathbf{K})$ and $V_{\kappa,fi}(\Delta\mathbf{K})$ have comparable magnitudes and the same sign, whereas optical modes are characterized by a displacement field where the atoms of the unit cell move one against the other and their contributions tend to cancel each other at distances as large as that of the turning point. This is especially true at long wavelengths where the center of mass is almost at rest and is a general and basic reason why HAS spectroscopy is less sensitive to optical modes.

As discussed in the Chap. 10, the intensity enhancement accompanying selective adsorption resonances has been successfully used to amplify the response of optical modes, but in this case the determination of eigenvectors from scattering amplitudes would be even more complicated. Since the expression in squared parenthesis in (7.58) is complex, its squared modulus for real displacement components will be independent of the sign of $\Delta\mathbf{K} = \mathbf{Q} + \mathbf{G}$ (forward or backward propagation). However, also for elliptical polarization in the sagittal plane, where the ratio of the vibrational amplitudes $u_{\mathbf{Q}\nu,\kappa x}/u_{\mathbf{Q}\nu,\kappa z}$ is imaginary (as occurs, e.g., for RWs), the orientation of the elliptical polarization changes sign with $\mathbf{Q}$ so as to preserve time-reversal invariance.

7.5.2 The Dynamic Factor for the Exponential Surface Interaction

When the integration of the differential reflection coefficient over the final state energies E_f, leading to the one-phonon angular reflection coefficient $d\Re^{(1)}/d\Omega_f$ (7.42), is restricted to a narrow range of energies in the vicinity of a surface phonon energy the inelastic intensity associated with that phonon is obtained. Using the DWBA expression for the coupling force as derived from the Zener-Jackson-Mott (ZJM) potential, (7.40) and (7.41), the restricted integration for a phonon of wavevector $\mathbf{Q}$ and energy ΔE gives

$$\left.\frac{d\Re^{(1)}}{d\Omega_f}\right|_{\mathrm{Q},\Delta E} = \frac{m^2}{8\pi^3\hbar^4\rho^2 \mathrm{A}} K(\Delta E) J_\nu(\mathrm{Q}, \Delta E), \tag{7.60}$$

where

$$K(\Delta E) = A_0^2 (1 + \Delta E/E_i)^{1/2} S^2(p_i, p_f) e^{-Q^2/Q_c^2} \tag{7.61}$$

is a *kinematic factor* for the ZJM theory, which mainly depends on the scattering potential and geometry and is only weakly dependent on phonon dynamics, whereas

$$J_\nu(\mathbf{Q}, \Delta E) = \mathrm{e}^{-2W} |n(\Delta E)| \left| \mathbf{Q} \cdot \mathbf{u}_{\mathrm{Q}\nu} - i\beta\ u_{\mathrm{Q}\nu,z} \right|^2 \tag{7.62}$$

is a *dynamic factor* explicitly depending on phonon energies and displacements. The kinematic factor $K(\Delta E)$ exhibits an exponential decay for increasing $\mathbf{Q}$ due to the cut off factor $exp(-Q^2/Q_c^2)$ and the softness factor $S^2(p_i, p_f)$ defined in (7.41). For processes involving acoustic surface phonons the limit $\mathbf{Q} \to 0$ implies $\omega_{\mathbf{Q}\nu} \to 0$, $p_f \to p_i$ and $S^2(p_i, p_f) \to 1$ (see Fig. 7.3), whereas for increasing $\mathbf{Q}$ the factor $S^2(p_i, p_f)$ decays exponentially as either $p_f^3 exp(-2p_f)$ or $p_i^3 exp(-2p_i)$ depending on whether $p_f > p_i$ or $p_f < p_i$, respectively. The dynamic factor $J_\nu(\mathbf{Q}, \Delta E)$ also decreases for increasing $\mathbf{Q}$ since the quantized square displacement is proportional to $\hbar/2M\omega_{\mathrm{Q}\nu}$ and the Bose-Einstein factor for small $\mathbf{Q}$ is $\approx kT/\hbar\omega_{\mathrm{Q}\nu}$. These terms can be partially compensated for by an increase with $\mathbf{Q}$ of the polarization vectors, see (3.16), of the surface acoustic phonons at the surface layer due to the rapid decrease of their penetration length. In this respect, however, there is an important difference between RW and L modes. In a first approximation, a linear (i.e., non-elliptical) polarization is expected for both the RW and the L modes along the z-direction (normal to the surface) and in the x-direction (parallel to $\mathbf{Q}$), respectively. According to (3.16) and (3.33) the amplitude of Rayleigh phonons at the surface layer can be approximated by

$$u_{Qz,RW} \cong \left(\frac{\hbar}{2M\omega_{\mathbf{Q},RW}}\right)^{\frac{1}{2}} \sin^{1/2}(Qa/2), \tag{7.63}$$

where a is the lattice constant and M the mass of the surface atoms. Since the dispersion curve of a Rayleigh phonon with frequency $\omega_{\mathbf{Q},RW}$ can be approximated by a sine law, (3.15), $u_{Qz,RW}$ is simply a constant. It is worth noting that at least in cubic crystals no longitudinal surface mode exists in the continuum limit $Q \to 0$. In this limit the longitudinal surface mode velocity tends at least linearly to the bulk LA velocity, and therefore $u_{Qx,L}$ decreases at least as fast as $\sin(Qa/2)$. Thus the L mode contribution to $J_\nu(\mathbf{Q}, \Delta E)$ according to (7.62) vanishes for $Q \to 0$. However, similarly to the RW, the L resonance reaches its maximum amplitude and localization at the zone boundary ($Q = \pi/a$), with a longitudinal displacement $(\hbar/2M\omega_{\pi/a,L})^{1/2}$. This however is smaller than for the RW since $\omega_{\pi/a,L} > \omega_{\pi/a,RW}$.

The dynamic factor $J_\nu(\mathbf{Q}, \Delta E)$ can be obtained experimentally for different surface phonons and provide useful information on surface interactions. As discussed in Sect. 8.5 the analysis of the dynamic factor is especially interesting in the case of metal surfaces.

7.6 Multiphonon Scattering and Temperature Effects

7.6.1 Mechanisms for Multiphonon Scattering

The assumption that the inelastic response of the surface to the probe atom occurs through the creation or annihilation of one-phonon at a time relies on the following four approximations:

(i) The force exerted by the probe atom on the surface atoms is independent of the surface atom displacements: only the first linear term is retained on the r.h.s. of (7.3) (*linear coupling*).

(ii) Only one collisional event occurs in each scattering process: multiple collisions are neglected and only the first term in the expansion of the transition matrix element, (7.20), is retained (*Born approximation and DWBA*). If intermediate bound states are included to account for inelastic bound state resonances (see second term in equation (7.20)), only one vertex is assumed to be inelastic.

(iii) The surface atom displacements are assumed to be much smaller than the probe atom wavelength, so that the latter behaves as a quantum probe (*quantum regime*). For large incident energies (short wavelengths) of the probe atoms the terms in the exponent of (7.31) like $(\Delta k_z u_{lz})^2 \cong (4mE_i \cos^2\theta_i/\hbar^2)u_{lz}^2$ may not be sufficiently less than one. This would prevent retaining only the term proportional to $\langle \mathbf{u}_l(0)\mathbf{u}_{l'}(t)\rangle$ in the power expansion, unless experiments are made at grazing incidence ($\theta_i \approx 90°$), so as to keep $(\Delta k_z u_{lz})^2 \ll 1$. Diffractive and inelastic scattering processes have been observed even with He atoms in the keV energy domain at grazing incidence, [64, 63, 74–76] which is opening another promising area in the study of surface excitations by helium atom scattering (Chap. 13).

(iv) Small displacements are also assumed in the *harmonic approximation* for lattice dynamics, so that in the expansion of the crystal potential energy, (3.18), only quadratic terms are retained.

If any of the above approximations is not satisfied inelastic scattering from more than one-phonon at a time can occur. Assumptions (i) and (ii) requiring surface mean-square displacements much shorter than the closest atom-surface and crystal interatomic distances, respectively are more easily fulfilled. In ordinary HAS experiments the approximation (iii) is the most critical one since He atom wavelengths are only an order of magnitude larger than, or even comparable to, the mean-square displacements of surface atoms. However, in the Born approximation the multiphonon contributions, when approximation (iii) is not satisfied, can be readily incorporated in the differential reflection coefficient to all orders by means of the Fourier transform convolution theorem.

7.6.2 Multiphonon Processes in the Distorted-Wave Born Approximation

The expansion into a power series of the exponential in (7.31) with respect to the displacement-displacement correlation function $<\mathbf{u}_l(0)\mathbf{u}_{l'}(t)>$ leads to

$$\frac{d^2\Re}{dE_f d\Omega_f} = \frac{m^2}{4\pi^2\hbar^5\rho^2 \mathrm{A}}\frac{k_f}{|k_{iz}|}\,|\mathbf{v}(\Delta\mathbf{k})|^2 \sum\nolimits_{ll'} \mathrm{e}^{i\Delta\mathbf{k}\cdot(\mathbf{r}_l-\mathbf{r}_{l'})} \times \int_{-\infty}^{\infty}\frac{dt}{2\pi}\,\mathrm{e}^{-i\omega t}\exp[\Delta\mathbf{k}\cdot <\mathbf{u}_l(0)\mathbf{u}_{l'}(t)> \cdot\Delta\mathbf{k} - 2W] \tag{7.64a}$$

$$= \frac{m^2}{4\pi^2\hbar^5\rho^2 \mathrm{A}}\frac{k_f}{|k_{iz}|}\,|\mathbf{v}(\Delta\mathbf{k})|^2 \mathrm{e}^{-2W}\sum_{n=0}^{\infty}\frac{(2W)^n}{n!}B_n(\Delta\mathbf{K},\omega), \tag{7.64b}$$

where the n-phonon spectral response function $B_n(\mathbf{Q},\omega)$ is defined by the convolution

$$B_n(\mathbf{Q},\omega) = \int_{-\infty}^{\infty} d\omega' \sum\nolimits_{\mathbf{Q}'} B_{n-1}(\mathbf{Q}',\omega')\,B_1(\mathbf{Q}-\mathbf{Q}',\omega-\omega'), \tag{7.65}$$

with

$$B_0(\mathbf{Q},\omega) \equiv \delta_{\mathbf{Q},0}\delta(\omega), \tag{7.66}$$

$$B_1(\mathbf{Q},\omega) \equiv \frac{1}{2W}\sum\nolimits_{\nu}\frac{\hbar|\Delta\mathbf{k}\cdot\mathbf{e}_{\mathbf{Q}\nu}(0)|^2}{2N_L M\omega_{\mathbf{Q}\nu}} \times [(n(\omega_{\mathbf{Q}\nu})+1)\delta(\omega-\omega_{\mathbf{Q}\nu}) + n(\omega_{\mathbf{Q}\nu})\delta(\omega+\omega_{\mathbf{Q}\nu})] \tag{7.67}$$

and the usual definitions $\Delta\mathbf{k} = (\Delta\mathbf{K}, \Delta k_z)$, $-\Delta\mathbf{K} = \mathbf{Q} + \mathbf{G}$ and N_L the number of surface unit cells. The index ν runs over all surface and bulk eigenvectors of parallel wavevector $\mathbf{Q}$, $\omega_{\mathbf{Q}\nu}$ and $\mathbf{e}_{\mathbf{Q}\nu}(0)$ being the respective angular frequencies and polarization vectors at the surface layer. As customary the phonon wavevectors $\mathbf{Q}$ are restricted to lie within the surface Brillouin zone. However *umklapp* combinations in the convolution (7.65) are possible with the extension $B_r(\mathbf{Q}' + \mathbf{G}) = B(\mathbf{Q}')$. In this derivation it has been assumed for simplicity that the surface is monoatomic with atoms of mass M, and the atom-surface interaction is restricted to the surface atoms. Then the Debye-Waller exponent is the same at all sites and is given by

$$2W = \sum\nolimits_{\mathbf{Q}\nu} [2n(\omega_{\mathbf{Q}\nu}) + 1] \frac{\hbar|\Delta\mathbf{k} \cdot \mathbf{e}_{\mathbf{Q}\nu}(0)|^2}{2N_L M \omega_{\mathbf{Q}\nu}}. \tag{7.68}$$

It is easily seen that the n-phonon spectral response functions are normalized to unity:

$$\int_{-\infty}^{\infty} d\omega \sum\nolimits_{\mathbf{Q}} B_n(\mathbf{Q}, \omega) = 1, \qquad \forall n. \tag{7.69}$$

The present derivation of the multiphonon contributions naturally holds for the DWBA provided that the dependence of W on $q_z \equiv \Delta k_z$ can be neglected in (7.39). This is the case for the approximate expressions of W which will be discussed in the next section.

In connection with approximation (ii), it was mentioned in the previous Section that Born-type approximations are probably inadequate for strongly corrugated surfaces. The exponentiated Born approximation (EBA), developed by Brenig et al. [7, 77, 78], Gumhalter et al. [79–83], represents a major extension of the DWBA, which proved successful in the treatment of multiphonon He atom scattering in the quantum regime (see Sect. 13.1) [80]. In this theory the exponential argument $[\Delta\mathbf{k} \cdot \langle\mathbf{u}_l(0)\mathbf{u}_{l'}(t)\rangle \cdot \Delta\mathbf{k} - 2W]$ in (7.64a) is replaced by a correlation function $C(t, \mathbf{R})$ which includes contributions from all powers of the interaction potential $\delta V(\mathbf{r}, t)$. Thus a similar convolution structure of the multiphonon spectrum may be obtained with the n-phonon processes weighed by a Poisson distribution, as expressed by (7.64b).

To deal with multiphonon scattering many authors [36, 77, 83–91] have used the powerful scattering-matrix (S-matrix) formalism, from which the semiclassical trajectory approximation can be conveniently derived. As shown by Manson in an excellent summary of multiphonon scattering theory [91], also the semi-classical trajectory approximation, in the quick-scattering limit, leads to the same (7.64b) with $v(\Delta\mathbf{k})$ replaced by the semiclassical scattering amplitude

$$\sigma_{fi} = \langle\mathbf{k}_f| \exp\left[-\frac{i}{\hbar}\int_{-\infty}^{+\infty} V(\mathbf{r}(s))ds\right] |\mathbf{k}_i\rangle, \tag{7.70}$$

where the integral over the static surface potential in (7.70) is performed along the time-dependent path $\mathbf{r}(s)$ of a classical trajectory bringing the atom from the initial state $|\mathbf{k}_i\rangle$ to the final state $|\mathbf{k}_f\rangle$.

In general a Poisson distribution of the phonon number indicates that phonons are generated in a coherent state [92, 93]—a feature common to other impulsive phonon generation phenomena like multiphonon inelastic neutron scattering [92, 94] and phonon sidebands of localized optical transitions in solids [95]. In the classical limit, occurring for large $2W$, i.e., at high temperature and/or strong coupling ($\Delta\mathbf{k}\cdot\mathbf{e}_{\mathbf{Q}\nu}(\hbar/2M\omega_{\mathbf{Q}\nu})^{1/2} \gg 1$), the Poisson distribution tends to the normal (Gaussian) distribution, i.e.

$$\mathrm{e}^{-2W}\frac{(2W)^n}{n!} \approx \frac{1}{\sqrt{4\pi W}}\mathrm{e}^{-(n-2W)^2/4W}, \qquad W \gg 1. \tag{7.71}$$

By setting $\Delta E \equiv n\varepsilon_0$, $\hbar\Omega_m \equiv 2W\varepsilon_0$ and $\alpha \equiv 1/(2\varepsilon_0\hbar\Omega_m)$, with ε_0 an average phonon energy, and taking for simplicity $B_n = N_L^{-1}\,\delta(\Delta E - n\varepsilon_0)$, the multiphonon spectrum in the classical limit turns out to be a series of equally spaced δ-functions. Its envelope should be, according to (7.71), a Gaussian function peaked at the average energy $\hbar\Omega_m$ with a FWHM = $2\sqrt{(\ln 2/\alpha)}$. Note that this is valid only in the low temperature limit, otherwise convolutions give all possible combinations of creation and annihilation processes. In this case the sum in (7.64b) yields a Gaussian around an average energy independent of temperature [91].

In addition to the coherent multiphonon peak, there must also be an incoherent multiphonon sideband associated with the diffuse elastic peak [90], so that for fitting purposes the entire multiphonon background may be chosen in the form

$$I_{multiph}(\Delta E) = a_{coh}\sqrt{\frac{\pi}{\alpha}}\exp[-\alpha(\Delta E - \hbar\Omega_m)^2] + a_{incoh}\sqrt{\frac{\pi}{\alpha'}}\exp[-\alpha'(\Delta E)^2], \tag{7.72}$$

where the amplitudes a_{coh} and a_{incoh}, the multiphonon peak energy $\hbar\Omega_m$ and the exponential coefficients α and α' are temperature-dependent fitting parameters. The coherent multiphonon background is subtracted from the total intensity in order to isolate the one-phonon contribution, so that its amplitude a_{coh} is obtained by excluding the terms n = 0,1 in the sum of (7.64b) and integrating over the frequency. This gives the temperature dependence of the coherent amplitude through that of the DW factor in the form

$$a_{coh}(T) \propto 1 - [1 + 2W(T)]\mathrm{e}^{-2W(T)}. \tag{7.73}$$

It should be noted that in general, as argued by Manson [91], the resulting coherent distribution is not exactly a Gaussian because $\hbar\Omega_m$ also depends on the energy transfer via $\Delta\mathbf{k}$ in (7.68).

The assumption of δ-like spectral functions is only valid for a set of equal independent oscillators (Einstein model) e.g., for weakly coupled adsorbed atoms at

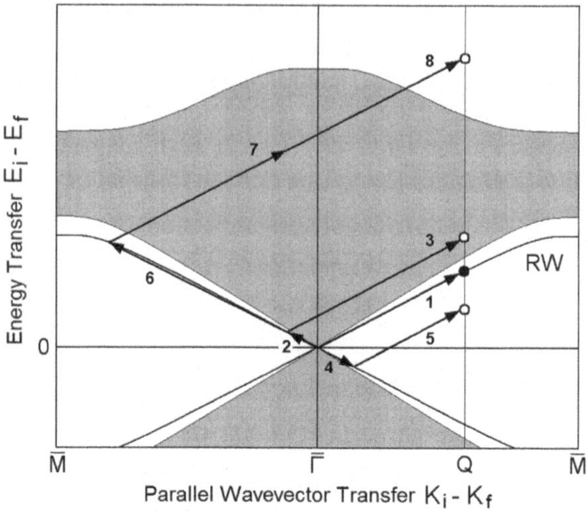

Fig. 7.9 Kinematics of multiphonon processes only involving Rayleigh waves (RW). For a fixed wavevector transfer $\mathbf{K}_i - \mathbf{K}_f = \mathbf{Q}$ only one creation process (●) involving one Rayleigh phonon (1) is possible, whereas there are infinite combinations of two (e.g., 2 + 3 and 4 + 5) or more (e.g., 6 + 7 + 8) Rayleigh phonons for the same wavevector transfer (○), thus giving a continuous energy loss ($E_i - E_f$) spectrum and a broadening of the one-phonon peak. The many-phonon energy loss spectrum normally extends well beyond the bulk one-phonon density (gray area). The two-phonon process 4 + 5 involves the annihilation of phonon 4, whereas in 2 + 3 both phonons are created. The latter is more probable than the former, thus giving an asymmetric broadening of the peak associated with the one-phonon process 1

low coverage (see Sect. 13.1). In general $B_1(\mathbf{Q},\omega)$, (7.67), at a fixed $\mathbf{Q}$ is a set of δ-functions, associated with the surface modes, and of continuous bands, associated with the surface-projected bulk phonon density (Chap. 3). Even neglecting the bulk phonon contributions, the convolutions in (7.65) lead to continuous spectral distributions. This is illustrated schematically in Fig. 7.9 which shows the kinematics of a few possible multiphonon processes involving only the Rayleigh branch. For a fixed wavevector transfer $\mathbf{K}_i - \mathbf{K}_f = \mathbf{Q}$ only one creation process (●) involving one single Rayleigh phonon (1) is possible, whereas there are infinite combinations of two (e.g., 2 + 3 and 4 + 5) or more (e.g., 6 + 7 + 8) Rayleigh phonons for the same wavevector transfer (○). The two-phonon process 4 + 5 involves the annihilation of phonon 4, whereas in the process 2 + 3 both RW phonons are created. Since the latter is more probable than the former, the broadening of the one-phonon peak (process 1) turns out to be asymmetric with a larger intensity on and a peak shift toward the bulk band side. The apparent shift of the RW peak into the bulk continuum is indeed observed at small wavevectors in soft materials like, e.g., Xe(111) [96] and KBr(001) [97] (see Sect. 11.1). All these processes give the multiphonon background a continuous energy loss spectrum, and a broadening of the one-phonon peak associated with the creation of the Rayleigh phonon 1.

The intensity of the multiphonon processes depends on temperature through the phonon occupation numbers in (7.67). The many-phonon energy loss spectrum normally extends well beyond the bulk one-phonon density (gray area) with an intensity which grows with temperature and incident energy. A remarkable example of this effect has been shown for Pt(111) [89] and will be discussed in Chapter 13.

The convolutions of a complex spectral function like that of any ordinary surface smooth out all the sharp features associated with surface modes, so that also in this case the multiphonon spectrum for $W \gg 1$ tends to an approximate Gaussian distribution. In all intermediate cases where W is not sufficiently large and some one-phonon features are still discernible, the actual structure of the spectral function $B_1(\mathbf{Q},\omega)$ plays an important role. In particular the contributions from surface-projected bulk phonon bands, whose integrated intensity is normally comparable to that of surface modes, cannot be neglected. A few examples of multiphonon HAS spectra observed for clean metal surfaces and adsorbates will be shown and discussed in Sect. 13.1.

7.6.3 Temperature Effects: The Debye-Waller Factor

A very important variable in inelastic phonon scattering is the surface temperature T_S, which in this section will be denoted simply as T. The temperature enters into the differential reflection coefficient 7.69 where it leads both to an *enhancement* of the inelastic scattering signal due to the Bose-Einstein (BE) factor and related convolutions $B_n(\mathrm{Q}, \omega)$, but also to an *attenuation,* described by the DW factor $\exp(-2W)$. Since the integrated scattering intensity fulfils unitarity, the actual role of the DW exponent is to share the scattering intensity between the inelastic multiphonon channels (fraction: $1 - \exp(-2W)$) and the elastic channel (fraction: $\exp(-2W)$).

The effects of the two competing mechanisms are shown in Fig. 7.10 [98]. As seen in Fig. 7.10b the Rayleigh peak intensity is expected to increase with temperature. Too high a surface temperature however is undesirable since it leads to an increase of the multiphonon background and to an anharmonic shift and broadening of the one-phonon peaks. The temperature dependence of the DW factor determines the conditions under which a one-phonon experiment must be carried out to avoid too large an inelastic contribution from multiphonon processes. Thus a preliminary phenomenological knowledge of the DW factor and its dependence on temperature and scattering parameters is particularly important in order to choose the optimal experimental conditions for the measurements of surface phonon dispersion curves [99, 100].

Similarly to the case of X-ray and neutron scattering, where the interaction with the lattice particles has a *very short range*, the scattering of He atoms from closed

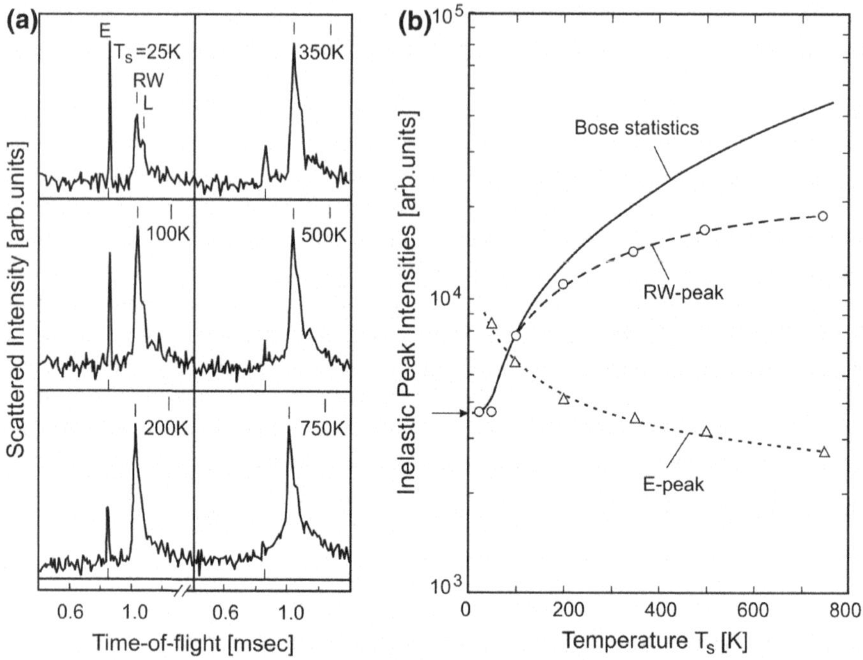

Fig. 7.10 The effect of temperature on one-phonon intensities. **a** HAS time-of-flight spectra on the phonon creation side for Pt(110) in the $\overline{\Gamma M}$ direction at different surface temperatures showing the evolution of the RW peak (together with an unresolved smaller peak from the surface longitudinal (L) mode) and the elastic diffuse peak (E). **b** The intensity of the RW peak, (open circles), increases less rapidly with temperature than the BE factor for creation processes $1+n(\hbar\omega) = [1-\exp(-\hbar\omega/kT)]^{-1}$ [see (7.12)] due to DW attenuation as indicated by the excellent agreement with theoretical predictions (broken line) [95]. The effect of DW attenuation is clearly visible in the intensity of the E peak (open triangles)

shell surfaces can be treated within the approximation of impulsive binary collisions. Under this condition the DW exponent W takes on the simple form given by (7.32), which can be approximated by $\approx \frac{1}{2}(\Delta k_z)^2 < u_z^2 >_s$, where $\Delta k_z = k_{iz} + |k_{fz}|$ is the normal momentum transfer and $<u_z^2>_s$ is the mean-square vertical displacement of the *surface* atoms at the temperature T. If the surface can be simply described by a hard corrugated surface with a dynamic corrugation $\delta D(\mathrm{R}, t)$, the DW exponent is given by (7.55).

At the low energies of interest in the present context, however, the derivation of the appropriate expression for the DW exponent for atom-surface collisions is by no means straightforward. Problems arise because the atom collisions are, by virtue of the relatively long range of the potential, no longer binary, nor are the collisions fast enough to be considered impulsive. The largest deviations from the binary collision model are found for conducting surfaces where the energy and momentum exchange between He and surface atoms occurs via the interposed conduction electrons, i.e., via the electron-phonon (e-p) interaction. Before the advent of the e-p theory of inelastic HAS, discussed in Chap. 8, a number of rather involved corrections had been introduced to account for these complications, as discussed at length, for example, by Levi and Suhl [99] and Meyer [100]. When particularly flat (uncorrugated) surfaces are considered as, e.g., the (110) surface of fcc metals, and only comparatively long wave (small **Q**) acoustic phonons are involved, the oscillations of the electronic surface charge, causing inelastic HAS scattering, follow adiabatically those of the surface atoms, so that the approximation of binary collisions can be used, with only the normal component $< u_z^2 >_S$ of the mean-square surface atom displacements involved in (7.68). This approximation for the Debye-Waller factor for a monoatomic surface with atoms of mass M is given by

$$2W \cong (\Delta k_z)^2 \sum\nolimits_{\mathbf{Q}\nu} [2n(\omega_{\mathbf{Q}\nu}) + 1] \frac{\hbar e^2_{\mathbf{Q}\nu z}(0)}{2N_L M \omega_{\mathbf{Q}\nu}} = (\Delta k_z)^2 \langle u_z^2 \rangle_s. \tag{7.74}$$

where Δk_z is the normal momentum transfer $\Delta k_z = k_{iz} + |k_{fz}|$ and $<u_z^2>_s$ is the mean-square vertical displacement of the *surface* atoms at the temperature T. The mean-square displacement can be conveniently rewritten as

$$\langle u_z^2 \rangle_s = \int_0^\infty d\omega \frac{\hbar}{2M\omega} [2n(\omega) + 1]\, D_{0,zz}(\omega), \tag{7.75}$$

where

$$D_{0,\alpha}(\omega) = \frac{1}{N_L} \sum\nolimits_{\mathbf{Q}\nu} |e_{\mathbf{Q}\nu\alpha}(0)|^2 \delta(\omega - \omega_{\mathbf{Q}\nu}) \tag{7.76}$$

is the surface-projected phonon DOS for the α displacement component. For some simple dynamical models and approximations $\langle u_z^2 \rangle_s$ can be calculated explicitly.

In the *Einstein model*, where all normal modes have the same frequency ω_E, $D^E_{0,z}(\omega) = \delta(\omega - \omega_E)$. Thus from (7.12):

$$\begin{aligned} \langle u_z^2 \rangle_E &= \frac{\hbar}{2M\omega_E} [1 + 2n(\hbar\omega_E)] \\ &= \frac{\hbar}{2M\omega_E} \coth \frac{\hbar\omega_E}{2k_B T} = \frac{\hbar}{2M\omega_E} \coth \frac{\theta_E}{2T}, \end{aligned} \tag{7.77}$$

where $\theta_E \equiv \hbar\omega_E / k_B$ is the Einstein temperature. Note that the *reduced mean-square displacement*, i.e., $\langle u_z^2 \rangle_E$ expressed in units of the square displacement quantum $\hbar/M\,\omega_E$ and plotted in Fig. 7.11a, is a universal function of the *reduced temperature* T/θ_E. The low- and high-temperature asymptotic expressions for $\langle u_z^2 \rangle_E$ are respectively given by

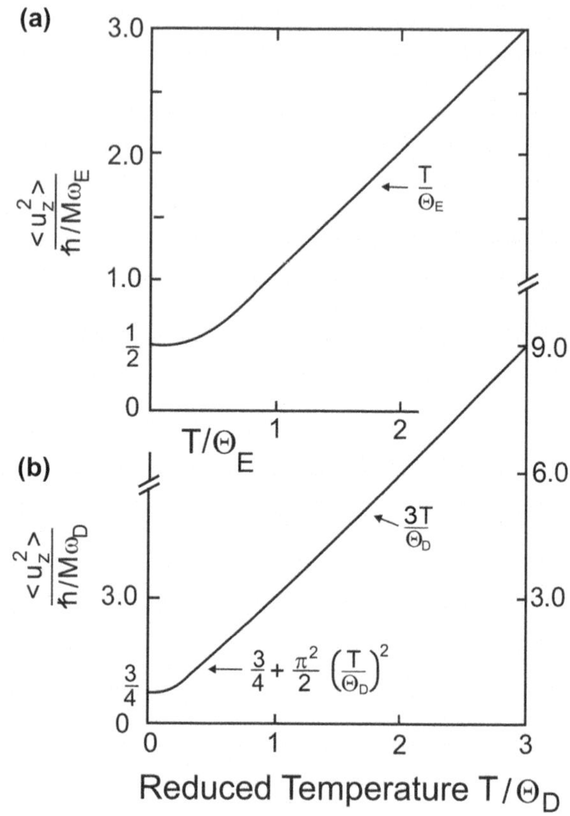

Fig. 7.11 The reduced mean-square displacement as a function of the reduced temperature. **a** For an Einstein (optical) phonon branch and (**b**) for a Debye (acoustic) phonon branch. The Einstein and Debye temperatures are related to the respective frequencies by $\theta_E = \hbar\omega_E/k_B$ and $\theta_D = \hbar\omega_D/k_B$. When referred to a surface plane, the effective surface Debye frequency ω_{Ds} and temperature Θ_{Ds} are used

$$\left\langle u_z^2 \right\rangle_E \cong \frac{\hbar}{M\omega_E} \cdot \begin{cases} \dfrac{1}{2} + \mathrm{e}^{-\hbar\omega_E/k_BT}, T \ll \theta_E \equiv \dfrac{\hbar\omega_E}{k_B} \\ \dfrac{T}{\theta_E}, \qquad T \gg \theta_E \end{cases}. \tag{7.78}$$

At high temperatures, when T is much larger than θ_E, the classical limit $\left\langle u_z^2 \right\rangle_E \cong k_BT/M\omega_E^2$ is recovered with the expected proportionality with T and cancellation of the Planck constant. In the low temperature limit, where k_BT becomes much smaller than the single phonon quantum $\hbar\omega_E$, the mean-square displacement tends exponentially to its zero point value $\hbar/2M\omega_E$, as a consequence of the phonon energy gap between 0 and $\hbar\omega_E$. Although unsuitable to fully represent the dynamics of a crystal, the Einstein model works well for a flat (dispersionless) optical phonon branch, as occurring, e.g., in weakly coupled molecular crystals, and to express its contribution to the total mean-square displacement. In this case M plays the role of a molecular effective mass.

In the *Debye model* the solid is approximated by an elastic continuum where each acoustic branch has a linear dispersion ($\omega = v_\alpha q,\ \ \alpha = \mathrm{TA}, LA$) with a constant phase velocity v_α equal to the respective speed of sound. The atomic structure of the solid is accounted for by that the sound wavelength cannot be shorter than twice the interatomic distance, which is equivalent to assigning a maximum vibrational frequency (Debye frequency). Even in isotropic materials there are different Debye frequencies for the LA and TA polarizations, but for many purposes it is enough to use a single effective Debye frequency ω_D for the total vibrational spectrum of the solid. Since the Debye model was originally formulated to explain the temperature dependence of the specific heat $c_V(T)$, the values of ω_D and of the related Debye temperature $\theta_D \equiv \hbar\omega_D/k_B$ reported in the literature for the various solids are generally those which give the best fit of the experimental $c_V(T)$. If no surface effects are considered so that all modes of the solid are plane waves, the mean-square displacement is independent of the lattice site and the surface-projected phonon DOS, (7.76), is just the DOS (normalized to unity) for the elastic continuum with a cut off at ω_D:

$$D^D_{0,z}(\omega) = \frac{3\omega^2}{\omega_D^3}\Theta(\omega_D - \omega), \tag{7.79}$$

where Θ is the Heaviside step function. The corresponding z-polarized surface mean-square displacement is then given by

$$\langle u_z^2 \rangle_D = \frac{3\hbar}{2M\omega_D^3}\int_0^{\omega_D} d\omega\ \omega \coth\frac{\hbar\omega}{2kT} \tag{7.80}$$

The integral, (7.80), cannot be performed analytically but can be expressed as a function of the reduced temperature T/θ_D

$$\langle u_z^2 \rangle_D = \frac{3\hbar}{M\omega_D}\left[\frac{1}{4} + \left(\frac{T}{\theta_D}\right)^2 \int_0^{\theta_D/T} \frac{xdx}{e^x - 1}\right]. \tag{7.81}$$

The low- and high-temperature asymptotic expressions for $\langle u_z^2 \rangle_D$ are respectively given by

$$\langle u_z^2 \rangle_D \cong \frac{\hbar}{M\omega_D} \cdot \begin{cases} \dfrac{3}{4} + \dfrac{\pi^2}{2}\left(\dfrac{T}{\theta_D}\right)^2, & T \ll \theta_D \leq 0.7 \\ \dfrac{3T}{\theta_D}, & T \gg \theta_D \end{cases} \tag{7.82}$$

In practice, as seen in Fig. 7.11b, the high temperature linear regime approximately starts already at $T/\theta_D \approx 0.7$, which will be assumed hereon as a

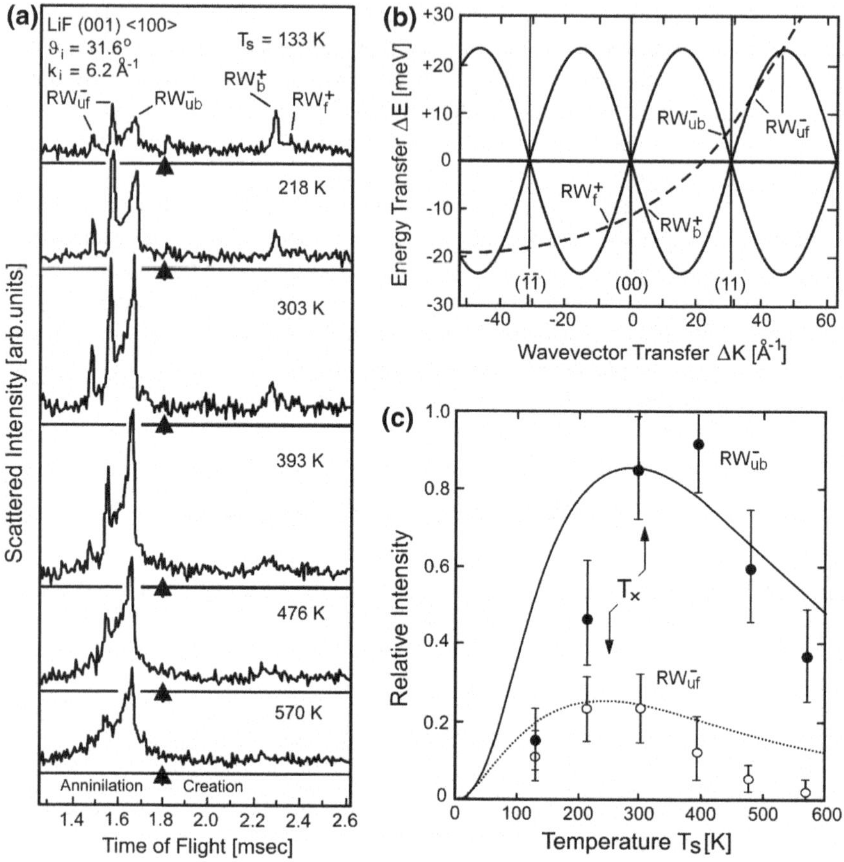

Fig. 7.12 Time-of-flight spectra and analysis illustrating the effect of temperature on single phonon peak intensities. **a** A series of HAS time-of-flight spectra from LiF(001) along the ⟨100⟩ direction with $k_i = 6.2$ Å^{-1} for an incident angle of 31.6° ($\vartheta_i + \vartheta_f = 90°$) recorded for different target temperatures [101] and **b** the corresponding scan curve (broken line) intersecting the RW dispersion curve in different Brillouin zones (BZ's) on both the creation and annihilation sides. The short black arrows in (**a**) indicate the position of the elastic peak. The backward creation peak RW$_b^+$, corresponding to a RW phonon in the first BZ, decrease with increasing temperature while the three peaks corresponding to the annihilation of two RW's in the forward direction (RW$_{\bar{u}f}^-$) and one RW in the backward direction (RW$_{\bar{u}b}^-$), all in the second BZ (*umklapp* processes), increase with increasing temperature. The forward-creation peak RW$_f^+$ is only barely visible at low temperature and has already disappeared above 200 K. At temperatures $T_s > 303$ K the RW$_{\bar{u}f}^-$ peaks rapidly lose intensity and are broadened by multiphonon effects, while the RW$_{\bar{u}b}^-$ peak starts decreasing above 393 K. **c** The plots of the intensities of the RW$_{\bar{u}b}^-$ peak and of the RW$_{\bar{u}f}^-$ peak at higher energy as a function of temperature show the crossover from the B-E quasi-linear increase to the DW exponential decrease, although the agreement with theoretical predictions (full and dotted lines) are only qualitatively reproduced. The cross-over temperature $T_\times$ is also indicated

conventional border between low and high-T regimes. At high temperatures $\langle u_z^2 \rangle_D \cong 3k_BT/M\omega_D^2$ is asymptotically linear in T, as for the Einstein model, whereas for vanishing temperatures $\langle u_z^2 \rangle_D$ tends quadratically to its zero point value, as expected for the acoustic modes with a linear dispersion. Note that the $T = 0$ and high-T expressions for $\langle u_z^2 \rangle_D$ coincide with those for $\langle u_z^2 \rangle_E$ if $\overline{\omega_E^{-1}}$ and $\overline{\omega_E^{-2}}$ are replaced by the corresponding moments (indicated by an upper line) of the Debye DOS $\overline{\omega^{-1}} = \frac{3}{2}\omega_0^{-1}$ and $\overline{\omega^{-2}} = 3\,\omega_0^{-2}$.

Although the temperature dependencies for the Einstein and Debye models (Fig. 7.11) look qualitatively similar, there is an important difference in the way the $T = 0$ value is reached, exponential in the Einstein model, quadratic in the Debye model. Moreover the reduced mean-square displacements in the linear high temperature region are a factor 3 larger for the Debye model than for the Einstein model due to the large contributions from the low frequency modes (absent in the Einstein model).

It is important to note that (7.76) is formally valid also when the surface acoustic phonon branches, notably the RW, are included. While the total DOS for the RW branch ($\omega = v_RQ$) is obtained by integrating over the 2D Brillouin zone and is therefore proportional to ω, the *surface-projected* DOS, (7.76), is proportional to ω^2 due to the presence of the square eigenvector, which increases with ω, (3.13). Thus (7.76) can be safely used to describe the actual surface-projected DOS including both surface and bulk phonons, by simply replacing the bulk ω_D with an effective *surface Debye frequency* ω_{Ds}. The extent to which ω_{Ds} differs from ω_D depends on the contribution of surface phonons to the total $\langle u_z^2 \rangle_s$. Actually this contribution is of the same order as that of the infinitely more numerous bulk modes, but the relative contributions from the RWs and the bulk modes depends on the degree of localization of the RWs, i.e. on the elastic constants of the solid (Chap. 3). A reasonable approximation is to assume that a surface atom moving in the normal direction feels only one half of the force constant acting on a bulk atom due to the reduced coordination. Thus $\omega_{Ds} \approx \omega_D/\sqrt{2}$ and at high temperatures the surface mean-square displacement $\langle u_z^2 \rangle_{D_s}$, which is inversely proportional to ω_{Ds}^2, is about twice as large as the bulk $\langle u_z^2 \rangle_D$.

As anticipated in Chap. 1 (Table 1.1), and shown by lattice dynamical calculations and molecular dynamics simulations (Table 7.1) [102–104] the normal component of mean-square displacement at the surface, $\langle u_z^2 \rangle_s$, at a high temperature is indeed up to a factor 2 larger than in the bulk. On the contrary the parallel component $\langle u_\parallel^2 \rangle_s$ remains comparable to the bulk value because both the longitudinal surface acoustic resonance and the shear-horizontal surface acoustic mode (see Chap. 3) tend to the respective band edges for $Q \to 0$ and cannot be distinguished from bulk modes. Effective surface Debye temperatures measured by either

Table 7.1 Bulk and surface mean-square vibrational amplitudes (in Å^2) for some metals at T = 300 K.

	Bulk $\langle u_z^2 \rangle$			Average surface $\langle u_z^2 \rangle_s$ [a]		
	Debye[b]	Lindemann[c]	Molecular dynamics[d]	Debye[e]	Lindemann[f]	Molecular dynamics[c]
Ag	0.0084	0.0087	0.0080	0.0178	0.0174	0.0166
Cu	0.0062	0.0060	0.0070	0.0104	0.0121	0.0129
Ni	0.0042	0.0042	0.0039	0.0058	0.0085	0.0084

Other data obtained from model and *ab initio* calculations for various surfaces are given in Table 1.1
[a]Values averaged over the three low index surfaces [100, 105]
[b]From (7.86) with bulk θ_D (Appendix B)
[c]From (7.90)
[d]From [104]
[e]From (7.86) with θ_{Ds} for the (001) surface from LEED data (Appendix B)
[f]From (7.91)

LEED or He atom scattering for a wide range of systems are summarized in Appendix B where they are compared with the corresponding bulk values θ_D.

The DW exponent for binary impulsive collisions can now be expressed within the Debye model introduced into (7.74). In the high-temperature limit:

$$2W \cong \frac{3k_B T}{M\omega_{Ds}^2}(\Delta k_z)^2, \qquad \frac{T}{\theta_{Ds}} > 0.7\,. \tag{7.83}$$

For impulsive collisions with a surface atom at rest $(\hbar\,\Delta k_z)^2$ is simply related to the energy transfer ΔE

$$(\hbar \Delta k_z)^2 = 2M\,\Delta E, \tag{7.84}$$

where ΔE is customarily estimated using the simple Baule formula [106]. This model assumes a hard sphere collision and considers the momentum and energy transfer along the line of centers [107]. According to the Baule formula the impulsive energy transfer between an incident gas atom of mass m and an isolated stationary free atom of mass M of the solid is given by

$$\Delta E/E_{iz} = 4\mu/(1+\mu)^2, \tag{7.85}$$

where $\mu = m/M$ and $E_{iz} = \hbar^2 k_{iz}^2/2m$ is the incident energy along the surface normal. For $\mu \ll 1$, as is usually the case for He probe atoms, (7.83) can now be written in the following two alternative forms

$$2W \cong \begin{cases} \dfrac{24\mu E_{iz} T}{k_B \theta_{Ds}^2} = \dfrac{12k_{iz}^2}{M\omega_{Ds}^2} k_B T, & \frac{T}{\theta_{Ds}} > 0.7 \\ \dfrac{6\mu E_{iz}}{k_B \theta_{Ds}} = \dfrac{3k_{iz}^2}{M\omega_{Ds}^2} k_B \theta_{Ds}, & T \to 0 \end{cases}. \tag{7.86}$$

Because of their simplicity these expressions for the Debye-Waller factor are very useful for estimating the effects of temperature on both elastic and inelastic HAS intensities, especially for insulator surfaces, where the scattering process is well described by two-body collisions [108].

7.6.4 Criteria for One-Phonon Scattering

As discussed above the DW exponent provides information on the probability for single phonon as well as n-phonon scattering. As seen in (7.64b), the n-phonon processes are weighed by the Poisson distribution

$$P(0 \to n) = \exp(-2W)\frac{(2W)^n}{n!}. \tag{7.87}$$

In general the experimental observation of a Poisson distribution is indicative of coherent scattering [92, 93]. Since the maximum of $P(0 \to n)$ with respect to $2W$ occurs at $2W = n$, the DW exponent gives the dominant number of phonons generated in the process. The one-phonon processes will dominate over the multiphonon processes when $2W \leq 1$. In the high temperature regime this occurs for

$$2W \cong \frac{24\mu E_{iz}T}{k\theta_{Ds}^2} < 1, \quad \frac{T}{\theta_{Ds}} > 0.7, \tag{7.88}$$

while in the low temperature limit the condition for dominant one-phonon processes is

$$2W \cong \frac{6\mu E_{iz}}{k\theta_{Ds}} < 1, \quad T \to 0, \tag{7.89}$$

which is clearly less stringent than the above high temperature condition. This means that one-phonon spectroscopy is better done at a low surface temperature.

At $2W = 1$ the one-phonon probability $1/\mathrm{e} \cong 0.368$ slightly exceeds the total multiphonon probability for $n > 1$, which is $1 - 2/\mathrm{e} \cong 0.264$. However, as discussed in Sec. 7.5, the relevant part of the one phonon spectrum mostly consists of sharp peaks associated with surface phonons whereas the multiphonon background is made of combination of bands and is in general rather uniform. For this reason the actual emergence of the one-phonon features over the multiphonon background depends on the overall spectrum of surface phonons and surface-projected bulk phonons.

Other more or less stringent conditions for the dominance of one-phonon processes can be found in the literature. For example, a much more stringent condition, $2W < 0.24$, was first proposed by Weare [60], also on the basis of an impulsive approximation. On the other hand a detailed model system study carried out by Lin, Adelman and Wolken [105] leads to the less stringent condition of $2W < 1.68$.

In order to satisfy the condition (7.88) for single phonon scattering from a surface composed of atoms of mass M and with a surface Debye temperature θ_{Ds} the parameters E_{iz} and T should be small and, for a given incident energy, also μ should be chosen as small as possible by choosing a light scattering particle. Among the possible light probes such as ^{3}He, ^{4}He, H_2 and H atoms, ^{4}He atoms have many experimental advantages, as will be discussed in Sect. 9.2, and therefore are widely used. Obviously for a given probe particle μ will be largest in scattering from surfaces consisting of light atoms. This effect, however, may be largely compensated for by the larger values of θ_{Ds} characteristic of surfaces made up of light atoms.

A simpler approximation for estimating the optimal conditions for single phonon scattering, which does not require an explicit use of the Debye model, can be obtained via the Lindemann criterion for melting [109]. The term $M\omega_{Ds}^2$, which appears in (7.83) and (7.86), can be considered to be equivalent to an effective force constant (stiffness) of the solid surface. Thus the stiffness rather than the Debye frequency (or temperature) of the surface, may be more appropriate for evaluating the best conditions for one-phonon processes. For example, the refractory 5d metals like W, Re, etc., all have a comparatively low Debye temperature due to their large atomic mass, but very large force constants and, consequently, a high stiffness. This allows, for example, for a HAS study of one-phonon processes up to very high temperatures, above 1000 K, as found for W(001) [48] and discussed Chapter 11.

The effective bulk force constant $\beta \equiv M\omega_D^2$ is empirically related to the melting temperature T_m through the Lindemann criterion [109] which states that the melting of a crystal sets in when the root mean-square displacement $<u^2>^{1/2}$ exceeds one tenth of the interatomic distance a_o. Thus, by means of (7.82), the bulk value of $<u_z^2> = <u^2>/3$ is approximately given by

$$\langle u_z^2 \rangle \cong \frac{a_o^2}{300}\frac{T}{T_m}, \qquad \frac{T}{\theta_D} \geq 0.7\,. \tag{7.90}$$

The validity utility of this approximation can be appreciated from the comparison shown in Table 7.1 among the Lindemann mean-square amplitudes, (7.90), and those from the Debye approximation, (7.82), and molecular dynamics simulations [103–105]. The data refer to both the bulk and an average over three low-index surfaces of Ag, Cu and Ni at room temperature, where the condition $T/\theta_D > 0.7$ is fulfilled. For the bulk a very good consistency is found among the values calculated with the three methods are in close agreement. For the surface the

Debye values for $<u_z^2>_s$ are obtained by using θ_{Ds} instead of θ_D. The agreement with the molecular dynamics results is only fair, however. A better agreement is obtained instead by simply setting $<u_z^2>_s$ equal to twice the Lindemann bulk values, (7.90). Thus the Lindemann formula, (7.90), can be adapted to the surface case in the form

$$\langle u_z^2 \rangle_s \cong \frac{a_o^2}{150} \frac{T}{T_m}, \qquad \frac{T}{\theta_D} \geq 0.7, \tag{7.91}$$

where T_m is the *bulk* melting temperature. With this the high temperature DW exponent becomes

$$2W \cong \frac{4mE_{iz}a_o^2}{75\hbar^2} \frac{T}{T_m} = \frac{2}{75} (a_o k_{iz})^2 \frac{T}{T_m}, \qquad \frac{T}{\theta_{Ds}} \geq 0.7\,. \tag{7.92}$$

The one-phonon condition $2W \leq 1$ appears to be fulfilled at any temperature $T < T_m$ for $a_o k_{iz} < 6$.

Some caution in using the Lindemann formula for the estimating of W is needed for those systems where surface melting occurs [48, 60, 101, 105, 108–110]. In this case an empirical Lindemann criterium linking $<u_z^2>_s$ to the surface melting temperature T_{ms} would be more appropriate. It should be noted that the condition (7.92) and the other expressions of the DW factor involve the normal component of the incident wavevector and therefore the choice of sufficiently large incidence angles also helps to keep the inelastic processes within the one-phonon regime. It is indeed intuitive that head-on collisions between He and surface atoms favor multiphonon processes and are surpressed at grazing incidence.

In general the possibility of observing single surface phonons at high temperatures requires not only a small DW exponent but also a small anharmonic broadening of the phonon peak [111], an aspect which is not included in the one-phonon criteria discussed so far. Both quantities, however, are proportional to the mean-square displacement of the surface atoms, and therefore, the highest temperature at which surface phonons can be distinctly observed may be substantially lowered by anharmonic broadening. Anharmonic broadening also increases dramatically with the phonon frequency due to the rapid increase with frequency of the number of possible decay channels (see Sect. 7.5). This is illustrated by the data presented in Fig. 7.12 showing the effect of the surface temperature on the experimental TOF spectra of He atom scattering from LiF(001) [112]. In Fig. 7.12a at 570 K only the lowest RW peaks on the annihilation side RW_{uf}^- and RW_{ub}^- survive. The two RW peak on the creation side RW_b^+ and RW_f^+ and the higher energy annihilation peak RW_{uf}^- are so broad at 476 K to be barely visible. Thus anharmonic broadening may lead to a more stringent condition for the observation of one-phonon peaks.

Finally it must be cautioned that the direct experimental verification of the various one-phonon conditions, discussed above, cannot be decided easily since it is difficult to clearly identify the multiphonon contributions in the TOF spectra. If the spectrum taken at T $\leq$ 303 K is somewhat arbitrarily attributed to pure single phonon scattering, (7.92), based on the Lindemann formula, with the kinematic data given in Fig. 7.12 and the LiF constants $a_o = 2.0$ Å and $T_m = 1115$ K (see Appendix B) yields the condition $2W \leq 0.82$ for dominant one-phonon processes. This is considerably less stringent than Weare's criterium, but consistent with $2\ W \leq 1$.

The TOF spectra for LiF(001) collected in Fig. 7.12a and the temperature dependence of the peak intensities show another interesting effect of the competition between the DW and the BE factors. The three peaks seen on the annihilation side correspond to two RW's propagating in the forward direction (RW_{uf}^{-}) and one in the backward direction (RW_{ub}^{-}), all with **Q** in the second Brillouin zone (Umklapp processes), whereas the RW peaks RW_f^+ and RW_b^+ correspond to normal forward and backward creation processes with **Q** in the first Brillouin zone, respectively. The RW_{ub}^{-} annihilation peak, corresponding to the lowest phonon frequency ω_{ub}, is weighed by a rapidly increasing BE factor $n(\hbar\omega_{ub}) \sim kT/\hbar\omega_{ub}$ and only at sufficiently high temperature, above 393 K, does the exponential decrease of the DW factor gain over the quasi-linear increase of the BE factor. The other two annihilation peaks (RW_{uf}^{-}) show a comparatively slower increase with temperature due to the larger frequencies in their BE factor, and the decrease induced by the DW factor starts already above 303 K. The crossover from the BE to the DW regimes is clearly seen in Fig. 7.12c. The larger annihilation peak intensities of the backward phonon (RW_{ub}^{-}) are favoured by the BE factor compared to the forward phonon of higher energy (RW_{uf}^{-}). In Fig. 7.12a the RW_b^+ creation peak shows in contrast a steady decrease, since the DW dominates over the BE factor, which is now $1 + n$ and is almost constant at low temperature where $n \ll 1$. The forward creation peak RW_f^+, at a higher phonon frequency, is only barely visible at low temperatures and has already disappeared above 200 K.

It is interesting to note that the behavior described above is practically independent of the parallel wavevector transfer because the quantity

$$|\Delta k_z|^2 = k_i^2[\cos\theta_i + (1 + \hbar\omega/E_i)^{1/2}\cos\theta_f]^2 \tag{7.93}$$

appearing in the DW exponent is only slowly dependent on the scattering geometry (for $\theta_i + \theta_f = 90°$ and $\hbar\omega \ll E_i$, $k_i < |\Delta k_z| \leq \sqrt{2}k_i$). Thus the crossover from the BE to the DW regime occurs at a temperature $T_\times$ (see Fig. 7.12c), where either the function $n\exp(-2W)$, for annihilation processes, or $(1+n)\exp(-2W)$, for creation processes, has a maximum. A simple analysis shows that for annihilation processes there is always a crossover temperature which increases with the phonon energy, whereas for creation processes no crossover occurs unless the phonon energy is much smaller than $k_B\theta_D$. When the phonon energy $\hbar\omega$ is much smaller than $k_B\theta_D$, the crossover temperature is such that $k_BT_\times \gg \hbar\omega$ and the classical limit $n(\hbar\omega) \gg 1$ holds so that creation and annihilation processes have approximately the same intensities and crossover temperatures. Thus the lack of a crossover for creation processes is a quantum effect.

Consistent with the above analysis, the experience gained in the laboratory has shown that it is best to study events near the zone origin having small $\hbar\omega(Q)$ with a low beam energy, while at the zone boundary it is best to concentrate on creation events using a high beam energy [113].

7.6.5 *Limitations of the Two-Body Collision Model*

The temperature effects on atom scattering from insulator surfaces have been discussed in some detail in the previous Sections since insulator closed shell surfaces provide the best example of two-body collisions, where the above derivation and discussion of the DW exponent is most appropriate. Despite the many experiments and theory on metal surfaces, the reinterpretation of the Debye frequency and temperature, and even of the surface atom mass *M* as effective surface quantities, may not be sufficient to account for the important deviations from the two-body collision model for conducting or even highly polarisable surfaces. As shown in the next chapter, for these systems the He atom-surface atom interaction occurs via the interposed electrons, and the DW exponent contains valuable information on the electron-phonon interaction [114] which is completely lost in the two-body collision approach.

References

1. L. Miglio, F. Quasso, G. Benedek, Surf. Sci. **136**, L9 (1984)
2. L. Miglio, F. Quasso, G. Benedek, J. Chem. Phys. **83**, 913 (1985)
3. V. Bortolani, A. Franchini, F. Nizzoli, G. Santoro, Phys. Rev. Lett. **52**, 429 (1984)
4. A. Liebsch, J. Harris, B. Salanon, J. Lapujoulade, Surf. Sci. **123**, 338 (1982)
5. M.S. Daw, M.I. Baskes, Phys. Rev. Lett. **50**, 1285 (1983)
6. M.S. Daw, M.I. Baskes, Phys. Rev. B **29**, 6443 (1984)
7. J.S. Nelson, E.C. Sowa, M.S. Daw, Phys. Rev. Lett. **61**, 1977 (1988)
8. P.R. Underhill, Surf. Sci. **200**, L441 (1988)
9. W.K. Jacobsen, J.K. Nørskov, M.J. Puska, Phys. Rev. B **35**, 7423 (1987)
10. P.D. Ditlevsen, J.K. Nørskov, J. El. Spectr. Rel. Phenom. **54**(55), 237 (1990)
11. P.D. Ditlevsen, J.K. Nørskov, Surf. Sci. **254**, 261 (1991)
12. F. Montalenti, M.I. Trioni, G.P. Brivio, S. Crampin, Surf. Sci. Lett. **364**, L595 (1996)
13. M.I. Trioni, F. Montalenti, G.P. Brivio, Surf. Sci. Lett. **401**, L383 (1998)
14. M.I. Trioni, S. Marcotulio, G. Santoro, V. Bortolani, G. Palumbo, G.P. Brivio, Phys. Rev. B **58**, 11043 (1998)
15. G.P. Brivio, M.I. Trioni, Rev. Mod. Phys. **71**, 231 (1999)
16. N.D. Lang, Phys. Rev. Lett. **46**, 842 (1981)
17. E. Zaremba, W. Kohn, Phys. Rev. B **13**, 2270 (1976)
18. E. Zaremba, W. Kohn, Phys. Rev. B **15**, 12769 (1977)
19. N. Esbjerg, J.K. Norskov, Phys. Rev. Lett. **45**, 807 (1980)

20. J. Harris, A. Liebsch, Phys. Rev. Lett. **49**, 341 (1982)
21. A. Liebsch, J. Harris, Surface Sci. **111**, 2721 (1981)
22. M.W. Cole, F. Toigo, Phys. Rev. B **31**, 727 (1985)
23. V. Celli in *Helium Atom Scattering from Surfaces*, E. Hulpke (Ed.), (Springer, Berlin, 1992), p. 25
24. P. Nordlander, J. Harris, J. Phys. C **17**, 1141 (1984)
25. Y. Chen, S.Y. Tong, K.P. Bohnen, T. Rodach, K.M. Ho, Phys. Rev. Lett. **70**, 603 (1993)
26. S.Y. Tong, Y. Chen, K.P. Bohnen, T. Rodach, K.M. Ho, Surf. Rev. Lett. **1**, 97 (1994)
27. K.P. Bohnen, K.M. Ho, Surf. Sci. Reports **19**, 99 (1993)
28. J.P. Senet, G. Benedek, J.P. Toennies, Europhys. Lett. **57**, 430 (2002)
29. G. Benedek, M. Bernasconi, V. Chis, E. Chulkov, P.M. Echenique, B. Hellsing, J. Peter, Toennies. J. Phys.: Cond. Matter **22**, 084020 (2010)
30. IYu. Sklyadneva, G. Benedek, E.V. Chulkov, P.M. Echenique, R. Heid, K.-P. Bohnen, J. P. Toennies, Phys. Rev. Lett. **107**, 095502 (2011)
31. G. Benedek, M. Bernasconi, K.-P. Bohnen, D. Campi, E.V. Chulkov, P.M. Echenique, R. Heid, IYu. Sklyadneva, J.P. Toennies, Phys. Chem. Chem. Phys. **16**, 7159 (2014)
32. L. Schiff, *Quantum Mechanics* (McGraw-Hill, New York, 1968)
33. B.H. Brandsen, C.J. Joachain, *Introduction to Quantum Mechanics* (Longman, New York)
34. N. Cabrera, V. Celli, J.R. Manson, Phys. Rev. Lett. **22**, 346 (1969)
35. J.R. Manson, V. Celli, Surf. Sci. **24**, 495 (1971)
36. V. Bortolani, A.C. Levi, Riv. Nuovo Cim. **9**, 1 (1986)
37. L. van Hove, Phys. Rev. **95**, 249 (1954)
38. V. Celli, N. Garcia, J. Hutchison, Surf. Sci. **87**, 112 (1979)
39. A.A. Maradudin, E.W. Montroll, G.H. Weiss, I.P. Ipatova, *Theory of Lattice Dynamics in the Harmonic Approximation*, Solid State Physics, Suppl. 3 (Academic Press, New York, 1971)
40. A.A. Maradudin, R.F. Wallis, L. Dobrzynski, *Handbook of Surfaces and Interfaces*, vol. 3 (Garland, New York, 1980)
41. D. Eichenauer, J.P. Toennies, J. Chem. Phys. **85**, 532 (1986)
42. C. Zener, Proc. Roy. Soc. (London) 40 (1932) 178 and 335
43. J.M. Jackson, N.F. Mott, Proc. Roy. Soc. (London) A **137**, 703 (1932)
44. G. Benedek, G. Seriani, Japan J. Appl. Phys. Suppl **2**(Pt. 2), 545 (1974)
45. G. Benedek, G. Boato, Europhys. News **8**, 5 (1977)
46. Y. Wang, B.H. Choi, N.L. Liu, Phys. Rev. B **43**, 7458 (1991)
47. B.H. Choi, R.T. Poe, J. Chem. Phys. **83**, 1330 (1985); *ibidem* **83**,1344 (1985)
48. H.-J. Ernst, E. Hulpke, J.P. Toennies, Phys. Rev. Lett. **58**, 1941 (1987)
49. N. Garcia, J. Ibanez, J. Solana, N. Cabrera, Surf. Sci. **60**, 385 (1976)
50. N. Garcia, Phys. Rev. Lett. **37**, 912 (1976)
51. N. Garcia, J. Chem. Phys. **67**, 897 (1977)
52. G. Benedek, N. Garcia, Ned. Tijdsch. Vacuumtechniek **16**, 344 (1978)
53. G. Benedek, N. Garcia Surf. Sci. **80**, 543 (1979). This work is very similar to that of Armand and Manson (AM), quoted at [53]. Here however the first order correction to the source function *(δf)* is included while AM produced an argument to disregard it. Indeed δf contributes two additional terms in the one-phonon cross section, the first being an inelastic correction to the diffraction amplitudes (required by unitarity), the second an incoherent contribution to the elastic diffuse scattering due to the thermal motion. Both terms, though physically meaningful, do not affect the shape of the one-phonon HAS spectrum and were therefore disregarded in AM theory as well as in all subsequent calculations based on the GR method
54. G. Benedek, in *Dynamics of Gas-Surface Interaction*, ed. by G. Benedek, U. Valbusa (Springer, Berlin, Heidelberg, 1982), p. 227
55. G. Benedek, N. Garcia, Surf. Sci. **103**, L143 (1981)
56. G. Armand, J.R. Manson, Surf. Sci. **80**, 532 (1979)
57. G. Benedek, R.B. Doak, J.P. Toennies, Phys. Rev. B **28**, 7277 (1983)

58. G. Benedek, N. Garcia, Le Vide Suppl. **201**, 818 (1980)
59. G. Benedek, G. Brusdeylins, R.B. Doak, J.P. Toennies, J. Physique (Paris) **42**, C6–793 (1981)
60. J.H. Weare, J. Chem. Phys. **61**, 2900 (1974)
61. G. Brusdeylins, R.B. Doak, J.P. Toennies, Phys. Rev. Lett. **44**, 1417 (1980)
62. A.C. Levi, Nuovo Cim. **54** B, 357 (1979)
63. A.C. Levi, G. Benedek, L. Miglio, G. Platero, V.R. Velasco, F. Garcia-Moliner, Surf. Sci. **143**, 253 (1984)
64. G. Platero, V.R. Velasco, F. Garcia-Moliner, G. Benedek, L. Miglio, Surf. Sci. **143**, 244 (1984)
65. G. Brusdeylins, R. Rechsteiner, J.G. Skofronick, J.P. Toennies, G. Benedek, L. Miglio, Phys. Rev. Lett. **54**, 466 (1985)
66. B. Dorner, *Coherent Inelastic Neutron Scattering in Lattice Dynamics* (Springer, Berlin, Heidelberg, 1982)
67. R.S. Leigh, B. Szigeti, V.K. Tewari, Proc. R. Soc. London Ser. A **320**, 505 (1971)
68. W. Cochran, Acta Crystallogr. Sec. A **27**, 556 (1971)
69. N.J. Chesser, J.D. Axe, Phys. Rev. B **9**, 4060 (1974)
70. H. Ibach, D.L. Mills, *Electron Energy Loss Spectroscopy and Surface Vibrations* (Academic Press, New York, 1982)
71. J.E. Yater, A.D. Kulkarni, F.W. de Wette, J.L. Erskine, J. Electr. Spectr. Rel. Phenom. **54/55**, 395 (1990)
72. J.L. Erskine, E.-J. Jeong, J. Yater, Y. Chen, S.Y. Tong, J. Vac. Sci. Technol., A **8**, 2649 (1990)
73. A. Glebov, W. Silvestri, J.P. Toennies, G. Benedek, J.G. Skofronick, Phys. Rev. B **54**, 17866 (1996)
74. A. Schuller, S. Wethekam, H. Winter, Phys. Rev. Lett. **98**, 016103 (2007)
75. J.R. Manson, H. Khemliche, P. Roncin, Phys. Rev. B **78**, 155408 (2008)
76. N. Bundaleski, H. Khemliche, P. Soulisse, P. Roncin, Phys. Rev. Lett. **101**, 177601 (2008)
77. W. Brenig, Z. Phys. B **36**, 81 (1979)
78. J. Böheim, W. Brenig, Z. Phys. B **41**, 243 (1981)
79. K. Burke, B. Gumhalter, D.C. Langreth, Phys. Rev. B **47**, 12852 (1993)
80. B. Gumhalter, K. Burke, D.C. Langreth, Surf. Rev. Lett. **1**, 133 (1994)
81. A. Bilić, B. Gumhalter, Phys. Rev. B **52**, 12307 (1995)
82. B. Gumhalter, Phys. Rep. **351**, 1 (2001)
83. W. Brenig, B. Gumhalter, J. Phys. Chem. B **108**, 14549 (2004)
84. J.L. Beeby, J. Phys. C **5**, 3438 (1972)
85. J.L. Beeby, J. Phys. C **5**, 3457 (1972)
86. R. Brako, D.M. Newns, Phys. Rev. Lett. **48**, 1859 (1982)
87. R. Brako, D.M. Newns, Surf. Sci. **117**, 42 (1982)
88. R. Brako, D.M. Newns, Surf. Sci. **123**, 439 (1982)
89. V. Celli, D. Himes, P. Tran, J.P. Toennies, Ch. Wöll, G. Zhang, Phys. Rev. Lett. **66**, 3160 (1991)
90. D. Himes, V. Celli, Surf. Sci. **272**, 139 (1992)
91. J.R. Manson, Comp. Phys. Comm. **80**, 145 (1994)
92. R.J. Glauber, Phys. Rev. **98**, 1692 (1955); Phys. Rev. **131**, 2766 (1963)
93. N. Terzi, in *Collective Excitations in Solids*, ed. by. B. Di Bartolo (Plenum, New York, 1983) pp. 149–181; G.M. Mazzucchelli, N. Terzi, Solid State Comm. **48**, 679 (1983)
94. L. van Hove, Phys. Rev. **95**, 249 (1954)
95. M.H.L. Pryce, in *Phonons in Perfect Lattices and in Lattices with Point Imperfections*, ed. by R.W.H. Stevenson (Oliver and Boyd, Edimburgh, 1966), p. 403
96. A. Giorgetti, N. Terzi, Solid State Comm. **39**, 635 (1981)
97. A. Šiber, B. Gumhalter, A.P. Graham, J.P. Toennies, Phys. Rev. B **63**, 115411 (2001)
98. K. Kern, Dissertation (University of Bonn, 1986)
99. A.C. Levi, H.G. Suhl, Surf. Sci. **88**, 221 (1979)

100. H.-D. Meyer, Surf. Sci. **104**, 117 (1981)
101. T. Rahman, private communication
102. S.K.S. Ma, F.W. de Wette, G.P. Alldredge, Surf. Sci. **78**, 598 (1978)
103. L. Yang, T.S. Raman, Phys. Rev. Lett. **67**, 2327 (1991)
104. L. Yang, T.S. Rahman, M.S. Daw, Phys. Rev. B **44**, 13725 (1991-II)
105. Y.W. Lin, S.A. Adelman, G. Wolken, Surf. Sci. **66**, 376 (1977)
106. B. Baule, Ann. der Physik **44**, 145 (1914)
107. For a derivation of the Baule formula see R.E. Stickney, Adv. Atomic and Mol. Physics 3 (1967) 143 Appendix
108. J.W.M. Frenken, P.M.J. Maree, J.F. van der Veen, Phys. Rev. B. **34**, 7506 (1986)
109. F. Lindemann, Phys. Z. **11**, 609 (1910)
110. H. Lowey, Phys. Rep. **237**, 249 (1994)
111. T.S. Rahman, J.D. Spangler, A. Al-Rawi, J. Phys. Cond. Matt. **14**, 4903 (2002)
112. G. Brusdeylins, R.B. Doak, J.P. Toennies, Phys. Rev. B **27**, 3662 (1983)
113. D. Eichenauer, U. Harten, J.P. Toennies, V. Celli, J. Chem. Phys. **86**, 3693 (1987)
114. J.R. Manson, G. Benedek, S. Miret-Artès, Chem. Phys. Lett. **7**, 1016–1021 (2016)

Chapter 8
Theory of Atom Scattering from Surface Phonon: The Role of Electron-Phonon Interaction

The effect of a concept-driven revolution is to explain old things in new ways. The effect of a tool-driven revolution is to discover new things that have to be explained.
(Frank Watson Dyson, Astronomy, 2014)

Abstract The forces that an He atom exerts on the atoms of a solid surface causing inelastic scattering have much in common with the interatomic forces that govern the dynamics of the lattice. The scattering theory, described in Chap. 7 based on two-body collisions, provides the correct interpretation of data for closed-shell surfaces, much as their lattice dynamics is well described by phenomenological interatomic potentials. The inelastic HAS experiments from metal surfaces, however, cannot be successfully described in the same way. In this Chapter the approaches developed for the lattice dynamics of metals, where the interatomic forces are mediated by free electrons, such as the Multipole Expansion (ME) and the Density Functional Perturbation Theory (DFPT) are introduced for the analysis of inelastic HAS intensities from metal surfaces. These theories have the important consequences that inelastic HAS intensities are directly proportional to the electron-phonon coupling strength for individual phonons and that HAS can detect deep sub-surface phonons. The propensity of HAS to excite certain phonons of conducting surfaces rather than others is found to depend on the electron-phonon interaction and the surface electron band structure.

8.1 Introduction

At metal surfaces, where the corrugation is negligible due to the smoothing effect of the free-electron density, the inelastic energy exchange with phonons occurs largely through the electrons at the surface. The displacement of the surface charge density by the impinging He atoms leads to time-dependent oscillations of the surface atomic displacements. In this case the inelastic response may be amplified by a large electron-phonon interaction, and is often associated with the appearance of

G. Benedek and J. P. Toennies, *Atomic Scale Dynamics at Surfaces*, Springer Series in Surface Sciences 63, https://doi.org/10.1007/978-3-662-56443-1_8

surface phonon anomalies. Section 8.2 describes a convenient platform for the calculation of the phonon-induced surface charge-density oscillations within the phenomenological framework of the Multipole Expansion method with its Pseudocharge (PC) parametrization. Section 8.3 then discusses the Density Functional Perturbation Theory (DFPT) which provides a suitable *ab initio* approach to describe the coupling between the electrons and the lattice atom motions. Sections 8.4 and 8.5 describe the selection rules and propensities in atom-surface inelastic scattering and the dependence of inelastic intensities on parallel momentum transfer, respectively. The ME method and DFPT developed for the dynamics of conducting surfaces have also been used for the analysis of HAS data from the surface of topological insulators [1, 2]. They also provide the appropriate tool for the analysis of inelastic HAS intensities, leading to the new concepts of *mode-lambda spectroscopy* and *quantum sonar effect* [3, 4].

8.2 Coupling of the He Atoms to the Electrons at Metal Surfaces

The inelastic scattering theory developed in Chap. 7 for closed shell neutral and ionic systems relies on the accuracy provided by pairwise additive potentials for describing the direct inelastic interactions of the He atom with the individual atoms of the crystal surface. For metal surfaces, however, the explicit contribution to the He atom-surface potential originating from the density of conduction electrons plays a major role and must be accounted for. The redistribution of conduction electrons at the surface of metals also called *Smoluchowski* smoothening [5] was investigated by Finnis and Heine [6] in the early seventies, and shown by Garcia and Soler [7] to produce a substantial softening of the He-surface potential as compared to that of ionic crystals. This effect is clearly shown in Fig. 6.7.

A general *ab initio* theory of the coupling forces between a He atom and the phonons at a metal surfaces was developed by Senet et al. [8]. The forces are directly derived from the static He-surface potential in the presence of conduction electrons and many-body forces, discussed in Sect. 6.6, from which some equations are repeated here for sake of clarity.

The force exerted upon an electron by a He atom impinging on metal surface at position **x** is derived from the following pseudopotential

$$V_{ps}(\mathbf{x},\mathbf{r}) = V_{ps,kin}(\mathbf{x},\mathbf{r}) + V_{ps,pol}(\mathbf{x},\mathbf{r}). \tag{8.1}$$

The first term $V_{ps,kin}$ is a short range potential which accounts for the increase in the kinetic energy of the surface electrons due to the additional presence of the He atom. The second term $V_{ps,pol}$ is a long range potential due to the induced electrostatic polarization of the atom. The dipolar double layer at a metal surface and its deformation due to the impinging He atom induce a polarization in the He atom itself, as shown in *ab initio* calculations by Petersen et al. [9]. The resulting weak long range electrostatic potential, in addition to the dispersion force, are not considered in this discussion. Moreover, Senet et al. [8] provide an argument that the potential $V_{ps,pol}$ largely screens the weak direct electrostatic potential between the polarized He atom and the metal ions. With this *Ansatz* the time-dependent part of the scattering potential, (7.2), which couples the He-atom with the phonon field, only involves the short-range part of the pseudopotential

$$\delta V(\mathbf{r},t) \cong \int d^3x\, V_{ps,kin}(\mathbf{x},\mathbf{r})\, \delta n(\mathbf{x},t), \tag{8.2}$$

where

$$\delta n(\mathbf{x},t) = \int d^3r' \chi(\mathbf{x},\mathbf{x}') \sum\nolimits_l \frac{\partial v_{ion}(\mathbf{x}')}{\partial \mathbf{r}_l} \cdot \mathbf{u}(l,t), \tag{8.3}$$

is the oscillating electron charge density $n(\mathbf{x}, t)$ induced by the phonon displacement fields $\mathbf{u}(l, t)$ of the l surface atoms and $\chi(\mathbf{x}, \mathbf{x}')$ is the nonlocal dielectric susceptibility of the electrons, as already introduced in (5.14). From the definition of $v_{ion}(\mathbf{x})$ as a sum of single ionic potentials $v_l(\mathbf{x} - \mathbf{r}_l)$ [(5.2) with $\mathbf{u}(l) = 0$] the resulting forces on the ions of the surface can be shown to be linearly related to the derivative of $n(\mathbf{x})$ with respect to the atomic displacements $\mathbf{u}(l)$ [8]

$$\frac{\partial n(\mathbf{x})}{\partial \mathbf{u}(l)} = \int d^3x' \chi(\mathbf{x},\mathbf{x}') \frac{\partial v_l(\mathbf{x}'-\mathbf{r}_l)}{\partial \mathbf{r}_l}. \tag{8.4}$$

By inserting (8.3) into (8.2) a direct proportionality between the perturbing potential and the phonon displacements $\mathbf{u}(l,t)$ is obtained, which can be expressed as

$$\delta V(\mathbf{r},t) = \sum\nolimits_l \frac{\partial V_{eff}(\mathbf{r},\mathbf{r}_l)}{\partial \mathbf{r}_l} \cdot \mathbf{u}(l,t), \tag{8.5}$$

where

$$V_{eff}(\mathbf{r},\mathbf{r}_l) = \int d^3x\, V_{ps,kin}(\mathbf{x},\mathbf{r}) \int d^3x' \chi(\mathbf{x},\mathbf{x}')\, V_l(\mathbf{x}'-\mathbf{r}_l) \tag{8.6}$$

are the *effective two-body potentials* between the He atom and the individual surface ions which were discussed in connection with (6.49). Their derivatives are the *effective two-body forces*. Such effective two-body potentials are non-central, however, as they incorporate the effects of many-body interactions. Their separate dependence on $\mathbf{r}$ and $\mathbf{r}_l$ and not on their difference does not permit the change of variable which leads to (7.27) and (7.28) and to the formulation in terms of displacement-displacement correlation functions. Note that this procedure is entirely analogous to that used in connection with (5.13), where an effective force constant matrix $R^* = R - TH^{-1}T^+$ was obtained. Exactly in the same way the substitution of $\delta n(\mathbf{x}, t)$ into (8.3) yields an effective two-body inelastic scattering potential.

In general $V_{ps,kin}(\mathbf{x}, \mathbf{r})$ also has a non-central character since it incorporates the effects of many-body forces, which may lead to subtle effects in the inelastic scattering intensities. For many purposes, however, one may take advantage of the short-range nature of $V_{ps,kin}$ and represent it in the local density approximation as

$$V_{ps,kin}(\mathbf{x}, \mathbf{r}) \cong A_n \delta(\mathbf{x} - \mathbf{r}), \tag{8.7}$$

where A_n is a semi-empirical constant. In this way the simple Esbjerg-Nørskov approximation [10], already introduced in (6.50), is obtained:

$$V(\mathbf{r}) = A_n n(\mathbf{r}). \tag{8.8}$$

The correct value A_n has been extensively discussed in the literature [11–14]. A value for He appropriate to most metal surfaces is $A_n = 364\,\text{eV}\,a_0^3$ [13]. According to (8.8) the He atoms at thermal energy (E_i = 60 meV) are already repelled by the comparatively weak electron density of about 10^{-4} $e^-/(\text{a.u.})^3$ which prevails at distances of about 3 Å above the top plane of atoms [15–17]. Although the potential $V(\mathbf{r})$, (8.8), was originally introduced to explain the diffraction intensities from surfaces, insertion of (8.7) into (8.2) shows that also the time-dependent part of the interaction potential $\delta V(\mathbf{r}, t)$, entering the inelastic scattering theory through (7.2), is approximately proportional to the oscillations of the surface electron density $\delta n(\mathbf{r}, t)$ resulting from the vibrations of the atomic cores. Therefore the inelastic HAS intensities are expected to depend on the phonon-induced modulation $\delta n(\mathbf{r}, t)$ of the surface electron density occurring at the position $\mathbf{r}$ of the He atom. This modulation is expressed by (7.3) with the derivatives of $V(\mathbf{r})$ replaced by the derivatives of $n(\mathbf{r})$ with respect to the atomic positions $\mathbf{r}_l$.

It should be noted, however, that at the turn-around-distance of about 3 Å the incident atom encounters only the most peripheral, less bound electrons. Thus only the loosely bound outermost electrons with energies close to the Fermi surface, and not all the electrons which concur to give the total charge density, are mostly involved in the transmission of the forces between He and atoms at the surface of a metal. This fact clearly appears in the *ab initio* calculations by Petersen et al. of the *static* potential for He and Ne on Rh(110) [9] and poses some limitation on the validity of the Esbjerg-Norskov potential. The *ab initio* treatment by Senet et al. draws similar conclusions for the perturbation [8]. The effective potentials $V_{eff}(\mathbf{r}, \mathbf{r}_l)$ defined by (8.6) lead to a dynamic surface corrugation (the change of surface profile due to the motion of surface ions) which is different, in general, from that produced by the bare ion potentials $V_l(\mathbf{r} - \mathbf{r}_l)$. The two potentials would be identical only if $V_{ps,kin}(\mathbf{x}, \mathbf{r}_a)$ and $\chi(\mathbf{x}, \mathbf{x}')$ are local, i.e., proportional to a δ-function. It is their non-locality which is responsible for a stronger electronic response between the surface ions, e.g., at bridge or hollow positions, compared to the response at the top position. Thus the above description of the electron-density modulation is consistent with the dynamic anti-corrugation effects which have been discussed in Sect. 6.6.3 [18–20]. A detailed discussion can be found in the review by Gumhalter [21].

8.3 Calculation of Inelastic HAS Intensities from Metal Surfaces

8.3.1 Application of the Multiple Expansion (ME) Method

The *ab initio* calculation of $\delta n(\mathbf{x}, t)$, (8.3), as a function of the ion displacements, although feasible within DFPT for common metal surfaces, can be surrogated by the ME method, especially when the method is used for the surface dynamics. As was shown in Chap. 5, the ME method is formally a transcription of DFPT on a local basis, and therefore the ME method [22] and its PC parametrization [23, 24] can also provide an efficient scheme to calculate the inelastic HAS intensities and account for the many-body inelastic interaction potential. The inelastic potential resulting from the modulation of the electron density produced by the vibrations of the surface atoms can be obtained directly from (8.2) and (8.3) by using the expression for the density oscillation $\delta n(\mathbf{x})$ derived from (5.14) and (5.15):

$$\delta V(\mathbf{r}, t) = \sum\nolimits_{l\lambda\Gamma} c_\Gamma(l\lambda, t) \int d^3x \, V_{ps,kin}(\mathbf{x}, \mathbf{r}) Y_\Gamma(\mathbf{x} - \mathbf{r}_{l\lambda}). \tag{8.9}$$

In principle the multipolar expansion of the charge density also encompasses the contributions from the direct potential between the He atoms and the tightly bound valence electrons, such as the *d* electrons of transition metal ions or the weakly polarizable shells in closed shell systems. The contributions of these terms to inelastic HAS intensities can be treated in much the same way as for the closed shell crystal surfaces discussed in the previous section. Here, however, only the short range repulsion between the He and the surface ion shells is included. As mentioned above in connection with (8.1), the direct interaction between the ion charges and the He induced dipole is screened out by conduction electrons through the potential $V_{ps,pol}(\mathbf{x}, \mathbf{r})$. In fact, the attractive part of the He metal-surface potential is almost entirely due to dispersion forces. By approximating $V_{ps,kin}$ with the Esbjerg and Nørskov potential, (8.9) reduces to

$$\delta V(\mathbf{r}, t) = A_n \delta n(\mathbf{r}, t) - A_n \sum\nolimits_{l\lambda\Gamma} c_\Gamma(l\lambda, t) Y_\Gamma(\mathbf{r} - \mathbf{r}_{l\lambda}). \tag{8.10}$$

The matrix elements of the perturbing potential $\delta V(\mathbf{r}, t)$ can be expressed by (7.21) with $\partial V(\mathbf{r}, t)/\partial \mathbf{r}_l$ replaced by $A_n Y_\Gamma(\mathbf{r} - \mathbf{r}_{l\lambda})$ and $\mathbf{u}_l(t)$ by $c_\Gamma(l\lambda, t)$. The essential physics can be captured in the simple Born approximation, discussed in Sect. 7.3. In this case the matrix elements of $\delta V(\mathbf{r}, t)$ coupling the He atom to a surface phonon of wavevector **Q** and branch index ν can be obtained as a Fourier transforms of (8.10),

$$\delta V_{fi} = A_n \sum\nolimits_{\Gamma\lambda} Y_{\Gamma\lambda}(\Delta\mathbf{k})\, c_\Gamma(\mathbf{Q}\nu,\lambda), \qquad (8.11)$$

where $c_\Gamma(\mathbf{Q}\nu, \lambda)$ are the multipolar components of the dynamical matrix eigenvectors (Sect. 5.3) and

$$Y_{\Gamma\lambda}(\Delta\mathbf{k}) \equiv \int d^3r\, e^{i\Delta\mathbf{k}\cdot\mathbf{r}} Y_\Gamma(\mathbf{r} - \mathbf{r}_\lambda), \qquad (8.12)$$

with $\Delta\mathbf{k} = \mathbf{k}_f - \mathbf{k}_i$ and $\mathbf{Q} = \mathbf{K}_f - \mathbf{K}_i - \mathbf{G}$. A **G**-vector has been added to **Q** to account for all possible *umklapp* processes. For low index metal surfaces, however, only inelastic processes with $\mathbf{G} = 0$ are observed due to the very small corrugation.

Following the arguments developed in Sect. 6.1, it is instructive to express the perturbation potential $\delta V(\mathbf{r}, t)$, (8.9), which has an intrinsic many-body character, as a sum of effective two-body interatomic potentials. This can be done by means of (5.12) which links $c_\Gamma(\mathbf{Q}\nu,\lambda)$ to the eigenvector $\mathbf{u}(\mathbf{Q}\nu)$ and gives, after substitution into (8.11),

$$\delta V_{fi} = -A_n \sum\nolimits_{\Gamma\lambda\alpha} Y_{\Gamma\lambda}(\Delta\mathbf{k}) \left(H^{-1}T^+\right)_{\Gamma\lambda\mathbf{Q}\alpha} u_\alpha(\mathbf{Q}\nu), \qquad (8.13)$$

where the Fourier transform of the electron response function is given by

$$\left(H^{-1}T^+\right)_{\Gamma\lambda,\mathbf{Q}\alpha} \equiv \sum\nolimits_L \left(\Gamma L\lambda\middle|H^{-1}T^+\middle|L'\alpha\right) e^{i\mathbf{Q}\cdot(\mathbf{R}_L - \mathbf{R}_{L'})}. \qquad (8.14)$$

As explained in Sect. 5.3, the matrix elements of T and H can in principle be derived from a density functional scheme and a suitable set of multipolar basis functions Y_Γ. Equation (8.13) indicates that the differential reflection coefficient, (7.18), is proportional to the squared elements of the electronic response function $H^{-1}T^+$. This by itself illustrates the close link existing between the microscopic theory of helium inelastic scattering and that of surface dynamics. In particular the inelastic HAS peak intensities contain important information also on the electron-phonon interaction, as will be shown in detail in the next section. A tutorial example of how this information can be extracted from experimental data will be discussed below in Sect. 8.4.

8.3.2 Implementation of the Pseudocharge PC Model

A parametrized PC model, which was successfully used to calculate the inelastic He scattering intensities for copper surfaces [23] is described. In addition to the interaction of the He atoms with the electron charge localized at the center of the surface ions, due largely to d electrons rigidly bound to the ion cores, the He interactions with a number N_p of PC's located at the midpoints $\mathbf{r}_{l\lambda}$ between surface nearest neighbor ions were also included. For the (111) surface of a *fcc* crystal (Fig. 8.1) there are $N_p = 3$ pseudocharges per surface unit cell.

As illustrated in Fig. 8.2, both dipolar deformations of the surface PC's, expressed by the components of the vector $\mathbf{c}_d(\mathbf{Q}\,\nu, \lambda)$, $\lambda = 1,2, \ldots N_p$, (see Fig. 5.2d), and the tetragonal PC quadrupolar deformation $c_q(\mathbf{Q}\,\nu, \lambda)$ ($A_{g,2}$ component in Fig. 5.2c).

As explained in Sect. 5.3, the coefficients $A_n Y_\Gamma(\mathbf{r} - \mathbf{r}_{l\lambda})$ are within the ME model the derivatives with respect to the PC coordinates $c_\Gamma(l\lambda, t)$ of an atom-PC potential $V_\lambda(\mathbf{r})$. Thus their Fourier transforms $A_n Y_{\Gamma\lambda}(\Delta\mathbf{k})$ may be written as $V_\lambda\,(\Delta\mathbf{k})\Delta\mathbf{k}$ with $V_\lambda(\Delta\mathbf{k})$ the Fourier transform of $V_\lambda(\mathbf{r})$. The inclusion of the PC dipolar and quadrupolar contributions in the coupling potential energy, expressed by the scalar product in (7.33b), is then achieved by assuming for the atom-PC potentials a Born-Mayer form $v_{BM}(r) = B_\lambda\, e^{-\beta r}$, (6.14). With a substitution similar to that of (7.35) the following expression is obtained for the dynamic work exerted on the lattice,

$$
\begin{aligned}
\mathbf{F}_{fi} \cdot \mathbf{u}_{\mathbf{Q}\nu} \rightarrow\ & e^{-W(\Delta\mathbf{k})} v(\Delta\mathbf{k})\,\Delta\mathbf{k} \cdot \mathbf{u}_{\mathbf{Q}\nu} \\
& + \sum_{\lambda=1}^{N_p} e^{-W_{\lambda,d}(\Delta\mathbf{k})} v_\lambda(\Delta\mathbf{k})[\mathbf{Q} \cdot \mathbf{c}_d(\mathbf{Q}\nu, \lambda) - i\beta\, c_{dz}(\mathbf{Q}\nu, \lambda)]\ e^{i\mathbf{Q}\cdot\mathbf{r}_\lambda} \\
& + \sum_{\lambda=1}^{N_p} e^{-W_{\lambda,q}(\Delta\mathbf{k})} v_\lambda(\Delta\mathbf{k})\left[-i\beta\, c_{q,3\xi^2-r^2}(Q\nu, \lambda)\right]\ e^{i\mathbf{Q}\cdot\mathbf{r}_\lambda},
\end{aligned}
\tag{8.15}
$$

where $W_{\lambda,d}(\Delta\mathbf{k})$ and $W_{\lambda,q}(\Delta\mathbf{k})$ are the Debye-Waller exponents constructed with the mean-square displacement of the PC dipolar and quadrupolar components.

The complex interplay of the three contributions to the scattering amplitude, i.e., the ion (rigid d-electron), dipolar and quadrupolar (deformable $sp\ d$ electrons)

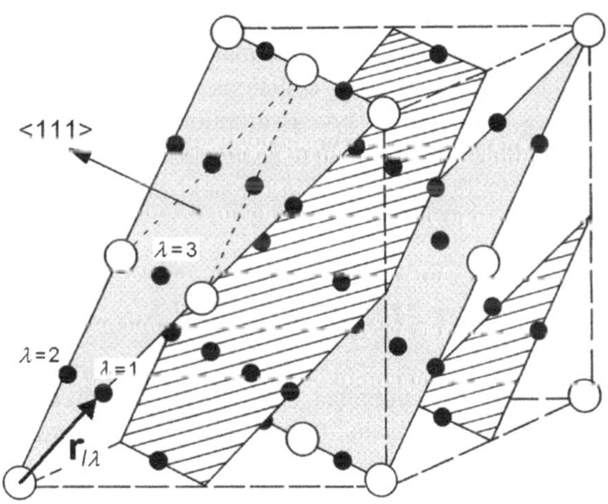

Fig. 8.1 Schematic diagram showing the location of pseudocharges in the vicinity of the (111) surface of a fcc metal. The (111) surface is projected into the cubic cell (broken lines), with the ions (open circles) and the pseudocharges (black circles) located at the midpoints $\mathbf{r}_{l\lambda}$ between nearest neighbor ions [23]. There are three pseudocharges (labelled by λ) in the surface unit cell (dashed lines)

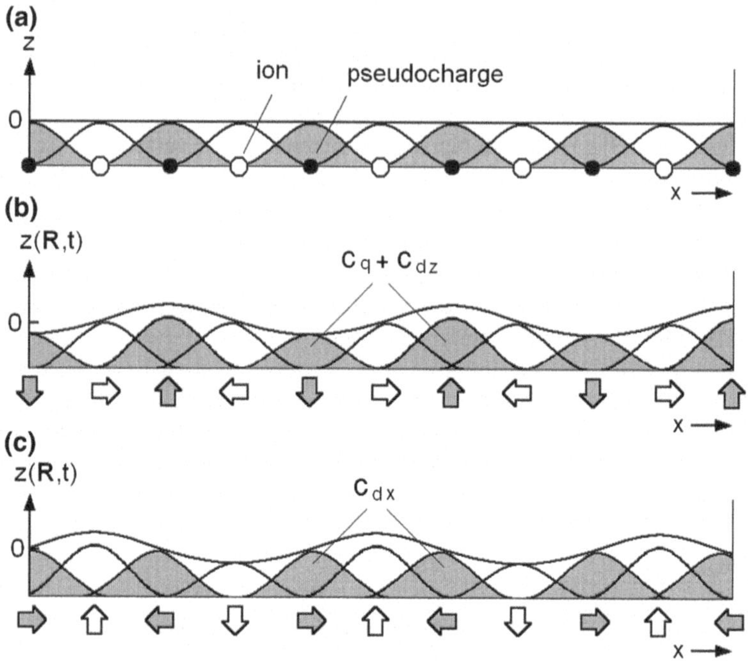

Fig. 8.2 Schematic diagram illustrating the pseudocharge (PC) description of the lattice dynamics at metal surfaces and the inelastic coupling of the He atoms to the pseudocharges. **a** The contributions of the charges on the ion cores and on the PC's add up to a flat undistorted surface. **b** The longitudinal ion displacements (open arrows), typical of the L resonance, produce a PC distortion of both quadrupolar (c_q) and *vertical* dipolar (c_{dz}) types (full arrows) and a dynamical surface corrugation $z(\mathbf{R}, t)$ with extrema at the bridge positions. **c** The shear-vertical ion displacements, typical of Rayleigh waves, produce a PC in-plane dipolar distortion c_{dx} and a dynamical surface corrugation with extrema at on-top positions

contributions, provides a framework within which it is possible to understand the anomalous intensities first observed on the silver and copper surfaces (see Sect. 11.3) [23]. This new formalism supplants the previous concepts of spring constants and two-body collisions of He atoms with individual atoms at metal surfaces [25]. The *ad hoc* assumptions in the PC semi-empirical theory find their full justification in the first principles DFPT treatment of surface dynamics and HAS intensities illustrated in the next Section and in Chap. 11.

8.3.3 Other Models for Metal Surfaces

It is interesting to note that after the replacement $\mathbf{c} = -H^{-1}T^{+}\mathbf{u}$, the coupling energy of (8.15) is proportional to $\mathbf{u}_{\mathbf{Q}\nu}$ and to some factor $S(\mathbf{Q})$ containing terms proportional to $\exp(\mathbf{Q} \cdot \mathbf{r}_\lambda)$, where $\mathbf{r}_\lambda$ are the basis vectors of the PC sub-lattice. Indeed

Santoro et al. [20] have shown that a factor $S(\mathbf{Q})$ can be chosen as an *ad hoc* structure factor reflecting the surface anti-corrugation, namely the surface electron density which actually responds to the He atom (see Fig. 8.8b). Their calculations of the inelastic HAS spectra for both Cu(111) and Cu(001), in which the He-ion forces for a soft corrugated potential, combined with a simple Born-von Karman surface dynamics (see Sect. 4.4), are multiplied by $S(\mathbf{Q})$, are in good agreement with the experiments, comparable to that obtained with the PC model. This is not surprising because the distribution of pseudocharges for the surfaces of noble metals [23] corresponds indeed to an anti-corrugated surface. This effect has been confirmed by DFT calculations for the Cu(111) surface by Jean et al. [26]. The anti-corrugation effect, first found in diffraction studies of metal surfaces and briefly discussed at the end of Sect. 6.6.3, has also been used to explain the large HAS inelastic intensities from LA resonances in Rh(111) [19, 23]. A subsequent analysis carried out by Santoro et al. [20] shows that the anomalously large inelastic HAS intensities of the anomalous longitudinal resonance can be attributed to a manifestation of the anti-corrugation effect.

A thorough comparison of the two methods is found in the review by Gumhalter [21]. It is important to note, however, that the response function $H^{-1}T^{+}$, appearing in (8.13) and (8.14), has a smooth dependence on $\mathbf{Q}$ only in the absence of Kohn anomalies and surface states at the Fermi level, as is the case in noble metals. In regions of the $\mathbf{Q}$-space next to Kohn anomalies, where the electron susceptibility and the He-electron forces increase rapidly, a pure geometric factor like $S(\mathbf{Q})$ is not expected to provide a good approximation.

Other related phenomenological models have been proposed to explain the HAS measurements on transition metals. Bortolani et al. [27] have represented the interaction between the He atoms and the surface atoms in the conventional way by a pairwise sum of potentials and calculated the phonon spectrum within the Born-von Karman force constant theory described in Chap. 3. In order to account for delocalized interactions with the electrons, the surface atoms are treated as oblate ellipsoids with the short axes normal to the surface plane. The use of these flattened potentials gives rise to a smoother surface potential for elastic scattering than with a pairwise summation of spherical potentials. The repulsive cores and attractive van der Waals terms are parameterized so that the calculated diffraction patterns agree well with HAS measurements [27]. To account for inelastic scattering, it is assumed that the flattened potentials move with the ion cores. Thus the motion of the conduction electrons between the ionic cores is effectively mimicked in a manner very similar to that of the pseudocharge model. A disadvantage of this *ad hoc* phenomenological approach is that more empirical parameters are required to describe not only the lattice dynamics but also the form of the flattened potentials. This approach has certain similarities to a theory introduced long ago by Quasso et al. for alkali halide surfaces [28, 29], which invoked quadrupole deformations. However, in the case of the simpler closed shell alkali halide crystals the quadrupolar polarizabilities could be obtained from previous quantum mechanical calculations and no *ad hoc* parameters were needed.

8.4 A First Principles Electron-Phonon Theory of Inelastic He Atom Scattering

8.4.1 The Role of the Electron-Phonon Coupling

The previous discussion of the role of electrons in transmitting collision energy to the phonons at the surface of free electron metal surfaces can easily be extended to semimetals, semiconductors and insulators with large atomic (or ionic) polarizabilities. The phonon-induced deformation of the surface electronic density in metals, as well as the polarization of surface atomic (ionic) shells in insulators are manifestations of what is generally termed *electron-phonon interaction*. Thus the inelastic scattering intensities and the range of the He atom interaction with the crystal atom displacements must be related to the electron-phonon interaction.

These concepts were already clear in the 80s after the first HAS measurements on noble metal surfaces [30, 31], and motivated a consistent theory of surface dynamics and inelastic scattering based on the ME method [22, 23, 32]. Only DFPT calculations could, however, fully account for the complexity of inelastic HAS spectra from metal surfaces [33–35] and ultrathin films [3]. Especially the HAS spectra from thin lead films revealed the surprising ability of HAS to detect subsurface phonons even localized several layers beneath the surface (termed *quantum sonar*) [3, 4]. The theory of inelastic HAS based on the electron-phonon interaction can be formulated as follows.

A phonon of parallel wavevector **Q**, branch index ν and energy $\varepsilon_{\mathbf{Q}\nu} \equiv \hbar\omega_{\mathbf{Q}\nu}$, causing the transition of an electron from a state $\psi_{\mathbf{K}n}(\mathbf{r})$ near the Fermi level of parallel wavevector **K** and band index n to an empty state $\psi_{\mathbf{K}+\mathbf{Q}n'}(\mathbf{r})$ of wavevector **K + Q** and band index n', induces a modulation of the charge density $\delta n_{\mathbf{K}n,\mathbf{Q}\nu}(\mathbf{r})$. Within Esbjerg-Nørskov approximation, (8.8), the matrix element between the He atom initial and final states of the inelastic scattering potential produced by the charge density variation is then given by

$$\begin{aligned}\delta V_{fi}(\mathbf{K}n, \mathbf{Q}\nu) &= A_n \langle f|\delta n_{\mathbf{K}n,\mathbf{Q}\nu}(\mathbf{r})|i\rangle \\ &= -\frac{A_n}{\varepsilon_{\mathbf{Q}\nu}} \sum\nolimits_{n'} g_{nn'}(\mathbf{K}, \mathbf{K}+\mathbf{Q}; \nu)\langle f|\psi^*_{\mathbf{K}n}(\mathbf{r})\psi_{\mathbf{K}+\mathbf{Q}n'}(\mathbf{r})|i\rangle.\end{aligned} \tag{8.16}$$

where $g_{nn'}(\mathbf{K}, \mathbf{K}+\mathbf{Q}; \nu)$ is the e-ph interaction matrix [36]. The product of the two electron wavefunctions on the r.h.s. is a non-diagonal element of the electron density matrix. It represents the charge density oscillation (CDO) produced by a surface phonon of wavevector **Q,** which admixes an electron state $(\mathbf{K}, n|$ with an electron state $|\mathbf{K} + \mathbf{Q}, n')$. A Feynman diagram representation of the scattering processes considered in (8.15) is shown in Fig. 8.3. As discussed in [3], the matrix element between final and initial atom states in (8.16), associated with the upper vertex of the diagram, turns out to be proportional to a function $I(\varepsilon_{\mathbf{Q}\nu})$ which, in a first approximation, only depends on the energy transfer, independently of whether

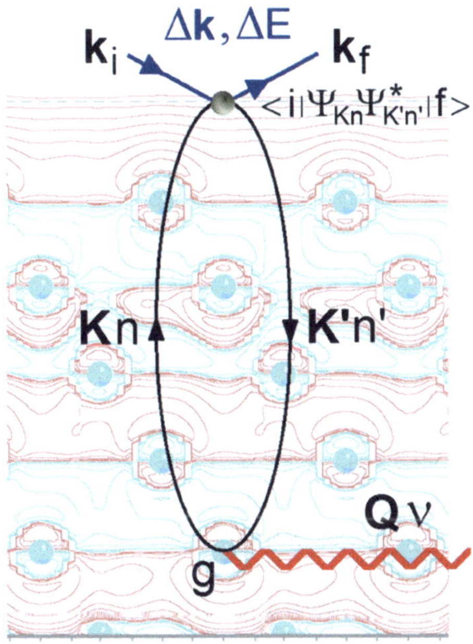

Fig. 8.3 Feynman diagram representing the scattering processes of (8.16). The diagram is superimposed on the contour lines of the phonon-induced charge density oscillations (for the color code see Fig. 8.4) extending over five atomic planes at a metal surface. The He atom is repelled by the surface charge density far away from the first atomic layer but can transmit energy and parallel momentum to a sub-surface phonon via a virtual electron-hole pair. The two vertex terms connected by Kn and K′n′ arrows correspond to the two factors in the r.h.m. of (8.16). The considerable depth (5 atomic layers in this picture) at which a phonon can be excited within the range of the e-ph interaction is the essence of the Quantum Sonar Effect

the electron transmits the energy to the nuclei via an intra- ($n = n'$) or an interband ($n \neq n'$) transition:

$$\langle f|\psi^*_{\mathbf{K}n}(\mathbf{r})\,\psi_{\mathbf{K}+\mathbf{Q}n'}(\mathbf{r})|i\rangle \cong \delta(\mathbf{K}_f - \mathbf{K}_i + \mathbf{Q})I(\varepsilon_{\mathbf{Q}\nu}). \tag{8.17}$$

This permits the factorization

$$\sum\nolimits_{\mathbf{K}n}\left|\sum\nolimits_{n'} g_{nn'}(\mathbf{K},\mathbf{K}+\mathbf{Q};\nu)\langle f|\psi^*_{\mathbf{K}n}(\mathbf{r})\psi_{\mathbf{K}+\mathbf{Q}n'}(\mathbf{r})|i\rangle\right|^2 \cong \tfrac{1}{2}N_F\varepsilon^3_{\mathbf{Q}\nu}I^2(\varepsilon_{\mathbf{Q}\nu})\lambda_{\mathbf{Q}\nu}, \tag{8.18}$$

where N_F is the density of electronic states at the Fermi level. The factor $\lambda_{\mathbf{Q}\nu}$, known as the *mode selected electron-phonon coupling constant* (or simply the *mode-λ*), gives the e-ph coupling for each individual surface phonon of parallel wavevector $\mathbf{Q}$ and branch index ν [37]. The average $\sum_{\mathbf{Q}\nu}\lambda_{\mathbf{Q}\nu}/3N$ over the $3N$ phonon modes of the solid defines the mean e-ph coupling constant λ, introduced in the BCS theory of superconductivity as the electron mass renormalization

$\Delta m_e/m_e$ at the Fermi level due to the e-ph interaction [38–40]. In metals Δm_e is always positive, and therefore λ is also known as the e-ph mass-enhancement parameter (or factor) [36, 41, 42].

The information of the phonon-induced CDO is instead contained in the factor $I^2(\varepsilon_{\mathbf{Q}\nu})$ in (8.18). In this way the DWBA inelastic transition rate, cfr. (7.33) can be written as

$$w(\mathbf{k}_f, \mathbf{k}_i) = K(\Delta E)\, \mathrm{e}^{-2W^{eff}(\mathbf{k}_f, \mathbf{k}_i)} \sum\nolimits_{\mathbf{Q}\nu} \lambda_{\mathbf{Q}\nu}\, \delta(\Delta E + \varepsilon_{\mathbf{Q}\nu}), \tag{8.19}$$

where the kinematic factor

$$K(\Delta E) \equiv |N_F \Delta E\, n(\Delta E)| [I(\Delta E) A_n]^2 \overset{T \gg \Delta E}{\longrightarrow} N_F [I(\Delta E) A_n]^2 kT \tag{8.20}$$

is found to be almost independent of the energy loss ΔE and proportional to the surface temperature in the high temperature limit. The squared matrix element of the coupling takes the general DWBA form (see Chap. 7), with

$$\left|\mathbf{F}_{fi} \cdot \mathbf{u}_{\mathbf{Q}\nu}\right|^2 = |N_F \Delta E| [I(\Delta E) A_n]^2 \mathrm{e}^{-2W^{eff}(\mathbf{k}_f, \mathbf{k}_i)} \lambda_{\mathbf{Q}\nu}. \tag{8.21}$$

The effective DW exponent $W^{eff}(\mathbf{k}_f, \mathbf{k}_i)$ accounts for surface potential fluctuations due to those of the surface charge density, either induced by the thermal motion of surface atoms or directly due to the thermal fluctuations of the Fermi gas. The latter is generally negligible, while the phonon contribution to $W^{eff}(\mathbf{k}_f, \mathbf{k}_i)$, mediated by the electron-phonon interaction, is discussed below in Sect. 8.4.3. Thus the above theory predicts that the inelastic scattering intensities are simply related, through (8.16–8.19), to the square phonon-induced surface charge density oscillations and not directly to the square displacements of the surface atoms as in the earlier theories reviewed in Chap. 7.

8.4.2 Mode-Selected Electron-Phonon Coupling Strengths from HAS Inelastic Intensities

The implications of the theory have been visualized by DFPT calculations of the actual CDOs at the surface of Cu(111). This is illustrated for the example of two HAS energy transfer spectra for the Cu(111) surface measured at two different scattering angles in the $\overline{\Gamma \mathrm{M}}$ direction shown in Fig. 8.4a, c. The spectra have been chosen to display the peaks associated with the Rayleigh wave (RW) and the longitudinal acoustic resonance S_3. In (a) the peaks correspond, via the scan curve, to the wavevector of the RW at $\overline{\mathrm{M}}/2$ while that of S_3 is at about one fourth of the zone ($\overline{\mathrm{M}}/4$). In (c) the wavevector of the RW is at the zone boundary ($\overline{\mathrm{M}}$-point) while that of S_3 is at about one half of the zone ($\overline{\mathrm{M}}/2$). In Fig. 8.4b the two large red

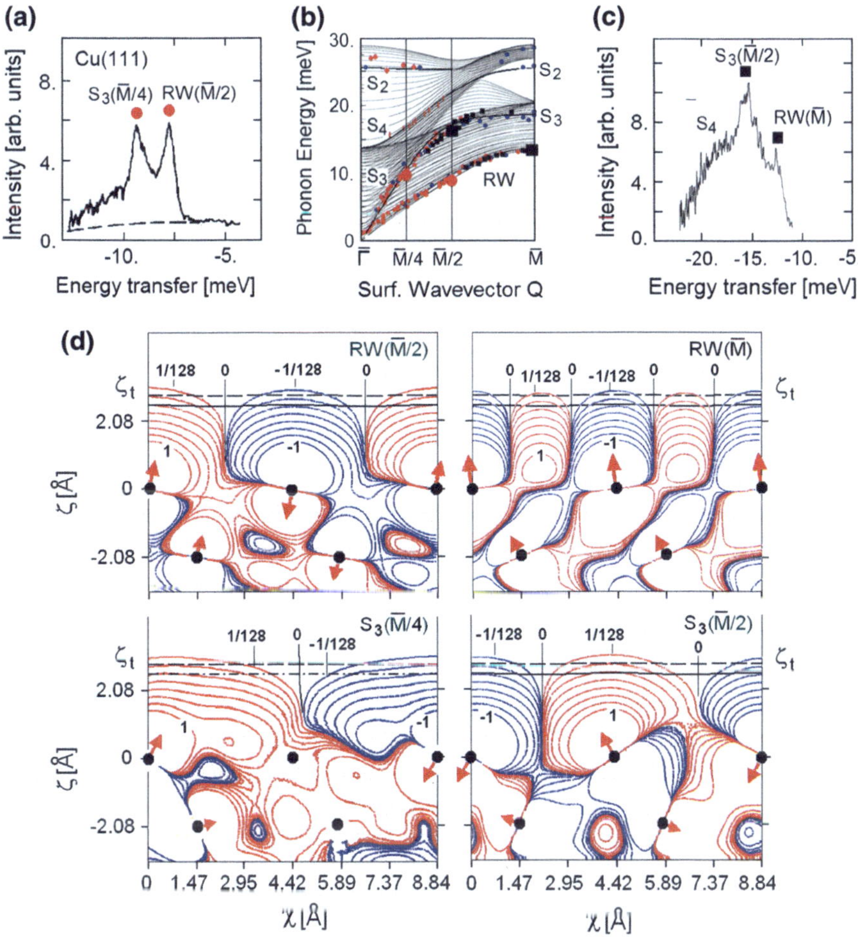

Fig. 8.4 Evidence for the electron-phonon mechanism of inelastic HAS from a metal surface. **a** HAS spectrum for the Cu(111) surface showing the peaks of the Rayleigh wave (RW) with wavevector at one half of the zone ($\overline{M}/2$) and the longitudinal acoustic resonance S_3 at about one fourth of the zone ($\overline{M}/4$). **b** The Cu(111) dispersion curves calculated with DFPT in the $\overline{\Gamma M}$ direction with the HAS and EELS experimental points (small symbols), including those of (**a**) and (**c**) (large symbols). **c** HAS spectrum for the RW at the zone boundary ($\overline{M}$-point) and S_3 at about $\overline{M}/2$. **d** DFPT calculations of the charge density oscillations (CDOs) produced by the RW and S_3 modes at the points in the energy transfer spectra in (**a**) and (**c**). The red arrows show the normalized amplitudes of the displacements of the atoms (black circles), while red and blue curved lines are the contour plots of the corresponding CDOs (in units of $2^{-n} \times 10^{-4}$ a.u. with $n = 0$ to 7; red for density increase, blue for density decrease). The broken line ζ_t marks the turning points for He atoms of incident energy of 35.3 meV and incident angle of 31°. The full line marks the turning point for normal incidence and energy of 63 meV [33, 34]

circles and the two black squares show the four surface phonons on the DFPT dispersion curves [33, 34]. While RW($\overline{M}$/2) and S_3($\overline{M}$/4) have about the same intensity, the RW($\overline{M}$) peak is much smaller than S_3($\overline{M}$/2).

In Fig. 8.4d the red and blue contours show the DFPT calculated CDOs associated with the RW and S_3 modes at the three points along the $\overline{\Gamma M}$ direction. The red arrows show the normalized amplitudes of the atomic displacements, while the curves correspond to the density distribution of the CDO (in units of $2^{-n} \times 10^{-4}$ a.u. with $n = 0$ to 7; red for density increase, blue for density decrease) [33, 34]. According to (8.16–8.19) the experimental intensities should be proportional to the square CDOs at the classical turning point of the scattered He atom (broken line ζ_t for an incident energy of 35.3 meV and incident angle of 31°). Thus the larger square CDOs of the S_3($\overline{M}$/2) of $\sim 2.0 \times 10^{-12}$ a.u. compared to that of the RW($\overline{M}$) of $\sim 1.0 \times 10^{-12}$ explains the larger HAS peak intensity of the S_3 mode in Fig. 8.4b. The more equal square CDOs for S_3($\overline{M}$/4) of $\sim 2.5 \times 10^{-12}$ a.u. compared to that of the RW($\overline{M}$/2) of $\sim 2.0 \times 10^{-12}$ is also consistent with the corresponding nearly equal peak intensities in Fig. 8.4a. Additional confirmation for the importance of the CDOs comes from a comparison of the lengths of the red arrows showing the atomic displacements for the RW and S_3 modes. According to the two-body collision model the intensity of the RW($\overline{M}$) inelastic peak should be much larger than that of the S_3 mode, in clear disagreement with the energy gain spectrum in Fig. 8.4c. The example illustrates the severe limitations of the two-body collision model when applied to metal surfaces.

A further consequence of the HAS electron-phonon theory is that in first approximation, the HAS peak intensities for individual phonons are proportional to the respective mode-selected e-p coupling constants, thus enabling a so-called *mode-λ spectroscopy* [3]. This also implies that inelastic HAS is sensitive to all atomic displacements that produce a sufficiently intense CDO at the surface. Although He atoms at thermal energies "tickle" the surface at about 3 Å above the first atomic layer (Fig. 8.4d), they can be sensitive to subsurface phonons via the CDOs as deep as the range of the electron-phonon interaction, which in many metals can extend over several layers.

This ability of inelastic HAS, termed *quantum sonar effect* and the related *mode-λ spectroscopy* found a first clear demonstration in Pb ultrathin films [3, 4]. Figure 8.5 shows some HAS spectra for Pb(111) films of 5 and 7 monolayers (ML) on a Cu(111) substrate, measured at different incident angles, are compared to the mode-λ spectrum calculated with DFPT for all phonons of the films having a sagittal polarization [3, 4].

The close correspondence between the measured phonon intensities and the respective *ab initio* mode-λ (further improved when the slowly varying prefactor, (8.21), is included) indicates that information on the electron-phonon coupling constants for individual surface phonons can be extracted directly from HAS inelastic intensities. The knowledge of the mode selected coupling constants $\lambda_{\mathbf{Q}\nu}$ provides detailed information on the phonons which are relevant or irrelevant for superconductivity.

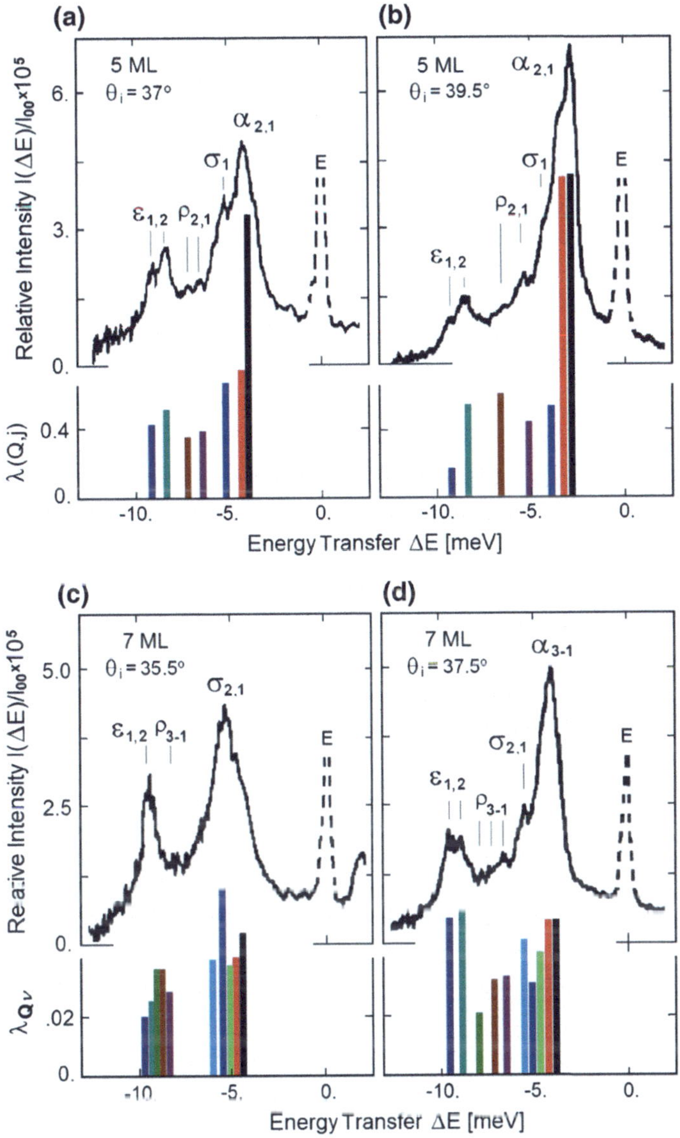

Fig. 8.5 Comparison of energy transfer HAS phonon peak intensities of sagittal polarization from Pb(111) films on a Cu(111) substrate with the respective *ab initio* mode selected e-ph coupling constants $\lambda_{Q\nu}$ [3, 4]. **a**, **b** Inelastic spectra for a 5-monolayer (ML) film measured at two closely spaced incident angles θ_i are compared with the corresponding calculated mode-lambdas $\lambda_{Q\nu}$ indicated by the colored bars. **c**, **d** The same for a 7-ML film. The colors of the $\lambda_{Q\nu}$ bars refer to different modes indicated by greek letters in the energy transfer spectra. The approximate peak proportionality to the respective mode selected electron-phonon coupling constants illustrates the *mode-λ spectroscopy*

For Pb(111) films the most important phonon appears to be the optical mode ε_2 which is localized at the interface with the substrate [3, 4]. This surprising finding fully illustrates the quantum sonar effect and the relevant information it can provide.

A similar analysis has been made for Bi(111), where the HAS scattering amplitude from the highest optical mode SV1 is comparatively far more intense than that for the ε_1 mode of Pb(111) films, although both modes are localized above the maximum bulk energy and restricted within the first atomic bilayer. According to the e-ph theory of inelastic HAS intensities, the optical first layer shear vertical (SV-1) mode of Bi(111) has a comparatively large mode-λ, a fact that has been associated with the enhancement of the e-ph interaction observed with ARPES at the Bi(111) surface [43], and the appearance of superconductivity in Bi crystallites and interfaces [44, 45], whereas bulk Bi is not superconducting.

8.4.3 The Electron-Phonon Coupling Strength from the HAS Debye-Waller Factor

Recently it has been recognized that the mean e-ph coupling constant λ can be directly estimated from the measurement of the HAS Debye-Waller factor $\exp\left[-2W^{eff}(\mathbf{k}_f, \mathbf{k}_i)\right]$, e.g., from the temperature dependence of the elastic intensities [46, 47]. It is expected that the constant λ at the surface as derived from HAS is somewhat different from the corresponding value in the bulk [73]. The effective Debye-Waller exponent $W^{eff}(\mathbf{k}_f, \mathbf{k}_i)$ in (8.19) and (8.21) requires special consideration for conducting surfaces. To a large extent mediated by the electron-phonon interaction, the thermal fluctuations affect the atom-surface potential. The fluctuations have two different sources: one is due to the intrinsic thermal fluctuation of the Fermi sea, which at ordinary surface temperatures can be neglected; the other, more important, is due to the charge density oscillations induced by the thermal vibrations of the crystal atoms via the e-ph interaction. As recently shown by Manson et al. [46–48] the latter contribution leads to the effective Debye-Waller exponent

$$W^{eff}(\mathbf{k}_f, \mathbf{k}_i) = 2N_F \sum\nolimits_{\mathbf{Q}\nu} \left[\frac{\Delta k_z}{2\kappa_F} \frac{A_n I(\varepsilon_{\mathbf{Q}\nu})}{(f|V^{eff}_{\mathbf{Q}\nu}(T,z)|i)}\right]^2 \varepsilon_{\mathbf{Q}\nu} \left[n(\varepsilon_{\mathbf{Q}\nu}) + \frac{1}{2}\right] \lambda_{\mathbf{Q}\nu}, \qquad (8.22)$$

where Δk_z is the normal wavevector change of the scattering atom ($2k_{iz}$ for specular elastic scattering), and $\kappa_F = \sqrt{2m_e^* \phi}/\hbar$ with ϕ the work function of the surface material and m_e* the effective mass of the electron. In (8.22) the sum over $\mathbf{Q}$ is restricted to the first BZ, and

$$V^{eff}_{\mathbf{Q}\nu}(T,z) = \frac{1}{2\pi}\int_{-\infty}^{+\infty} dk\, e^{ikz}\, e^{-W_{\perp\nu}(k)} V_{\mathbf{Q},k} \tag{8.23}$$

with $V_{\mathbf{Q},k}$ the 3D Fourier transform of He-surface-atom potential $V(\mathbf{r})$. The 2D Fourier component of the effective potential $V^{eff}_{\mathbf{Q}\nu}(T,z)$ depends on the temperature via the normal component of the ordinary Debye-Waller exponent approximately cancel with the weakly energy-dependent function $W_{\perp\nu}(k) = \frac{1}{2}k^2 < u^2_{z\nu} >$, with $< u^2_{z\nu} >$ the surface atom mean square displacement for the ν-th mode in the z-direction. The analogous 2D Fourier transform of an exponential repulsive potential was already discussed in Sect. 6.4.2. For the case that $\beta \gg Q$ the effective potential $V^{eff}_{\mathbf{Q}\nu}(T,z)$ depends weakly on $\mathbf{Q}$. The condition $\beta \gg Q$ is generally fulfilled for $\mathbf{Q}$ within the first BZ, as can be seen in Table 6.5. In this case the matrix element of $V^{eff}_{\mathbf{Q}\nu}(T,z)$ can be approximated by its $Q = 0$ value [46] thereby approximately canceling the weakly energy-dependent function $A_n I(\varepsilon_{\mathbf{Q},\nu})$ in the numerator of (8.22) (cfr. (8.20)). This gives for $k_B T \gg \Delta E$ [46]

$$W^{eff}(\mathbf{k}_f, \mathbf{k}_i) \approx N_F \left(\frac{k_{iz}}{\kappa_F}\right)^2 \lambda k_B T, \tag{8.24}$$

where $\lambda \equiv (1/3N)\sum_{\mathbf{Q}\nu} \lambda_{\mathbf{Q}\nu}$, with $3N$ the number of phonons of the solid, is the (surface) e-ph coupling constant. This turns out to be directly proportional to the DW exponent, and can therefore be directly determined from DW measurements, once the multiplying factor is known [46]. For a simple estimation, N_F may be replaced by the Fermi level density of states per atom of a free electron gas, $3Zk_B m^*_e/\hbar^2 k_F^2$, with Z the number of conduction electrons per atom and k_F the Fermi wavevector. When surface electronic states occur at the Fermi level, a surface Fermi wavevector k_{Fs} should be used in order to account for the fact that such surface states extend into the vacuum further than bulk states and are mainly responsible for repelling low energy He atoms. With this approximation, valid at sufficiently high temperatures and for specular scattering, it is found that

$$2W^{eff}(\mathbf{k}_f, \mathbf{k}_i) \approx 6Z \frac{k_B T}{\phi} \frac{k^2_{iz}}{k_F^2} \lambda, \tag{8.25}$$

with $E_{iz} = \hbar^2 k^2_{iz}/2m = E_i \cos^2\theta_i$. As a consequence λ can be estimated from the logarithmic slope of the specular intensity as a function of the surface temperature

$$\lambda \approx \frac{-d \ln I}{dT} \frac{\phi}{6Zk_B} \frac{k_F^2}{k^2_{iz}}. \tag{8.26}$$

Figure 8.6 shows in a logarithmic scale the experimental relative HAS specular intensities $I(T)/I(0)$ plotted as functions of the surface temperature for Bi(111) [49] and Sb(111) [50]. The observed linear dependence of $\ln[I(T)/I(0)]$ on temperature indicates that zero point quantum effects are negligible in the measured temperature range and therefore $-\ln[I(T)/I(0)]$ is just $2W(T)$.

Table 8.1 lists a comparison of λ derived from the DW temperature dependence of HAS specular peaks with literature values for five conducting surfaces. It appears that, despite the crude approximations, the values of λ extracted from HAS measurements are in satisfactory agreement with those found in the literature. This is especially true if one considers that λ, as known from other sources, may vary over a fairly wide range depending on the experimental method. Note that at this level of approximation the difference between the nominal surface λ derived from HAS DW data with (8.26) and the bulk λ from other sources may not be so significant. The important parameter in (8.26) is the square Fermi wavevector k_F^2 relative to that of the surface electron gas which contributes most to the He-surface potential at the turning point, cfr. (8.17). For the surfaces of Bi, Cu, Sb and W in Table 8.1, the values of k_F associated with relevant surface bands cutting the Fermi level have been extracted from the quoted literature [51, 52]. Ultrathin films like 4 ML-Pb (111)/Cu(111), whose thickness is less than the range of electron-phonon interaction, present the interesting case that all the electronic states of the quantum well are involved and therefore the use of the bulk k_F appears to be more appropriate. For example, in the case of copper, measurements on epitaxial nanometric films yield λ of the order of 0.1 [53], whereas photoemission data from the Cu(110) surface [54, 55] yield $\lambda = 0.23 \pm 0.02$ for the surface states at $\overline{\mathrm{Y}}$. The value derived from HAS data with $k_F = 0.25$ Å^{-1} for the surface band at $\overline{\mathrm{Y}}$ [56] is $\lambda = 0.15$, intermediate to the above experimental values but closer to the Cu bulk values reported in Allen's review ($\lambda = 0.11$–0.14) [42]. Because of the quantum sonar effect, the information provided by HAS may not exclusively come from surface states. On the other hand for W(001) the value of λ derived from HAS data with k_F from calculated surface states [57] is substantially larger than the bulk value [42]. Bulk W is a superconductor with a critical temperature T_c of only 15 mK, but T_c rapidly grows in W thin

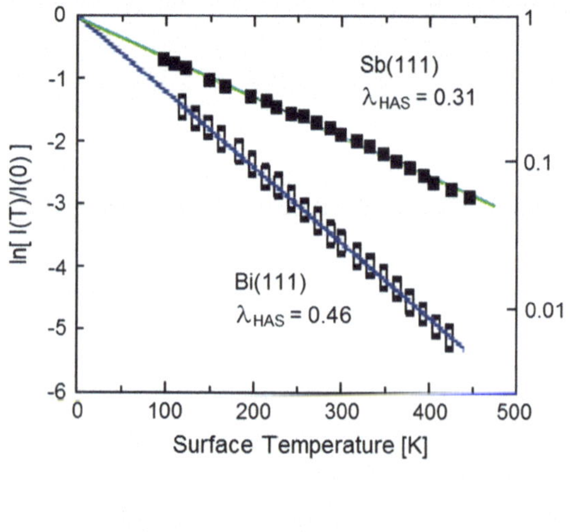

Fig. 8.6 Intensity of the HAS specular peak as a function of the surface temperature relative to the values extrapolated to 0 K for Bi(111) [49] and Sb(111) [50]. The mass-enhancement factors λ are derived from the slopes of the logarithm of the relative intensities

Table 8.1 The e-ph coupling constants λ derived from the temperature dependence of the HAS specular intensity for some conducting surfaces are compared with the corresponding values from other sources. The parameters entering (8.26) and the values of λ as known from the literature are also listed. From [46]

Surface	$-dlnI/dT$ (K^{-1})	k_{iz}^2 ($Å^{-2}$)	Z	ϕ (eV)	k_F ($Å^{-1}$)	λ HAS	λ Other
Bi(111)	0.0115[a]	16.8[a]	5	4.23[b]	0.65[c]	0.46	0.60[d]
Cu(110)	0.0017[e]	6.20[e]	1	4.48[f]	0.25[g]	0.15	0.093[h], 0.23[i]
4ML-Pb(111)/Cu (111)	0.0028[j]	22.3[j]	4	4.25[k]	1.71[l]	0.76	1.01[m]
Sb(111)	0.0056[n]	22.8[n]	5	4.56[o]	0.80[p]	0.31	0.27[q]
W(001) 1x1	0.0041[r]	26.3[r]	6	4.32[s]	1.09[t]	0.39	0.28[u]

[a][49] [e][59] [j][60] [n][50] [r][61]
[b][62] [f][63] [k][64] [o][65] [s][66]
[c][67] [g][56] [l][68] [p][69] [t][57]
[d][70] [h][53] [m][4] [q][71] [u][42]
[i][54, 55]

films [58]. This probably explains the larger e-ph coupling constant found with HAS for W(001).

8.5 Propensities for Scattering from Metal Surfaces

8.5.1 *Difference in Propensities Between Insulator and Metal Surfaces*

The predictions of Sect. 7.5 of the previous Chapter which were derived for insulators disagree with most HAS data from conducting surfaces, where the L resonance is regularly found to have an amplitude much larger than predicted on the basis of pairwise He-surface atom potentials. The propensity to excite certain surface modes rather than others depends strongly on the interactions mediated by conduction electrons. Figure 8.7 illustrates the difference for the case of Cu(001) in the [100] direction at $T = 150$ K. In Fig. 8.7 some representative HAS spectra for the Rayleigh and anomalous longitudinal surface modes are plotted as functions of the energy transfer together with the corresponding dispersion curves and intensities as functions of the momentum transfer within the first Brillouin zone [72]. The spectra in Fig. 8.7a show a dramatic increase of the L-resonance intensity with respect to that of the RW for increasing phonon energies and surface wavevectors. The surface wavevector dependence of the intensities of the two modes shown in Fig. 8.7c is obtained from the respective scan curves in Fig. 8.7b. The more gradual decrease of the L mode intensity with increasing ΔK compared to the RW, and for wavevectors larger than 0.8 $Å^{-1}$ the equal L and RW intensities differ from the predictions of Sect. 7.5. According to the dynamic factor in (7.62), and the

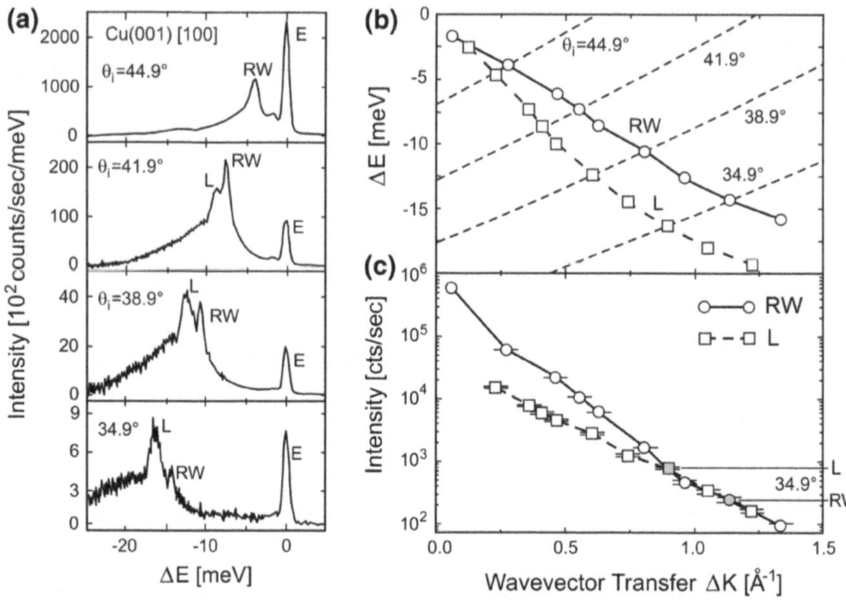

Fig. 8.7 Typical inelastic peak intensities measured for a metal surface illustrated for the [50] azimuth of the Cu(001) surface (E_i = 40.4 meV, T = 150 K). **a** HAS energy-loss spectra, from which the frequencies and intensities of the RW and L modes are obtained after subtraction of the multiphonon background, for a few selected incidence angles. **b** The dispersion curves measured by HAS of the Rayleigh waves (RW) and surface longitudinal modes (L). The dashed lines are scan curves corresponding to the spectra. **c** The measured intensities (open circles and squares) as functions of the parallel momentum transfer [72]. The L mode, as measured, e.g., at 34.9°, is apparently much more intense than the RW mode due to its smaller wavevector. In reality the L mode intensity decreases for increasing ΔK less fast than that of the RW, and at the same wavevectors larger than 0.8 $Å^{-1}$ the L and RW intensities are equal

discussion in Sect. 7.52, the repulsive BM parameter β in (7.62) is in general larger than π/a and Q_c (cfr. Table 6.5) and $u_{Qz,RW} > u_{Qx,L}$. Thus the L intensity should remain considerably smaller than that of the RW even at large Q. Moreover for $Q \rightarrow 0$ the L intensity should die out as predicted by (7.62), a trend which is not observed at small values of Q in the HAS data reported in Fig. 8.7b.

In summary, the predictions of (7.60–7.62) disagree with most HAS data from conducting surfaces, where the L resonance is regularly found to have an amplitude much larger than predicted on the basis of pairwise additive He-surface atom potentials. The propensity to excite certain surface modes rather than others depends strongly on the interactions mediated by conduction electrons. This has been quantitatively established with the *ab initio* method discussed in Sect. 8.3. A qualitative understanding can however be gained from simple arguments based on the PC model discussed below.

8.5.2 Pseudocharge Effects at Conducting Surfaces

In the PC model the deformation of the surface potential profile may not faithfully follow the displacements of the atoms of the surface layer. In the HCS model, Sect. 7.4, the phonon displacement components are weighted by the functions $\partial D_o(\mathbf{R})/\partial \mathbf{r}_\kappa$. The surface corrugation $D_o(\mathbf{R})$ depends on the surface charge density and the manner in which the latter is changed by the atomic displacements can be derived from the PC model. The PC contributions as given by (8.15) are formally similar to the square brackets of (7.58), but now the determination of phonon eigenvectors from scattering amplitudes depends on the relationship between the PC coordinates $c_\Gamma(\mathbf{Q}\nu,\lambda)$ and the atomic displacements. In many metals the coupling to the longitudinal modes is dramatically enhanced thanks to the vertical PC deformation induced by the longitudinal compressional motion of the surface ions, especially at the zone boundary [24].

The additional degrees of freedom introduced by the PCs can be explicitly included in the formalism of (7.60–7.62), by writing the dynamic factor, (7.62), as

$$J_\nu(\mathbf{Q},\Delta E) = |n(\Delta E)| \big|\mathbf{Q}\cdot\mathbf{u}_{\mathbf{Q}\nu} - i\beta\, u_{\mathbf{Q}\nu,z} + (\mathbf{Q}\cdot\mathbf{c}_{d,\mathbf{Q}\nu} - i\beta\, c_{d,\mathbf{Q}\nu z})\exp(iQa/2) - i\beta\, c_{q,\mathbf{Q}\nu}\exp(iQa/2)\big|^2, \tag{8.27}$$

where $\mathbf{c}_{d,\mathbf{Q}\nu}$ and $c_{q,\mathbf{Q}\nu}$ are the PC dipolar displacement vector and the tetragonal component of the PC quadrupolar deformation, respectively (cf. Sects. 5.3 and 8.3). One single PC has been considered in the unit cell, located at the midpoint of the interatomic distance a. In this case the dynamic factors for the RW and L modes are given by

$$RW:\quad J_{RW} = |n(\hbar w_{Q,RW})| \left[\beta^2 u^2_{Qz,RW} - 2\beta\, Q u_{Qz,RW} c_{d,Qx,RW}\sin(Qa/2) + Q^2 c^2_{d,Qx,RW}\right], \tag{8.28}$$

$$L:\quad J_L = |n(\hbar w_{Q,RW})| \left[\beta^2 c^2_{Qz,L} + 2\beta\, Q u_{Qx,L} c_{Qz,L}\sin(Qr_0/2) + Q^2 u^2_{Qx,L}\right], \tag{8.29}$$

where $c_{Qz,L} \equiv c_{d,Qz,L} + c_{q,Q,L}$ incorporates the dipolar and quadrupolar contributions to the PC deformation along z (see Fig. 8.2). Due to their microscopic nature both $c_{Qz,L}$ and $c_{d,Qx,L}$ decrease for $Q \to 0$ at least as fast as $\sin(Qr_0/2)$. Thus for not too large $Q = \Delta K$ and not too soft potentials so that $Q \ll \beta$, only the leading terms proportional to β^2 can be retained in (8.28) and (8.29).

Under these conditions the longitudinal motion of surface atoms can produce inelastic scattering of helium atoms in as much as it induces a multipolar deformation of the pseudocharges and a dynamic corrugation of the otherwise flat surface electron density. Thus in this model the intensity of the inelastic scattering from the longitudinal surface branch is a measure of the phonon-induced electron density oscillation. For example, as shown below, the combined measurements of the anomalous softening $\Delta\omega^2_{\mathbf{Q},L}$ of the surface longitudinal phonon branch with respect to the bulk LA band edge (see, e.g., the S_3 branch in Fig. 3.1c, d) and of its intensity with respect to the RW intensity at the same **Q** can provide in principle useful information on the surface electron susceptibility and electron-phonon interaction.

The matrix relationship

$$c_{Qz,L} = -H^{-1}T^{+}u_{Qx,L}, \tag{8.30}$$

derived from (5.12), is now used in a one-dimensional form, where H^{-1} and $T^{+} \equiv T$ are no longer matrices but pure numbers expressing the electron susceptibility (in units of Z^2_{PC}/v_c, with Z_{PC} the PC charge and v_c the unit cell volume) and the electron-phonon interaction at a wavevector Q, respectively. Equation (8.29) follows from the argument that the PC deformations normal to the surface are strongly coupled by symmetry reasons to the longitudinal modulation of the interatomic distance between two adjacent surface atoms (Fig. 8.2). Thus using a $\sin(Qa/2)$ dependence for $u_{Qx,L}$ and the high-T approximation for the BE factor $|n(\hbar\omega_{Qv})| \approx kT/\hbar\omega_{Qv}$, one obtains from (8.28) and (8.29) the following expression for the ratio of the dynamic factors

$$\frac{J_L}{J_{RW}} \approx \left(\frac{\omega_{Q,RW}}{\omega_{Q,L}}\right)^2 \left(\frac{T}{H}\right)^2 \sin^2\frac{Qa}{2}, \tag{8.31}$$

with $(\omega_{Q,RW}/\omega_{Q,L})^2 \approx 0.5$. Since the factor $K(\mathbf{Q}, \Delta E)$ in (7.60) and (7.61) is about the same (within 10%) for the RW and L modes at the same Q all over the surface Brillouin zone, (8.31) can be identified with the experimental intensity ratio. This ratio as derived from the experimental data for Cu(001), shown in Fig. 8.7b, is roughly fitted by ${}^3/_2 \sin^2(Qa/2)$ (Fig. 8.8), which yields $T/H \cong \sqrt{3}$.

If the anomalous softening $\Delta\omega^2_{Q,L}$ is entirely attributed to the electron susceptibility via e-ph interaction, it can be easily calculated from the expression of the effective dynamical matrix, (5.13):

$$M\Delta\omega^2_{Q,L} \approx -T^2/H. \tag{8.32}$$

Hence one obtains the electron-phonon interaction and the inverse electron susceptibility as functions of Q:

$$\frac{T}{M} \cong \sqrt{3}\frac{H}{M} \approx -\frac{1}{\sqrt{3}}\Delta\omega^2_{Q,L}. \tag{8.33}$$

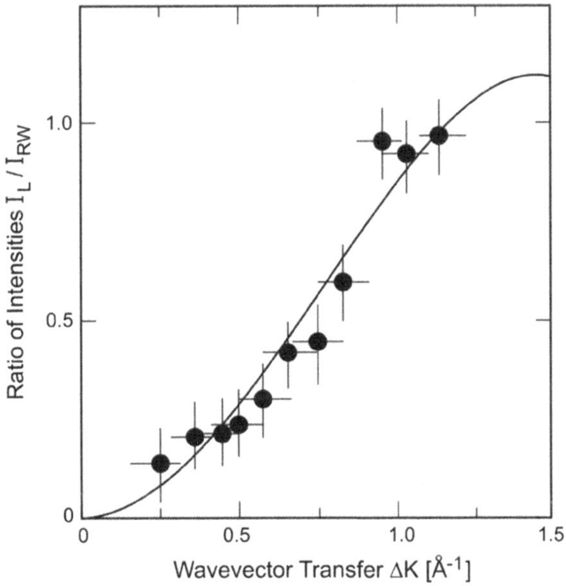

Fig. 8.8 The ratio I_L/I_{RW} of the longitudinal-to-Rayleigh wave scattering intensities from Fig. 8.7c (open circles) and a sine-square fit (full line).

From the dispersion curves of Cu(001) along the [100] direction (Chap. 11), $\Delta\omega^2_{Q,L}$ can be fitted by $(22.3\text{ meV})^2 \sin^3(Qa/2)$. Incidentally the constant value of the response function $-H^{-1}T^{+}$, appearing in (8.30) and (8.13), supports in this case the structure factor approximation proposed by Santoro et al. for the calculation of the He-phonon forces [18].

The present analysis, though highly qualitative, has nevertheless a pedagogical value as it further illustrates how relevant microscopic interactions, like those expressed by (8.33), can be estimated from the HAS measurement of surface phonon dispersion curves and their scattering intensities via a PC model analysis.

8.5.3 Propensities in Inelastic HAS Scattering from Conducting Ultrathin Films

As a concluding remark an interesting propensity property is illustrated for the first principle theory presented in Sect. 8.4. In this case single electronic wavefunctions are explicitly involved in the energy and momentum exchange between the probe atom and the phonon gas. Consider the special situation in which the electronic wavefunctions can be approximated by plane waves also in the normal (ζ) direction (cfr. Fig. 8.4) so as to have the form $\psi_{\mathbf{K}n}(\mathbf{R},\zeta) = (N_s N_3)^{-1/2}\exp(i\mathbf{K}\cdot\mathbf{R} - ik_n\zeta)$, with N_s the number of surface atoms, N_3 the number of atomic layers, and n labelling the different bands intersecting the Fermi level. This approximation may apply to ultrathin films of thickness N_3d, with d the interlayer spacing and a small N_3, having a finite set of wavelike quantum well states in the ζ direction. Then the e-ph interaction matrix takes the simple form [3]

$$g_{nn'}(\mathbf{K},\mathbf{K}+\mathbf{Q};\nu) = \mathbf{q}_{nn'}\cdot\xi(\mathbf{q}_{nn'}\nu)\,V(\mathbf{q}_{nn'}), \tag{8.34}$$

where $V(\mathbf{q}_{n,n'})$ is the pseudopotential Fourier transform with $\mathbf{q}_{nn'} \equiv (\mathbf{Q}, k_n - k_{n'})$, and

$$\xi(\mathbf{q}_{nn'}\nu) \equiv \sum\nolimits_{l_3} \mathbf{u}(\mathbf{Q}\,\nu, l_3)\exp[i(k_n - k_{n'})l_3 d]\exp\left[-\frac{1}{2}\left\langle|\mathbf{q}_{nn'}.\mathbf{u}_{l_3}|^2\right\rangle\right]. \tag{8.35}$$

Note that the inelastic electron scattering process creating or annihilating one-phonon has its own DW factor, where $\mathbf{u}_{l_3}$ is any displacement of the l_3-th layer and <…> indicates the usual thermal average. The scalar product in (8.34) determines the propensity of the phonon electronic transitions to couple with either SV or L components of the phonon modes of the film: (a) pure SV modes couple only to interband transitions ($n \neq n'$); (b) *bulk-like* L modes, i.e., modes with a wavelike displacement field $\mathbf{u}(\mathbf{Q}\nu, l_3)$ with respect to l_3, couple only to intraband transitions ($n = n'$); c) surface L modes couple preferentially to intraband transitions ($\mathrm{q}_{nn} = 0$), due to the DW attenuation of interband processes ($\mathrm{q}_{nn'} \neq 0$). Moreover SH modes along mirror symmetry directions do not couple to any electronic transition since the electron-phonon interaction cannot break the selection rules established for planar inelastic phonon scattering from surfaces with pairwise additive potentials. It should be noted, however, that the surface phonon modes, away from symmetry points where their group velocity is zero, have in general a sagittal polarization, so that the above arguments indicate more the propensities than the selection rules. The important conclusion is however that inelastic HAS intensities from conducting surfaces actually depend on the features of the band structure and response functions of the surface electrons.

References

1. X. Zhu, L. Santos, R. Sankar, S. Chikara, C. Howard, F.C. Chou, C. Charmon, M. El-Batanouny, Phys. Rev. Lett. **107**, 186102 (2011)
2. X.T. Zhu, L. Santos, C. Howard, R. Sankar, F.C. Chou, C. Chamon, M. El-Batanouny, Phys. Rev. Lett. **108**, 185501 (2012)
3. I. Yu. Sklyadneva, G. Benedek, E.V. Chulkov, P.M. Echenique, R. Heid, K.-P. Bohnen, J. P. Toennies, Phys. Rev. Lett. **107**, 095502 (2011)
4. G. Benedek, M. Bernasconi, K.-P. Bohnen, D. Campi, E.V. Chulkov, P.M. Echenique, R. Heid, IYu. Sklyadneva, J.P. Toennies, Phys. Chem. Chem. Phys. **16**, 7159–7172 (2014)
5. R. Smoluchowski, Phys. Rev. **60**, 661 (1941)
6. M.W. Finnis, V. Heine, J. Phys. F (Metal Phys.) **4**, L37 (1974)
7. N. Garcia, J.M. Soler, Surface Sci. **126**, 689 (1983)
8. P. Senet, J.P. Toennies, G. Benedek, Europhys. Lett. **57**, 430 (2002)

9. M. Petersen, S. Wilke, P. Ruggerone, B. Kohler, M. Scheffler, Phys. Rev. Lett. **76**, 995 (1996)
10. N. Esbjerg, J.K. Norskov, Phys. Rev. Lett. **45**, 807 (1980)
11. J. Harris, A. Liebsch, Phys. Rev. Lett. **49**, 341 (1982)
12. A. Liebsch, J. Harris, Surface Sci. **111**, 2721 (1981)
13. M.W. Cole, F. Toigo, Phys. Rev. B **31**, 727 (1985)
14. V. Celli, in *Helium Atom Scattering from Surfaces*, ed. by E. Hulpke, (Springer, Berlin, 1992), p. 25
15. N.D. Lang, Phys. Rev. Lett. **46**, 842 (1981)
16. E. Zaremba, W. Kohn, Phys. Rev. B **13**, 2270 (1976)
17. E. Zaremba, W. Kohn, Phys. Rev. B **15**, 12769 (1977)
18. G. Santoro, A. Franchini, V. Bortolani, Phys. Rev. Lett. **80**, 2378 (1998)
19. G. Santoro, A. Franchini, V. Bortolani, Nuovo Cim. D **20**, 1055 (1998)
20. G. Santoro, A. Franchini, V. Bortolani, D.L. Mills, R.F. Wallis, Surf. Sci. **478**, 99 (2001)
21. B. Gumhalter, Phys. Rep. **351**, 1 (2001)
22. C.S. Jayanthi, H. Bilz, W. Kress, G. Benedek, Phys. Rev. Lett. **59**, 795 (1987)
23. C. Kaden, P. Ruggerone, J.P. Toennies, G. Zhang, G. Benedek, Phys. Rev. B **46**, 13509 (1992)
24. C. Kaden, P. Ruggerone, J.P. Toennies, G. Benedek, Nuovo Cim. **14D**, 627 (1992)
25. V. Bortolani, A. Franchini, F. Nizzoli, G. Santoro, Phys. Rev. Lett. **52**, 429 (1984)
26. N. Jean, M.I. Trioni, G.P. Brivio, V. Bortolani, Phys. Rev. Lett. **92**, 013201 (2004)
27. V. Bortolani, D. Cvetko, F. Floreano, A. Franchini, A. Lansi, A. Morgante, M. Peloi, G. Santoro, F. Tommasini, T. Zambelli, A.F. Bellman, J. Elect. Spectr. Rel. **64**(65), 671 (1993)
28. L. Miglio, F. Quasso, G. Benedek, Surf. Sci. **136**, L9 (1984)
29. L. Miglio, F. Quasso, G. Benedek, J. Chem. Phys. **83**, 913 (1985)
30. R.B. Doak, U. Harten, J.P. Toennies, Phys. Rev. Lett. **51**, 578 (1983)
31. U. Harten, J.P. Toennies, Ch. Wöll, Faraday Disc. Chem. Soc. **80**, 137 (1985)
32. C. Kaden, P. Ruggerone, J.P. Toennies, G. Zhang, G. Benedek, Nuovo Cim. D **14**, 627 (1992)
33. V. Chis, B. Hellsing, G. Benedek, M. Bernasconi, E.V. Chulkov, J.P. Toennies, Phys. Rev. Lett. **101**, 206102 (2008)
34. V. Chis, B. Hellsing, G. Benedek, M. Bernasconi, E.V. Chulkov, J.P. Toennies, Phys. Rev. Lett. **103**, 069902 (2009)
35. G. Benedek, M. Bernasconi, V. Chis, E. Chulkov, P.M. Echenique, B. Hellsing, J. Peter, Toennies. J. Phys.: Cond. Matter **22**, 084020 (2010)
36. G. Grimwall, *The Electron–Phonon Interaction in Metals* (North-Holland, New York, 1981), Chap. 3
37. IYu. Sklyadneva, A. Leonardo, P.M. Echenique, S.V. Eremeev, E.V. Chulkov, J. Phys.: Condens. Matter **18**, 7923 (2006)
38. W.L. McMillan, Phys. Rev. **167**, 331 (1968)
39. P.B. Allen, M.L. Cohen, Phys. Rev. **187**, 525 (1969)
40. J. Bardeen, *Nobel Lecture, December 11, 1972*, in ed. by S. Lundqvist, Nobel Lectures. Physics 1971–1980 (World Sci. Publ. Co., Singapore, 1992), p. 54
41. P.B. Allen, Phys. Rev. B **36**, 2920 (1987)
42. P.B. Allen, in *Handbook of Superconductivity*, ed. by C.P. Poole, Jr. (Academic Press, New York, 1999) Chap. 9, Sec. G, pp. 478–483
43. A. Tamtögl, P. Kraus, M. Mayrhofer-Reinhartshuber, D. Campi, M. Bernasconi, G. Benedek, W.E. Ernst, Phys. Rev. B **87**,035410 (2013) and **87**, 159906 (2013) (E)
44. B. Weitzel, H. Micklitz, Phys. Rev. Lett. **66**, 385 (1991)
45. F. Muntyanu, A. Gilewski, K. Nenkov, A. Zaleski, V. Chistol, Solid State Commun. **147**, 183 (2008)
46. J.R. Manson, G. Benedek, Salvador Miret-Artès, J. Phys. Chem. Lett. **7**, 1016 (2016)
47. G. Benedek, S. Miret-Artès, J.P. Toennies, J.R. Manson, J. Phys. Chem. Lett. **9**, 76 (2018)
48. J.R. Manson, G. Benedek, S. Miret-Artès, Surface Sci. Rep. (2017)
49. M. Mayrhofer-Reinhartshuber, A. Tamtögl, P. Kraus, K.H. Rieder, W.E. Ernst, J. Phys. Cond. Matt. **24**, 104008 (2012)

50. A. Tamtögl, M. Mayrhofer-Reinhartshuber, P. Kraus, W.E. Ernst, Surf. Sci. **617**, 225 (2013)
51. W.R. Datars, J. Van der Koy, IBM J. Res. Dev. **8**, 247 (1964)
52. R. Herrmann, V.S. Edelman, Zh. Eksp. Tear. Fiz. **53**, 1563 (1967); Engl. Transl.: Sov. Phys. JETP, **26**, 901 (1968)
53. X. Shen, Y.P. Timalsina, T.-M. Lu, M. Yamaguchi, Phys. Rev. B **91**, 045129 (2015)
54. P. Straube, F. Pforte, T. Michalke, K. Berge, A. Gerlach, A. Goldmann, Phys. Rev. B **61**, 14072 (2000)
55. E.W. Plummer, J. Shi, S.-J. Tang, E. Rotenberg, S.D. Kevan, Prog. Surf. Sci. **74**, 251–268 (2003)
56. S.S. Tsirkin, S.V. Eremeev, E.V. Chulkov, Phys. Rev. B **84**, 115451 (2011)
57. K.E. Smith, G.S. Elliott, S.D. Kevan, Phys. Rev. B **42**, 5385 (1990)
58. W. Li, J.C. Fenton, Y. Wang, D.W. McComb, P.A. Warburton, J. Appl. Phys. **104**, 093913 (2008)
59. J. Lapujoulade, J. Perreau, A. Kara, Surf. Sci. **129**, 59–78 (1983)
60. G. Zhang, Ph.D. Dissertation, University of Göttingen, 1990; Max-Planck-Institut für Strömungsforschung Report No. 102 (1991)
61. H.-J. Ernst, E. Hulpke, J.P. Toennies, Phys. Rev. B **46**, 16081 (1992)
62. Ch. Bronner, P. Tegeder, Phys. Rev. B **87**, 035123 (2013)
63. P.O. Gartland, S. Berge, B.J. Slagsvold, Phys. Rev. Lett. **28**, 738 (1972)
64. H.B. Michaelson, J. Appl. Phys. **48**, 4729 (1977)
65. J. Speight, in *Lange's Handbook of Chemistry*, 16th ed, vol. 1 (McGraw Hill, Boston, 2004), p. 132
66. E. Wimmer, A.J. Freeman, M. Weinert, H. Krakauer, J.R. Hiskes, A.M. Karo, Phys. Rev. Lett. **48**, 1128 (1982)
67. Y. Ohtsubo, L. Perfetti, M.O. Goerbig, P. Le Fèvre, F. Bertran, A. Taleb-Ibrahimi, New J. Phys. **15**, 033041 (2013)
68. B.J. Hinch, C. Koziol, J.P. Toennies, G. Zhang, Europhys. Lett. **10**, 341 (1989)
69. K. Sugawara, T. Sato, S. Souma, T. Takahashi, M. Arai, T. Sasaki, J. Magn. Magn. Mat. **310**, 2177 (2007)
70. P. Hofmann, Prog. Surf. Sci. **81**, 191 (2006)
71. D. Campi, M. Bernasconi, G. Benedek, Phys. Rev. B **86**, 075446 (2012)
72. G. Benedek, J. Ellis, N.S. Luo, A. Reichmuth, P. Ruggerone, J.P. Toennies, Phys. Rev. B **48**, 4917 (1993)
73. J. Kröger, Rep. Prog. Phys. **69**, 899–969 (2006)

Chapter 9
Experimental Methods of HAS Surface Phonon Spectroscopy

These experiments are of interest not only because of their confirmation of the predictions of quantum mechanics, but also because they introduce the possibility of applying atom diffraction to investigations of the atomic constitution of surfaces. A beam of atomic hydrogen, for example, ... has a range of wavelengths of the right magnitude ... centering around 1 Å, and a complete absence of penetration of these waves will ensure that the effects observed arise entirely from the outermost atomic layer.

(T. H. Johnson, Phys. Rev. 37 (1931) 847,
about diffraction of H, H_2 and He from LiF)

Abstract Helium Atom Scattering (HAS) and Electron Energy Loss Spectroscopy (EELS) are the major experimental surface scattering techniques which have provided during the last decades information on the energies and dispersion curves of surface phonons. The projectiles, either a He atom or an electron are inelastically diffracted in a coherent fashion from the phonon wave oscillations at the surface of the target. In this chapter the kinematics, the intensities, the apparatus and the resolution of HAS experiments are discussed in detail, with quantitative comparisons with EELS experiments.

9.1 Kinematics of Surface Phonon Spectroscopy

He atoms with energies of about 20 meV have a wavelength of about 1 Å, thus comparable with the size of atoms and their interatomic distances in a solid (cf. Fig. 1.8). This represents an ideal range for measuring surface phonon dispersion curves [1, 2]. On the other hand, electrons in Electron Energy Loss Scattering (EELS) experiments, by virtue of their factor 7350 smaller mass, attain comparably small wavelengths at incident kinetic energies almost four orders of magnitude larger than those of He atoms, i.e., in the range of 100–300 eV. In contrast to the more usual *dipole scattering*, where electrons have incident energies of less than about 10 eV and essentially probe optical excitations of very small wavevectors [3],

G. Benedek and J. P. Toennies, *Atomic Scale Dynamics at Surfaces*, Springer Series in Surface Sciences 63, https://doi.org/10.1007/978-3-662-56443-1_9

this mode of EELS operation is called *impact scattering*. Because of the comparatively high energy of the electrons in both types of EELS experiments several collisions may occur with the ion cores in the vicinity of the surface. At least one collision is needed to turn the electron around so that it can leave the surface. At incident energies of 200–300 eV the penetration of the elastically reflected electrons is only about two or three layers. At the much lower energies of the strongly interacting He atoms there is, of course, no penetration at all. Thus, He atom scattering is one of the few surface techniques—and the only diffractive probe—which is sensitive exclusively to the outermost layer and is absolutely non-destructive.

Because of the different particles and energies involved in the two techniques the experimental methods used to detect the inelastic energy changes are quite different. In EELS the energy of the incident electrons is selected using an electrostatic sector field. A similar sector field serves to analyze the distribution of final energies after scattering. Such techniques are obviously not applicable to neutral He atoms. Thanks to the inherent narrow velocity distributions, energy selection of the incident beam of He atoms from a high pressure jet source is not required. The final energies are determined by measuring the time-of-flight (TOF) distribution (see Fig. 1.3). In both types of experiments the incident energies are chosen so that only single phonons processes dominate the spectra.

Although the energies are very different the kinematics of one-phonon scattering events are the same in both cases. As illustrated in Fig. 1.3, the kinematics are determined by the conservation laws of the total energy and parallel (surface) wavevector of the probe particle and the lattice. They are expressed by (1.2) and (1.3), which are reproduced here for convenience:

$$\Delta E \equiv \frac{\hbar^2}{2m}\left(k_f^2 - k_i^2\right) = -\hbar\omega(\mathbf{Q}), \tag{9.1}$$

$$\Delta \mathbf{K} \equiv \mathbf{K}_f - \mathbf{K}_i = -\mathbf{G} - \mathbf{Q}, \tag{9.2}$$

where $\mathbf{k}_f$ and $\mathbf{k}_i$ are the final and incident wavevectors of the probe particle, and $\mathbf{K}_f$ and $\mathbf{K}_i$ are the respective components projected on the surface plane (Fig. 9.1). Here $\mathbf{G}$ is a surface reciprocal lattice vector which is added in order to allow for the possibility of a diffraction process being coupled with the surface phonon wavevector $\mathbf{Q}$ (*Umklapp* process). It is recalled that $\mathbf{Q}$ is only defined within the first surface Brillouin zone. Events with $\hbar\omega(\mathbf{Q}) > 0$ and $\hbar\omega(\mathbf{Q}) < 0$ are referred to as phonon creation and annihilation scattering processes, respectively.

By way of introduction it is instructive to first consider the various vectors and defining planes involved in a general non-planar scattering experiment (Fig. 9.1). In most experiments the kinematics and consequently the interpretation of the experiments is greatly simplified in comparison to Fig. 9.1 by locating $\mathbf{k}_i$ and $\mathbf{k}_f$ in a common plane normal to the surface. This is called a *planar scattering* configuration, which was shown already in Fig. 1.3. In this geometry conservation of parallel momentum requires that the surface wavevector transfer $\Delta\mathbf{K}$ also must lie

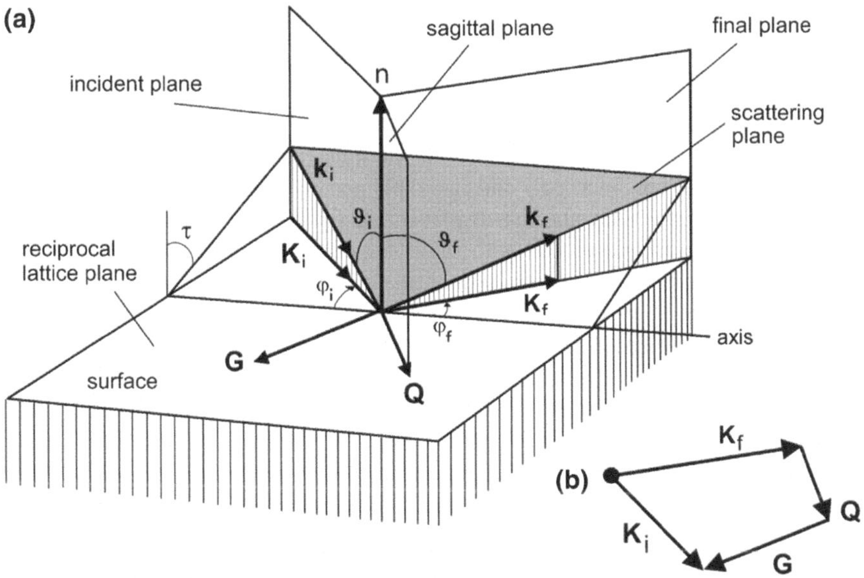

Fig. 9.1 Schematic diagram for the scattering of a monoenergetic beam of incident particles from a crystal surface in a general *non-planar scattering* geometry. **a** The particles are scattered from the *incident plane*, defined by the incident wavevector $\mathbf{k}_i$ and the surface normal **n**, into the *final plane*, defined by the final wavevector $\mathbf{k}_f$ and the surface normal. The scattering is said to be non-planar when the incident and final planes do not coincide. The wavevectors $\mathbf{k}_i$ and $\mathbf{k}_f$ define the *scattering plane*, which for non-planar scattering is not normal to the surface but forms an angle τ (*tilt angle*) with **n**. The intersection of the scattering plane with the surface defines the scattering plane *axis*. The incident and final particle directions are defined by the incident and final angles, ϑ_i and ϑ_f, and the azimuthal angles φ_i and φ_f, respectively. The latter are referred to the *axis* and have a sign (in the figure $\varphi_i < 0$, $\varphi_f > 0$). For planar scattering $\varphi_i = \varphi_f = 0$; however an azimuthal angle may be used to define the orientation of the scattering plane with respect to a high symmetry direction of the surface. **b** In an inelastic process in which a single surface phonon is annihilated or created the parallel surface component $\mathbf{K}_i$ of the incident wavevector equals the sum of the parallel component $\mathbf{K}_f$ of the final wavevector, plus the phonon wavevector **Q** and, in *Umklapp* processes, plus or minus an additional reciprocal surface lattice vector **G**. All these vectors lie in the reciprocal lattice plane coincident with the surface. Another important plane is the *sagittal plane*, which is defined by **Q** and **n**. In the frequently used *planar scattering* geometry the incidence and final planes coincide ($\varphi_i = \varphi_f = 0$), and the scattering plane is normal to the surface ($\tau = 0$). For planar inelastic scattering the sum **Q** + **G** must also lie in the scattering plane, and for a scattering plane along a high symmetry direction of the surface, **Q** and **G** must both belong to this plane, which coincides with the sagittal plane

in this plane. However (see Sect. 9.3) it is not necessary in general that the vectors **G** and **Q** also lie in the scattering plane, as illustrated in Fig. 9.8. If the scattering plane, or simply $\Delta\mathbf{K}$, is aligned parallel to a high symmetry direction, i.e. along one of the short reciprocal lattice vectors **G**, then also **Q** will be in the scattering plane, which will then coincide with the sagittal plane defined in Fig. 9.1.

9.2 Planar Scattering

In planar scattering along a high symmetry direction the scattering plane, besides coinciding with the sagittal plane, can also be a mirror symmetry plane for the crystal. In this case phonons have a definite sagittal or shear-horizontal (SH) polarization (see Sect. 3.9) and, as explained in Chap. 8, the incident particles can only couple with phonon displacement fields which are symmetric with respect to the sagittal plane, i.e., with sagittally polarized phonons. SH modes are anti-symmetric with respect to the sagittal plane and therefore cannot be excited in this scattering geometry.

In both HAS and EELS planar scattering experiments the energy distribution of the scattered particles is measured for an optimal fixed incident energy and certain incident (ϑ_i) and final (ϑ_f) scattering angles with respect to the surface normal and a suitable azimuthal angle $\varphi_i = \varphi_f$. Note that in planar scattering the surface wavevectors $\mathbf{K}_i$ and $\mathbf{K}_f$ are aligned along the scattering plane axis (see Fig. 9.1) so that $\varphi_i = \varphi_f = 0$. In this case however an azimuthal angle is used to define the orientation of the scattering plane with respect to a high symmetry direction. From the experimentally observed energy transfer and the incident and outgoing wavevectors, the energies $\hbar\omega$ and wavevectors $\mathbf{Q}$ of single phonons are determined via the conservation equations (9.1) and (9.2).

For the interpretation of the HAS experiments it has proven useful to consider the so-called *scan curves*, which are illustrated by the dashed lines in Fig. 9.3. Each of the scan curves connect all possible values of $\Delta E = -\hbar\omega$ and $\Delta\mathbf{K} = -(\mathbf{Q}+\mathbf{G})$ which are accessible for a given incident energy E_i and set of incident ϑ_i and final ϑ_f scattering angles [4]. They are given by the equation

$$\frac{\Delta E}{E_i} + 1 = \frac{\sin^2\vartheta_i}{\sin^2\vartheta_f}\left(1 + \frac{\Delta K}{K_i}\right)^2, \tag{9.3}$$

where $\Delta\mathbf{K}$ is now along the scattering plane axis and is written as ΔK. For HAS it is experimentally of great advantage and simplification to arrange the incident and scattered beam directions such that the sum of the angles $\vartheta_i + \vartheta_f$ is fixed at some total angle Θ_{SD}. This has the advantage that it is only necessary to rotate the sample and both the bulky and massive source and detector chambers do not need to be moved (Fig. 9.2a). As the crystal is rotated both ϑ_i and ϑ_f change simultaneously since $\vartheta_f = \Theta_{SD} - \vartheta_i$. The scan curves shown in Fig. 9.3 correspond to the usual $\Theta_{SD} = 90°$ arrangement for scattering from Ag(111) at $E_i = 17.5$ meV. The sinusoidal curves in Fig. 9.3 represent the Rayleigh wave dispersion curves in an extended zone scheme. The intersections of the scan curves with the dispersion curves indicate the $(\Delta K, \Delta E)$ for which inelastic peaks can be expected. These intersections may occur for either positive or negative ΔE and ΔK according to whether phonons are created or annihilated in the forward or backward direction. The quadrants of the $(\Delta K, \Delta E)$ plane where the four possible processes occur are illustrated in Fig. 9.4.

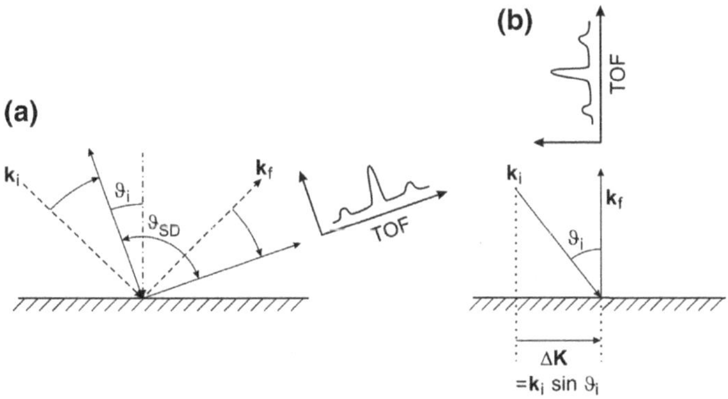

Fig. 9.2 Wavevector diagram illustrating constant angle and constant wavevector transfer scattering. **a** In the more customary total constant angle $\vartheta_i + \vartheta_f = \Theta_{SD}$ experimental arrangement instead of rotating $\mathbf{k}_i$ and $\mathbf{k}_f$ as shown, it is more convenient to keep the angle between $\mathbf{k}_i$ and $\mathbf{k}_f$ fixed in the laboratory and to rotate the crystal about an axis perpendicular to the plane of $\mathbf{k}_i$ and $\mathbf{k}_f$ and passing along the crystal surface. **b** Vector diagram showing the incident and final wavevectors in a constant **Q** scan geometry. This arrangement is, however, rarely used

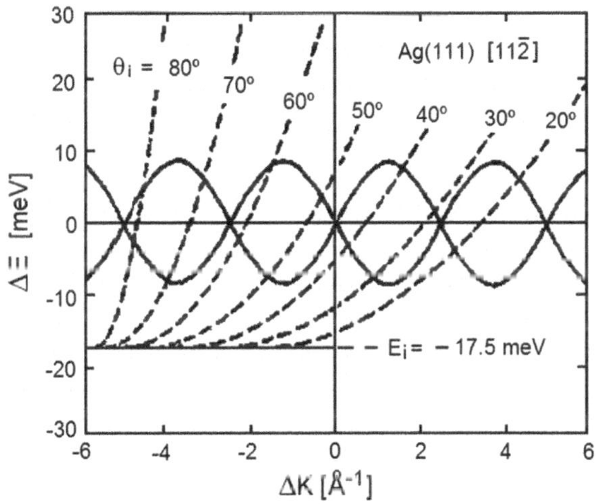

Fig. 9.3 Extended zone diagram showing some typical scan curves (broken lines) labeled by the incident angles for He atom scattering at an incident energy of E_i = 17.5 meV (k_i = 5.8 $Å^{-1}$) from a Ag(111) surface. The angle between incident and final beams is $\Theta_{SD} = 90°$. The intersections of the scan curves with the RW dispersion curves (full lines), reproduced over several BZs in the $[11\bar{2}]$ symmetry direction for both creation and annihilation processes, give the energy and wavevector of the phonons probed for each incident angle. Scan curves which pass through the nodes of the dispersion curves at ΔE = 0 correspond to the elastic diffraction peaks

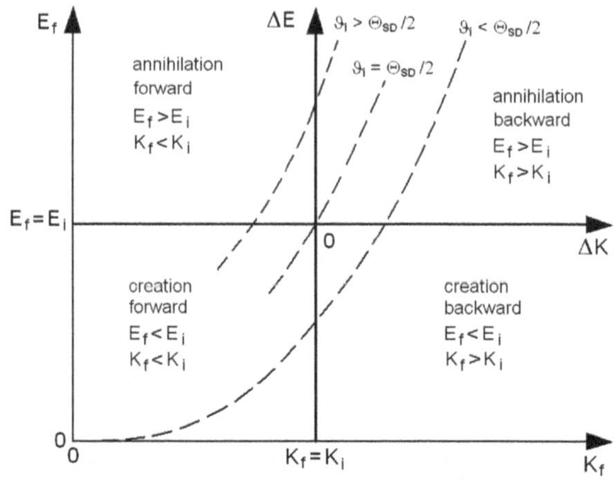

Fig. 9.4 Four different quadrants in the (ΔK, ΔE) plane for creation and annihilation of a phonon travelling in either the forward or backward direction with respect to the $\mathbf{K}_i$ of the incident atom. The scan curves (broken lines) for planar scattering defined by (9.3) cut three of the four quadrants for either $\vartheta_i < \Theta_{SD}/2$ or $\vartheta_i > \Theta_{SD}/2$. The scan curve for specular reflection ($\vartheta_i = \Theta_{SD}/2$) passes through the origin. The vertex of each scan curve parabola is at the origin of the corresponding (K_f, E_f) coordinates

For the usual $\Theta_{SD} = 90°$ planar scattering geometry the implementation of a scan curve for interpreting experimental TOF spectra is illustrated in Fig. 9.5. Figure 9.5a shows one of the He atom TOF distributions taken from an entire series of similar spectra measured in one of the early studies of surface phonons in LiF (001) [5]. The incident angle was set at $\vartheta_i = 64.2°$ ($\vartheta_f = 25.8°$) and the beam energy was $E_i = 19.0$ meV. The same TOF spectrum is converted into an energy transfer spectrum in Fig. 9.5b. Figure 9.5c displays the region of an extended phonon energy wavevector diagram with $\Delta K < 0$, which is probed by this particular TOF spectrum. The entire diagram has been rotated by 90° so that the phonon energies $\hbar\omega$ coincide with the energy transfer scale in Fig. 9.5b. The scan curve calculated from (9.3) for the scattering angles used in this particular experiment is shown by the dashed line curve. The intersections of the energy transfer peaks in Fig. 9.5b with the scan curve determine the locations of these peaks in the extended zone diagram. The projections on the ΔK axis provides the corresponding ΔK-values for each of the measured peaks. The ΔK-values in this particular experiment are centered around $\Delta K = -3.1$ Å^{-1} which is just the reciprocal lattice vector for this symmetry direction. Thus in this particular experiment all the phonons are processes in the vicinity of the corresponding diffraction peak (*Umklapp* processes).

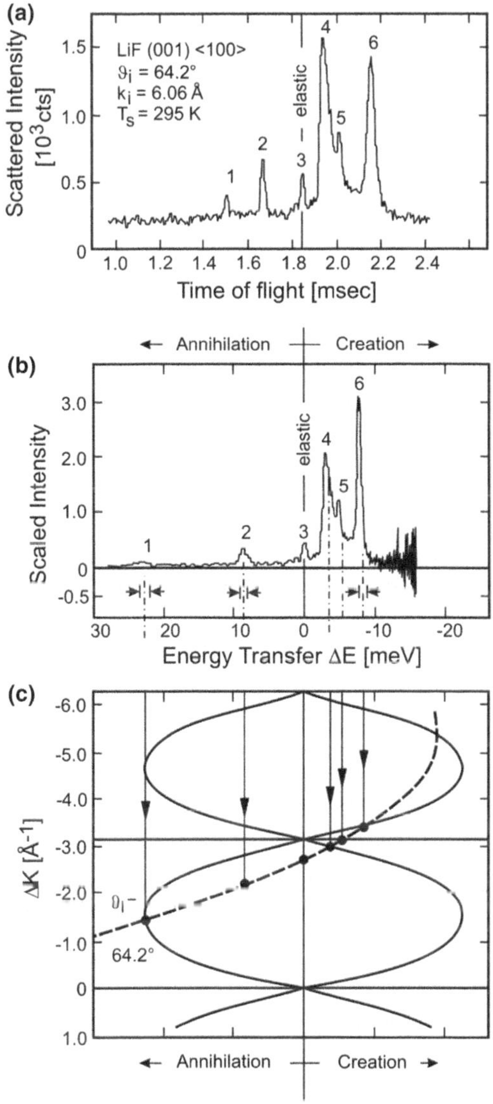

Fig. 9.5 The interpretation of measured time of flight spectra is illustrated for He atom single-phonon inelastic scattering from LiF(001) along the [100] azimuth with $\Theta_{SD} = 90°$ [5]. **a** The original measured TOF spectrum is transformed to an energy scale in (**b**). **c** The locations of the maxima are interpreted with the aid of a scan curve (----) for $\vartheta_i = 64.2°$ in an extended phonon energy wavevector diagram which has been rotated by 90°. The solid line curves (—) show the LiF(001) Rayleigh surface phonon dispersion curve. From the intersections of the energy loss peak locations with the scan curve the corresponding momentum transfers are determined. Consequently the peaks 1, 4 and 6 correspond to an inelastic coupling with the Rayleigh mode, peak 3 is due to elastic incoherent scattering from defects and peak 5 is due to a coupling to bulk phonons. Peak 2 is a *spurion* (see Sect. 9.9)

The example shown in Fig. 9.5 illustrates the rich information provided by HAS TOF spectra. First of all it is apparent that it is kinematically possible to observe both backward and forward phonons also for *Umklapp* processes involving one reciprocal lattice vector. The RW creation processes (peaks 4 and 6) are the most intense features, while the annihilation process (peak 1) is quite weak, though still quite sharp in the TOF spectrum. Such weakness is due to the larger values of phonon frequency and momentum, the former giving a smaller Bose factor, the latter a larger cut off effect, as explained in Chap. 7. According to this argument also peak 6 should be smaller than peak 4, but it is known that the former is resonance-enhanced [6]. The scattering intensity between peaks 4 and 6 contains information on the continuous bulk phonon density projected onto the surface plane, discussed in Sect. 3.4. Under favorable resolution and kinematical conditions certain features of the bulk phonon density can also be resolved. In this case kinematics indicates that peak 5 is due to a resonance-enhanced bulk phonon density response. As discussed in Chap. 10 by a fine tuning of a resonance the resulting enhancement can be exploited to detect weakly coupled phonons, provided the locations of the bound states are precisely known [6].

Peak 3, occurring at $\Delta E = 0$, corresponds to incoherent diffuse elastic scattering from surface defects. This incoherent intensity provides valuable information on the concentration and nature of surface defects, such as step edges or isolated atoms or molecules. Measurements of the small energetic broadening of the incoherent elastic peak have been used to gain detailed information on the dynamics of microscopic diffusion of atoms and molecules (see Sect. 13.2). Finally, the other remaining peak 2 is attributed to a process in which the elastic scattering of an incident atom with an energy deviating from the average incident energy E_i appears like an ordinary inelastic process. The kinematics of such spurious events, also called *spurions*, makes it possible to distinguish them from genuine inelastic processes, as is discussed in detail in Sect. 9.9. Thus in an actual experiment, involving up to perhaps a hundred TOF spectra for a given incident beam energy over a wide range of incident and final angles, a large amount of information can be acquired, not only on the phonon dispersion curves, but also on resonances and surface defects.

The expression for the scan curve in planar scattering (9.3) illustrates an important difference between electrons and He atoms. For electrons, because of the larger incident energies, the relative energy change $\Delta E/E_i$ in (9.3) is typically 10^{-3}–10^{-4} so that ΔK is nearly independent of the phonon energy $\hbar\omega$. This is not true for He atoms where $\hbar\omega$ is comparable to E_i. Another way of understanding this difference is in terms of the much larger group velocity *dE*/*dk* of electrons compared to He atoms. Thus for electrons the measurements are carried out under conditions where ΔK is virtually independent of $\hbar\omega$ and is simply fixed by choosing the incident and final scattering angles. In HAS it is in principle possible to also achieve this simplification by adopting a constant **Q**-scan geometry as commonly practiced in analogous scattering experiments with neutrons. Such a geometry suitable for HAS is illustrated in Fig. 9.2b where $\vartheta_f = 0°$. Then (9.3) simplifies considerably and the surface projection of the incident wavevector $\Delta K = k_i \sin\vartheta_i$ is independent of $\hbar\omega$ and equal to **Q** (if **G** is zero). In order to scan ΔK, without changing the magnitude of the incident wavevector k_i, the incident angle must be scanned. Since, however, ϑ_f is fixed this requires that Θ_{SD} is continuously varied over a wide range of angles. As will be discussed in Sect. 9.4, although possible in some of the HAS apparatus this mode of operation can be cumbersome, because of the rather bulky source and detector vacuum chambers. To our knowledge the constant **Q**-scan mode of operation has not been used yet to any extent in HAS experiments.

If the apparatus is arranged in such a way that other angles Θ_{SD} can be accessed, then a large portion of the total $(\vartheta_i, \vartheta_f)$ space (Fig. 9.6), including the conditions needed for a constant **Q**-scan can be made available. In Fig. 9.6 the constant Θ_{SD} scan lines are diagonal straight lines extending from the upper right to the bottom left. They intersect the other diagonal line $\vartheta_i = \vartheta_f$ corresponding to specular scattering at right angles. The solid line curves show the positions of the diffraction peaks (m, n) for scattering from LiF(001) at $E_i = 85.3$ meV. Thus virtually the

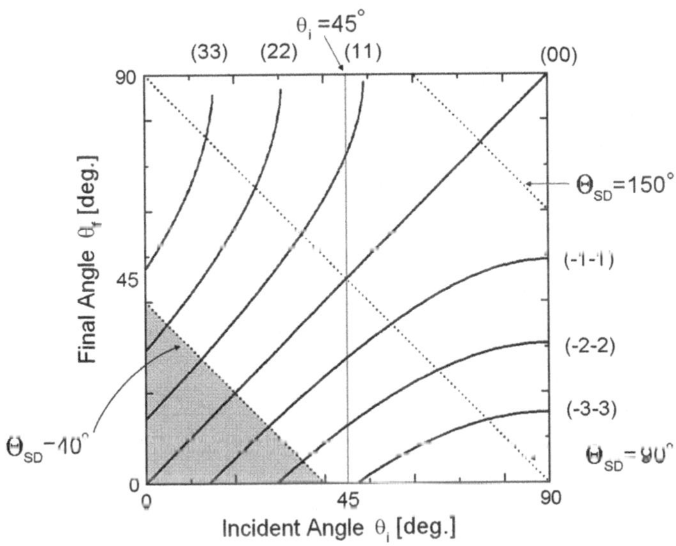

Fig. 9.6 Schematic diagram showing the (θ_i, θ_f) plane illustrating the range of incident and final scattering angles which can be accessed for a fixed scattering angle Θ_{SD}. In the diagram a fixed Θ_{SD} corresponds to a diagonal line as illustrate for Θ_{SD} = 90° and 150° (dotted lines). The solid curved lines illustrate the angles at which the elastic diffraction peaks are expected for the example of scattering of He atoms from LiF(001) along the [100] direction with k_i = 12.8 Å (E_i = 85.3 meV). Most variable angle apparatus are limited to $\Theta_{SD} \geq 40°$, in which situation the shaded region of angles cannot be accessed

entire $(\vartheta_i, \vartheta_f)$ space and all phonon wavevectors between the diffraction peaks, except for a small corner in the bottom left depending on the smallest angle Θ_{SD} available with the particular apparatus, can be covered.

9.3 Non-sagittal and Non-planar Scattering

Although in most experiments the simple planar scattering along crystal mirror symmetry planes is used, other geometries may have special advantages. When $\Delta\mathbf{K}$ is away from a high symmetry direction of the surface Brillouin zone, *umklapp* inelastic processes are possible in which the phonon wavevector $\mathbf{Q}$ is not aligned with $\Delta\mathbf{K}$ (*non-sagittal* scattering) (Fig. 9.7). This condition can be realized either in a planar scattering geometry away from a mirror symmetry plane (Fig. 9.8a) or in a non-planar scattering geometry (Fig. 9.8b). In both cases the final and/or initial states of the entire (atom plus crystal) system have no definite mirror symmetry and both SH and sagittal polarized modes can be excited. It is in principle always possible, however, to choose a scattering geometry so that the wavevector $\mathbf{Q}$ varies along a high symmetry direction (Fig. 9.7b). This opens up the following several possibilities for detecting SH phonons even along high symmetry directions.

In the first method, exploiting the configuration of Fig. 9.8a, the crystal is rotated around the normal to the surface (*azimuthal* rotation) so that the scattering plane is away from a symmetry direction along one of the reciprocal lattice vectors $\mathbf{G}$, and *umklapp* processes are measured with both $\mathbf{G}$ and $\mathbf{Q}$ out of the scattering plane as shown in Fig. 9.7b. In this case, in order to measure phonons with the wavevector $\mathbf{Q}$ along a symmetry direction normal to $\mathbf{G}$, $\Delta\mathbf{K} = \mathbf{K}_f - \mathbf{K}_i$ has to be increased together with the azimuth angle φ so that

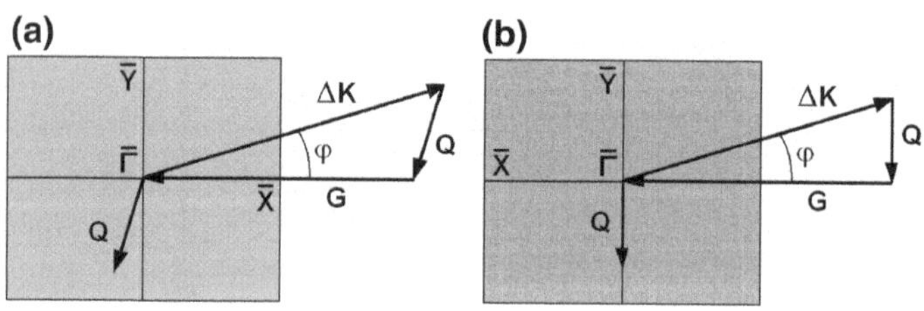

Fig. 9.7 Configurations in reciprocal lattice space for detecting SH polarized phonons. **a** An *umklapp* inelastic process in which $\Delta\mathbf{K}$ is away from a high symmetry direction of the surface Brillouin zone (gray area) and the phonon wavevector $\mathbf{Q}$ is not aligned with $\Delta\mathbf{K}$ (*non-sagittal* scattering). This condition can be realized either in a planar scattering geometry away from a mirror symmetry plane or in a non-planar scattering geometry. **b** In both cases the scattering geometry can be chosen so as to keep $\mathbf{Q}$ along a high symmetry direction ($\overline{\Gamma\mathrm{Y}}$), where surface phonons have a definite sagittal or SH polarization. Both types of phonons can be excited in this configuration

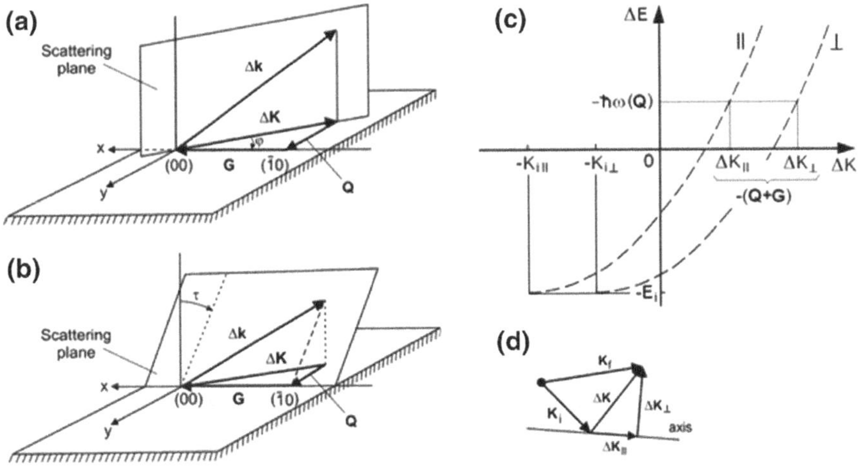

Fig. 9.8 Non-planar scattering for measuring shear-horizontal phonons. **a** In-plane scattering, with the scattering plane azimuthally rotated with respect to a high symmetry direction to sample the phonon displacement field of a SH *umklapp* phonon. **b** Out-of-plane scattering where the scattering plane with the axis oriented along a symmetry direction is tilted away from the normal to the surface to achieve a similar phonon geometry [7]. **c** Scan curves (broken lines) for non-planar scattering of He atoms of incident energy E_i and incident parallel wavevector $\mathbf{K}_i = (K_{i\parallel}, K_{i\perp})$, where $K_{i\parallel}$ and $K_{i\perp}$ are the components of $\mathbf{K}_i$ parallel and perpendicular to the scattering plane axis. For a given energy transfer $\Delta E = -\hbar\,\omega(\mathbf{Q})$ and scattering geometry the scan curves give simultaneously the two components $\Delta K_\parallel$ and $\Delta K_\perp$ of the wavevector transfer $\Delta\mathbf{K}$ with respect to the scattering plane axis (**d**) and from these values the phonon wavevector **Q** can be obtained after subtraction of a **G** vector in order to bring **Q** into the first BZ

$$|\Delta\mathbf{K}| = G/\cos\varphi, \quad \text{and} \quad Q = G\tan\varphi. \tag{9.4}$$

Such an experiment was first published in 1983 [5] and subsequently exploited for the observation of shear-horizontal modes [7, 8].

Alternatively the scattering plane, with the axis oriented along a symmetry direction, is tilted away from the normal as shown in Fig. 9.8b. If the projection of $\Delta\mathbf{k} = \mathbf{k}_f - \mathbf{k}_i$ onto the scattering plane axis equals a **G** vector, the phonon wavevector **Q** is along a symmetry direction normal to **G** and its magnitude increases with the tilt angle τ

$$Q = (k_{fz} - k_{iz})\tan\tau. \tag{9.5}$$

This method has the advantage that **Q** can be varied along a symmetry direction without changing $\mathbf{k}_f - \mathbf{k}_i$. Such experiments have been first reported using EELS [9–11] and subsequently also using HAS [7, 12].

For the general case of *non-planar scattering* illustrated in Figs. 9.1 and 9.8b, the relationship between ΔE and $\Delta\mathbf{K}$ is expressed by two equations

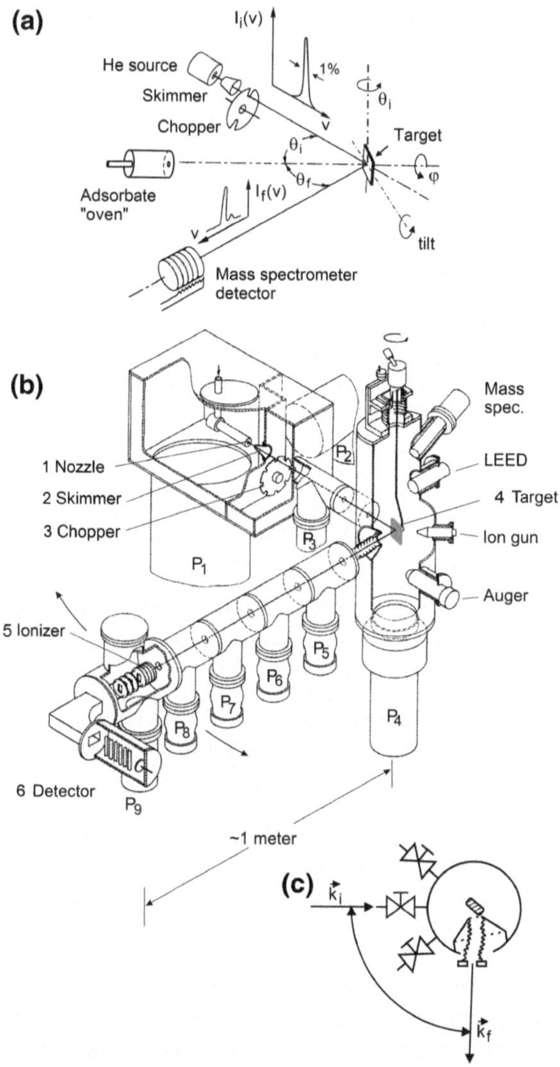

Fig. 9.9 The Göttingen HAS apparatus HUGO II. **a** Schematic view illustrating the geometry, and the velocity distributions of the incident and final He beams. **b** The nearly monoenergetic helium atom beam is emitted from the nozzle (1) and after scattering from the target surface (4) a part of the scattered intensity enters the time-of-flight chambers (P_5–P_8) and, after ionization (5) is detected by the mass spectrometer (6). Rotation of the sample and/or the detector arm allows the angular distribution of scattered He atoms to be mapped out. For time-of-flight studies the beam is pulsed by passing it through a rapidly rotating chopper disc with a few equally spaced narrow slits (3). P_1–P_9 are differential pumping stages which provide for a reduction of the helium partial pressure in the detector chamber with respect to the source chamber by over 12 orders of magnitude. Some of the other important apparatus components are described in the text. **c** Top view of the scattering geometry

$$\frac{\Delta E}{E_i} + 1 = \frac{\sin^2 \vartheta_i}{\sin^2 \vartheta_f} \left(\frac{\cos \varphi_i}{\cos \varphi_f} + \frac{\Delta K_{\|}}{K_i \cos \varphi_f} \right)^2 = \frac{\cos^2 \vartheta_i}{\cos^2 \vartheta_f} \left(1 + \frac{\Delta K_{\perp}}{K_i \sin \varphi_i} \right)^2, \quad (9.6)$$

where $\Delta K_{\|}$ and $\Delta K_{\perp}$ are the components of $\mathbf{\Delta K}$ parallel and perpendicular to the scattering plane axis, respectively. For a fixed scattering geometry the two right hand sides of (9.6) describe in the (ΔK, ΔE) plane two parabolic scan curves with respect to the components $\Delta K_{\|}$ and $\Delta K_{\perp}$ of the parallel wavevector transfer (Fig. 9.8c, d). Thus for each value of the energy transfer $\Delta E = -\hbar\omega(\mathbf{Q})$ the scan curves give simultaneously the two components $\Delta K_{\|}$ and $\Delta K_{\perp}$ and therefore $\mathbf{Q}$, after subtraction of a suitable $\mathbf{G}$ vector so as to bring $\mathbf{Q}$ into the first Brillouin zone.

9.4 He Atom Scattering (TOF-HAS) Apparatus

Helium has a number of important advantages over other atoms and molecules for surface scattering experiments. For one it is extremely inert and is not expected to stick to any surface at temperatures greater than about 5–10 K. Free jet expansions, which are generally used to produce intense molecular beams, show particularly advantageous properties in the case of helium [13]. Helium because of its weak interaction potential has the lowest of all heats of vaporization and consequently still has a sizeable vapor pressure of 1 bar at 4.2 K. For this reason beams with very low energies of less than meV can be made with He. More important, since the very weak He–He intermolecular potential just barely supports a bound state of the diatomic He_2 molecule at low temperatures the scattering cross section is greatly enhanced by a *zero energy resonance* [14]. As a consequence the He–He scattering cross section in the limit of 0 K increases dramatically from the room temperature value of about 30 Å^2 to about 269,000 Å^2 which is the largest cross section for collisions of ground state atoms. This has a big effect on the velocity distribution of the adiabatically expanded gas which is driven by collisions. Thus as the *ambient* translational temperature associated with the velocity fluctuations within the gas jet, which is moving in the laboratory system, drops in temperature the cross sections which serve to convert the random fluctuations into forward directed motion increase sharply resulting in ultra-low *ambient* temperatures approaching $T_\infty = 10^{-3}$ K [14, 15]. At the same time these same weak interactions also inhibit cluster formation which, because of the concomitant heat release, can spoil the expansion and, moreover, can contaminate the atom beam with clusters. These two effects, giant cross sections and reduced clustering, explain the extremely low *ambient* translational temperatures in the moving coordinate system of the beam and the resulting very narrow velocity half-widths of about $\Delta v/v \approx (T_\infty/T_o)^{1/2} \approx 1\%$ [13, 16].

Another important advantage of helium is that it can be especially sensitively detected with a mass spectrometer because of the very low background pressure which can be achieved with helium. For this same reason helium is extensively used in commercial leak detectors. Finally, because of its low mass and the possibility to go to low source temperatures and beam energies, scattering conditions especially favorable for the dominant excitation of only single phonons are easily achieved with helium. Single- and multi-phonon scattering are discussed in detail in Sect. 7.6 .

9.4.1 The Göttingen Time-of-Flight (TOF-HAS) Apparatus

A schematic diagram of an apparatus, named HUGO II, which was used from about 1980 to 2005 in Göttingen for He atom inelastic surface scattering [1, 17, 18] is shown in Fig. 9.9. A photograph of the same apparatus is in Fig. 9.10. The characteristic parameters and operating conditions of HUGO II are listed in Table 9.1. The following discussion provides a description of the important components designated by numbers in Fig. 9.9.

Fig. 9.10 Photograph of the Göttingen HAS apparatus HUGO II. The rectangular source chamber with a large diffusion pump is seen on the left. The large 20 cm diameter aluminum pipe in the upper left corner leads to a Roots blower pump in the room above. The scattering chamber is slightly to the right of the center of the picture. The time-of-flight tube is directed towards the right foreground. The detector magnet and electron multiplier chamber are seen at the end of the flight tube. The flight tube and the vacuum pumps for the pumping stages P_5–P_9 (Fig. 9.9) are mounted on rigid frame which can rotate around the target for experiments in which Θ_{SD} is varied. The orange rotating pumps seen below the detector arm are on wheels so that they can move with the detector. They are connected to the rigid frame by rubber tubing or bellows to avoid vibrations that would affect the detector position

Table 9.1 Characteristic parameters and operating conditions of the apparatus shown in Figs. 9.9 and 9.10

Source	Nozzle orifice diameter d, nominal	10 μm
	Nozzle stagnation pressure	20–200 bar
	Nozzle stagnation temperature	40–400 K
	He beam velocity	644–2037 m/s
	Relative velocity spread (FWHM) $\Delta v/v$	0.01
	Estimated intensity	3×10^{19} atom/sr s
Skimmer	Entrance diameter	0.62 mm
	Length: entrance to base	37 mm
	Outside/inside full angles	32°/25°
	Orifice-to-skimmer distance	10 mm
Chopper	Double slit disk, slit diameter	150 mm
	Double slit disk, slit width	1.5 mm
	Disk rotational frequency	≈300 Hz
	Shutter function (FWHM), effective	10.6 μs
Dimensions	Source-target distance	501 mm
	Target-detector distance	1202 mm
	Chopper-detector distance	1635 mm
	Fixed source-target-detector angle Θ_{SD}	90°
	Incident beam full angular spread, nominal	0.81°
	Angle subtended by detector from target, nominal	0.42°
	Angular resolution, (00) peak, nominal (scanning θ_i)	0.42°
	Angular resolution, (00) peak, measured (scanning θ_i)	(=0.2°)
Pressures	Target chamber P_5, total pressure	
	Base	1×10^{-10} mbar[a]
	Operation, beam chopped	2×10^{-9} mbar
	Operation, beam unchopped	1×10^{-7} mbar
	Target chamber P_5, He partial pressure	
	Operation, beam chopped	2×10^{-9} mbar
	Operation, beam unchopped	1×10^{-7} mbar
	Detector chamber P_9, total pressure	
	Base	$\approx 10^{-11}$ mbar
	Operation	2×10^{-10} mbar
	Detector chamber P_9, He partial pressure	
	Base	$\approx 10^{-15}$–10^{-14} mbar
	Operation, beam chopped, specular	$\approx 10^{-12}$ mbar
	Operation, beam unchopped, specular	$\approx 10^{-10}$ mbar

[a] 1 mbar = 100 Pa

A nearly monoenergetic beam of He atoms ($\Delta v/v \sim 1\%$) is formed by expanding 99.9999% pure helium gas from a source with a temperature T_o, which typically is about 80 K, and a high pressure $P_o \approx 100$ atm through a small thin walled orifice (1) with a diameter of typically ≈5–10 μm into a vacuum chamber evacuated by a large vacuum pump with a pumping speed greater than about 10^3 l/s. A conical thin walled skimmer (2) with an opening of typically 0.5 mm diameter and a total internal angle of typically 30° and an external of about 35° serves to both collimate the beam, while at the same time reducing the gas flow into the next vacuum chamber. In the course of the adiabatic expansion the gas is cooled to the low translational temperatures and sharply peaked velocity distributions. Such highly expanded free jet expansions are usually characterized by their *speed ratio*

$$S = 1.67\, v/\Delta v, \tag{9.7a}$$

where $\Delta v = 2.35\ (5kT_\infty/m)^{1/2}$ is the velocity half-width and T_∞ is the ambient temperature inside the fully expanded gas jet [19–21]. For an *ideal* gas, such as He at not too high pressures, the most probable velocity is given by conservation of enthalpy:

$$v_{mp} = (5kT_0/m)^{1/2}. \tag{9.7b}$$

As with most light atoms, the mostly probable He atom beam velocity is about 30% greater than the most probable velocity in the source chamber [19, 20]. In the example introduced above $S = 167$ and $T_\infty = 7 \times 10^{-3}$ K. As discussed in Chap. 14 much higher speed ratios can be achieved but require a more complex apparatus.

By simply changing the temperature of the source from, say, 300–40 K, the most probable beam velocity v_{mp} can be varied from about 2000 m/s. ($E_i \approx 60$ meV, $k_i = 11$ Å, de Broglie wavelength $\lambda = 0.5$ Å) down to about 700 m/s. ($E_i \approx 8$ meV, $k_i = 4$ Å^{-1}, $\lambda = 1.5$ Å), respectively. For this range of incident beam energies the energy half-widths are typically 2% of the beam energy or between about 1 and 0.2 meV, respectively [13, 16]. Lower beam energies down to about 1 meV have been reported [22] using even lower source temperatures. Unless the source pressure is reduced, however, cluster formation, which in helium occurs at very low source temperatures below about 30 K, leads to an increase in the relative energy width and, ultimately, to a loss of atom intensity [21].

In most modern apparatus the energy distribution of the scattered atoms is measured by the time-of-flight technique. This technique, also used in HUGO II, requires that the incident beam is mechanically chopped into short pulses of about 10 μs by a rotating disc chopper (3) with between 2 and 8 narrow slits of about 0.5 mm equally spaced on the perimeter of the disc rotating with typically 100–300 Hz. The time of arrival of the scattered atoms at the electron impact ionizer (5) of a mass spectrometer detector is measured and the count rate for each time bin is recorded usually with the aid of a small personal computer or dedicated electronic equipment.

The scattered beam is usually detected by electron bombardment ionization and subsequent mass analysis with a mass spectrometer, which in the Göttingen apparatus is a 90° magnet. With conventional electron beam ionizers the probability of converting each incoming atom into a single ion count is only of the order of

10^{-5}–10^{-6} [23]. The effective length of the Brink-type [24] open-cage detector developed and optimized in Göttingen is about 2 cm [25]. The short length is important for the TOF resolution (see Sect. 9.8). The low detection efficiency is largely limited by space charge effects. As the electron beam density in the ionization region is increased the resulting negative space charge works against an efficient extraction of the positive ions [26]. A great improvement in sensitivity is possible if the electrons are stabilized by a coaxial magnetic field [27, 28]. Such an ionizer has a much longer ionizer of 6–10 cm, and therefore is not suitable for high resolution time-of-flight energy analysis. In the case of spin echo measurements discussed in Sect. 9.4.3, where the length is of no importance, a magnetic field stabilized ionizer has been developed which has a sensitivity of several orders of magnitude greater [29].

The detector cannot distinguish the He atoms in the beam from those in the residual gas in the detector chamber. In order to keep the diffuse helium background partial pressure in the mass spectrometer detector chamber, as low as possible nine differentially pumped vacuum stages each of which is indicated by P_i (i = 1–9) in Fig. 9.9 are introduced along the beam path from the source to detector in order to reduce the effusive transport of helium gas from the relatively high pressure of typically 10^{-3} mbar ($\cong$0.1 Pa) in the source chamber (P_1) into the detector chamber (P_9). In the apparatus of Fig. 9.9 a helium partial pressure as low as $\sim 10^{-15}$ mbar, which corresponds to a particle density of less than 10 helium atoms/cm^3, was achieved. The overall reduction in the helium pressure between the source chamber P_1 and the detector chamber P_9 thus amounts to more than 12 orders of magnitude! The remaining very low background ion signal of about 5–10 counts/s places a lower limit on the smallest detectable beam ion signal of about the same magnitude. To achieve such a low background the detector chamber must be operated under ultra-high vacuum conditions with a total pressure below about 10^{-11} mbar. The straight-through signal of the highly collimated ($\Delta\vartheta \approx 10^{-2}$ rad) incident beam arriving on the target is typically 10^8–10^9 ion counts/s which corresponds to a neutral He atom flux of about 10^{13}–10^{14} atoms/s. Thus measurements are possible in which the scattered beam signals varies over a range of up to 8 or 9 orders of magnitude.

In the apparatus shown in Figs. 9.9 and 9.10 both Θ_{SD} and ϑ_f (for a given ϑ_i) can be varied by rotating the detector arm which is attached to the target chamber by a flexible bellows (see Fig. 9.9c). The bellows allow the angles to be continuously changed within a maximum angular range of $\pm 20°$ [17]. A wider range of scattering angles can be accessed by rotating the target chamber (P_4 in Fig. 9.9) with respect to the incident beam. In the apparatus shown in Figs. 9.9 and 9.10. This is achieved by disconnecting the target chamber from the incident beam line and rotating the entire target chamber to access another entrance port of the target chamber [17]. In most of the experiments Θ_{SD} was fixed at 90°. The smallest angle Θ_{SD} which can be accessed with HUGO II is $\Theta_{SD} = 50°$. Other variable Θ_{SD} apparatus have smallest angles of 35° [30] and 40° [31] and can in principle operate in the constant **Q** configuration (Fig. 9.2b) for sufficiently large $\Delta\mathbf{K}$, e.g., around a diffraction peak.

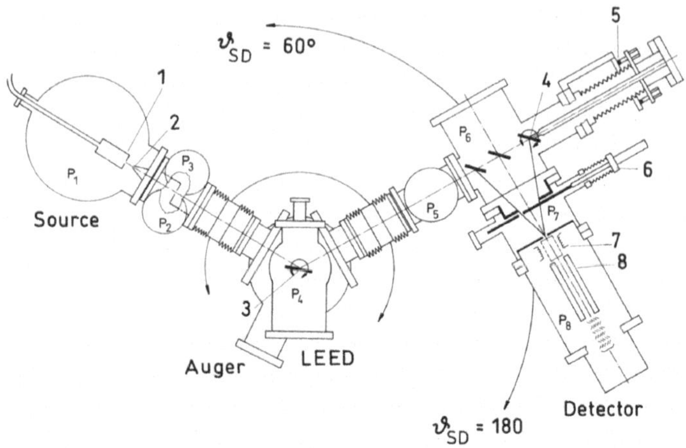

Fig. 9.11 Top view of a He atom scattering apparatus featuring a crystal diffractometer for energy analysis. P_1–P_8 are the differential pumping stages which provide for a reduction of the helium partial pressure in the detector chamber. The nearly monoenergetic He atom beam from the nozzle (1) passes through the skimmer (2) and is scattered from the target (3). The manipulator (5) positions an echelette grating (4) for diffractive analysis of the beam scattered from the target. The diffracted beam from the grating is detected by a stationary detector with ionizer (7) and a quadrupole mass spectrometer (*8*). The slit on the manipulator (6) moves synchronously with the grating (4). Only the source chamber is stationary and both the target chamber and the diffractometer-detector arm can be rotated continuously with respect to each other so that ϑ_{SD} can be varied between 60°–180°. (adapted from Schief et al. [32])

The single crystal target sample (4 Target in Fig. 9.9) is positioned on a precision manipulator which provides for rotations around and translations along all three Cartesian coordinates. The manipulator construction adopted by the Göttingen group is described in a compendium on surface science experimental techniques [33]. The total vacuum in the target chamber (P_4) is typically in the 10^{-10}–10^{-11} mbar ($\cong 10^{-9}$ Pa) range without the He beam incident on the target. With the helium beam on the He partial pressure typically rises to 10^{-8} mbar ($\cong 10^{-6}$ Pa). The target chamber is usually equipped with a mass spectrometer for partial pressure analysis and with an ion gun for sputter cleaning the sample. LEED and Auger equipment are also usually available for characterizing the crystal surface structure and the elemental cleanliness, respectively. An equally detailed description of a similar apparatus also featuring different ports and bellows for fine tuning the scattering angles has been described by Doak [30].

9.4.2 Other HAS Apparatus

Several similar TOF-HAS apparatus with a flexible bellows and rotatable detector arm have been described in the literature [30, 34–37]. A very similar apparatus with a fixed $\Theta_{SD} = 90°$ angle has been used by the Jülich group [31]. Other HAS machines have been designed so as to reduce the distance l between the nozzle and target. Since the beam density decreases according to $(d/l)^2$, where d is the nozzle diameter, such a more compact construction has the advantage of providing a higher density on the target, if all other factors are the same. In one fixed-angle apparatus l could be reduced from the 50 cm in HUGO II to only 15.3 cm [38]. In a more recent design l was reduced to 20 cm [39]. Another compact apparatus has been described by the Genoa group of Tatarek and Bracco [37, 40]. The compactness, however, necessitates compromises with other components, which make the operation of the apparatus more cumbersome and may also compensate for part of the gain.

In an unconventional TOF-HAS apparatus the vacuum problems associated with detection by electron beam ionization is circumvented by electron beam excitation of the scattered He atoms into a long lived 2^3S metastable state, which can be directly detected with an electron multiplier [41]. A similar technique has been applied to the detection of hydrogen atoms [42]. In the most recent application the hydrogen atoms were excited by laser pulses [43]. An early TOF apparatus specifically designed to operate with Ne atoms, prior to the advent of high speed ratio He atom beams, dates back to 1976 [44].

In several of the HAS apparatus the simple single chopping time-of-flight technique described here has been replaced by the pseudo-random chopping technique [45–48]. In this technique the transmission is increased by replacing the few regularly spaced narrow slits on the chopper disc by a series of closely spaced slits with a pseudo-random distribution of widths. The measured time-of-flight distribution is a convolution of this pseudo-random sequence and the velocity distribution of the scattered atoms. After a computer deconvolution of the measured distribution with the known sequence the sought after velocity distribution is extracted. Since the average width of the slits and intervening slots is about the same, the duty cycle (transmission) with this technique is about 50% compared to only about 1% for single pulse time of flight. A gain in signal-to-noise ratio, which can be as much as a factor 100 under favorable circumstances, is, however, only expected when the detector background is comparable to or larger than the signal [45]. Generally this is not the case in a helium apparatus with a large number of differential pumping stages as in the apparatus shown in Fig. 9.9. Advantages are, however, expected for an apparatus with fewer pumping stages or in experiments with other beam particles such as H_2 or Ne, where the intrinsic partial pressure in the detector is much more difficult to suppress and can be much larger than for helium. The pseudo-random chopping technique has the disadvantage that the desired time-of-flight spectrum is only obtained after a deconvolution, which with modern computers is carried out online in the course of the measurements. Moreover, the precision depends sensitively on the accuracy in the position of the slits on the chopper disc and the stability of the rotor frequency. Therefore this technique places higher requirements on these apparatus components than in the single chopping method.

Another method for carrying out time-of-flight experiments employs the use of an electromechanically pulsed nozzle beam source [49]. This method was applied to surface scattering in one of the first surface phonon scattering experiments by Feuerbacher and Willis [50, 51]. In this experiment Ne atoms, which inherently do not have the same sharp velocity distributions as He were used. The energy resolution was only 12%. The technology has since been greatly improved [52, 53] and the method is extensively used in other areas of chemical physics, such as for the production of large cold helium droplets [54]. Several reports of high speed ratios in He atom pulsed beams have recently been reported [55, 56] and in one instance pulsed beams were used to achieve the largest ever measured speed ratios of a He atom beam of $S \geq 1000$ [57]. In only one case was a pulsed beam source used for scattering of He atoms from a surface, albeit from a moving surface with the aim to create a very slow atom beam [55].

Another scheme for measuring inelastic scattering, which dates back to 1932 [58], uses a single crystal surface, typically LiF(001), as a grating monochromator to both select and analyze the momentum of the scattered atoms. An analogous double crystal diffractometer technique is the standard method for inelastic neutron scattering determinations of bulk phonon dispersion curves [59–61]. In the early 1970s this technique was further developed by Mason and Williams [62, 63] and successfully used by them in a large number of pioneering surface scattering experiments [64–68] including the first study of surface phonons on clean Cu(001) in 1981 [69]. In their apparatus a single diffractometer is used for energy transfer analysis. In many of its properties a crystal diffractometer has features which are similar to those of the time-of-flight technique for energy analysis of the scattered atoms. For instance, in both cases the experimental energy resolution decreases with increasing final beam energy. This is not a serious problem since the energy half-width of the nozzle beam increases roughly proportionally to the beam energy anyway. The biggest advantage of the crystal diffractometer as described by Mason and Williams [62] is that the apparatus can be made very compact. Probably the main disadvantage is the greater mechanical complexity since in addition to the target manipulator a precise rotational manipulation of the spectrometer crystal is required.

Kern's group in 1992 proposed using a stepped metal surface as an echelette grating for He atom beams [70]. Later in 1996 they reported the realization of this idea using a Pt (997) crystal surface which was operated in a low-pressure oxygen atmosphere ($p(O_2) = 5.0 \times 10^{-9}$ mbar) at 750 K to maintain a clean surface over several weeks [32]. Figure 9.11 shows an elegant apparatus based on this concept using a grating to analyze the final velocities [71]. This apparatus has two bellows for changing the incident and final scattering angles and can cover a continuous range of ϑ_{SD} between 180° and 60° without breaking the vacuum.

For more details on technical aspects of modern TOF-HAS spectrometers the reader is referred to some classical reviews [72, 73] and some papers on He atom beam sources [74] and on detectors [75]. Factors affecting the resolution of a HAS-TOF spectrometer have been analyzed in [76] and reviewed in [30]. They are discussed in detail in Sect. 9.8 at the end of the chapter.

9.4.3 Spin Echo ^{3}He Atom Scattering (SE-HAS) Apparatus

Another scheme for studying inelastic processes using helium atoms, which also derives from previous innovative developments in neutron scattering is the spin echo method [77, 78]. This technique has been successfully adapted to ^{3}He atoms by two groups, first in 1995 by the group of DeKieviet in Heidelberg [79, 80], and about ten years ago in Cambridge, UK [81–84]. The Heidelberg apparatus was restricted to incident beam energies of about 3 meV, which is too low for accessing the full range of parallel wavevector transfer needed for measuring the full surface phonon dispersion curves. On the other hand, the apparatus had a spectacular energy resolution of <0.2 μeV which made it possible to measure the gravitational potential of a beam of atoms [79].

In the Cambridge ^{3}He spin echo (SE-HAS) apparatus the incident beam energy was increased to about 8 meV, which is sufficient for measuring the full dispersion curve of surface acoustic modes, at least in soft materials. Compared to a time-of-flight apparatus with an energy resolution of the order of 0.1 meV, the SE-HAS apparatus has a much better resolution of several μeV making it particularly suitable for quasi-elastic HAS (QHAS) studies of surface diffusion discussed in Sect. 13.2

The SE-HAS technique can be understood using either a classical description, which is particularly useful for understanding elastic studies (bound state resonances) and inelastic scattering experiments, or using a semi-classical description which is convenient for quasi-elastic scattering studies. Within the classical approach the basic idea is to use the Larmor precession of the nuclear spin of a polarized ^{3}He atom beam as a "clock" instead of the actual time of flight.

Figure 9.12 presents a schematic diagram of the apparatus. Before "starting" the clock the ^{3}He atoms are first spin polarized in one of the possible $m_s = \pm ½$ states. In the Cambridge apparatus this is achieved in a carefully designed 30 cm long hexapole magnet [82] in which the $m_s = +½$ state atoms are focused and the $m_s = -½$ atom defocused and removed from the beam. The selected atoms are then polarized so that they are lined up in the y-direction perpendicular to the scattering plane before they enter an 81 cm long solenoid. The small nuclear magnetic moment requires very large magnetic fields of about 1 and 0.1 T in the polarizer and in the solenoid, respectively. The clock is started when the atoms enter a solenoid with a homogeneous longitudinal magnetic field in which the nuclear magnetic moment of ^{3}He ($\mu(^3\mathrm{He}) \cong -0.00158\mu_B$, with μ_B the Bohr magneton) precesses several thousand times before arriving at the target. After scattering the atoms enter a second identical solenoid with exactly the same magnetic field but with a reversed polarity to that of the first field. This field serves to "unwind" the polarization acquired in the first solenoid. Finally, the resulting "difference" in polarization of the beam leaving the second solenoid is determined by passing the beam through another hexapole field which serves as a spin analyzer. The analyzer transmits atoms according to the cosine of their classical spin phase Φ with respect to the original y-polarization direction so that the signal detected is proportional to

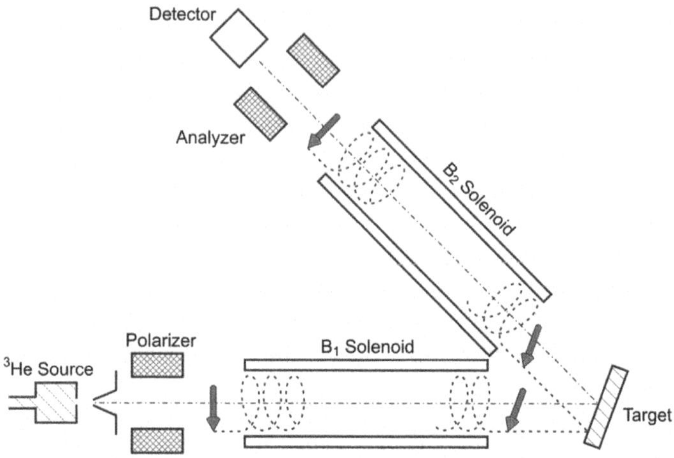

Fig. 9.12 Schematic diagram illustrating the classical functioning of a ^{3}He spin echo spectrometer. The ^{3}He atoms formed in a free jet expansion are first polarized by passing them through a magnetic hexapole magnet. In a longitudinal magnetic field directed along the beam path the nuclear spins precess several thousand times before the atoms arrive at the target. After scattering from the surface the atoms enter a second identical, but oppositely poled magnetic field in which the spin precesses in the opposite direction. The final difference in polarization is analyzed by a hexapole field mounted in front of the detector. The small differences in the precession phase provide a very precise information on the change in velocity after scattering

$P_y\langle\cos\Delta\Phi\rangle$. The large velocity resolution results from the small difference in the phase of precession of the ^{3}He atoms which is a small fraction of a precession period out of a total of several thousand precessions in the second solenoid.

For elastic scattering experiments the current in the second solenoid is set equal to zero. The phase the atoms accumulate through the first magnetic field is given by [83]

$$\Phi_1 = \frac{\gamma}{\bar{v}} \int_0^L B dl \cong \gamma \frac{m_3 \lambda}{h} BL, \tag{9.8}$$

where γ is the gyromagnetic ratio of ^{3}He $\left(\gamma = 204 \times 10^6\,\text{rad}\,\text{s}^{-1}\,\text{T}^{-1}\right), \bar{v}$ is the average beam velocity, which is about 700 m/s at E_i = 8 meV, the optimal energy of the Cambridge machine, λ is the corresponding de Broglie wavelength, and m_3 the ^{3}He atom mass. The magnetic field is about B = 0.1 T and its length about L = 1 m. With these parameters ϕ_1 equals 3×10^4 rad. This corresponds to about 5×10^3 precessions of the spin. In an elastic scattering experiment, the measured polarization can be written as a function of the de Broglie wavelengths of the incident beam, λ_1, [85].

$$P_y(K_1) \propto \int_{\lambda_1} I(\lambda_1)[\cos 2\pi(K_1\lambda_1)]d\lambda_1, \tag{9.9}$$

where $K_1 = (\gamma m_3/2\pi h)B_1L$ is an experimentally controlled variable proportional to the strength of the first magnetic field. Thus, by scanning K_1 and performing an inverse Fourier transform, the wavelength distribution $I(\lambda_1)$ of the scattered beam can be reconstructed with a wavelength resolution given by $\Delta\lambda = \frac{1}{2K_{max}}$, where K_{max} denotes the K_1 value obtained with the strongest magnetic field. The non-linear relation between energy and wavelength results in an energy resolution that is a function of energy which improves with decreasing energy. For the typical energy of the Cambridge apparatus (8 meV) and the maximal field strength, the resolution ΔE is approximately 2 μeV. In practice the energy distribution in elastic scattering measurements is limited by the signal-to-noise ratio and the spectra is often reconstructed with energy resolutions of only about a few tens of μeV. Since the velocity spread of the beam does not limit the resolution in an elastic scattering experiment, in these experiments the velocity spread of the incident beam is often intentionally broadened, by reducing the pressure behind the nozzle, to simultaneously measure a wide range of bound state resonances.

In this way the Cambridge group have explored in great detail the elastic selective adsorption resonances of ^{3}He from LiF(001) over a broad range of energies and scattering angles [86, 87]. To reduce the measuring time the resolution was chosen to be 30 μeV, which is less than the natural half-width of the resonances. Another impressive demonstration is provided by the high resolution investigation of the selective adsorption resonances in scattering from c(2×2) CO molecules on Cu(001) [88]. A close coupling calculation of the scattering resonances made it possible to compare the resonance energies and angles with a He atom-surface potential based on a potential for an isolated He–CO interaction [89] and a surface potential derived from experimental HAS scattering data [90]. The high resolution of these new experiments made it possible to arrive at an improved He atom-adsorbate surface potential, which appears to be presently the most accurate potential of this type.

In the application of the spin echo method to inelastic scattering it is necessary to take full account of the velocity distributions of the incident and scattered beams. In this situation the polarization in the y-direction is given by [84, 85]

$$P_y(K_1,K_2) \propto \int_{\lambda_1}\int_{\lambda_2} I(\lambda_1,\lambda_2)[\cos 2\pi(K_1\lambda_1 + K_2\lambda_2)]d\lambda_1 d\lambda_2 \tag{9.10}$$

where $K = (\gamma m_3/h)BL$ accounts for the magnetic field setting in each of the solenoids 1 and 2 and $I(\lambda_1, \lambda_2)$ denotes the wavelength intensity matrix. It is defined as the product of the incident beam velocity distribution and the scattering matrix $S(\lambda_1 \rightarrow \lambda_2)$.

By measuring the polarization over a sufficient range of K_1 and K_2 the wavelength intensity matrix can be obtained from the two dimensional Fourier transform defined by (9.10). In the λ_1, λ_2-plane elastic events correspond to sharp diagonal lines. Their sharpness results from the excellent resolution while the length of the lines depends on the width of the velocity distributions prior to and after scattering. Inelastic events appear on a diagonal with a slope which depends on whether a creation or an annihilation event is involved. Since these experiments require time consuming scans over the K_1, K_2-space other more expedient schemes have been introduced to obtain information on selected regions of the intensity matrix [84, 85].

The wavelength intensity matrix method has, as of spring 2018, only been applied in a few instances to surface phonon inelastic scattering. A pioneering study was reported in the review by Alexandrowicz and Jardine [85]. As shown in Fig. 9.13 by a sophisticated tuning of the magnetic field in the second solenoid it was possible to reconstruct a phonon energy spectrum of Cu(001). The Rayleigh peak had a FWHM of only 70 μeV, but the *L* mode and the diffuse elastic peak have a width comparable with a TOF-HAS experiment. More recently, Kole et al. [84] reported a full two dimensional Fourier transform study of the Rayleigh mode of Cu(111) at an incident angle of -1° off-specular. The separation between the energy loss and elastic peak could be measured with an accuracy of 50 μeV despite

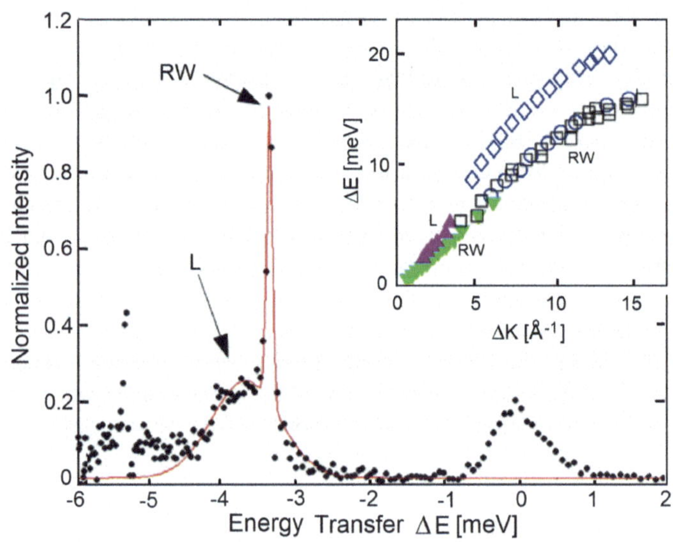

Fig. 9.13 A spin echo measurement of the inelastic energy spectrum for scattering from Cu(001) [85]. The reconstruction of the K_1, K_2-space, (9.10) was optimized to measure the Rayleigh mode RW with a resolution of 70 μeV at FWHM. The longitudinal mode (L) and the diffuse elastic peak E are broadened by the simplified reconstruction which was optimized for the Rayleigh mode. The inset compares the spin echo data points (solid colored triangles) with previous TOF-HAS measurements of the Rayleigh mode (circles) and of the longitudinal mode (diamonds) and with EELS (squares) measurements

the incident beam energy spread of 700 μeV. A similar study has also been reported for the scattering from graphene on Ni(111) [91].

The high resolution of SE-HAS spectroscopy makes it especially suitable for the study of surface phonons in comparatively soft very interesting materials like topological insulators. An example is shown in Fig. 9.14, where in this case the Fourier analysis of the complex spin polarization oscillations was used to determine the energy transfer spectrum under given kinematical conditions in $Bi_2Te_3(111)$. In this material the Rayleigh wave is a pseudo-surface mode (Chap. 3). Since the observed peak FWHM is much wider than the actual instrumental resolution it can be assigned to the true natural width.

Another impressive demonstration of the spin echo method for inelastic scattering is a recent study of the phason dispersion of misfit dislocations on the $23x\sqrt{3}$ reconstructed Au(111) surface [93]. This recent impressive demonstration was made possible by the greater sensitivity of an improved ionization detection system

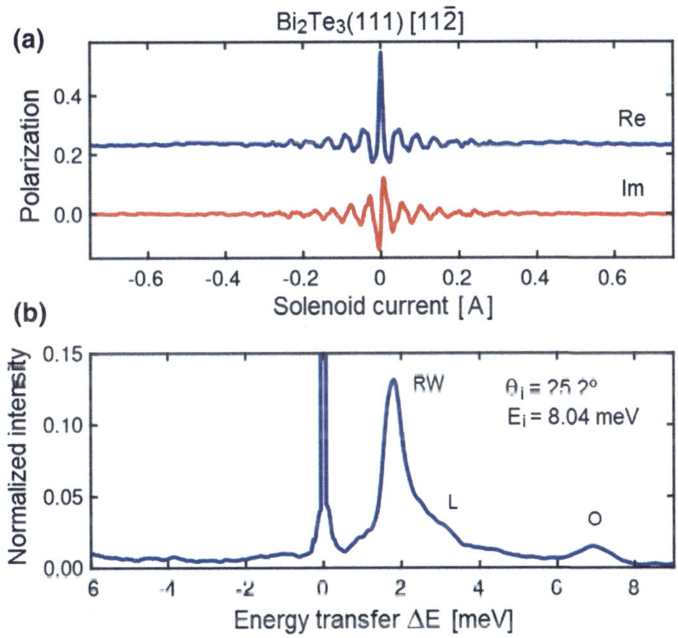

Fig. 9.14 A typical ^{3}He spin echo measurement of inelastic scattering from the topological insulator surface $Bi_2Te_3(111)$ along the $[11\bar{2}]$ symmetry direction at 115 K [92]. **a** The real and imaginary components of the complex polarization. The oscillations are due to phonon events which become evident after Fourier transforming and converting to the energy transfer scale. **b** The very sharp peak at $\Delta E = 0$ is due to diffuse elastic scattering, and the phonon modes are detected, under the indicated kinematic conditions, on the positive energy transfer (annihilation) side. The most prominent phonon peak corresponds to the Rayleigh wave (RW), the broader peak at 6.8 meV (O) is an optical surface resonance and the shoulder at 3 meV corresponds to a longitudinal acoustic resonance (adapted from [92])

[29]. For the first time the acoustic and optical dispersion relations of a phason could be measured at energies in the range of between 40 and 500 μeV with a resolution of about 100 μeV. These experiments are discussed in more detail in Sect. 11.5.1.

The major application of the spin echo technique has been quasi-elastic helium atom scattering (QHAS) investigations of surface diffusion. In this case the magnetic fields B_1 and B_2 are set equal and information on the diffusion process is obtained by varying their field strengths. The method is discussed in more length in Sect. 13.2 where also the large number of results obtained by the Cambridge group up to the time of writing is surveyed. Briefly, in the conventional TOF-QHAS method [94] also discussed in Sect. 13.2 the broadening of the elastic peak $\Delta\hbar\omega$ is measured over a wide range of parallel wavevector transfers $\Delta\boldsymbol{K}$ at low coverages of the adsorbed molecules. From the data the scattering matrix $S(\Delta\boldsymbol{K},\omega)$ is obtained directly. $S(\Delta\boldsymbol{K},\omega)$ is the Fourier transform of the time-dependent pair correlation $G(\boldsymbol{R},t)$ which contains the full information on the diffusing ensembles. In the spin echo method, called SE-QHAS, the time-dependent change in the phase of the scattered atoms produced by the evolving ensemble on the surface is measured directly. This has the advantage that it enables the direct measurement of the time domain Fourier transform of $G(\boldsymbol{R},t)$.

Despite the great progress made by the Cambridge group the spin echo apparatus presents a number of imposing technical challenges. Because of the many vacuum stages and large overall dimensions of the apparatus, the intensities tend to be less than with the conventional TOF apparatus. To compensate for this a new type of detector with a greatly lengthened ionization region of about 6 cm, compared to the short 5 mm ionizer in a typical TOF apparatus, was developed for the Heidelberg apparatus [95] and recently implemented in the Cambridge machine [29]. Moreover, very great precision and advanced machining technology are required in the construction of the hexapoles and the precession magnets [96]. Also the use of the much more expensive rare isotope ^{3}He requires a special gas handling and recycling system to keep losses to a minimum. Another disadvantage is the limitation to low beam energies. As mentioned above the carefully optimized Cambridge apparatus can handle beams with energies up to only about 8 meV. Another problem results from the broader inherent velocity distribution of ^{3}He supersonic expansions due to the fermionic nature of the atoms and leads to a reduced resolution in K-space compared to an apparatus with ^{4}He as the beam gas. Another less severe complication results from the requirement of well-defined homogeneous magnetic fields in the target chamber. Thus experiments on magnetic targets such as Ni and Fe may require special precautions. Finally, for analyzing the data sophisticated methods are required which are in the process of being further optimized. Despite these complications the spin echo method holds great promise for the future.

Since the resolution of a spin echo apparatus is to a large extent independent of the beam velocity, experiments can be envisaged with other probe particles with magnetic moments, like H atoms, metastable orthohelium ^{4}He * (2^3S_1), alkali atoms as well as o-H_2 and p-D_2, or even molecules like ortho-water (o-H_2O), ammonia, methane or acetylene. The successful application of the method to the ortho-water (o-H_2O), was recently demonstrated with the new spin echo apparatus at Technion, Haifa [97, 98]. Most recently the same group have demonstrated the method for controlling the initial hyperfine aligned quantum states of ortho-H_2 (J=1) molecules in scattering from stepped and flat Cu surfaces [99].

9.5 Electron Energy Loss Scattering (EELS) Apparatus

In principle the electron scattering apparatus used in phonon studies is similar to the standard electron energy loss (EELS) spectrometers which have been frequently described in the literature [3, 100, 101]. Figure 9.15 shows a schematic diagram of a commercially available EELS spectrometer which has been developed over the years mainly by Ibach and his colleagues in Jülich, Germany [101]. This apparatus is much more compact than the HAS machines making it possible to mount it on a single flange and insert it into a standard vacuum chamber for in situ analysis. As with the HAS apparatus the target chamber can also be equipped with LEED and

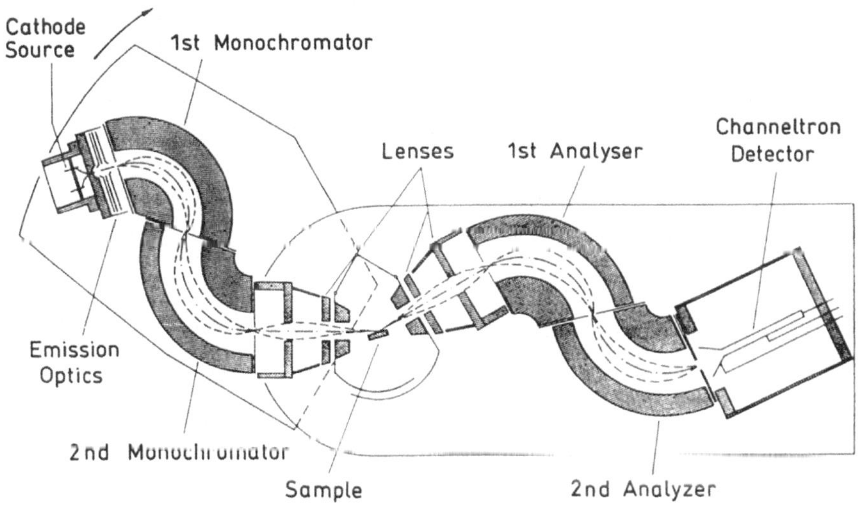

Fig. 9.15 A typical electron energy loss spectrometer (EELS). The electrons from a hot cathode emission system pass through a two stage electrostatic 127° monochromator and are focused onto the target. The scattered electrons are then focussed into a second two stage electrostatic 127° monochromator for energy analysis and detected by a channeltron. The electron trajectories in the cathode system, the energy dispersive systems and the lens systems have to be matched to each other for the optimum performance of the entire system. It is customary to mount the sample on a carousel manipulator. Thus the sample can be prepared and characterized in another position of the carousel with LEED and Auger electron spectroscopy before being placed into the EELS spectrometer (adapted from [101])

Auger. Since the late 1980s a considerable improvement in the energy resolution became possible and measurements with an overall resolution of less than about 1 meV in the dipole regime ($E_i \sim 3$ eV) are now standard [102, 103]. Because of space-charge limitations, however, the current j invariably falls off with increasing energy resolution $\Delta E_{1/2}$ (FWHM) as $j \propto \Delta E_{1/2}^{5/2}$.

As noted at the beginning of this chapter for phonon studies with electrons much higher kinetic energies of between 50 and 300 eV are used compared to the standard EELS machine operating in the dipole regime. At these high energies, referred to as the impact regime, the usual dipole selection rules no longer apply since the interaction is impulsive in a somewhat similar way as with He atoms. The actual resolution achieved depends quite sensitively on the reflected electron signal which in turn depends on the effective reflection coefficient. At $E_i = 100$ eV a resolution of 5–7 meV was achieved in the early phonon experiments [101, 104–109]. In [110] the ultimate limit in resolution for phonon studies is estimated to be 1.5 meV. A resolution of about 1 meV in energy and 10^{-3} Å^{-1} in parallel momentum has recently been achieved in a high resolution EELS (HREELS) apparatus [111].

The range of scattering angles probed is typically between $135° - \vartheta_f$ and $90° - \vartheta_f$ with ϑ_f between 55° and 65°. On average the count rates in electron scattering are much the same for zone boundary phonons as in the helium atom experiments [1] and are typically about 10–40 counts/s per meV energy channel width. The available electron currents, which are severely limited by space charge, are smaller than the He atom flux by about a factor 10^5–10^6. This is fully compensated, however, by the 100% detection efficiency for electrons compared to the 10^{-5}–10^{-6} efficiency of the mass spectrometer used to detect He atoms. For a more detailed comparison of the relative merits of HAS and EELS see the discussion in [1]. In actual practice the electron scattering signal is strongly influenced by interferences which depend on the mode excited and the experimental conditions (see next section and Fig. 9.17b).

9.6 Comparison of HAS and EELS Experiments

Figure 9.16 shows a comparison of some typical energy loss spectra measured by He atom [112] and electron scattering [107], both on the Ni(001) surface. The spectra from the two different experiments have been chosen to correspond to about the same points on the dispersion curve. Note that because of the different nature of the scan curves the electron energy loss distributions are measured for a fixed value of **Q**, whereas the He atom spectra are for a fixed incident angle. In the HAS energy loss spectrum each of the inelastic peaks corresponds to different values of **Q** as determined by the particular scan curve (see Sect. 9.1). For this reason the **Q**-values are listed next to the corresponding inelastic peaks. In both sets of spectra well resolved peaks are seen for energy transfers equal to zero corresponding to diffuse elastic scattering. In the case of the HAS spectra the incoherent elastic peaks are

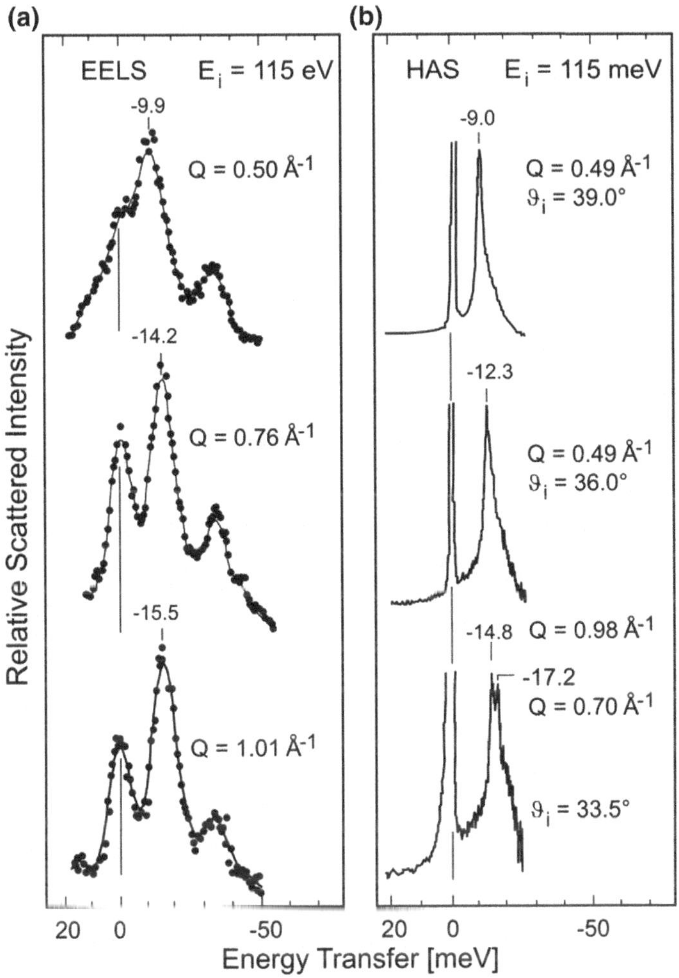

Fig. 9.16 Side-by-side comparison of typical electron energy loss spectra [107] with helium energy loss spectra derived from time-of-flight distributions [112]. In both cases the energy loss peaks are due to surface phonons of clean Ni(001) along the $\overline{\Gamma X}[110]$ direction of about the same phonon energy. **a** In the electron scattering experiment the incident electron beam energy is 115 eV and final angle ϑ_f is 55.1° and different Q values are sampled by adjusting ϑ_i. **b** In the helium atom scattering experiment the total angle between incident and final beam is fixed at $\vartheta_{SD} = 90°$, the beam energy is 40.4 meV and ϑ_i is varied

much larger relative to the inelastic peaks than in the EELS spectra as a result of the greater sensitivity of the non-penetrating He atoms to even small concentrations of defects on the surface. This side-by-side comparison of the two sets of spectra illustrates not only the greater energy resolution of the He atom scattering experiments for low energy surface vibrations, but also the greater sensitivity of EELS for high energy vibrational modes.

Figure 9.17 presents a comparison of a logarithmic plot of inelastic intensities as a function of phonon wavevector for the two techniques. In the case of electron scattering the intensities show a seemingly erratic behavior with wavevector which strongly depends on the incident energy and scattering angles [113, 114]. This is attributed to constructive and destructive interference accompanying the multiple scattering of the electrons from the several layers below the surface [114]. These interferences have been found to provide a useful technique for enhancing special modes and suppressing others and for identification purposes. The He scattering intensities, on the other hand, are characterized by a smooth but sharp drop off over several orders of magnitude with increasing energy transfer [115, 116]. This decrease in intensity is due to several factors relating to the coupling of the He atoms to the surface vibrations, as discussed in Chap. 7.

A comparison of the typical intensities occurring in atom and electron inelastic scattering experiments for the incident, specular elastic and inelastic peaks from either zone center or zone boundary surface phonons is given in Table 9.2. Despite the much lower detection efficiency for He atoms with respect to electrons, the counting rate is about two orders magnitude larger for HAS than for EELS in the long wave acoustic region, and is about the same for zone boundary phonons. As

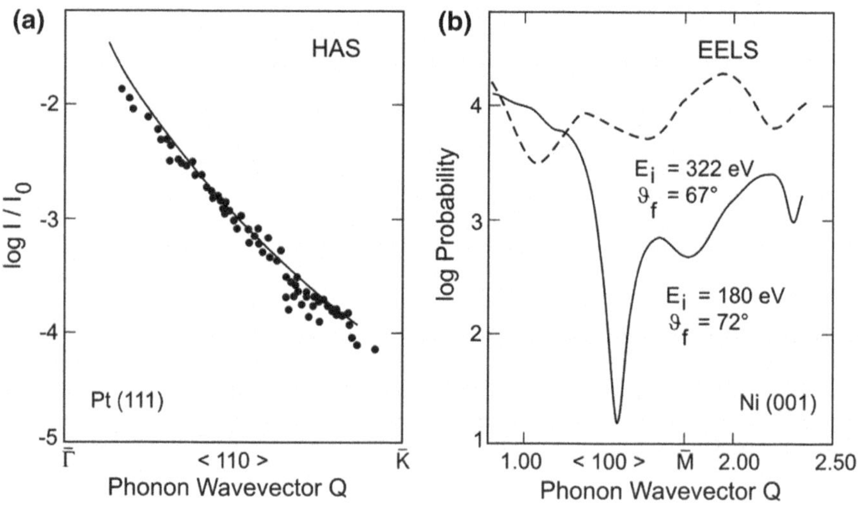

Fig. 9.17 Comparison of the inelastic intensity measured in He atom scattering (HAS) [115] with the inelastic intensities of electron energy loss scattering (EELS) measured for two different energies and angles [113] as a function of the phonon wavevector. **a** The drop off in the He scattering intensities is largely due to the rapidly decreasing inelastic reflection coefficient with increasing wavevector [116]. **b** In the case of electron scattering the intensities show an erratic behavior which is due to interferences associated with the multiple scattering [113, 114]

Table 9.2 Comparison of intensities in He atom and electron scattering from surface phonons

	He atoms	Electrons[b]
Incident onto target	3×10^{14} atoms/s[a]	10^9 electrons/s
In specular peak (I_{00})		
– Smooth surface	1×10^{14} atoms/s	10^5 electrons/s
– Corrugated surface	1×10^{12} atoms/s	10^5 electrons/s
Inelastic peaks		
– Near zone origin (10^{-3} of I_{00})	1×10^9 atoms/s	10 electrons/s
– At zone boundary (10^{-5} of I_{00})	1×10^7 atoms/s	10 electrons/s
Detection efficiency	10^{-5}–10^{-6}	1
Count rate		
– Near zone origin	10^3 counts/s	10 counts/s
– At zone boundary	10 counts/s	10 counts/s
Background count rate at zone boundary	0.3 counts/s[c]	0.3 counts/s

[a]Assuming a beam flux on 3×10^{19} atoms/s ster. and a target area of 1 mm^2 for the apparatus described in Table 9.1
[b]Vattuone et al. [111]
[c]The total background count rate is typically 10 counts/s. It is distributed uniformly over the entire time-of-flight spectrum and typically only a small fraction amounting to about 3×10^{-2} is at the time of the inelastic peak

Table 9.3 Relative fraction I/I_0 (in percent) of electrons of the incident beam which emerge from the surface after interacting with the n-th layer [111]. The angles are chosen to correspond to a parallel momentum transfer $\Delta K = 1.78$ Å^{-1}. λ is the mean free path of the electrons in the crystal

Scattering conditions					I/I_0		
E_i (eV)	ϑ_i (degree)	ϑ_f (degree)	λ (Å)	Penetration (Å)	$n = 2$ (%)	$n = 3$ (%)	$n = 4$ (%)
100	28	55	4	5.0	37	14	5
100	37	72	4	7.8	14	2	0.3
180	44	72	6	8.0	26	7	1.8
220	36	55	8	5.2	52	28	15

shown in Table 9.3 the penetration depth and the mean free path of electrons in the impact regime conditions allow the electrons to probe several atomic planes beneath the surface. Thus EELS provides in-depth information about surface dynamics, although disentangling it from the complicated effects of energy- and momentum-dependent electron surface interaction, with, in addition, a less favorable resolution, can be a formidable task. The analysis of HAS time-of-flight data also can be demanding, as discussed below in Sect. 9.10. The detailed analysis of the TOF spectra, aimed at extracting intrinsic dynamical properties of the surface, is discussed in the next sections.

9.7 Intensity Figure of Merit of a HAS Apparatus

The actual intensities observed in a HAS surface scattering experiment and the achievable resolution depend on a number of factors which are influenced by the parameters of the apparatus. Obviously, as for all types of spectrometers, an increase in the energy resolution leads to a drastic reduction in the signal and, consequently, an optimal compromise between resolution and intensity has to be made. This is an especially important consideration in the case of HAS surface phonon experiments since, as seen in Fig. 9.17a, the inelastic intensity of one-phonon Rayleigh events falls off by several orders of magnitude as the phonon wavevector increases. Thus the trade off between energy resolution and intensity also depends on which part of the dispersion curve is of interest.

The scattering signal $\dot{N}$ (in ions per second at the detector) is obtained by integrating the differential reflection coefficient $d^2R/dE_f d\Omega_f$, see (7.33), over the velocity distribution of the incident beam $f(\mathrm{v}_i)$, the interval of final energies dE_f centered about $E_f = \hbar^2 k_f^2/2m$, and the scattering solid angle $d\Omega_f$ (see Fig. 7.1):

$$\dot{N} = \int\limits_{\Delta \mathbf{v}_i} \int\limits_{\Delta E_f} \int\limits_{\Delta \Omega_f} n_i f(\mathbf{v}_i) A_i\, \alpha\, \varepsilon_{\det} \frac{d^2R(\mathbf{v}_i, E_f, \Omega_f)}{dE_f d\Omega_f} d\mathbf{v}_i dE_f d\Omega_f \tag{9.11}$$

Equation (9.11) also accounts for the dependence upon n_i, the particle density in the incident beam at the target, as well as on the effective target area struck by the incident beam, which depends on the area of the beam A_i and an important geometric factor α, which is discussed below. In (9.11) $\varepsilon_{\det}$ is the fraction of the atoms which pass into the detector and are ionized and detected by the electronics. $\varepsilon_{\det}$ also is a measure of the overall detection efficiency and is typically between 10^{-6} and 10^{-5}.

From Sect. 7.2 it is recalled that the differential reflection coefficient depends on a number of intrinsic factors related to the nature of surface phonons and, especially, on their wavevector and polarization as well as on their coupling to the probe particle. It contains all the information on the dynamical aspects of the microscopic interaction.

Before analyzing the factors determining the optimum trade off referred to above, it is expedient to first discuss the factor α. This factor accounts for unavoidable geometrical constraints imposed either by the detector or the incident beam geometries, which can dramatically affect the effective area of the surface, which is probed in the scattering. In general the factor α is given by [1]

$$\alpha = \min\left(1, \frac{A_f'}{A_i'}\right), \tag{9.12}$$

where $\min(1, A'_f/A'_i)$ denotes the smaller of the two quantities in parenthesis. The areas A'_i, A'_f are defined in Fig. 9.18a and some typical values of α are shown in Fig. 9.18b.

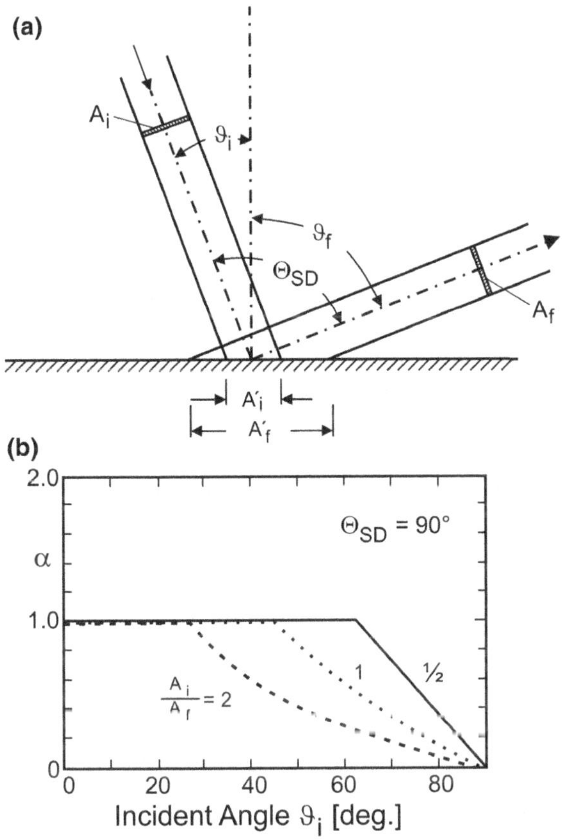

Fig. 9.18 Schematic diagram illustrating the geometrical reduction in the effective surface area in a surface scattering experiment for $\Theta_{SD} = 90°$. **a** In the geometry shown all the particles scattered from the exposed surface can be "seen" by the detector since $A'_i < A'_f$ and $\alpha = 1$. If for the same geometry the indices i and f are exchanged then $A'_i > A'_f$ and the exposed surface seen by the detector is reduced and $\alpha < 1$. **b** The factor α, which defines the relative reduction in scattered intensity is plotted as a function of the incident angle. It also depends on the ratio of the areas of the incident and final beams A_i and A_f, as well as on the incident and final scattering angles and, of course, on Θ_{SD}

In the optimization of the scattered signal for a given differential reflection coefficient it is useful to consider the following quantity, which follows directly from (9.11).

$$\frac{\dot{N}}{\left\langle \frac{d^2R}{dE_f d\Omega_f} \right\rangle} = n_i\, \bar{v}_i\, A_i\, \alpha\, \varepsilon_{\text{det}} \Delta E_f \Delta\Omega_f, \tag{9.13}$$

where $\left\langle \frac{d^2R}{dE_f d\Omega_f} \right\rangle$ denotes the mean value of the reflection coefficient averaged over the incident energy distribution of the incident beam and over the angular and energy resolution of the detector, with half-widths of $\Delta\Omega_f$ and ΔE_f, respectively. $\bar{v}_i$ is the average incident beam velocity. Thus the left hand side is simply the signal per reflection coefficient. The right hand side of (9.13) describes the overall efficiency or the *figure of merit* of a given apparatus. Thus (9.13) emphasizes the importance of a large incident particle density n_i on the surface, a large effective illuminated area $A_i \cdot \alpha$ and a large value of ε_{det}, implying an efficient detector. Moreover, (9.13) accounts for the proportionality of the signal on the resolution via the factors ΔE_f and $\Delta\Omega_f$. As discussed in the next section many different apparatus parameters determine the actual size of ΔE_f and $\Delta\Omega_f$. If these parameters are chosen to optimize the signal as discussed in the next section then the following general and somewhat simplified result has been derived [117]:

$$\frac{\dot{N}}{\left\langle \frac{d^2R}{dE_f d\Omega_f} \right\rangle} \propto B_i\, \varepsilon_{\text{det}} \left(\Delta E_f\right)^5 \quad \text{for} \quad \Delta E_f > \Delta E_i, \tag{9.14}$$

where ΔE_i is the half width of the incident beam energy. B_i is the brightness of the source (the number of particles per second per steradian emitted by the source) and, for the forward direction, is given by

$$B_i \cong c\, n_0\, A_0\, \bar{v}/\Delta\Omega_i = c\, n_0 \bar{v}\, L_{ST}^2 (A_0/A_i), \tag{9.15}$$

where the constant c depends on the source conditions and is of the order of unity, A_0 is the source area ($A_0 = \pi d^2/4$, where d is the orifice diameter) and $\Delta\Omega_i = A_i/L_{ST}^2$ is the solid angle subtended by the target, with L_{ST} the distance from source to target as illustrated in Fig. 9.19. Since the brightness is essentially determined by the total flux through the nozzle it is proportional to the size of the vacuum pump in the source chamber. The other efficiency determining factor ε_{det} has over the years, for electron impact detectors with a short ionizer as required for a good resolution in a TOF experiment, reached the optimum value of $\varepsilon_{\text{det}} \cong 10^{-6}$–$10^{-5}$, which is limited essentially by space charge effects. It is presently about the same for the best commercial and home-made detectors. Thus the crucial factor in (9.14) is the strong fifth power dependence of the apparatus figure of merit on the energy resolution ΔE_f. The huge effect of ΔE_f underlines the extreme care which must be taken in designing and operating an apparatus in order to be able to carry out measurements with the highest possible sensitivity and resolution.

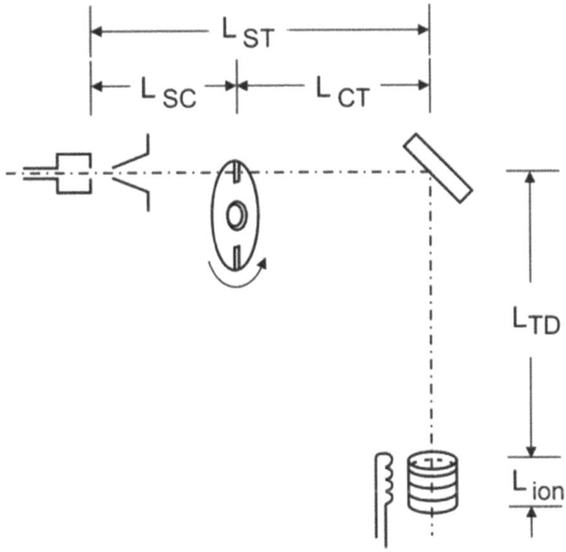

Fig. 9.19 Schematic diagram of a typical TOF-HAS apparatus ($\Theta_{SD} = 90°$) showing some important distances referred to in the text

9.8 Resolution of a HAS Apparatus

As illustrated in Fig. 9.5 the raw data, usually in the form of time-of-flight spectra, are customarily transformed from a time-of-flight scale to an energy transfer abscissa. This has two effects on the appearance of the spectra. First of all, the change in scale leads to a non-linear distortion in the distribution. From conservation of energy it is straight forward to show that for a spectral feature corresponding to an energy transfer $\hbar\omega$ the apparatus defined time-of-flight resolution Δt leads to an effective energy spread ΔE_f given by [117, 118]

$$\frac{\Delta E_f}{E_i} = 2\left(1 + \frac{\hbar\omega}{E_i}\right)^{3/2} \left|\frac{\Delta t}{t_{el}}\right|, \tag{9.16}$$

where t_{el} is the time-of-flight for elastic scattering. According to (9.16) for extreme creation events ($\hbar\omega < 0$), $\Delta E_f / E_i$ approaches zero and accordingly the resolution is very good. For annihilation processes ($\hbar\omega > 0$) the upper limit on $\hbar\omega$ is not only determined by the phonon spectrum, but may also be limited by the kinematics and the surface temperature. According to (9.16) for extreme annihilation events ($\hbar\omega \gg E_i$) or low incident beam energies the effective energy spread may become unacceptably large.

Secondly, the Jacobian of the transformation from TOF to energy transfer leads to a distortion of the intensity distribution as a function of the energy transfer $\hbar\omega$:

$$f_\omega(\omega) = f_t[t(\omega)]\left|\frac{dt(\omega)}{d\omega}\right|, \tag{9.17}$$

where $f_t[t(\omega)]$ is the experimental time-of-flight distribution. The Jacobian is given by

$$\left|\frac{dt(\omega)}{d\omega}\right| = \frac{\hbar t_{TD}^3}{m L_{TD}^3}, \tag{9.18}$$

where t_{TD} is the inelastic time-of-flight and L_{TD} is the distance from the target to the detector ionizer (see Fig. 9.19). Both of these effects can be seen by comparing Fig. 9.5a with Fig. 9.5b.

The apparatus energy resolution ΔE_f can be estimated by considering the four contributions to the total time-of-flight of the atoms from the chopper to the detector and their effect on the total energy spread which can be expressed in terms of the half-widths of four basic components:

$$\Delta E_f = \left(\Delta E_{chop}^2 + \Delta E_{CT}^2 + \Delta E_{det}^2 + \Delta E_{en}^2\right)^{1/2}. \tag{9.19}$$

The first three terms describe the energy spread resulting from the non-ideal apparatus components, whereas the last term is inherent to any inelastic surface phonon experiment. The first term ΔE_{chop}, is given by:

$$\Delta E_{chop} \cong 2E_f(\Delta t_{chop}/t_{CT}), \tag{9.20}$$

where Δt_{chop} is assumed to be equal to the pulse time. The second term ΔE_{CT} accounts for the additional energy dispersion resulting from the spread in the arrival times of the atoms at the target:

$$\Delta E_{CT} \cong 2E_f(\Delta t_{CT}/t_{TD}), \tag{9.21}$$

where Δt_{CT}, the spread in the flight time between chopper and target, depends on the velocity spread Δv_i of the incident beam:

$$\Delta t_{CT} \approx (\Delta v_i/v_i)\, t_{CT}. \tag{9.22}$$

Thus, since $t_{CT} = L_{CT}/v_i$, ΔE_{CT} is directly proportional to the distance between chopper and target (Fig. 9.19). The third term ΔE_{det} accounts for the smearing resulting from the finite length L_{ion} of the detector ionizer:

$$\Delta E_{det} \cong 2E_f(\Delta t_{ion}/t_{TD}), \tag{9.23}$$

Table 9.4 Expressions for the individual energy spread contributions to the overall energy resolution (9.19) in detecting a phonon of energy $\hbar\omega(\Delta K)$ using a HAS time-of-flight apparatus

Contributing factor	Contribution to ΔE_f
Chopper pulse length[a]	$\Delta E_{chop} = 2\left(\frac{2}{m}\right)^{1/2}\frac{(E_i+\hbar\omega)^{3/2}}{L_{TD}}\frac{T_{chop}}{nf}$
TOF dispersion between chopper and target	$\Delta E_{CT} = \frac{L_{CT}}{L_{TD}}\frac{(E_i+\hbar\omega)^{3/2}}{E_i^{3/2}}\Delta E_i$
Finite length of ionizer	$\Delta E_{det} = 2\,(E_i+\hbar\omega)\left(\frac{L_{ion}}{L_{TD}}\right)$
Kinematic smearing[b, c]	$\Delta E_{en} = E_i\left[\left(\frac{C_1+C_2}{2}\right)^2\left(\frac{\Delta E_i}{E_i}\right)^2 + C_2^2\Delta\vartheta_i^2 + C_3^2\Delta\vartheta_f^2\right]^{1/2}$

[a] m is the mass of the atom, T_{chop} is the chopper transmission, equal to $n(w_{slit}/2\pi R)$, n is the number of slits on the chopper disk, w_{slit} is the width of the chopper disk slits which are assumed to be equal to the beam width, R is the effective radius of the chopper at the slit and f is the chopper rotational frequency

[b] Expressions for the coefficients C_1, C_2 and C_3 have been derived and are presented in [117]. These coefficients depend in a complicated manner on the apparatus parameters and have to be calculated at the local intersection of the scan curve with the dispersion curve

[c] $\Delta\vartheta_i$ and $\Delta\vartheta_f$ are the angular half-widths (FWHM) of the incident and scattered beams

where $\Delta t_{ion} = L_{ion}/v_f$ is the spread in point-in-time at which ionization occurs. Each of the three terms discussed so far depends on the various lengths along the beam path in the apparatus.

Explicit expressions for estimating each of these terms for a given apparatus and experiment as a function of incident energy E_i, phonon energy $\hbar\omega$ and apparatus parameters are summarized in Table 9.4. Examination of these terms indicates that it is advantageous to minimize L_{ST}, (9.15), L_{CT} and L_{ion} and maximize L_{TD}. However, vacuum and pumping considerations and the need for accessibility limit the former and the loss of signal, which is proportional to L_{TD}^{-2}, limits increasing the latter.

The last term in (9.19), ΔE_{en}, called the *energy term*, is more complicated. As mentioned above it contains those terms which would apply independent of the technique used to analyse for the energy loss. In addition to the inherent spread in energy of the incident beam ΔE_i and the resulting dispersion in passage time from the target to the detector, the loss in effective resolution resulting from the *kinematic* smearing of the energy transfer, denoted by $\Delta\hbar\omega$ makes an important contribution:

$$\Delta E_{en} \cong \Delta E_{TD} = \Delta E_i + \Delta\hbar\omega. \tag{9.24}$$

The term $\Delta\hbar\omega$ depends on the angle between the scan curve and the phonon dispersion curve at the point at which they cross and has been examined in detail by Smilgies and Toennies [117]. At the crossing point the phonon energy must satisfy two equations, one expressing the scan curve, which for simplicity is referred to planar scattering, (9.3), and is written here as

$$\hbar\omega = S[k_i, \vartheta_i, \vartheta_f, \Delta K(\hbar\omega)], \tag{9.25}$$

where $\Delta K(\hbar\omega)$ is the inverse of the dispersion relation and is a multi-valued function to account for *Umklapp* processes. The dependence of $\hbar\omega$ on the apparatus parameters k_i, ϑ_i and ϑ_f cannot be obtained analytically and requires a numerical calculation based on the chain rule for differentiation

$$\Delta\hbar\omega = \frac{\partial S}{\partial k_i}\Delta k_i + \frac{\partial S}{\partial \vartheta_i}\Delta\vartheta_i + \frac{\partial S}{\partial \vartheta_f}\Delta\vartheta_f + \frac{\partial S}{\partial \Delta K}\frac{\partial \Delta K}{\partial \hbar\omega}\Delta\hbar\omega. \tag{9.26}$$

The final results for ΔE_{en} as a function of the important experimental parameters are also summarized in Table 9.4. Note that only the energy term ΔE_{en} depends explicitly on the angular half-widths of the incident and scattered beams. Their effects on the other terms are much smaller and have been neglected.

In the discussion so far the emphasis has been on the energy resolution. Alternatively, the apparatus resolution can also be expressed in terms of the wavevector resolution $\Delta(\Delta K)$. For this quantity the corresponding expressions which take account of the same smearing contributions have also been derived [117]. The results can be conveniently expressed by means of the kinematic term $\Delta(\Delta K)_{kin}$ and the apparatus length term $\Delta(\Delta K)_{app}$.

$$\Delta(\Delta K) = \left\{[\Delta(\Delta K)]^2_{kin} + [\Delta(\Delta K)]^2_{app}\right\}^{1/2}, \tag{9.27}$$

where the first term is given by

$$\Delta(\Delta K)_{kin} = \left|\frac{\partial\hbar\omega}{\partial\Delta K}\right|^{-1}\left[\left(\frac{C_1}{2}\right)^2\left(\frac{\Delta E_i}{E_i}\right)^2 + C_2^2\Delta\vartheta_i^2 + C_3^2\vartheta_f^2\right]^{1/2}. \tag{9.28}$$

The coefficients C_1, C_2, C_3 in (9.28) are the same as those used in the calculation of ΔE_{TD} (see Table 9.4). The three terms which depend on beam path lengths are given by

$$\Delta(\Delta K)_{app} = k_f \sin\theta_f \left\{\left(\frac{(2/m)^{1/2}E_f}{2\pi f R}\frac{w_{slit}}{L_{TD}}\right)^2 + \left[\frac{1}{2}\frac{L_{CT}}{L_{TD}}\left(\frac{E_f}{E_i}\right)^{1/2}\frac{\Delta E_i}{E_i}\right]^2 + \left(\frac{L_{ion}}{L_{TD}}\right)^2\right\}^{1/2}. \tag{9.29}$$

Figure 9.20 shows an example of an analytical calculation of the energy resolution for the apparatus of Fig. 9.9 in the investigation of Rayleigh phonons on the Ag(111) surface at E_i = 17.5 meV in the vicinity of the specular peak [117]. For this particular apparatus the three main contributions to the squared half-width of the energy loss peak ΔE_f^2 come from (1) the finite length of the chopper pulse

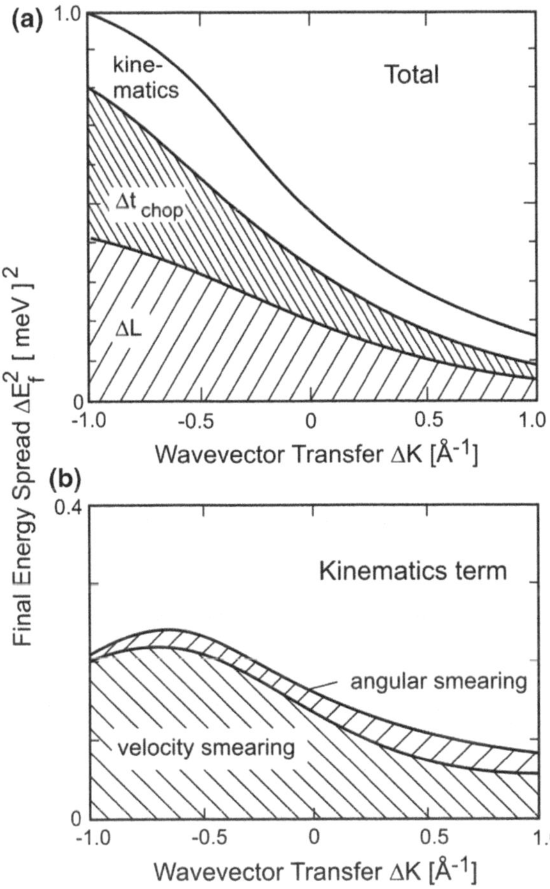

Fig. 9.20 The contributions to the half-widths of the inelastic peaks for the apparatus shown in Fig. 9.9 are plotted as a function of the wavevector transfer [117]. The contributions apply to both the experimentally preferred annihilation-backward and creation-forward branches. **a** The relative contributions of the three major terms contributing to the total half-width. **b** The contributions to the energy term (9.26) from the incident beam energy half-widths ΔE_i via ΔE_{CT} and ΔE_{en} and from the angular half-widths via ΔE_{en} (see Table 9.4) are shown on an expanded scale. The calculations are for the Ag(111) surface along the $[11\bar{2}]$ azimuth [117] and operating conditions of $E_i = 17.5$ meV for the $\Theta_{SD} = 90°$ apparatus with parameters listed in Table 9.1

Δt_{chop}, (2) the finite length of the ionizer L_{ion}, and (3) an effective kinematic smearing term. The latter term includes ΔE_{en} and ΔE_{CT} since both depend on ΔE_i. The above three terms are displayed as a function of the parallel momentum transfer ΔK in Fig. 9.20a. Figure 9.20b shows that the major contribution to the kinematic term comes from the finite velocity spread and only a relatively small contribution is due to the angular spread. From Fig. 9.20a it is seen that in this particular experiment the relative contributions from the chopper smearing, detector length smearing and the effective kinematics smearing are all nearly the same. This implies that with regard to these three terms the transmission of the apparatus for a given resolution is nearly optimal. However a closer examination of the kinematics contribution suggests that an improvement in the sensitivity (figure of merit) with only a small decrease in the overall resolution could be achieved if, in fact, the angular resolution would have been reduced by increasing the solid angle of acceptance of the detector. The calculations indicate, furthermore, that the effective total squared energy resolution is best for extreme creation (small ϑ_i, $\Delta K > 0$) events where it is less than about 0.40 meV. For extreme annihilation events (large ϑ_i, $\Delta K < 0$), on the other hand, the squared energy resolution is considerably worse and approaches a value of 1.0 meV. In most experiments the relative wavevector resolution $\Delta K/K$ is usually comparable to $\Delta E_f/E_i$.

The quantitative results on resolution discussed here have been checked by a complete Monte Carlo simulation of a typical apparatus [117]. This simulation avoids the assumption implicit in the use of (9.19) that the various contributions are treated independently. The Monte Carlo simulation yields results for the total energy resolution which closely follow the trends of Fig. 9.20.

These considerations on resolution have been confirmed in a number of experiments. Figure 9.21 shows a series of time-of-flight spectra transformed to an energy loss scale taken along the [112] direction of the Al(111) surface measured for a range of different incident angles [117]. Aluminum is one of the few surfaces which are dominated by a sharp Rayleigh mode (see Sect. 11.3.2). The scan curves corresponding to the series of spectra plotted in Fig. 9.21 are shown in Fig. 9.22. With the aid of the scan curve a number of features discussed previously can be identified. At $\vartheta_i = 43.0°$ the peak marked 1 is much sharper than the features 2 and 3 since for 1 the scan curve and dispersion curves cross at nearly right angles. Under these conditions the kinematic smearing is minimized. On the contrary the peak marked 2 is broadened because of the small angle between the scan curve and dispersion curve (Fig. 9.22) so that the conditions are near to those for kinematic focussing discussed in Chap. 10. The crossing angle is more favourable for the peak 3, which, however, is much weaker because of the larger momentum transfer and also, to a minor extent, because of its smaller thermal population.

Usually, as illustrated in these measurements, the conditions for observing intense peaks are most favourable for creation ($\Delta E < 0$) processes. Often, if the surface temperature is sufficiently high, inelastic spectra are also readily obtained for annihilation ($\Delta E > 0$) processes in both forward ($\Delta K < 0$) and backward ($\Delta K > 0$) directions. With further decreasing incident angles the intersections of the scan curves with the Rayleigh dispersion curve move more and more to the backward creation side and the corresponding spectra (Fig. 9.21) develop broad tails extending to larger energy losses. These tails are attributed to the excitation of modes within the bulk bands which are present in this region.

In Figs. 9.21 and 9.22 the forward (3) and a backward (4) creation of the same zone boundary Rayleigh phonon can be directly compared. As appears from the ordinate scales of Fig. 9.21, the *absolute* peak intensities of the two processes are

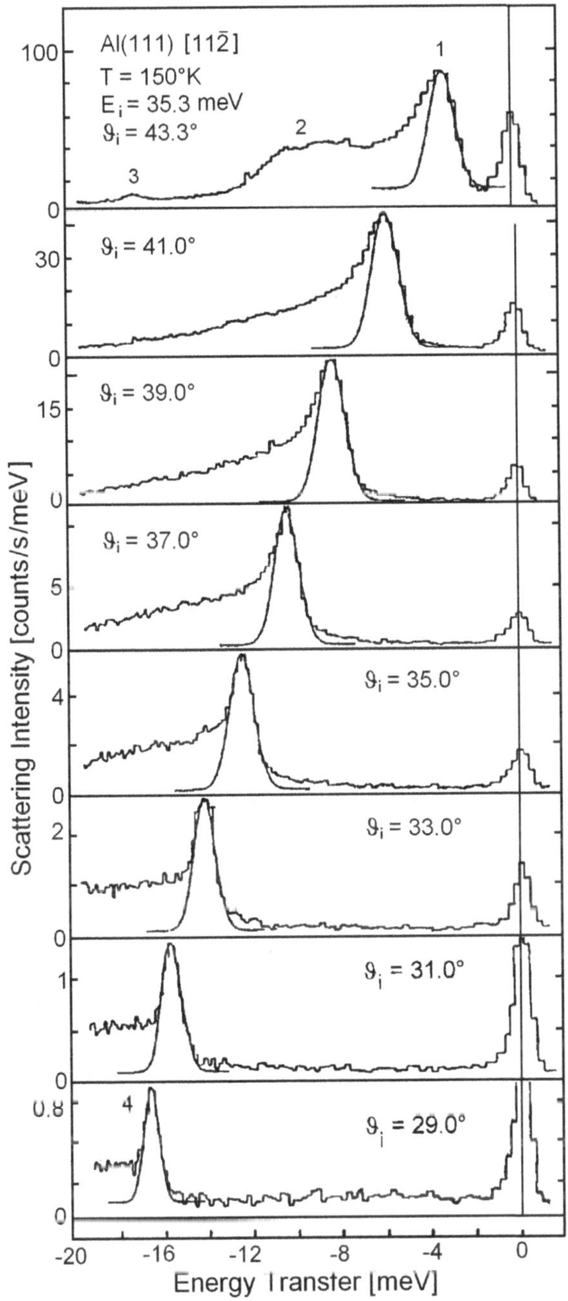

Fig. 9.21 Creation-forward Rayleigh wave energy transfer peaks based on HAS time-of-flight spectra from Al(111) [11$\bar{2}$] as a function of incident angle (T_s = 150 K, E_i = 35.3 meV, k_i = 8.23 Å^{-1} Θ_{SD} = 90° [117]). The labels 1–4 in the spectra for ϑ_i = 43.0° and 29.0° refer to the corresponding scan curves shown in Fig. 9.22. Note the increase in resolution with decreasing angle corresponding to increasing Q

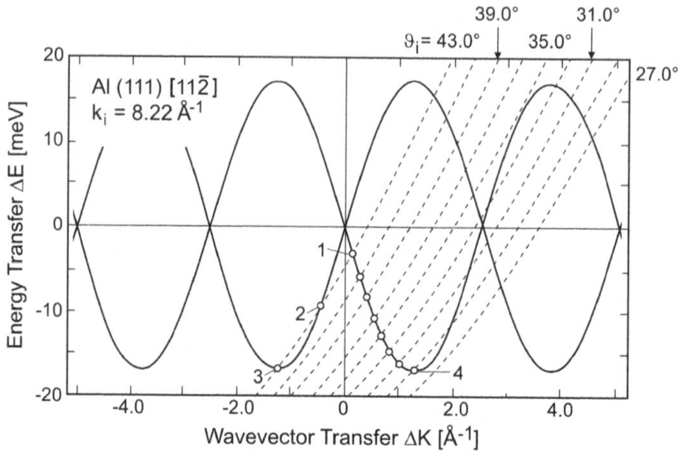

Fig. 9.22 Scan curves and Rayleigh dispersion curves in an extended zone diagram showing the intersections (small circles) which correspond to the peaks observed in the series of TOF measurements of Fig. 9.21. Note that the TOF spectra of Fig. 9.21 do not show *Umklapp* features corresponding to dispersion curve scan curve interactions beyond $\Delta K \cong 1.4$ Å^{-1}. This is expected since for the Al(111) surface the first order diffraction peak intensities are very small ($\sim 10^{-4}$) compared to the specular peak. The labels 1–4 mark the intersections of the scan curves for $\vartheta_i = 43.0°$ and 29.0° with the Rayleigh dispersion curves in the first BZ in correspondence with the respective peaks of the TOF specta in Fig. 9.21

about 4 and 0.8 counts/s/meV, respectively, i.e., the forward creation peak is about five times more intense than the backward creation peak. The intersections occurring beyond $\Delta K \cong 1.4$ Å^{-1}, i.e., in the second BZ and corresponding to *Umklapp* processes (Fig. 9.22), do not correspond to any feature above the noise level in the TOF spectra (Fig. 9.21). This is expected since for the Al(111) surface the first order diffraction peak intensities are very small ($\sim 10^{-4}$) compared to the specular peak. It should be noted, however, that the zone boundary phonons created in processes 3 and 4 have zero group velocity and are therefore standing waves neither moving forward nor backward. Thus process 4 can be viewed as an *Umklapp* equivalent of process 3 with the intensity reduced by a factor five. Clearly for a surface as flat as that of Al(111) a much larger drop in intensity occurs for *Umklapp* processes involving phonons of smaller wavevector.

The change in resolution of the apparatus with increasing ΔK, in the same experiment and for $\Delta K > 0$, can also be seen in Fig. 9.21 by noting the trend in the peak width with decreasing scattering angle. At $\vartheta_i = 43°$ the peak 1 has a half-width (FWHM) of 1.3 meV which drops to 0.75 meV at $\vartheta_i = 29°$. In Fig. 9.23 the experimental peak half-widths are shown to agree with the calculated half-widths. The divergence of the peak half-width predicted at $\Delta K = 1.5$ Å^{-1} is due to kinematic focussing (Sect. 10.2.1) which occurs for $\vartheta_i = 28°$ just beyond the crest of the Rayleigh phonon dispersion curve (see Fig. 9.22).

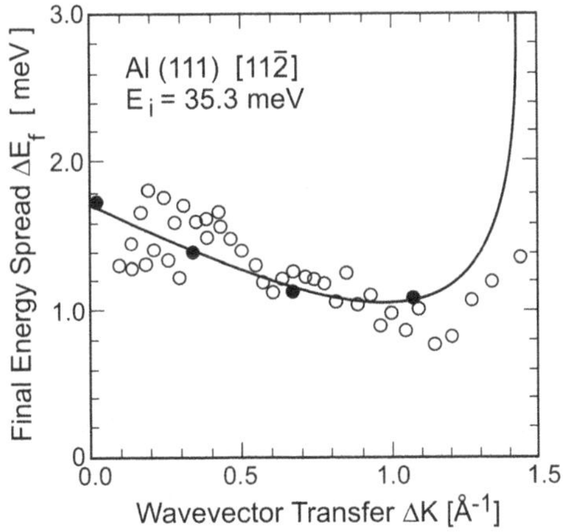

Fig. 9.23 Comparison of experimental Rayleigh phonon peak half-widths for Al(111) [11$\bar{2}$] and E_i = 35.3 meV (O) (see Fig. 9.21) with calculations based on (9.19) and Table 9.4 (—) and a full Monte Carlo simulation (●) [117]

A very good resolution of 0.2 meV for $\Delta K = Q > 0$ was achieved on a Cu(111) surface at a He beam energy of 9.2 meV [119]. A more comprehensive discussion of the resolution determining factors presented here can be found in the original paper by Smilgies and Toennies [117], and from a somewhat different perspective in [5] and in the excellent review by Doak [118]. A more detailed theoretical and computational analysis has been presented by Vollmer [120].

9.9 Deceptons and Spurions

One problem encountered in early He phonon studies using the time-of-flight technique originated from the presence of a weak unexpected component in the incident beam with a broad Maxwellian velocity distribution in addition to the much larger desired component with a sharp velocity distribution. The small component is difficult to detect with ordinary techniques and was only discovered by accident in the first inelastic scattering studies of He from LiF(001) [5, 121, 124] through the appearance of unexpected peaks that appeared to be from an inelastic event, which could however not be assigned as such. Since these peaks are easily confused with the true inelastic structures they have been dubbed *deceptons* (sometimes *phonyons*) [5, 122]. Deceptons are the surface scattering counterpart of *spurions* (which shall be considered a synonym) observed in inelastic neutron triple-axis spectrometry of crystal lattices [123]. Decepton peaks are explained by considering an additional component with a broad velocity distribution in the incident beam. Since the incident wavevectors k_i^* of the atoms of this component differ significantly from the nominal wavevector k_i their elastic diffraction peaks appear at angles removed from the nominal diffraction peak of the major sharp velocity component. Thus corresponding to each value of the wavevector k_i^* there is an elastic diffraction peak at a specific final angle ϑ_f^* which satisfies the Bragg condition:

$$k_i^* (\sin \vartheta_f^* - \sin \vartheta_i) = nG, \tag{9.30}$$

where G is a reciprocal lattice vector and n is the diffraction order. Since the narrow collimating slits permit only a narrow range $d\vartheta_f$ and a corresponding narrow range dk_i^* to arrive at the detector additional sharp maxima appear in the TOF spectra with an *apparent* energy transfer equal to the difference between the nominal k_i and the unexpected component wavevector k_i^*. An example of a "spurion" is peak 2 in Fig. 9.5. Thus although the sharp peaks are due to atoms which, in fact, are elastically diffracted they appear, because of their different velocities and low intensities, to have been involved in an inelastic process.

In other words, the apparatus serves as a grating monochromator where the corrugated single crystal surface is the grating. Since the spurions are produced solely by diffraction they are expected to occur for a corrugated surface and then only in the vicinity of the diffraction peaks. Evidence that spurions can also be caused by multiple terraces and other surface defects has also been reported [122].

The spurion positions can be predicted by noting that incident atoms with a wavevector k_i^*, not equal to the nominal wavevector k_i, will appear as if they had been inelastically scattered with a final wavevector k_f^* given by [5]

$$k_f^* = k_i^* \left[\frac{L_{CD}}{L_{TD}} \left(\frac{k_i^*}{k_i} - 1 \right) + 1 \right]^{-1}, \tag{9.31}$$

where L_{CD} and L_{TD} are the chopper-detector and target-detector path lengths, respectively (see Fig. 9.19). This expression accounts for the fact that the observed apparent time delay is attributed to particles which were incorrectly assumed to have changed their velocities at the target. From the conservation equations the scattered atoms with k_f^* are erroneously assigned to the following false frequencies ω^* with wavevector ΔK^*

$$\hbar\omega^* = (\hbar^2/2m)(k_f^{*2} - k_i^2), \tag{9.32}$$

$$\Delta K^*/k_i = (k_f^*/k_i^*) \sin \vartheta_f - \sin \vartheta_i. \tag{9.33}$$

To obtain an effective spurion dispersion curve, ϑ_f and ϑ_i are eliminated from (9.33) via the Bragg condition (9.30).

Figure 9.24 shows the result of an evaluation of the intensities of spurion peaks observed in scattering from LiF(001) [1]. The broad velocity distribution appears to be Maxwellian. This suggests that it is very likely due to a beam of He atoms produced by effusion of the unavoidable background gas in the nozzle-skimmer region. Even though the relative intensity of this broad distribution amounts to only about 5×10^{-4} of the major component the effect is comparable to that of the true small single phonon intensities. Spurions are not known to occur in EELS.

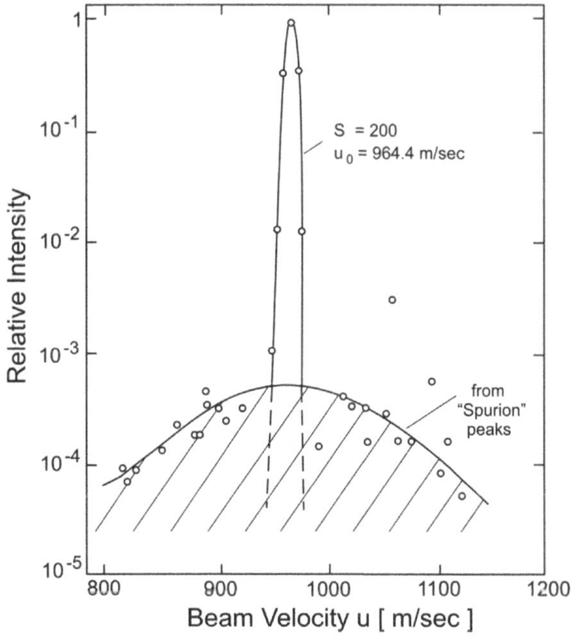

Fig. 9.24 Spurious intensity contribution to the velocity distribution of the incident beam as determined from a TOF analysis of "spurion" inelastic peaks from LiF(001) [1]. The broad spurious velocity distribution is attributed to a diffuse beam originating upstream of the beam chopper. The source temperature was $T_0 = 80$ K. Note that the intensity is plotted on a log scale and extends over more than four orders of magnitude

Another type of spurion, which is related to an artifact in the time-of-flight electronics, designated the *wraparound* TOF peak, has been described by Doak [118].

9.10 Procedures and Data Analysis

9.10.1 A Wealth of Valuable Information from a Complex Experimental Method

HAS surface phonon spectroscopy is conceptually similar to the inelastic neutron scattering (INS) spectroscopy for bulk phonons. However the analysis and interpretation of HAS TOF spectra is far more complicated than that of INS spectra. The latter normally show a few well resolved peaks which can be readily assigned, by means of a simple kinematic analysis, to single bulk phonons of defined energy and wavevector. There is much more than that in a HAS TOF spectrum, due to the following three different reasons. (i) The intrinsic complexity of surface phonon spectra, consisting of peaks due to surface localized modes and resonances on top of a structured continuum of surface-projected bulk modes (Chaps. 3–5). (ii) The complexity of the He-surface interaction: while neutrons collide directly with the atomic nuclei of the solid, the interaction of He atoms with the massive part of the surface atoms only occurs via the interposed electrons, which generally gives the

He-surface interaction a many-body character (Chaps. 6–8). (iii) The same interposed electrons exert on the He atoms dispersion forces and therefore an attractive surface potential with a number of bound states. Thus flying-by He atoms have their own complex dynamics with bound state resonances and various focusing effects which may largely enhance the inelastic scattering intensities (Chap. 10).

Due to all these complications, the development of HAS spectroscopy required an iterative feed-back with *ad hoc* theoretical simulations of the scattering cross section and surface lattice dynamics. In this way the surface phonon dispersion curves of a large amount of different solid surfaces, adsorbates and ultrathin films could be measured with a comparatively high resolution (see Appendix A). The considerable care that must be exercised in inelastic HAS experiments is however largely rewarded by the wealth of information which can be extracted. Besides the dispersion curves of surface phonons, information can be obtained on their individual electron-phonon interaction (Chap. 8), their role in various surface dynamical processes in adsorbates and thin films (Chap. 12), and possibly other surface low-energy elementary excitations (Chaps. 12 and 13).

9.10.2 Importance of Angular Distributions

In general it is possible to determine the phonon dispersion curves directly from the experimental HAS spectra with the aid of the conservation equations (9.1) and (9.2). This procedure can meet, however, several pitfalls and lead to erroneous results, due to the physical reasons enumerated above as well as to instrumental reasons like, for example, the presence of unresolved nearby lying peaks or the apparatus smearing effects discussed above in Sect. 9.8. For this reason it is important to follow a protocol where the relevant features of the He-surface interaction potential are determined first. The HAS spectrometer, once the crystal surface has been prepared and characterized in situ with the equipment installed in the scattering chamber (Fig. 9.9), allows to collect a rather complete information on the surface structure (periodicity, corrugation) and He-surface interaction potential (bound states) from angular distributions, before measuring the inelastic scattering spectra with the TOF method. The large amount of information which can be extracted from the angular distributions and TOF spectra in combination with various theoretical methods is schematically illustrated in Fig. 9.25.

High-resolution HAS angular distributions, providing sharp specular and diffraction peaks, allow for a precise determination of the surface lattice parameters and alignment. A Fourier analysis permits to evaluate the surface corrugation from the diffraction amplitudes (Sect. 6.1.2). The growth regime of ultrathin metal films, their thickness and related quantum size effects (see Sect. 12.3) can be monitored from the time dependent oscillations of the specular and diffraction peaks. The dependence of diffraction peaks, besides revealing possible surface phase transitions (Sects. 11.3.4, and 11.4) provides the DW factor, i.e., a preliminary

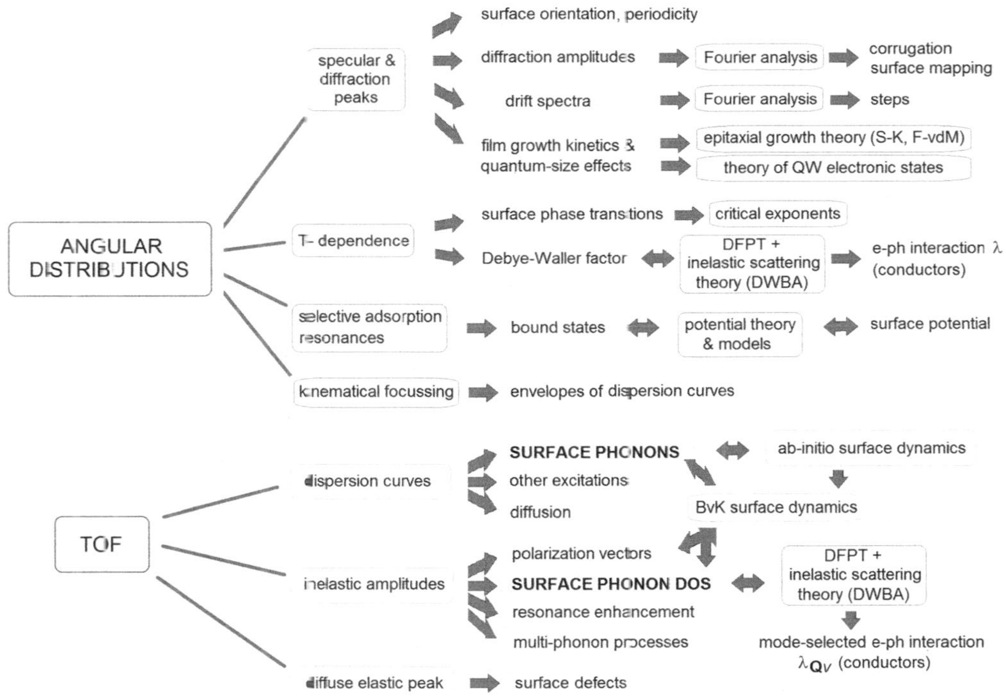

Fig. 9.25 Computational pathways showing the rich information that can be obtained from HAS angular distributions and TOF spectra when used in combination with theoretical simulations. The latter include the Fourier analysis of diffraction and drift spectra, the theory of epitaxial growth (Frank-van der Merwe (F-vdM) vs. Stranski-Krastanov (S-K) regimes) and surface phase transitions, the theory of electronic Quantum Well (QW) states involved in thin film quantum size effects, the theory and models of He-surface potentials for the calculation of bound states, the surface dynamics, either with the Born-von Kármán (BvK) force constant models or *ab initio* methods (e.g., DFPT), and the theory of inelastic scattering with the methods discussed in Chaps. 7 and 8. Double arrows indicate possible iterations of parametrized simulations (e.g., a BvK fit of HAS data with adjustable force constants) up to convergence

information about the best conditions for the observation of single-phonon scattering processes. As discussed in Chap. 8, a DFPT calculation of the phonon-induced charge density oscillations combined with a DWBA treatment permits to extract from the DW factor for conducting surfaces information on the electron-phonon coupling.

Most of the numerous weak features observed in the angular distributions away from the specular and diffraction peaks (see, e.g., Figs. 10.4 and 10.5) are assigned to series of bound state resonances by means of simulations with a suitable He-surface potential (Sect. 10.1). Other features correspond to various focusing effects, discussed in Sect. 10.2, involving surface phonons. Thus determining the kinematic conditions of inelastic resonances and focusing effects is a necessary step before starting the analysis of TOF spectra. A useful procedure is to mark on each TOF spectrum the expected positions of inelastic resonances and kinematic focusing, in order to distinguish real surface phonon peaks from local resonance enhancements of the bulk phonon DOS. On the other hand a precise control of the resonance kinematic parameters permits to tune bound state resonances on weak phonon features in order to enhance their response (see Sect. 10.2).

9.10.3 Analysis of Time-of-Flight (TOF) Data

After an accurate study of the angular distributions, the measurements and interpretation of TOF spectra can be undertaken, possibly in combination with a reliable scattering theory and a fast simulation of the surface dynamics based on a BvK force constant model (Sect. 3.3) so as to determine the set of force constants which provide the best fit of both the dispersion curves and inelastic scattering intensities. As discussed in Chap. 11 in the early analysis of TOF spectra from metal surfaces, the simultaneous BvK fit of the dispersion curves and the inelastic peak intensities resulted in an unphysically large force constant softening at the surface. This led to new theoretical approaches to the surface dynamics of metals which properly incorporate the electronic degrees of freedom (Sects. 5.2–5.5). Semi-empirical methods like the PC and the EA models introduce additional electronic degrees of freedom, similarly to the various bond charge and shell models for semiconductors and insulators, so as to reduce the dynamical matrix to that of a BvK model with effective force constants. In this way comparatively fast surface dynamics simulations can be performed also for conducting surfaces for an iterative comparison with the TOF data.

In general a set of force constants which provides a best fit of the bulk dispersion curves and possibly the bulk elastic constants is used in a first trial calculation. These coupling constants depend on the lattice dynamical theory used and, for example, may consist of several two- and three-body force constants as in the Born-von Karman theory, or effective force constants as in the multipole expansion method discussed Chap. 5. Force constant calculations do not require large computer facilities and can be implemented directly online with the data analysis. From the solution of the dynamical matrix the surface phonon dispersion curves and polarization vectors for this set of coupling constants are determined (Chap. 3). Then a scattering calculation based, e.g., on the distorted-wave-Born approximation in the case of helium atom scattering, is used to obtain the inelastic reflection coefficient for one-phonon events as given for example by (7.33) with the effective two-body potential of the form surveyed in Chap. 6. For potentials with an attractive well which allows for bound states, resonance processes may have to be considered in the simulations (see Chap. 10). In the final step the inelastic reflection coefficients are averaged over the apparatus resolution. Computational techniques accounting for the instrumental and spurious effects, which arise from the various apparatus components and can severely affect inelastic spectra, have been discussed and analyzed by Smilgies and Toennies [117] and Doak [30] (see Sects. 9.6 and 9.8). The sequence of calculations described above is then repeated with modified coupling constants for the surface layers and for modified potential parameters until a best fit of the time-of-flight distributions is achieved. This iteration procedure is based on simple models for surface dynamics and simple approximations for the surface potential and the scattering theory. The recent progress of *ab initio* surface dynamics calculations with a high predicting power and similar advances in the quantum simulations of inelastic scattering have superseded the iteration method, at least for the simplest surfaces. In particular DFPT provides the surface phonon dispersion curves as well as the surface charge density oscillations (CDOs) for some selected phonons within a reasonable CPU time. As illustrated in Sect. 8.4.2, the DFPT calculation of the phonon-induced CDOs, by providing an accurate description of the energy gain spectra permitted to establish a direct proportionality between the inelastic HAS intensities and the mode selected electron-phonon coupling strength.

The last line of the diagram shown in Fig. 9.25 emphasizes the important role of the diffuse elastic peak (cf. Fig. 9.5) which appear in the TOF spectra. The intensity of this peak is extremely sensitive to the quality of the surface and increases sharply with decreasing surface quality and thus provides important information on the accretion of surface contaminants as well as on the formation of defects. By constantly monitoring the elastic peak so that it is as small as possible it is possible to retain only TOF measurements performed under optimal surface conditions.

References

1. J.P. Toennies, in *Surface Phonons*, ed. by F.W. deWette, W. Kress, Springer Series Surface Science, vol. 21 (Springer, Berlin, Heidelberg, 1991)
2. G. Brusdeylins, R.B. Doak, J.P. Toennies, Phys. Rev. Lett. **44**, 1417 (1980)
3. H. Ibach, D.L. Mills, *Electron Energy Loss Spectroscopy and Surface Vibrations* (Academic Press, New York, 1982)
4. G. Benedek, Phys. Rev. Lett. **35**, 234 (1975)
5. G. Brusdeylins, R.B. Doak, J.P. Toennies, Phys. Rev. B **27**, 3662 (1983)
6. D. Evans, V. Celli, G. Benedek, J.P. Toennies, R.B. Doak, Phys. Rev. Lett. **50**, 1854 (1983)
7. A. Glebov, W. Silvestri, J.P. Toennies, G. Benedek, J.G. Skofronick, Phys. Rev. B **54**, 17866 (1996)

8. G. Benedek, R. Gerlach, A. Glebov, G. Lange, W. Silvestri, J.G. Skofronick, J.P. Toennies, Vuoto Sci. Tecn. **25**, 26–29 (1996)
9. J.E. Yater, A.D. Kulkarni, F.W. de Wette, J.L. Erskine, J. Electron Spectrosc. Relat. Phenom. **54**(55), 395 (1990)
10. J.L. Erskine, E.-J. Jeong, J. Yater, Y. Chen, S.Y. Tong, J. Vac. Sci. Technol. A **8**, 2649 (1990)
11. M. Balden, S. Lehwald, H. Ibach, A. Ormeci, D.L. Mills, Phys. Rev. B **46**, 4172 (1992)
12. R.B. Doak, Superlattices Microstruct. **7**, 201 (1990)
13. J.P. Toennies, K. Winkelmann, J. Chem. Phys. **66**, 3965 (1977)
14. R.E. Grisenti, W. Schöllkopf, J.P. Toennies, G.C. Hegerfeldt, T. Köhler, M. Stoll, Phys. Rev. Lett. **85**, 2284 (2000)
15. J. Wang, V.A. Shamamian, B.R. Thomas, J.M. Wilkinson, J. Riley, C.F. Giese, W.R. Gentry, Phys. Rev. Lett. **60**, 969 (1988)
16. G. Brusdeylins, H.-D. Meyer, J.P. Toennies, K. Winkelmann, in *Progress in Astronautics and Aeronautics*, vol. 51, ed. by J.L. Poetter (AIAA, New York, 1977), p. 1047
17. G. Lilienkamp, J.P. Toennies, J. Chem. Phys. **78**, 5210 (1983)
18. F. Hofmann, Dissertation, University of Göttingen, 1995
19. J.B. Anderson, in *Molecular Beams and Low Density Gas Dynamics*, ed. by P.P. Wegener (Dekker, New York, 1974) pp. 1–74
20. D.R. Miller, in *Atomic and Molecular Beam Methods*, vol. 1, ed. by G. Scoles (Oxford, New York, 1988) pp. 14–53
21. L.W. Bruch, W. Schöllkopf, J.P. Toennies, J. Chem Phys. **117**, 1544 (2002)
22. J.C. Mester, E.S. Meyer, M.W. Reynolds, T.E. Huber, Z. Zhao, B. Freedman, J. Kim, I.F. Silvera, Phys. Rev. Lett. **71**, 1343 (1993)
23. Y.T. Lee, J.D. McDonald, P.R. LeBreton, D.R. Herschbach, Rev. Sci. Instrum. **40**, 1402–1408 (1969)
24. G.O. Brink, Rev. Sci. Instrum. **37**, 857–860 (1966)
25. B. Samelin, Diplomarbeit, Max Planck Institut für Strömungsforschung and University of Göttingen, 1993
26. R. Weiss, Rev. Sci. Instrum. **32**, 397–401 (1961)
27. M. DeKieviet, D. Dubbers, M. Klein, U. Pieles, C. Schmidt, Rev. Sci Instr. **71**, 2015 (2000)
28. A. R. Alderwick, A. P. Jardine, H. Hedgeland, D. A. MacLaren, W. Allison, J. Ellis, Rev. Sci Instr. **79**, 123301 (2008)
29. D. M. Chisnall, Dissertation, University of Cambridge, 2012
30. R.B. Doak, in *Atomic and Molecular Beam Methods*, vol. 2, Chap. 14, ed. by G. Scoles (Oxford University Press, 1992), pp. 384–443
31. R. David, K. Kern, P. Zeppenfeld, G. Comsa, Rev. Sci. Instrum. **57**, 2771 (1996)
32. H. Schief, V. Marsico, K. Kuhnke, K. Kern, Surf. Sci. **364**, L631 (1996)
33. J.T. Yates Jr., *Experimental Innovations in Surface Science* (Springer, New York, 1998), p. 4
34. B. Gans, S.F. King, P.A. Knipp, D.D. Koleske, S.J. Sibener, Surf. Sci. **264**, 81 (1992)
35. L. Niu, D.D. Koleske, D.J. Gaspar, S.J. Sibener, J. Chem. Phys. **102**, 9077 (1995)
36. Y. Ekinci, J.P. Toennies, Phys. Rev. B **72**, 205430 (2005)
37. L. Pedemonte, A. Gussoni, R. Tatarek, G. Bracco, Rev. Sci. Instrum. **73**, 4257 (2002)
38. D. Cvetko, A. Lausi, A. Morgante, F. Tommasini, K.C. Prince, M. Sastry, Meas. Sci. Technol. **3**, 997 (1992)
39. T. Miyake, E.S. Gillman, I. Oodake, H. Petek, Jpn. J. Appl. Phys. **36**, 4531 (1997)
40. L. Pedemonte, R. Tatarek, G. Bracco, Rev. Sci. Instrum. **74**, 4404 (2003)
41. K.M. Martini, W. Franzen, M. El-Batanouny, Rev. Sci. Instrum. **58**, 1027 (2002)
42. L. Schnieder, K. Seekamprahn, F. Liedeker, H. Steuwe, K.H. Welge, Faraday Discuss. **91**, 259 (1991)
43. O. Bünermann, H. Jiang, Y. Dorenkamp, A. Kandratsenka, S.M. Janke, D.J. Auerbach, A. M. Wodtke, Science **350**, 1346 (2015)
44. P. Scherb, E.M. Hörl, Rev. Sci. Instrum. **47**, 1511 (1976)
45. H.D. Meyer, *Bericht 137/1972* (Max-Planck-Institut für Strömungsforschung, Göttingen, 1972)
46. G. Comsa, R. David, B.J. Schumacher, Rev. Sci. Instrum. **52**, 789 (1981)

47. G. Rotzoll, J. Phys. E (Sci. Instrum.) **15**, 708 (1982)
48. D.J. Auerbach, in *Atomic and Molecular Beam Methods*, vol. 1, Chap. 14, ed. by G. Scoles (Oxford University Press, 1988), p. 369ff
49. W.R. Gentry, Rev. Sci. Instrum. **49**(5), 595 (1978)
50. B. Feuerbacher, R.F. Willis, Phys. Rev. Lett. **47**, 526 (1981)
51. B. Feuerbacher, in *Dynamics of Gas-Surface Interactions*, ed. by G. Benedek, U. Valbusa (Springer, Heidelberg, 1982), p. 263
52. U. Even, J. Jortner, D. Noy, N. Lavie, J. Chem. Phys. **112**, 8068 (2000)
53. U. Even, EPJ Techn. Instrum. **2**, 17 (2015)
54. D. Pentlehner, R. Riechers, B. Dick, A. Slenczka, U. Even, N. Lavie, R. Brown, K. Luria, Rev. Sci. Instrum. **80**, 043302 (2009)
55. A. Libson, M. Riedel, G. Bronshtein, E. Nareviscius, U. Even, M.G. Raizen, New J. Phys. **8**, 77 (2006)
56. W. Christen, J. Chem. Phys. **139**, 024202 (2013)
57. J. Wang, V.A. Shamamian, B.R. Thomas, J.M. Wilkinson, J. Riley, C.F. Giese, W.R. Gentry, Phys. Rev. Lett. **60**, 696 (1988)
58. I. Estermann, R. Frisch, O. Stern, Z. Physik **73**, 348 (1932)
59. B.N. Brockhouse, A.T. Stewart, Phys. Rev. **100**, 756 (1955)
60. B.N. Brockhouse, S. Hanteder, H. Stiller, in *Interaction of Radiation with Solids*, ed. by Strumane et al. (North-Holland, Amsterdam, 1963)
61. B. Dorner, *Coherent Inelastic Neutron Scattering in Lattice Dynamics* (Springer, Berlin, Heidelberg, 1982)
62. B.F. Mason, B.R. Williams, Rev. Sci. Instrum. **43**, 375 (1972)
63. B.F. Mason, B.R. Williams, Rev. Sci. Instrum. **49**, 897 (1978)
64. R.E. Subbarao, D.R. Miller, J. Vac. Sci. Technol. **9**, 808 (1972)
65. J.M. Horne, D.R. Miller, Phys. Rev. Lett. **41**, 511 (1978)
66. B.R. Williams, J. Chem. Phys. **55**, 3220 (1971)
67. B.R. Williams, J. Chem. Phys. **56**, 1895 (1972)
68. B.F. Mason, B.R. Williams, J. Chem. Phys. **75**, 2199 (1981)
69. B.F. Mason, B.R. Williams, Phys. Rev. Lett. **46**, 1138 (1981)
70. K. Kuhnke, E. Hahn, R. David, P. Zeppenfeld, K. Kern, Surf. Sci. **272**, 118 (1992)
71. H. Schief, Ph.D. thesis, Ecole Polytechnique Fédérale de Lausanne, 1995
72. W. Raith, Adv. At. Mol. Opt. Phys. **12**, 281 (1976)
73. D.J. Auerbach, For a review on time of flight spectroscopy of atoms and molecules, in *Atomic and Molecular Beam Methods*, vol. I, ed. by G. Scoles (Oxford, New York, 1994) p. 362
74. J. Braun, P.K. Day, J.P. Toennies, G. Witte, E. Neher, Rev. Sci. Instrum. **68**, 3001 (1997)
75. K. Kuhnke, K. Kern, R. David, G. Comsa, Rev. Sci. Instrum. **65**, 3458 (1994)
76. D.M. Smilgies, J.P. Toennies, Rev. Sci. Instrum. **59**, 2185 (1998)
77. F. Mezei, Int. J. Mod. Phys. B **7**, 2885 (1993)
78. F. Mezei, C. Pappas, T. Gutberlet (eds.), *Neutron Spin Echo Spectroscopy*, Lecture Notes in Physics (Springer, Berlin, 2002)
79. M. DeKieviet, D. Dubbers, C. Schmidt, D. Scholz, U. Spinola, Phys. Rev. Lett. **75**, 1919 (1995)
80. M. DeKieviet, D. Dubbers, M. Klein, C.H. Schmidt, M. Skrzipczyk, Surf. Sci. **377**, 1112 (1997)
81. P. Fouquet, A.P. Jardine, S. Dworski, G. Alexandrowicz, W. Allison, J. Ellis, Rev. Sci. Instrum. **76**, 053109 (2005)
82. S. Dvorski, G. Alexandrowicz, P. Fouquet, A.P. Jardine, W. Allison, J. Ellis, Rev. Sci. Instrum. **75**, 1963 (2004)
83. A.P. Jardine, H. Hedgeland, G. Alexandrowicz, W. Allison, J. Ellis, Prog. Surf. Sci. **84**, 232–379 (2009)
84. P.R. Kole, A.P. Jardine, H. Hedgeland, G. Alexandrowicz, J. Phys.: Cond. Matter **22**, 304018 (2010)
85. G. Alexandrowicz, A.P. Jardine, J. Phys.: Condens. Matter **19**, 305001 (2007)
86. A.P. Jardine, S. Dworski, P. Fouquet, G. Alexandrowicz, D.J. Riley, G.Y.H. Lee, J. Ellis, W. Allison, Science **304**, 1790 (2004)

87. D.J. Riley, A.P. Jardine, S. Dworsky, G. Alexandrowicz, J. Ellis, W. Allison, J. Chem. Phys. **126**, 104702 (2007)
88. D.J. Riley, A.P. Jardine, G. Alexandrowicz, H. Hedgeland, J. Ellis, W. Allison, J. Chem. Phys. **128**, 154712 (2008)
89. L.D. Thomas, W.P. Kraemer, G.H.F. Diercksen, Chem. Phys. **51**, 131 (1980)
90. J. Ellis, K. Hermann, F. Hoffmann, J.P. Toennies, Phys. Rev. Lett. **75**, 886 (1975)
91. A. Tamtögl, E. Bahn, J. Zhu, P. Fouquet, J. Ellis, W. Allison, J. Phys. Chem. **119**, 25983 (2015)
92. A. Tamtögl, D. Campi, M. Bremholm, E.M.J. Hedegaard, B.B. Iversen, M. Bianchi, P. Hofmann, J. Ellis, W. Allison, To be published (2017)
93. E.M. McIntosh, P.R. Kole, M. El-Batanouny, D.M. Chisnall, J. Ellis, W. Allison, Phys. Rev. Lett. **110**, 086103 (2013)
94. A.P. Graham, W. Silvestri, J.P. Toennies, *Surface Diffusion: Atomistic and Collective Processes* (Plenum Press, New York, 1997), pp. 565–580
95. M. DeKieviet, D. Dubbers, M. Klein, U. Pieles, C. Schmidt, Rev. Sci. Instrum. **71**, 2015 (2000)
96. A.P. Jardine, P. Fouquet, J. Ellis, W. Allison, Rev. Sci. Instrum. **72**, 3834 (2001)
97. T. Kravchuk, M. Reznikov, P. Tichonov, N. Avidor, Y. Meir, A. Bekkerman, G. Alexandrowicz, Science **331**, 319 (1911)
98. P.-A. Turgeon, P. Ayotte, E. Lisitsin, Y. Meir, T. Kravchuk, G. Alexandrowicz, Phys. Rev. A **86**, 062710 (2012)
99. O. Godsi, G. Corem, Y. Alkoby, J.T, Cantin, R. Krems, M. Somers, J. Meyer, G.-J Kroes, T. Maniv, G. Alexandrowicz, Nature Commun. **8**, 15357 (2017). https://doi.org/10.1038/ncomms
100. H. Ibach, J. Vac. Sci. Technol., A **5**, 419 (1987)
101. H. Ibach, *Electron Energy Loss Spectrometers*, vol. 63, Springer Series in Optical Sciences (Springer, Berlin, 1991)
102. H. Ibach, J. Electr. Spectr. Related Phenom. **64–65**, 819 (1993)
103. H. Ibach, M. Balden, S. Lehwald, J. Chem. Soc. Faraday Trans. **92**, 4771 (1996)
104. H. Ibach, Phys. Rev. Lett. **24**, 1416 (1970)
105. H. Ibach, J. Vac. Sci. Technol. **9**, 713 (1971)
106. H. Ibach, Phys. Rev. Lett. **27**, 253 (1971)
107. S. Lehwald, J.W. Szeftel, H. Ibach, T.S. Rahman, D.L. Mills, Phys. Rev. Lett. **50**, 518 (1983)
108. M. Balden, S. Lehwald, E. Preuss, H. Ibach, Surf. Sci. **307–309**, 1141 (1994)
109. SPECS GmbH, Berlin, Germany. http://www.specs.de
110. H. Ibach, J. Electron Spectrosc. Relat. Phenom. **64/65**, 819 (1993)
111. L. Vattuone, M. Rocca, L. Savio, in *Surface Science Techniques*, Springer Series in Surface Sciences, vol. 51, ed. by G. Bracco, B. Holst (Springer, Berlin, Heidelberg, 2013), Chap. 17, p. 499
112. R. Berndt, J.P. Toennies, Ch. Wöll, J. Electron Spectrosc. **44**, 183 (1987)
113. M. Rocca, H. Ibach, S. Lehwald, T.S. Rahman, in *Structure and Dynamics of Surfaces I*, ed, by W. Schommers, P. von Blanckenhagen (Springer, Berlin, 1987), p. 245
114. B.M. Hall, D.L. Mills, Phys. Rev. B **34**, 8318 (1986)
115. V. Bortolani, A. Franchini, G. Santoro, J.P. Toennies, Ch. Wöll, G. Zhang, Phys. Rev. B **40**, 3524 (1989)
116. D. Eichenauer, U. Harten, J.P. Toennies, V. Celli, J. Chem. Phys. **86**, 3693 (1987)
117. D.M. Smilgjes, J.P. Toennies, Rev. Sci. Instrum. **59**, 185 (1988)
118. R.B. Doak, in *Atomic and Molecular Beam Methods*, vol. 2, ed. by G. Scoles (Oxford University Press, New York, 1992), pp. 384–443
119. U. Harten, J.P. Toennies, Ch. Wöll, Faraday Disc. Chem. Soc. **80**, 137 (1985)
120. R. Vollmer, Dissertation, University of Göttingen, 1991, Max-Planck-Institut für Strömungsforschung, Göttingen, Bericht, 4/1992
121. G. Brusdeylins, R.B. Doak, J.P. Toennies, Phys. Rev. Lett. **46**, 437 (1981)
122. E.A. Akhadov, T.W. Trelenberg, S.A. Safron, J.G. Skofronik, D.H. Van Winkle, F.A. Flaherty, W. Theis, Phys. Rev. B **67**, 113406 (2003)
123. G. Shirane, S.M. Shapiro, J.M. Tranquada, Neutran scattering with a triple-axis spectrometer: basic techniques (Cambridge University Press 2004)
124. Allison, R. F. Willis, M. Cardilo, Phys. Rev. B **23**, 6824 (1981)

Chapter 10
Resonances and Critical Kinematic Effects

Atoms moving along the surface with the right energy and in the right direction may be diffracted so as to leave the surface with positive energy and thus be evaporated. This is a new mechanism of evaporation which has not been previously expected.

J. E. Lennard-Jones and A. F. Devonshire, Nature 1936

Abstract The resonant interaction of the He atom with the bound states of the atom-surface potential can have a great influence on the elastic and inelastic intensities. Examples from experiment are used to illustrate how bound state resonances can be used to detect weakly coupled surface phonons by means of resonance enhancement. In addition there are a variety of focusing and surfing effects, arising from a coupling between the motion of the atomic projectile and that of the surface phonons. Several different focusing phenomena called *critical kinematic effects* are characterized by various tangency conditions among scan, resonance and phonon dispersion curves. Kinematical focusing, in particular, allows to determine envelopes of the surface phonon dispersion curves without resorting to time-of-flight measurements. Focused resonances provide a possible mechanism for an atom beam monochromator, while the surfing effect, allowing trapped atoms to ride a Rayleigh wave and leading to very sharp resonances, corresponds to a type of atomic polaron.

10.1 Bound State Resonances

In one of the first He atom diffraction experiments Otto Robert Frisch and Otto Stern in 1933 observed a number of anomalous dips in the broad diffraction peaks of He scattered from LiF (001) as shown in Fig. 10.1 [2]. They also found that by rotating the crystal the specular peak intensity showed a number of unexpected sharp minima and maxima. At the time they were not able to provide an explanation, but their careful documentation of the anomalies enabled Lennard-Jones and

G. Benedek and J. P. Toennies, *Atomic Scale Dynamics at Surfaces*, Springer Series in Surface Sciences 63, https://doi.org/10.1007/978-3-662-56443-1_10

Devonshire in 1936 [1] to conclude that the anomalies resulted from the resonant trapping by diffraction of the atoms into the bound states of the atom-surface potential where they were incoherently scattered and therefore lost from the elastic scattered intensity. Because the anomalies only occur at special angles they called the new phenomenon *selective adsorption* (SA). Since their discovery in 1933 [2] and the explanation by Lennard-Jones and Devonshire [1] selective adsorption resonances have been extensively studied since they provide the most sensitive and most direct probe of the bound states of the atom-surface potential. Since SA involves diffraction, a sufficient corrugation of the surface is necessary. Results are presently available for a wide variety of corrugated insulator surfaces [3] and also on metal surfaces with sufficient corrugation as in the case of higher index stepped surfaces [4]. Selective adsorption resonances have not been observed on the smooth low index metal surfaces because of their very small corrugation. Since 1981 several new types of resonances involving resonant *inelastic* processes, in which the bound states play an important role, have been found. After a discussion of the elastic selective adsorption resonances the different types of inelastic resonances will be discussed below.

The simplest effect of the attractive potential is to accelerate the atoms towards the repulsive barrier and then, after they have been reflected, to decelerate the atoms as they leave the surface. This effect can be simply incorporated in the previous treatment based on the HCS model as suggested by Beeby [5] and discussed by Celli and Evans in the framework of the DWBA [6]. According to the theoretical

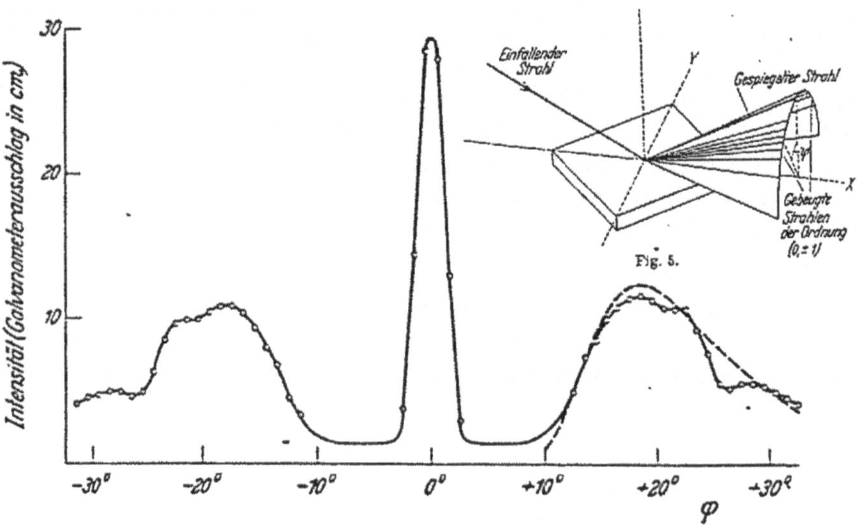

Fig. 10.1 The first experimental evidence for selective adsorption comes from the angular distribution reported by Frisch and Stern in 1933 for scattering of He atoms from LiF(001) [2]. The insert shows how the angle φ was measured. The indentations at 20° and 26° in both of the broad first order diffraction peaks were later explained by Devonshire and Lennard-Jones in 1936 as due to selective adsorption [1]. The broadening of the diffraction peaks is attributed to the Maxwellian velocity distribution of the incident beam. Taken directly from [2]

analysis of Beeby the effect of the attraction can be approximately accounted for by adding the well depth energy to the incident kinetic energy. This simple correction, however, neglects the bound vibrational states of the atom at the surface (Fig. 10.1) which, as mentioned above, under special conditions can affect the scattering.

10.1.1 Atom-Surface Potentials and Bound States

The selective adsorption peaks provide even today the best direct probe of the bound state energies of atoms and molecules on the surface. The comparison with the experimental energies ε_n derived from bound state resonances can be used to establish the most reliable model for the atom-surface potential. For some of the potential models discussed in Chap. 6 the bound state energies are given by analytical expressions [3, 7]. The simplest of the various model potentials is the two-parameter *9-3 Lennard-Jones* potential [8, 9] which derives from a pairwise summation of the Lennard-Jones atom-atom potential in the continuum limit of the solid. It is given by

$$V_{LJ}(z) = \frac{D}{2}\left[\left(\frac{z_m}{z}\right)^9 - 3\left(\frac{z_m}{z}\right)^3\right], \tag{10.1}$$

where D is the atom-surface potential well depth and z_m the distance from the surface of the potential well (see Sect. 6.3). This potential enjoys great favor because its long range z^{-3} behavior agrees with theory. Moreover its bound state energies can be calculated analytically within the Wentzel-Kramers-Brillouin (WKB) approximation. They are given by [8]

$$\varepsilon_n \simeq -D\left[1 - \frac{n+1/2}{L}\right]^6, \quad L = \frac{2.55\, z_m}{\pi\hbar}(2mD)^{1/2}, \tag{10.2}$$

where m is the mass of the adsorbed atom. In general attractive potentials decreasing faster than $-1/z$ admit only a finite number of bound state levels. In the present case the maximum of n is the integer part of L − ½ (Fig. 10.2).

Another popular model is the 3-parameter Morse potential which combines two exponentials [3]:

$$V_M(z) = D\left[e^{-2\kappa(z-z_m)} - 2e^{-\kappa(z-z_m)}\right], \tag{10.3}$$

where κ is a reciprocal range parameter. As compared to the $\mathbf{G} = \mathbf{0}$ component of the two-body Morse potential given by (6.42) the potential (10.3) is an approximation valid for $\beta z \ll 1$, where β (= 2 κ) is the range parameter of (6.42). In any case the Morse potential has the disadvantage that it does not have the proper z^{-3} long range dependence, but incorporates an exponential repulsion, which is a better representation of the short

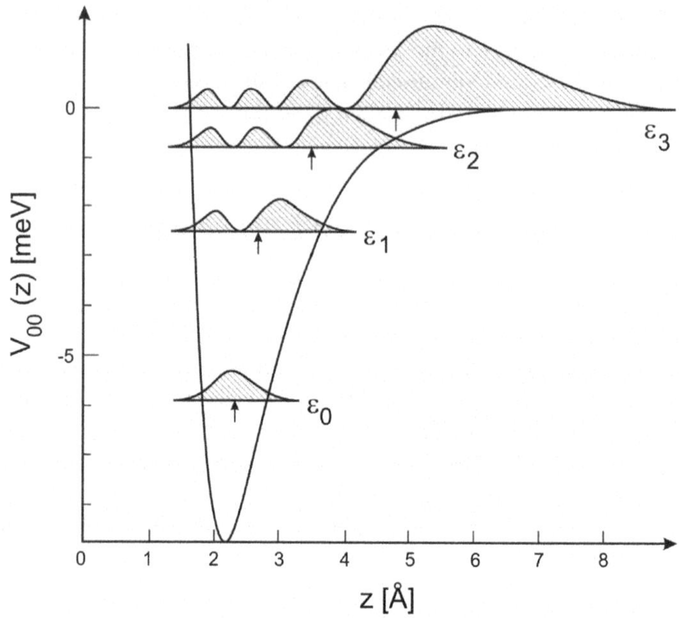

Fig. 10.2 Schematic diagram showing the **G** = 0 (surface averaged) He-LiF (001) potential $V_{00}(z)$ as a function of z, the distance from the first surface layer of atoms. The four known bound vibrational levels ε_n ($n = 0 - 3$), and the estimated probability distributions (not normalized) are shown. The arrows indicate the estimated expectation values of z for each of the bound states. With increasing vibrational level the average distance of the bound atoms from the surface increases (adapted from [10])

range interatomic Pauli repulsion (see Sect. 6.4.2). The WKB bound state energy levels of the Morse potential (10.3) are given by

$$\varepsilon_n = -\left[\sqrt{D} - \left(\hbar\kappa/\sqrt{2m}\right)(n + 1/2)\right]^2 \tag{10.4}$$

with $n = 0, 1, 2 \ldots \leq \left[\left(\sqrt{2mD}/\kappa\hbar\right) - 1/2\right]$ [8].

Several other potentials, which provide a better description of the experimental bound states have been introduced in the past. One of these is the 4-parameter Variable Exponent Potential [11]:

$$V_{VEP} = D\left[\left(1 + \frac{\kappa(z - z_e)}{p}\right)^{-2p} - 2\left(1 + \frac{\kappa(z - z_e)}{p}\right)^{-p}\right]. \tag{10.5}$$

This potential has the disadvantage that its energy levels can only be approximated for which several expressions have been reported [12].

The 5-parameter Shifted Morse Potential given by

$$V_{SM}(z) = (D'/D)V_M(z) - D'\Delta, \tag{10.6}$$

has more flexibility. Its bound state energy levels are correspondingly more complicated:

$$\varepsilon_n = -D'(1+\Delta) + \hbar\kappa[2D'/m]^{1/2}(n+1/2) - (\hbar^2\kappa^2/2m)(n+1/2)^2. \tag{10.7}$$

The well depth in this potential is $D = D'(1+\Delta)$, where the parameter Δ is fitted to the bound states [13]. Several other models have been introduced and have been surveyed by Hoinkes [3] and more recently by Kraus et al. [7].

10.1.2 Elastic Bound State Selective Adsorption (SA) Resonances

In the process of selective adsorption the incident atom undergoes diffraction into one of the bound states of the potential. Since the particle can only enter the bound state by diffraction the process involves one of the **G**-components of the surface potential. Moreover since the bound state levels have discrete energies ε_n, the incident energies E_i and surface wavevector $\mathbf{K}_i$ (or incident angle) must satisfy the following resonance condition.

$$E_i = E_n(\mathbf{K}_i) \cong -|\varepsilon_n| + \frac{\hbar^2(\mathbf{K}_i+\mathbf{G}')^2}{2m}. \tag{10.8}$$

Thus by conservation of energy the free particle kinetic energy for directed motion along the surface $\hbar^2(\mathbf{K}_i+\mathbf{G}')^2/2m$ is greater than the total energy $E_n(\mathbf{K}_i)$ of the incident beam by the energy $\varepsilon_n = -|\,\varepsilon_n\,|$ gained in entering the bound state.

Equation (10.8) assumes that the kinetic energy $\hbar^2(\mathbf{K}_i+\mathbf{G}')^2/2m$ in the bound state is sufficiently large that the effects of the periodic corrugation become negligible and that the Bloch wavefunction reduces to a free particle state. In this approximation the number and the energies of the bound state levels ε_n only depend on the $\mathbf{G} = 0$ component of the periodic surface potential. Equation (10.8) can also be obtained via the factorization of (7.25) into the wavefunction for the component of the motion normal to the surface, and the Bloch wavefunction for the parallel component of the motion along the surface. By virtue of Bloch's theorem all $E_n(\mathbf{K}_i+\mathbf{G}')$ for any surface reciprocal lattice vector $\mathbf{G}'$ involved in the diffraction into the bound state are also energy eigenvalues. Any redundancy is eliminated by restricting $\mathbf{K_i}$ within the first Brillouin zone (BZ). In general the free particle band structure consists of an infinite set of intersecting parabolas $E_n(\mathbf{K}_i+\mathbf{G}')$ as given by (10.8).

Figure 10.3 shows the classical atom trajectory in a selective adsorption process. An atom in the incident state $\mathbf{k}_i = (\mathbf{K}_i, k_{iz})$ with energy E_i impinging on a corrugated surface is diffracted *elastically* into a bound state by picking up a surface lattice vector $\mathbf{G}'$. The parallel kinetic energy is increased from initially $\hbar^2\mathbf{K}_i^2/2m$ to $\hbar^2(\mathbf{K}_i+\mathbf{G}')^2/2m$ at the expense of the perpendicular energy, which is decreased by the same amount from $\hbar^2k_{iz}^2/2m$ to the bound state energy $-|\varepsilon_n|$. The trapped atom moves along the surface while at the same time oscillating in the bound state in the z-direction. The same corrugation which lead to the trapping also enables the atom to leave the surface elastically by giving back either the same $\mathbf{G}'$-vector and joining the elastically reflected beam, or by another $\mathbf{G}''$-vector and joining the $\mathbf{G}''-\mathbf{G}'$ diffracted beam. The final parallel wavevector of the ejected atom is then given by

$$\mathbf{K}_f = \mathbf{K}_i + \mathbf{G}' - \mathbf{G}''. \tag{10.9}$$

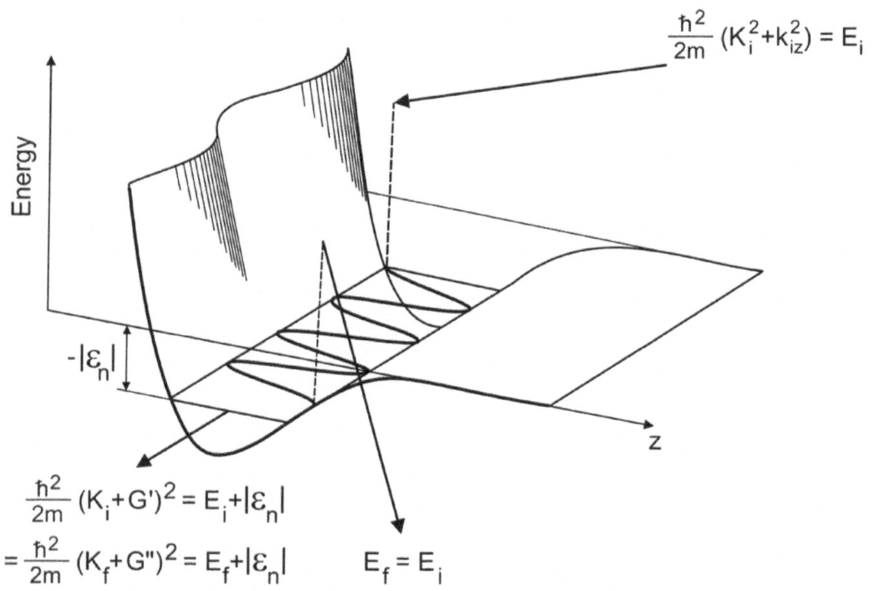

Fig. 10.3 Classical trajectories involved in the selective adsorption process in which an atom impinging on a corrugated surface with energy E_i is diffracted *elastically* into a bound state by picking up a surface lattice vector $\mathbf{G}'$. The parallel kinetic energy is increased from initially $\hbar^2\mathbf{K}_i^2/2m$ to $\hbar^2(\mathbf{K}_i+\mathbf{G}')^2/2m$ at the expense of the perpendicular energy, which is decreased by the same amount from $\hbar^2k_{iz}^2/2m$ to the bound state energy $-|\varepsilon_n|$. For a given incident energy E_i this process is possible only at resonance for special incident angles. The trapped atom moves along the surface while at the same time oscillating in the bound state in the z-direction. By the same corrugation which lead to the trapping the atom has a finite probability for leaving the surface elastically by giving back either the same $\mathbf{G}'$-vector and joining the elastically reflected beam, or by another $\mathbf{G}''$-vector and joining the $\mathbf{G}''-\mathbf{G}'$ diffracted beam. The quantum interference between the direct reflection (diffraction) and the reflection (diffraction) through a bound state yields resonant features in the intensity I_{00} of the specular beam

In this way the conservation of momentum is fulfilled simultaneously for both the incident and the final states. Conservation of energy for the elastic resonances ($E_i = E_n(\mathbf{K}) = E_f$) yields

$$k_i^2 = (k_i \sin\theta_i + G'_{\parallel})^2 + G'^2_{\perp} - \frac{2m}{\hbar^2}|\varepsilon_n| \tag{10.10a}$$

$$= (k_f \sin\theta_f + G''_{\parallel})^2 + G''^2_{\perp} - \frac{2m}{\hbar^2}|\varepsilon_n| = k_f^2, \tag{10.10b}$$

where the indices $\parallel$ and $\perp$ label, respectively, the parallel and perpendicular components of the vectors $\mathbf{G}'$ and $\mathbf{G}''$ relative to the vectors $\mathbf{K}_i$ and $\mathbf{K}_f$.

Figure 10.4 provides a convenient illustration of (10.8) in a diagram for in-plane scattering in (E, K) space, first proposed by Vittorio Celli [14]. The left side shows the bound states and the atom-surface potential. On the right side of the diagram Fig. 10.4 the difference between a given incident energy E_i and incident parallel wavevector K_i, and the parabolas corresponds to the perpendicular component of the incident energy. Accordingly the difference between the parabolas and the origin for a given bound state corresponds to the parallel energy. The shadowed area on the right side of Fig. 10.4 covers all possible translational states between the boundaries $K_i = 0$ (normal incidence) and the maximum $K_i = \sqrt{2mE_i}/\hbar$ (grazing incidence). Each point in the shaded area also corresponds to an incident angle $\theta_i = \arcsin[\hbar K_i/\sqrt{2mE_i}]$. According to (10.8) if the sum of K_i and a reciprocal lattice vector $\mathbf{G}'$ equal the wavevector $\mathbf{K}$ in a bound state of energy $E_n(\mathbf{K}) = E_i$, a resonant elastic selective adsorption can take place, and a feature will be observed in the angular distribution at the angle θ_i.

It should be noted that the same diagram Fig. 10.4 also describes the reverse elastic process in which an atom in a bound state of positive energy $E_n(\mathbf{K})$ is diffracted into a final translational state of energy E_f via the exchange of some reciprocal lattice vector $-\mathbf{G}^n$ (cf. Figs. 10.3 and 10.5b). When the corrugation of the periodic surface potential is switched on and no longer can be neglected, the intersections of the parabolas $E_n(\mathbf{K})$ with the G-vectors and the associated Brillouin zone boundaries transform into avoided crossings with the opening of gaps at the BZ boundaries and the G-vectors. The opening of the gaps for the bands for the n = 0, 1 bound states shown in Fig. 10.4 depends on the corrugation amplitude as has been predicted in several studies [3, 15] and observed experimentally [16, 17]. However, (10.8) is a useful approximation in the present context since, as mentioned above, all gaps induced by corrugation decrease with increasing bound state energy, since the atom spends most of the time far from the surface (e.g., >2.4 Å in Fig. 10.1). Moreover small deviations from the free particle parabolic bands can always be represented by parabolic free particle bands, (10.8), provided the atom mass m is replaced by an effective mass $m^*(K)$ [6, 18].

Figure 10.4 also illustrates the particular case of an inelastic resonance in LiF (001) reported in Fig. 10.6 with E_i = 19.4 meV and at a value of $\mathbf{K}_i$ = 3.78 Å^{-1} corresponding to $\theta_i = 38.9°$ (red lozenge) in the [100] direction. Under these conditions selective adsorption into the $n = 0$ state of energy $E_0(\mathbf{K}) = E_i$ is possible via the in-plane $\mathbf{G}'$ vector (1, 1). In the experiment shown in Fig. 10.6 the

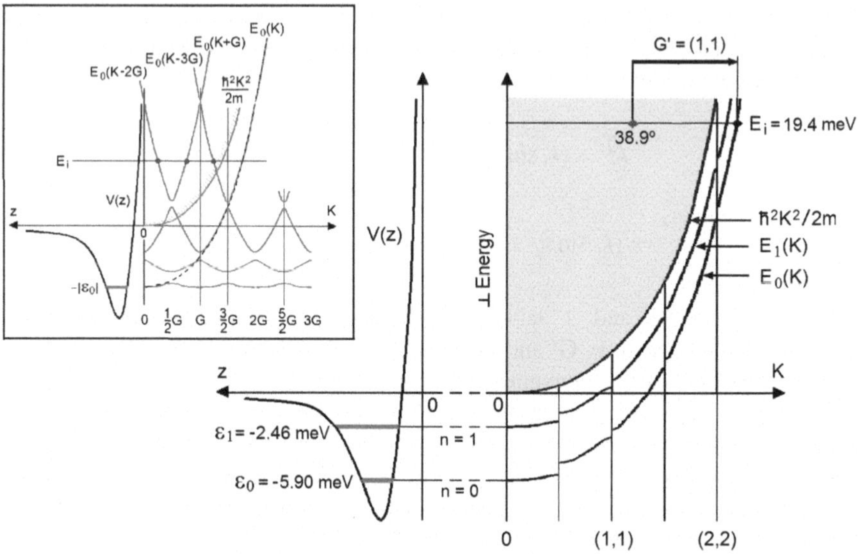

Fig. 10.4 Celli diagram for the energies and wavevectors involved in in-plane scattering from LiF (001) [14]. The left side shows the potential and the bound states. On the right side the energies $E_0(\mathbf{K})$ and $E_1(\mathbf{K})$ of a He atom in the lowest bound states $-|\varepsilon_0|$ and $-|\varepsilon_1|$ in the surface potential well $V(z)$ of LiF(001) are shown as functions of the parallel wavevector **K** along the [100] direction. The band structure (inset) of the atom in a bound state (here $n = 0$) shows avoided crossings at the band intersections, which accounts for the gaps appearing in the extended zone plot of $E_n(\mathbf{K})$. The difference between a given incident energy and the parabolas corresponds to the perpendicular component of the incident energy at the particular incident parallel wavevector $\mathbf{K}_i$. Accordingly the difference between the parabolas and the origin for a given bound state corresponds to the parallel energy. All possible incident He atom translational states fall within the shadowed area limited by the $\mathbf{K} = 0$ axis (normal incidence) and the parabola $\hbar^2\mathbf{K}^2/2m$ (grazing incidence). Scan curves (not shown) for a specific incident angle are parabola-like curves which can be displayed in the diagram. The bands, plotted in the extended zone representation, show gaps at the symmetry points of the Brillouin zone centers, corresponding to the G vectors (1, 1) and (2, 2), and the intermediate zone boundaries due to the periodic corrugation of the potential of the surface. The diagram illustrates the conditions for the *inelastic* resonances in Fig. 10.6 observed at an incident energy $E_i = 19.4$ meV. The red dot corresponds to an incident angle of 38.9° where the parallel wavevector $\mathbf{K}_i$ is exactly less than the wavevector **K** of the lowest bound state of energy $E_0(\mathbf{K})$ by the surface reciprocal lattice vector $\mathbf{G}' = (1, 1)$. A simple geometrical construction allows to represent also out-of-plane resonances

resonance is followed by an *inelastic* process so that the feature appears at an angle different from the diffraction peak and stands out against the small background resulting from direct inelastic processes. There are several other resonances, not shown in Fig. 10.4, but clearly seen in the angular distribution of Fig. 10.6 such as that at $\theta_i = 58.0°$ via the out-of-plane $\mathbf{G}'$ (1, 0) vector. Since in this case the atoms are inelastically scattered out of the scattering plane and lost in the forward direction the feature appears as a minimum. Figure 10.8 shows Celli diagrams for several resonant processes discussed next.

The SA resonances as discussed above can only occur if the surfaces are sufficiently corrugated. For this reason they have, up to the present, been mostly observed on insulator surfaces and on layered semimetals. The low index surfaces of metals are generally so smooth that the diffraction coupling to the bound states is too weak to lead to observable effects. However, similar phenomena have been seen in HAS studies of stepped metal surfaces [4, 19–22] or in inelastic scattering of molecules (e.g., H_2 or D_2) which are somewhat more sensitive to the small corrugation than He atoms on ordinary metal surfaces [23]. The level of accuracy depends on how well the band structure of the bound states can be reconstructed from the observed resonances. At low incident energies, especially for values of E_i around the gaps of $E_n(\mathbf{K})$, the avoided crossings of the energy bands can be well resolved from the maps of polar and azimuthal incident angles at which resonances occur [15, 24]. The experiments on elastic SA resonances and their theoretical interpretation in terms of potentials, carried out up to 1980, have been extensively reviewed by Hoinkes [3].

Simple diagrams for the elastic resonances are illustrated in the upper half of Fig. 10.5. Figure 10.5a shows a diagram for simple non-resonant elastic specular or diffractive scattering and Fig. 10.5b shows a diagram for the *elastic SA resonances* discussed above. These diagrams are especially useful for characterizing the many different *inelastic resonances* summarized in the second row in Fig. 10.5. These are discussed in more detail in the following sections.

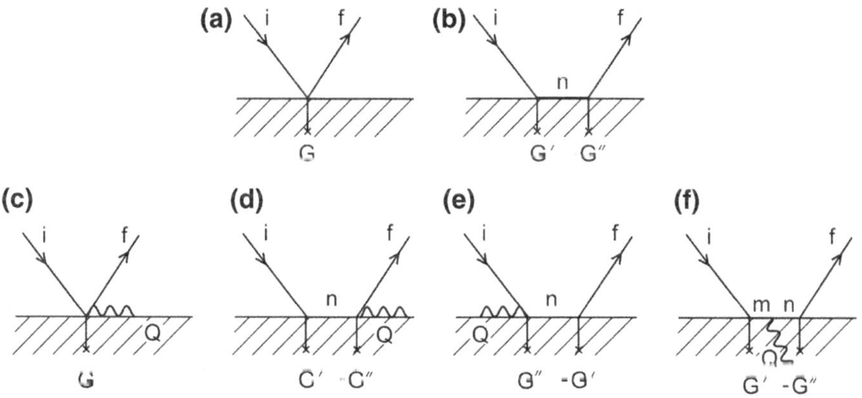

Fig. 10.5 Diagrams showing the different types of possible elastic and inelastic resonances. The elastic (**a**, **b**) and inelastic (**c**–**f**) atom-surface scattering processes can be either direct (**a**, **c**), or involve one bound state n, as in elastic (**b**) and inelastic (**d**, **e**) resonances, or two bound states m and n (**f**). The resonance in (**d**) involves initially selective adsorption followed by phonon-assisted desorption (SA-PAD) and that in (**e**) phonon-assisted adsorption followed by selective desorption (PAA-SD). In (**f**) after selective adsorption into a bound state the atom undergoes a phonon transition into another bound state from which it selectively desorbs. Resonant features in the final beam intensity are produced by the interference of a bound state scattering channel with the respective direct scattering channel for both elastic (**b**) or inelastic (**d**–**f**) processes

10.1.3 Inelastic Bound State Resonances

The first indirect evidence for inelastic resonances was reported in the pioneering experiments by Williams in 1971 [25]. Later the groups of Cantini and Tatarek and Hoinkes and Wilsch [26–28] also observed unexpected features in the angular distributions which could not be explained by elastic bound state resonances. With the advent of high resolution inelastic scattering the true nature of these resonances could be examined in detail. In 1981 Brusdeylins et al. reported high resolution time-of-flight measurements which confirmed directly the role of the inelastic process in explaining the sharp peaks observed in the angular distributions [29].

Figure 10.6a shows the angular distributions measured in an experiment on LiF (001) in the [100] direction with an incident wavevector of $k_i = 6.03$ Å^{-1} and a surface temperature of 300 K [30]. All the features between the diffraction peaks are due to inelastic processes. The downward arrows for $\theta_i < 45°$ mark sharp maxima assigned to in-plane (1,1) selective adsorption inelastic resonances. The upward arrows for $\theta_i > 45°$ mark the depletion of inelastic intensity following trapping into out-of-plane (10) resonances. Figure 10.6b displays the TOF spectrum measured at nearly the same energy at incident angles $\theta_i = 32.7°$ and 38.9° corresponding to the two in-plane $\mathbf{G}' = (1,1)$, $n = 0,1$ SA resonances, respectively [29]. The Celli diagram in Fig. 10.4 corresponds to the resonance at $\theta_i = 38.9°$. For comparison two other nearby out-of-resonance TOF spectra at $\theta_i = 30.7°$ and 34.7° are also shown. While the latter two spectra are simply due to a direct inelastic scattering from the surface, the former two spectra are from atoms which were coupled to a bound state. Their large intensities in Fig. 10.6a are from direct inelastic scattering, with an additional contribution from a SA resonant adsorption into a bound state followed by an inelastic desorption process.

The in-plane resonances giving large maxima in the angular distribution at 32.7° and 38.9° also produce a general amplification of their TOF spectra with respect to the non-resonant spectra, as clearly seen in Fig. 10.6b. Since in this process the resonating bound state and the incident state have the same energy and the same parallel momentum up to a reciprocal lattice vector $\mathbf{G}'$, the same phonon spectrum involved in the direct inelastic scattering is also effective in the inelastic desorption, bringing the atom into the same final states, provided the same vector $\mathbf{G}'$ is given back to the surface lattice on desorption. Thus no selectivity is expected in the desorption process, unless one single surface phonon, e.g., a RW, dominates in the phonon DOS along the scan curve.

For the resonance at 32.7° phonon annihilation processes are mostly involved whereas for the resonance at 38.9° phonon creation processes dominate the inelastic desorption. Since the total energy of the atom in the bound state is equal to the incident energy, the weights of annihilation and creation processes depend on where the scan curve cuts the phonon spectrum, and of course also on the surface temperature. The intensity of TOF peaks are proportional to the corresponding Bose-Einstein factor (7.12), thus increasing with the surface temperature, with a faster increase for the annihilation processes [30]. A similar temperature

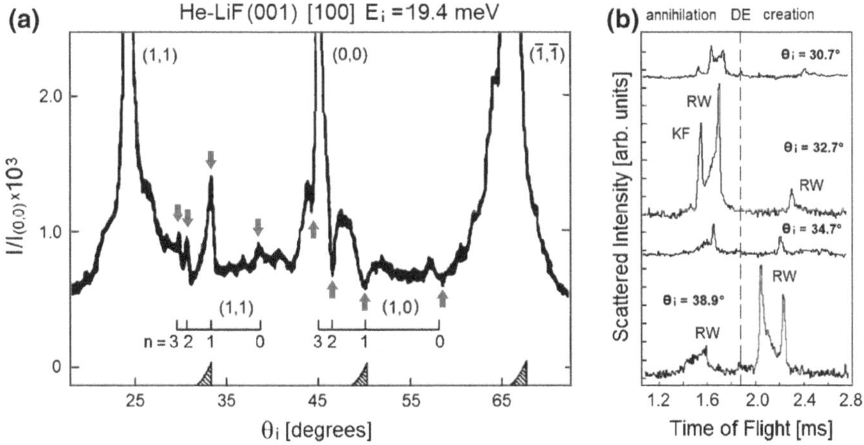

Fig. 10.6 Inelastic resonances in the angular and time-of-flight distributions of He atoms scattered from LiF(001) at E_i = 19.4 meV (k_i = 6.03 Å^{-1}), measured at the surface temperature of 293 K in a 90° constant total angle apparatus [30]. **a** The downward arrows for $\theta_i < 45°$ mark sharp maxima assigned to in-plane (1,1) selective adsorption phonon-assisted desorption resonances (SA-PAD) with the bound states n. The upward arrows for $\theta_i > 45°$ mark out-of-plane (1, 0) SA-PAD resonances which lead to a loss of intensity. Note that the intensities of the sharp features are about only a factor 10^{-4} of the specular peak and that not all the features are identified. **b** Comparison of time-of-flight (TOF) spectra measured at nearly the same incident beam energy at nearby non-resonant angles of $\theta_i = 30.7°$ and 34.7° with the TOF spectra at angles $\theta_i = 32.7°$ and 38.9° corresponding to the n = 1 and n = 0 SA-PAD resonances, respectively [29]. The larger intensity of the latter spectra illustrates the resonant amplification of the phonon spectrum. Note that at $\theta_i = 32.7°$ the RW peak at 1.7 ms is enhanced by the bound state resonance, whereas the RW peak at 1.5 ms is additionally enhanced by kinematical focussing (KF) (Sect. 10.2.1)

dependence is expected for the intensity of resonance peaks in angular distributions. Note, however, that the particularly large intensity of the **G**′ = (1, 1), n = 1 resonance, Fig. 10.6a, is due to a concomitant *kinematical focussing* (see Sect. 10.2.1) at the same incident angle.

For out-of-plane resonances like the **G**′ = (1, 0) series in Fig. 10.6a the resonance feature is normally a minimum because the motion of the atom in the bound state is oblique with respect to the incident direction. Also the phonon involved in the desorption process is likely an out-of-plane phonon. In this case the atoms are resonantly trapped into final directions other than that of direct scattering, and a minimum is observed.

In the experiment shown in Fig. 10.6a the energy and angular resolution were much greater than in earlier studies and consequently the widths Γ of the peaks could be used to establish the lifetimes $\tau_n = \hbar/\Gamma$ in the different resonant bound states (Table 10.1). As expected they are largest for the high lying states which spend most of their time far from the surface and least for the strongly bound states. From the beam velocities the distance travelled along the surface before a surface phonon was annihilated are also listed.

Table 10.1 Lifetimes and distances travelled by the He atom along the surface calculated from the widths of the SA-PAD resonances shown in Fig. 10.5 along the [100] direction on LiF [29]

Bound state n	Lifetime τ (10^{-12} s)	Distance travelled L (Å)
0	7	80
1	11	120
2	17	170
3	60	560

The process discussed above, where the atom is *elastically* and selectively diffracted into a bound state and then exits after exchanging energy *inelastically* to leave the surface, will be referred to as *selective adsorption and phonon-assisted desorption* (SA-PAD). It is illustrated schematically in Fig. 10.5c.

In later HAS experiments on LiF(001) the other peaks in Fig. 10.6a not identified as initial state SA resonances were attributed to *phonon-assisted adsorption* followed by *selective desorption (PAA-SD)* [10]. In this case the atoms are trapped into a bound state by creating or annihilating a phonon. These peaks can be identified experimentally by the fact that they appear always at the same final angles θ_f in angular distribution measurements in which the total angle Θ_{SD} was varied between 50° and 115° [10]. The corresponding time-of-flight spectra indicate that the features are associated with the creation of surface phonons.

For inelastic processes involving a single surface phonon of energy $\hbar\omega_{\mathbf{Q}\nu}$ and parallel wavevector $\mathbf{Q}$, the conservation of parallel momentum for both resonances is the same, and given by:

$$\mathbf{K}_f = \mathbf{K}_i + \mathbf{G}' - \mathbf{G}'' - \mathbf{Q}. \tag{10.11}$$

The conservation of energy can be written for *initial-state resonances* (leading to SA-PAD) as

$$\begin{aligned} k_i^2 &= (K_i + G'_{\|})^2 + G'^2_{\perp} - \frac{2m}{\hbar^2}|\varepsilon_n| \\ &= (K_f + G''_{\|} + Q_{\|})^2 + (G''_{\perp} + Q_{\perp})^2 - \frac{2m}{\hbar^2}|\varepsilon_n| = k_f^2 + \frac{2m}{\hbar}\omega_{\mathbf{Q}\nu}, \end{aligned} \tag{10.12a}$$

or for *final state resonances* (leading to PAA-SD) as

$$\begin{aligned} k_i^2 - \frac{2m}{\hbar}\omega_{\mathbf{Q}\nu} &= (K_i + G'_{\|} - Q_{\|})^2 + (G'_{\perp} - Q_{\perp})^2 - \frac{2m}{\hbar^2}|\varepsilon_n| \\ &= (K_f + G''_{\|})^2 + G''^2_{\perp} - \frac{2m}{\hbar^2}|\varepsilon_n| = k_f^2, \end{aligned} \tag{10.12b}$$

where $Q_{\|}$ and $Q_{\perp}$ are the phonon wavevector components parallel and normal to the direction of $\mathbf{K}_i$ in the surface plane, respectively. The first line of (10.12b) defines the n-th bound state scan curve. If the final state of the inelastic process is a bound state, the atom in this state is moving horizontally along the surface. Thus the final angle is 90° and sin 90° = 1. The kinetic energy of the atom is increased by the absolute value of the bound state energy plus/minus the phonon energy, while the wavevector transfer is that of the phonon plus the G vector needed to get into the bound state. Thus the "resonance equation" is nothing else than the scan curve for the bound state as a final state.

In phonon-assisted adsorption the initial state is given and for each final bound state there is a scan curve (*bound state scan curve*) which links the energy to the parallel wavevector of all possible phonons involved in the phonon-assisted adsorption. The intersections of the bound state scan curves with the ordinary scan curve for the direct inelastic scattering into the final direction give the specific phonons which have helped the atom to enter the right bound state so as to be elastically diffracted into the same final state as for direct inelastic scattering. In this case a *resonance enhancement* in the TOF spectrum only occurs for that specific phonon. When changing the incident angle, the intersection eventually moves across the continuum of the surface-projected phonon DOS, and no specific feature would be observed in the angular distribution. However at the angles where the intersections meet the dispersion curve of a surface localized phonon, e.g., the RW, the PAA-SD process is strongly amplified yielding sharp SD features in the angular distributions, as reported in [10, 30]. As discussed in Sect. 10.1.4, resonance enhancement is particularly pronounced when surface localized phonons are involved, and has been exploited for measuring with HAS the dispersion curves of high energy surface optical phonons in LiF(001) (Fig. 11.3) [31].

The SA-PAD and the PAA-SD resonances are schematically illustrated through Celli diagrams in Fig. 10.5d, e, respectively. From the diagrams it appears that the PAA-SD resonance is the time reverse of SA-PAD. Strictly speaking, however, phonon creation is not the time-reverse of phonon annihilation, since their probabilities are different due to the BE factor. In the formulation of the thermodynamic response functions there a complex time $\tilde{t} = t + i\hbar/kT$ so that the displacement time dependence $e^{i\omega\tilde{t}} = e^{i\omega t}e^{-\hbar\omega/kT}$ has its own Boltzmann factor. The complex time reversal $\tilde{t} \to -\tilde{t}$ is equivalent to changing $\hbar\omega$ into $-\hbar\omega$, and this converts $n(\hbar\omega)$ for annihilation into $n(\hbar\omega) + 1$ for creation and viceversa. Thus time reversal is re-established.

In 1982 an especially intense resonance feature was discovered in He scattering from LiF(001) [32]. In these experiments the inelastic intensities were measured over a wide range of both incident and final scattering angles for different total scattering angles Θ_{SD}. Figure 10.7 shows the resonance peak which entirely by chance was first found at the usual fixed angle $\Theta_{SD} = 90°$. The large peak, which the authors dubbed *supernova* because of the unusually large intensity, appeared only at special incident and final scattering angles of $\theta_i^* = 40°$ and $\theta_f^* = 50°$. This implies that (10.12a) and (10.12b) are fulfilled simultaneously resulting in a *double resonance*. Equation (10.12a) indicates diffraction into a bound state $-|\varepsilon_m|$, while (10.12b) corresponds to diffraction out of another bound state $-|\varepsilon_n|$ (Fig. 10.5f). Thus the atom must have interacted with a phonon in order to jump from one bound state to the other. In the particular experiment the atom was diffracted into the $n = 2$

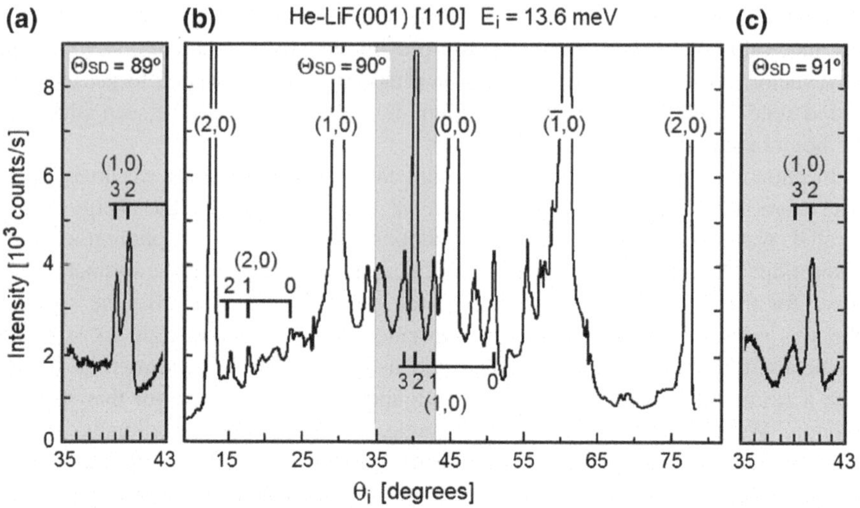

Fig. 10.7 HAS angular distribution measured as a function of incident angle θ_i for three values of Θ_{SD} from LiF(001) at E_i = 17.6 meV along the [110] direction illustrating the *double resonance supernova* peak at $\theta_i = 40°$ [32]. **a** For $\Theta_{SD} = 89°$ the angular distribution in the range $\theta_i = 35°$ to $\theta_i = 43°$ shows a small peak at 40^0. **b** Overview of the entire angular distribution for $\Theta_{SD} = 90°$ with the diffraction peaks and the assignment of selective adsorption phonon-assisted desorption (SA-PAD) resonances. The resonance at $\theta_i = 40°$ corresponding to the n = 2 bound state has an increased intensity. **c** For $\Theta_{SD} = 91°$ the intensity at $\theta_i = 40°$ drops off again. The strong dependence on Θ_{SD} indicates that both the incident and final angles have special selected values of $\theta_i = 40°$ and $\theta_f = 50°$. The final angle is also a FIR angle (see Sect. 10.2.1)

state and created a phonon enabling it to drop down to the n = 0 bound state from which it departed by diffraction. In a double resonance processes the phonon frequency is equal to

$$\hbar\omega_{\mathbf{Q}\nu} = |\varepsilon_n| - |\varepsilon_m| + \frac{\hbar^2}{2m}[(\mathbf{K}_i + \mathbf{G}')^2 - (\mathbf{K}_f + \mathbf{G}'')^2], \tag{10.13}$$

where $\mathbf{Q}$ is implicitly included via (10.11). The transition between the two resonances is allowed by the exchange of a phonon, and the coupling of a trapped atom to a surface phonon is strongly enhanced by the increased interaction time. A phenomenon similar to the *supernova,* was already identified in 1981 by Cantini and Tatarek in He atom scattering from graphite (0001) through a very careful analysis of the angular distributions [28].

As discussed in the next section such a double resonance involves two intermediate steps and is therefore a process of higher order in the transition matrix expansion (7.20), third term on the right. It is expected to yield an intense feature in the angular distribution as it may be viewed as a final state resonance which is reinforced by an initial state resonance.

Examples of Celli's diagrams for the bound state processes described above and in Figs. 10.4, 10.5 and 10.7 are shown in Fig. 10.8. First it is noted that the specular peak corresponds to $i = f$, which is just a single point in the Celli diagram (not shown in Fig. 10.8). Diagram (a) illustrates an elastic diffraction involving the lattice reciprocal vector **G**. Diagram (b) shows a SA process in which by resonant diffraction with **G**′ a bound states is accessed, which subsequently is left via diffraction with reciprocal lattice vector **G** to a final state f. Diagram (c) illustrates two possible phonon transfer processes, $i \rightarrow f$, corresponds to phonon creation, while $i \rightarrow f_u$ describes *umklapp* phonon creation. Diagrams (d) and (e) correspond, respectively, to an initial-state resonance involving a **G**-vector followed by a phonon-assisted desorption (SA-PAD), and the reverse process, phonon-assisted adsorption followed by a final state resonance (PAA-SD). Diagram (f) represents a double inelastic resonance involving two bound states s and s'. It is called a *supernova* because of the large resonance enhancement [32].

10.1.4 Theory of Inelastic Intensities at Resonances

In order to account for the effects of inelastic resonances within the DWBA theory, intermediate bound states have to be included in the sum appearing on the right hand side of (7.20). In principle the sum over the state index s includes both bound ($s = m, n, \ldots$) and free atom ($s = j$) intermediate states. An approximate solution of (7.20) providing T_{fi} beyond the Born approximation, so as to account for resonant processes involving one or two bound states, can be obtained with a procedure described by Celli [6] in the form

$$\begin{aligned} T_{fi} \cong V_{fi} &+ \sum_{m} V_{fm}(E_i^{tot} - E_m^{tot} - t_{mm})^{-1} V_{mi} \\ &+ \sum_{m,n} V_{fm}(E_i^{tot} - E_m^{tot} - t_{mm})^{-1} V_{mn}(E_i^{tot} - E_n^{tot} - t_{nn})^{-1} V_{ni}, \end{aligned} \tag{10.14}$$

where m and n run over the bound states, and

$$t_{nn} = \sum_{j \neq n} (E_i^{tot} - E_j^{tot} - i\delta)^{-1} |V_{nj}|^2 \tag{10.15}$$

is the self-energy of the n-th bound state, with j running over the free atom states. The real part of t_{nn} determines a shift of the resonance position due to the coupling of the bound state with the free atom states, while the imaginary part is related to the resonance line width

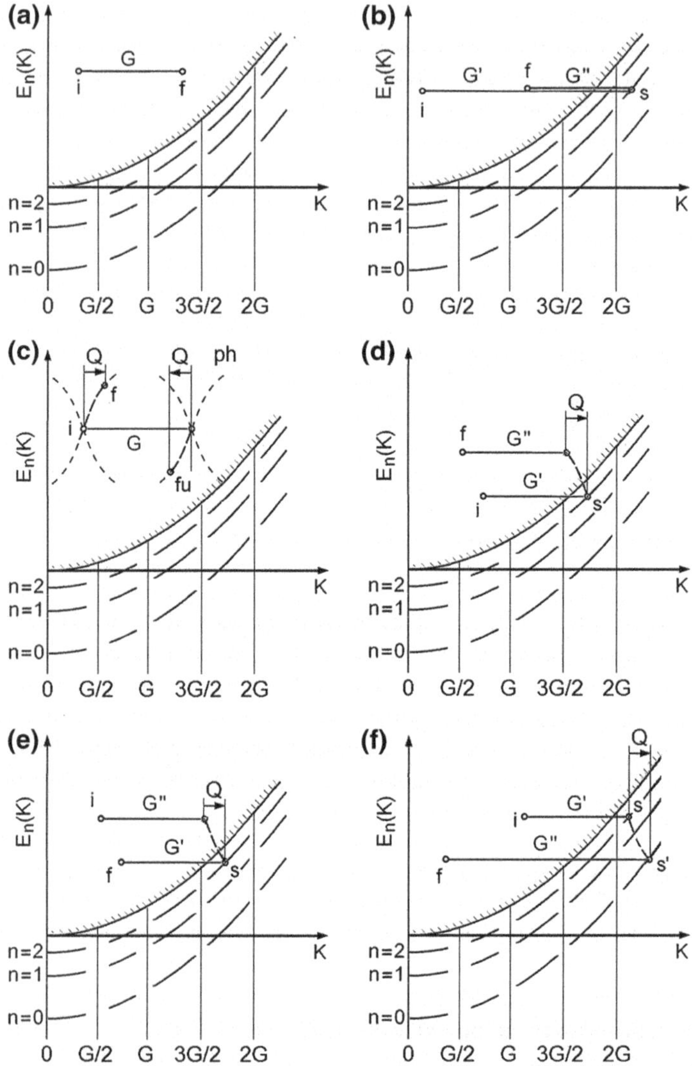

Fig. 10.8 Celli diagrams showing various elastic and inelastic scattering processes. The solid line curves show the bound state energies as a function of the parallel momentum, described in Fig. 10.4, for the lowest three bound states (n = 0, 1, 2). The diagrams **a–f** describe the kinematics of scattering processes from an initial (i) to a final (f) state corresponding to the processes **a–f** depicted in Fig. 10.3, respectively. Diagrams **a** and **b** represent a direct elastic diffraction and an elastic diffraction into a bound state at the point s, which corresponds to *selective adsorption*. Diagram **c** describes a direct ($i \rightarrow f$) and an umklapp ($i \rightarrow f_u$) inelastic process, with the broken lines representing the dispersion curves of Rayleigh phonons. Diagram **d** shows an SA-PAD resonance involving a **G**-vector. Diagram **e** is the same as **d** for a final state inelastic (PAA-SD) resonance. Diagram **f** represents a double inelastic resonance involving two bound states s and s', which is called a *supernova* [32]

$$\Gamma = -2\mathrm{Im}(t_{nn}) = 2\pi \sum_{j \neq n} \delta(E_i^{tot} - E_j^{tot}) \left| V_{nj} \right|^2 \tag{10.16}$$

as well as to the bound state lifetime $\tau_n = \hbar/\Gamma$.

The inclusion of bound states in the calculation of the differential reflection coefficient, (7.18), may yield important effects such as the resonant enhancement (or suppression) of certain inelastic processes occurring under the kinematic conservation conditions given by (10.11–10.13). The time dependent correlation function $< T_{fi}(0)T_{fi}^{+}(t) >$ in (7.18) now contains, besides the direct scattering term of order $O(V^2)$, two terms of order $O(V^3)$, corresponding to initial- and final state resonances with one bound state, and three $O(V^4)$ terms involving two bound states. Two of these contain V_{mn} and yield therefore a double resonance as e.g. a supernova with either $m \rightarrow n$ or $n \rightarrow m$, while the third $O(V^4)$ term is a mere sequence of two ordinary resonances. Since $E_i^{tot}(= E_f^{tot})$, E_j^{tot} and $E_{m,n}^{tot}$ denote the *total* energy of the particle-crystal system for the atom in either a free or a bound state, the energy differences appearing in (10.14–10.15) may include all possible inelastic processes (creation or annihilation of phonons). Thus the resonances affect the intensity of the scattered atoms and from the incident energy and scattering angle, which determines $\boldsymbol{K}$, the bound state energies ε_n can be determined.

In the case of an elastic resonance the intensity feature may either be a maximum or a minimum depending on whether the interference between the directly scattered part of the incident wave and the part of the wave which passed through the bound state (Fig. 10.3) is constructive or destructive, respectively. Since the resonant channels through a bound state are allowed only at special angles, whereas the directly reflected channel is continuously allowed at all angles, the modulation of the peak intensity, as a function of the incident angle or energy, displays the typical Fano resonance shape [33–35].

The calculation of *inelastic* resonance intensities requires the numerical solution of a large set of close coupling equations which must be sufficiently large to assure convergence. For this reason such calculations are an order of magnitude larger in size than those based on the DWBA without bound states. Since the inelastic amplitudes involving bound states depend sensitively on both the repulsive and attractive parts of the atom-surface potentials they provide a discriminating test of these potentials.

This is illustrated in Fig. 10.9 for inelastic scattering from LiF(001). In this calculations Evans et al. [36] extended Celli, Garcia and Hutchison's formalism to account for inelastic processes with an HCS model potential [37]. As seen in Fig. 10.9a, c, e the calculations reproduce only in part the strong $n = 1$ resonance enhancement of the RW at -3.45 Å^{-1} while the $n = 2$ resonance enhancement of the bulk phonon response at -3.2 Å^{-1} is largely overestimated. On the other hand, with a close-coupling calculation based on the more realistic Tang and Toennies semi-*ab initio* potential model [38], described in Chap. 6. Eichenauer and Toennies [39, 40] obtained an excellent fit of the $n = 1$ inelastic resonance at -3.2 Å^{-1} (Fig. 10.9b, d, f). This comparison explains the extreme sensitivity of inelastic bound state resonances on the atom-surface potentials. This is due to the fact that an atom trapped in a bound state probes both the repulsive and attractive parts of the surface potential, as well as their derivatives (determining the coupling strength to phonons) and their corrugation.

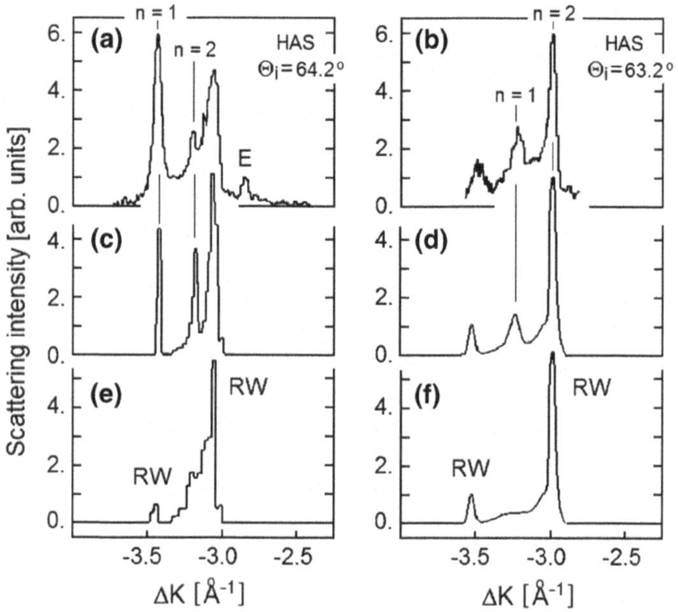

Fig. 10.9 Comparisons showing the sensitivity of the calculations of resonances in inelastic HAS spectra on the potential. The experimental resonances are shown together with calculations for scattering from LiF(001) along the [001] ($\overline{\Gamma M}$) direction as a function of the parallel momentum transfer $\Delta\mathbf{K}$ for a 90° scattering geometry ($\vartheta_f = 90 - \vartheta_i$) at an incident wavevector $k_i = 6.06$ Å^{-1} [36]. The corresponding energies can be calculated from the scan curves (Fig. 9.3). **a** The experimental TOF spectrum at incident angle $\vartheta_i = 64.2°$ is compared to the spectra calculated for a HCS potential using the S-matrix formalism with (**c**) and without (**e**) the inclusion of bound state resonances. The weak RW at $\Delta K = -3.45$ Å^{-1} is seen to be strongly enhanced by the resonance. **b** After a small shift of the incident angle to $\vartheta_i = 63.2°$ the experimental $n = 1$ resonance in (**a**) is shifted to -3.2 Å^{-1}. **d, f** Comparison with close-coupling calculations by Eichenauer and Toennies [39, 40] based on the more realistic pairwise Tang-Toennies potential [38] with (**d**) and without (**f**) the inclusion of bound state resonances. In this case the inclusion of the $n = 1$ resonance reproduces the strong enhancement of the bulk phonon density

10.2 Critical Kinematic Effects

A number of interesting kinematic effects can occur when the scan curves are tangent to the phonon dispersion curves or resonance curves. In other cases the resonance curves may be tangent to the phonon dispersion curves. Figure 10.10 illustrates the various possibilities for critical kinematic effects in the $(\hbar\omega, \mathbf{Q})$-plane. For convenience these are shown only for creation processes in the lower left hand quadrant of Fig. 9.4 and thus are inverted with respect to Figs. 9.3 and 9.4. The conditions for in-plane inelastic bound state resonances due to phonon-assisted adsorption (PAA) are given by the first line of (10.12b). In Fig. 10.10 these PAA resonance curves for the different bound states $n = 0, 1, 2, \ldots$ are also represented by a set of parabolas (blue curves denoted by r_n in Fig. 10.10) in the $(\hbar\omega, \mathbf{Q})$-plane. A triple intersection of the scan curve (SC) and a resonance (r_n) curve with a surface phonon dispersion curve (e.g., RW) yields the strong resonance enhancement (RE in Fig. 10.10a), like those described in Fig. 10.9, panel (a) for $n = 1$ and (b) for $n = 2$. A resonance enhancement of the bulk phonon response is also observed when the intersection of the scan curve with a resonance curve occurs in the surface-projected bulk phonon DOS (denoted by $\mathrm{r_n}$ in Fig. 10.10a) The same resonances are described in Fig. 10.9, panel (a) for $n = 1$ and (b) for $n = 1$.

When any pair of these curves are tangent important focusing phenomena can occur which belong, from the mathematical point of view, to the general class of *critical kinematic* (CK) effects [42–44]. The three possible tangency cases, for *kinematic focusing* (KF), for *focused inelastic resonance* (FIR) and *surfing* are illustrated in Fig. 10.10b–d in the $(\hbar\omega, \mathbf{Q})$-plane and are discussed in detail in the following subsections.

10.2.1 Kinematical Focusing (KF)

The critical kinematic effect depicted in Fig. 10.10b and mentioned in connection with Fig. 10.6 (small shaded triangles at the bottom of Fig. 10.6a) is known as *kinematical focusing* (KF) and was first predicted in 1975 [45]. In many cases there are special incident angles for each given incident energy at which the scan curve is tangent to the surface phonon dispersion curve for either creation or annihilation processes. Figure 10.10b illustrates this condition for creation of a surface RW phonon, while Fig. 10.11 provides an experimental demonstration of KF on the annihilation side. The TOF spectra measured on the NaF(001) surface (Fig. 10.11a) show for $\vartheta_i = 32.0°$ two large RW peaks, while at a slightly larger angle of 32.5° the peak at about 1.84 ms has suddenly disappeared. As seen in Fig. 10.11b at $\vartheta_i = 32.0°$ the scan curve is exactly tangent to the Rayleigh dispersion curve on the annihilation side in the second Brillouin zone, thus explaining the strong peak at

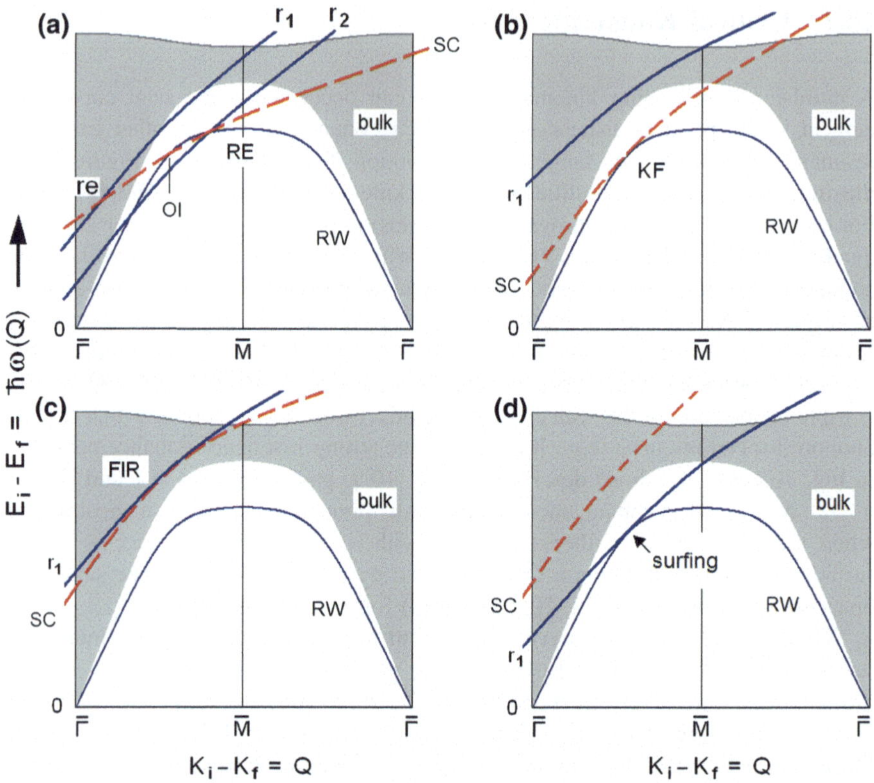

Fig. 10.10 Diagrams illustrating the energy loss as a function of the parallel wavevector transfer in various HAS critical kinematic effects involving the creation of a phonon of energy $\hbar\omega(Q)$ and wavevector Q. The *scan curve* SC represents the kinematically allowed energy loss as a function of the parallel momentum transfer in experiments with a given scattering geometry and incident energy (9.3). The intersections of the scan curve with a surface phonon dispersion curve (RW) and the bulk phonon continuum (grey areas) determine the features of the time-of-flight spectra. The intersections of the resonance curves r_1 and r_2 with the phonon spectrum give the energy loss as a function of the parallel wavevector transfer for phonon-assisted adsorption (PAA) into different bound states. **a** The intersection of SC with the RW surface phonon represents an ordinary inelastic process (OI), whereas the triple intersection of SC and r_2 with the RW branch determines a resonance enhanced (RE) inelastic process, re indicates a resonant enhancement of a bulk phonon. **b** The tangency of a scan curve with a phonon branch causes a kinematical focusing (KF) yielding a peculiar saw-tooth peak in the angular distribution (see Fig. 10.6a). **c** The tangency of SC with r_1 inside the bulk phonon continuum gives a focused inelastic resonance (FIR). **d** The tangency of r_1 with a surface phonon branch leads to the surfing phenomenon [41]

1.84 ms. In contrast at $\vartheta_i = 32.5°$ no intersection occurs between the scan curve and the RW dispersion curve and the additional peak is absent [46]. A similar effect can be seen in Fig. 10.6b for LiF(001). At $\vartheta_i = 30.7°$ there are three peaks

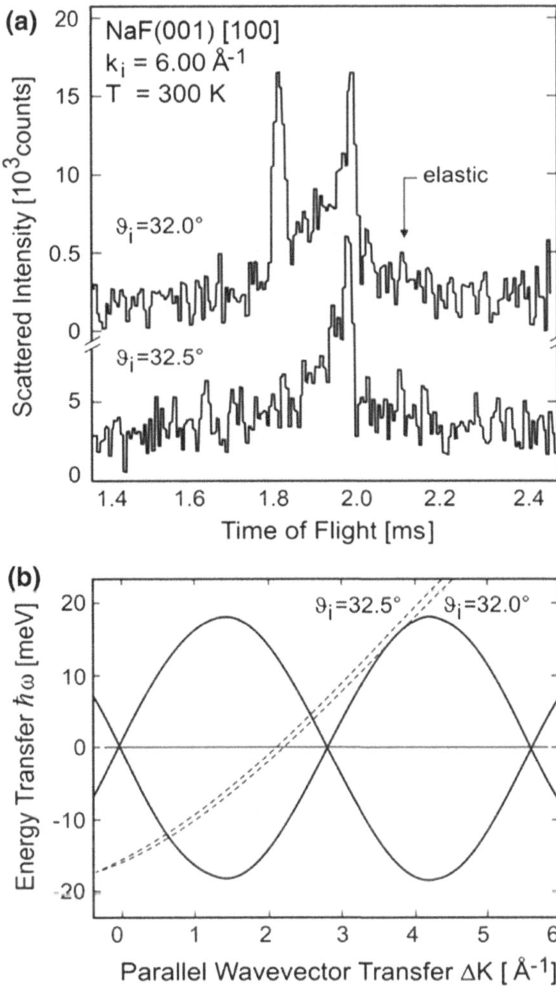

Fig. 10.11 **a** The kinematic focusing effect is illustrated with two experimental TOF spectra of He atoms scattered from NaF(001) along the [100] azimuth ($k_i = 6.0\,\text{Å}^{-1}$, $T_s = 300$ K) at two nearby incident angles of $\vartheta_i = 32.0°$ and 32.5° for a 90° scattering geometry ($\vartheta_f = 90 - \vartheta_i$). The intense annihilation peak at 1.84 ms measured at 32.0° corresponds to the scan curve shown in (**b**) which is tangent to the Rayleigh dispersion curve. At $\vartheta_i = 32.5°$ a corresponding peak does not appear in the TOF spectrum since the corresponding scan curve no longer crosses the Rayleigh dispersion curve [46]

associated with three intersections of the scan curve with the RW branches on the annihilation side; at $\vartheta_i = 32.7°$ two of these intersections coalesce into a tangency point giving rise to the single intense peak marked KF in Fig. 10.6b. The coalescence conditions are very sharp as seen by the disappearance at $\vartheta_i = 34.7$ where only one RW peak is observed.

The coalescence of two peaks at the tangency point, occurring when $\theta_f = \overline{\theta}_f$, yields an inverse square root singularity at the corresponding angle in the angular distribution [45, 46]:

$$\frac{d\mathcal{R}}{d\Omega} \propto \left[\frac{\tan \theta_f}{(\overline{w}_s - \overline{w}_d)(\overline{\theta}_f - \theta_f)} \right]^{1/2}, \tag{10.17}$$

where $\overline{w}_s$ and $\overline{w}_d$ are the second derivatives with respect to ΔK of the scan curve and the dispersion curve at the tangency point, respectively. The singularity can be particularly strong if also the two curvatures are alike. However the features of KF in the angular distributions from corrugated surfaces are often obscured by the tails of intense diffraction peaks and the concurrence of bound state resonances. These disturbances, however, are greatly reduced by rotating the scattering plane away from a high symmetry direction by a small amount, which does not appreciably affect the phonon dispersion and the KF feature. Under such circumstances it is even possible to use the KF effect to determine the surface phonon dispersion curve without resorting to TOF measurements. This is illustrated by the angular distribution measured in a fixed angle apparatus ($\Theta_{SD} = 90°$), also from NaF(001), shown in Fig. 10.12. The incident angles at which the scan curves are tangent to the RW dispersion curve are listed in Fig. 10.12a for creation and annihilation processes in different Brillouin zones. When the crystal was rotated by 5° away from the [100] direction (Fig. 10.12b) the sharp features due to the KF effect for both annihilation (A) and creation (C) processes become prominent and can be easily identified in the angular distribution. Other sharp minima, indicated by red arrows, are instead the signatures of surfing, discussed in Sect. 10.2.3. From the corresponding scan curves folded back into the first Brillouin zone an envelope of the RW dispersion curve is obtained as shown in Fig. 10.12c. The results were found to be in excellent agreement with TOF measurements along the [100] direction.

For heavier atoms (e.g., Ne) the scan curves have a much smaller slope than for He and the tangency points occur near the *van Hove singularities* of the surface phonon density. Generally van Hove singularities are expected to appear at the zone-boundary symmetry points and, for the optical modes, also at the $\overline{\Gamma}$ point [45]. They can contribute to a further enhancement of KF singularities and under favorable conditions even optical Lucas modes can yield detectable features [45, 47].

A particularly interesting focusing effect involves a bunch of scan curves of slightly different incident wavevectors but fixed incidence angle which converge to a spot around a single point of the ($\Delta\mathbf{K}$, ΔE)-plane (Fig. 10.13). When the spot is tuned along a surface phonon dispersion curve a large intensity of the inelastic signal can be obtained even for a limited energy resolution of the incident beam. This particular kind of kinematic focusing has been discussed by Miret-Artés [42] within the unifying framework of the critical kinematic effects [43, 44] and has been called *inelastic focusing* (IF). From the condition $\partial\hbar\omega/\partial k_i = 0$ applied to the scan curve, (9.3), at constant ϑ_i one finds that a bunching of the scan curves occurs at a wavevector transfer given by

$$\Delta K = k_i \cos 2\vartheta_i / \sin \vartheta_i. \tag{10.18}$$

This effect was first clearly observed in NaCl(001) [48]. The example reproduced in Fig. 10.13a shows a bunch of scan curves for the same incident angle of 37.2° but a range of incident wavevectors varying from 4.45 to 5.93 Å^{-1}, as well as an isolated scan curve for 7.5 Å^{-1}. They cut the RW dispersion curves in three

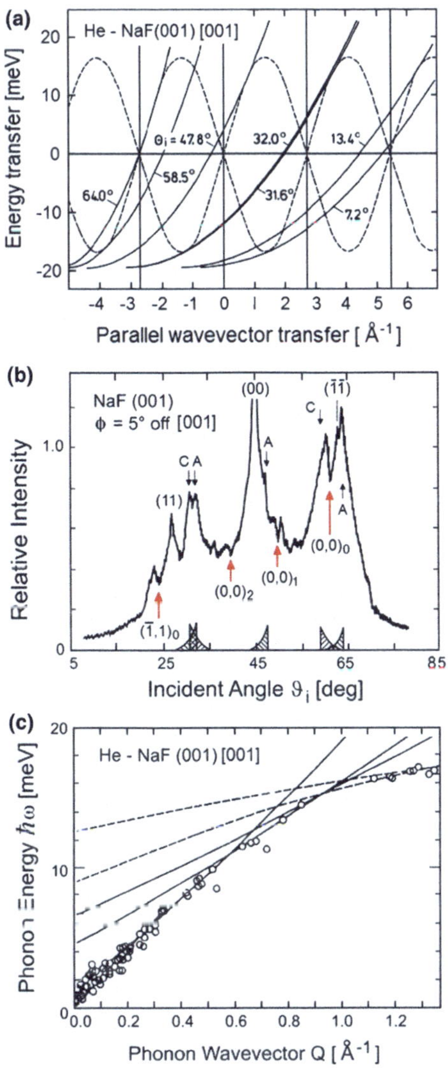

Fig. 10.12 The measurement of a RW dispersion curve based only on the KF effect. **a** The scan curves for an incident wavevector $k_i = 6.0\,\text{Å}^{-1}$ and the special incident angles at which they are tangent to the RW dispersion curve of NaF(001) in the [100] direction for creation and annihilation processes in various Brillouin zones. **b** HAS angular distribution measured from NaF(001) at an azimuth $\phi = 5°$ slightly removed from the [100] direction with a $\Theta_{SD} = 90°$ apparatus at an incident energy $E_i = 18.8$ meV ($k_i = 6.0\,\text{Å}^{-1}$). Each of the small peaks at the angles marked by an "A" (phonon annihilation) and a "C" (phonon creation) correspond to incident angles at which the inelastic kinematic focussing (KF) effect occurs. **c** Scan curves corresponding to the KF angles for phonon annihilation (continuous line) and creation (broken lines) folded into the irreducible part of the surface Brillouin zone for positive frequency. The resulting envelope of the RW dispersion curve compares well with the experimental points (o) derived from the TOF spectra [46]

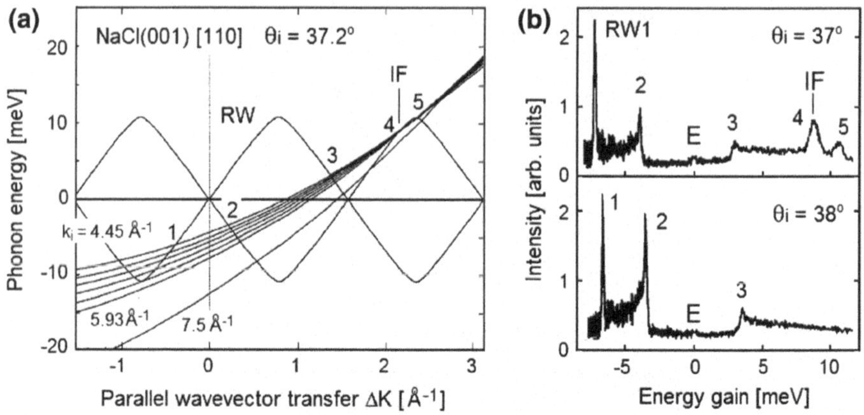

Fig. 10.13 Scan curves and TOF spectra illustrating the *inelastic focusing* (IF) effect [33]. **a** The scan curves for a range of incident wavevectors varying from 4.45 to 5.93 $Å^{-1}$ (as well as an isolated one for 7.5 $Å^{-1}$) at $\vartheta_i = 37.2°$ (for a 90° scattering geometry $\vartheta_f = 90 - \vartheta_i$) cross the RW dispersion curve of NaCl(001) along [110] nearly tangent at two nearby points 4 an 5. **b** The resulting inelastic focusing (IF) near $\Delta K = 2.2\,Å^{-1}$ leads to two large peaks in the energy gain spectrum. At the nearby angle $\theta_i = 38°$ no intersection occurs and the peaks 4 and 5 disappear. The IF amplification is particularly prominent because annihilation processes of zone boundary phonons in the second Brillouin zone are generally expected to be very weak

Brillouin zones along the $[1\bar{1}0]$ direction on both the creation (1, 2) and annihilation (3, 4, 5) sides. The scan curves for k_i between 4.45 and 5.93 $Å^{-1}$ lead to inelastic focusing (IF) near $\Delta K = 2.2\,Å^{-1}$ in an approximate KF condition. A large amplification of the response from the RW phonons 4 and 5 is clearly seen in the corresponding energy gain spectra (Fig. 10.13b). At = 38° the bunch of scan curves no longer intersects the RW branch around 2.2 $Å^{-1}$ and the peaks 4 and 5 disappear. It is important to note that the IF amplification is particularly significant because the annihilation of zone boundary phonons in the second Brillouin zone is expected to be improbable. Indeed, under similar conditions an ordinary KF without IF like that in Fig. 10.12a for $\theta_i = 13.4°$ does not produce any appreciable feature in the angular distribution (Fig. 10.12b).

10.2.2 Focused Inelastic Resonances (FIR)

Another interesting critical kinematic effect is the *focussed inelastic resonance* (FIR) [48, 49], illustrated in Fig. 10.10c. It occurs whenever a scan curve is tangent to a resonance curve within the region in the (Q, ω) plane where surface phonons are found. The final angle at which a FIR intensity enhancement can be expected is given by

$$\tan^2 \theta_f^{FIR} = (2m|\varepsilon_n|/\hbar^2 - G''^2)/G_{\|}''^2. \quad (10.19)$$

It has the remarkable property of being independent of the incident energy and incident angle and therefore also independent of the final energy. An example can be seen in Fig. 10.7. By coincidence the final angle of the supernova double resonance in Fig. 10.7 also happens to be at a *focused inelastic resonance* (FIR) angle.

The final energy at a FIR peak, given by

$$E_f^{FIR} = \frac{\hbar^2 G_{\|}^2}{2m} \tan^2 \theta_f^{FIR} / \cos^2 \theta_f^{FIR}, \quad (10.20)$$

is the same for all the atoms involved in the FIR scattering. In other words if the incident beam is not exactly monochromatic but has a certain velocity distribution, under FIR conditions a continuum of surface phonons is available to provide the right energy and parallel momentum to cast the atoms, through an inelastic resonance process, into the same final state. Thus the FIR effect provides a possible mechanism for the realization of an atom beam monochromator [49]. The intensity could be comparable to that of a diffraction peak if the FIR peak occurs in association with a *supernova* double resonance (Fig. 10.7b).

In principle the FIR effect is also applicable to the study of *electronic image states* in large-period stepped metal surfaces, as discussed in [49]. In general FIR is expected to occur for a particle scattered from a surface which supports bound surface states where the particles can move freely along the surface. In the inelastic scattering of molecules also the rotational excitations can be involved in matching the FIR conditions. Rotationally mediated focused inelastic resonances have been reported for D_2 scattering from the Cu(100) surface [50]. In general a low index metal surface does not support any inelastic resonance in helium atom scattering due to the lack of surface corrugation, but with molecules a rotationally mediated focusing process is possible.

10.2.3 Surfing

The third tangency condition leading to a focusing effect is called *surfing* [33]. In this case the tangency is between an inelastic resonance curve for a given bound state n and a surface phonon dispersion curve (specifically that of a RW), as depicted in Fig. 10.9d. Under this focusing condition an atom is trapped into a bound state through an inelastic process, Fig. (10.14), by creating a RW, as in PAA-SD and travels along the surface with the same group velocity and direction of the RW (*parallel surfing*, Fig. 10.14a), or at least with the component of the velocity along the RW direction equal to the RW group velocity (*oblique (tube-riding) surfing*, Fig. 10.14b) [41].

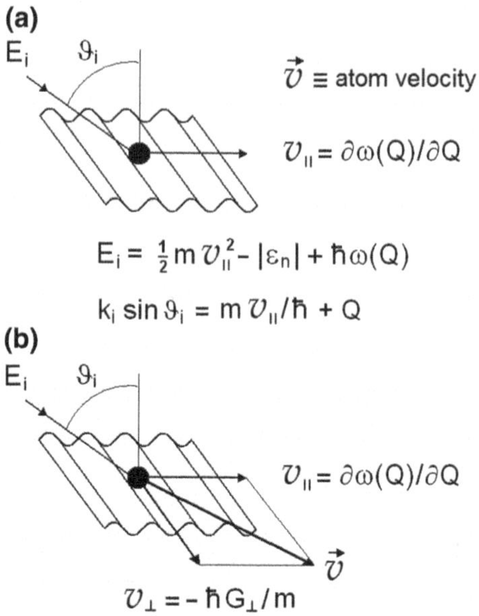

Fig. 10.14 Schematic diagram illustrating two different surfing conditions. **a** The atom impinging onto a surface is selectively adsorbed into a bound state of energy $-|\varepsilon_n|$ via the excitation of a surface phonon of energy $\hbar\omega(Q)$ and wavevector Q as in PAA-SD. For special values of the incident angle ϑ_i or energy E_i the atom and the surface phonon wave have the same group velocity and travel together, with the atom trapped in a trough of the surface phonon wave. **b** The inelastic selective adsorption may also occur with the concurrent exchange of a surface reciprocal lattice vector **G**. Under surfing conditions with a **G**-vector normal to the phonon wavevector a oblique (*tube-riding*) surfing occurs with the atom running sideways in a trough of the surface phonon wave. The symbols are explained in the text [41]

A natural question is whether at the incident angle and energy at which atoms are able to enter into a surfing state there is any signature in the angular distribution of the scattered atoms. In surfing as in selective adsorption the bound state is involved in a resonant process, i.e., in a virtual transition from the incident beam to the bound state and back into the final beam. The interference between the direct scattering channel and the one through the bound state will give either a maximum or a minimum in the angular distribution depending on the phase shift. In the present case the scan curve does not cross, in general, the tangency point (Fig. 10.10d) and therefore the surface atoms will be lost from the scattered beam. Thus the surfing process will show up as a dip in the angular distribution. The effect of surfing was first identified in a reexamination of experimental angular distributions of NaF(001) at an azimuthal angle of 5° off the [100] direction [41].

The matching of velocities can be calculated from (10.12) by solving for the phonon energy (with $Q_{||} = Q$):

$$\hbar\omega_{RW}(Q) = E_i + |\varepsilon_n| - \frac{\hbar^2}{2m}\left[(k_i \sin\theta_i - G_{\|} - Q)^2 + G_{\perp}^2\right]. \tag{10.21}$$

From the derivative with respect to Q, it is seen that the tangency condition requires that the RW group velocity at the tangency point $Q = Q_s$ fulfils the equation

$$\upsilon_{RW}(Q_s) \equiv \frac{\partial\omega_{RW}(Q_s)}{\partial Q} = \frac{\hbar}{m}(K_i - Q_s - G_{\|}). \tag{10.22}$$

The r.h.m. of this equation is just the component of the He atom velocity along the RW direction:

$$\upsilon = (\upsilon_{\|}, \upsilon_{\perp}) = \frac{\hbar}{m}(K_i - Q_s - G_{\|}, G_{\perp}). \tag{10.23}$$

The equality $\upsilon_{RW}(Q_s) = \upsilon_{\|}$ has the interesting consequence that in a phonon-assisted adsorption process, where the phonon is *created* in the *forward direction*, the atom and the surface phonon travel together or, in other words, the atom is carried along the surface by the phonon, trapped in a potential minimum of the phonon wave. It should be noted that, unlike the case of elastic selective adsorption, (10.22) can be fulfilled also for **G** = 0, provided the phonon energy is sufficiently large. Besides the direct process, *umklapp* processes involving a non-zero **G** are also possible. Among the *umklapp* processes, the one for **G** = (0, $G_{\perp}$) corresponds to the atom carried by the surface phonon in a trough of the dynamic corrugation and rapidly moving along the trough in a direction normal to the phonon wavevector (Fig. 10.14b). Thus the surfing atom may also exhibit some carving and tube-riding abilities as an experienced surfer riding ocean waves.

The surfing conditions can be easily fulfilled under the usual HAS experimental conditions. For a mass m like that of a He atom, a typical $k_i \approx 10$ Å^{-1} and **G** = 0 the atom velocity given by (10.23) is of the order of a few 10^3 m/s, just like the RW velocity in most crystal surfaces. The matching of the two velocities leads to the formation of *an atom-phonon bound state*, where the atom travels along the surface "dressed" by a phonon cloud. This generally implies a renormalization of the atom effective mass ($m \rightarrow m^*$), similarly to the electron mass renormalization occurring in the polaron state (electron-phonon bound state) [51].

A special case occurs when the scan curve intersects the resonance and dispersion curves at the tangency point. In this case the surfing atoms may selectively resonantly desorb into the final state and a maximum can be observed in the angular distribution (*focused surfing* (FS)) [48]. In Fig. 10.15 these different situations are illustrated with HAS angular distributions taken at different incident wavevectors varying within restricted ranges and at a surface temperature of 300 K for the [110] direction on NaCl(001). The sequences of deep minima marked with small vertical red bars in Fig. 10.15a and b and labeled by the respective $(G_{\|}, G_{\perp})_n$ are found to

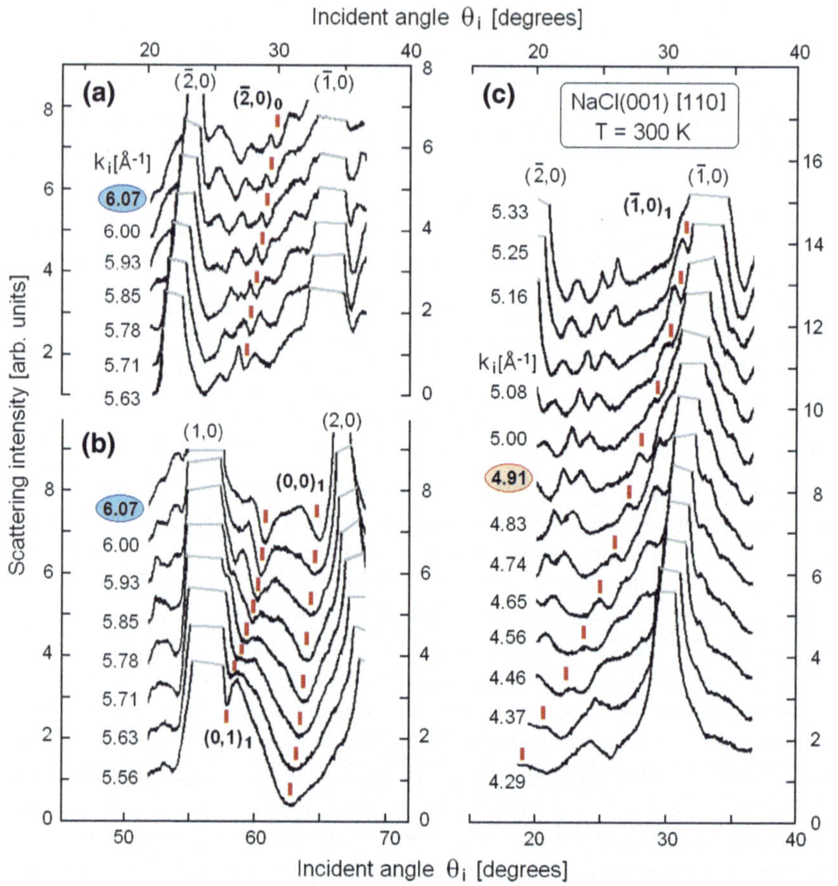

Fig. 10.15 Angular distributions of HAS from NaCl(001) in the [110] direction measured at room temperature for different incident wavevectors k_i and incident angles. **a** For k_i between 5.63 and 6.07 $Å^{-1}$ and angles between the diffraction peaks $(\overline{2}, 0)$ and $(\overline{1}, 0)$. **b** For the same values of k_i in the angular region between diffraction peaks $(1, 0)$ and $(2, 0)$. **c** Same as (**a**) for smaller incident wavevectors. The red short vertical bars mark sequences of features assigned to surfing states labeled by the corresponding $(G_{\|}, G_{\perp})_n$. Surfing states are characterized by dips, but the sequence in (**c**) meets the triple intersection condition for k_i between 4.5 and 5 $Å^{-1}$ and the dips evolve into maxima for $k_i < 5$ $Å^{-1}$ [41]. The resonance and scan curves for the encircled values of k_i are shown superimposed to the RW dispersion curves in the corresponding panels of Fig. 10.16 (adapted from [48])

be associated with direct (**G** = 0) and tube-riding surfing in the $n = 1$ bound state ($(0, 0)_1$ and $(0, 1)_1$ in Fig. 10.15b, respectively) and with parallel surfing ($(\overline{2}, 0)_0$ in Fig. 10.15a). The sequence of features $(\overline{1}, 0)_1$ seen in the angular distributions of Fig. 10.15c is also assigned to parallel surfing on the $n = 1$ bound state, but in this case the minima observed for $k_i > 5$ gradually convert to maxima for $k_i < 5$ due to

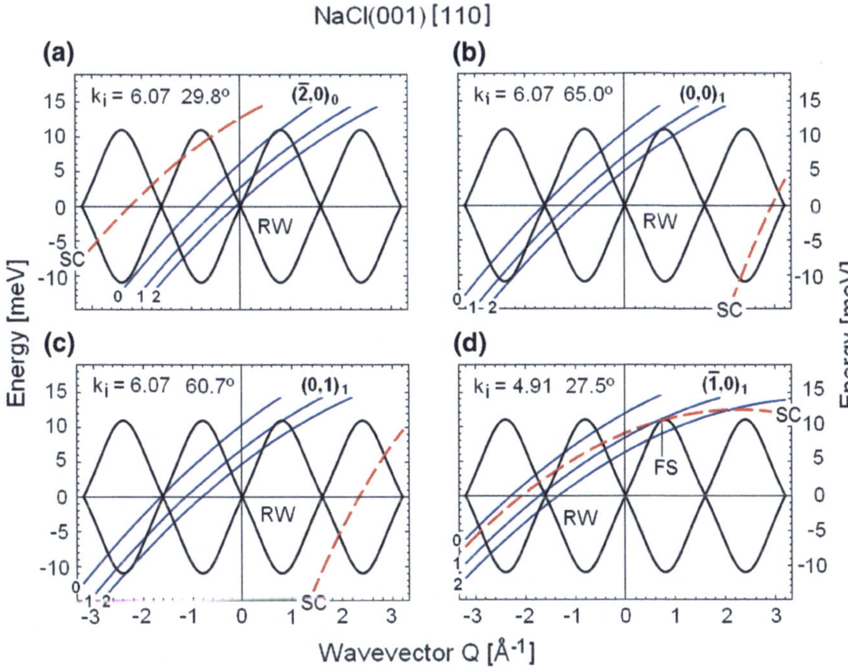

Fig. 10.16 The resonance curves for the lowest bound state levels (n = 0, 1, 2) and the HAS scan curve (SC) superimposed upon the RW branches for NaCl(001). As in Fig. 10.10, the phonon energies are reversed compared to the usual atom energy-gain vs. wavevector-gain plots adopted in Chap. 9. The incident wavevectors k_i (in Å^{-1}), corresponding to those encircled in Fig. 10.15 and incident angles (degrees) indicated in each panel all fulfil the surfing condition, i.e., a tangency between a resonance curve (labeled by the corresponding **G**-vector and bound state index) and the RW dispersion curve, and yield surfing features in the angular distributions (red bars in Fig. 10.15). **a** The resonance curve for n = 0 (cfr. Fig. 10.15a) is tangent to the RW curve on both the creation side in the first BZ and annihilation side in the second BZ, but only the creation process supports surfing. **b** The **G** = (0, 0), n = 1 resonance curve, which is tangent to the RW curve on the creation side, provides an example of direct phonon-assisted selective adsorption (cfr. Fig. 10.15b). **c** Same as (**b**) for the out-of-plane bound states. **d** The intersection of the scan curve (SC) with the tangency point on the n = 1 resonance curve, converts the surfing dip into a maximum (*focused surfing* (FS)) (cfr. Fig. 10.15c)

the approaching FS triple intersection of the scan curves with the resonance and RW dispersion curves.

The attribution to surfing to the above sequences is based on the analysis of the resonance, scan and RW dispersion curves in the $(\Delta E, \Delta K)$-plane. Figure 10.16a–d shows selected diagrams, one for each of the four surfing series of Fig. 10.15 at selected incident wavevectors (encircled) and corresponding resonance incident angles (red bars). In Fig. 10.16a the $n = 0$, $\mathbf{G} = (\bar{2}, 0)$ in-plane resonance curve, labelled by $(\bar{2}, 0)_0$ (cfr. Fig. 10.15a), is tangent to the RW on both the creation side (in the first BZ) and the annihilation side (in the second BZ), but only the creation process can lead to surfing. In Fig. 10.16b the $(0, 0)_1$ resonance curve (cfr. Fig. 10.15b), is tangent to a direct phonon creation-assisted selective adsorption into the $n = 1$ bound state. The $(0, 0)_2$ resonance curve, although also tangent to the RW, is on the annihilation side and surfing is not possible. Similarly the out-of-plane resonance curves for **G** = (0, 1), shown in Fig. 10.16c, give an oblique surfing series $(0, 1)_1$ (cfr. Fig. 10.14b), only for $n = 1$. In Fig. 10.16d the $(\bar{1}, 0)_1$ in-plane resonance is at values of k_i and θ_i with a triple intersection, which corresponds to the special condition of *focused surfing* (FS) and therefore to a maximum in the angular distribution (cfr. Fig. 10.15c).

The surfing phenomenon has some additional interesting consequences, to date only based on theoretical arguments but not yet supported by experimental studies. The reader can find a detailed discussion on surfing in the recent review of [41]. As mentioned in Chap. 1, travelling surface acoustic waves (SAWs) generated by a transducer at a semiconductor surface can transport a charge carrier. A similar application based on the surfing effect, where surface adsorbed atoms are carried by a surface wave, is conceivable, with the important difference that in the case of surfing the surface wave is generated by the atoms themselves in the phonon-assisted selective adsorption process.

Bound state inelastic resonances and kinematic effects provide interesting information not only on the surface phonon dispersion curves as in the case of kinematic focussing but also complementary information on the surface bound states. Only alkali halide surfaces, because of their large corrugation were discussed in this section. Similar effects are however expected to arise in the case of corrugated layered materials like graphite [28, 37, 57–59], as well as of high index metal and semiconductor surfaces, or of stepped surfaces. Because of the important role of electron-phonon interaction in HAS from these surfaces interesting novel effects can be expected.

References

1. J.E. Lennard-Jones, A.F. Devonshire, Nature **137**, 1069 (1936)
2. R. Frisch, O. Stern, Z. Phys. **84**, 430 (1933). The first clue for the presence of unexpected minima in the angular distributions was reported in 1930 by I. Estermann, O. Stern, Z. f. Phys. **61**, 95 (1930). See page 116 therein
3. H. Hoinkes, Rev. Mod. Phys. **52**, 993 (1980)
4. J. Perreacu, J. Lapujoulade, Surf. Sci. **122**, 341 (1982)
5. J.L. Beeby, J. Phys. C: Solid State Phys. **4**, L359 (1971)
6. V. Celli, D. Evans, in *Dynamics of Gas-Surface Interaction*, ed. by G. Benedek, U. Valbusa (Springer, Berlin, 1982), p. 2
7. P. Kraus, M. Mayrhofer-Reinhartshuber, Ch. Gösweiner, F. Apolloner, S. Miret-Artes, W.E. Ernst, J. Chem. Phys. (2014)
8. M.W. Cole, T.T. Tsong, Surf. Sci. **69**, 325 (1977)
9. W.A. Steeele, *The Interactions of Gases with Solid Surfaces*(Chapter 3) (Pergamon, Oxford, 1974)
10. G. Lilienkamp, J.P. Toennies, J. Chem. Phys. **78**, 5210 (1983)
11. L. Mattera, F. Rosatelli, C. Salvo, F. Tommasini, U. Valbusa, G. Vidali, Surf. Sci. **93**, 515 (1980)

12. L. Mattera, C. Salvo, S. Terreni, F. Tommasini, Surf. Sci. **97**, 158 (1980)
13. C. Schwartz, M.W. Cole, J. Pliva, Surf. Sci. **75**, 1 (1978)
14. V. Celli in *Dynamics of Gas-Surface Interaction*, ed. by G. Benedek, U. Valbusa (Springer, Berlin, 1982), p. 29
15. H. Chow, E.D. Thompson, Surf. Sci. **59**, 225 (1976)
16. W.E. Carlos, G. Derry, D.R. Frankl, Phys. Rev. B **19**, 3258 (1979)
17. D.A. Wesner, D.R. Frankl, Phys. Rev. B **24**, 1798 (1981)
18. A.C. Levi, H.G. Suhl, Surf. Sci. **88**, 221 (1979)
19. J.R. Manson, G. Mantovani, G. Armand, in *Dynamics on Surfaces*, ed. by A. Pullman, J. Jortner, A. Nitzan, B. Gerber (D. Reidel Publ. Co., Dordrecht, 1984), p. 23
20. L. Niu, D.J. Gaspar, S.J. Sibener, Science **268**, 847 (1995)
21. L. Niu, D.D. Koleske, D.J. Gaspar, S.J. Sibener, J. Chem. Phys. **102**, 9077 (1995)
22. D.J. Gaspar, A.T. Hanbicki, S.J. Sibener, J. Chem. Phys. **109**, 6947 (1998)
23. J. Lapujulade, J. Perreacu, Physics Scripta **T4**, 138 (1983)
24. G. Boato, P. Cantini, C. Guidi, R. Tatarek, G.P. Felcher, Phys. Rev. B **20**, 3957 (1979)
25. B.R. Williams, J. Chem Phys. **55** (1971)
26. P. Cantini, G.P. Felcher, R. Tatarek, Phys. Rev. Lett. **37**, 606 (1976)
27. H. Frank, H. Hoinkes, H. Wilsch, Surf. Sci. **63** (1977)
28. P. Cantini, R. Tatarek, Phys. Rev. B **23**, 3030 (1981)
29. G. Brusdeylins, R.B. Doak, J.P. Toennies, J. Chem. Phys. **75**, 1784 (1981)
30. G. Brusdeylins, R.B. Doak, J.P. Toennies, Phys. Rev. B **27**, 3662 (1983)
31. G. Bracco, M. D'Avanzo, C. Salvo, R. Tatarek, F. Tommasini, Surface Sci. **189**(190), 684 (1987)
32. G. Lilienkamp, J.P. Toennies, Phys. Rev. B **26**, 4752 (1982)
33. U. Fano, Phys. Rev. **121**, 1866 (1961)
34. A. Bohm, *Quantum Mechanics: Foundations and Applications* (Chap. XVIII. 9) (Springer, Berlin, 1979)
35. J.R. Manson, V. Celli, D. Himes, Phys. Rev. B **49**, 2782 (1994)
36. D. Evans, V. Celli, G. Benedek, J.P. Toennies, R.B. Doak, Phys. Rev. Lett. **50**, 1854 (1983)
37. V. Celli, N. Garcia, J. Hutchison, Surf. Sci. **87**, 112 (1979)
38. K.T. Tang, J.P. Toennies, J. Chem. Phys. **80**, 3726 (1984)
39. D. Eichenauer, J.P. Toennies, in *Dynamics on Surfaces*, ed. by B. Pullman, J. Jortner, A. Nitzan, B. Gerber (Reidel, Dordrecht, 1984), p. 1
40. D. Eichenauer, J.P. Toennies, J. Chem. Phys. **85**, 532 (1986)
41. G. Benedek, P.M. Echenique, J.P. Toennies, F. Traeger, J. Phys.: Cond. Matter **22**, 304016 (2010) and 359801(C)
42. S. Miret-Artés, Surf. Sci. **294**, 141 (1993)
43. M. Hernández, S. Miret-Artés, P. Villarreal, G. Delgado-Barrio, Surf. Sci. **274**, 21 (1992)
44. S. Miret-Artés, Surf. Sci. **339**, 205 (1995); ibid. 366 (1996) L735
45. G. Benedek, Phys. Rev. Lett. **35**, 234 (1975)
46. G. Benedek, G. Brusdeylins, J.P. Toennies, R.B. Doak, Phys. Rev. B **27**, 2488 (1983)
47. G. Benedek, L. Miglio, G. Seriani, in *Helium Atom Scattering from Surfaces*, ed. by E. Hulpke (Springer, Heidelberg, 1992), p. 207
48. G. Benedek, R. Gerlach, A. Glebov, G. Lange, S. Miret-Artes, J.G. Skofronick, J.P. Toennies, Phys. Rev. B **53**, 11211 (1996)
49. G. Benedek, S. Miret-Artes, Surf. Sci. **339**, L935 (1995)
50. M.F. Bertino, S. Miret-Artés, J.P. Toennies, G. Benedek, Phys. Rev. B **56**, 9964 (1997)
51. D. Pines, *Elementary Excitations in Solids* (W.A. Benjamin, 1963). See also: N.N. Bogoliubov, *Polaron Theory: Model Problems* (Gordon & Breach Science Publ., 2000)
52. D. Schmicker, Dissertation, Georg-August University, Göttingen (1992)
53. A. Glebov, V. Senz, J.P. Toennies, G. Gensterblum, J. Appl. Phys. **82**, 2329 (1997)

54. A. Gelbvo, V. Senz, J.P. Toennies, P. Lambin, G. Gensterblum, P. Senet, A.A. Lucas, Phys. Rev. B **56**, 9874 (1997)
55. B. Winter, M. Faubel, Chem. Rev. **106**, 1176 (2006)
56. J.A. Faust, G.N. Nathanson, Chem. Soc. Rev. **45**, 3609 (2016)
57. N. García, W.E. Carlos, M.W. Cole, V. Celli, Phys. Rev. B **21**, 1636 (1980)
58. J. S. Hutchison, Phys. Rev. B **22**, 5671 (1980)
59. J. Hutchison, V. Celli. Surface Sci. **93**, 263 (1980)

Chapter 11
Experimental Results: Surface Phonons

Measure what is measurable, and make measurable what is not so.

Galileo Galilei

Abstract In this chapter some exemplary high resolution HAS experiments on surface phonon dispersion curves of single crystal surfaces based largely on the TOF technique are reviewed. A selection is made of significant results for insulator, semiconductor and conducting surfaces, with special attention to the HAS results for metals, semimetals, and layered materials, where evidence is found for important effects of, for example, electron-phonon coupling and the related observation of Kohn anomalies, anomalous longitudinal branches and deep subsurface modes. Studies on complex surfaces with large-period unit cells due to reconstruction or high index orientation are also reviewed. Here also the comparatively high resolution of HAS permits to observe low energy dynamical excitations and their dispersion. Each section is meant as an introduction to the specific phenomena and issues characteristic to each of these different surface types, with special consideration to the questions solved and the new theoretical issues raised. This chapter is complemented by Appendix A, which contains a comprehensive list of the available HAS experimental data on phonon dispersion curves and of the theoretical studies for the different classes of materials, as of early 2018 which update previous reviews: [1, 2] (all systems), [3, 4] (layered compounds), [5, 6] (semiconductors), [7] (metals) and Landot-Börnstein Chapters [8–10, 484].

11.1 Insulators: Nearly Ideal Surfaces

11.1.1 Rare Gas Crystals: Xe(111)

The solid rare gas surfaces were the first systems for which complete model calculations of the surface phonon dispersion curves have been made [12, 13]. In view of the weak interactions of these closed shell systems it is not surprising that the

G. Benedek and J. P. Toennies, *Atomic Scale Dynamics at Surfaces*, Springer Series in Surface Sciences 63, https://doi.org/10.1007/978-3-662-56443-1_11

bulk phonon dispersion curves of all rare gas solids, including both helium isotopes are well described by simple third nearest neighbor force constant models [14]. Although the monolayer, bilayer and trilayer rare gas films on flat metal surfaces, were among the first to be studied by HAS [15–20], only a few experiments are available for somewhat thicker films on Ag(111) [21–23] and Au(111) [24]. In these early experiments evidence was obtained for the existence of the lowest Sezawa wave, whose dispersion curve starts at finite energy at zero wavevector (organ-pipe mode) and evolves into the Rayleigh wave of the solid rare gas surface at wavevectors much larger than the inverse film thickness. This is discussed in some detail in Sect. 12.3.

Xe is the only rare gas system for which thick layer measurements are available. Šiber et al. [25] reported an extensive HAS investigation of 160 layer thick Xe crystals grown *in situ* on Pt(111). This system was chosen as a benchmark for testing the transferability of gas phase two-body Xe-Xe potentials to the calculation of phonon dispersion curves and gas phase two-body He-Xe potentials for calculating the inelastically scattered intensities. From the sharp regular diffraction patterns the epitaxially grown surface could be identified as a regular close packed (111) surface with a lattice constant of 4.33 Å, which is the same as for the monolayer on the same substrate [26]. Figure 11.1 displays both some typical measured time-of-flight spectra for the two main symmetry directions (panels (a, b)), as well as the resulting experimental surface phonon dispersion curves (panel (c), open circles). The solid line curves show for comparison the theoretical Born-von Kármán (BvK) force constant model dispersion curves calculated from the semi-empirical gas phase two-body Xe-Xe potential of Dham et al. [27, 28]. The overall agreement is very good for the three acoustic modes RW, L and SH, and provides the first clear evidence for the transferability of gas phase potentials to simulations of surface phonons. Especially interesting are the few points in the $\bar{\Gamma}\bar{K}$ direction along the SH-curve. For planar scattering the excitation of this mode is symmetry forbidden in the $\overline{\Gamma M}$ direction, but the breaking of the mirror symmetry along the $\overline{\Gamma K}$ direction endows the SH mode with a small vertical component thereby making the mode HAS-active.

A similar breakdown of the selection rules allowing for the observation of the SH mode was observed for monolayers of Ar, Kr and Xe on Pt(111) by Bruch et al. [29–32]. Since a self standing monolayer would have a mirror symmetry along both $\bar{\Gamma}\bar{K}$ and $\bar{\Gamma}\bar{M}$ they attributed this to a small misalignment of about 1–2 degrees. Along $\bar{\Gamma}\bar{K}$, however, they attributed the mirror symmetry break primarily to the monolayer-substrate interaction, its effect being expected to be much bigger than that of a small misalignment.

The major deviation from the theory is a nearly dispersionless mode at about 4.3 meV, indicated by an X, found both for the $\bar{\Gamma}\bar{K}$ and $\bar{\Gamma}\bar{M}$ directions. This mode was first thought to be due to defects but since the 4.3 meV energy is greater than predicted for the S-mode (vertical stretch vibrations of an adsorbate) whose frequencies are about 3 meV. Thus another mechanism involving two phonons was proposed. According to an analysis by Šiber and Gumhalter [33] the simultaneous

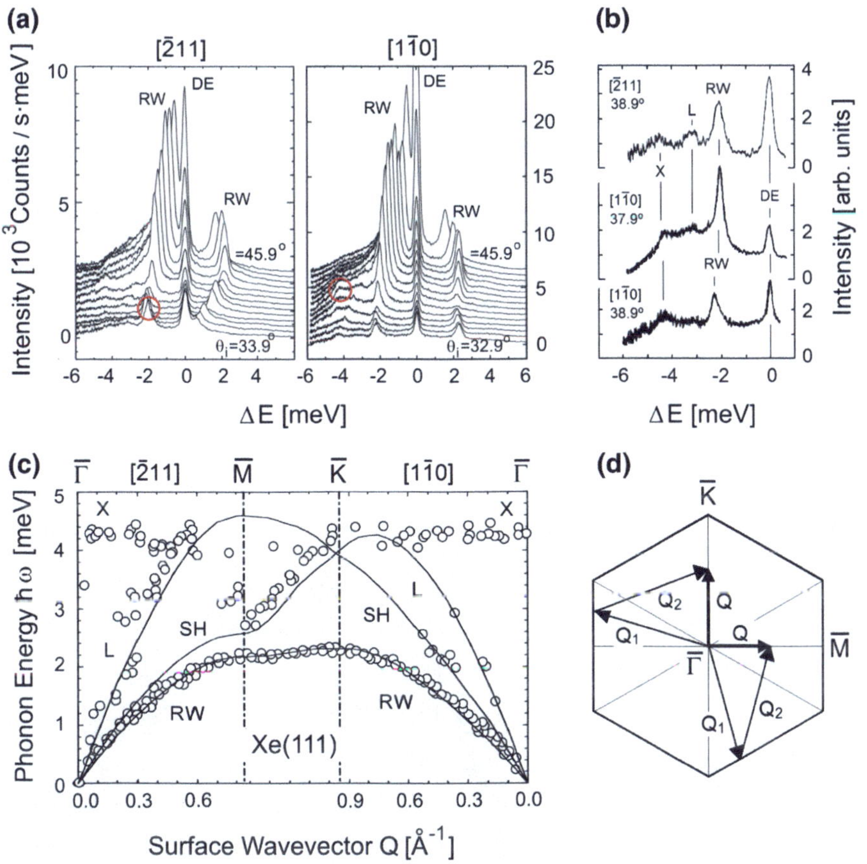

Fig. 11.1 Overview of HAS measurements of the dispersion curves of an epitaxially grown 160 layer film of Xe atoms on a Pt(111) substrate [25]. **a** Some typical time-of-flight spectra for the two main symmetry directions $\overline{\Gamma M} \equiv [\bar{2}11]$ (incident energy E_i = 10.5 meV) and $\overline{\Gamma K} \equiv [\bar{1}10]$ (E_i = 10.2 meV) and a surface temperature of 40 K. The main inelastic features on both creation and annihilation sides of the diffuse elastic (DE) peak are the Rayleigh waves (RW). **b** The three spectra plotted on an enlarged scale show, besides the RW and DE peaks, distinct inelastic features at larger energy transfers (L, X). At these incident angles the RW is near the zone boundary. **c** The resulting dispersion curves for the RW, L and SH acoustic branches folded back onto the first BZ (open circles) agree with the a Born-von Kármán force constant calculation (continuous curves) [25]. **d** The X branch was originally attributed to two-phonon excitations involving $\overline{MK}$ zone edge phonons according to wavevector combinations $\mathbf{Q}_1 + \mathbf{Q}_2 - \mathbf{Q}$ for any $\mathbf{Q}$ along the two symmetry directions [33]

excitation of two $\overline{MK}$ zone edge Rayleigh waves with almost oppositely directed wavevectors is favored by the coupling matrix elements and kinematics. As illustrated in Fig. 11.1d along the $\overline{\Gamma K}$ direction, there are infinite combinations of zone edge RW phonons of wavevectors $\mathbf{Q}_1$ and $\mathbf{Q}_2$ giving the same two-phonon state of wavevector **Q**. Whereas only zone edge wavevector combinations $\mathbf{Q}_1 + \mathbf{Q}_2 = \mathbf{Q}$ with $|\mathbf{Q}_1| = |\mathbf{Q}_2|$ give a two-phonon state with **Q**. The fact that the X branch is observed with similar intensity in both directions (Fig. 11.1b) suggests therefore some other origin, as discussed in the following.

Recently it was observed [34] that surprisingly the experimental dispersion curves for Xe(111) are in excellent agreement with those of Cu(111) calculated with a simple BvK force constant model [35], but with the energies reduced by about a factor 6 (Fig. 11.2a). This surprising correspondence has been confirmed by a new calculation for a 38-layer Xe(111) slab based on the same BvK force constant model in which the force constants were fitted to the experiment curves (Fig. 11.2b) [34]. The comparison shows that the experimental flat X-mode branch, originally attributed to two-phonon processes, actually corresponds to the subsurface shear vertical optical resonance S_2 localized in the second layer. This assignment is further supported by a DFPT calculation of the surface charge density oscillations

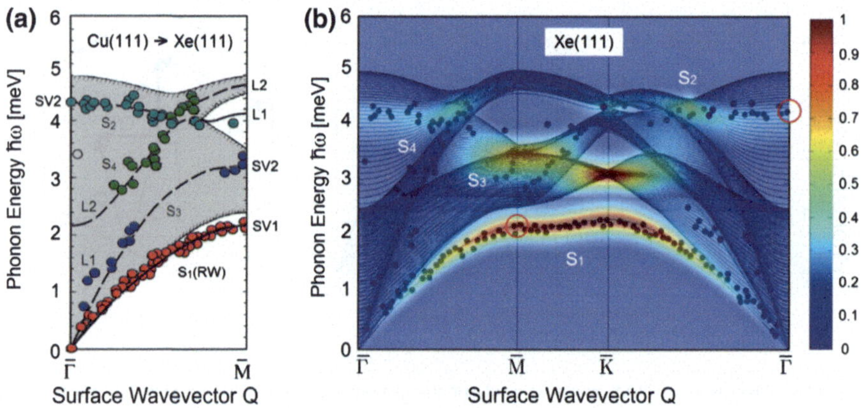

Fig. 11.2 Comparison of the experimental surface phonon dispersion curves (points) of Xe(111) in the $\bar{\Gamma}\bar{M}$ direction [25] with the calculated dispersion curves of Cu(111) [36]. **a** The Cu (111) energy axis has been reduced by a factor of about 6 so as to fit the dynamics of Xe(111). Full lines: localized surface modes; broken lines: surface phonon resonances; gray area: surface projected bulk phonon bands; S_1: Rayleigh wave (RW); S_2: Wallis mode; S_3: longitudinal acoustic resonance; S_4: longitudinal Lucas mode; SV1 (SV2): shear vertical polarization in the first (second) surface plane; L1(L2): longitudinal polarization in the first (second) surface plane. The excellent agreement with experiment suggests a surprising similarity between the surface dynamics of a noble metal and that of a highly polarizable noble gas solid. **b** The full set of HAS data are superimposed on the surface phonon densiy of states of SV polarization (see color code) projected onto the first and second atomic layer calculated with a Born-von Kármán force constant model for a 38-layer Xe(111) slab [34]. Theory predicts that the S_2 and S_4 optical resonances have their largest amplitude on the second layer below the surface as in Cu(111). The red circles indicate the mode S_2 at $\bar{\Gamma}$, which is localized in the second layer, and S_1 (RW) at $\bar{M}$, localized on the first layer, respectively. As seen also in Fig. 11.1a, c they have comparable HAS intensities. Thus the experimental flat X-mode branch, originally attributed to two-phonon processes, is in fact a subsurface shear vertical optical resonance S_2 localized in the second layer (adapted from [34])

induced by the shear vertical displacements of the RW (S_1) at $\bar{M}$, and by the subsurface S_2 phonon at $\bar{\Gamma}$. It appears that the two modes have comparable amplitudes, in agreement with the experimental amplitudes (Fig. 11.2b), after the Bose-Einstein factor has been divided out, thus providing a further example of the quantum sonar effect (see Sect. 8.4). A similar effect is found in Pb nanofilms (see Figs. 8.3 and 8.4).

Especially surprising is the observation of this effect in an insulator surface. As explained in [34] Xe atoms are highly polarizable and the surface shell oscillations, pushed up and pulled down by the motion of the second layer atoms, mimic the effect of the pseudocharges which allow for the observation of subsurface modes in Cu(111) and Pb(111) thin films (see Sect. 11.3.2). Thus the possibility of directly measuring the surface electron-phonon coupling constants from inelastic HAS intensities, originally demonstrated for conducting surfaces, can be extended to highly polarizable insulating surfaces, as well as oxides and heavier chalcogenides.

11.1.2 Ionic Crystals: The Alkali Halides

The first HAS measurements of a surface phonon dispersion curve in 1980, carried out by Brusdeylins et al. [37–39] for the LiF(001) surface, stimulated much of the work reported in this monograph. The choice of LiF(001) was motivated largely by experimental considerations since the surface is easily prepared by cleavage in the laboratory and then remains clean in the apparatus over long-periods even in a relatively poor vacuum of only 10^{-7} mbar. Moreover, at that time the alkali halide surfaces were the only ones for which realistic calculations were available and which were readily accessible for experimental investigation. The calculations were based on either the shell model or the breathing shell model (BSM) in the framework of the slab method [40–45] or the GF method [46–50], respectively. With the advent of the HAS experiments these theoretical methods could be tested for the first time and they were found to be in overall good agreement with the HAS data [51, 52]. Figure 11.3 shows a comparison between experimental and theoretical dispersion curves for this important prototype system. The initial experimental observation of a 10% lower Rayleigh mode frequency near the $\overline{M}$-point zone boundary subsequently led to various theoretical improvements including elastic relaxation [53–56], changes in the surface electronic structure and polarizability [57, 58], and many-body effects [54–56, 58]. All these mechanisms, however, have a similar effect so that a clear-cut interpretation of the $\overline{M}$-point RW softening in LiF (001) has, surprisingly, not yet been given.

The LiF surface also provided the first surface with which to test the theory for the inelastic differential reflection coefficient. Because of the largely ionic nature of

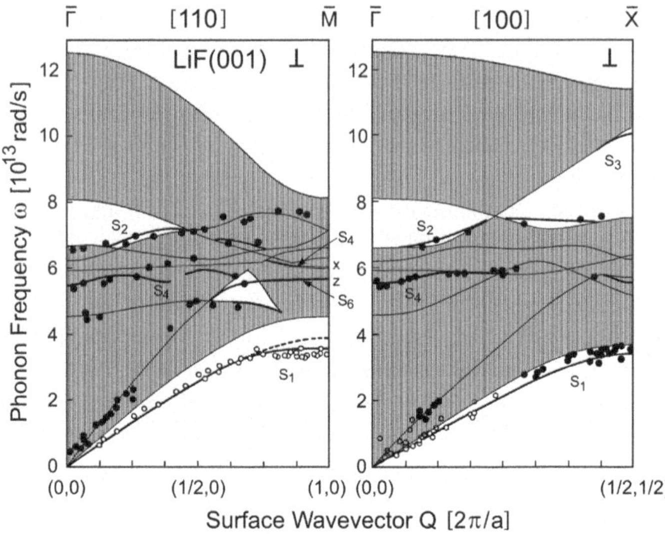

Fig. 11.3 Comparison of calculated and experimental surface phonon dispersion curves of LiF (001) with sagittal polarization (⊥) along the $\overline{\Gamma M} \equiv [100]$ and $\overline{\Gamma X} \equiv [110]$ directions. The surface wavevector Q is expressed in units of $2\pi/a$, where a is the lattice constant. The calculated curves based on the Green's function method and the breathing shell model for lattice dynamics are in agreement [57]. Heavy lines are surface modes; shaded areas correspond to bulk bands projected onto the surface. The curves have been calculated with an 8% increase in the surface F^- ion polarizability [57]. This effect is largest for the RW frequency near the $\overline{M}$-point. The broken line near the $\overline{M}$-point shows the RW dispersion for the ideal case where the polarizability of F^- ions at the surface is the same as in the bulk. The He scattering data are from [39] (open circles) and from [52] (full circles)

the bonding with a covalent contribution of less than about 5% in the lattice [59], the assumption of pairwise additivity of the separate two-body potentials between the He atoms and the Li^+ and F^- ions is well justified making this another ideal model system. Figure 11.4 shows a typical comparison of an experimental energy transfer spectrum with DWBA calculations based on various potential models [60]. The best agreement is achieved with an atom-surface potential, which is obtained from a sum over two-body potentials, based on sophisticated semi-empirical methods developed for gas phase atom-atom interactions [61] (see Sects. 6.2–6.5).

These initial experiments and comparisons with reliable theoretical models provide evidence that indeed the TOF spectra are largely dominated by single phonon events and that at least for He atoms with incident kinetic energies of less than about 40–60 meV multiphonon events, which are not included in these calculations, can be largely neglected. Moreover they confirm the theoretical expectation that bulk phonons play a subordinate role compared to the surface modes [51]. The calculations presented in Fig. 11.4 show indeed that the Rayleigh wave peaks (histograms) rather than the bumps of the bulk phonon density (dashed lines) account for the main features of the experimental spectrum.

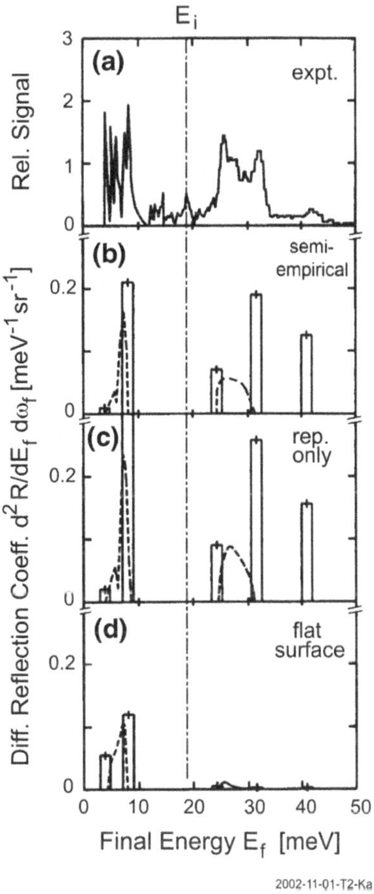

Fig. 11.4 Comparison of a typical experimental energy transfer spectrum for LiF (001) [100] with three calculations for different potential models [61]. **a** The HAS spectrum measured for $E_i = 18.6$ meV at $T_s = 300$ K [61]. **b–d** Calculated energy transfer spectra for inelastic scattering from the Rayleigh wave surface mode (histograms) and the additional contributions from bulk phonons (dashed line curves) (**b**) for the corrugated semi-empirical potential described in Fig. 6.4 of Chap. 6; **c** without the attractive terms in the potential; **d** without the corrugation terms in the Fourier expansion but with the attractive part included. The comparisons reveal the dominant effect of the corrugated repulsive potential

Under the conditions of this particular experiment, the same model potential in which, however, the attractive dispersion terms in the interaction potential were neglected altogether (Fig. 11.4c) also provides nearly equally good agreement with the inelastic intensities. This was also found in another systematic analysis carried out for the same surface, but using the Hard Corrugated Surface (HCS) model [51]. However as seen in Fig. 11.4d the complete neglect of the repulsive surface corrugation, which is particularly large in the case of LiF, leads to an unsatisfactory description of the experimental data. The dominant effect of the repulsive potential has been found to hold also for the low index metal surfaces and appears to be a universal phenomenon. As discussed in Chap. 6 most low index metal surfaces have only a negligible corrugation and usually can also be neglected altogether.

In the course of these early scattering experiments the bound state resonances, discussed at some length in Sect. 10.1, were found to have a big influence on the phonon intensities (see for example Fig. 10.6). The phenomenon of bound state resonances, originally called *selective adsorption*, was discovered experimentally in angular distributions in the early 1930s by Otto Stern's group in Hamburg [62, 63], but the correct explanation was provided a few years later in the famous series of theoretical publications by Devonshire, Lennard-Jones and Strachan in England (see also Chap. 2) [64–70]. This effect has been extensively studied [71, 72] and exploited for determining the bound state energies in the surface potential [73–75]. In 1981 Brusdeylins et al. [73] and later Lilienkamp and Toennies [76] observed for the first time small intensity maxima in the weak inelastic background in their angular distributions, which, from an analysis of the corresponding TOF spectra [51], could be assigned to inelastic resonances. The various inelastic processes involving bound state resonances are discussed in the previous Chap. 10.

No clear evidence of surface optical modes could be gleaned from these first He scattering TOF studies of surface phonons on LiF(001). This was expected since the neutral He atoms cannot couple directly but only weakly to the anti-phase (dipolar) optical mode motion of the ions at the surface. Peaks ascribed to optical modes were first found for NaF(001) in HAS time-of-flight spectra at higher energies of about 100 meV [77, 78]. The price paid by raising the beam energy is, however, an increase in the multiphonon background. This is illustrated by the comparison shown in Fig. 11.5 between two inelastic HAS spectra from NaF(001) measured at different incident energies E_i and angles θ_i for a surface temperature of 300 K [78]. At E_i = 30 meV and $\theta_i = 31.5°$ the one-phonon scattering dominates the spectrum in (a), as appears from the comparison with the calculated spectrum in (b). Whereas as seen in (c) at a high incident energy and angle (E_i = 104 meV, $\theta_i = 66.9°$) a large multiphonon background, dotted line in (c), arises. In the latter case, however, the comparison with the calculated spectrum in (d) allows to assign the weak features to the optical modes S_4 and S_2 [78], whereas at lower incident energy (a, b) only the Rayleigh (S_1) and folded (S_8) modes are observed.

Since the multiphonon background should be almost structureless, it is generally accounted for by a smooth interpolation of the intensity on both sides of the sharp single-phonon peaks. After subtraction from the measured spectrum the remaining features can then be assigned to one-phonon events. A theoretical justification for such a treatment of the multiphonon background, discussed in Sect. 7.6, was first given by Celli et al. [80].

The transition from single-phonon to multiphonon scattering has been studied theoretically by Manson in collaboration with several experimental groups [81–84] for a wide range of different surfaces, and by Gumhalter [85]. In LiF(001) the optical surface modes, which are stiffer and have weaker HAS intensities than in NaF(001), could also be identified by exploiting the enhancement due to bound state inelastic resonances, as shown in a benchmark experiment carried out at Genoa by Bracco et al. (full circles in Fig. 11.3) [52].

Skofronick and coworkers, at the Florida State University, have found and analyzed in a number of the heavier alkali halides additional phonon dispersion features such as the folded Rayleigh modes discussed in Sects. 3.1 and 3.7.3 [86–91], as well as surface modes which appear in the gap between the optical and the acoustic bulk bands [88, 89, 92]. An interesting example is KBr(001), shown in Fig. 11.6. The HAS time-of-flight spectra measured for this surface along the [110]

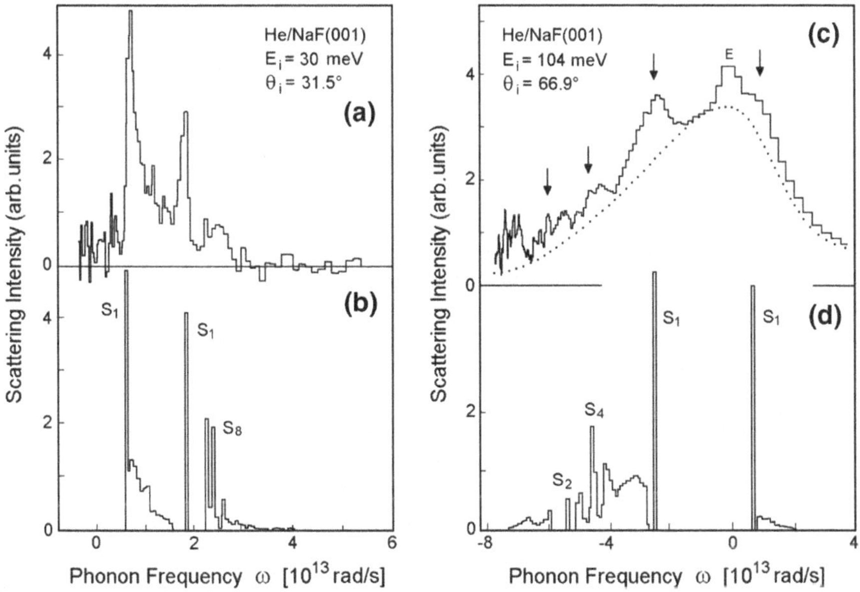

Fig. 11.5 Comparison between inelastic HAS spectra from the NaF(001) surface at 300 K measured along [79] for a 90° scattering geometry at two different incident energies E_i and angles θ_i with calculations based on the eikonal approximation for one-phonon excitation [78]. **a** At E_i = 30 meV and θ_i = 31.5° the calculated one-phonon contributions S_1 and S_8 in (**b**) account for a large part of the experimental spectrum. **c** At the larger incident energy (E_i = 104 meV) and angle (θ_i = 66.9°) the experimental spectrum is dominated by a large multiphonon background (dotted line). The comparison with the calculated spectrum in (**d**) allows to assign distinct features to the optical modes S_4 and S_2 [78], whereas at the lower incident energy (**a**, **b**) only the Rayleigh (S_1) and folded (S_8) modes are observed. The label E in (**c**) indicates the diffuse elastic peak

and [100] symmetry directions (Fig. 11.6a, c) show several inelastic peaks (S_j) on both the creation and annihilation side [89]. The corresponding points in the (ω, Q) diagram (Fig. 11.6b), when compared with the theoretical surface phonon dispersion curves (full lines), are seen to correspond to the acoustic modes (S_1, S_6) as well as to the sagittally polarized optical surface modes (S_2, S_4) [90]. The Wallis mode S_2 falls in the gap between the optical and acoustic bulk bands and well below the sagittal Lucas mode S_4, which is a special feature of potassium halides discussed in [92]. As a consequence of its pronounced surface localization (cf. Sect. 3.7), the S_2-mode gives a comparatively strong peak in the TOF spectra.

Another intriguing feature of KBr(001) is the observation of the folded Rayleigh mode (S_8) (see Table 3.1) which is normally expected in quasi-monoatomic (*nearly isobaric*) crystals such as NaF, MgO, KCl and RbBr [91]. De Wette et al. [93] were the first to point out that the near equality of ionic masses is a sufficient but not necessary condition for the existence of a folded Rayleigh branch. As shown by these authors there is indeed another possible mechanism responsible for the

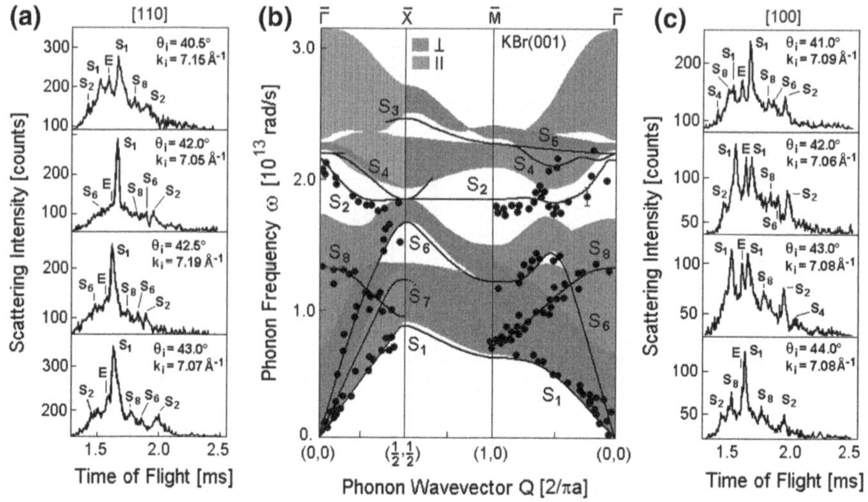

Fig. 11.6 Overview of HAS measurements of the dispersion curves of the KBr (001) surface. **a**, **c** Some typical time-of-flight for the two main symmetry directions. Besides the diffuse elastic peak E several inelastic peaks corresponding to surface phonons (S_j) are observed. **b** The corresponding points are superimposed on the theoretical surface-projected bulk phonon densities-of-state with sagittal (⊥) and shear-horizontal (||) polarization and the theoretical surface phonon dispersion curves (full lines). The surface wavevector Q is expressed in units of $2\pi/a$, where a is the lattice constant (adapted from [89, 90])

appearance of this branch, which consists in the hybridization (i.e., polarization exchange) between the acoustic and optical modes in certain directions of the bulk Brillouin zone. It was subsequently recognized that the same mechanism is also responsible for the anomaly observed in the optical surface mode S_2 at one half of the $\bar{\Gamma}\bar{M}$ direction where the gap narrowing indicates an avoided crossing between an optical and an acoustic bulk branch [90].

The complete observation of all the expected surface modes with sagittal polarization for many alkali halides at the time demonstrated the potential of He atom scattering TOF spectroscopy and also provided the first confirmation of the theory of surface dynamics and the atom-surface scattering theory. In the ionic insulator crystals, the experimentally determined phonon dispersion curves agreed in most cases with the theoretical predictions, even in some significant details as seen for the case of KBr(001). This appears to be related to the fact that in a lattice made up of closed shells the tightly bound electrons closely follow the ion cores during the surface vibrations. Electronic effects are largely restricted to dynamical ion polarization which is quite precisely accounted for by the shell models used in surface dynamics. Yet the problem of the change of ion polarizabilities and related many-body effects at the surface still awaits a full clarification even in insulators, particularly in the chalcogenides. The oxygen ion O^{2-}, for example, only exists in ionic oxides due to the stabilizing effect of the crystal field in the bulk and has a reduced stability at the surface, resulting in a large polarizability [94, 95].

11.1.3 Ionic Crystals: AgBr(001)

The silver halides are another important class of ionic crystals, in which bulk vibrations are largely affected by ion polarizabilities, notably the quadrupolar polarizability of open shell silver ions due to sd hybridization [14, 96, 97]. The conjecture that this effect may be more pronounced at the surface is one of the reasons for studying surface phonons in silver halides. Another reason is the importance of the silver halides in the photographic process [98]. Because of their inherent sensitivity to radiation their surfaces cannot be studied with either photons or electrons, the latter having the additional disadvantage of producing an electrostatic charge at the surface. Their surface phonons are also of some interest in the present context since, whereas the alkali halides have a nearly 100% ionic character, the degree of ionic bonding in the case of AgBr is estimated to be between 75.7 and 84% [36]. The surfaces are difficult to prepare. Because of their inherent softness the crystals cannot be easily cleaved, even at low temperatures and, moreover, they cannot be polished. The experiments reported here were made possible by combining several HAS techniques, which for these systems are especially ideal because of the non-destructive gentle nature of the interaction of the neutral atoms with this ultra-sensitive surface [99]. Thick films with up to 100 layers could be grown *in situ* by vapor deposition on NaCl(001) substrates. This was possible since the surface smoothness could be monitored and the deposition conditions optimized via the reflectivity of the specular peak and the diffraction patterns measured during growth.

The AgBr surface exhibits a number of anomalous features. The lattice has a comparatively large coefficient of thermal expansion [100]. The corrugation amplitude, as estimated from the diffraction intensities using the eikonal approximation and assuming a corrugated hard wall potential (cfr. Sect. 6.2 and Fig. 6.4), is only $\xi_0 = 0.16 \pm 0.005$ Å [99] making it smoother than expected for a purely ionic lattice in agreement with the stronger covalent bonding. Moreover after a prolonged heat treatment the AgBr(001) surface exhibits an unusual reconstruction to a (2×2) phase on cooling to below ~60 K [99].

Figure 11.7 shows some typical time-of-flight spectra and the measured dispersion curves for the two main symmetry directions [101]. Unfortunately calculations for direct comparison have not yet been reported. The Rayleigh mode (S_1) with energies of 3.7 and 3.5 meV at the $\overline{M}$- and $\overline{X}$ points, respectively, are consistent with the neutron scattering bulk TA mode frequency of 3.8 meV measured at an equivalent point in the corresponding Brillouin zone. The mode marked S_8 in Fig. 11.7b at first glance appears to be a folded Rayleigh mode. This mode, however, can hardly be attributed to the isobaric mechanism since the Ag^+ ion is about 25% heavier than the Br^- ion. A more likely explanation is based on the hybridization mechanism, discussed above for KBr(001), which can be justified by the unusual nature of the bulk dynamics of AgBr. Fischer et al. in 1972 demonstrated long ago, by means of a quadrupolar shell model (QSM), that the tetragonal component of the quadrupolar polarizability of the Ag^+ ion plays an important role

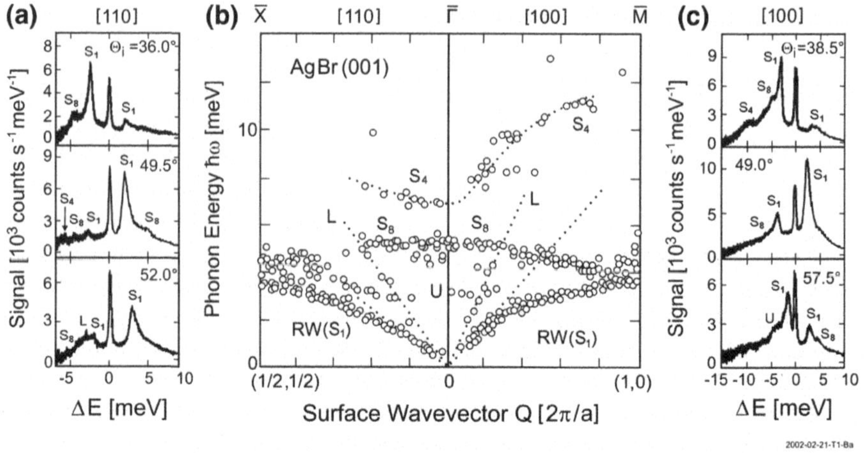

Fig. 11.7 Overview of HAS measurements of the dispersion curves of an epitaxially grown 10 layer AgBr(001) crystal surface on a NaCl(001) substrate [101]. **a**, **c** Some typical energy transfer spectra for the two main symmetry directions measured with $E_i = 39$ meV at $T_s = 100$ K. **b** The resulting dispersion curves folded back onto the first BZ. The surface wavevector Q is expressed in units of $2\pi/a$, where a is the lattice constant. The dotted lines in (**b**) are guides to the eye

in the bulk dynamics of silver halides and has to be included for a satisfactory description of the experimental dispersion curves [96]. The large quadrupolar deformability of the silver ions explains the inversion of the transverse mode TO-TA eigenvectors near the L-point of the bulk Brillouin zone as experimentally confirmed [102]. Moreover, the AgBr bulk dispersion curves exhibit large TO-TA and LO-LA hybridizations in the middle of the [110] and [100] directions, respectively [14]. Thus the occurrence of the folded Rayleigh mode in AgBr(001) can also be viewed as an effect of hybridization and, ultimately, of the large quadrupolar polarizability of the Ag^+ ion.

The HAS spectra at energy transfers above 7 meV in Fig. 11.7 reveal another unexpected mode. This weak peak is attributed to an optical mode, which is tentatively assigned to the quasi-longitudinal Lucas mode S_4 (Fig. 11.7b), although the lower edge of the surface-projected TO band at $\overline{\Gamma}$, is significantly higher at 10.4 meV [14]. Such a large softening of a surface mode below the edge of the corresponding bulk band is rather unusual and again can be attributed to the large surface quadrupolar polarizability of Ag^+. This enables the Ag^+ ion to deform when the neighboring Br^- ions move against it, thus reducing the mutual repulsion. The role of the quadrupolar polarizability in determining the high mobility of Ag^+ and the photographic process has been further highlighted by *ab initio* calculations [97]. The pronounced diffusivity of Ag^+ implies the formation at the surface of many defects such as vacant silver sites, adatoms and clusters. The additional phonon branch observed in the low energy region around 3 meV (U in Fig. 11.7b) is thought to be associated with some surface defect, since it cannot be assigned to any known surface phonon.

11.2 Semiconductors: Surface Phonons and Structure

11.2.1 Si(111) 2×1

The lattice dynamics of semiconductors are of considerable interest because of the highly covalent bonds holding atoms together in these substances [105, 106]. The technological relevance of semiconductors and their surfaces is another motivation for the full characterization of their dynamical properties. The neutron measurements of bulk phonons [107, 108] stimulated the development of several different models, of which the bond charge model (Sect. 5.2) has been particularly useful also for describing the surface phonons. Thus both the bulk and surface phonons of the semiconductors have been a testing ground for advanced semi-empirical and first principle methods in lattice dynamics (see Chap. 5). Compared to the alkali halides the surface dynamics of elemental and compound semiconductors is made more complex by the ubiquitous spontaneous reconstruction of their surfaces. This is due to the intrinsic instability of the half-filled orbitals associated with the broken (dangling) bonds produced when the surface is initially formed. For example the native Si(111) surface reconstructs spontaneously on cleavage to a 2×1 structure (Fig. 11.8) [109–112]. At $T_s \geq 500$ K this surface then reconstructs irreversibly to produce the stable 7×7 structure [113].

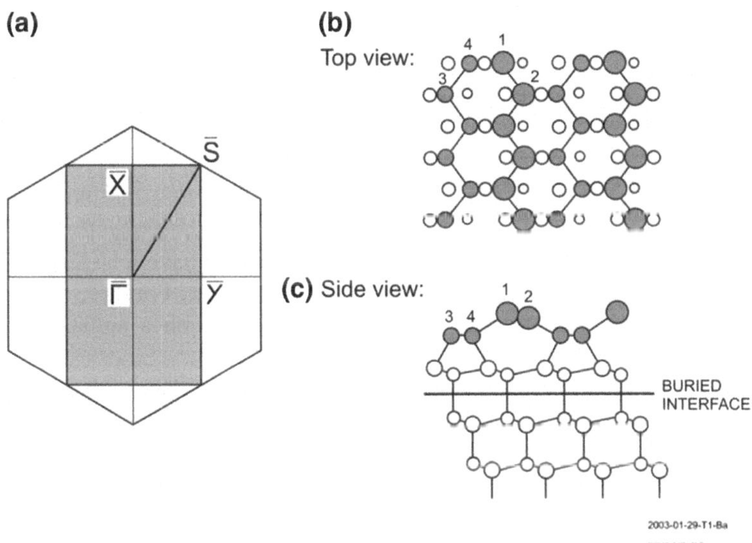

Fig. 11.8 The surface Brillouin zone and structures of the ideal Si(111) 2×1 reconstructed surface. **a** The shaded region shows the Brillouin zone of the 2×1 reconstructed surface. **b** Top view of the 2×1 reconstructed surface as obtained from the projection onto a $(1\bar{1}0)$ plane normal to the surface. **c** Side view showing the alternating sequence of 7-fold and 5-fold atomic rings, according to Pandey's reconstruction model [111, 112]. Atoms 1 and 2, here shown in the positive buckled (*positive modified Pandey*, +MP) configuration, belong to the surface π-bonded chain and atoms 3 and 4 to the groove between two neighboring chains

The first surface phonon studies of Si(111) were carried out by Harald Ibach with EELS in the low energy dipole coupling regime and led to the discovery of an intense optical mode at 55 meV on Si(111) 2×1 [114, 115]. Since silicon is fully covalent and cannot sustain any macroscopic Fuchs and Kliever surface mode (Sect. 3.8), this observation is regarded as the first experimental evidence of a genuine microscopic optical surface phonon.

The first helium atom TOF investigations of the semiconductor surfaces were reported for Si(111) 2×1 [116] and GaAs(110) [79, 117] in 1986 and 1987 by Harten et al. In both systems a flat nearly dispersionless surface phonon branch was found with frequencies of about 10 meV. This branch, seen for Si(111) 2×1 in Fig. 11.9, corresponds roughly to a folded Rayleigh branch resulting from the reconstruction, which leads to a doubling of the surface unit cell with respect to the bulk. Despite this seemingly simple origin the quantitative explanation has challenged theoreticians no less than the interpretation of the optical mode at 55 meV discovered on the same Si(111) (2×1) surface earlier by EELS and mentioned above [114].

The alternating sequence of five- and seven-atom rings occurring in the π-bonded 2×1 reconstructed surface [111, 112] requires a theoretical treatment with a fairly large unit cell. As seen in Fig. 11.8b, c, even assuming that the lattice structure immediately below the 5- and 7-fold rings remains unaffected by the surface reconstruction, the surface unit cell includes already eight atoms. Even the simplest surface model requires in this case a surface perturbation extending at least down to the fifth atomic layer. Because of this deep penetration of the surface reconstruction, it has proven useful to introduce the concept of a buried interface separating the surface atomic layers, affected by the reconstruction, from the bulk layers which are assumed to remain unaffected (Fig. 11.8c). This makes it possible to define the number of degrees of freedom and surface phonon branches which are affected by the perturbation. With respect to the ideal surface the more compact five-atom rings are expected to give stiffer modes whereas the expanded seven-fold rings will have softer modes. Moreover, some modes may have a larger amplitude below the surface near the buried interface than at the surface. For these reasons it is especially important that the slab calculations are based on a sufficiently thick sample.

The available experimental data for the surface phonon dispersion curves of Si (111) 2×1, shown in Fig. 11.10, deviate significantly from the predictions for the ideal surface [104, 118, 119]. The flat branch at 10.5 meV and the 55 meV optical mode have been studied by a variety of theoretical techniques such as the tight-binding method [120–122], the bond charge model [103], the Car-Parrinello quantum molecular dynamics [123, 124] and density functional theory methods [125–127]. The bond-charge model (BCM) calculation for a 22 layer slab [103], based on Weber's model for bulk dynamics [128–130] adapted to the unusual surface geometry, is in good agreement with both and the optical mode at 55 meV at $\bar{\Gamma}$ and the flat 10 meV (Fig. 11.9). The almost isotropic dipolar activity of the former mostly arises from the *vertical* motion of atoms in the subsurface layers (mode $I_z(00{+}{+})$ as

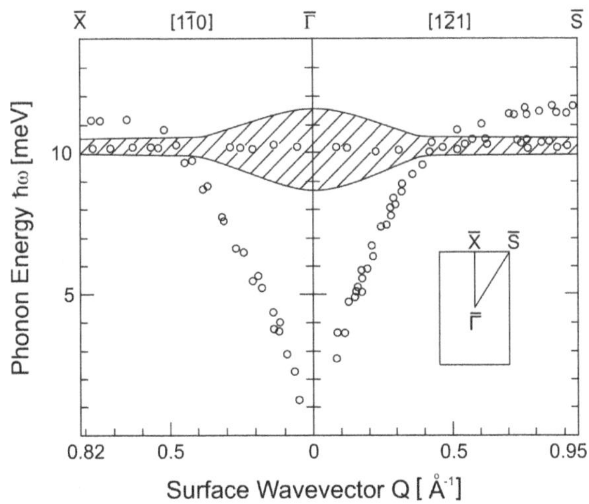

Fig. 11.9 The surface phonon dispersion curves of Si(111) 2×1 measured by HAS in the low-frequency part of the spectrum in two symmetry directions. The surface Brillouin zone is shown in the inset [115]. The hatched area of the flat branch at 10.5 meV gives the width (FWHM) of the measured peaks. The additional width at $Q \leq 0.4$ Å^{-1} is due to a coupling to the bulk phonons

illustrated in the left margin of Fig. 11.10. In this mode the chain atoms are practically at rest, whereas the pair of groove atoms 3 and 4 oscillates vertically against the anti-phase motion of the two next atomic layers underneath. The bond charge redistribution in the surface region endows this mode with a non-negligible dipole moment. The fact that the 55 meV mode observed with EELS is not a chain mode was confirmed as early as 1997 by a DFPT calculation [125] (Fig. 11.11) and later by the first principle analysis by Patterson et al. [127] of the infrared conductivity spectrum of Si(111)2×1 in both the positively buckled (+MP) and and negatively buckled (−MP) configurations. Buckling refers to the difference in the vertical positions of the two adjacent chain atoms (1 and 2 in Figs. 11.8c, 11.10 and 11.11). In the theory of Patterson et al. the strongest optical-active mode localized on the chain (x(+−−+) in Fig. 11.10) is predicted at 50.4 meV, well below the 55 meV interface mode $I_z(00{+}{+})$ in essential agreement with the BCM calculations.

With the aid of the BCM theory it was also possible to gain some information on the structure of the surface from the HAS measurements. For example, the calculations indicate that the small splitting of the 10.5 meV mode observed experimentally at the zone boundary along the $\overline{\Gamma S}$ direction (Fig. 11.9) is a sensitive measure of the buckling of the surface chains resulting from the predominant π - bonding. The best fit of the measurements by Harten et al. [116], based on the BCM, suggests that the buckling is positive (atom 1 lies above atom 2) and amounts to 0.20 Å, in agreement with density functional structure calculations [133, 134], but significantly smaller than the value of 0.30 Å previously deduced from fits to LEED data [135]. The suitability of BCM for this structural analysis relies on the fact that the bond angles enter the model as adjustable parameters whereas the bond lengths are kept fixed. The relationship between buckling and zone boundary phonons revealed by the BCM analysis has been confirmed by the TB calculations of the structure and zone boundary phonon frequencies for the III-V (110) surfaces by Duke et al. [136–138]. Also the DFPT calculations performed by Zilztsperger

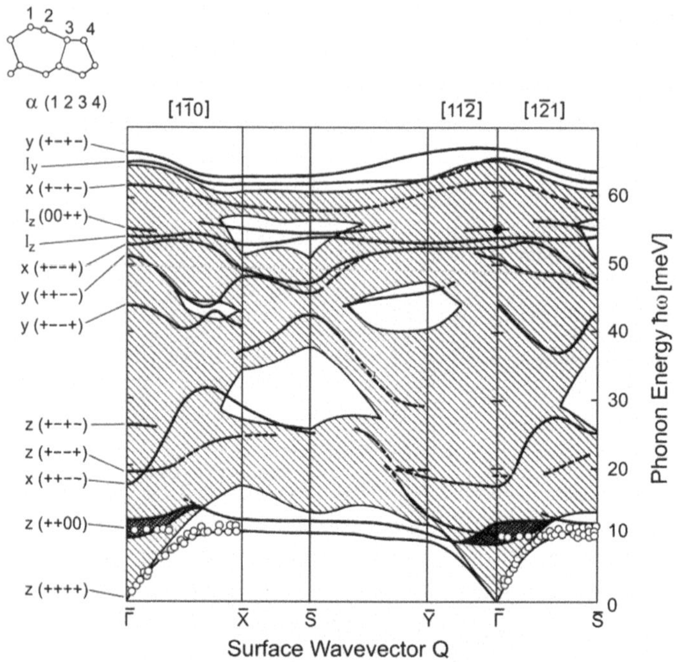

Fig. 11.10 Comparison of bond charge model (BCM) calculations of surface phonon dispersion curves for the positively buckled (+MP) reconstructed Si(111) 2×1 surface [103] with the results of HAS (open circle) [116] and EELS (filled circle) experiments [114, 131, 132]. The calculations predict a broad resonance at about 10 meV (darker shaded regions). The left margin shows the displacements of the four surface atoms (inset, top left), and polarizations α for the zone center surface modes are indicated by their orientations (+ or −) or their being at rest (o) with respect to the direction α, where α = x is parallel to the chain, α = y is in the surface plane but normal to the chain and α = z is normal to the surface plane. The three modes designated on the left column as I_y and I_z are localized at the internal interface between the reconstructed surface region and the inner unperturbed bulk

et al. [125] of the Si(111) 2×1 dispersion curves for both a positive and a reversed (negative) buckling (Fig. 11.11). Their calculations predict correctly the 10.5 meV mode but only outside the bulk band, where it becomes strongly localized on the π-bonded chain. They also predict a 20% stiffening in the negative buckled phase. Whereas the experiment indicates a perfectly flat branch down to Q = 0, as often found for weakly bound surface defects (see Chap. 12), the BCM calculation yields a fairly large dispersion of the resonance inside the band amounting to about 2.5 meV. The flat 10.5 meV mode can be viewed as resulting from the folding of the RW dispersion curve due to the 2×1 reconstruction, although heavily affected by the extended surface perturbation. In this picture the zone edge splitting (∼1.5 meV in experiment) increases with the extent of the reconstruction.

An important theoretical prediction, on which the BCM and *ab initio* calculations agree, is the large number of surface phonon branches, which amount to

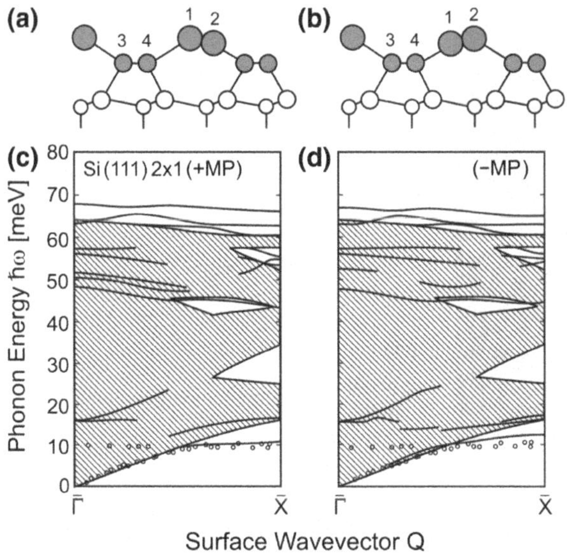

Fig. 11.11 Density functional perturbation theory calculations of the surface phonon dispersion curves of Si(111) 2×1 in the $\bar{\Gamma}\bar{X}$ direction for a positive and negative buckled surface. **a**, **c** The structure and the dispersion curves of the positive buckled (+MP) surface, respectively. **b**, **d** The same for the negative buckled (-MP) surface (adapted from Zitzlsperger et al. [125]). The experimental HAS data circles are the same as in Fig. 11.10 [116]. Note the close correspondence (average deviation 2.1%) of the surface modes at Q = 0 (the doublet at 17 meV, the quintet between 40 and 60 meV and the triplet above 60 meV) with those calculated with BCM (Fig. 11.10), except for the 10.5 and 26 meV resonances, not resolved in DFPT calculation. DFPT predicts correctly the 10.5 meV mode outside the bulk band, where it becomes strongly localized on the π-bonded chain, and predicts a 20% stiffening in the negative buckled phase

sixteen including the gap modes. This number is consistent with the fact that at least eight atoms are affected by the surface perturbation. All calculations predict two surface phonon branches above the bulk phonon bands, which are associated with anti-phase stretching of the bonds between neighbor atoms in the chains and grooves and are stiffened by the surface charge redistribution.

11.2.2 Si(111) 7×7

The Si(111) 7×7 reconstructed surface presents a formidable experimental challenge [139] because of the much larger unit cell and the structural complexity of the dimer-adatom stacking-fault (DAS) model introduced by Takayanagi et al. [140]. The STM images and the top and side views of this model are shown in Fig. 11.12. The smallest set of surface atoms involved in the reconstruction includes 12 adatoms, 72 rest atoms and 18 dimer atoms, which makes altogether 102 atoms. Also

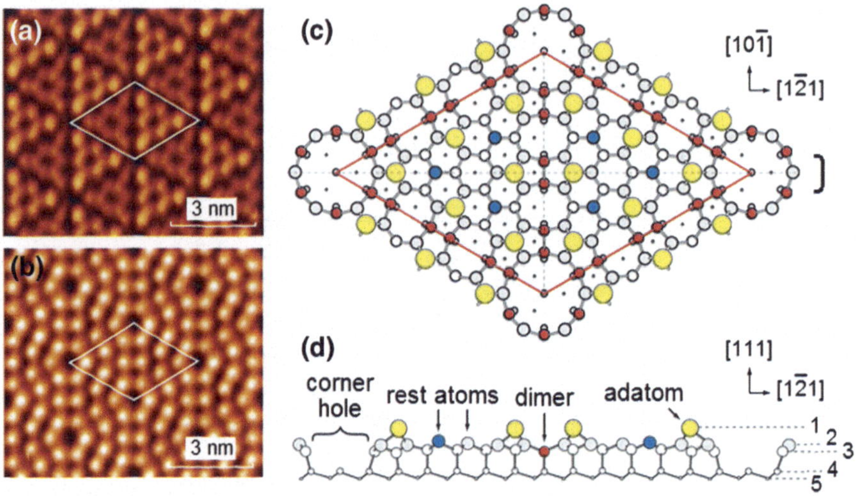

Fig. 11.12 The dimer-adatom stacking-fault (DAS) model of the Si(111) 7×7 reconstructed surface [139, 140]. **a**, **b** STM images; **c** top and **d** side views of the reconstructed surface. The unit cell (lozenge) contains 12 adatoms, 72 rest atoms in two layers (30 in the lower layer (open circles) and 42 in the upper layer, of which 6 (blue circles) are isolated and 36 are linked to adatoms (grey circles)) and 18 atoms forming 9 dimers, which makes altogether 102 atoms. The side view (**d**) encompasses the three central rows within the bracket on the left side in (**c**). The surface unit cell encompasses 49 unit cells of an ideal (111) bulk substrate (adapted from *thesaurus.rusnano.con/wiki/article14156*, from Takayanagi et al. [140])

in this case the buried interface can be located between the 5th and the 6th atomic layer (Fig. 11.12d). The dynamical perturbation matrix necessarily includes also the 98 atoms of the 7×7 = 49 bulk unit cells of the ideal (111) substrate, which the dimers and rest atoms are linked to, and reaches therefore the dimension of 600 × 600.

Because of the extensive folding and bunching of the modes in a narrow region of reciprocal lattice space, EELS measurements [141] only reveal some unresolved features at 25–33 meV and at 70.8 meV. These were interpreted in terms of vibrations localized on the adatoms and atoms in the second and third layers beneath the adatoms [142, 143]. The HAS time-of-flight (TOF) spectra converted to an energy transfer scale, shown in Fig. 11.13, on the other hand, reveal a rich mode structure [144] with resolved phonon peaks with energies up to about 16 meV. Due to the high density of diffraction peaks and the large corrugation felt by the helium atoms, several of the inelastic peaks are however due to spurions (Sect. 9.9), which are marked in Fig. 11.13 by an S.

Figure 11.14a shows the HAS angular distributions and Fig. 11.14b the surface phonon dispersion curves extracted from TOF measurements, some of which are shown in Fig. 11.13. For direct comparison the former have been plotted on the same momentum transfer abscissa. For comparison with the experiment the same BCM force constants which provided a good fit of the H-terminated Si(111)1×1

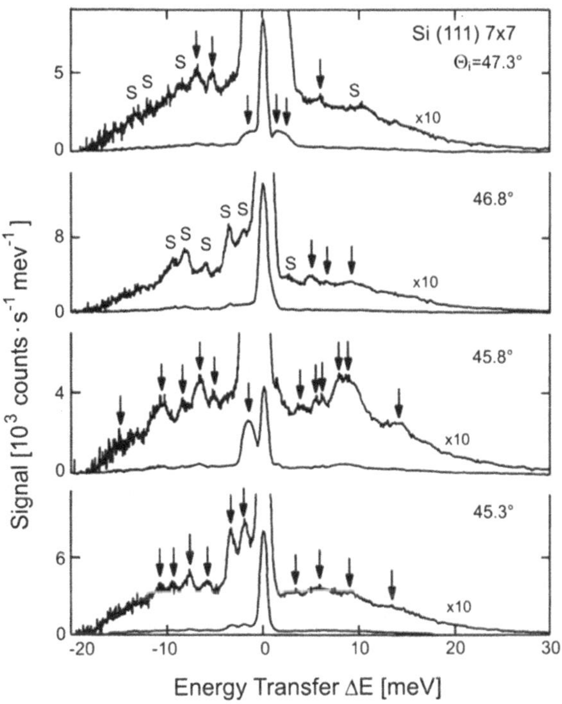

Fig. 11.13 Four typical HAS time-of-flight spectra (transformed to an energy transfer scale) from the Si (111) 7×7 surface for closely spaced incident angles along the $[1\bar{2}1]$ direction with $E_i = 28.0$ meV; $T_s = 300$ K. The arrows indicate the positions of single phonon peaks. Spurions are marked with an *S* [144]

surface [145], which is considered to be close to the ideal termination of the silicon surface, were used to calculate the Rayleigh wave dispersion curve. The solid lines in Fig. 11.14b show the dispersion curves after they had been folded to correspond to the 7×7 surface. This simple choice of force constants for an ideal termination was made in lieu of any previous information on the effect of the 7×7 reconstruction on the forces between the Si atoms in the surface region. Because of the high density of lattice points, umklapp phonons from nearby off-axis lattice vectors, which are not along the symmetry direction chosen in the experiment, also make a significant contribution to the TOF spectra. This explains why some of the dispersion curves (Fig. 11.14b) do not go to zero and also explains some of the measured points between about 6 and 16 meV.

It is interesting to see in Fig. 11.14b that the simple model of a folded ideally terminated surface provides a reasonable description of the results at phonon energies in the lower part of the spectrum. This correspondence suggests that the low frequency phonon branches of the macroscopic modes are dominated by the collective vibrations of a large number of atoms extending deeply into the crystal and therefore are not greatly affected by the complicated surface structure [146].

More information on the surface forces can be derived from the phonons with energies above about 6 meV which penetrate less into the bulk, and therefore are more sensitive also to the motions of the adatoms, which are a unique feature of the

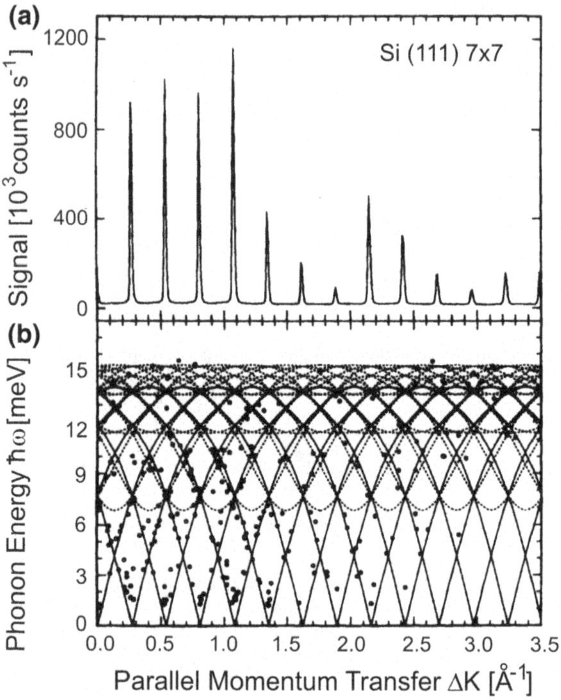

Fig. 11.14 Comparison of the HAS angular distributions and dispersion curves from the Si(111) 7×7 surface along the $[1\bar{2}1]$ azimuthal direction with BCM calculations [144]. **a** The angular distribution (transformed to a parallel momentum transfer scale) for an incident wavevector $k_i = 7.3$ Å^{-1} ($E_i = 28.0$ meV) and a surface temperature $T_s = 300$ K. **b** The measured phonon energies on the same parallel momentum transfer scale. The solid curves show the predicted dispersion curves based on force constants which provide a good fit of the ideal termination, but folded to take account of the 7×7 structure. The reciprocal lattice vector is $G = 0.27$ Å^{-1}. The dashed curves show the projection along the symmetry direction of phonon dispersion curves calculated for nearby reciprocal lattice points of the folded structure

complicated 7×7 unit cell [147–149]. Figure 11.15a shows a compilation of all the HAS dispersion curve data along the symmetry directions folded into the irreducible BZ [144]. The solid line and dashed curves have the same meaning as in Fig. 11.14b. Some of the RW folding points at the $\overline{\Gamma}$-point correspond to Raman-active modes and compare quite well with the Raman scattering data by Liebhaber et al. [150] (large red dots). Clearly a slightly stiffer force constant should have been chosen for fitting the folded-sine curve to match the Raman and HAS data points above 16 meV, which are in essential agreement.

In the high energy part of the acoustic spectrum the data points are, however, too much dispersed for a meaningful fit, and could only be analyzed by evaluating the total phonon density of states (DOS) obtained from several hundred experimental points. The experimental DOS shown in Fig. 11.15b agrees reasonably well with

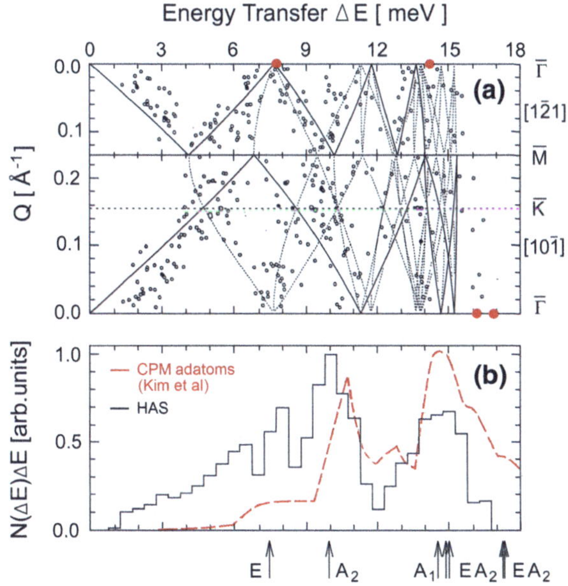

Fig. 11.15 Dispersion curves and density of states of the Si(111) 7×7 surface folded into the reduced 7×7 surface Brillouin zone. **a** The small circles are the HAS data [144] and the large red dots mark the experimental Raman frequencies measured by Liebhaber et al. [150]. The theoretical curves are obtained by reproducing the Si(111) 1×1 RW branch for all possible in-plane (full lines) and nearby out-of-plane (broken lines) **G**-vectors of the 7×7 reconstructed surface [12, 144]. **b** The black line histogram is the density of states (DOS) derived from the HAS data points. The dashed red line curve is the theoretical DOS projected onto the adatom coordinates parallel to the surface calculated by Kim et al. [146] with the Car-Parrinello method (CPM). The qualitative agreement suggests that the longitudinal adatom modes dominate the spectrum. The arrows below part (**b**) indicate the lowest Q = 0 frequencies obtained by Stich et al. [148, 149] from *ab initio* molecular dynamics calculations. The labels A_1, A_2 and E refer to the irreducible representations of the C_{3v} point group

the DOS projected onto the components of the adatom coordinates in the plane of the surface as calculated by Kim et al. with the Car-Parrinello method [146] (Fig. 11.15b, dashed red line curve). This suggests a propensity of He atoms to be inelastically scattered by the longitudinal adatom modes. Stich et al. [148, 149], using *ab initio* molecular dynamics combined with the multiple signal classification (MUSIC) algorithm for the spectral analysis, were able to predict the locations of the zone center phonons for the different symmetry irreducible representations of the C_{3v} group. Their results, shown by the arrows at the bottom of Fig. 11.15b, are in rather good agreement with the experimentally observed peaks of the DOS. The A_1 and E and EA modes are also in reasonable agreement with the Raman points (red dots). The predicted position of the lowest A_2 (rotational Raman inactive) mode at about 10 meV corresponds well to the largest feature of the experimental DOS, although the major contribution to the DOS comes in general from zone-boundary modes. An *ab initio* calculation of the Si(111) 7×7 surface dynamics by Liu et al. [147] suggests an important role of the rest atoms (see Fig. 11.12). Their lowest SV modes are at about 7, 11 and 16 meV, the latter peak being much stronger than the other two.

It should be noted that the out-of-phase SV motion of the rest atoms next to an adatom is strongly coupled to the adatom longitudinal (L) motion. The propensity of He atoms to be inelastically scattered by the adatom L vibrations is mostly due to the large surface corrugation (Fig. 11.12a, b), but also the dynamical redistribution of the adatom-rest-atom bonding charge via electron-phonon interaction (see Chap. 8) may play a significant role.

These studies show how much can be learned about the dynamics of a surface of great technological importance and complexity such as Si(111)7×7 from the combination of high resolution HAS with *ab initio* DFT calculations.

11.2.3 GaAs(110)

A more favorable situation occurs on the GaAs(110) surface where the six-fold rings remain intact at the surface, unlike Si(111) 2×1 where the original bulk structure is restructured into a sequence of 5- and 7-fold rings. The structure of the GaAs(110) surface is shown in Fig. 11.16a, b. It is now well established that the surface relaxation is mainly characterized by a bond-length-conserving rotation of the surface chains by a tilt angle of about 30° [151, 152], with the As atoms shifted upwards above the ideal (110) plane and the Ga atoms shifted downwards toward the bulk. The early studies by EELS in the dipole scattering regime revealed the expected strong peak associated with the surface phonon polariton (Fuchs-Kliever (FK) mode) at about 36 meV [153, 154]. On the basis of the bulk phonon data [14] the FK mode for GaAs is in fact expected at 34.9 meV.

The HAS dispersion curves measured later by two groups in 1987 are shown in Fig. 11.17a [155, 156]. Semi-empirical tight-binding [136–138], BCM [157, 158, 159] (Fig. 11.17a) and calculations based on first principle methods [5, 6, 165–169] (Fig. 11.17b) all reproduce the experimental data, particularly the flat low lying 10 meV branch from HAS [155–157] and EELS [153, 160, 161] measurements. The 10 meV branch also occurs in Si(111) 2×1, but in contrast to silicon, this branch lies well above the zone boundary Rayleigh wave (S_1), since the reconstruction and the associated seven-fold ring softening are absent. In GaAs as with silicon the calculated surface phonon frequencies depend strongly on the geometrical arrangement of the surface atoms. This situation allowed for a determination of the surface crystallography from an accurate fit of the dispersion curves, in good agreement with previous LEED data [170, 171].

EELS data in the impact scattering regime besides confirming the HAS data in the low part of the spectrum, provide evidence for two surface optical branches at about 23 and 37 meV, which are well reproduced by the first principle calculations (Fig. 11.17b). The calculations also predict another bunch of surface optical branches at around 30 meV, similar to those predicted in Si(111) 2×1 between 40

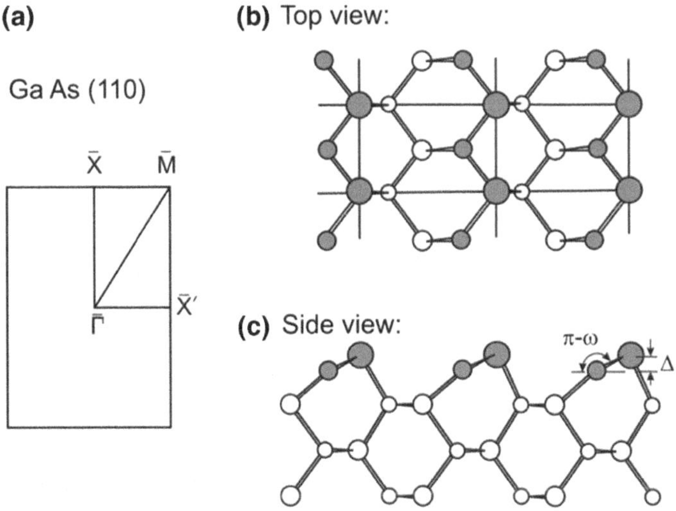

Fig. 11.16 Structure and zone diagram of the GaAs(110) surface. **a** The surface Brillouin zone. **b** The top view of the lattice showing the Ga and As atoms as small and large circles, respectively. **c** Side view projected onto a $(1\bar{1}0)$ plane normal to the surface indicating the bond-length-conserving rotation of the surface chains by a tilt angle of $\omega \cong 30°$ resulting in an upward shift of the As atoms and a downwards shift of the Ga atoms which are vertically displaced by about $\Delta = 0.83$ Å [152]

and 60 meV (Fig. 11.10), which however have not been detected by EELS. Although no reconstruction occurs for the GaAs surface, very likely the 30 meV modes are associated with a substantial relaxation occurring in the first layers below the surface chains. The 37 meV surface branch above the maximum bulk frequency, found by EELS, as well as in first principle calculations is very similar to the 55 meV mode of Si(111) 2×1. This suggests that it may be associated with a surface relaxation and a consequent bond stiffening occurring between the surface atoms in the chains and grooves which are mostly affected by the surface charge redistribution. In connection with the 37 meV mode it is interesting to note that at low electron energies in the dipolar scattering regime EELS detects the macroscopic FK mode at 36 meV [151, 172], while EELS in the impact scattering regime reveals a flat branch of microscopic surface modes at just 1 meV above the FK mode.

Unlike the FK modes, microscopic surface modes are very sensitive to the surface geometry and their spectroscopy may help to determine details of the structure like the tilt angle ω (Fig. 11.16b). Similarly to the case of Si(111) 2×1, the gaps at the zone boundary point $\bar{X}$ between the S_8 and RW energies and between S_3 and S_8 energies (Fig. 11.17a) depend crucially on the chain geometry, notably on the two most important parameters involved in surface relaxation: the tilt angle and the interchain distance. For example, a bond-length-conserving decrease of the tilt angle, from 31° to 29°, enlarges the $S_3 - S_8$ gap by 1.1 meV, whereas a decrease of the interchain distance as small as 2% with a fixed tilt angle reduces the

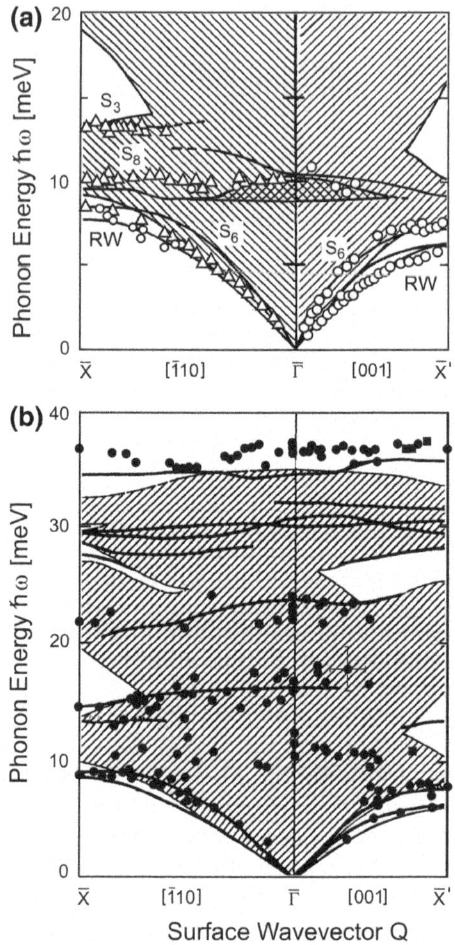

Fig. 11.17 Comparison of experimental surface phonon dispersion curves of the unreconstructed GaAs(110) surface with theory. **a** HAS data [155, 157], (open circles) and [156] (open triangles), are compared with a BCM calculation [157] (continuous line curves labelled according to Table 1.1). The cross-hatched area shows the width of the strong resonance at ~10 meV. **b** EELS data [153, 160, 161] (closed circles), which show evidence for a surface phonon branch localized above the maximum bulk frequency, are compared with a full first principles CP calculation by Di Felice et al. [161]

RW energy by 1 meV [157]. Both changes are large enough to cause appreciable disagreement with experiment. The BCM theoretical analysis by Santini et al. shows, indeed, that the geometry derived from the LEED analysis by Duke et al. [170] provides the best agreement with the HAS dispersion curves.

11.3 Metals: Effects of the Electron-Phonon Interaction

11.3.1 Simple Alkali and Alkaline Earth Metal Surfaces

The naturally grown crystals of alkali metals are very soft and cannot be easily prepared for surface studies. Instead epitaxial thin films several nanometers thick are grown by atom beam deposition on suitable metal substrates such as Cu(111) or

Cu(001). In this way the surfaces of thin films with up to about 30 layers of the alkali metals Li [162], Na [163], K [164, 173], Rb [174], Cs [175, 176] have been studied by the HAS technique. Because of the comparatively low phonon frequencies of the fairly soft materials, which makes it necessary to carry out the measurements at low surface temperatures of typically 25–50 K, the quality of the data is not as good as for the other metals. In all cases, except for Cs(110), only the Rayleigh mode is found, with no evidence for the longitudinal resonance. This was originally attributed to the nearly free electron spill-out which makes the surface practically insensitive to the longitudinal displacements of the surface atoms. More important, however, is the comparatively small electron-phonon interaction in alkali metals [177] which makes the inelastic scattering of He atoms from the surface longitudinal modes quite weak. It should be noted that early *ab initio* calculations did not predict any longitudinal acoustic resonance in Na(110) [178, 179] whereas they are predicted by the more recent EA calculations by Wilson and Riffe [180].

Cesium represents an interesting case because its bulk lattice dynamics was never measured experimentally by the inelastic scattering of neutrons due to their high absorption cross section. Whether Cs belongs to the simple metal homology class has been long debated in the literature [181, 182]. The epitaxial growth of Cs at low temperature on noble metal (111) surfaces occurs layer by layer (Frank-van der Merwe growth regime) [175], but the absence of any diffraction peak in the HAS angular distribution makes it impossible to determine experimentally the crystallographic indices and the orientation of the surface with respect to that of the substrate. This information can however be obtained from the analysis of inelastic HAS data. Figure 11.18a, b display the bulk and surface zone diagrams of bcc Cesium. Figure 11.18c shows a typical series of TOF spectra shown on the energy transfer scale for a 25 ML film measured at a fixed incident energy of 10 meV, a surface temperature of 40 K and different incident angles θ_i above the specular (final angle $90°-\theta_i$, planar scattering). The surface could be assigned to the (110) surface and bcc stacking since the dispersion curves calculated for the (111) and (100) surfaces did not agree at all with the experimental data in any other direction for both bcc and fcc stackings [176]. The calculations also made it possible to assign the symmetry direction to be along the $[1\bar{1}0]$ direction of the Pt(111) surface, corresponding to the $\bar{\Gamma}\bar{\mathrm{H}}$ direction of the Cs(110) surface (Fig. 11.18b). In the energy transfer spectra shown in Fig. 11.18c the main features on both creation and annihilation sides of the diffuse elastic (DE) peak could be assigned to the Rayleigh wave (RW). At the thickness of 25 ML the film Sezawa waves (see Sect. 12.3) form a broad background, which is no longer distinguishable from the surface-projected bulk density and therefore only the genuine surface modes like the RW and the longitudinal (L) resonance are detected. As seen in the two selected parts of the spectra marked in yellow in Fig. 11.18c the longitudinal resonance is quite weak unlike the anomalous L resonance in the transition and noble metals, where it can be more intense than the RW. This indicates that in Cs, as probably in all alkali metals, the L resonance makes only a modest contribution to the

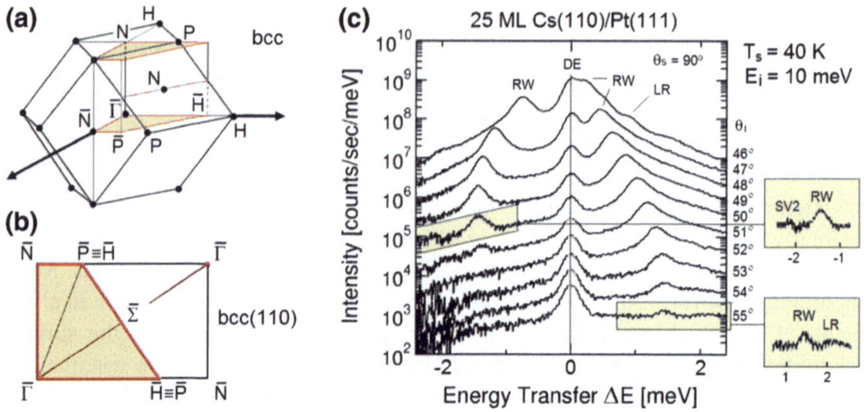

Fig. 11.18 Brillouin zone diagrams and HAS energy transfer spectra from a 25 ML film of Cs on Pt(111) [176]. **a** The bcc Brillouin zone with an octant (bases shadowed) of the equivalent (110)-slab adapted prismatic zone. **b** The corresponding irreducible quadrant of the bcc(110) surface Brillouin zone (shadowed) and a quadrant in the next zone showing the equivalence of the zone edge points $\bar{P}$ and $\bar{H}$. **c** A series of inelastic HAS energy transfer spectra from a 25 ML Cs(110) film on a Pt(111) substrate for an incident energy of 10 meV, a surface temperature of 40 K, and different incident angles above the specular angle of 45° [175, 176]. The scattering plane is oriented along the surface symmetry direction $\bar{\Gamma}\bar{H}$. The compressed energy scale of the yellow insets highlights the weaker features of the 50° and 55° spectra associated with 2nd-layer SV2 and the L resonances (LR), respectively

electron-phonon interaction, with a mode-lambda $\lambda_{\mathbf{Q},\mathrm{LR}}$ (see Sect. 8.4) which is an order of magnitude smaller than $\lambda_{\mathbf{Q},RW}$.

Figure 11.19 shows a comparison of the experimental Rayleigh wave (RW) dispersion curves and also the few points associated with the surface longitudinal resonance with the DOS of both SV1 and L1 from recent DFPT calculations for a 11 layer slab [175, 176]. Creation (top part) and annihilation (bottom part) processes are plotted separately. The left side shows the comparison with HAS data for a 25-layer film at a surface temperature of 40 K (black lozenges) and at 100 K (red lozenges). At 100 K the RW dispersion curve shows some softening with respect to the 40 K data, indicating an appreciable anharmonic effect. The right side of Fig. 11.19 shows the data for a thicker film at a surface temperature of 40 K. Although the data for thicker films are somewhat stiffer towards the zone boundary than for the 25 ML film, experiment and theory are in good agreement. The calculations reveal interestingly that the L resonance extends over four layers and therefore is excited via a long range electron-phonon interaction at the surface. Such a comparatively deep penetration into the bulk is related to the small separation of the L resonance from the bulk edge, due in turn to the weak e-ph interaction.

In the group of alkaline earth metal surfaces only Ba(0001) films of a few overlayers have been investigated so far by means of inelastic HAS [183]. The surfaces (0001) and $(10\bar{1}0)$ of Be have been the object of experimental studies with

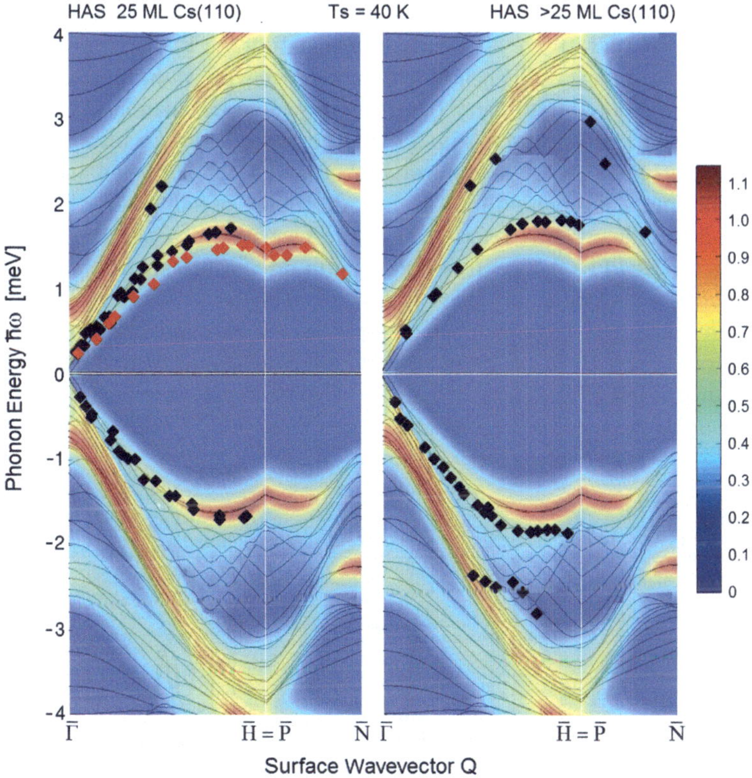

Fig. 11.19 DFPT calculated density of phonon states of Cs(110) films for both the SV1 and L1 polarizations are compared with HAS dispersion curves for a 25-layer film (left side) at a surface temperature of 40 K (black lozenges) and at 100 K (red lozenges) and for a thicker film (right side) at 40 K. The top halves show the experimental points for creation and the bottom halves for annihilation events [175, 176]

EELS [184–187], and of force constant [184–187] Green's function [189] and *ab initio* calculations [188, 190, 191, 193], but no HAS investigation has been reported to date.

11.3.2 Aluminum and Lead Surfaces

The inelastic HAS spectra of the less ideal free electron metals such as Al and Pb present more interesting features. Because of its sp nature, aluminum is after the alkali and alkaline earth metals the most nearly ideal free electron metal. Its low index surfaces can be prepared without difficulty and have been investigated by HAS [192, 194–196]. Some typical TOF spectra for the (110), (001) and

(111) aluminum surfaces are shown in Fig. 11.20 [192, 197]. These spectra, with the exception of Al(110) are all clearly dominated by the Rayleigh peak.

By taking advantage of the nearly jellium-like behavior of the electrons, Eguiluz and coworkers were able to carry out a first principle calculation [192] for comparison with the TOF spectra in Fig. 11.20. They used a surface response function for interacting electrons in a uniform background [198–200] and modelled the electron-ion interaction with a local Heine-Abarenkov pseudopotential [201, 202]. In Chap. 5 this theory was referred to as the Density Response Pseudopotential Perturbation (DR-PPP) theory. It is important to remark that the calculations were first made for a comparatively thick slab of as many as 17 atomic layers, further widened by adding a large number of inner layers. A multiparameter Born-von Kármán force constant fit of the dispersion curves was also carried out and compared with the DR-PPP theory and with the experiment, in which central force constants up to the tenth nearest neighbors and three-body forces up to the second

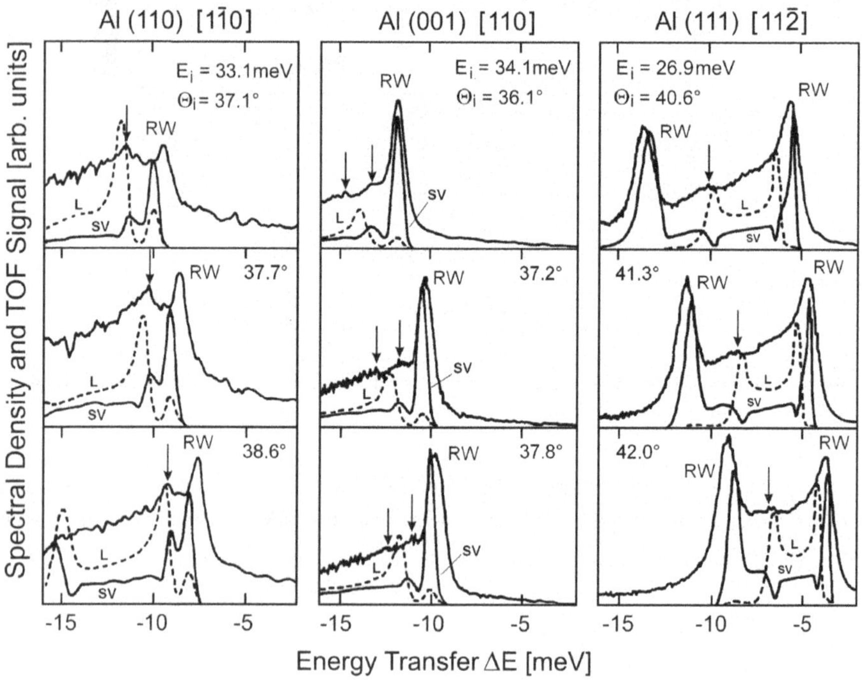

Fig. 11.20 Comparison of some representative experimental TOF spectra plotted as a function of the energy transfer for all three of (110), (100) and (111) surfaces of Al with the phonon densities of states calculated by the DR-PPP method along the scan curve, for SV polarized modes (solid lines) and L polarized modes (dashed lines) [192]. Each panel is labeled by the azimuth, incident angle Θ_i. and incident energy. The arrows mark the location of the weak resonances in the experiments, which could be identified with the help of the calculations

nearest neighbors were included [197]. Both approaches lead to very similar dynamical matrices, thereby providing a convincing vindication of the Born-von Karman procedure for the surface phonons, as long as a sufficiently large number of nearest neighbor shells and parameters are included.

With the aid of the theory it was possible to assign the large intensity of the energy transfer tails to the left of the main RW peak on the (110) surface and eventually extending up to the RW peak on the (111) surface to the bulk continuum. The DR-PPP calculations indicate further that the bulk continuum receives a major contribution from the longitudinal modes (broken lines in Fig. 11.20) and that the weak resonances (indicated by arrows) arising above the continuum can mostly be attributed to the surface longitudinal resonance. Their comparatively small intensity implies that the surface charge density is only weakly affected by the longitudinal (in-plane) displacement of the surface atoms and that their frequencies are not significantly softened by the electron density response. In this respect there is an important difference between nearly free electron metals and noble and transition metals, where in the latter, as seen in Sect. 11.3.3, the large HAS intensity from the longitudinal modes is accompanied by a considerable frequency reduction of the longitudinal resonance.

In Fig. 11.21 for the (001) and (111) surfaces there is overall good agreement, whereas on the (110) surface the calculated RW frequencies lie about 10% above the experimental ones. The experimental points associated with the weak resonances are in reasonable agreement with the dispersion curves predicted for the longitudinal resonance (L) and another weak resonance of quasi-shear vertical (SV) character. It is interesting that the combined analysis of the DR-PPP and the semi-empirical Born-von Kàrmàn calculations [197] indicate that the weak longitudinal resonance in aluminum is not related to a reduction of a single force constant as found for the noble metals, but rather is due to the cumulative effect of changes in the entire force constant field extending over several layers in the vicinity of the surface [203]. This was interpreted as indicating that the changes in force constants were mediated by conduction electrons over an extended region of the surface.

The previous analysis has been largely confirmed by a calculation based on the density functional perturbation theory (DFPT), also for a 17-layer slab [204]. Figure 11.22 shows the ion core phonon density of states for Al(001) as a function of the phonon energy and wavevector along the $\overline{\Gamma X}$ direction for the shear-vertical and longitudinal components of the ion cores projected on the first surface layer (SV-1 and L-1, respectively) and on the second surface layer (SV-2 ans L-2, respectively). For comparison the experimental points (red circles), attributed to the quasi-SV and quasi-L polarizations, are plotted on the respective SV-1 and L-1 density of states. The RW amplitude is the main feature in SV-1 and shows the typical sine-like increase towards the X-point. The agreement of the calculated RW dispersion with HAS data is nearly perfect. Moreover the points at higher energy in SV-1 correspond to a much weaker ridge, as already found in the DR-PPP calculation (Fig. 11.22). As appears in L-1, the longitudinal component of the RW is

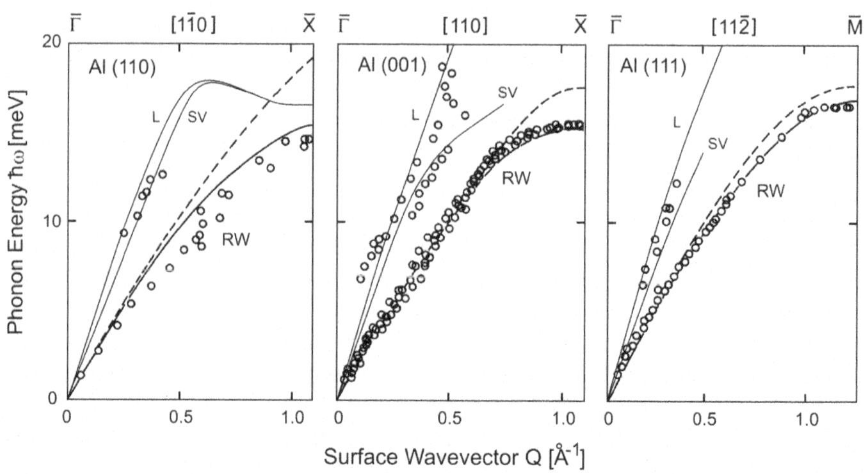

Fig. 11.21 Comparison of HAS experimental (open circles) and DR-PPP calculated (solid line curves) dispersion curves for the (110), (100) and (111) surfaces of Al [192]. The lower edge of the transverse bulk phonon continuum is indicated by the dashed line curves. The experimental points correspond to the Rayleigh mode (RW) and the weak L and SV resonances, identified in Fig. 11.20

very small, indicating a small vibrational ellipticity. The main ridge feature in L-1 starting from the origin at $\bar{\Gamma}$ and ending into a gap mode at the $\bar{X}$-point, also corresponds very well to the experimental points. This is a clear confirmation that the longitudinal acoustic resonance occurs also for *sp*-metal surfaces. The nature of the resonance, however, is not trivial. The longitudinal amplitudes in the second layer, Fig. 11.22b (L-2) reveal an avoided crossing between the longitudinal acoustic resonance and a surface optical branch running between 20 meV at $\bar{\Gamma}$ to 17 meV at $\bar{X}$. This longitudinal optical resonance and the SV optical resonance at around 34 meV have the largest amplitude in the second layer, but almost no signature in the first layer. This provides additional evidence that the surface perturbation involves several layers, as correctly concluded in the original analysis by Gaspar et al. [192] and Franchini et al. [197].

The other less ideal free electron metal Pb is even more interesting. Lead becomes superconducting at 7.193 K and shows conspicuous electron-phonon anomalies in the dispersion curves of bulk phonons [206]. Because of the softness of lead and the difficulty to prepare a surface from a bulk crystal the (111) surface was prepared by depositing thick films on a Cu(111) substrate [207–209]. Figure 11.23 shows a comparison of the HAS surface phonon data [210] with a DFPT calculation for Pb(111) based on an accurate minimization of the potential energy [211]. The equilibrium configuration of the Pb(111) surface, calculated with the inclusion of the spin-orbit coupling (SOC), shows a contraction of the first interlayer spacing of −4.5%, and a +1.95% expansion of the second interlayer

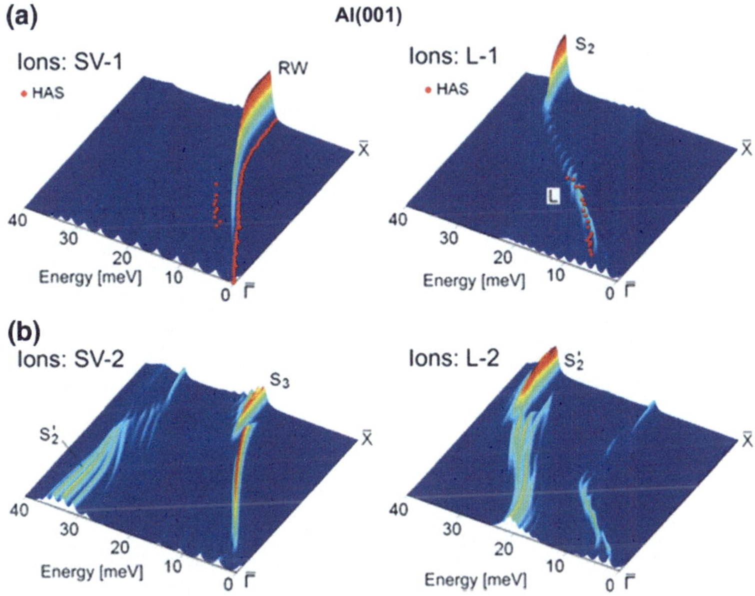

Fig. 11.22 Comparison of HAS surface phonon dispersion curves (red circles) [192] with a DFPT calculation of the Al(001) phonon density of states (DOS) of the ion cores along the $\overline{\Gamma X}$ wavevector direction [204, 205]. **a** The ion DOS projected onto the first surface layer for shear-vertical (SV-1) and longitudinal (L-1) polarizations. **b** The ion DOS projected onto the second layer for shear-vertical (SV-2) and longitudinal (L-2) polarizations. The main features of the first layer density of states are the RW and the longitudinal resonance, whereas the second layer densities (SV-2, L-2) show two surface optical resonances of SV and L polarization and a complex hybridization (non-crossing) pattern with the surface acoustic modes

spacing. This is consistent with LEED experimental data of $-3.5 \pm 1\%$ and $+0.5 \pm 1.4$, respectively [212]. This relaxation leads to a partial decoupling of the first bilayer from the rest of the semi-infinite lattice, which can explain the flat phonon branch ε_1 above the bulk phonon maximum (Fig. 11.23). Altogether the theory is in essential agreement with the experiment and reveals interesting transfers of intensity from the RW to upper branches when approaching the zone boundaries. This appears to be another effect of the surface bilayer contraction. The study of ultrathin Pb films on Cu(111), discussed in Chap. 12, where the surface bilayer spacing oscillates in sign with the film thickness, better illustrates this interesting manifestation of the strong electron-phonon coupling leading to a *quantum size effect*, first discovered by Hinch et al. in 1989 [209].

A large stiffening of the RW and a branch ε_1 slightly above the surface-projected bulk spectrum [213] as effects of the surface relaxation are also predicted in Tl(0001) by a DFPT calculation with SOC [214], although the first interlayer contraction is in comparison smaller than in Pb(111). Thallium, however, is like lead an elemental superconductor with a comparatively large electron-phonon interaction in the bulk, and an appreciable enhancement (15%) at the surface.

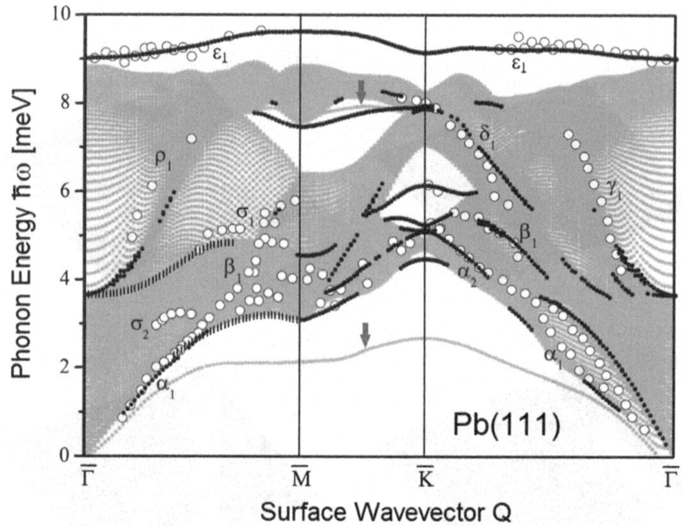

Fig. 11.23 Comparison of HAS surface phonon dispersion curves for the Pb(111) surface (open circles) [210] of a thick (≈50 ML) film with a DFPT calculation including spin-orbit coupling (dotted lines for surface branches, grey areas for the surface projected bulk bands) [214]. The calculated RW and ε_1 branches for the unrelaxed surface (light gray lines marked by arrows) are much softer than the corresponding experimental branches. This is attributed to the surface contraction of the first interlayer spacing, which amounts to −4.5% (−4.7% with no spin-orbit coupling) [214]

11.3.3 Noble Metals

11.3.3.1 Experimental Results

As mentioned in the introductory Chapter the very first TOF-HAS measurements for a metal surface were carried out in 1983 for the close packed Ag (111) surface [215]. This surface was chosen for the exploratory study since it is known to be relatively inert and is characterized by only a modest relaxation ($\sim -2.5\%$) [216] so that its structure does not differ appreciably from that inside the bulk. Moreover, at the time the calculated dispersion curves had been provided by two groups, prior to publication, based on bulk force constants [217, 218] with very similar results so that a direct comparison with theory would be possible. Figure 11.24a, c shows some typical time-of-flight spectra for the two main symmetry directions. The two sharp inelastic peaks of comparable intensity came as a surprise. Whereas the one with the lower frequency agreed nicely with the predicted Rayleigh mode, the second mode was found to lie in a region where no surface mode was predicted. Its energy was significantly above the Rayleigh mode and below the bulk LA mode (see Fig. 11.24b) and the surface longitudinal (L) resonance for the ideal (111) surface which is expected to follow the lower edge of the LA bulk mode

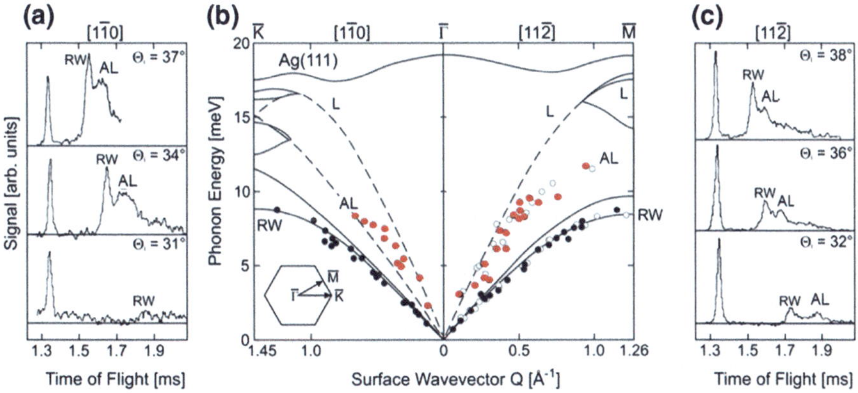

Fig. 11.24 Overview of the first HAS surface phonon measurements of the dispersion curves of a metal surface Ag (111). **a**, **c** Typical TOF spectra for the two main symmetry directions [215]. **b** Comparison of the HAS dispersion curves with calculations based on a BvK model with bulk force constants [217, 218]. While the experimental dispersion curve of the Rayleigh waves (RW) is well reproduced by the simple model, the experimental data associated with the inelastic peaks at higher energy (red dots in (**b**), denoted as AL in (**a**) and (**c**)) deviate substantially, in both the [11$\bar{2}$] and [1$\bar{1}$0] symmetry directions, from the lower edge of the longitudinal acoustic bulk band (L), and has been termed the anomalous longitudinal (AL) resonance, or S_3 resonance (cfr. Table 3.1). For an ideal surface, a longitudinal acoustic *surface mode* is not expected to be peeled-off from the longitudinal acoustic bulk edge except in the gap around the $\bar{M}$-point. This anomalous mode has been subsequently found for the other noble metal surfaces [219], and then recognized as a distinctive feature of practically all metal surfaces. The open circles are measurements from a 1000 Å thick epitaxial grown film on mica

[217]. The deviation is clearly seen in both [1$\bar{1}$0] directions in the dispersion curves in Fig. 11.24b.

Originally it was conjectured that the new mode was indeed the surface L mode which, however, was substantially shifted to lower frequencies. For this reason this feature was referred to as the Anomalous Longitudinal (AL) resonance and classified as S_3 (cfr. Table 3.1). The AL resonance was subsequently also found in HAS measurements of Cu(111) [219, 220]. Then in 1988 it was independently confirmed in one of the first EELS experiments in the high energy impact regime on the Cu(111) surface along the $\overline{\Gamma M}$ direction. Figure 11.25 shows some representative TOF spectra and the full set of dispersion curves of Cu(111) from HAS [219, 220] and EELS measurements [35, 221]. A new more recent analysis of TOF-HAS spectra provides further data for the surface optical resonances S_4 (Lucas mode) and S_2 (Wallis mode), which are plotted in Fig. 8.4 for the $\overline{\Gamma M}$ direction [222].

The HAS results on the surface phonons for Cu(001) [223] and Ag(001) [224] provide additional support for the interpretation presented above. Some representative TOF spectra and the full set of dispersion curves for Cu(001) are shown in Fig. 11.26. An interesting finding is the large anisotropy of the AL resonance peak

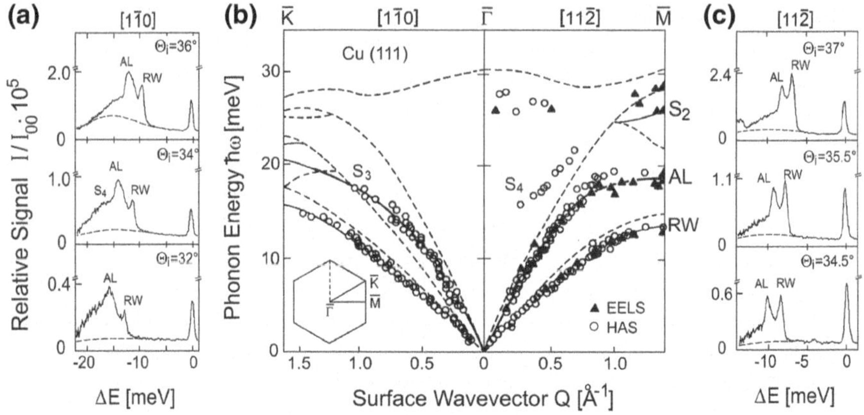

Fig. 11.25 Overview of HAS and EELS measurements of the dispersion curves of Cu(111). **a** and **c** Some typical HAS energy loss spectra for the two main symmetry directions. The dashed lines in the energy loss spectra indicate the interpolated multiphonon background. **b** The dispersion curves from HAS [219, 220] and EELS [35] measurements. The dotted lines in the dispersion curves delineate the bulk band edges whereas the solid lines show the best fit dispersion curves based on the PC model [220] discussed in Sect. 5.3. Note that along [1$\bar{1}$0] the AL peak is more intense than the RW peak

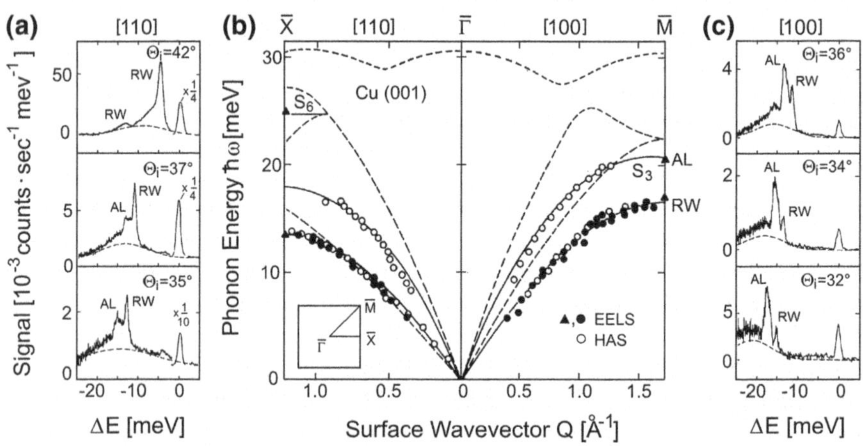

Fig. 11.26 Same as Fig. 11.25 but for Cu(001). The HAS data are from [223] and the EELS data are from [225–228]. The calculated dispersion curves in (**b**) are best fits based on the PC model [223]. The bulk band edges of sagittal polarization are shown by broken line curves. **c** Note that along [100] the AL peak is much more intense than the RW peak

intensity that is relatively weak in the $\overline{\Gamma X}$ direction, but in the $\overline{\Gamma M}$ direction it is as much as five times more intense than the Rayleigh wave!

Similar differences in the relative intensities for the two azimuths are also present in Ag(001) [224] but they are less pronounced than for the Cu(001) surface. No set of short range surface force constants within the framework of the Born-von

Karman model, no matter how strongly modified, could be found to explain these results on either Cu(001) or Ag(001). Surprisingly, whereas EELS has been able to detect the AL mode on Cu(111), at least for the $\overline{\Gamma M}$ direction, there is no reported EELS evidence for the very strong AL mode on Cu(001) [225–228] and Ag(001) [228–230], except at the $\overline{X}$- and $\overline{M}$-points where the EELS experiments do provide evidence for the longitudinal S_6 and S_3 modes, respectively, on both Cu(001) (Fig. 11.29b) and Ag(001) [228].

The Au(111) surface, which exhibits a long-period $23\times\sqrt{3}$ reconstruction [231–235] is a unique case among the noble metal surfaces. As seen in Fig. 11.27 the experimental dispersion curves of Au(111) exhibit only a split RW mode and there is no clear evidence for the AL resonance. Indeed reconstruction squeezes 24 surface atoms into a 23 atom period along the $[1\bar{1}0]$ direction with an average 4% compression [236], causing a general stiffening of nearest neighbor radial force constants between surface atoms instead of the usual softening responsible for the ordinary AL resonance. Thus the interpretation of the additional branch just above the RW in both symmetry directions as a particularly soft longitudinal resonance would require an unphysical softening of more than 50% of the force constants [237] or of the interatomic potential between surface atoms [238, 239] with respect to the bulk. The apparent splitting of the RW is more likely due either to the two inequivalent possible domain orientations of the reconstructed surface, or to the removal of the sagittal plane mirror symmetry, thus allowing for the observation of the SH acoustic branch in all directions.

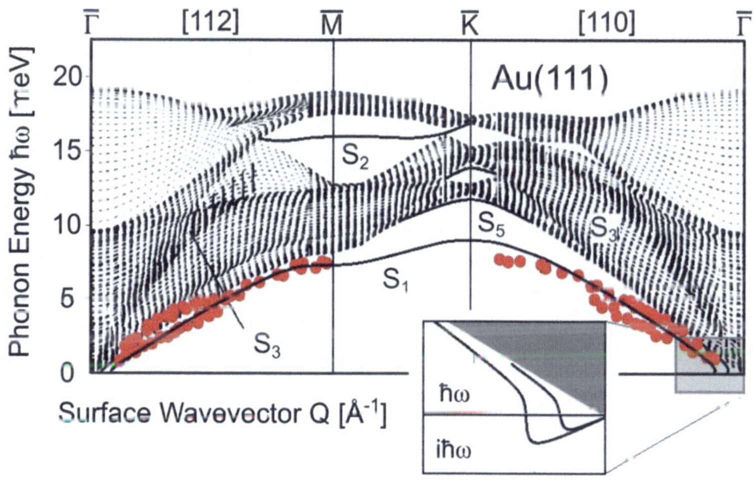

Fig. 11.27 HAS measurements of the dispersion curves (red dots) of surface phonons for Au (111) [219] are compared with a PC model calculation [240]. The well known instability of the ideal Au(111) surface, leading to a long-period $23\times\sqrt{3}$ reconstruction (see Sect. 11.5.1) [236] is reflected in the imaginary RW and SH frequencies occurring at very small wavevectors (inset)

11.3.3.2 Pseudocharge Model Calculations

The first pseudocharge (PC) model calculations from 1987 were based on the multipole expansion and predicted the L resonance (S_3) for the unreconstructed Au (111) surface at the usual place [240], but revealed an instability of the RW and SH acoustic branches at very small wavevectors in the $[1\bar{1}0]$ direction (Fig. 11.27, inset), consistent with the known surface reconstruction [236]. Since the origin of the reconstruction is essentially electronic, as discussed below in Sect. 11.5, this confirms that the multipole expansion at the basis of the PC model contains the essential features of the electron-phonon interaction and provides a tool for complex surfaces where DFPT calculations are still too time-expensive [241–243]. The recent results on the dispersion curves of the acoustic phasons and optical phasons of the $23\times\sqrt{3}$ reconstructed Au(111) surface are described in Sect. 11.5.2.

The interpretation of the anomalously intense LA mode in Ag(111) and the HAS observations of an even more intense AL resonance on the Cu(001) [244] and Ag (001) [245] surfaces stimulated a considerable amount of theoretical activity and controversy. The review by Heid and Bohnen [7] contains a comprehensive account and discussion of the *ab initio* studies devoted to the surface dynamics of metals, including noble metals, with a comparison to former semi-empirical approaches. In an effort to finally resolve the problem the group of Bortolani and Santoro starting in the mid-eighties carried out a series of careful Born-von Kármán fits. These authors for the first time also paid special attention to fitting the intensities in the experimental TOF spectra. Using an impact model for the coupling of the He atoms to the surface atoms they were compelled to use six force constants to fit the dispersion curves [218, 246]. Their first layer intraplanar nearest neighbor radial force constants $\beta_{\|}$, in particular, had to be considerably reduced compared to the bulk values by 69% for Cu(111) [246], 52% for Ag(111) [237] and 70% for Au (111) [237, 247]. This strong reduction was attributed to the surface electron spill-out which leads to a weakening of the sp-d hybridization and the lateral bonding in the surface layer. This initial interpretation of the HAS data in terms of extreme reductions in the force constants was, however, subsequently questioned by Hall, Mills et al. [35, 221]. Surprisingly these authors could explain the location of the anomalous longitudinal mode in the $\overline{\Gamma M}$ direction found by EELS for Cu (111) by using a much simpler force constant model with only a moderate 15% reduction of $\beta_{\|}$. They made no attempt, however, to fit the HAS intensities. This apparent contradiction between the HAS and EELS interpretations was called the *Bortolani-Mills* paradox [248]. Since the large force constant reduction required by Bortolani et al. resulted from their attempt to explain both the frequencies and the HAS scattering intensities this raised questions about how to properly describe the AL mode and/or the mechanism by which He atoms couple to the surface.

The *Bortolani-Mills* paradox, attracted considerable attention and a number of different explanations were proposed. In 1987 Nelson et al. [249] observed in their EA calculations an avoided crossing between two longitudinal polarized modes along $\overline{\Gamma M}$, similar to that shown for Al(001) in Fig. 11.22(L-2). One of these originates at the $\overline{\Gamma}$-point with a finite frequency and ends up at the $\overline{M}$-point as the S_2 mode, while the other lies at the lower edge of the longitudinal bulk band. According to their theory, the avoided crossing enhances the longitudinal polarization of the lower mode, which they ascribe to the L-mode seen in HAS. Subsequently Ditlevsen and Nørskov in 1990 [250] proposed, on the basis of the effective medium theory, that the strong longitudinal peak is in fact due to a strong second layer transverse polarized mode, which they found in their calculations for Cu(111) and also for the other low index surfaces of copper, Ni(111) and even for Al(111) [231]. Then in 1993 Chen et al. [232, 251] reported first principle calculations based on a frozen phonon determination of surface interplanar force constants at high symmetry points. Through an interpolation procedure exploiting the knowledge of bulk force constants, the entire surface phonon dispersion curve was obtained. For the $\overline{\Gamma K}$ direction on Cu(111) and Ag(111) their results are in good agreement with the EA calculations by Nelson et al. [249]. Along the $\overline{\Gamma M}$ direction these new first principle calculations reveal an unexpected z-polarized mode labeled by them as R_1, which lies above the Rayleigh mode at about the position of the anomalous L-mode observed with HAS. Subsequent DFPT calculations of the surface phonon dispersion curves of Cu(001) [7, 252] reproduced quite well the anomalous L branch (S_3) in the $\overline{\Gamma M}$ direction, where the corresponding HAS peaks are most intense, but give no indication of any AL resonance in the other direction $\overline{\Gamma X}$. At about the same time Raouafi et al. [253] calculated the surface dynamics of the Cu(001) surface and a few other vicinal surfaces of copper on the basis of a semi-empirical potential similar to the Rosato et al. tight-binding second moment (TBSM) potential [254]. The resulting dispersion curves for Cu(001) are in close agreement with the HAS data and show the L resonance in both symmetry directions. The one along $\overline{\Gamma X}$ has an appreciable surface localization only within a restricted portion of the $\overline{\Gamma X}$ direction, in agreement with the weakness of the L peak in this direction (Fig. 11.26a), and around the $\overline{X}$-point. However, in none of these theoretical works were quantitative calculations carried out of the HAS peak intensities for direct comparison with the TOF spectra.

The importance of the electron charge polarization accompanying lattice vibrations in possibly providing a simultaneous explanation of the AL-mode and its HAS intensity was first pointed out by Heinz Bilz at a small seminar in Stuttgart where the first HAS results on Ag(111) were presented in 1983. This stimulated the application of Phil Allen's multipole expansion (ME) method, originally devised for the lattice dynamics of transition metals with Kohn anomalies [255], to the problem of surface phonons in noble metals (Sect. 5.3) [240]. In the ME method the electron-mediated force constants are accounted for in a local basis picture, suitable to noble and transition metals. Calculations first performed in 1987 by Jayanthi et al. [240] with this method in its phenomenological version (the so-called pseudocharge (PC) model) provided an excellent fit of the dispersion curves of both RW and L branches for Cu(111) and Ag(111).

In the PC model version (Sect. 5.3) adopted by Kaden et al. for copper [220] besides the direct *nn* ion-ion force constant A, the surface atoms are coupled via PC dipolar and quadrupolar polarizations by the force constants pairs B,C, and H, F, involving, as described in Fig. 5.2, the first and second *nn* ion displacements, respectively. By defining an effective radial *nn* force constant for a cubic lattice as $A_{eff} \equiv \frac{1}{4}a(c_{11} + 2c_2)$, where a is the lattice constant (3.615 Å for Cu), c_{11} and c_{12} are elastic constants (c_{11} = 11.68 × 10^{11} N/m^2 and c_{12} = 1.21 × 10^{11} N/m^2 for Cu [256]), and using the relationships between the elastic constants and the model force constants [220, 257], it is found:

$$A_{eff} \equiv A + \frac{1}{2}B + 3H - \frac{1}{2}(B + 2H)S^2, \tag{11.1}$$

where $S \equiv (C - 3F)/(B + 2H)$. In copper A = 28.6 N/m, B = 1.1 N/m, H = 2.5 N/m, C = 7.8 N/m, F = 2.4 N/m [220] and therefore S = 0.098 and A_{eff} = 36.6 N/m. With this model it was possible to reproduce very well both the RW and AL dispersion curves by merely increasing the dipolar and quadrupolar polarizabilities of the electronic charge distributed around the bridge positions between the surface ions (pseudocharges), with no change in the direct ion-ion force constant A. The changes in surface polarizabilities were shown to originate from a change of C and H while B and F remain unaffected. In Cu(111) these changes amount to ΔH = −1.6 N/m, ΔC = 3.2 N/m [220] and

$$\Delta A_{eff} = (3 + S^2)\Delta H - S\Delta C = -5.13\ \mathrm{N/m}. \tag{11.2}$$

Although ΔC is twice as large in absolute value as ΔH, the small value of S has a consequence that the surface perturbation of the effective *nn* radial force constant ΔA_{eff} mostly derives from the 1st neighbor quadrupolar term. Thus ΔA_{eff} is about −14% of A_{eff}, in remarkable agreement with the more recent DFPT calculations discussed below [257]. A basic requirement of a non-central many-body force constant model like the PC model is to account for the deviation from the Cauchy relation $c_{12} = c_{44}$ [258], which is expressed by

$$c_{12} - c_{44} = \frac{2F}{a}\frac{3C - 2F}{B + 2H}. \tag{11.3}$$

The r.h.m. is equal to 0.41 × 10^{11} N/m^2, in good agreement with the experimental value of l.h.m., 0.46 × 10^{11} N/m^2 [259]. Thus the non-central forces and their surface perturbation mostly originate from the PC quadrupolar polarizability terms, in agreement with the multipole expansion (ME) formulation by Jayanthi et al. [240]. As discussed in Chap. 5, the ME theory is a local basis representation of DFPT, of which the PC model is a simple parametrization. Thus the multipolar polarization of the electron density induced by the atomic displacements is a way to introduce electron-phonon interaction in lattice dynamics. Thus the anomalies observed in the phonon dynamics of metal surfaces are direct manifestations of the electron-phonon interaction.

The above PC model calculations by Kaden et al. [220] led the authors to question the assumption at the basis of all the previous calculations of the inelastic differential reflection coefficient, including those of Bortolani et al. [218, 237, 246, 247] that the surface phonons are excited by impulsive collisions between the He atoms and the individual surface atoms (see Sect. 7.4). This can be understood from Fig. 11.28a which shows the densities of surface phonon states (DOS) of sagittal polarization projected on the *pseudocharge* coordinates in the $\overline{\Gamma M}$ symmetry direction [220]. The pseudocharge DOS is dominated by the AL (S_3) resonance with a smaller density at the location of the RW mode. This DOS distribution agrees quite nicely with the energy transfer spectra for selected azimuths found on Cu(111) and Cu(001). For comparison the *ion* DOS projected on the first surface

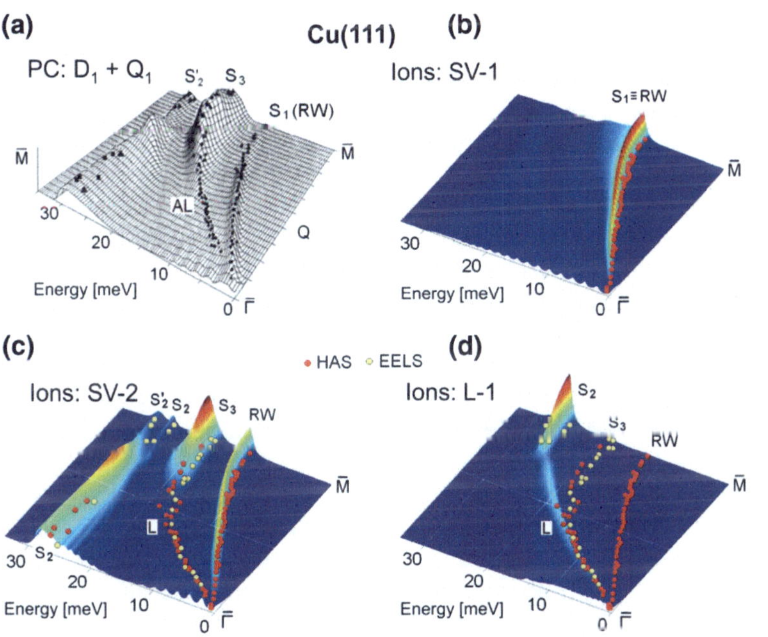

Fig. 11.28 The pseudocharge density of states (DOS) at the surface of Cu(111) calculated with the PC model are compared with the ion DOS calculated with DFPT (colored frames). The points show the HAS (red dots) [244, 245] and EELS (yellow circles) [35, 221] experimental points. **a** PC model DOS calculations with dipolar and quadrupolar contributions of the first layer. The dots show the HAS and EELS experiments. **b** First layer ion DOS of SV polarization, which in the impact model would determine the RW excitation probability. **d** First layer ion DOS of L polarization which in the impact model would determine the AL resonance excitation. **c** Second layer ion DOS with SV polarization. The DOS for the SV motion of the ions in the second layer (SV-2) is very similar to the DOS for the first layer PC polarization. This is confirmed by other DFPT calculations of the surface charge density deformations produced by the SV motion of the underlying ions

layer for both SV and L polarizations calculated with DFPT [222] are shown in Fig. 11.28b, d, respectively. The DFPT DOS agree well with the earlier early EA calculations of Nelson et al. [249], as well as with the PC based calculations of the ion DOS. The DFPT ion DOS projected on the first surface layer, which determine the excitation in the impact model, are dominated by the SV Rayleigh wave (RW) and by the gap mode S_2, whereas at the position of the AL resonance they exhibit only a negligible state density. The first layer and second layer SV ion density of states shown in Fig. 11.28b, d can however explain the intensities of the EELS signal, since the electrons in the impact regime are only scattered by the motion of the ions in the first few layers. Indeed EELS reveals the S_2 and S_2' modes as well as the AL resonance near the zone boundary, though with only a small intensity. Surprisingly, the second layer SV ion density of states appears to be remarkably similar to the PC density of states in the first layer. This led to the conclusion that perhaps the second layer motion could induce the charge density oscillations at the surface. This coordinated motion was then confirmed by the more extended DFPT calculations shown in Fig. 8.4. There the actual time dependent amplitudes of the motion of the charge distributions and the ions in the first few layers are shown and the influence of the second layer ion motions on the surface charges is clearly apparent.

The advantage of the PC formulation is that it provides a convenient way for the calculation of the inelastic intensities coming from the interaction of the He atoms with the surface PCs (see Sect. 8.3). The results are illustrated in Fig. 11.29 where the TOF spectra from Cu(111) shown in Fig. 11.25a, c and the TOF spectra from Fig. 11.26a, c after subtraction of the multiphonon background are compared with calculations based on the PC model for coupling either to both the PC's and the ions (full line) or to only the ions (dotted line). The dramatic effect of the coupling of the atoms to the PC oscillations at the AL resonance is clearly apparent especially for the $[1\bar{1}0]$ direction in Cu(111) and in the [100] direction in Cu(001).

It is presently possible to carry out first principle calculations of the inelastic HAS amplitudes where the surface charge density oscillations are consistently derived from a DFPT treatment of surface dynamics (see Chap. 8, Fig. 8.3) [11]. In this respect DFPT is regarded as the most powerful *ab initio* method, and the PC model based on the ME method as an efficient parametrization of DFPT (cfr. Sects. 5.3 and 5.5). The accurate *ab initio* Hartree-Fock calculations for the interface Ag/Mg(001) carried out by Heifets et al. [260] provide independent confirmation of the PC picture. These authors demonstrated that a complex charge redistribution occurs in the metal layer at the interface, characterized by large quadrupole moments as well as an electron density transfer towards bridge and hollow positions between the nearest and the next nearest Ag atoms. Incidentally this calculation, in using an exact non-local treatment of the exchange interaction between atoms, goes beyond LDA. Thus it appears that such non-local approaches may be needed in future first principle treatments aiming at an accurate description of the ubiquitous anomalous L branch and the related intense HAS signal in metals.

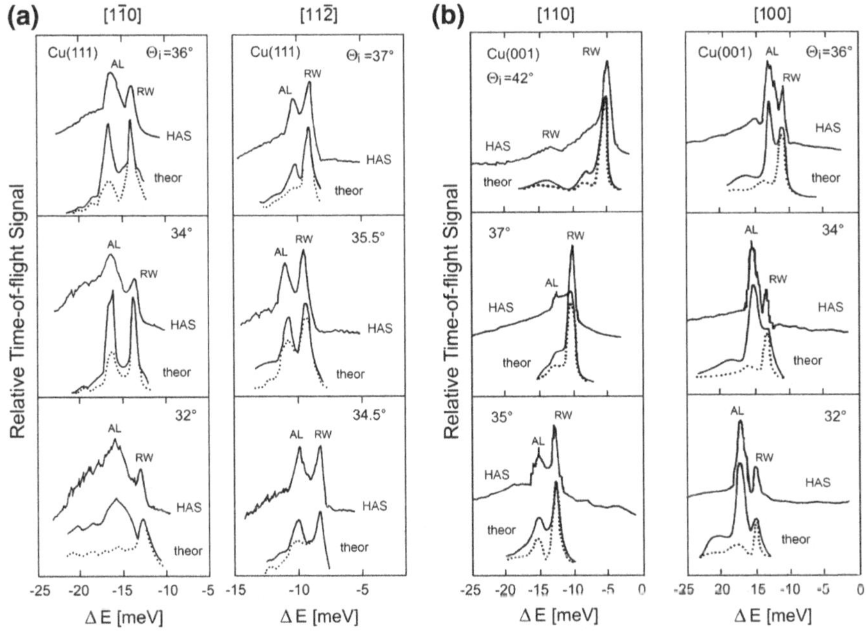

Fig. 11.29 Comparison of some typical HAS energy loss spectra for the Cu(111) and Cu(001) surfaces with calculations based on the PC model showing the contribution from the coupling to the PC electrons. **a** Cu(111) HAS spectra from Fig. 11.25 for the symmetry directions [1$\bar{1}$0] and [11$\bar{2}$] of Cu(111) (upper full lines) after subtraction of the multiphonon background and spectra based on the PC model [220] with inclusion of the coupling to the PC electrons (lower full line curves) and with only a coupling to the ions (dotted line curves). **b** Same for Cu(001) and symmetry directions [110] and [100] [223]. Note that the intensity of the AL resonance dominates over that of the RW along both the [1$\bar{1}$0] direction of Cu(111), and also along [100] direction of Cu (001)

11.3.4 Transition Metals

A strong longitudinal mode has also been seen in a number of transition metals including Ni(001) [261], Fe(110) [262], Pt(111) [263, 264] and Rh(111) [265, 266] and the refractory metals Mo and W which are discussed in the next subsection. Ni (001) was the first metal system studied by impact EELS [267–275] in 1983. It was subsequently investigated by HAS [276] in 1987 and thus it is the first system studied by both techniques (see Sect. 9.6). Figure 11.30b presents a comparison of the dispersion curves for the two high symmetry azimuths from both techniques. For orientation the bulk band edges, the surface modes and resonances obtained from a single force constant Born-von Kármán calculation are also shown. Panels (a) and (c) show typical HAS TOF spectra for the two symmetry directions. In two of the spectra the respective scan curves are shown in Fig. 11.30b (dashed-dotted

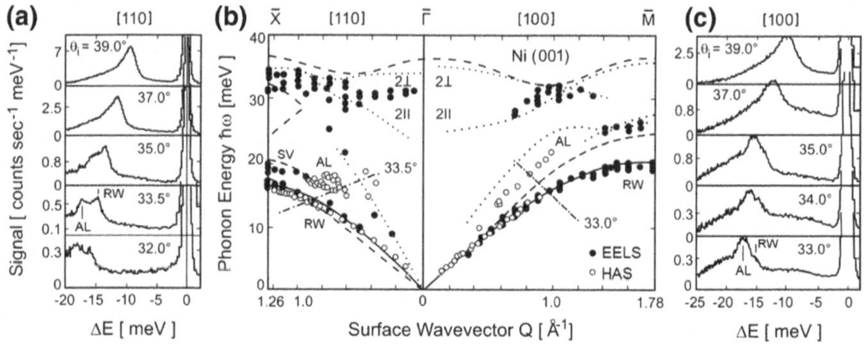

Fig. 11.30 Comparison between surface phonon dispersion curves of clean Ni(001) measured by EELS [267–275] (filled circle) and by helium atom scattering [276] (open circle). **a**, **c** Typical HAS energy loss distributions. **b** Comparison of the experimental dispersion curves with calculations for a single nn force constant model showing the bulk band edges (dashed line), the Rayleigh wave (solid line) and resonances (dotted line). The symbols 2⊥ and 2II designate second layer transverse and second layer longitudinal polarized modes. The scan curves (dot-dashed lines) for two incident angles in the two symmetry directions are also shown. The intersections with the RW and L branches correspond to the peaks in the respective energy loss spectra. In these experiments the cut off at large momentum and energy transfers prevented the observation of the SV, 2⊥ and 2|| branches by HAS

lines). In both cases the spectra exhibit a second peak denoted L at energies greater than that of the Rayleigh peak. Along the [100] direction the intensity of the second peak is considerably larger than that of the Rayleigh peak. This peak is attributed to a downward shifted, intense longitudinal resonance. Again, as with Cu(001), EELS does not indicate a significant peak at the same place in the dispersion curve for either direction as found with HAS. In Cu(001), it is recalled, the anomalous longitudinal resonance was also much larger relative to the Rayleigh peak along the same [100] direction. It is interesting, moreover, to see that the EELS measurements do reveal some resonances above 30 meV close to the zone boundary in both directions. The points at about 26 meV at the $\overline{\mathrm{M}}$-point are attributed to second layer vibrations. It is also instructive to compare the experimental EELS and HAS data in the $\overline{\Gamma \mathrm{X}}$ direction of Ni(001) with the DFPT calculation for Al(001) shown in Fig. 11.22. The surface phonon structure appears to be almost the same for the two metals, even as regards the energy scale. The Ni(001) EELS data correspond quite well to the main features of both the first- and second-layer shear vertical densities SV-1 and SV-2. Similarly the Ni(001) HAS data, like Al(001), correspond to the main features of both SV and L components the first layer DOS (SV-1 and L-1), with the data points above the RW branch restricted inside the first half of the Brillouin zone. In Ni(001) the branch marked SV in Fig. 11.30b, and seen in Al (001) (Figs. 11.20 and 11.21), corresponds closely to the R_1 mode reported by Tong et al. [251].

The observation of an anomalous longitudinal mode on Ni(001) and on all the other transition metals so far investigated is further confirmation of the close link between the presence of sp-d hybridization and the pseudocharge coupling mechanism. Thus the longitudinal anomaly is a general feature of noble and transition metals, and probably of all metal surfaces as shown above for Cs(110) (Fig. 11.18) and Al(001) (Fig. 11.22(L-2)).

Another manifestation of the coupling of phonons to electrons in metals are the local anomalies in the dispersion curves called Kohn anomalies which are well known for bulk phonons. Generally, these anomalies occur for longitudinal phonons since they are able to excite e-h pairs due to the alternating local compressions and expansions of the neighboring lattice atoms. The electrons tend to deplete the compressed regions (leaving holes behind), and to accumulate in the lower density regions. In the process each electron exchanges energy and momentum with the phonon field. The ability of phonons to excite e-h pairs, i.e., by the *electron-phonon coupling* [e.g., the matrix T in (5.8)], is expressed by the matrix element between initial and final electronic states of the derivative of the potential with respect to the phonon displacements. The size of the electronic response, which is given by the *electron susceptibility* χ[e.g., the matrix H^{-1} in (5.16)] is an integral over all possible electronic transitions weighted by the respective Fermi probabilities and is therefore a function of ΔE, ΔK and temperature. Under nesting conditions there is a multitude of electrons which couple to the phonon field and consequently a strong coupling between phonons and e-h excitations, leading to large electronic oscillations in resonance with the atomic motion (*electron-phonon resonances*). At the values of ΔK where such nesting occurs, χ can become very large. The resulting spill out of the electrons from the regions of lattice compression, will substantially weaken the repulsive forces between ions arising from the electronic kinetic energy and Pauli repulsion. The resulting dips in the longitudinal phonon frequencies near the nesting wavevectors are known as Kohn anomalies [277–280].

Since the reduced dimensionality at the surface should enhance surface Kohn anomalies, in the early studies of metal surface phonons the question was raised whether Kohn anomalies may affect the dispersion curves of Rayleigh waves or surface resonances, also in crystals which do not show anomalies in the bulk phonons. The search was naturally addressed to transition metals, some of which are known to have Kohn anomalies in the bulk and are expected to have unfilled d-bands with surface states near the Fermi energy [281]. On the clean transition metal surfaces discussed so far, besides the unexpected observation of an anomalous longitudinal resonant branch, no anomalies were found in the measured dispersion curves, consistent with the regularity of the corresponding bulk dispersion curves. Thus it came somewhat as a surprise when the first clear evidence for Kohn-type anomalies was found in the RW branches of Pt(111).

This system was extensively studied by He atom scattering by three groups [263, 282, 283] at about the same time in 1985. Figure 11.31 displays the high precision HAS dispersion curves for the Pt(111) surface from the group in Göttingen. They reveal the same two predominant dispersion curves found on the other (111) metal surfaces discussed so far. It is interesting to see from the illustrative TOF spectrum in the inset of Fig. 11.31 that the longitudinal mode along the $\overline{\Gamma \mathrm{K}}$ direction is once

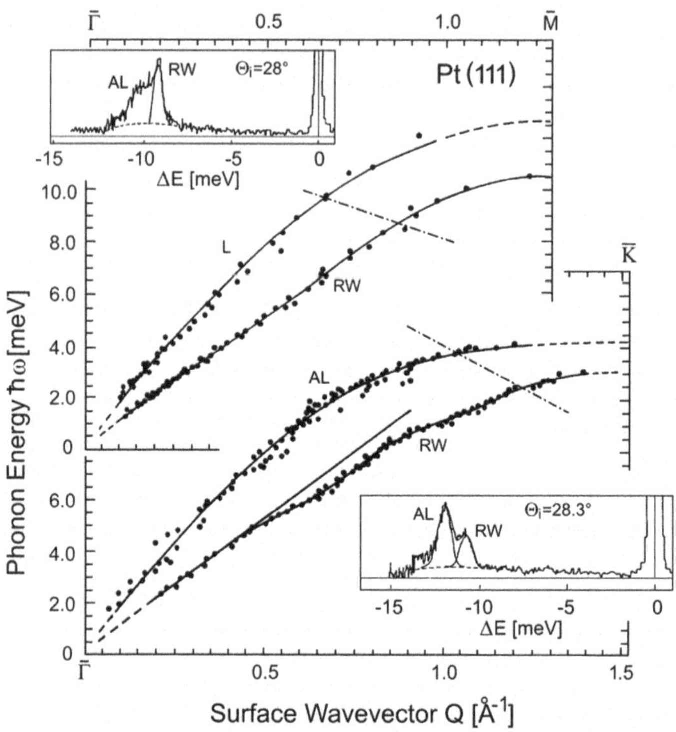

Fig. 11.31 Helium atom scattering measurements of surface phonon dispersion curves for Pt(111) [263] at beam energies between 10 and 32 meV. The target temperature in the $\overline{\Gamma M}$ measurements was 400 K, and in the $\overline{\Gamma K}$ measurements 160 K. The solid line curves show the best-fit with a nine-term Fourier expansion for the longitudinal resonance (L) and Rayleigh (RW) mode for the $\overline{\Gamma M}$ direction (upper two curves) and for the $\overline{\Gamma K}$ direction (bottom two curves). The straight line fit of the $\overline{\Gamma K}$ RW data corresponds to a group velocity of 1590 m/s, which is slightly less than the bulk TA velocity for the equivalent direction. The insets show two HAS time-of-flight spectra (converted into an energy-loss scale) for the two scan curves represented by the dot-dashed lines. The Gaussian fits of the L and RW peaks are shown after subtraction of the multiphonon background (broken lines)

more also much more intense than the Rayleigh mode, as in Cu(111) (see Fig. 11.25a). Through several improvements in technique over the earlier work the group in Göttingen could fit the data to determine the exact shape of the Rayleigh dispersion curve with root mean square deviations of only about 0.09 meV (90 μeV) [263]. Along the $\overline{\Gamma K}$ direction it was therefore possible to discern two small broad dips at about 0.65 and 1.05 Å^{-1} in the otherwise regular shape of the RW dispersion curve which were attributed to Kohn anomalies (Sect. 5.3). In the $\overline{\Gamma M}$ direction a single weak dip was discernible at about 0.7 Å^{-1}.

To highlight the anomalies in the $\overline{\Gamma K}$ direction, the experimental group velocity, shown in Fig. 11.32a, was determined by fitting a nine-term Fourier expansion to the Rayleigh mode dispersion curve in Fig. 11.31. The maximum depth of an anomaly in the dispersion curve approximately corresponds to an intersection of the actual group velocity with the ideal group velocity (broken line in Fig. 11.32a), which is close to an inflexion point of the group velocity curve. Indeed two well-defined anomalies stand out at 0.65 and 1.05 $Å^{-1}$ (arrows in Fig. 11.32a). For comparison the group velocities of the corresponding bulk mode along the equivalent direction are also shown in Fig. 11.32b. The anomalies at the surface are much stronger and sharper than the two bulk anomalies of the transverse phonon branch at 0.9 and 1.9 $Å^{-1}$. These bulk anomalies can be mimicked by a Born-von Kármán model calculation (Chap. 3), provided that a sufficient number of force constants is included in the expansion [264]. The corresponding calculated surface phonon dispersion curves are in good agreement with the observed surface anomalies identifying them as a projection of the bulk anomaly [264]. The observed shift to lower wavevectors of the surface phonon anomalies with respect to the positions in the bulk seen in Fig. 11.32 is also consistent with theoretical studies of surface phonon anomalies in other systems such as the refractory superconductor TiN(001) [284] and the layered CDW wave system 2H-$TaSe_2$(0001) [285] (see Sect. 11.4).

This finding was questioned on the basis that early angle-resolved photoemission studies [287] produce Fermi contours at the Fermi level which do not support the existence of surface phonon anomalies. However the subsequent improvements in ARPES resolution and measurements on platinum [256], as well as in DFT calculations [448] corroborated the early HAS observation of Kohn anomalies in the RW dispersion curves of Pt(111). The inset of Fig. 11.32 shows a portion of the $\overline{\Gamma K}$ surface band structure near the Fermi level of Pt(111) calculated by Dal Corso [286] with the local density fully relativistic projector-augmented-wave (LDA-FR-PAW) method. The resulting structure, notably the strong resonances S_9, S_9' and the gap band S_3 ending near the $\overline{M}$-point as a strong resonance, is in good agreement with the ARPES data by Frantzeskakis et al. [256] showing intense spots in the surface-projected DOS near the Fermi level crossing points of the three resonances and around the electron pocket at $\overline{K}$ (vertical lines in the inset of Fig. 11.32). The transitions at the Fermi level between surface DOS spots marked in the inset by horizontal heavy lines are in good correspondence with the observed anomaly wavevectors of 0.65 and 1.05 $Å^{-1}$. The calculations by Dal Corso also attributed the weak anomaly at 0.7 $Å^{-1}$ in the $\overline{\Gamma M}$ direction to an *umklapp* transition across $\overline{M}$ between the S_9 resonances at the Fermi level. This is considered as conclusive confirmation for Kohn anomalies observed in the RW dispersion curves of Pt(111).

Further important experimental evidence in support of the coupling of phonons to electrons and the PC model come from four HAS studies of the effect of hydrogen on the phonon intensities on Ni(001) [276], Pt(111) [288], Rh(111) [288–290] and W(001) [291]. In all four cases there is no change in the symmetry of the

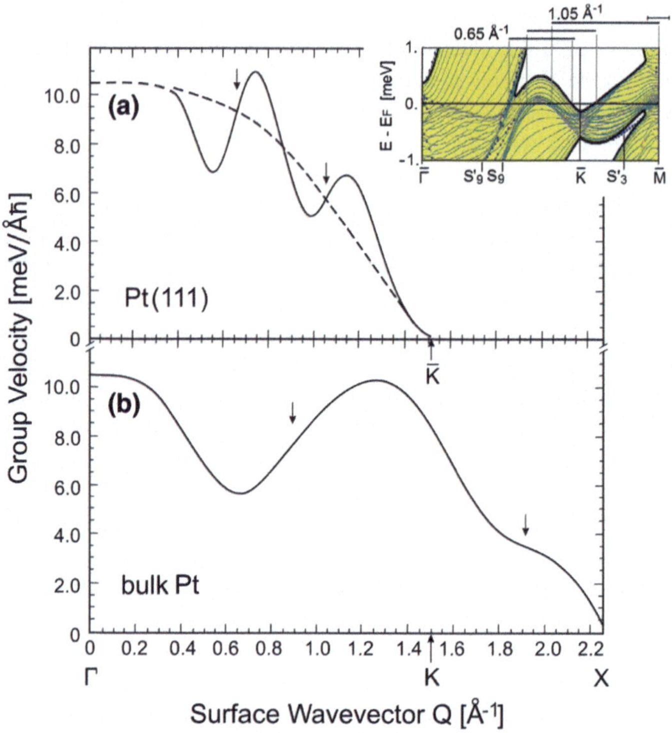

Fig. 11.32 Comparison of the experimental results for the surface and bulk group velocities in platinum: **a** Rayleigh mode group velocity as a function of phonon wavevector from the best fit Fourier expansion of the measured Rayleigh dispersion curve along the $\overline{\Gamma K}$ direction shown in Fig. 11.31. The arrows indicate the inflexion points with positive derivative, which approximately correspond to the intersections of the ideal group velocity with the ideal group velocity curve (broken line) and give the positions of Kohn anomalies. The inset shows an *ab initio* surface band structure of Pt(111) in the $\overline{\Gamma K}$ direction near the Fermi level [448] with possible Fermi level transitions corresponding to the observed Kohn anomaly wavevectors, between intense DOS spots associated with S_9, S_9' and S_3' surface electronic resonances. The error bar for the anomaly wavevectors corresponds to the anomaly FWHM. **b** Bulk group velocity calculated from a best fit Fourier expansion of the experimental transverse bulk mode measured by neutron scattering for an equivalent bulk direction and polarization [263]

surface lattice and only a small relaxation of the metal surface. Moreover since the vibrational modes for the hydrogen-metal vibrations have frequencies much greater than 50 meV, they cannot be excited by He atoms of lower incident energies. On Ni (001), Pt(111) and W(001) the longitudinal resonance is found to disappear entirely and only the Rayleigh mode remains. Its frequency is usually significantly shifted to higher frequencies so that it agrees with that calculated with bulk force constants. In the case of Rh(111) a longitudinal peak still can be seen after hydrogenation but it is significantly weaker in intensity than observed on the clean surface. These effects and the increase in the diffraction peak intensities indicating a greater corrugation, correlate nicely with the observed increase in work function [292] on these same systems. They are also in qualitative agreement with an *ab initio* calculation of the changes of the electronic density of states induced by hydrogen atom adsorption on Pd(001) [293]. Thus the previously highly polarizable delocalized electrons on the clean surfaces appear to be localized into bands which are associated with the strong chemical bond to the hydrogen atom adsorbate.

11.3.5 The Clean (001) Surfaces of Tungsten and Molybdenum

More striking Kohn-like anomalies were found on the (001) surface of another refractory material, the bcc transition metal W [294–296]. This surface probably exhibits the best known example of a nearly perfect two-dimensional phase transition on a clean metal surface driven by a soft surface mode [297]. The structural changes have been extensively studied using a wide variety of techniques including LEED [298–300], high energy ion scattering [301], and X-ray diffraction [302]. As shown in Fig. 11.33, in the phase transition the top layer atoms, which exhibit a (1×1) structure at temperatures above about 500 K, undergo significant lateral shifts in their positions as the temperature is reduced well below the transition temperature at $Tc \approx 280$ K. The resulting zigzag chain structure in the low temperature phase can be seen in Fig. 11.33a. In going through the transition the reciprocal lattice vectors are reduced in size and are rotated by 45° (Fig. 11.33c). The remarkable feature of this phase transition is that it is largely restricted to the top layer in which the lateral displacements are much greater by about a factor four than in the second layer [303, 304] as illustrated schematically in Fig. 11.33b.

Figure 11.34 summarizes some representative HAS results on the angular distributions [305] and the surface phonon dispersion curves [294–296] for three different temperatures well above (T_s = 1200 K), just above (T_s = 500 K), and well below (T_s = 220 K) the phase transition. For comparison in Fig. 11.34a, b the corresponding measurements are shown for the H atom 1×1 saturated surface [291]. At temperatures above 500 K the clean surface exhibits the ideal 1×1 structure as evidenced by sharp first order HAS diffraction peaks which were even seen at temperatures as high as 1200 K (Fig. 11.34c, e). As the temperature drops below 280 K both LEED [306] and HAS [296] (Fig. 11.34g) find the $(\sqrt{2}\times\sqrt{2})$ R45° structure of the reconstructed surface corresponding to zig-zag chains along the $\overline{\Gamma M}$ direction as illustrated in Fig. 11.33a.

Even at temperatures as high as 1900 K sharp single phonon inelastic features were still found in the TOF spectra [296]. This is the highest temperature at which single phonons have been observed. As discussed in Sect. 7.6 the possibility of observing surface phonons at high temperatures requires a small Debye-Waller

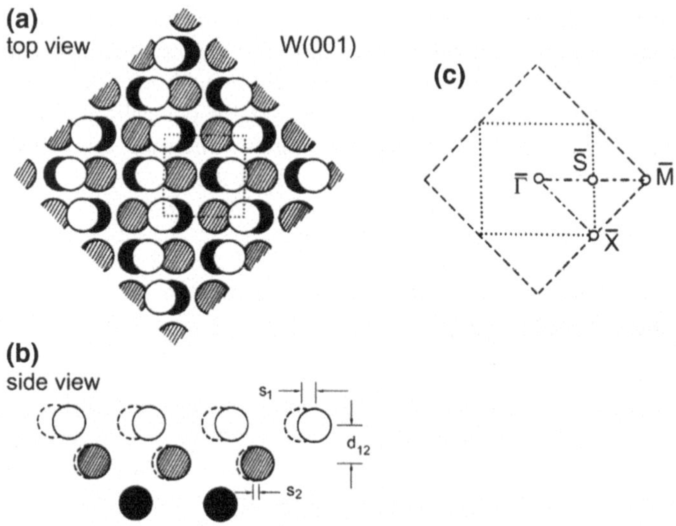

Fig. 11.33 Schematic diagram showing the changes accompanying $(\sqrt{2}\times\sqrt{2})$ R45° reconstruction of the W(001) surface at surface temperatures below 280 K [298–305]. **a**, **b** In the reconstruction the atoms of the top layer (white circles) are displaced from their equilibrium positions by the amount S_1 to form diagonal chains. The atoms of the second layer (shadowed circles) are much less displaced (S_2), while in the third layer (black circles) the atoms remain practically in the bulk positions. At high temperatures ($T_s > 300$ K) the top layer atoms occupy average positions above the third layer atoms. **c** The reciprocal lattice of the two phases are indicated by the larger square (dashed lines) and the smaller 45°-rotated square (dotted lines) corresponding to the high temperature and low temperature phases, respectively

exponent and a small anharmonic broadening of the phonon peak. According to (7.92), one-phonon processes prevail up to temperatures which are generally larger than the surface Debye temperature but smaller than 37.5 $T_m/(a_o k_{iz})^2$, where T_m is the bulk melting temperature, k_{iz} the incident normal wavevector, and a_0 the interatomic distance. In this respect tungsten is an ideal case since it combines the highest meltimg temperature (T_m = 340 K) with fairly soft phonons due to the large atomic mass, as indicated by the comparatively low bulk and surface Debye temperatures, $\Theta \cong 280$ K and $\Theta_S \cong 200$ K, respectively (Appendix B).

As seen in Fig. 11.34d at the high temperature of 1200 K on the clean W(001) surface, there is clear evidence for an anomalous mode with a phonon energy of only about 6–7 meV which changes little with Q in the range 0.4–1.2 Å^{-1} [296]. On the basis of theoretical simulations [303], the experimental analysis [307] and the common wisdom about the origin of Kohn anomalies, this mode appears to be longitudinally polarized. As opposed to the longitudinal mode on the fcc Cu(111) and Cu(001) surfaces discussed above, however, on W(001) the anomalous

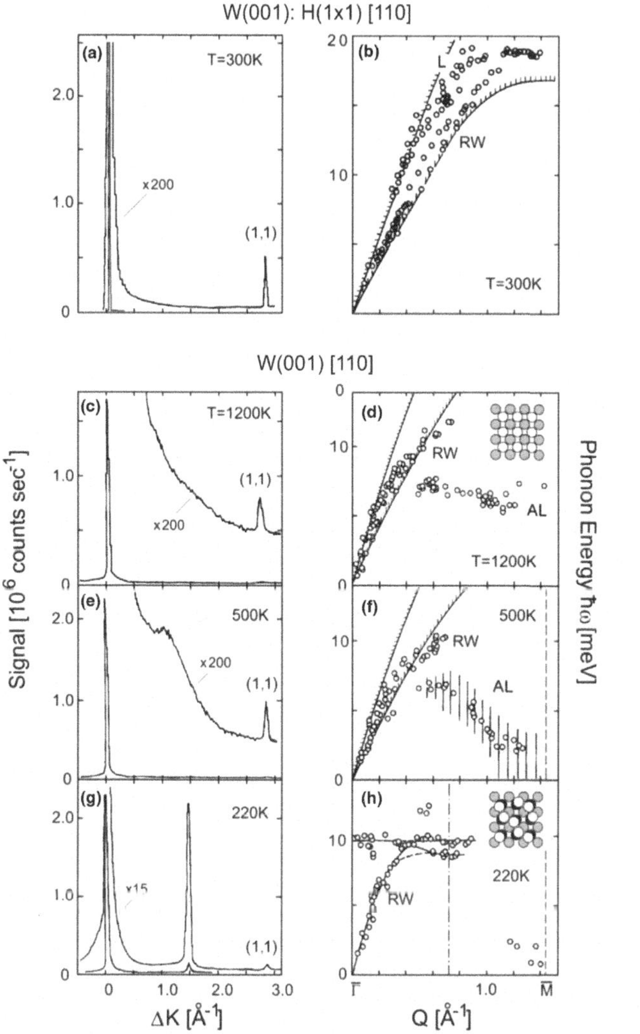

Fig. 11.34 HAS measurements on the W(001) and W(001): H 1×1 surface. **a** Angular distribution for the 1×1 H atom covered surface at T = 300 K [291]. **b** The corresponding surface phonon dispersion curves. **c**, **e**, **g** Angular distributions along the [110] ($\overline{\Gamma M}$) azimuth of the clean W(001) surface at three different temperatures [294–296]. **d**, **f**, **h** The corresponding surface phonon dispersion curves. The insets in **d**, **h** illustrate the structure of the 1200 K unreconstructed and the 220 K ($\sqrt{2} \times \sqrt{2}$)R45° reconstructed surfaces of W(001), respectively

longitudinal mode (AL) lies below the Rayleigh mode. On lowering the temperature further this mode decreases significantly in frequency and already at T_s = 500 K approaches zero within the half width of the energetically broad peaks at $Q \sim 1.2$ Å^{-1}. At temperatures between 500 and 220 K a broadening of the features in the HAS TOF spectra is observed (Fig. 11.34f). This correlates with the disorder found in several experiments [301, 302]. Then at T_s = 220 K, below the transition to the commensurate phase, indicated by the appearance of a sharp 1/2 order diffraction peak, a new system of dispersion curves is found which is symmetric with respect to the new Brillouin zone boundary at $Q = 0.7$ Å^{-1} (Fig. 11.34h).

This remarkable temperature behavior, which at that time had not been seen for any other system, clearly indicates a structural phase transition driven by a soft mode in agreement with the now classic theories of Cochran [308, 309] and Anderson [310]. At the Q-value at which the frequency approaches zero the force constants between adjacent atoms are dramatically softened and the longitudinal displacements accompanying the reconstruction (see Fig. 11.33a, b) increase in amplitude. Then at the phase transition the atoms switch over to their new positions. As seen in Fig. 11.34h below the transition temperature the reconstructed surface is characterized by two regular phonon branches originating from the folding of the dispersion curves with respect to the new Brillouin zone boundary [311]. The upper branch is an optical mode with the same polarization as the original RW branch.

Additional information on the electronic mechanism driving the soft mode comes from the angular distributions shown in Fig. 11.34e. At T_s = 500 K an unexpected broad but distinct peak appears at about $Q = 1.00$ Å^{-1} which is not seen at higher temperatures. With a further lowering of the temperature this peak becomes more pronounced and shifts towards larger Q-values converging on a value of $Q = 1.4$ Å^{-1} which is the position of the diffraction peak of the reconstructed phase which was first observed by Salanon and Lapujoulade [305] in 1986 and later by another group [312]. A careful TOF analysis of this unexpected peak shows that it is due entirely to elastic scattering and therefore cannot come from kinematical focusing (Sect. 10.2.1) nor from a bound state resonance (Sect. 10.1) [296]. The appearance and behavior of this peak is explained by assuming that the helium atoms are scattered from a periodic modulation of the surface electron density associated with an incipient incommensurate charge CDW.

The important role of electrons on this phase transition is confirmed by examining the phonons of the hydrogen saturated surface in Fig. 11.34b. The hydrogen saturated W(001) surface forms a 1×1 structure so that the surface periodicity is the same as that of the unsaturated surface just as with the other transition metals discussed earlier in this section. In Fig. 11.34a, the first order first order diffraction peak (1,1) has much the same amplitude as for the clean surface above the phase transition but is somewhat sharper. Note also that the background intensity between the specular and first order diffraction peak is considerably reduced. This decrease is largely due to a reduction in the phonon inelastic intensities and is consistent with a stiffening of the phonons on the saturated surface. As seen in Fig. 11.34b the longitudinal mode is shifted upwards by more than 10 meV and now appears above the Rayleigh mode as a surface resonance as in the other metals discussed so far.

The comparison with calculated phonon dispersion curves indicate that the phonons of the hydrogen saturated surface are very similar to those for an ideal bulk termination [291]. Thus as in the other transition metals the H-atoms have the effect of stabilizing the surface making it similar to the bulk.

The above observations provide strong evidence that the reconstruction is driven by a CDW mechanism very similar to that introduced to explain the reconstruction on the quasi-two-dimensional layered compounds such as $TaSe_2$ and TaS_2 [313–315], which are discussed in Sect. 11.4. For a CDW structure to occur large nearly parallel nesting regions of the electronic states separated by $2k_F$ must be present at the Fermi surface in an extended zone scheme. Although early photoelectron studies [316, 317] appear to have ruled out such nesting, a better resolved photoelectron experiment [318] indicates that the nesting is sufficient. In this picture electronic states sharply localized in K-space are involved implying a delocalization in real space. This interpretation is supported by the HAS observation of the weak peak at about $\Delta K \sim 1$ Å^{-1} already at temperatures as high as 500 K (Fig. 11.34e). As mentioned above and seen in Fig. 11.34g, this peak shifts to larger ΔK with decreasing temperature and below the phase transition goes over to the half-order diffraction peak at $\Delta K = 1.4$ Å^{-1}. Despite several attempts this shifted peak could not be found in especially dedicated LEED [319] and X-ray scattering experiments [302]. This supports the interpretation that the He atoms are diffracted from the weak electronic corrugations produced by the CDW at the surface [294] whereas electrons and X-rays are diffracted by the atom cores.

A similar strongly anomalous temperature behavior of phonon dispersion curves has been measured with HAS for Mo(001) [320], but there are significant differences in the behavior of the HAS [321], LEED [322] and X-ray [323] diffraction patterns compared to W(001), and the electronic mechanism driving the reconstruction appears to be more complex.

11.3.6 The Hydrogen Saturated (110) Surfaces of Tungsten and Molybdenum

An even more striking example of Kohn anomalies was subsequently observed with HAS in 1992 on the hydrogen saturated W(110) and Mo(110) surfaces [324, 325]. In view of the stabilizing effect of hydrogen in all the previous metals these observations came as a surprise. The dense packed (110) surfaces of the bcc refractory metals W and Mo, which are easily prepared by merely flashing to 2600 K, were among the earliest surfaces to be studied by LEED [326]. The clean surfaces show no evidence for either a reconstruction nor appreciable relaxation which is estimated to be less than 2% [327]. In 1986 Chung et al. [328] proposed that at a $\theta > 0.5$ ML coverage with hydrogen the top layer is shifted in the $[1\bar{1}0]$ direction so that each hydrogen atom is surrounded by three top layer atoms and is centered above a second layer atom (see Fig. 11.35b). First principle calculations confirmed that the hydrogen atoms occupy the three-fold hollow sites on W(110) but found only a very small top layer shift of less than 0.05 Å [329, 330]. The same group found that the hydrogen saturated Mo(110) surface does not exhibit the same "top layer shift"-reconstruction [331].

Figure 11.36 shows a series of HAS TOF spectra measured on the hydrogen saturated surface along the $\overline{\Gamma H}$ direction by Hulpke and Lüdecke [324, 332, 333]. In addition to the Rayleigh and longitudinal modes (open and closed circles), which appear at about the same positions as for the clean surface, there is another set of modes, indicated by circles with crosses which exhibit a precipitous fall-off in energy over a narrow range of wavevectors. The importance of these anomalous features becomes clear from an examination of Fig. 11.37 in which the surface phonon dispersion curves for the clean surface are compared with those of the hydrogen saturated surface in a reduced zone plot. There contribute to the sharp minimum in the lowest (RW) dispersion curve along the $\overline{\Gamma H}$ direction at a wavevector of $Q_s = 0.95$ Å^{-1} where it drops down to a frequency of only about 2 meV. At the same wavevector there is also evidence for a second series of points corresponding to the AL resonance at an energy of about 15 meV with only a small dip in the anomalous region. This anomaly is found to be largest in the $\overline{\Gamma H}$ direction and then weakens as the crystal is rotated towards $\overline{\Gamma S}$ [333] where it has about the same wavevector (Fig. 11.39) and is still quite pronounced. No anomaly is observed along the $[1\bar{1}0]$ direction, which, is the direction of the top layer-shift induced by hydrogen coverage (Fig. 11.35b, c). In contrast the clean W(110) surface does not exhibit any unusual behaviour aside from the ubiquitous AL

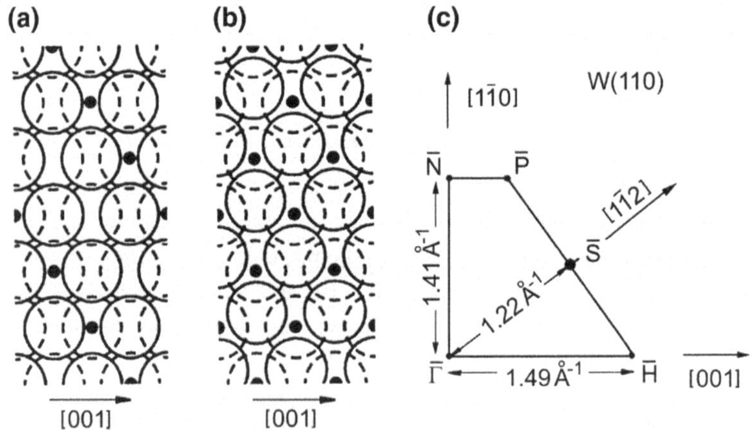

Fig. 11.35 The structure and Brillouin zone of the hydrogen covered W(110) surface. **a** At low coverage ($\theta < 0.5$ ML) the hydrogen atoms (●) occupy bridge sites. **b** At saturation the top layer is shifted and the hydrogen atoms now occupy hollow three-fold sites. **c** The irreducible surface Brillouin zone in the same orientation as in (**a**) and (**b**). The $\overline{\Gamma N}$ distances are 1.41 and 1.42 Å^{-1} and for $\overline{\Gamma H}$ 1.49 and 1.50 Å^{-1} for W(110) and Mo(110), respectively

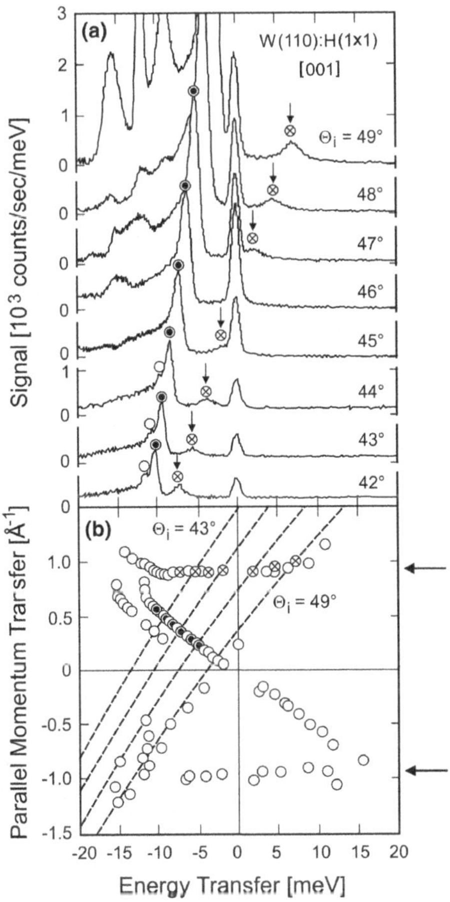

Fig. 11.36 HAS energy transfer spectra for the W (110):H(1×1) surface. **a** Series of spectra for several different incident angles ($\Theta_{SD} = 101.16°$) measured with an incident wavevector of $k_i = 8.56$ Å^{-1} ($E_i = 38$ meV) and a surface temperature $T_s = 150$ K [324, 332, 333]. The filled and open circles mark the Rayleigh and longitudinal mode, respectively, and the circles with crosses mark the lower branch of the sharp Kohn-like anomaly. **b** Corresponding points in an extended energy transfer and parallel wavevector transfer diagram

resonance. Apart from the narrow anomalous region it has about the same dispersion curve as for the hydrogen covered surface.

The HAS data for the Mo(110):H(1×1) surface show almost exactly the same behavior with the sharp anomaly now positioned at $Q_s = 0.89$ Å^{-1} (Fig. 11.38a) [325, 333]. Moreover for both hydrogen saturated W(110) and Mo(110) surfaces a broad elastic peak is seen in the angular distribution at the same value of Q_s at which the phonon anomaly is seen [325, 335]. On the other hand, no anomaly has been reported for the hydrogen saturated Cr(110) surface [336, 337], although Cr is in the same column of the periodic system.

The discovery of unusually sharp anomalies in the H-saturated W(110) and Mo (110) surfaces led to an extensive EELS study by Balden et al. [341]. The EELS data, also included in Fig. 11.37, are able to reproduce most of the HAS data points outside the region of the anomalies. However at $Q_s = 0.93$ Å^{-1} the EELS data do not follow the other HAS points of the sharp anomaly extending down to about

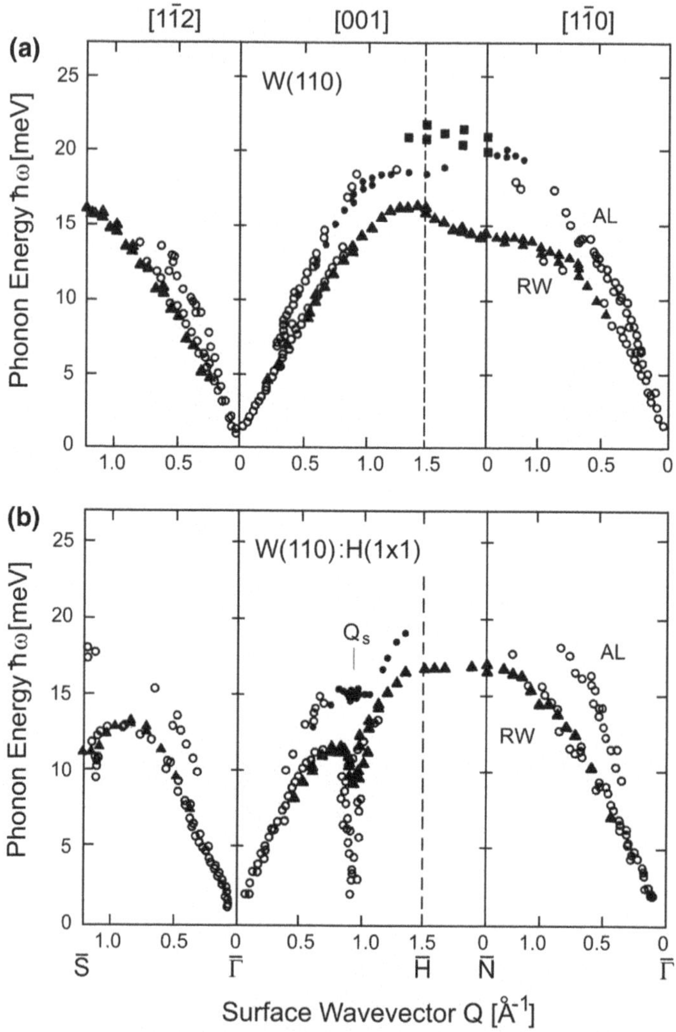

Fig. 11.37 Comparison of EELS (filled triangle, filled circle) [334] and HAS (open circle) [324, 332, 333] measured dispersion curves. **a** For the clean W(110) surface. **b** For the hydrogen saturated W(110):H(1×1) surface. The few second layer modes measured by EELS are indicated by squares

2 meV, only showing a smaller dip with a minimum at about 8 meV. On the other hand, it is seen from Fig. 11.37a that for the clean W(110) surface there is generally good agreement between EELS and HAS. Another intriguing aspect of the sharp anomaly is that it is most strongly developed along a direction perpendicular to that of the previously postulated top layer shift. Further experiments confirm the same phonon behavior with D_2, thereby ruling out that the anomalous features could be

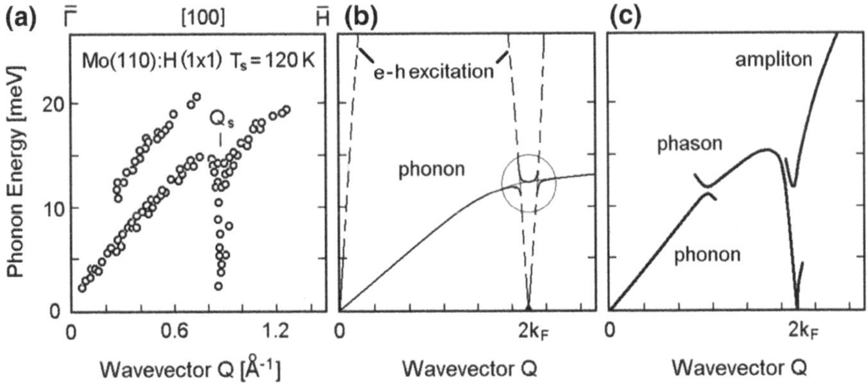

Fig. 11.38 Comparison of the anomaly on the Mo(110):H(1×1) surface with two models. **a** HAS dispersion curves showing a very sharp anomaly at $Q = Q_s$ [325, 333]. **b** This is attributed to electron-hole (e-h) pair excitations of a one-dimensional free electron gas at $2k_F$ [338], according to which a non-crossing between the e-h excitation curve and the phonon branches (encircled) is expected as seen in experiment. **c** The coupling between the two branches at $2k_F$ leads to the formation of a charge density wave (CDW), whose spectrum consists of a "phason" and an "ampliton" branch (adapted from Giuliani and Tosatti [339, 340])

due to some type of unusual low frequency excitation of the adsorbate [341]. Rather the fact that the phonon anomaly is accompanied by the appearance of a peak in the angular distribution of the *elastic* intensity at the same values of Q_s [324, 325] is indicative of an incipient charge-density wave with a $2k_F = Q_s$. Indeed, the intensity of this peak depends strongly on temperature just as for the analogous peak observed in W(001), with the important difference, however, that for W(110):H (1×1) as well asfor Mo(110):H(1×1) no structural phase change is observed.

The interpretation of the anomalies based on the Kohn mechanism was initially ruled out since the position of the anomalies was not consistent with possible nesting wavevectors in the electronic surface states, measured for both hydrogen saturated W(110) [342] and Mo(110) [343] and were too large by about 40%. The value of $2k_F$ for surface states is, however, much more precisely determined from the position of sharp anomalies than from the extrapolation of the broad angle-resolved photoemission features associated with surface states. Indeed the phonon anomalies show the unmistakable features of the well known Kohn anomalies in quasi-one-dimensional conductors such as seen in $K_2[Pt(CN)_4]$ $Br_{0.3}$ $3H_2O$ (KPC) [344–346] and this is *per se* a strong argument in favor of the electron-phonon mechanism.

As shown schematically in Fig. 11.38b the electron-hole (e-h) excitations of a one-dimensional free electron gas (1DEG) have a nearly verticle parabolic dispersion curve, which intersects the phonon dispersion curve at two points very close to $2k_F$, where k_F is the 1DEG Fermi wavevector. The sharp anomaly in the phonon branch originates from an avoided crossing behavior. As Q approaches $Q_s = 2k_F$ the phonon rapidly converts into an e-h excitation with a sharp drop of the frequency. The oscillations of the surface atoms become negligibly small, while those of the surface electron density are amplified. It should be noted that the Kohn anomaly mechanism normally involves longitudinal phonons. For this reason the upper mode observed at Q_s with both EELS and HAS is attributed to a shallow anomaly of the RW, while the deep anomaly is associated with the longitudinal mode which in these systems lies below the RW. Of course, a shallow anomaly is expected to retain most of the original phonon-like character and is therefore observed with both techniques, whereas the deep anomaly acquires a strong e-h pair character and is invisible to EELS. The fact that also the RW shows a weak anomaly at $2k_F$ is attributed to a small longitudinal component which is weakly hybridized with the e-h pair excitations. This interpretation explains why HAS, by virtue of its unique sensitivity to electronic density fluctuations, shows a much deeper anomaly than EELS. If the frequency (energy) of the anomaly were to vanish then the electon gas would stabilize into a CDW with a lattice period $\sim\pi/k_F$, yielding additional diffraction peaks at $\sim 2k_F$. The stabilization in the broken symmetry configuration is actually signalled by the small gap.

Giuliani and Tosatti have shown theoretically that the excitation spectrum of a CDW-distorted one-dimensional system consists of what they have denoted as a phase mode (*phason*) and an amplitude mode (*ampliton*, also called *amplitudon*) illustrated in Fig. 11.38c [339]. There it is seen that the predicted phason and ampliton dispersion branches are strikingly similar to the observed HAS and EELS anomalies, respectively [340].

These unexpected observations led to a theoretical analysis beyond the present phenomenological frameworks [330, 340, 347]. An alternative explanation, in which an involvement of the hydrogen atoms was invoked, was suggested by Balden et al. [334]. This interesting model is not completely satisfactory since it places the anomaly at 1.0 Å^{-1} instead of 0.93 Å^{-1}, the difference being larger than the resolution of the HAS apparatus. Moreover, the model cannot explain the sharpness of the phonon minimum, nor the differences between the HAS and EELS data, and the lack of any observable isotope effect. The most convincing theoretical approach comes from the first principles calculations of Kohler et al. [338]. Using a carefully optimized full-potential linearized augmented plane-wave (FLAPW) program they calculated the local density of electron-hole excitations in K-space [342, 338]. Figure 11.39 shows a comparison of an experimental search for the anomalies as a function of the azimuthal angle extending from the $\overline{\Gamma\text{H}}$ to the $\overline{\Gamma\text{N}}$ direction. Both the predicted nestings along the $\overline{\Gamma\text{H}}$ and $\overline{\Gamma\text{S}}$ directions are observed but the experiment failed to detect the predicted nesting along the $\overline{\Gamma\text{N}}$ direction. Similar nestings at about the positions of the observed phonon anomaly are also predicted for the Mo(110) hydrogen saturated surface [329].

The first *ab initio* dynamical calculation in which the coupling between the electronic and surface atom degrees of freedom is fully accounted for was carried out by Bungaro et al. [349] using a special form of DR theory described briefly in Sect. 5.3. For the H-saturated W(110) surface this calculation correctly predicts the phonon-like branch with the shallow anomaly of the RW at the right position and

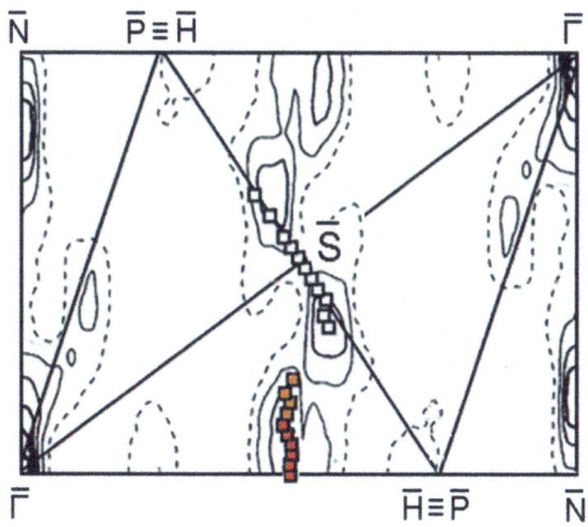

Fig. 11.39 Comparison of the position of the HAS anomalies on the W(110):H(1×1) surface [348] with the calculated local density of electron-hole pair excitations at the Fermi surface indicated by the closed loops [338]. The depth of the anomaly decreases from the square points, colored in red, to the orange squares above the red points, while the white squares correspond to the weakest anomalies [348]. Overall there is good correspondence between the locations predicted by theory and those found in experiment

confirms the mechanism of hybridization between the phonon and electron-hole branches around $2k_F$ and the associated plasmon-like motion of the hydrogen atoms. The deep anomaly, however, was not predicted because in this calculation, performed in the *adiabatic approximation*, the spectrum of e-h excitations cannot be obtained. The theoretical approach used by Bungaro et al. [349] has, however, all the requisites to solve this problem provided that non-adiabatic effects can be included in some way. A phenomenological description of the W(110):H(1×1) and Mo(110):H (1×1) anomalies based on a simplified non-adiabatic PC model is found in [350].

The demonstration that inelastic HAS can probe also collective low energy electronic excitations such as surface phasons, amplitudons, and acoustic surface plasmons, has opened up new horizons in the THz spectroscopy of interesting materials like topological insulators, 2D-superconductors, etc. These new developments are discussed in Chap. 13.

11.4 Layered Crystals

Layered crystals are an important class of materials because of their inherent quasi two dimensional nature. These substances are characterized by weak van der Waals interactions between the individual layers and much stronger intralayer bonds. They were recognized early to be interesting with regard to their surface dynamical properties both as pure materials [3, 315, 351–354] as well as intercalation compounds [355]. Owing to the weak interlayer coupling the dynamical properties of the surface are expected to be very similar to those of bulk layers and thus the surface phonon dispersion curves are expected a priori to be nearly identical to those of the bulk. The weak interlayer interaction has also the effect that the surface-projected bulk phonon bands associated with intralayer modes are comparatively narrow and separated by sizeable gaps. On this basis HAS and EELS studies of layered crystals were initially considered to provide information on their bulk properties [356]. This could be especially important when the thick monocrystalline specimens required for neutron studies can not easily be grown or are not naturally available. For UHV surface studies layered crystals have the advantage that clean surfaces are easily produced cleaving in vacuum. This is achieved by using either the sticky tape peeling technique in an argon atmosphere or in UHV, or by activating in UHV a lever glued to the surface prior to mounting in UHV [3].

11.4.1 Graphite and Graphene

Graphite is probably the best known of the layered crystals because of its inertness and resulting cleanliness and the fact that it is commercially available in the exfoliated form with very large surface areas. As a consequence it has been considered as the ideal substrate for neutron scattering studies of physisorption and two-dimensional phase transitions [357–359]. Early extensive neutron studies gave only a partial description of the bulk phonons for graphite in the acoustic region [360, 361]. Complementary data from IR reflectance and Raman scattering experiments provide the frequencies of optical modes with their displacement patterns at the zone origin [362].

The surface vibrations of graphite were first modeled theoretically in 1981 by de Rouffignac et al. [351] and later by Alldredge et al. [352] with slab calculations. As mentioned above, these calculations provided a quantitative basis to the expectation that the RWs and all other surface modes are extremely close to the edge of the respective bulk bands, well beyond the limits of ordinary HAS resolution, but probably within that of ^{3}He spin echo spectroscopy. The early slab calculations, as well as the bulk calculations by Maeda et al. [363], Al-Jishi et al. [364, 365] and Grüneis et al. [366] based on more refined force constant models extending up to 4th nearest neighbors (4NNFC) with several fitting parameters, still show some significant discrepancies compared to the experiments, especially in the zone boundary regions. This stimulated further theoretical studies with models based on the Valence Force Field (VFF) concept [367], like the calculations by Aizawa et al. [368], Siebentritt et al. [355], and those based on the bond charge model (BCM) [370, 373].

The first HAS experimental studies of surface phonons in graphite were reported in 1981 by the Genoa group [353, 371] who based their dispersion curves on the angular positions of inelastic bound state resonances. Figure 11.40 presents the more recent HAS TOF data points measured along the symmetry direction $\overline{\Gamma M}$ in the low energy region [3, 372], where an unusual upward currature of the SV transverse (*flexural*) mode is observed. For comparison the theoretical dispersion curves showing the surface-projected ZA phonon bands (shadowed areas) from a 6-parameter BCM calculation by Benedek and Onida [373] are also shown. There is a good correspondence of the ordinary RW branch (black full and open squares) with the lower edge of the ZA bulk band of SV polarization, while the data points associated with the longitudinal acoustic modes (full circles) are substantially above the bulk LA branch, though approaching it for increasing wavevector. This apparent deviation occurring at small wavevectors from the theoretical LA branch is attributed to the unfavorable contributions of the forward (Q>0) and backward (Q<0) LA phonons. In general the separation of surface phonon branches from the respective bulk bands, as well as the band width of the surface-projected optical bulk phonon branches are too small for being resolved on the energy scales of Fig. 13.40. This was clearly shown in the early calculations by de Rouffignac et al. [351, 352] and has the simple consequence that in weakly bound layered crystals the phonon dispersion curves measured with surface sensitive probes like EELS and HAS agree very well with the *bulk* phonon branches as obtained from neutron and optical data.

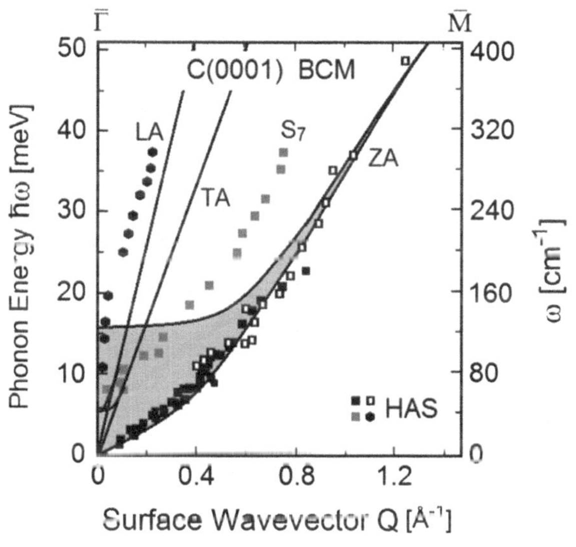

Fig. 11.40 Comparison of HAS experimental and BCM calculated theoretical surface phonon dispersion curves of the (0001) surface of graphite along the $\overline{\Gamma M}$ direction [372]. In addition to the ordinary RW along the lower edge of the ZA bulk band (black full and open squares), data points (full circles) approaching the LA branch are observed. Their deviation from the theoretical LA branch can be attributed to the unresolved contribution of the forward (Q>0) and backward (Q<0) LA phonons. The additional unexpected branch labeled S_7 (grey squares) has been recently confirmed by HAS measurements in graphene/Ru(0001) [374] and is believed to be some kind of collective electronic excitation

An intriguing feature of inelastic HAS experiments on graphite is the occurrence of an unexpected branch, labeled S_7 in Fig. 11.40b (grey squares), whose HAS intensity can be comparable to that of Rayleigh waves [372]. The following explanations have been proposed for this unexpected branch: (1) An (unphysical) surface softening of the in-plane shear force constants [372] can be ruled out, since it is symmetry-forbidden for planar scattering in the $\overline{\Gamma M}$ direction. (2) An effect of surface defects [375] can be ruled out by the observation of an extremely small diffuse elastic peak. (3) Identifying it as the upper edge of the ZA bulk band [376] which can be ruled out since it was not found in subsequent accurate EELS measurements (see Fig. 11.42). Thus the S_7 branch presently is likely to be an intrinsic feature of graphite, a conclusion which has been recently supported by inelastic HAS measurements in graphene/Ru(0001) [374], which also show a S_7 branch (Fig. 11.43). Since theory predicts that the S_7 mode can not be a phonon branch and He atoms are primarily scattered by surface conduction electrons, the S_7 modes are probably some kind of low energy collective electronic excitation. It lies just in the range of the low energy plasmons observed in graphite with HREELS by Jensen et al. [377], Laitenberg and Palmer [378], and Portail et al. [379], and with STS by Vitali, Wirtz et al. [380, 381].

Figure 11.41 shows a compilation of the data from EELS [362, 384, 386, 387, 388] and IXS studies [382, 383], which give a complete picture of the dispersion

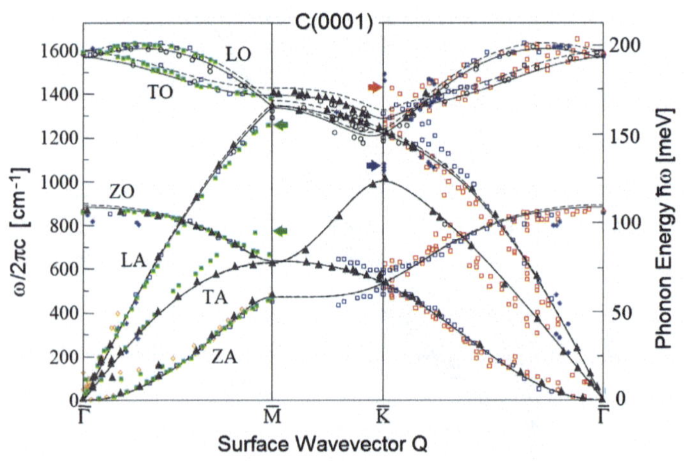

Fig. 11.41 Experimental dispersion curves of the (0001) graphite surface. The data points are from IXS (black full triangles [382], black open circles [383]), HREELS (green full squares [375], red open squares [355], blue open squares [384]), double-resonance Raman scattering (DRRS, blue crosses [366]), infrared absorption (open red triangles [385]), and neutron scattering data (orange open diamonds [361]). The arrows mark the end points of the TA and LA branches at $\bar{M}$, and of TA and TO branches at $\bar{K}$ as obtained from HREELS and DRRS which show significant differences to the IXS data. The *ab initio* dispersion curves from DFT-LDA (dashed line) and DFT-GGA (solid line) calculations by Wirtz and Rubio [381] are shown for comparison (adapted from [381])

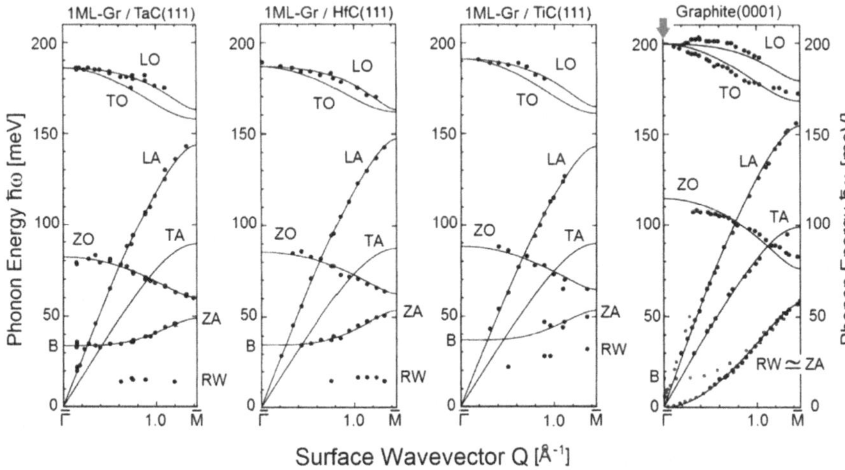

Fig. 11.42 HREELS phonon dispersion curves of graphene on the (111) surface of tantalum carbide, hafnium carbide and titanium carbides along the $\overline{\Gamma M}$ symmetry direction (dots) are compared with graphite (adapted from Aizawa et al. [386]). Charge transfer to the substrate not only softens the intralayer phonon branches with respect to graphite, but also removes the Kohn anomaly of the LO branch of graphene at $\bar{\Gamma}$ (downward arrow). On the other hand the stiffening of the B mode at $\bar{\Gamma}$ indicates that graphene interacts with the substrate considerably stronger than within graphite

curves up to the maximum phonon energy at about 200 meV. For comparison also the latest *ab initio* dispersion curves by Dubay and Kresse (DFT-LDA) [389] (dashed line) and by Wirtz and Rubio (DFT-GGA) [381] (solid line) are shown for comparison. Very similar results have been obtained by the early *ab initio* density functional calculations by Pavone et al. (DFPT) [390] and the DFT-LDA calculations by Maultzsch et al. (DFT GGA) [383]). The extensive analysis by Wirtz and Rubio demonstrates that similar results can also be obtained by fitting the 4NNFC model to DFT-GGA [381]. As indicated by the horizontal arrows in Fig. 11.41, there are, however, significant discrepancies between IXS and HREELS data, and also with double-resonance Raman scattering (DRRS) results reported in [366].

The 6-parameter BCM also provides a very good fit of EELS, HAS and DRRS, thanks to a realistic description of the dynamical role of bonding charges [373]. The BCM adapted to sp^2-bonded systems also provides an expedient method to deal with curvature-dependent force constants occurring, e.g., in fullerenes [391–393], and in general in curved sp^2 materials, for which a general updated version of the valence force field model was proposed [387].

Due to the presence of virtually massless Dirac fermions at the Fermi level around the Dirac $\overline{K}$ points, graphite was recognized already in the 1980s as an ideal benchmark system for the study and simulation of fundamental relativistic quantum field effects in two dimensions at the accessible scale of condensed matter physics [394, 395]. In the last decade the experimental and theoretical studies of graphite monolayers, under the name of *graphene* (Gr), have received a great boost due to the expectation that this (quasi-)ideal 2D material, with its unique transport properties [396–399], will have a large impact not only on fundamental science, but also in several areas of nanotechnology, like flexible electronics, optoelectronics, spintronics, sensors, nanomedicine, renewable energy, etc. [400].

The deposition and growth of graphene have been investigated since the early sixties [401–405]. The phonon dispersion curves have been extensively studied in the 1990s on various substrates by means of HREELS [368, 386, 405–411] and recently reviewed by Nika and Balandin [412]. Figure 11.42 shows the result of the seminal HREELS study of a graphene on various carbide substrates by Aizawa et al. [386]. The ZO, TO and LO intralayer modes of the graphene are considerably softened with respect to those of graphite which are also shown in Fig. 11.42. Charge transfer from the substrate alters the interatomic radial force constant and the two shear force constants within the monolayer. On the other hand, the vertical oscillations of graphene against the substrate (labeled B in Fig. 11.42) are strongly stiffened with respect to the equivalent breathing vertical mode of the graphite surface bilayer. This indicates that the interaction of graphene with those substrates is much stronger than that between two neighboring layers in graphite. The stronger dispersion forces exerted by the substrate as well as the charge transfer, which compensates the work function difference between graphene and the substrate, are the main causes of the stiffening. The shift of the graphene Fermi level away from the Dirac points at $\overline{\mathrm{K}}$, produced by the charge transfer to the carbide substrates also has the effect of removing the Kohn anomaly in the LO branch of graphene at $\overline{\Gamma}$ (downward arrow in Fig. 11.42). HREELS also has sufficient penetration to detect the Rayleigh waves of the first substrate layer, indicated by the RW in Fig. 11.42.

Figure 11.43 illustrates a recent detailed HAS study of Gr/Ru(0001) [374, 413]. The rumpled surface is shown in the top part of Fig. 11.43a. The side view of the hexagonal supercell extends over 11 lattice distances of Ru(0001) and 12 lattice spacings of graphene (Gr) lattice distances in both principal directions. The graphene sheet accommodates the commensurate 12/11 phase by forming a circular dome 1.15 Å high. Figure 11.43a shows the corresponding satellite peaks in the HAS angular distribution along the $[10\bar{1}0]$ azimuth of graphene on both sides of the specular peak (0,0) and of the first diffraction peaks of graphene, Gr(1,0) and Gr $(\bar{1},0)$. These satellites are associated with the so called Moiré pattern and labeled m(n,0); accordingly the Moirè peaks m(11,0) and m(12,0) coincide with the first diffraction peaks of Ru and Gr, respectively. The red points in Fig. 11.43b show the HAS dispersion curves of Gr/Ru(0001) in the Gr BZ along the $\overline{\Gamma\mathrm{M}}$ direction. The ZA and LA branches of Gr agree quite well with those of pure graphite and extrapolate to the same value of 15 meV at B (see Figs. 11.40, 11.41 and 11.42). This is indicative of a fairly weak interaction of Gr with the Ru(0001) substrate, consistent with the fact that an appreciable portion of the graphene layer is not in contact with the substrate due to rumpling. It is interesting to note that at about ~ 800 K the larger thermal expansion of ruthenium permits Gr to accommodate in the 12/11 commensurate phase without rumpling, thus improving the Gr contact with the surface and raising the B point up to about 25 meV [408].

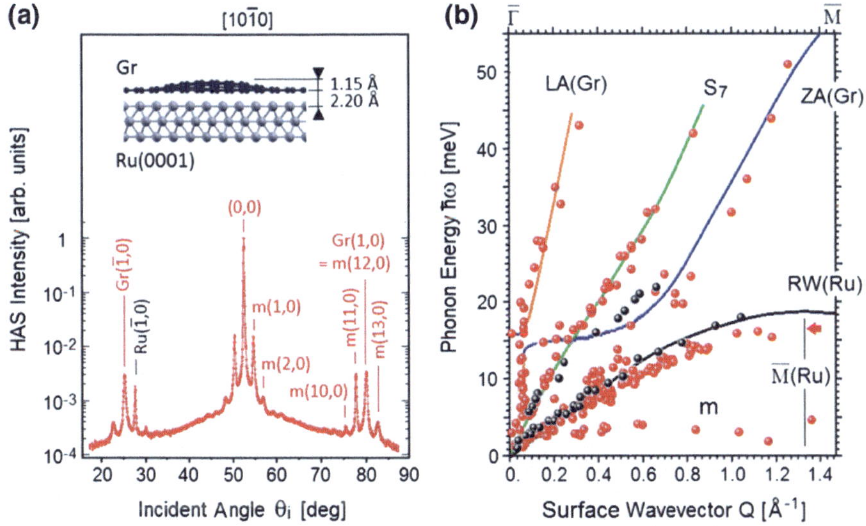

Fig. 11.43 HAS measurements of graphene (Gr) on Ru(0001) [374, 413]. **a** HAS angular distribution along the $[10\bar{1}0]$ azimuth. The diagram at the top is a side view of the hexagonal supercell which extends over 11 Ru and 12 Gr lattice spacings in both principal directions. The Gr sheet adjusts to the commensurate 12/11 phase by forming a circular dome 1.15 Å high in each supercell. The supercell diffraction peaks, forming the so called Moiré (m) pattern, appear as satellites of the specular (0,0) and Gr (1,0), $(\bar{1}, 0)$ diffraction peaks. The Moirè peaks m(11,0) and (12,0) coincide with the first diffraction peaks of Ru and Gr, respectively. **b** HAS dispersion curves (red points) of Gr/Ru(0001) in the Gr BZ along the $\overline{\Gamma M}$ direction. The $\overline{M}$-point of the Ru(0001) BZ is marked by a vertical line. HAS detects not only the expected ZA and LA branches and the same S_7 branch (green curve) observed in graphite (Fig. 11.40) but also the RW of Ru(0001) down to the smallest energies. The low energy branch m can very likely be attributed to phasons related to the large index (12/11) commensurate phase. The black curves and black points show for comparison the RW and the LA modes of the clean Ru(0001) surface (adapted from [374])

HAS can also detect the RW of the Ru(0001) substrate down to the smallest energies thanks to surface charge density oscillations above the graphene sheet induced by the oscillations of the underlying Ru atoms. As shown by the black points and full line in Fig. 11.43 the Ru(0001) RW dispersion curve agrees with that of the clean surface, except for a softening with increasing Q due to the loading effect of the Gr layer. The loading effect at the $\overline{M}$-point of the Ru(0001) BZ (marked in Fig. 11.43b by a vertical line) reduces the RW energy by a factor $M_{Ru}^{1/2}/(M_{Ru} + 2M_C)^{1/2} = 0.90$ (red arrow), where M_{Ru} and M_C are the atomic mass of ruthenium and carbon, respectively. There are two other intriguing aspects in the HAS data from Gr/Ru(0001). One is the observation of the same S_7 branch which was originally found in graphite (Fig. 11.40), and is thought to be of electronic origin, since it has no counterpart in HREELS, nor can it be associated with any known phonon branch. The other aspect is the low energy branch labeled "m" in Fig. 11.43b, which was tentatively attributed to vertical oscillations of the domes [414]. It seems more likely that these oscillations actually belong to the upper ZA (Gr) branch, whereas the low energy "m" branch, as suggested by the DFPT analysis [374], is more plausibly assigned to phason excitations related to the large index (12/11) commensurate phase. In many respects this branch is similar to that predicted [238, 239] and observed with ^{3}He spin echo spectroscopy for the long-period $23 \times \sqrt{3}$ reconstruction of the Au(111) surface [415] (see Sect. 11.5).

The intrinsic instability of a perfectly flat isolated monolayer against a long-wave rumpling [398], discussed in connection with Fig. 11.43, explains why graphene is a *quasi-two-dimensional* film and not an ideal free standing 2D film. In the latter case the parabolic dispersion of the flexural acoustic branch (ZA) would have a vanishing energy at Q=0 as seen in graphite Fig. 11.43. The physical properties of a supported graphene monolayer depend very much on the interaction with the substrate, which should be as weak as possible. This interaction is complicated by the different periodicity of graphene with respect to the substrate, which may either allow for a perfectly flat monolayer (e.g., Gr/Ni(111) [410]), or impose a commensurate superstructure with a possible periodic rumpling of the graphene sheet as illustrated in Fig. 11.43 for Gr/Ru(0001) [374, 413]. In all these cases the graphene-substrate interaction has the greatest effect on the dispersion of acoustic phonons in the lowest part of the spectrum, which can be investigated best with HAS.

11.4.2 Pnictogen Semimetals

The stable rhombohedral forms of the pnictogen semimetals Bi, Sb, and As have in common a layered structure formed by the stacking of (111)-oriented atomic bilayers, with stronger, partially covalent bonds within the bilayer and weaker bonds between adjacent bilayers. Despite the comparatively weak interlayer bonding, the electronic structure and the electron-phonon properties of the surface bilayer are rather different from those of the bulk. The presence of spin-split surface bands crossing the Fermi level leads to the formation of narrow electron and hole pockets at the symmetry points and along the symmetry directions of the surface Brillouin zone with relevant topological properties [416– 418]. The concentration of Fermi level electrons around well-defined pocket wavevectors has effects similar to nesting and favors the formation of charge density waves and associated elementary excitations such as phasons and amplitons (see Sect. 11.3.3). Moreover the corresponding enhancement of the electron-phonon interaction may yield Kohn anomalies in the surface phonon branches as well as in those of the phasons and amplitons.

Figure 11.44a illustrates these unusual features in the surface electronic band structure of Sb(111), derived from a DFT calculation including spin-orbit coupling. [418]. The two spin-split surface bands, eventually crossing at $\bar{\Gamma}$, confer to the Sb (111) the character of a weak topological insulator and lead to the formation at the Fermi level of electron pockets around the $\bar{\Gamma}$ and $\overline{M}$ symmetry points (downward arrows) and a hole pocket in the $\overline{\Gamma M}$ direction (upward arrow). Figure 11.44b shows the DFPT calculated surface phonon dispersion curves for Sb(111) without the inclusion of the spin-orbit interaction for a 24-layer thick slab [418].

Figure 11.45 compares the HAS measured surface phonon experimental points on Sb(111) for two different surface temperatures [419] with the phonon branches with the largest DOS in Fig. 11.44b. Despite the large scatter of the data the theory provides a reasonable description of the HAS data plotted in Fig. 11.45 [419]. It is particularly interesting to note that the second bilayer modes (SV-3 and L-3, with index 3 refering to the third atomic layer) appear to have an amplitude comparable to the amplitude in the first bilayer (SV-1 and L-1, with index 1 refering to the surface atomic layer). This provides a further example of the quantum sonar effect. The HAS data at 170 K show, in addition to the acoustic and optical surface phonon branches, two more low energy branches (crosses) of weaker intensity. Since they cannot be assigned to vibrations of the clean surface they have been tentatively assigned to phasons (P) and amplitons (A) [419]. The latter branches, as well as the L-1 acoustic phonon branch show what appears to be a Kohn anomaly near the $\overline{K}$ symmetry point, while a similar softening seems to affect the optical surface branch L-3 at $\bar{\Gamma}$.

A similar set of surface phonon branches have been measured with HAS on Bi (111) [420–422]. The experimental data points are plotted in Fig. 11.46a on top of the dispersion curves calculated with DFPT without spin-orbit coupling [422]. The intensity of the phonon branches corresponds to the phonon-projected density onto the first and second bilayers, according to the same color code of Fig. 11.45. As already observed for Sb(111), the inelastic scattering intensities from phonons in the first or second bilayer are comparable, as shown in Fig. 11.46b. There the intensities of a HAS energy transfer spectrum for one particular incident angle are assigned to the various modes at different layers at or near the surface. The DFPT calculated phonon-induced surface charge density oscillations are very similar to the calculations discussed in Sect. 8.4.2 for Cu(111). As in the case of Cu(111) the subsurface modes, localized in the upper atomic layer of the third layer (SV-3 and L-3), which is the second bilayer, produce surface charge density oscillations comparable to those of the first-bilayer phonons SV-1 and L-1, thereby explaining the observed HAS intensities.

The HAS measurements for Bi(111) do not show any additional branch associated with phasons or amplitons. A Kohn anomaly in the RW is seen at $\overline{M}$, which is quite pronounced at 123 K (Fig. 11.46a), but is almost absent at room temperature [422]. This is similar to the Kohn anomaly observed in Sb(111) near the $\overline{K}$-point at 170 K (Fig. 11.45).

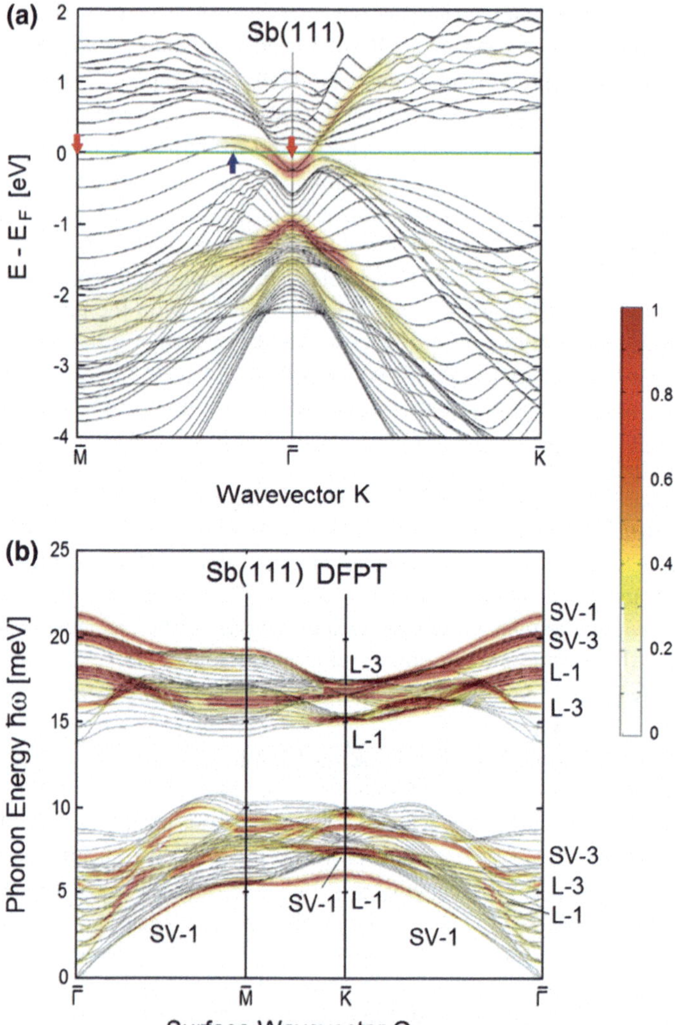

Fig. 11.44 Electronic band structure and surface phonon dispersion curves of the Sb(111) surface. **a** DFT calculated electronic bands for a 12-layer slab including the spin-orbit interaction The color code indicates the amplitude of electronic bands at the surface bilayer. The surface spin-split bands, crossing at the zone center form two electron pockets at $\overline{\Gamma}$ and $\overline{M}$ (downward arrows) and a hole pocket in the $\bar{\Gamma}\bar{M}$ direction (upward arrow) [418]. **b** DFPT phonon dispersion curves calculated for a 24-layer Sb(111) slab without spin-orbit interaction The color code indicates the surface-projected density of the modes. SV-1 and SV-3 label shear-vertical modes with the largest amplitude in the first and the upper layer of the second bilayer, respectively, and similarly L-1 and L-3 for the longitudinal surface modes [418]

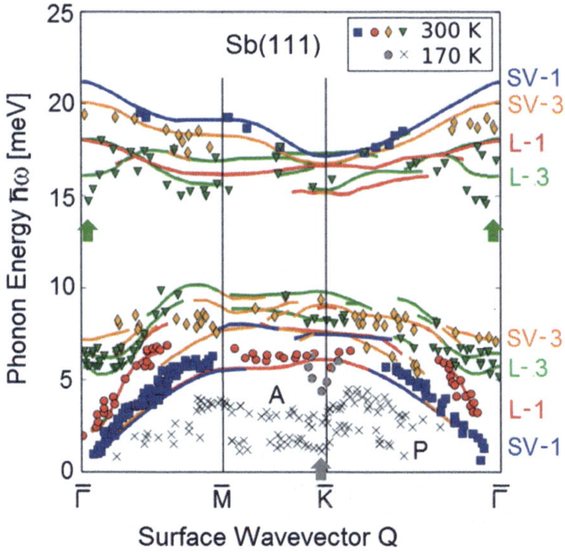

Fig. 11.45 HAS phonon dispersion curves of Sb(111) measured at 170 and 300 K [419] are compared with the phonon branches with the largest DOS of the DFPT surface phonon dispersion curves in Fig. 11.44b [418]. In addition to the usual surface phonon dispersion curves, at 170 K HAS provides evidence for two low energy branches (×), which cannot be ascribed to any phonon branch of the clean surface and are interpreted as the dispersion curves of phason (P) and ampliton (A) modes [419]. The upward arrows indicate Kohn-type anomalies. The Kohn anomalies affect the surface phonon dispersion of L-1 as well as the P and A branches near the $\overline{K}$-point

The spin-orbit coupling is quite important in Bi(111) and was shown to give a sizeable contribution also to the electron-phonon interaction [423]. However all recent DFPT calculations of the surface phonon dispersion curves of Bi(111) with the inclusion of the spin-orbit coupling [423–425] agree reasonably well with experiment in the acoustic region, but do give somewhat softer optical phonons than observed in the experiment. This is explained by the fact that the van der Waals forces which are comparatively strong in bismuth are neglected. They are estimated to have a compensating effect over the softening produced by spin-orbit coupling [425]. This explains why DFPT calculations without spin-orbit and van der Waals corrections perform nevertheless quite well.

11.4.3 Layered Semiconductor Chalcogenides GaSe and InSe

The layered metal chalcogenides have been studied for a long time because of their various important transport and optical properties of technological interest. In the last decade they have undergone a renaissance, similar to graphite, stimulated by

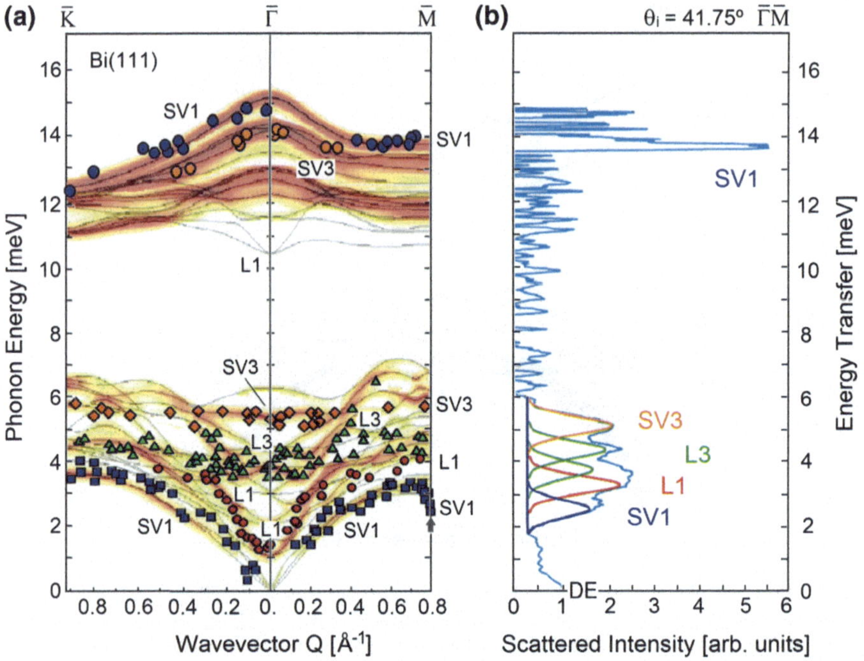

Fig. 11.46 Phonon dispersion curves and intensity distributions of the Bi(111) surface. **a** The HAS phonon data at 296 and 103 K are superimposed on the dispersion curves calculated with DFPT without spin-orbit coupling [422]. The labels of the surface phonon branches and the color code have the same meaning as in Fig. 11.45b. The upward directed small arrow at Q = 0.8 indicates a Kohn anomaly seen only at 103/123 K. **b** HAS energy transfer intensity distribution measured at an incident energy of 16.15 meV and incident angle of 41.75° along the $\overline{\Gamma M}$ direction for a 91.5° planar scattering geometry and a surface temperature of 123 K. Besides the sharp peaks associated with the optical surface modes, SV-1 and SV-3, all the expected acoustic surface modes in the first and second bilayer are resolved with comparable intensities

the discovery of several high-temperature superconductors [426, 427] and topological insulators [428, 429] within their large family.

The insulating layered chalcogenides GaSe(0001) and InSe(0001) have been extensively studied using HAS since 1986 and with various theories. GaSe is an ideal candidate for a pilot study since the earlier bulk studies [369, 388, 430–433] indicated that, unlike layered transition metal dichalcogenides, it had no phase transitions. This was also expected from its electronic structure, which is that of a semiconductor [434–436]. Figure 11.47 shows the structure of the ε-polytype of GaSe, which is similar to that of many of the layered dichalcogenides. Sheets of Ga dimers are sandwiched between two similar sheets of Se atoms to form a layer consisting of $(GaSe)_2$ dimers, each of which is held together by strong covalent Ga-Ga bonds lined up perpendicular to the planes [431]. The Se atoms and Ga dimers are hexagonally close packed in their respective sheets with the same periodicity. The Ga and Se atoms within the layer are held together by strong

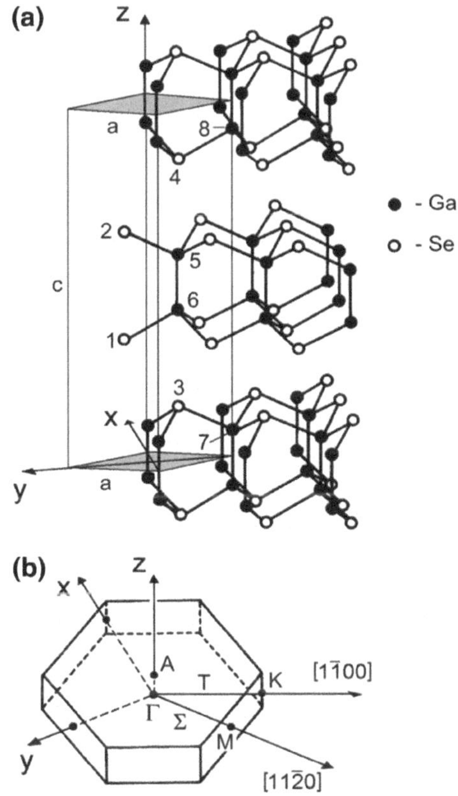

Fig. 11.47 Structure and unit cell of the hexagonal ε-polytype of GaSe showing the interlayer bonding. **a** The unit cell with dimensions a = 3.75 Å and c = 15.9 Å contains the atoms labeled 1–8 [3, 356]. **b** The three-dimensional Brillouin zone diagram and the high symmetry surface directions

covalent bonds whereas the bonding between layers is of the weak van der Waals type [431–433]. Different relative positions of the layers with respect to each other within a stack lead to various polytypes characterized by different unit cells. The experiments were made on the ε-polytype structure shown in Fig. 11.47.

Presumably because of the difficulty to grow thick monocrystalline samples the neutron phonon data are not copious, being limited to the $\overline{\Gamma A}$ and $\overline{\Gamma M}$ direction [369, 388, 430]. Moreover only a few model calculations of the bulk phonon dispersion relations with the force constants fitted to the neutron data and Raman frequencies have been carried out [14, 430]. The first HAS measurements, which were reported in 1986 [356], were made on crystals grown by standard techniques. These were cleaved in an argon atmosphere and then mounted in vacuum. The measurements were 13 years later repeated for GaSe and extended to InSe with samples grown *ex-situ* using molecular beam epitaxy (MBE) techniques on the hydrogen terminated Si(111) 1×1 surface with a thickness of about 14 layers [437]. These more recent improved HAS energy transfer spectra and dispersion curve measurements are displayed in Fig. 11.48a–c for GaSe (0001) and (d–f) for InSe (0001). At least five different surface phonon modes, labeled S_1, S_2, S_3, S_6, and S_8, are clearly resolved in the energy transfer spectra on both surfaces. Surprisingly the

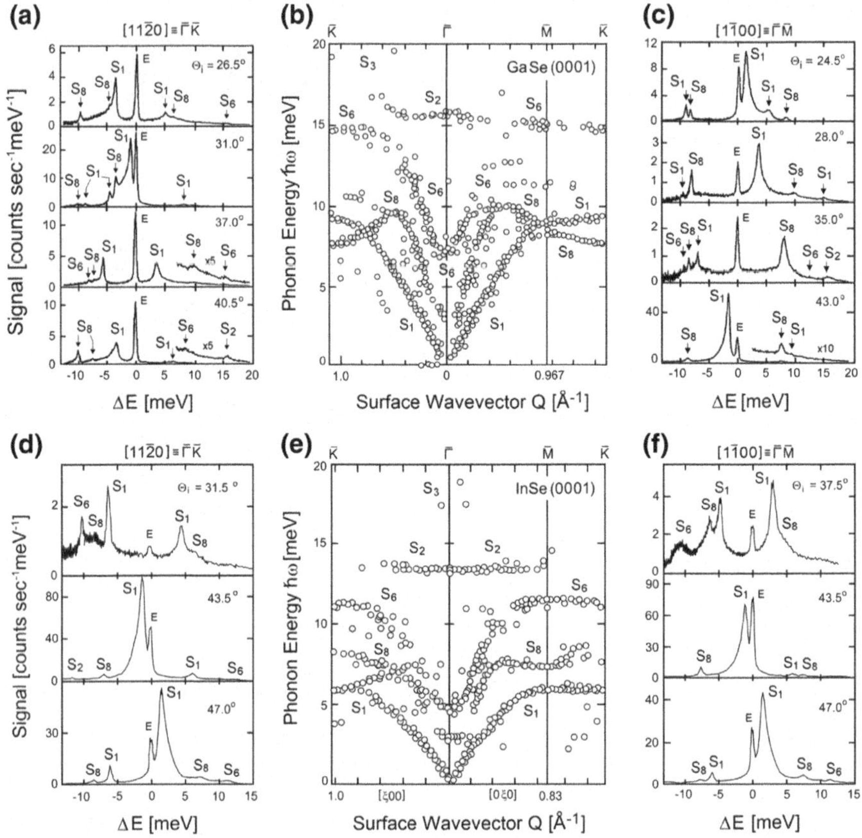

Fig. 11.48 HAS energy transfer spectra and surface phonon data for a 14-layer MBE grown GaSe (0001) and InSe(0001) films on Si(111):H(1×1). **a**, **c** energy transfer spectra measured for GaSe at E_i = 20 meV and T_s = 300 K **b** Dispersion curves derived from spectra in (**a**) and (**c**). For comparison analogous data are shown in panels (**d–f**) for InSe (adapted from [437])

S_8 mode in GaSe lies *below* the Rayleigh mode S_1, at both the $\overline{M}$-and $\overline{K}$-points by 1 and 2 meV, respectively.

These dispersion curves appear to be much different than the surface-projected phonon bands derived from the bulk data [369]. According to a phenomenological analysis based on the DLC model [356, 438–440] the surface modes S_1, S_2, S_3, S_6 could be reasonably well assigned from their proximity to the corresponding bulk modes, whereas the mode S_8 could not be assigned to any bulk band unless by invoking an anomalous softening of the first transverse optical mode at short wavelengths. A similar situation occurs for InSe(0001) (Fig. 11.48e). The puzzle was solved by the DFPT calculations for GaSe and InSe by Schröder et al. [437]. Figure 11.49a, c shows comparisons of their calculated dispersion curves with the HAS data. The agreement is reasonably good but the calculations differ sharply from the early fits of the neutron experimental points and related assignments proposed by Jandl and Brebner [369, 430]. The S_8 branch is then confirmed as intrinsic by the DFPT analysis and has its origin in the bulk electronic response.

Figure 11.49b shows the inelastic electron tunneling spectra (IETS) of a thin 10–20-layer slab of GaSe squeezed between two superconducting lead contacts [441, 442]. The IETS data, which provide a direct measurement of the density of states is in good agreement both with the GaSe and also with the InSe dispersion curve. To emphasize the similarity of the two systems the latter are plotted on an ordinate scale in the ratio 4/3 approximately corresponding to the square root of the cation mass ratio to compensate for the greater mass of the indium cation. The largest peak of the IETS profile (Fig. 11.49a, b) corresponds to the energy range 8–10 meV where the S_8 branch is concentrated over a large portion of the BZ demonstrated. This analysis demonstrated that DFPT can be highly predictive as regards surface phonons and can settle controversial interpretations. Moreover the original conjecture that for layer crystals HAS spectroscopy can do a better job than neutrons for bulk phonons received here a strong support, at least for insulators and semiconductors. The weak interlayer forces, which should make the surface layer very similar to the bulk layers, can also be responsible for the insurgence of peculiar low-dimensional effects such as an enhancement of electron phonon interaction. In this case a factor two reduction of the interlayer forces occurring at the surface may have dramatic effects. The transition metal chalcogenides discussed next, indeed, do show noticeable departures from the bulk phonon behavior associated with the occurrence of a CDW phase transition.

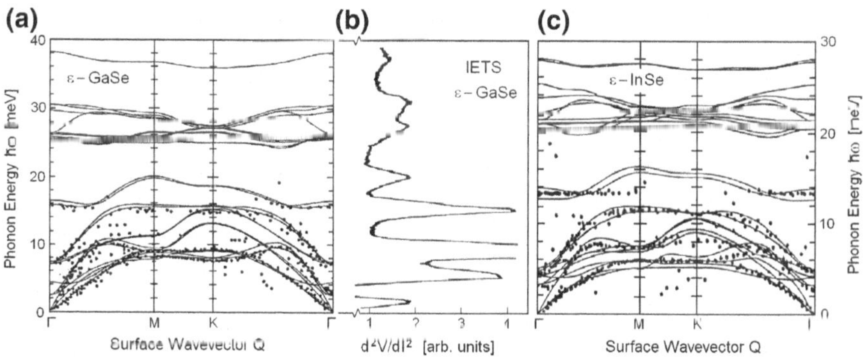

Fig. 11.49 Comparison of the surface phonon data points from HAS for ε-GaSe(0001) (**a**) and ε-InSe(0001) (**c**) with *ab initio* DFPT calculations [437]. The few data point falling outside the phonon spectral region are attributed to inelastic bound state resonances. To show the similarities between the dispersion curves of the two systems the InSe dispersion curves are plotted on an ordinate scale in the ratio 4/3, approximately corresponding to the square root of the cation mass ratio to compensate for the greater mass of the In cation. **b** The inelastic electron tunneling spectrum (IETS) of a Pb–GaSe–Pb tunneling junction measured by Yamaguchi et al. [441, 442]. The peaks of the IETS spectrum agree with the flat branches occurring at the BZ boundary or around the BZ center, which mostly contribute to the total phonon density. There is also good agreement with the rescaled ε-InSe phonon branches

11.4.4 Layered Metallic Chalcogenides $TaSe_2$ and TaS_2

The layered dichalcogenides 2H-$TaSe_2$ and 1T-TaS_2 have been extensively studied in the bulk using neutrons [434, 435, 443] and high resolution electron microscopy [444] since they provide opportunities to investigate charge-density wave (CDW)-induced structural phase transitions in two dimensions. From these extensive structural studies several mechanisms are now thought to be involved. As discussed for clean and H-covered tungsten surfaces (Sects. 11.3.5 and 11.3.6), Fermi surface nesting leads to singularities in the dielectric response function at certain wavevectors $2k_F$. This leads to a gap at the Fermi level together with a modulation of the electronic charge-density in real space which can be either commensurate or incommensurate with the crystal lattice geometry. At the same time, these divergences affect the phonon dispersion curves via the electron-phonon interaction leading at low temperatures to a softening of the phonons at $2k_F$ [434]. With HAS it is possible to observe not only the structural changes accompanying the phase transitions, but also to study the phonon softening and search for differences with respect to the bulk dispersion curves. The early HAS studies on some layered transition metal dichalcogenides and on the layered high-T_c superconductors were reviewed in 1992 by Skofronick and Toennies [3].

Figure 11.50 shows the crystal structure of the 2-H polytype of $TaSe_2$. The hexagonal unit cell (Fig. 11.50b) contains a dimer consisting of two Ta ions and four Se ions [443, 445, 446] and spans two layers [435]. The Se–Ta–Se subunits form strongly bonded layers which are only weakly bonded to the adjacent layers. The weak bond connects the cleavage planes (0001) and because of the strong anisotropy in the bond structure, each layer is expected to behave as a nearly two-dimensional system, similar to graphite. Different stacking arrangements of the layers lead to different polytypes with different unit cells and slightly different electronic and structural properties. Figure 11.50a, c shows the surface structure and the Brillouin zone with the high symmetry directions. At temperatures below T_o = 123 K a CDW leads to a change in the structure [446] and on further cooling at least three different structural phases appear. Moreover, the transition temperatures and occurrence of certain phases are observed to depend upon whether the crystal is cooled or heated. These structural transitions have been extensively studied at the surfaces of both $TaSe_2$ and TaS_2 using AFM and STM [313, 447, 448] and helium atom diffraction [3].

Figure 11.51 provides an overview of the HAS dispersion curves of 2H-$TaSe_2$(0001) at T_s = 140 K, well above the CDW transition temperature at T_0 = 123 K [285, 449]. The optical branches, which were not found in previous neutron studies, were first determined by HAS. The bulk bands with polarization in the sagittal plane, as obtained from a DLC fit of the data [450], are represented by the vertical hatched regions. The two acoustic bands were directly obtained from fitting neutron data [435], whereas the optical bands were fitted to the HAS [449] and Raman [451, 452] data. Even well above T_c, the RW branch (continuos line)

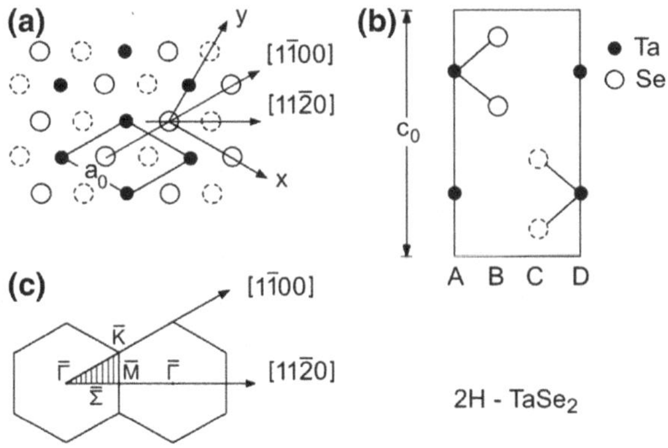

Fig. 11.50 Crystal structure and surface Brillouin zone diagram of 2H-$TaSe_2$.
a Atomic arrangement as seen from the surface. **b** Side view of the unit cell. **c** The high symmetry directions and the irreducible part of the surface Brillouin zone (after [445])

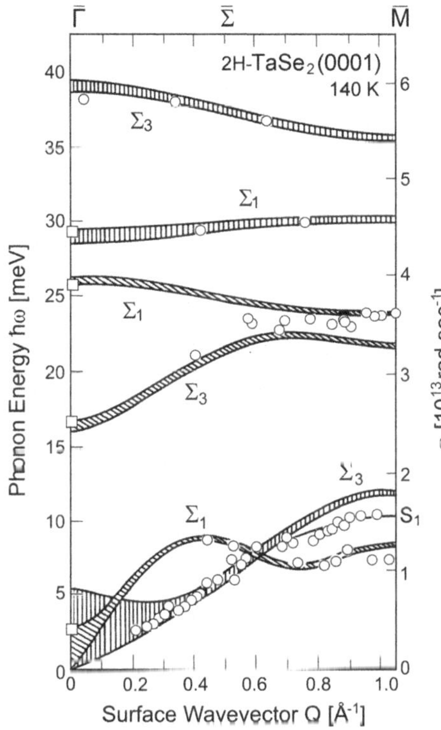

Fig. 11.51 HAS phonon measurements for 2H-$TaSe_2$(0001) along the $\bar{\Gamma}\bar{M} \equiv \bar{\Sigma}$ direction at 140 K (open circles) [449, 453] and Raman scattering data (open squares) [451, 452] are compared with a dispersive linear chain (DLC) fit [450] which included previous neutron data. Hatched areas correspond to the surface-projected bulk phonon bands for predominantly x- (oblique hatching) or z- (vertical hatching) polarization. The Σ_3 and Σ_1 modes have TA and LA character, respectively, S_1 is the Rayleigh mode

shows a clear softening at the zone boundary below the Σ_3 bulk band. The good fit with the DLC model of the entire set of experimental data including the HAS data for the optical branches [314, 450] provided insight not only into the inter- and intra-layer interactions but also on the perturbation needed to reproduce the zone boundary softening of the Rayleigh wave (S_1 in Fig. 11.51).

As shown in Fig. 11.52 at temperatures much below 140 K the Rayleigh wave anomaly in 2H-$TaSe_2$(0001) is shifted to smaller wavevectors with respect to the corresponding bulk anomaly [453]. The Rayleigh branch softens anomalously only in the fairly narrow temperature range between 70 and 140 K at one-half of the reduced Brillouin zone with the largest softening at around 110 K. A comparison with the bulk modes is shown in Fig. 11.52a where the room-temperature bulk dispersion curves Σ_1 and Σ_3 measured by Moncton et al. [435] are shown along $\overline{\Gamma M}$. At 140 K the HAS Rayleigh mode (S_1) dispersion curve almost coincides with that of the transverse bulk mode Σ_3, except for some softening at the zone boundary. The bulk anomaly, which occurs at $Q_b = 0.70$ Å^{-1} in the longitudinal polarized Σ_1 branch, exhibits a strong temperature dependence and reaches its deepest value at the bulk transition temperature (123 K). As seen in Fig. 11.53 the temperatures of the bulk and the surface phonon anomalies are also distinctly different and no obvious direct correlation can be established between the two effects.

According to the Kohn anomaly mechanism, the wavevector shift of the anomaly indicates that a surface electronic state cuts the Fermi surface at a wavevector different from that of the bulk. Actually, angle-resolved photoemission measurements by Smith and Traum [454] show, in the region just below the Fermi

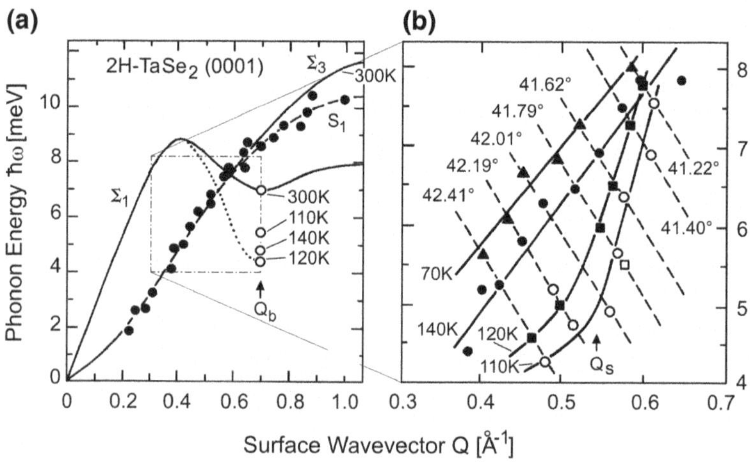

Fig. 11.52 HAS experimental data for the Rayleigh mode of 2H-$TaSe_2$ along $\overline{\Gamma M}$ [453] **a** Rayleigh mode (S_1) measurements (solid points and dashed line) at 140 K are compared with the bulk Σ_1-mode energy (open circles) at $Q_b = 0.7$ Å^{-1} at four different temperatures. The dotted line indicates the deepest bulk anomaly at 123 K. The solid line curves are the bulk TA (Σ_3) and LA (Σ_1) modes [435]. **b** HAS Rayleigh mode dispersion curves for 4 temperatures over a limited wavevector range, revealing the anomalous softening at $Q_s = 0.54$ Å^{-1} (filled circles = 140 K; filled squares = 120 K; open squares = 115 K; open circles = 110 K; triangles = 70 K). The interpolating solid lines are guides to the eye. The dashed lines indicate the HAS scan curves at different incident angles [449, 453]. The surface anomaly has its minimum at 110 K compared to 123 K in the bulk

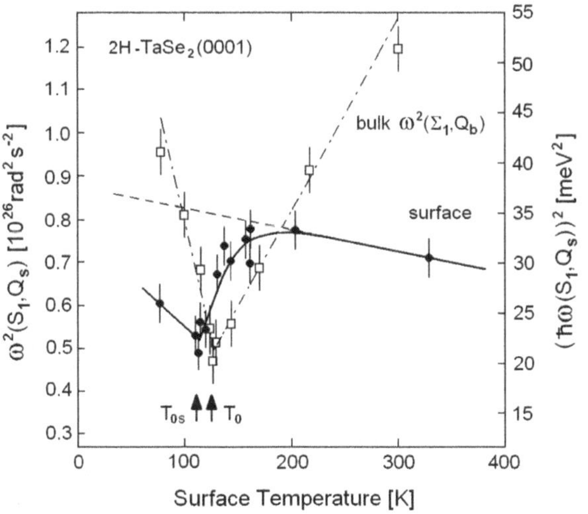

Fig. 11.53 Comparison of the temperature dependence of the Rayleigh (S_1) surface phonon and the Σ_1 bulk anomalies in 2H-$TaSe_2$(0001). The HAS data (filled circles) is for $Q_s = 0.54$ Å^{-1} [453] and the neutron data (open squares) [435] is for $Q_b = 0.70$ Å^{-1}. The largest softening of the squared phonon frequency occurs at $T_{0s} = 110$ K for the surface S_1 mode and at $T_0 = 123$ K for the bulk Σ_1 mode

surface, a band derived from tantalum d-states moving downward for increasing Q. Because of the band flatness, the smallest softening, induced by the surface, would appreciably shift the band Fermi wavevector to a smaller value. The effect observed here has a striking similarity to that predicted in the (001) surface of TiN [284]. In this case the change in the range of the electron-phonon interaction was attributed to the modified coordination at the surface as discussed in the next paragraph. Since in layered crystals like $TaSe_2$ the metal atom which provides an extra free electron to the conduction band has the same coordination at the surface as in the bulk, the explanation given for TiN(001) is not expected to apply to $TaSe_2$.

Another intriguing mechanism determining the position of certain phonon anomalies has been proposed by Bilz and coworkers [455, 456] for layered perovskites. They suggested that the position of the anomaly is given by the wavevector of nonlinear periodic waves (*periodons*) coupled to the phonons. Such waves originate from a nonlinearity in the electron-phonon interaction. Their period vector determined by the exponent of the nonlinear term is an integer fraction of a **G**. For example, a cubic term in the electron-phonon force yields an anomaly at **G**/3, of the kind occurring, e.g., in the layered perovskite K_2SeO_4 [457]. In the absence of inversion symmetry, as occurs at crystal surfaces, also even-power electron-phonon terms may be induced and cause anomalies at even fractions of **G**, as happens, for example, in $TaSe_2$(0001) and TiN(001). Even-power electron-phonon terms, however, do not seem to be strong enough to induce a superstructure with period four, corresponding to the half-zone anomaly, since ordinary (3×3) superstructures due to the bulk CDW are also detected at the surface of 2H-$TaSe_2$ by He atom diffraction [458]. In general shifted anomalies may be viewed as a manifestation of competing symmetries, in this case not only between the symmetry of the lattice (represented by **G**) and that of the electron gas (represented by $2\mathbf{k}_F$) but also between those of the surface and the bulk. Since the surface anomaly is peaked at the striped-incommensurate transition temperature of 2H-$TaSe_2$ rather than at the bulk CDW T_C [458], one could consider the intrinsic tendency of the surface to adopt a different periodicity as responsible for the internal stress yielding discommensuration.

TaS_2 is perhaps the most interesting among the transition metal dichalcogenides exhibiting charge-density waves. It has been most intensively studied because of its complex phase diagram shown in Fig. 11.54 [445, 446, 459, 460]. On lowering the temperature, the 1T normal phase develops into an aligned incommensurate structure below 543 K ($1T_1$ phase), then below 351 K into a rotated nearly commensurate phase ($1T_2$ phase). Finally below 183 K, as shown in Fig. 11.55a, TaS_2 converts into a fully commensurate CDW induced structure ($1T_3$ phase) with a $(\sqrt{13}\times\sqrt{13})$R13.9° superstructure, in which the hexagonal axes are rotated by 13.9° with respect to those of the normal phase. The higher transition temperature of 183 K compared to $TaSe_2$ for onset of the commensurate $1T_3$ CDW-phase suggests a much stronger electron-phonon coupling. This is presumably the reason why Boato, Cantini and Colella [461], in an important pioneering He diffraction study, were able to observe satellite diffraction peaks associated with the CDW at T_s = 20 K. The peak intensities were comparable to those of the Bragg peaks for

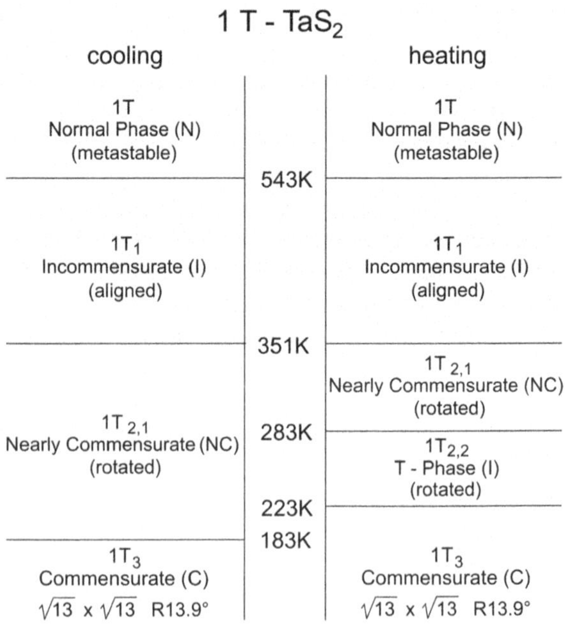

Fig. 11.54 Phase diagram of the 1T polytype of TaS_2 [445, 446, 459, 460]

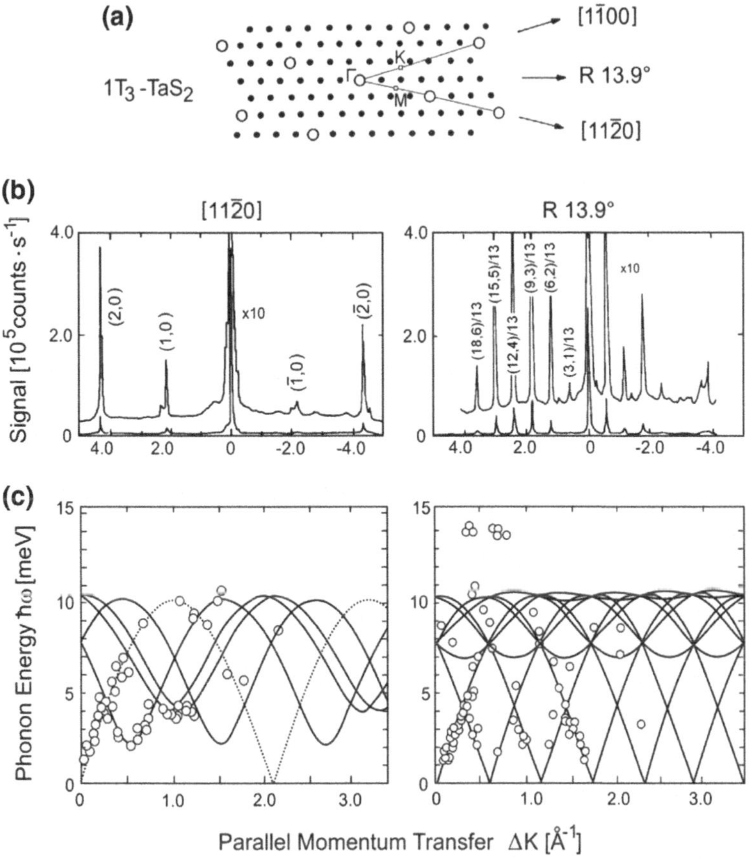

Fig. 11.55 Structure, HAS diffraction patterns and phonon dispersion curves measured at $T = 120$ K along two symmetry directions of the $1T_3$-TaS_2 (0001) surface. **a** Schematic view of the normal and CDW lattice structure for $1T_3$-TaS_2 (0001) showing in reciprocal space the normal phase G-vectors (open circles) and the reciprocal vectors of the CDW superstructure (filled circles). The symmetry points of the normal phase BZ and the corresponding symmetry directions in both three-index and four-index notations are also shown. **b** The corresponding HAS diffraction patterns. **c** HAS dispersion curves of the Rayleigh mode for a symmetry direction of the normal lattice (left frame) and of the CDW lattice (right frame), which, as shown in (**a**), is rotated by 13.9° from the [1$\bar{1}$00] direction [466]. The experimental points (open circles) are fitted by single sinusoidal components (dashed line curve in the left frame) with the superlattice reciprocal vectors. Note that the diffraction peaks in the 13.9° direction (**c**) are all due to the CDW and are even more intense than the peaks due to lattice atoms measured along the [1$\bar{1}$00] and [11$\bar{2}$0] directions [3, 4]

the atomic surface lattice. They estimated that the CDW-induced surface deformation was more than a factor 7 greater than in $TaSe_2$ [462]. By contrast X rays [463] and electrons [464] show comparatively weak CDW satellite peaks. Presumably because of the non-availability of sufficiently large crystals for inelastic neutron scattering, experimental information on bulk phonon dispersion curves

[465] is rather limited and restricted to room temperature. Only the acoustic branches along the c-axis and the LA and TA_1 branches (the latter polarized nearly parallel to the c-axis) along the Σ-direction could be measured.

Figure 11.55b, c show the HAS diffraction patterns and phonon data [3, 4, 466] measured at $T_s = 120$ K along two symmetry directions of the $1T_3$-TaS_2 (0001) surface. Along the $[11\bar{2}0]$ direction the sharp intense diffraction peaks arise solely from the atomic lattice. The measured points at small parallel momentum transfer (left frame of Fig. 11.55c) follow the expected sine-law behavior near the origin, but with increasing momentum exhibit additional characteristic minima which are attributed to dispersion curves with their origins at the CDW lattice. This is confirmed by measurements along the R13.9° direction (right frame of Fig. 11.55c, which is a symmetry direction for the CDW lattice. In this direction the dispersion curves show a regular behavior with a periodicity corresponding to the CDW **G**-vector. A simple analysis based on a sine-law fit [466] gives convincing evidence of phonons induced by the CDW, which arise from the folding of the crystallographic Brillouin zone into that of the commensurate $\sqrt{13} \times \sqrt{13}$ CDW superstructure. Phonons associated with practically all the 13 different **G**-vectors of the CDW contained in the crystallographic Brillouin zone are observed along the three directions investigated.

These HAS observations of anomalously folded Rayleigh modes associated with different CDW **G**-vectors can be explained by assuming that the He atoms are inelastically scattered from the phonon-induced modulation of the surface electronic density [466]. At wavevectors close to a CDW reciprocal **G**-vector, the small phonon induced displacements of the ions produce a relatively large dynamical modulation of the electronic charge-density to which the He atoms respond. Thus the Rayleigh phonons induce a modulation of the electronic density associated with the CDW. The signatures of such CDW-phonon coupling are the gaps which appear at the folding points in the left frame of Fig. 11.55c.

The above explanation has been confirmed by a lattice dynamical calculation for $1T_3$-TaS_2 in which the CDW-lattice coupling is accounted for in the framework of the PC model [4, 466]. As shown in Fig. 11.56b the CDW is mimicked by a sublattice of six pseudocharges located around the Ta ions at each David-star center in the $\sqrt{13} \times \sqrt{13}$ supercell. In the simplest version of the model it is sufficient to couple the PC′s to the neighbor metal and chalcogen ions through the radial force constants.

It is important finally to point out that in 1T-TaS_2 (0001) the strong coupling to the CDW is mediated by the *transverse* polarized Rayleigh waves [466], whereas the electron-phonon interaction in most metallic systems involves *longitudinal* phonons. In the light of the above analysis, the experimental HAS data for 1T-TaS_2(0001) is the first clear observation of the phonon branches in a quasi-incommensurate CDW crystal. This was made possible because the He atoms are scattered from the outermost electron density, whereas neutrons are scattered by the atomic nuclei and electrons in electron energy loss spectroscopy (EELS) are mainly scattered by the ion cores.

Unlike $TaSe_2$, the HAS data on 1T-TaS_2(0001) do not show any Kohn-like surface phonon anomaly, nor any deviation of the RW from the TA branch obtained by neutron scattering, at least within the experimental errors. This comes as a surprise

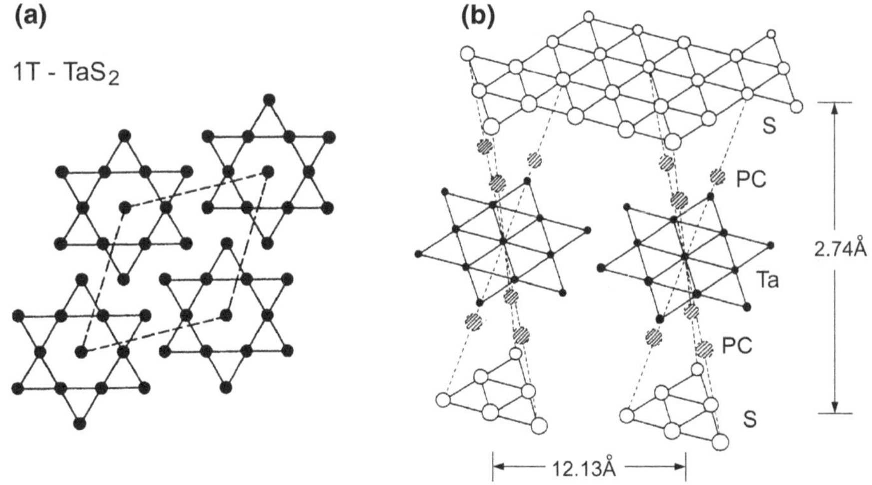

Fig. 11.56 Schematic diagram showing the structure inside a 1T-TaS_2 crystal. **a** Unit cell (dashed line) of the sheet of Ta ions in the presence of the commensurate $\sqrt{13}\times\sqrt{13}$ charge-density wave in the $1T_3$ phase. **b** Location of the pseudocharges (PC) between sheets of sulphur (S) and tantalum ions illustrating how an octahedral cage of sulphur ions encloses one Ta ion. In the pseudocharge model the Ta ions at the David-star centers are surrounded above and below by six pseudocharges (shaded circles) also forming a compressed octahedron

since the CDW-lattice coupling is larger in TaS_2 and the surface effects claimed for $TaSe_2$ should be enhanced in TaS_2. This could, however, be due to the fact that the available HAS experiments on TaS_2 were made at temperatures much below T_C, in a region where the CDW is fully stabilized and any possible competition between surface and bulk symmetries is definitely won by the bulk, whereas in $TaSe_2$ the measurements were made in the vicinity of T_C.

Another chalcogenide surface whose CDW transition is meeting a renewed interest is 2H-$NbSe_2$(0001). Its surface phonon dispersion curves were first investigated by HAS together with $TaSe_2$(0001) in the late 80s, though much less systematically [467]. At a surface temperature of 160 K, the HAS data show a shallow dip in the RW dispersion curve at about 2/3 of $\overline{\Gamma M}$ where also the bulk anomaly occurs. More recent studies with inelastic X-ray spectroscopy (IXS) by Murphy et al. [468, 469] and Weber et al. [470, 471] reveal a strong temperature dependent dip of the phonon frequency to zero at the CDW transition. These results are supported by first principle calculations [470, 471], though previous DFPT calculations by Calandra et al. [472] show that the large anomaly driving the CDW instability shifts from $^2/_3$ in the bulk to $^1/_2$ of $\overline{\Gamma M}$ in a single trilayer. This shift is consistent with that observed with HAS in 2H-TaS_2(0001) (Fig. 11.52), although the softening for the surface trilayer of a semiinfinite crystal is much weaker than that predicted for the single trilayer. The shift of the anomaly in the surface phonon branches does not seem to be a consequence of the crystal layer structure, since it was first predicted for the (001) surface of the cubic superconductor TiN(001) [473, 474]. Rather it results from the change in the electron-phonon coupling strength at the surface associated with the emergence of surface states. Similarly to what occurs in 2H-TaSe2 (0001) (Fig. 11.53) the critical temperature of the surface CDW transition in 2H-$NbSe_2$(0001) is, slightly shifted with the respect to the bulk T_c [475], although in the opposite direction.

11.4.5 Layered Pnictogen Chalcogenides

The heavier pnictogen chalcogenides (P_2X_3 with P = Bi, Sb and X = Te, Se), have in the past been intensively studied because of their thermoelectric properties. In the last decade they have attracted renewed interest as topological insulators in connection with the related surface properties. Despite their layered structure, organized in an alternate XPXPX stacking in quintuple layers (QLs) along the [111] direction of a rhombohedral lattice, the surface properties may be rather different from those of the bulk. Their special features are the presence of electron and hole pockets at the Fermi level, spin-split bands due to the large spin-orbit coupling, and the electron-phonon interaction involving surface Dirac electrons, similar to those considered in Sect. 11.4.2 for Sb and Bi. As a result they have stimulated much research on both the experimental and theoretical side. Due to their comparatively low phonon energies, Bi_2Se_3(111) and Bi_2Te_3(111) are particularly suitable to be studied with HAS [241–243]. As an example Fig. 11.57a shows the surface phonon dispersion curves of Bi_2Se_3(111) measured with HAS by Zhu et al. [241]. The data points are shown on top of DFPT phonon branches for 3 QLs calculated along the $\overline{\Gamma M}$ and $\overline{MK}$ directions without spin-orbit interaction [476]. There is substantial agreement between experiment and the theoretical surface branches S_1, S_3 and S_4, while near the zone center the data points deviate strongly from the S_2 branch, showing a narrow dip (marked by the arrow) in all wavevector directions. Because of its isotropic nature Zhu et al. attributed the anomaly to the interaction of the S_2 optical phonons with the Dirac electrons [241, 242], which they could fit with the pseudocharge (PC) model (see Sect. 5.3) in agreement with a calculation within the random-phase approximation (RPA) (Fig. 11.57b). Recent *ab initio* calculations of the mode selected electron-phonon interaction in these materials [477, 478] actually show that the largest values at which anomalies may occur are for the higher optical phonons above 15 meV, not observed so far with HAS. The calculated mass-enhancement factor λ for Bi_2Te_3(111) is however rather small, hardly exceeding 0.1 [478], in agreement with HAS measurements [542]. A more recent SE-HAS study of Bi_2Te_3(111) surface phonons (Fig. 9.14) compares very favorably with DFPT calculations [543], showing however a possible anomaly in the LA phonon rather than in the low energy optical branch S_2. It is important to remark that in these materials the Dirac-cone Fermi wavevector and in general the surface electronic structure at the Fermi level are very sensitive the amount of surface defects and doping.

Altogether these seminal HAS studies of the topological insulators Bi_2Se_3(111) and Bi_2Te_3(111) demonstrate once more that HAS, unlike other surface spectroscopies, has the appropriate resolution in the meV domain and the special sensitivity to investigate important manifestations of the electron-phonon interaction.

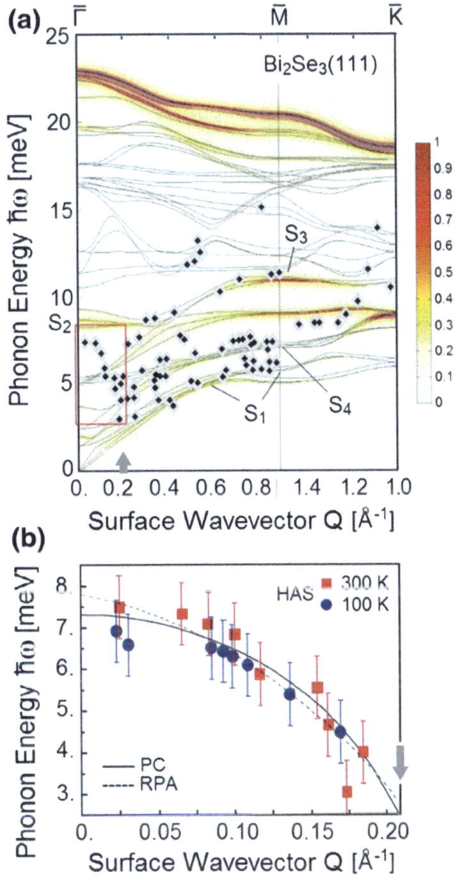

Fig. 11.57 Comparison of HAS surface phonon measurements on $Bi_2Se_3(111)$ [241] with theory. **a** The HAS data are superimposed on a DFPT calculation without spin-orbit interaction, with the SV component of the DOS projected onto the surface-layer [476]. The deep anomaly at 0.2 $Å^{-1}$ near $\bar{\Gamma}$ (rectangle and arrow), observed in all symmetry directions in $Bi_2Se_3(111)$ as well as in $Bi_2Te_3(111)$ [243], is attributed to the interaction of the optical surface phonon S_2 with Dirac fermions. **b** The anomaly on an expanded scale at two different temperatures, fitted with the pseudocharge model (PC, see Sect. 5.3) and compared to a random-phase approximation (RPA) calculation [242]

11.5 Complex Long-Period Surfaces

11.5.1 Stepped Surfaces

In systems with large unit cells the diffraction patterns become more closely spaced and the detection and resolution of phonon dispersion curves places severe demands on the experiment. Stepped surfaces are of special interest since they provide simple model systems with well-defined defects, which because of the high-order periodicity are accessible to study with diffraction techniques [479, 480]. Pioneering theoretical studies on the dynamics of stepped surfaces were reported in the late 1970s and early 1980s by several groups [481–484]. The first HAS study of surface phonons on a stepped surface was reported in 1990 by Lock, Toennies and Witte for Al(221) [479]. Previous EELS phonon experiments on vicinal surfaces [485] provided only data points near the center of the surface Brillouin zone. The HAS measurements of Lock et al. revealed a rich phonon structure including a new surface mode propagation along the step edges, in qualitative agreement with theoretical predictions [479]. Lattice dynamical calculations for stepped surfaces are especially difficult because of the large unit cells leading to very large dynamical matrices [482]. To circumvent such difficulties Lock et al. set up the dynamical matrix analytically by means of computer algebra starting from a force constant model for the dynamics of bulk aluminum. In a related theoretical study Knipp demonstrated the applicability of the Green's function method for stepped surface dynamics [486].

Subsequently in 1995 Witte et al. reported an extensive series of HAS experiments on the surface phonon dispersion curves of stepped metal surfaces [487]. Witte et al. chose the Cu(211) and (511) stepped surfaces since the terraces have different packings of (111) and (100), respectively, but nearly the same widths of 6.25 and 6.63 Å, respectively. Surfaces of copper have the advantage over aluminum that they do not roughen at temperatures below about 700 K [488] and highly ordered samples can be prepared. Figure 11.58a, b illustrate the structure and Brillouin zone of the Cu(211) surface. The HAS dispersion curves are shown in Fig. 11.58c. There it is seen that the Rayleigh mode (S_1) could be followed over nearly the entire Brillouin zone along azimuths parallel and perpendicular to the step edge directions. For both the Cu(211) and (511) surfaces backfolding of the Rayleigh mode was observed along the direction perpendicular to the step edges, while parallel to the step edges, a step localized mode, designated E_1 in Fig. 11.58c, with optical character and a weak dispersion was also found. On Cu(511), an additional longitudinal mode and an energy gap for the Rayleigh mode at the zone boundary, whereas on Cu(211), an additional shear horizontal mode, designated S_4 in Fig. 11.58c, could be identified. The coupling to this (otherwise forbidden) shear horizontal mode is facilitated by the tilt of the lattice planes with respect to the scattering plane. The low phonon energy of the S_4 mode, even below the S_1 mode, at the zone boundary results from the relaxation of the step atoms. Surprisingly, despite the longitudinal anomalies observed for the corresponding low index surfaces, the experimental data could be well explained by theoretical slab calculations in the framework of a single force constant model [487].

At about the same time Niu et al. presented a similar study of the Ni(997) surface which consists of (111) terraces and (100) step edges [489, 490]. They also observed a number of surface modes some of which were attributed to step edge localized displacements. By treating the surface with oxygen they were also able to produce steps with double the height of the ordinary (977) surface steps. These results have been analyzed theoretically by Mele and Pykhtin [491], using a continuum GF theory and by Witte and Toennies [492] using a single force constant slab calculation. The latter calculation provided evidence for a $\approx$20% softening of the observed step mode and of the terrace Rayleigh mode which are related to each other by the step-induced folding.

The great interest in stepped surfaces for various nanotechnologies has stimulated much theoretical work on their dynamical properties, which is generally based on semi-empirical approaches due to the large extension of the unit cells. The embedded atom (EA) model has proven to be very efficient for the vicinal surfaces

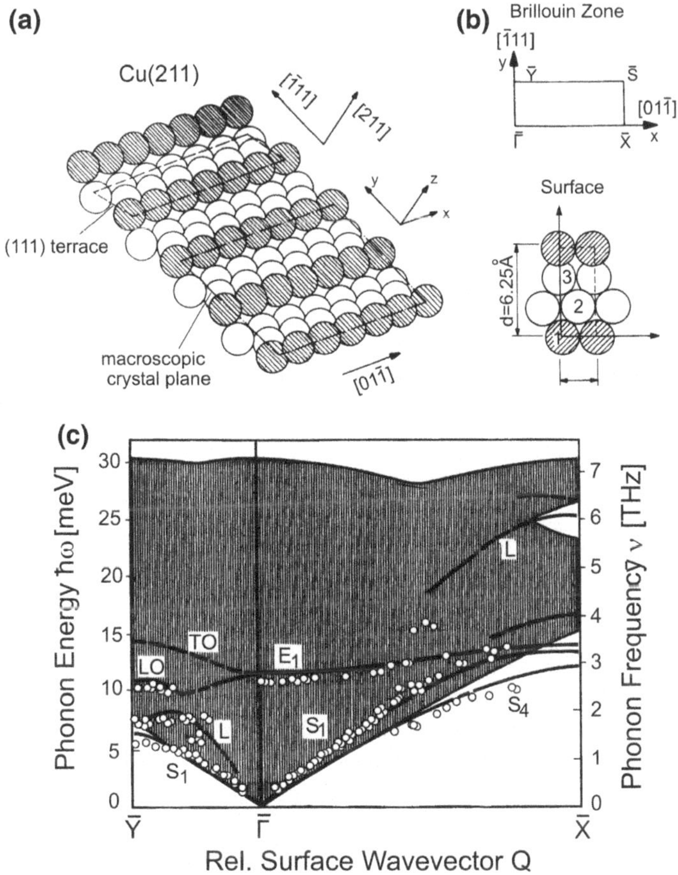

Fig. 11.58 Unit cell and HAS dispersion curves of the Cu(211) surface. **a** Perspective ball model showing the terrace atoms (white circles) and the step edge atoms (hatched circle). **b** Surface Brillouin zone and dimensions of the Cu(211) surface. **c** Experimental surface phonon dispersion curves (open circles) measured along the $\overline{\Gamma Y}$ direction perpendicular to the step edges and the $\overline{\Gamma X}$ direction parallel to the steps and folded back into the first Brillouin zone, are compared to the calculated phonon dispersion curve (bold lines). The calculations were performed for a slab with 216 layers with a single nearest neighbor radial force constant without any relaxation. The surface projected bulk phonon bands are represented by the hatched region (adapted from [487]). A ball model simulation of the surface phonons of stepped surfaces is available as a video at https://youtu.be/LTOLkhVEIGk

of metal surfaces [493–496], as well as the tight-binding second-moment (TBSM) method [497] and the combination of the EA model with the Green function method developed by Kara et al. [498, 499]. For an overview of the theory the reader is referred to the excellent review by Kara and Rahman [499].

11.5.2 Reconstructed Surfaces

Complex reconstructions with even larger periods can be investigated by a combination of high resolution experiments and state-of-the-art calculations. Examples have been given in Sect. 11.2 for the reconstructed (111) surfaces of silicon and germanium. In general any extensive reconstruction is characterized by prominent surface features, e.g., a different topological structure of the atomic rings, like in Si (111)2×1, or the presence of adatoms, like in Si(111)7×7 and Ge(111)2×8. These features in many cases lead to the appearance of important gaps in the phonon dispersion curves which separate the resulting higher frequency modes which are localized in the reconstructed region from the deeply penetrating low frequency modes, as in the ideal surface. Typical examples are the 7-fold rings in Si(111)2×1 or the adatoms in Si(111)7×7 and Ge(111) 2×8 which are weakly coupled to the other surface atoms.

Another example of a HAS study of a complex system is the investigation of the high-T_c superconductor $Bi_2Sr_2CaCu_2O_8$ compound (BISSCO) [385]. This layered compound is especially suited for such an investigation since fairly large (2 mm 2 mm), albeit very thin (0.1 mm) crystals are available and are quite stable in vacuum [500]. In addition to being structurally complex, BISSCO(001) exhibits an unusual $(1\times n)$ reconstruction where n varies between 4.7 and 5 both at the surface and in the bulk. The surface phonon spectrum of the BISSCO(001) surface, revealed four apparently dispersionless surface modes, as well as an unusual similarity between the surface dispersion curves measured in different directions on the same surface. Once the complex folding is included the flat dispersion curves were qualitatively explained by a shell model calculation as a direct consequence of the anisotropic superstructure at the surface [385]. This is the first example in which the effects of an anisotropic superstructure on surface phonons were observed.

The advent of ^{3}He spin echo (SE-HAS) spectroscopy with its unprecedented resolution has made it possible to investigate the low energy elementary excitations occurring in very large-period surfaces such as, e.g., the $23\times\sqrt{3}$ reconstructed Au (111) surface [415]. In this structure, illustrated in Fig. 11.59a, 24 surface Au atoms are squeezed along the direction into a 23-atom period of the bulk layers. According to the original HAS structure analysis by Harten et al. [236] this is made possible by a periodic set of misfit dislocations yielding alternating domains with surface atoms in fcc sites separated by domains with atoms in hcp sites. Accordingly there are 23 equivalent ways to accommodate the surface structure on top of the underlying lattice, with a very small energy barrier between the neighboring minima and a phase variable expressing the actual position of the surface layer [238, 239]. The small reciprocal lattice vectors **g** corresponding to the large-period along the $[1\bar{1}0]$ direction, besides yielding a few satellite diffraction peaks around the specular peak [236], allow for replicas of the RW branch having their origins at $\pm n\mathbf{g}$ with $n = 1, 2, 3, \ldots$. Figure 11.59b shows three typical energy transfer spectra reported by McIntosh et al. [415]. Because of the greatly improved energy resolution altogether 3 or 4 peaks could be resolved, which in a TOF measurement would be assigned to the

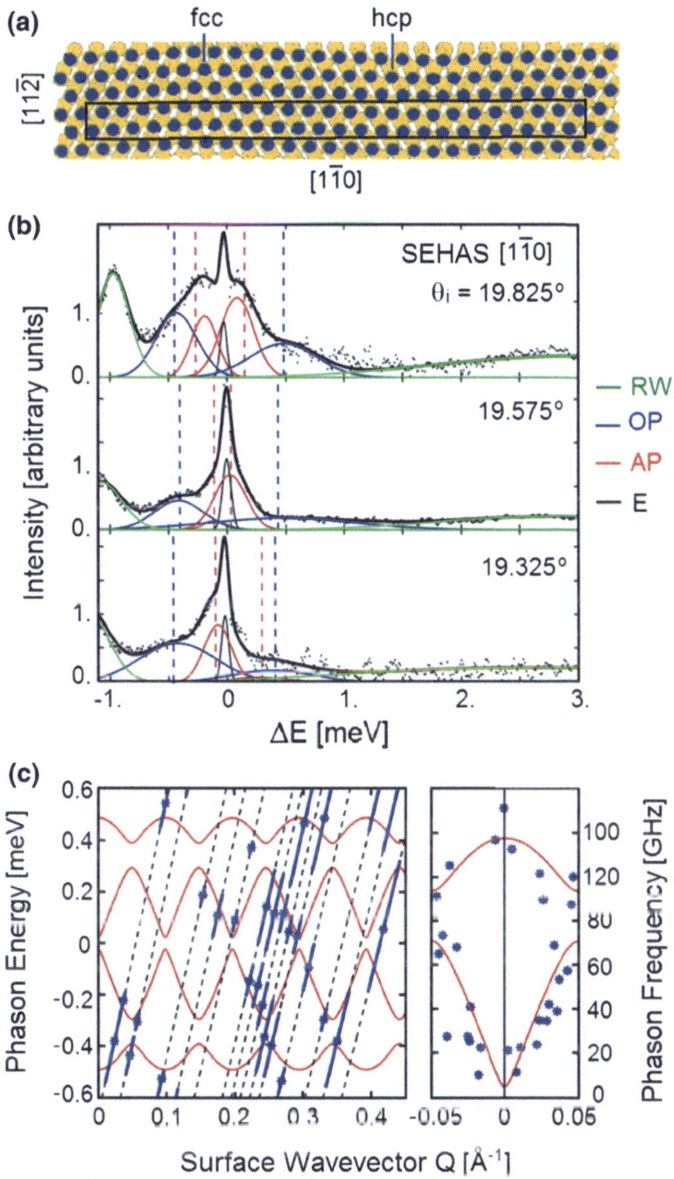

Fig. 11.59 [3]He spin echo (SH-EAS) inelastic scattering from the acoustic and optical phasons of the $23 \times \sqrt{3}$ reconstructed Au(111) surface [415]. **a** The 24 atoms are squeezed within 23 lattice spacings in the $[1\bar{1}0]$ direction of the second layer. The atom positions in the resulting serpentine vary continuously from the hcp to the fcc sites of the substrate. **b** Inelastic [3]He spin echo (SE-HAS) energy transfer spectra from Au(111) along [118] at three incidence angles θ_i differing by only 0.25° for a $\Theta_{SD} = 44.4°$ scattering geometry and incident energy of 8.05 meV. In addition to the Rayleigh wave (RW, green Gaussian) and diffuse elastic (DE, black Gaussian) peaks, features corresponding to acoustic (AP, red Gaussian) and optical phasons (OP, blue Gaussian) are identified. **c** The measured dispersion curves of surface AP and OP agree well with earlier MD predictions (full lines) [238, 239]

elastic scattering from defects. Since the apparatus was optimized to have the highest resolution at small creation energy transfers (see Chap. 9), the Rayleigh phonons (green line curves) are not so well resolved. The SE-HAS data are plotted in Fig. 11.59c together with a theoretical fit based on a sine-Gordon soliton lattice model. The long-period perturbation mentioned above opens up gaps at each intersection of the RW replicas, included those occurring at zero energy, thus producing the set of dispersion curves shown in Fig. 11.59c, each having a period **g**. The excitations of the lowest branch are designated as *acoustic phasons* (APs), those of the upper branch as *optical phasons* (OPs). It is easily verified that the group velocity of the APs away from the gaps is just the RW velocity, here ~7 meV Å. The OPs, as nicely depicted in Fig. 1 of [415], correspond to a counter-phase motion of fcc site and hcp site atoms producing density amplitude oscillations at the dislocation lines. Thus the OPs can also be described as *amplitons* (or amplitudons), already encountered in Sects. 11.3.5 (Fig. 11.38) and 11.4.2 (Fig. 11.45).

Figure 11.59 shows the first measurements of AP and OP dispersion curves of the $23 \times \sqrt{3}$ reconstructed Au(111) surface, which were made possible by the high resolution of SE-HAS spectroscopy. The clear detection of these very low energy modes demonstrates the new possibilities made available with SE-HAS spectroscopy.

It is important to recall that ^{3}He like ^{4}He interacts with the surface excitations of a metal via the surface electron density far from the first atomic plane. The instability leading to the $23 \times \sqrt{3}$ reconstruction essentially occurs in the surface electron system at the Fermi level leading to a CDW coupled to the atomic displacements via the electron-phonon interaction. In this respect He atom spectroscopies represent an ideal tool for investigating the structure of surface CDWs and their elementary excitations, as already pointed out for graphene on Ru(0001), pnictogens, and chalcogenides.

11.5.3 Quasicrystal Surfaces

Quasicrystals, discovered by Shechtman et al. in 1984 [501], are characterized by an infinite-range orientational order but no translational symmetry [502–505]. This may occur, e.g., with the occurrence of 5-fold or 10-fold rotational symmetry axes, which are incompatible with the translational symmetry. Mathematically quasicrystals may in general be viewed as three-dimensional cuts of lattices which are periodic in spaces with more than three dimensions. Thus the quasicrystal directions and planes also for the surface can be labelled with sets of more than three Miller indices.

Figure 11.60a shows the HAS spots derived form the diffraction pattern for the (00001) surface of *d*-AlNiCo, where *d* denotes decagonal. The spots exhibit a perfect decagonal symmetry [506, 507], which does not permit any periodicity in the reciprocal space. Nevertheless in the two independent symmetry directions

[10000] and [00$1\bar{1}$0] prominent diffraction peaks (labeled B, C, D, E) are found with parallel wavevector which can be viewed as quasi-**G**-vectors. Figure 11.60b, c show the HAS dispersion curves of the d-AlNiCo(00001) surface along the symmetry directions [10000] and [00$1\bar{1}$0]. The dispersion curves of the RW have a maximum at the wavevectors corresponding to one half of the 0B and 0D distances, respectively. Thus the latter act as boundaries of a quasi-BZ. It should be noted, however, that the RW branches would not go to zero at the quasi-**G**-vectors due to lack of translational order. Similar results have been obtained for the (100000) surface of the icosahedral *i*-AlPdMn quasicrystal, with the RW dispersion measured along the symmetry directions [00$1\bar{1}\bar{1}$1] and [000$\bar{1}\bar{1}$0] [506, 507].

Surprisingly out of the infinite set of reciprocal lattice vectors the data in Fig. 11.60 appear to indicate that there are distinct wavevectors corresponding to one half of the 0B and 0D distances. In an effort to understand this Theis et al. [507] carried out a series of simple one-dimensional spring model calculations of the inelastic HAS cross section from a 2D rhombic Penrose tiling of edge d = 2.44 Å. The edge length was chosen to represent the AlNiCo(00001) surface, while the force constant-to-mass ratio *f/M* has been taken the same for all edges. Figure 11.61b shows the simulated dispersion curve for a direction lying in the (00001) plane. The dispersion curves exhibit a sequence of quasi-**G**-vectors of which the shortest one (2.67 Å^{-1}) corresponds to an effective linear chain distance 2π/2.67 = 2.35 Å, which turns out to be only slightly less than the tile edge length. The deviation from the apparent wavevectors seen in the experiment is related to the Fibonacci sequence assumed in the model calculation. An interesting result of the simulation is represented by the intense minima (arrows) in the quasi-dispersion curves, superimposed on the weaker parabolic dispersion curves of an effective quasi-periodic linear chain. These minima are similar to those observed in inelastic

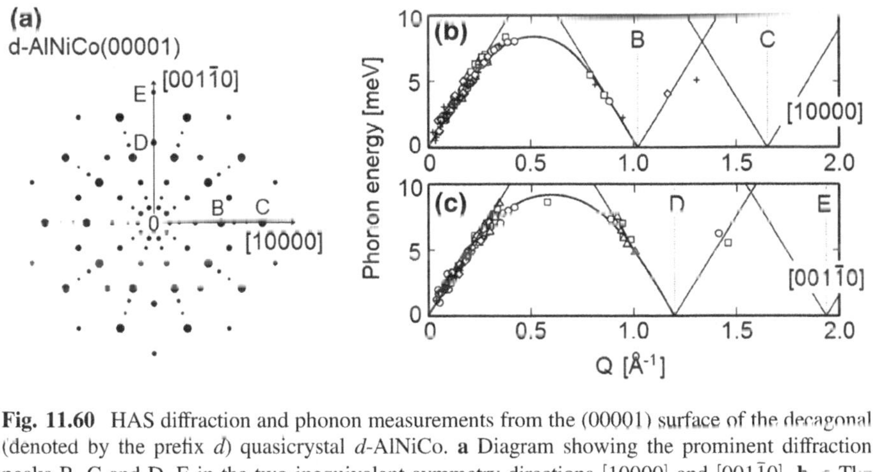

Fig. 11.60 HAS diffraction and phonon measurements from the (00001) surface of the decagonal (denoted by the prefix *d*) quasicrystal *d*-AlNiCo. **a** Diagram showing the prominent diffraction peaks B, C and D, E in the two inequivalent symmetry directions [10000] and [00$1\bar{1}$0]. **b**, **c** The RW dispersion curve along the two symmetry directions as measured by HAS (adapted from Sharma et al. [506] and Theis et al. [507])

neutron scattering from liquids as indicative of some short range order or disorder on a length scale larger than *d*. The minima clearly enhance the DOS with respect to that expected for a linear chain or a Debye model in the same spectral region. A similar phenomenon is well known to occur in the lattice dynamics of crystal defects [508, 509].

11.5.4 Disordered Surfaces

The quasicrystal simulations shown in Fig. 11.61b and the lattice dynamics of crystal defects are characterized by the localization of the phonons at wavelengths which are short enough to discern the topological disorder. Space localization also means spectral localization, a concept that can be directly applied to the so-called boson peak. The first studies by HAS of the phonons of the vitreous silica (v-SiO_2) surface were reported in a series of papers by Steurer et al. [510–513]. Figure 12.62 a shows a typical TOF-HAS spectrum converted to the energy transfer scale from the surface of vitreous silica (v-SiO_2) [512]. In addition to the diffuse elastic (DE) peak, the boson peaks for creation and annihilation processes are observed at −4 and +4 meV for a surface temperature T_s = 137.2 K. Their energies are practically independent of the wavevector transfer, but slowly increase with temperature up to about 5 meV at 368.5 K. If it is assumed that in analogy with the simulation of Fig. 11.61b there are two minima at different energies contributing to the excess

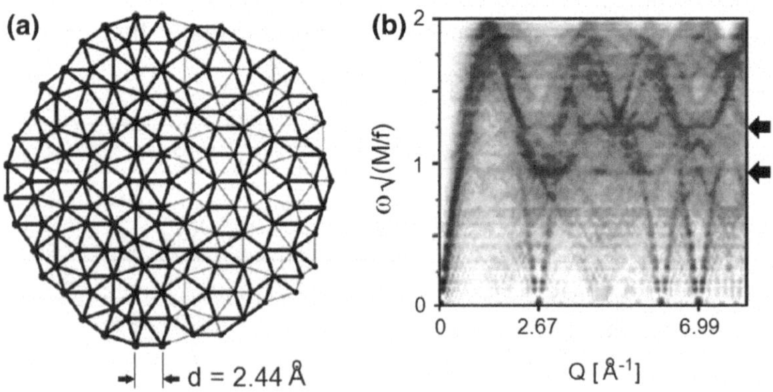

Fig. 11.61 Simulation of the inelastic HAS dispersion curves from a 2D rhombic Penrose tiling by Theis et al. [507] **a** Rhombic Penrose tiling of edge length d = 2.44 Å models a 2D quasicrystal similar to the AlNiCo(00001) surface. **b** A simulation of the HAS dispersion curves along a direction lying in the (00001) plane for a dynamical model with identical force constant-to-mass ratio (*f/M*) for each bond. The first quasi-**G**-vector (2.67 $Å^{-1}$) corresponds to an effective lattice distance 2π/2.67 = 2.35 Å, only slightly smaller than *d*. The quasi-**G**-vector is larger than found in the experiment. The arrows indicate minima in the quasi-dispersion curves, similar to those occurring in liquids, which contribute maxima in the total DOS above that expected for a Debye model (adapted from [507])

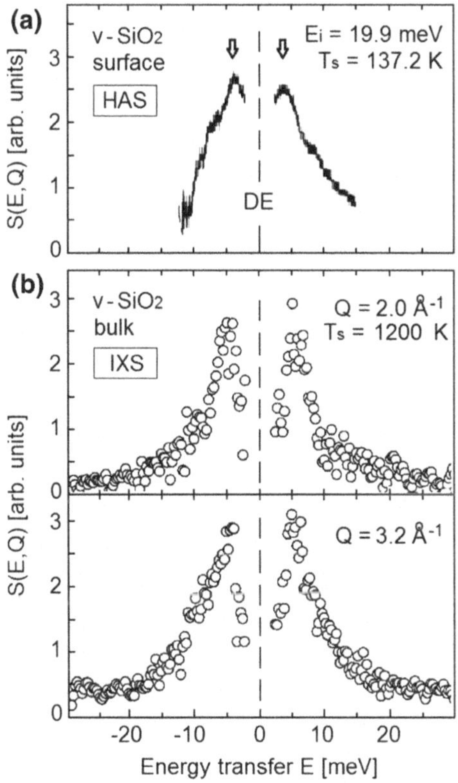

Fig. 11.62 Energy transfer spectra by HAS and Inelastic X-ray scattering (IXS) from the amorphous surface of vitreous silica (v-SiO_2) **a** TOF-HAS energy transfer spectrum showing the boson peaks at about 4 meV on the creation and annihilation side [512]. The boson peak at $T_s = 137.2$ K is practically independent of the wavevector transfer Q but shifts with the surface temperature up to ~ 5 meV at $T_s = 368.5$ K [511]. **b** IXS spectra from v-SiO_2 for two different values of the wavevector transfer, measured at 1200 K in order to enhance the inelastic response, is also independent of the wavevector transfer [514]

DOS, the upward shift of the boson peak with temperature can be attributed to the increasing population of the upper localized phonon density of states.

Figure 11.62 compares the HAS energy transfer spectra with inelastic X-ray scattering (IXS) spectra on the same surface v-SiO_2 by Pilla et al. [514]. The X-ray measurements were performed at high temperature (1200 K) in order to increase the phonon population and the corresponding inelastic intensity. In the examples shown in Fig. 11.62b for two different values of the wavevector transfer Q, the inelastic peaks for creation and annihilation are found at ± 5 meV and are also independent of Q. This energy may then be considered as the high temperature limit, implying that the boson peak energy measured with HAS must saturate above a certain surface temperature. This supports the model of two (or more) spectral regions of localized phonon states as depicted in Fig. 11.61b. IXS is indeed becoming an alternative powerful tool to investigate the dynamics of amorphous surfaces, as demonstrated also by the recent study on amorphous selenium by Scopigno et al. [515].

There appears to be some confusion about the concept of the boson peak, arising from different interpretations of the experimental observations. Localization of phonon excitations due to disorder is a key ingredient. Gurevich et al. [516] suggest a universal mechanism for the boson peak formation in glasses in which anharmonicity drives an instability in a system of quasi-localized but interacting oscillators, leading to a restructuring of the phonon spectrum with the emergence of a peak in $g(\omega^2)/\omega^{d-1}$, where $g(\omega^2)$ is the DOS and d the lattice dimension. Some other simple mechanism within harmonic theory is suggested by the weak dependence on temperature of the peaks discussed in connection with Fig. 11.62 and their persistence up to high temperatures. It has been well established in the sixties that defects and disorder in harmonic lattices may induce low frequency resonant features in the phonon spectrum with large DOSs above the Debye spectrum [508, 509]. According to Chumakov et al. [517] (see also the views of Zorn [518]) the boson peak is simply due to a van Hove singularity of the TA branch broadened by disorder. This explanation, however, although consistent with a purely harmonic origin, does not seem to apply to the case of SiO_2 since the lowest van Hove singularity in the ordered version, α-quartz, is around 10 meV [519]. Very likely what is currently termed "boson peak" encompasses different mechanisms, some related to disorder-induced modifications of the harmonic phonon spectrum, others, more fascinating, involving some kind of phase transition leading to a restructuring of the excitation spectrum [516, 520, 521].

References

1. F.W. de Wette, W. Kress (eds.), *Surface Phonons*, Springer Ser. Surf. Sci., vol. 21 (Springer, Berlin, Heidelberg, 1991)
2. W. Kress, in *Surface Phonons*, ed. by F.W. deWette, W. Kress, Springer Ser. Surf. Sci., vol 27 (Springer, Berlin, 1991)
3. J.G. Skofronick, J.P. Toennies in *Surface Properties of Layered Structures*, ed. by G. Benedek (Kluver, Amsterdam, 1992) p. 151
4. G. Benedek, F. Hofmann, P. Ruggerone, G. Onida, L. Miglio, Surf. Sci. Rep. **20**, 1 (1994)
5. J. Fritsch, C. Eckl, P. Pavone, U. Schröder, Fest-körperprobleme. Adv. Solid State Phys. **36**, 135 (1997)
6. J. Fritsch, U. Schröder, Phys. Rep. **309**, 209 (1999)
7. R. Heid, K.-P. Bohnen, Phys. Rep. **387**, 151 (2003)
8. R.F. Wallis, S.Y. Tong, *Surface Phonons*, in Landolt-Börnstein, New Series III, vol. 24: *Physics of Solid Surfaces*, ed. by G. Chiarotti (Springer V., Berlin, Heidelberg, 1994/1995) Part b, p. 434
9. V. Celli, *Interaction of Atoms with Surfaces*, in Landolt-Börnstein, New Series III, vol 24: *Physics of Solid Surfaces*, ed. G. Chiarotti (Springer V., Berlin, Heidelberg, 1994/1995) Part c, p. 278

10. M. Rocca, *Surface Phonon Dispersion*, in Landolt-Börstein, New Series III/42.A2/4.5 (Springer V., Heidelberg, Berlin, 2002)
11. G. Benedek, D. Campi, J.P. Toennies, in *Physics of Solid Surfaces*, Subvol. A, Landolt-Börnstein, *Numerical Data and Functional Relationships in Science and Technology*—New Ser., Subvol. 45A, ed. by P. Chiaradia, G.F. Chiarotti, Chap. 10, (2015), pp. 572–646
12. R.E. Allen, G.P. Alldredge, F.W. de Wette, Phys. Rev. B **4**, 1648 (1971)
13. R.E. Allen, G.P. Alldredge, F.W. de Wette, Phys. Rev. B **4**, 1661 (1971)
14. H. Bilz, W. Kress, *Phonon Dispersion Relations in Insulators* (Springer, Berlin Heidelberg, 1976)
15. B.F. Mason, B.R. Williams, Phys. Rev. Lett. **46**, 1138 (1981)
16. B.F. Mason, B.R. Williams, Surf. Sci. **111**, 609 (1981)
17. B.F. Mason, R. Caudano, B.R. Williams, J. Chem. Phys. **77**, 562 (1982)
18. B.F. Mason, B.R. Williams, Surf. Sci. **130**, 295 (1983)
19. B.F. Mason, B.R. Williams, Surf. Sci. **139**, 173 (1984)
20. K.D. Gibson, S.J. Sibener, Phys. Rev. Lett. **55**, 1514 (1985)
21. K.D. Gibson, S.J. Sibener, B.M. Hall, D.L. Mills, J.E. Black, J. Chem. Phys. **83**, 4256 (1985)
22. K.D. Gibson, S.J. Sibener, Faraday Discuss. Chem. Soc. **80**, 203 (1985)
23. K.D. Gibson, S.J. Sibener, J. Chem. Phys. **88**, 7862 (1988)
24. B.F. Mason, B.R. Williams, Surf. Sci. **177**, L925 (1986)
25. A. Šiber, B. Gumhalter, A.P. Graham, J.P. Toennies, Phys. Rev. B **63**, 115411 (2001)
26. A. Šiber, B. Gumhalter, J. Braun, A.P. Graham, M.F. Bertino, J.P. Toennies, D. Fuhrmann, Ch. Wöll, Phys. Rev. B **59**, 5898 (1999)
27. A.K. Dahm, A.R. Allnatt, W.J. Meath, R.A. Aziz, Mol. Phys. **67**, 1291 (1989)
28. A.K. Dahm, W.J. Meath, A.R. Allnatt, R.A. Aziz, M.J. Slaman, Chem. Phys. **142**, 173 (1990)
29. L.W. Bruch, A.P. Graham, J.P. Toennies, Mol. Phys. **95**, 579 (1998)
30. L.W. Bruch, A.P. Graham, J.P. Toennies, J. Chem. Phys. **112**, 3314 (2000)
31. F.Y. Hansen, L.W. Bruch, J. Chem. Phys. **127**, 204708 (2007)
32. L.W. Bruch, R.D. Diehl, J.A. Venables, Rev. Mod. Phys. **79**, 1381 (2007)
33. A. Šiber, B. Gumhalter, J.P. Toennies, Phys. Rev. Lett. **83**, 1375 (1999); Vacuum **54**, 315 (1999)
34. D. Campi, M. Bernasconi, G. Benedek, J.P. Toennies, J. Phys. Chem. C **119**, 14579 (2015)
35. D.M. Hall, D.L. Mills, M.H. Mohamed, L.L. Kesmodel, Phys. Rev. B **38**, 5856 (1988)
36. T. Yaminishi, T. Kanashiro, Y. Michihiro, Y. Kishimodo, T. Ohno, J. Phys. Soc. Jpn. **64**, 643 (1995)
37. G. Brusdeylins, R.B. Doak, J.P. Toennies, Phys. Rev. Lett. **44**, 1417 (1980)
38. G. Brusdeylins, R.B. Doak, J.P. Toennies, Phys. Rev. Lett. **46**, 437 (1981)
39. G. Brusdeylins, R.B. Doak, J.P. Toennies, Phys. Rev. B **27**, 3662 (1983)
40. T.S. Chen, G.P. Alldredge, F.W. de Wette, R.E. Allen, Phys. Rev. Lett. **26**, 1543 (1971)
41. T.S. Chen, G.P. Alldredge, F.W. de Wette, Solid State Commun. **10**, 941 (1972)
42. T.S. Chen, G.P. Alldredge, F.W. de Wette, Phys. Lett. **40A**, 401 (1972)
43. T.S. Chen, F.W. de Wette, G.P. Alldredge, Phys. Rev. B **15**, 1167 (1977)
44. G.P. Alldredge, Phys. Lett. **41A**, 291 (1972)
45. V.L. Zoth, G.P. Alldredge, F.W. de Wette, Phys. Lett. **47A**, 247 (1974)
46. G. Benedek, Phys. Status Sol. **B58**, 661 (1973)
47. G. Benedek, Surf. Sci. **61**, 603 (1976)
48. G. Benedek, L. Miglio, in *Ab Initio Calculation of Phonon Spectra*, ed. by J.T. Devreese, V. E. van Doren, P.E. Van Camp (Plenum, New York, 1983)
49. L. Miglio, G. Benedek, in *Structure and Dynamics of Surfaces II*, ed. by W. Schommers, P. von Blanckenhagen (Springer, Heidelberg, 1987), p. 35
50. G. Benedek, L. Miglio, in *Surface Phonons*, ed. by F.W. deWette, W. Kress, Springer Ser. Surf. Sci., vol. 27 (Springer, Berlin, Heidelberg, 1991), p. 37

51. G. Benedek, R.B. Doak, J.P. Toennies, Phys. Rev. B **28**, 7277 (1983)
52. G. Bracco, M. D'Avanzo, C. Salvo, R. Tatarek, F. Tommasini, Surf. Sci. **189**(190), 684 (1987)
53. F.W. de Wette, W. Kress, U. Schröder, Phys. Rev. B **32**, 4143 (1985)
54. W. Kress, F.W. de Wette, A.D. Kulkarni, U. Schröder, Phys. Rev. B **35**, 5783 (1987)
55. F.W. de Wette, A.D. Kulkarni, U. Schröder, W. Kress, Phys. Rev. B **35**, 2467 (1987)
56. F.W. de Wette in *Surface Phonons*, ed. by W. Kress, F.W. de Wette (Springer Verlag, Heidelberg, 1991), p. 67
57. G. Benedek, G.P. Brivio, L. Miglio, V.R. Velasco, Phys. Rev. B **26**, 497 (1982)
58. E.R. Cowley, J.A. Barker, Phys. Rev. B **28**, 3124 (1983)
59. H. Witte, E. Wölfel, Rev. Mod. Phys. **30**, 51 (1958)
60. D. Eichenauer, J.P. Toennies, J. Chem. Phys. **85**, 532 (1986)
61. V. Celli, D. Eichenauer, A. Kaufhold, J.P. Toennies, J. Chem. Phys. **86**, 2504 (1985)
62. I. Estermann, O. Stern, Z. Phys. **61**, 95 (1930)
63. R. Frisch, O. Stern, Z. Phys. **84**, 430 (1933)
64. J.E. Lennard-Jones, C. Strachan, Proc. Roy. Soc. (London) A **150**, 442 (1935)
65. C. Strachan, Proc. R. Soc. (London) A **150**, 456 (1935)
66. J.E. Lennard-Jones, A.F. Devonshire, Proc. R. Soc. (London) A **156**, 6, 29 (1936); Nature **137**, 1069 (1936)
67. A.F. Devonshire, Proc. R. Soc. (London) A **156**, 37 (1936)
68. J.E. Lennard-Jones, A.F. Devonshire, Proc. R. Soc. (London) A **158**, 242, 253 (1937)
69. A.F. Devonshire, Proc. R. Soc. (London) A **158**, 269 (1937)
70. C. Strachan, Proc. R. Soc. (London) A **158**, 591 (1937)
71. H. Chow, E.D. Thompson, Surf. Sci. **59**, 225 (1976)
72. G. Vidali, G. Ihm, H.-Y. Kim, M.W. Cole, Surf. Sci. Rep. **12**(4), 133–182 (1991)
73. G. Brusdeylins, R.B. Doak, J.P. Toennies, J. Chem. Phys. **75**, 1784 (1981)
74. W. Steele, Chem. Rev. **93**, 2355–2378 (1993)
75. G. Benedek, A., Glebov, W. Silvestri, J.G. Skofronick, J.P. Toennies, Surf. Sci. Lett. **381**, L540 (1997), and **406**, L621 (1998)
76. G. Lilienkamp, J.P. Toennies, J. Chem. Phys. **78**, 5210 (1983)
77. G. Brusdeylins, R. Rechsteiner, J.G. Skofronick, J.P. Toennies, G. Benedek, L. Miglio, Phys. Rev. Lett. **54**, 466 (1985)
78. A.C. Levi, G. Benedek, L. Miglio, G. Platero, V.R. Velasco, F. Garcia-Moliner, Surf. Sci. **143**, 253 (1984)
79. P. Santini, L. Miglio, G. Benedek, U. Harten, P. Ruggerone, J.P. Toennies, Phys. Rev. B **42**, 11942 (1990)
80. V. Celli, D. Himes, P. Tran, J.P. Toennies, Ch. Wöll, G. Zhang, Phys. Rev. Lett. **66**, 3160 (1991)
81. G. Armand, J.R. Manson, Phys. Rev. Lett. **53**, 1112 (1984)
82. G. Armand, J.R. Manson, in *Dynamics on Surfaces*, ed. by B. Pullman, J. Jortner, A. Nitzan, B. Gerber (Reidel, Dordrecht, 1984), p. 59
83. J.R. Manson, J.G. Skofronick, Phys. Rev. B **47**, 12890 (1993)
84. F. Hofmann, J.P. Toennies, J.R. Manson, Surf. Sci. Lett. **349**, L184 (1996)
85. B. Gumhalter, Phys. Rep. **351**, 1 (2001)
86. G. Chern, W.P. Brug, S.A. Safron, J.G. Skofronick, J. Vac. Sci. Technol. **A7**, 2094 (1989)
87. W.P. Brug, G. Chern, J. Duan, S.A. Safron, J.G. Skofronick, G. Benedek, J. Vac. Sci. Technol. **A8**, 2632 (1990)
88. G. Chern, J.G. Skofronick, W.P. Brug, S.A. Safron, Phys. Rev. B **39**, 12828 (1989)
89. J. Duan, W.P. Brug, G.G. Bishop, G. Chern, S.A. Safron, J.G. Skofronick, Surf. Sci. **251** (252), 782 (1991)
90. S.A. Safron, G. Chern, W.P. Brug, J.G. Skofronick, G. Benedek, Phys. Rev. B **41**, 10146 (1990)
91. G. Benedek, L. Miglio, G. Brusdeylins, J.G. Skofronick, J.P. Toennies, Phys. Rev. B **35**, 6593 (1987)

92. G. Benedek, F. Galimberti, Surf. Sci. **71**, 87 (1978). *Erratum:* Surf. Sci. **118**, 713 (1982)
93. F.W. de Wette, W. Kress, U. Schroeder, Phys. Rev. B **33**, 2835 (1986)
94. A.J. Martin, H. Bilz, Phys. Rev. B **19**, 6593 (1979)
95. Y. Li, D.C. Langreth, M.R. Pederson, Phys. Rev. B **55**, 16456 (1997)
96. K. Fischer, H. Bilz, R. Haberkorn, W. Weber, Phys. Status Solidi (b) **54**, 285 (1972)
97. C. Fu, K.S. Song, Phys. Rev. B **59**, 2529 (1999)
98. T.H. James (ed.), *The Theory of Photographic Process* (Macmillan, New York, 1977)
99. A.L. Glebov, J.P. Toennies, F. Träger, Phys. Rev. Lett. **82**, 4492 (1999)
100. B.R. Lawn, Acta Cryst. **16**, 1163 (1963)
101. A.L. Glebov, V. Panella, J.P. Toennies, Phys. Ref. B **60**, 2046 (1999)
102. B. Dorner, W. von der Osten, W. Bühren, J. Phys. C **9**, 723 (1976)
103. L. Miglio, P. Santini, P. Ruggerone, G. Benedek, Phys. Rev. Lett. **62**, 3070 (1989)
104. W. Ludwig, Japan J. Appl. Phys., Suppl. **2**, Pt. 2, 879 (1974)
105. J.C. Phillips, Phys. Rev. **166**, 832 (1968)
106. R.M. Martin, Chem. Phys. Lett. **2**, 268 (1968)
107. B.N. Brockhouse, Phys. Rev. Lett. **2**, 256 (1959)
108. J. Kulda, D. Strauch, P. Pavone, Y. Ishii, Phys. Rev. B **50**, 13347 (1994)
109. K.C. Pandey, Phys. Rev. Lett. **47**, 1913 (1981)
110. K.C. Pandey, Phys. Rev. Lett. **49**, 223 (1982)
111. T.F. Heinz, M.M.T. Loy, W.A. Thompson, Phys. Rev. Lett. **54**, 63 (1985)
112. R.M. Feenstra, W.A. Thompson, A.P. Fein, Phys. Rev. Lett. **56**, 608 (1986)
113. D. Hanneman, Rep. Prog. Phys. **50**, 1045 (1987)
114. H. Ibach, Phys. Rev. Lett. **27**, 253 (1971)
115. H. Ibach, D. L. Mills, *Electron Energy Loss Spectroscopy and Surface Vibrations* (Academic Press, 1982)
116. U. Harten, J.P. Toennies, Ch. Wöll, Phys. Rev. Lett. **57**, 2947 (1986)
117. U. Harten, J.P. Toennies, Europhys. Lett. **4**, 833 (1987)
118. W. Goldhammer, W. Ludwig, W. Zierau, C. Falter, Surf. Sci. **141**, 139 (1984)
119. W. Ludwig, Festkörperprobleme **29**, 107 (1989)
120. O.L. Alerhand, E.J. Mele, Phys. Rev. Lett. **59**, 657 (1987)
121. O.L. Alerhand, E.J. Mele, Phys. Rev. B **37**, 2536 (1988)
122. A. Mazur, J. Pollmann, in *Phonons 89*, ed. by S. Hunklinger, W. Ludwig, G. Weiss (World Scientific, Singapore, 1990), p. 943
123. F. Ancillotto, W. Andreoni, A. Selloni, R. Car, M. Parrinello, Phys. Rev. Lett. **65**, 3148 (1990)
124. A. Ancillotto, W. Andreoni, A. Selloni, R. Car, M. Parinello, in *Phonons 89*, ed. by S. Hunklinger, W. Ludwig, G. Weiss (World Scientific, Singapore, 1990), p. 931
125. M. Zitzlsperger, R. Honke, P. Pavone, U. Schröder, Surf. Sci. **377**, 108 (1997)
126. J. Pollmann, P. Krüger, A. Mazur, M. Rohlfing, Adv. Solid St. Phys. **42**, 189 (2002)
127. C.H. Patterson, S. Banerjee, J.F. McGilp, Phys. Rev. B **84**, 155314 (2011)
128. W. Weber, Phys. Rev. Lett. **33**, 371 (1974)
129. W. Weber, Phys. Rev. B **15**, 4789 (1977)
130. K.C. Rustagi, W. Weber, Solid State Comm. **18**, 1027 (1976)
131. Del Pennino, M.G. Betti, C. Mariani, S. Nannarone, I. Abbati, L. Braicovich, A. Rizzi, Surf. Sci. **189/190**, 689 (1987)
132. Del Pennino, M.G. Betti, C. Mariani, S. Nannarone, C.M. Bertoni, I. Abbati, A. Rizzi, Phys. Rev. B **39**, 10380 (1989)
133. J.E. Northrup, M.L. Cohen, Phys. Rev. Lett. **49**, 1349 (1982)
134. J.E. Northrup, M.L. Cohen, J. Vac. Sci. Technol. **21**, 333 (1982)
135. F.J. Himpsel, P.M. Marcus, R. Tromp, I.P. Batra, M.R. Cook, F. Jona, H. Liu, Phys. Rev. B **30**, 2257 (1984)
136. Y.R. Wang, C.B. Duke, Surf. Sci. **205**, L755 (1988)
137. C.B. Duke, Y.R. Wang, J. Vac. Sci. Technol. B **7**, 1027 (1989)
138. T.J. Gedin, J.P. La Femina, C.B. Duke, J. Vac. Sci. Technol. B **9**, 2282 (1991)

139. G. Binnig, H. Rohrer, C. Gerber, E. Weibel, Phys. Rev. Lett. **50**, 120 (1983)
140. T. Takayanagi, Y. Tanshiro, M. Takahashi, S. Takahashi, J. Vac. Sci. Technol. **A3**, 1502 (1985) and Surf. Sci. **164**, 367 (1985)
141. W. Daum, H. Ibach, J.E. Müller, Phys. Rev. Lett. **59**, 1593 (1987)
142. X.-P. Li, P.B. Allen, J.Q. Broughton, Phys. Rev. Lett. **61**, 243 (1988)
143. Y.-P. Li, P.B. Allen, J.Q. Broughton, Phys. Rev. B **38**, 3331 (1988)
144. G. Lange, J.P. Toennies, P. Ruggerone, G. Benedek, Europhys. Lett. **41**, 647 (1998)
145. U. Harten, J.P. Toennies, Ch. Wöll, L. Miglio, P. Ruggerone, L. Colombo, G. Benedek, Phys. Rev. B **38**, 3305 (1988)
146. J. Kim, M. Yeh, F.S. Khan, J.W. Wilkins, Phys. Rev. B **52**, 14709 (1995)
147. L. Liu, C.S. Jayanthi, S.Y. Wu, Phys. Rev. B **68**, 201301 (2003)
148. I. Stich, K. Terakura, B.E. Larson, Phys. Rev. Lett. **74**, 4491 (1995)
149. I. Stich, J. Kohanoff, K. Terakura, Phys. Rev. B **54**, 2642 (1996)
150. M. Liebhaber, U. Bass, P. Bayersdorfer, J. Geurts, E. Speiser, J. Räthel, A. Baumann, S. Chandola, N. Esser, Phys. Rev. B **89**, 045313 (2014)
151. H. Lüth, R. Matz, Phys. Rev. Lett. **46**, 1652 (1981)
152. R.M. Feenstra, J.A. Stroscio, J. Tersoff, A.P. Fein, Phys. Rev. Lett. **58**, 1192 (1987)
153. L.H. Dubois, G.P. Schwartz, Phys. Rev. B **26**, 794 (1982)
154. Y.-M. Chang, L. Xu, H.W.K. Tom, Chem. Phys. **251**, 283 (2000)
155. U. Harten, J.P. Toennies, Europhys. Lett. **4**, 833 (1987)
156. R.B. Doak, D.B. Nguyen, J. Electron Spectr. **44**, 205 (1987)
157. P. Santini, L. Miglio, G. Benedek, U. Harten, P. Ruggerone, J.P. Toennies, Phys. Rev. B **42**, 11942 (1990)
158. H.M. Tütüncu, G.P. Srivastava, Surf. Sci. **377**, 304 (1996)
159. H.M. Tütüncu, G.P. Srivastava, J. Phys. Chem. Sol. **58**, 685 (1997)
160. U. del Pennino, M.G. Betti, C. Mariani, Surf. Sci. **211**(212), 557 (1989)
161. N. Niehaus, W. Mönch, Phys. Rev. **B50**, 11750 (1994-II)
162. B. Flach, Dissertation, University of Göttingen 1999, Max-Planck-Institut für Strömungsforschung, Report 1/2000, 1999
163. G. Benedek, J. Ellis, A. Reichmuth, P. Ruggerone, H. Schief, J.P. Toennies, Phys. Rev. Lett. **69**, 2951 (1992)
164. D. Fuhrmann, E. Hulpke, W. Steinhögl, Phys. Rev. B **57**, 4798 (1998)
165. J. Fritsch, P. Pavone, U. Schröder, Phys. Rev. Lett. **71**, 4194 (1993)
166. R. Di Felice, A.I. Shkrebtii, F. Finocchi, C.M. Bertoni, G. Onida, J. Electr. Spectr. Rel. Phenom. **64/65**, 697 (1995)
167. W.G. Schmidt, F. Bechstedt, G.P. Srivastava, Phys. Rev. B **52**, 2 (1995)
168. W.G. Schmidt, F. Bechstedt, G.P. Srivastava, Surf. Sci. **352**, 83 (1996)
169. C. Eckl, J. Fritsch, P. Pavone, U. Schröder, Surf. Sci. **394**, 47 (1997)
170. C.B. Duke, S.L. Richardson, A. Paton, A. Kahn, Surf. Sci. **127**, L135 (1983)
171. C.B. Duke, Festkörperprobleme /Adv. Solid State Phys. **33**, 1 (2017)
172. See, e.g., W. Mönch, *Semiconductor Surfaces and Interfaces* (Springer, Berlin Heidelberg, 1993), p. 96
173. H. Range, Dissertation (University of Göttingen, 1995 and Max-Planck-Institut für Strömungsforschung Göttingen, Bericht 3/1995), 1995
174. B. Flach, E. Hulpke, W. Steinhögl, Surf. Sci. **412**, 12 (1998); phonons are reported in B. Flach, Diplomarbeit (University of Göttingen, 1998; unpublished), 1998
175. E. Hulpke, J. Lower, A. Reichmuth, Phys. Rev. B **53**, 13901 (1996-II)
176. D. Campi, M. Bernasconi, G. Benedek, A.P. Graham, J.P. Toennies, Phys. Chem. Chem. Phys. **19**, 16358–16364 (2017)
177. P.B. Allen, in *Handbook of Superconductivity*, ed. by C.P. Poole, Jr. (Academic press, New York, 1999) Ch. 9, Sec. G, pp. 478–483
178. T. Rodach, K.P. Bohnen, Phys. Rev. B **35**, 3446 (1985)
179. T. Rodach, K.P. Bohnen, K.M. Ho, Surf. Sci. **209**, 481 (1989)
180. R.B. Wilson, D.M. Riffe, J. Phys.: Condens. Matter **24**, 335401 (2012)

181. D.L. Price, K.S. Singwi, M.P. Tosi, Phys. Rev. B: Condens. Matter Mater. Phys. **2**, 2983 (1970)
182. V. Ramamurthy, M. Satishkumar, Physica, **111B**, 297 (1981)
183. S. Bartholmei, P. Fouquet, G. Witte, Surf. Sci. **473**, 227 (2001)
184. J.B. Hannon, E.W. Plummer, J. Electr. Spectr. Rel. Phenom. **65**(65), 683 (1993)
185. J.B. Hannon, E.J. Mele, E.W. Plummer, Phys. Rev. B **53**, 2090 (1996)
186. E.W. Plummer, J.B. Hannon, Prog. Surf. Sci. **46**, 149 (1994)
187. P. Hofmann, E.W. Plummer, Surf. Sci. **377**, 330 (1997)
188. K. Pohl, J.H. Cho, K. Terakura, M. Scheffler, E.W. Plummer, Phys. Rev. Lett. **80**, 2853 (1998)
189. F. Kwasniok, Surf. Sci. **329**, 90 (1995)
190. M. Lazzeri, S. de Gironcoli, Phys. Rev. Lett. **81**, 2096 (1998); Surf. Sci. **402–404**, 715 (1998)
191. M. Lazzeri, S. de Gironcoli, Surf. Sci. **454**, 442 (2000)
192. J.A. Gaspar, A.G. Eguiluz, M. Gester, A. Lock, J.P. Toennies, Phys. Rev. Lett. **66**, 337 (1991)
193. I.Yu. Sklyadneva, R. Heid, P.M. Echenique, K.-P. Bohnen, E.V. Chulkov, Phys. Rev. B **83**, 195437 (2011)
194. J.P. Toennies, Ch. Wöll, Phys. Rev. Rapid Comm. **36**, 4475 (1987)
195. A. Lock, J.P. Toennies, Ch. Wöll, V. Bortolani, A. Franchini, G. Santoro, Phys. Rev. B **37**, 7087 (1989)
196. M. Gester, D. Kleinhesselink, P. Ruggerone, J.P. Toennies, Phys. Rev. B **49**, 5777 (1994)
197. A. Franchini, V. Bortolani, G. Santoro, V. Celli, A.G. Eguiluz, J.A. Gaspar, M. Gester, A. Lock, J.P. Toennies, Phys. Rev. B **47**, 4691 (1993)
198. A.G. Eguiluz, Phys. Rev. Lett. **51**, 1907 (1983)
199. A.G. Eguiluz, Phys. Rev. B **31**, 3303 (1985) and Phys. Scr. **36**, 651 (1987)
200. A.A. Quong, R.F. Wallis, A.A. Maradudin, J.A. Gaspar, A.G. Eguiluz, in *Proceedings of the 3rd International Conference on Phonon Physics*, ed. by S. Hunklinger, W. Ludwig, G. Weiss (World Sci., Singapore, 1990), p. 934
201. A.G. Eguiluz, Phys. Rev. B **35**, 5473 (1987)
202. A.G. Eguiluz, A.A. Maradudin, R.F. Wallis, Phys. Rev. Lett. **60**, 309 (1988)
203. F. Ercolessi, E. Tosatti, M. Parrinello, Surf. Sci. **177**, 314 (1986)
204. V. Chis, B. Hellsing, G. Benedek, M. Bernasconi, J.P. Toennies, J. Phys. C **19**, 305011 (2007)
205. G. Benedek, M. Bernasconi, V. Chis, E.V. Chulkov, P.M. Echenique, B. Hellsing, J. P. Toennies, J. Phys.: Condens. Matter **22**, 084020 (2010)
206. B.N. Brockhouse, T. Arase, G. Caglioti, K.R. Rao, A.D.B. Woods, Phys. Rev. **128**, 1099 (1962)
207. G. Zhang, Dissertation (University of Göttingen, 1991 and Max-Planck-Institut für Strömungsforschung Göttingen, Bericht 102/1991)
208. J. Braun, Dissertation (University of Göttingen, 1997 and Max-Planck-Institut für Strömungsforschung Göttingen, Bericht 11/1997)
209. D.J. Hinch, C. Koziol, J.P. Toennies, G. Zhang, Europhys. Lett. **10**, 341 (1989)
210. Braun, J., Ruggerone, P., Zhang, G., Toennies, J.P., Benedek, G, Phys. Rev. B **79**, 205423 (2009)
211. Sklyadneva, I.Y., Heid, R., Bohnen, K.-P., Echenique, P.M., Chulkov, E.V.: J. Phys.: Condens. Matter **24**, 104004 (2012)
212. Y.S. Li, F. Jona, M.P. Marcus, Phys. Rev. B **43**, 6337 (1991)
213. T.G. Worlton, R.E. Schmunck, Phys. Rev. B **3**, 4115 (1971)
214. I.Y. Sklyadneva, R. Heid, K.-P. Bohnen, P.M. Echenique, G. Benedek, E.V. Chulkov, J. Phys. Chem. A **115**, 7352 (2011)
215. R.B. Doak, U. Harten, J.P. Toennies, Phys. Rev. Lett. **51**, 578 (1983)
216. P. Satiris, H.C. Lu, T. Gustafsson, Phys. Rev. Lett. **72**, 3574 (1994)
217. G. Armand, Solid State Commun. **48**, 261 (1983)

218. V. Bortolani, A. Franchini, F. Nizzoli, G. Santoro, Phys. Rev. Lett. **52**, 429 (1984)
219. U. Harten, J.P. Toennies, Ch. Wöll, Faraday Discuss. Chem. Soc. **80**, 137 (1985)
220. C. Kaden, P. Ruggerone, J.P. Toennies, G. Zhang, G. Benedek, Phys. Rev. B **46**, 13509 (1992)
221. M.H. Mohamed, L.L. Kesmodel, B.M. Hall, D.L. Mills, Phys. Rev. B **37**, 2763 (1988)
222. V. Chis, B. Hellsing, G. Benedek, M. Bernasconi, E.V. Chulkov, J.P. Toennies, Phys. Rev. Lett. **101**, 206102 (2008). Erratum: **103**, 069902 (2009)
223. G. Benedek, J. Ellis, N.S. Luo, A. Reichmuth, P. Ruggerone, J.P. Toennies, Phys. Rev. B **48**, 4917 (1993)
224. N. Bunjes, N.S. Luo, R. Ruggerone, J.P. Toennies, G. Witte, Phys. Rev. B **50**, 8897 (1994)
225. W. Wuttig, R. Franchy, H. Ibach, Solid St. Comm. **57**, 445 (1986)
226. W. Wuttig, R. Franchy, H. Ibach, Z. Physik B **65**, 71 (1986)
227. L.L. Kesmodel, M.L. Xu, S.Y. Tong, Phys. Rev. B **34**, 2010 (1986)
228. Y. Chen, S.Y. Tong, J.S. Kim, L.L. Kesmodel, T. Rodach, K.P. Bohnen, K.M. Ho, Phys. Rev. B **44**, 11394 (1991)
229. Y. Chen, S.Y. Tong, M. Rocca, P. Moretto, U. Valbusa, K.P. Bohnen, K.M. Ho, Surf. Sci. **250**, L389 (1991)
230. M. Rocca, U. Valbusa, Nuovo Cim. **15D**, 493 (1993)
231. P.D. Ditlevsen, J.K. Nørskov, Surf. Sci. **254**, 261 (1991)
232. Y. Chen, S.Y. Tong, K.P. Bohnen, T. Rodach, K.M. Ho, Phys. Rev. Lett. **70**, 603 (1993)
233. C.E. Bach, M. Giesen, H. Ibach, T.L. Einstein, Phys. Rev. Lett. **78**, 4225 (1997)
234. A.R. Sandy, S.G.J. Mochrie, D.M. Zehner, K.G. Huang, D. Gibbs, Phys. Rev. B **43**, 4667 (1991)
235. S.M. Driver, T. Zhang, D.A. King, Angew. Chem. Int. Ed. Engl. **46**, 700 (2007)
236. U. Harten, A.M. Lahee, J.P. Toennies, Ch. Wöll, Phys. Rev. Lett. **54**, 2619 (1985)
237. V. Bortolani, G. Santoro, U. Harten, J.P. Toennies, Surf. Sci. **148**, 82 (1984)
238. R. Ravelo, M. El-Batanouny, Phys. Rev. B **40**, 9574 (1989)
239. R. Ravelo, M. El-Batanouny, Phys. Rev. B **47**, 12771 (1993)
240. C.S. Jayanthi, H. Bilz, W. Kress, G. Benedek, Phys. Rev. Lett. **59**, 795 (1987)
241. X. Zhu, L. Santos, R. Sankar, S. Chikara, C. Howard, F.C. Chou, C. Chamon, M. El-Batanouny, Phys. Rev. Lett. **107**, 186102 (2011)
242. X.T. Zhu, L. Santos, C. Howard, R. Sankar, F.C. Chou, C. Chamon, M. El-Batanouny, Phys. Rev. Lett. **108**, 185501 (2012)
243. C. Howard, M. El-Batanouny, R. Sankar, F.C. Chou, Phys. Rev. B **88**, 035402 (2013)
244. G. Benedek, J. Ellis, N.S. Luo, A. Reichmuth, P. Ruggerone, J.P. Toennies, Phys. Rev. B **48**, 4917 (1993)
245. N. Bunjes, K.S. Luo, P. Ruggerone, J.P. Toennies, G. Witte, Phys. Rev. B **50**, 8897 (1994)
246. V. Bortolani, A. Franchini, G. Santoro, in *Electronic, Dynamic Quantum Structural Properties of Condensed Matter*, ed. by J.T. Devreese, P.E. van Camp (Plenum, New York, 1985), p. 401
247. G. Santoro, A. Franchini, V. Bortolani, U. Harten, J.P. Toennies, Ch. Wöll, Surf. Sci. **183**, 180 (1987)
248. J.P. Toennies, Superlattices Microstruct. **7**, 193 (1990)
249. J.S. Nelson, E.C. Sowa, M.S. Daw, Phys. Rev. Lett. **61**, 1977 (1988)
250. P.D. Ditlevsen, J.K. Nørskov, J. El Spectr. Rel. Phenom. **54**(55), 237 (1990)
251. S.Y. Tong, Y. Chen, K.P. Bohnen, T. Rodach, K.M. Ho, Surf. Rev. Lett. **1**, 97 (1994)
252. A. dal Corso, Phys. Rev. B **64**, 235118 (2001)
253. F. Raouafi, C. Barreteau, M.C. Desjonquères, D. Spanjaard, Surf. Sci. **507**, 748 (2002)
254. V. Rosato, M. Guillopé, B. Legrand, Phil. Mag. A **59**, 321 (1989)
255. P.B. Allen, Phys. Rev. B **16**, 5139 (1977)
256. E. Frantzeskakis, S. Pons, A. Crepaldi, H. Brune, K. Kern, M. Grioni, Phys. Rev. B **84**, 245443 (2011)

257. In Ref. [220] A_{eff} was defined as in Ref. [217] by $ac_{11}/2$. With this definition, however, A_{eff} is not purely radial and was found to decrease at the surface by about 30%. It should not be compared with the surface change of the nn radial force constant obtained with other methods.
258. H. Bilz, D. Strauch, R.K. Wehner, *Light and Matter Id*, Encyclopedia of Physics, vol. XXV/2d, ed. by L. Genzel (Springer, Berlin-Heidelberg, 1984)
259. N.W. Ashcroft, N.D. Mermin, Solid State Physics (Brooks Cole, 1976) 1st ed.
260. E. Heifets, YuF Zhukovskii, E.A. Kotomin, M. Causá, Chem. Phys. Lett. **283**, 395 (1998)
261. R. Berndt, J.P. Toennies, Ch. Wöll, J. Electron. Spectr. **44**, 183 (1987)
262. G. Benedek, J.P. Toennies, G. Zhang, Phys. Rev. Lett. **68**, 2644 (1992)
263. U. Harten, J.P. Toennies, Ch. Wöll, G. Zhang, Phys. Rev. Lett. **55**, 2308 (1985)
264. V. Bortolani, A. Franchini, G. Santoro, J.P. Toennies, Ch. Wöll, G. Zhang, Phys. Rev. B **40**, 3524 (1989)
265. G. Witte, H. Range, J.P. Toennies, Ch. Wöll, J. Elect. Spectr. Rel. Phenom. **64**(65), 715 (1993)
266. J.P. Toennies, G. Witte, Ch. Wöll, Surf. Sci. **323**, 228 (1995)
267. S. Lehwald, J.W. Szeftel, H. Ibach, T.S. Rahman, D.L. Mills, Phys. Rev. Lett. **50**, 518 (1983)
268. S. Lehwald, J.M. Szeftel, H. Ibach, T.S. Rahman, G.L. Mills, Surf. Sci. **138**, L 123 (1984)
269. M.L. Xu, B.M. Hall, S.Y. Tong, M. Rocca, H. Ibach, S. Lehwald, J.E. Black, Phys. Rev. Lett. **54**, 1171 (1985)
270. S. Lehwald, J.M. Szeftel, H. Ibach, T.S. Rahman, G.L. Mills, Surf. Sci. **171**, 632 (1986)
271. M. Rocca, H. Ibach, S. Lehwald, M.L. Xu, B.M. Hall, Y.S. Tong, in *The Structure of Surface*, ed. by M.A. van Hove, S.Y. Tong (Springer, Berlin, Heldelberg, 1985)
272. S. Lehwald, M. Rocca, H. Ibach, T.S. Rahman, J. Electr. Spectr. Rel. Phenom. **38**, 29 (1986)
273. M. Rocca, H. Ibach, S. Lehwald, T.S. Rahman, in *Structure and Dynamics of Surfaces I*, ed. by W. Schommers, P. von Blanckenhagen (Springer, Berlin, Heidelberg, 1987), p. 245
274. E.-J. Jeong, J.L. Erskine, Rev. Sci. Instr. **60**, 3139 (1989)
275. J.L. Erskine, E.-J. Jeong, J. Vac. Sci. Technol. A **8**, 2649 (1990)
276. R. Berndt, J.P. Toennies, Ch. Wöll, J. Electron. Spectr. **44**, 183 (1987)
277. E.J. Woll Jr., W. Kohn, Phys. Rev. **126**, 1693 (1962)
278. A.M. Afanas'ev, Y.M. Kagan, Soviet Phys. JETP **16**, 1030 (1963)
279. J. Friedel, Nuovo Cim., Suppl. **7**, 287 (1968)
280. A. Larose, B.N. Brockhouse, Can. J. Phys. **54**, 1819 (1976)
281. See for example A. Zangwill, *Physics at Surfaces* (Cambridge University Press, Cambridge, 1988), page 51 ff
282. K. Kern, R. David, R.L. Palmer, G. Comsa, T.S. Rahman, Phys. Rev. B **33**, 4334 (1989)
283. D. Neuhaus, F. Joo, B. Feuerbacher, Surf. Sci. Lett. **165**, L90 (1986)
284. G. Benedek, M. Miura, W. Kress, H. Bilz, Phys. Rev. Lett. **52**, 1907 (1984)
285. G. Benedek, G. Brusdeylins, C. Heimlich, L. Miglio, J.G. Skofronick, J.P. Toennies, R. Vollmer, Phys. Rev. Lett. **60**, 1037 (1988)
286. G. Benedek, L. Miglio, J.G. Skofronick, G. Brusdeylins, C. Heimlich, J.P. Toennies, J. Vac. Sci. Technol. A **5**, 1093 (1987)
287. W. Di, K.E. Smith, S.D. Kevan, Phys. Rev. B **45**, 3652 (1992)
288. V. Bortolani, A. Franchini, G. Santoro, J.P. Toennies, Ch. Wöll, G. Zhang, Phys. Rev. B **40**, 3524 (1989)
289. G. Witte, H. Range, J.P. Toennies, Ch. Wöll, J. Elect. Spectr. Rel. Phenom. **64**(65), 715 (1993)
290. J.P. Toennies, G. Witte, Ch. Wöll, Surf. Sci. **323**, 228 (1995)
291. H.J. Ernst, E. Hulpke, J.P. Toennies, Ch. Wöll, Surf. Sci. **262**, 159 (1992)
292. K. Christmann, Surf. Sci. Repts. **9**, 1 (1988)
293. D. Tomanek, S.G. Louie, C.-T. Chan, Phys. Rev. Lett. **57**, 2594 (1986)
294. H.-J. Ernst, E. Hulpke, J.P. Toennies, Phys. Rev. Lett. **58**, 1941 (1987)
295. H.-J. Ernst, E. Hulpke, J.P. Toennies, Europhys. Lett. **10**, 747 (1989)
296. H.-J. Ernst, E. Hulpke, J.P. Toennies, Phys. Rev. B **46**, 16081 (1992)

297. A. Fasolino, E. Tosatti, Phys. Rev. B **35**, 4264 (1987)
298. D.A. King, Phys. Scr. **T4**, 34 (1983)
299. P. Estrup, in *Chemistry and Physics of Solid Surfaces V*, ed. by R. Vanselow, R. Howe, Springer Ser. Chem. Phys. vol. 35 (Springer, Berlin, Heidelberg, 1984), p. 205
300. H. Landskron, N. Bickel, K. Heinz, G. Schmidtlein, K. Müller, J. Phys.: Condens. Matter **1**, 1 (1989)
301. I. Steensgaard, K.G. Purcell, D.A. King, Phys. Rev. B **39**, 897 (1989)
302. I.K. Robinson, A.A. MacDowell, M.S. Altman, P.J. Estrup, K. Evans-Lutterodt, J.D. Brock, R.J. Birgenau, Phys. Rev. Lett. **62**, 1294 (1989)
303. C.Z. Wang, A. Fasolino, E. Tosatti, Europhys. Lett. **7**, 263 (1988); Phys. Rev. B **37**, 2116 (1988)
304. L.P. Roelofs, T. Ramseyer, L.L. Taylor, D. Singh, H. Krakauer, Phys. Rev. B **46**, 9147 (1989)
305. B. Salanon, J. Lapujoulade, Surf. Sci. Lett. **173**, L613 (1986)
306. J.F. Wendelken, G.C. Wang, Phys. Rev. B **32**, 7542 (1985)
307. E. Hulpke, D.-M. Smilgies, Phys. Rev. B **42**, 9203 (1990-I)
308. W. Cochran, Phys. Rev. Lett. **3**, 412 (1959)
309. W. Cochran, Adv. Phys. **9**, 387 (1960)
310. P.W. Anderson, in proceeding of the All-Union Conference on the Physics of Dielectrics, ed. by G.I. Skanavi, Acad. Sci. USSR, Moscow, 290 (1958)
311. H.J. Schulz, Phys. Rev. B **18**, 5756 (1978)
312. E.K. Schweizer, C.T. Rettner, Surf. Sci. Lett. **208**, L29 (1989)
313. R.V. Coleman, Z. Dai, W.W. McNairy, C.G. Slough, C.W. Wang in *Surface Properties of Layered Structures*, ed. by G. Benedek (Kluver, Amsterdam, 1992), p. 27
314. E. Meyer, D. Anselmetti, R. Wiesendanger, H.-J. Güntherodt, F. Lévy, H. Berger, Europhys. Lett. **9**, 695 (1989)
315. G. Benedek, F. Hofmann, P. Ruggerone, G. Onida, L. Miglio, Surf. Sci. Rep. **20**, 1 (1994)
316. J.C. Campuzano, D.A. King, C. Somerton, J.E. Inglesfield, Phys. Rev. Lett. **45**, 1649 (1980)
317. M.I. Holmes, T. Gustafsson, Phys. Rev. Lett. **47**, 443 (1981)
318. K.E. Smith, G.S. Elliot, S.D. Kevan, Phys. Rev. B **42**, 5385 (1990)
319. E. Bauer, private communication
320. E. Hulpke, D.-M. Smilgies, Phys. Rev. B **40**, 1338 (1989-I)
321. D.-M Smilgies, E. Hulpke, Phys. Rev. B **43**, 1260 (1991)
322. R.S. Daley, T.E. Felter, M.L. Hildner, P.J. Estrup, Phys. Rev. Lett. **70**, 1295 (1993)
323. D.-M. Smilgies, P.J. Eng, I.K. Robinson, Phys. Rev. Lett. **70**, 1291 (1993)
324. E. Hulpke, J. Lüdecke, Phys. Rev. Lett. **68**, 2846 (1992)
325. E. Hulpke, J. Lüdecke, Surface Sci. **287**(288), 837 (1993)
326. V.V. Gonchar, Yu.M. Kagan, O.V. Kanash, A.G. Naumovets, A.G. Fedorus, Zu. Eksp. Teor. Fiz. **84**, 246 (1983) [Engl. transl.: Sov. Phys. JETP 57, 142 (1983)]
327. N.S. Luo, B. Legrand, Phys. Rev. B **38**, 1728 (1988)
328. J.W. Chung, S.C. Ying, P.J. Estrup, Phys. Rev. Lett. **56**, 749 (1986)
329. B. Kohler, P. Ruggerone, S. Wilke, M. Scheffler, Phys. Rev. Lett. **74**, 1387 (1995)
330. B. Kohler, P. Ruggerone, M. Scheffler, E. Tosatti, Z. Phys, Chem. **197**, 193–202 (1996)
331. M. Altman, J.W. Chung, P.J. Estrup, J.M. Kosterlitz, J. Prybyla, D. Saha, S.C. Ying, J. Vac. Sci. Technol. **A5**, 1045 (1987)
332. E. Hulpke, J. Lüdecke, Surf. Sci. **272**, 289 (1992)
333. E. Hulpke, J. Lüdecke, J. Electr. Spectr. Rel. Phen. **64**(65), 641 (1993)
334. M. Balden, S. Lehwald, E. Preuss, H. Ibach, Surf. Sci. **307–309**, 1141 (1994)
335. L.H. Germer, J.W. May, Surf. Sci. **4**, 452 (1966)
336. H. Köster, Diplom thesis (Göttingen, 1993, unpublished)
337. J. Lüdecke, Dissertation, University of Göttingen (1994), Max-Planck-Institut für Strömungsfoschung, Report 15 (1994)
338. B. Kohler, P. Ruggerone, S, M. Scheffler, Phys. Rev. B **56**, 13503 (1997-II)

339. G. Giuliani, E. Tosatti, Nuovo Cim. B **47**, 135 (1978); and in *Quasi-one Dimensional Conductors*, ed. by S. Baricic, A. Bjelis, J.R. Cooper, B: Leontic (Springer, Berlin, 1979), p. 191
340. E. Tosatti, in *Electronic Surface and Interface States on Metallic Systems*, ed. by E. Bertel, M. Donath (World Sci., Singapore, 1995), p. 67
341. M. Balden, S. Lehwald, H. Ibach, D.L. Mills, Phys. Rev. Lett. **73**, 854 (1994)
342. R.H. Gaylord, K.H. Jeong, S.D. Kevan, Phys. Rev. Lett. **62**, 2036 (1989)
343. K.H. Jeong, R.H. Gaylord, S.D. Kevan, Phys. Rev. B **39**, 2973 (1989)
344. B. Renker, H. Rietschel, L. Pintschovius, W. Gläser, W. Bruesch, D. Kuse, M.J. Rice, Phys. Rev. Lett. **30**, 1144 (1973)
345. W. Gläser, Festkörperprobleme XIV, O. Madelung, ed. (Verlag Vieweg u. Sohn, Braunschweig, 1974), p. 205
346. A textbook theoretical treatment of electron-phonon driven anomalies and lattice instabilities in low-dimensional systems can be found in: H. Böttger, *Principles of the Theory of Lattice Dynamics* (Physik-Verlag, Weinheim, 1983)
347. P. Ruggerone, B. Kohler, S. Wilke, M. Scheffler, in *Electronic Surface and Interface States on Metallic Systems*, ed. by E. Bertel, M. Donath (World Scientific, Singapore, 1995), p. 113
348. B. Flach, Dissertation (University of Göttingen, 2000 and Max-Planck-Institut für Strömungsforschung Göttingen, Bericht 1/2000)
349. C. Bungaro, S. de Gironcoli, S. Baroni, Phys. Rev. Lett. **77**, 2491 (1996)
350. G. Benedek, M. Pardo, J.P. Toennies, Theory of inelastic atom scattering from surface electron-hole and plasmon excitations, in *Highlights on Spectroscopies of Semiconductors and Nanostructures*, ed. by G. Guizzetti, A.C. Andreani, F. Marabelli, M. Patrini, Conf. Proc. vol. 94 (SIF, Bologna, 2007) pp. 151–167
351. E. de Rouffignac, G. Alldredge, F.W. de Wette, Phys. Rev. B **23**, 4208 (1981)
352. G.P. Alldredge, E. de Rouffignac, B. Firey, F.W. de Wette, Phys. Rev. B **29**, 3712 (1984)
353. P. Cantini, in *Dynamics of Gas Surface Interactions*, ed. by G. Benedek, U. Valbusa (Springer, Heidelberg, 1982), p. 283
354. G. Benedek (ed.), *Surface Properties of Layered Structures* (Kluver, Dordrecht, 1992)
355. S. Siebentritt, R. Pues, K.-H. Rieder, A.M. Shikin, Phys. Rev. B **55**, 7927 (1997)
356. G. Brusdeylins, R. Rechsteiner, J.G. Skofronick, J.P. Toennies, G. Benedek, L. Miglio, Phys. Rev. B **34**, 902 (1986)
357. S.K. Sinha, in *Methods of Experimental Physics*, ed. by D.L. Price, K. Sköld, vol. 23B (1987), p. 1
358. H. Godfrin, H.J. Lauter, *Progress in Low Temperature Physics*, ed. by W.P. Halperin, vol. XIV (Elsevier, Amsterdam, 1995), p. 213
359. W. Bruch, M.W. Cole, E. Zaremba, *Physical Adsorption: Forces and Phenomena* (Clarendon Press, Oxford, 1997), Chap. 6
360. G. Dolling, B.N. Brockhouse, Phys. Rev. **128**, 1120 (1962)
361. R. Nicklow, N. Wakabayashi, H.G. Smith, Phys. Rev. B **5**, 4951 (1972)
362. L.J. Brillson, E. Burstein, A.A. Maradudin, T. Stark, in *The Physics of Semimetals and Narrow Gap Semiconductors*, ed. by D.L. Carter, R.T. Bate (Pergamon Press, Oxford, 1971), p. 187
363. M. Maeda, Y. Kuramoto, C. Horie, J. Phys. Soc. Jpn. Lett. **44**, 337 (1979)
364. R. Al-Jishi, G. Dresselhaus, Phys. Rev. B **26**, 4514 (1982)
365. R.A. Jishi, L. Venkataraman, M.S. Dresselhaus, G. Dresselhaus, Chem. Phys. Lett. **209**, 77 (1993)
366. A. Grüneis, R. Saito, T. Kimura, L.C. Cancado, M.A. Pimienta, A. Jorio, A.G. Souza Filho, G. Dresselhaus, M.S. Dresselhaus, Phys. Rev. B **65**, 155405 (2002)
367. R. Tubino, L. Piseri, G. Zerbi, J. Chem. Phys. **56**, 1022 (1972)
368. T. Aizawa, R. Souda, S. Otani, Y. Ishizawa, C. Oshima, Phys. Rev. Lett. **64**, 768 (1990)
369. S. Jandl, J.L. Brebner, Can. J. Phys. **52**, 2454 (1974)
370. G. Benedek, F. Hofmann, P. Ruggerone, G. Onida, L. Miglio, Surf. Sci. Rep. **20**, 1 (1994)
371. G. Boato, P. Cantini, C. Salvo, R. Tatarek, S. Terreni, Surf. Sci. **114**, 485 (1982)

372. G. Benedek, G. Brusdelins, C. Heimlich, J.P. Toennies, U. Valbusa, Surf. Sci. **114**, 549 (1986). Additional unpublished data points (open squares in Fig. 11.36(b)) obtained with a higher energy beam of 73 meV have been added
373. G. Benedek, G. Onida, Phys. Rev. B **47**, 16471 (1993)
374. D. Maccariello, D. Campi, A. Al Taleb, G. Benedek, D. Farìas, M. Bernasconi, R. Miranda, Carbon **93**, 1–10 (2015)
375. C. Oshima, T. Aizawa, R. Souda, Y. Ishizawa, Y. Sumiyoshi, Solid State Commun. **65**, 1601 (1988)
376. J.L. Wilkes, R.E. Palmer, R.F. Willis, J. Electron. Spect. Rel. Phenom. **44**, 355 (1987)
377. E.T. Jensen, R.E. Palmer, W. Allison, J.F. Annett. Phys. Rev. Lett. **66**, 492 (1991)
378. P. Laitenberger, R.E. Palmer, Phys. Rev. Lett. **76**, 96 (1992)
379. M. Portail, M. Carrere, J.M. Layet, Surf. Sci. **433–435**, 863 (1999)
380. L. Vitali, M.A. Schneider, K. Kern, L. Wirtz, A. Rubio, Phys. Rev. B **69**, 121414 (2004)
381. L. Wirtz, A. Rubio, Solid State Comm. **131**, 141–152 (2004)
382. M. Mohr, J. Maultzsch, E. Dobardžić, S. Reich, I. Milošević, M. Damnjanović, A. Bosak, M. Krisch, C. Thomsen, Phys. Rev. B **76**, 035439 (2007)
383. J. Maultzsch, S. Reich, C. Thomsen, H. Requardth, P. Ordejòn, Phys. Rev. Lett. **92**, 075501 (2004)
384. H. Yanagisawa, T. Tanaka, Y. Ishida, M. Matsue, E. Rokuta, S. Otani, C. Oshima, Surf. Interf. Anal. **37**, 133 (2005)
385. U. Paltzer, D. Schmicker, F.W. de Wette, U. Schröder, J.P. Toennies, Phys. Rev. B **54**, 11989 (1996)
386. T. Aizawa, R. Souda, S. Otani, Y. Ishizawa, C. Oshima, Phys. Rev. B **42**, 11469 (1990), and **43**, 12060 (1991) (E)
387. C. Lobo, J.L. Martins, Z. Phys. D **39**, 159 (1997)
388. J.L. Brebner, S. Jandl, B.M. Powell, Solid State Commun. **13**, 1555 (1987)
389. O. Dubay, G. Kresse, Phys. Rev. B **67**, 035401 (2003)
390. P. Pavone, R. Bauer, K. Karch, O. Schütt, S. Vent, W. Windl, D. Strauch, S. Baroni, S. de Gironcoli, Phys. B **219**, **220**, 439 (1996)
391. G. Onida, G. Benedek, Europhys. Lett. **18**, 403 (1992) and **19**, 343 (1992) (Erratum)
392. S. Sanguinetti, G. Benedek, Phys. Rev. B **50**, 15439 (1994)
393. D.W. Snoke, M. Cardona, S. Sanguinetti, G. Benedek, Phys. Rev. B **53**, 12641 (1996)
394. G.W. Semenoff, Phys. Rev. Lett. **53**, 2449 (1984)
395. G.W Semenoff, Phys. Scr., T146, 014016 (2012) and references therein
396. K.S. Novoselov, A.K. Geim, S.V. Morozov, D. Jiang, Y. Zhang, S.V. Dubonos, L.V. Grigorieva, A.A. Firsov, K.S. Novoselov, Science **306**, 666 (2005)
397. Y.B. Zhang, Y.W. Tan, H.L. Stormer, P. Kim, Nature **438**, 2012005 (2005)
398. A.K. Geim, K. S Novoselov, Nature Mat. **6**, 183 (2007)
399. K.S. Novoselov, A.K. Geim, S.V. Morozov, D. Jiang, M.I. Katsnelson, I.V. Grigorieva, S.V. Dubonos, A.A. Firsov, Nature **438**, 197 (2006)
400. Graphene Flagship Annual Report 2015 (EU 2016), http://graphene-flagship.eu/project/Pages/Annual-reports.aspx
401. H.P. Boehm, A. Clauss, G. Fischer, U. Hofmann, in *Proceedings of 5th Conference on Carbon* (Pergamon Press, 1962), p. 73
402. W.M. Hess, L.L.Ban, Novelco Reporter **13**, 102 (1966) and in *Proceedings of 6th International Congress on Electron Microscopy (Kyoto)*, vol. I, p. 569
403. A.E. Karu, M. Beer, J. Appl. Phys. **37**, 2179 (1966)
404. Z. Hu, D.F. Ogletree, M.A. Van Hove, G.A. Somorjai, Sur. Sci. **180**, 433 (1987)
405. C. Oshima, A. Nagashima, J. Phys.: Cond. Mat. **9**, 1 (1997)
406. T. Aizawa, Y. Hwang, W. Hayami, R. Souda, S. Otani, Yoshio Ishizawa, Surf. Sci. **260**, 311 (1992)
407. Y. Hwang, T. Aizawa, W. Hayami, S. Otani, Y. Ishizawa, S.-J. Park, Surf. Sci. **271**, 299 (1992)
408. M.-C. Wu, Q. Xu, D.W. Goodman, J. Phys. Chem. **98**, 5104 (1994)

409. A.M. Shikin. D. Farìas, K.H. Rieder, Europhys. Lett. **44**, 44 (1988)
410. A.M. Shikin, D. Farias, V.K. Adamchuk, K.-H. Rieder, Surf. Sci. **424**, 155 (1999)
411. A. Politano, A.R. Marino, G. Chiarello, J. Phys.: Condens. Matter **24**, 104025 (2012)
412. D.L. Nika, A.A. Balandin, Rep. Prog. Phys. **80**, 036502 (2017)
413. M. Al Taleb, D. Farìas, J. Phys.: Condens. Matter **28**, 103005 (2016)
414. D. Maccariello, A. AlTaleb, F. Calleja, A.L. Vazquesz des Parga, P. Perna, J. Camarero, E. Gnecco, D. Farias, R. Miranda, Nano Lett. **16**, 2–7 (2015)
415. E.M. McIntosh, P.R. Kole, M. El-Batanouny, D.M. Chisnall, J. Ellis, W. Allison, Phys. Rev. Lett. **110**, 086103 (2013)
416. P. Hofmann, Prog. Surf. Sci. **81**, 191 (2006)
417. G. Bian, T. Miller, T.-C. Chiang, Phys. Rev. Lett. **107**, 036802 (2011)
418. D. Campi, M. Bernasconi, G. Benedek, Phys. Rev. B **86**, 075446 (2012)
419. P. Kraus, M. Mayrhofer-Reinhartshuber, A. Tamtögl, W.E. Ernst, private communication, to be published
420. A. Tamtögl, M. Mayrhofer-Reinhartshuber, N. Balak, W.E. Ernst, K.H. Rieder, J. Phys.: Condens. Matter **22**, 304019 (2010)
421. M. Mayrhofer-Reinhartshuber, A. Tamtögl, P. Kraus, K.H. Rieder, W.E. Ernst, J. Phys.: Condens. Matter **24**, 104008 (2012)
422. A. Tamtögl, P. Kraus, M. Mayrhofer-Reinhartshuber, D. Campi, M. Bernasconi, G. Benedek, W.E. Ernst, Phys. Rev. B **87**, 035410 (2013)
423. V. Chis, G. Benedek, P.M. Echenique, E.V. Chulkov, Phys. Rev. **87**, 075412 (2013)
424. M. Alcántara Ortigoza, I.Y. Sklyadneva, R. Heid, E.V. Chulkov, T.S. Rahman, K.-P. Bohnen, P.M. Echenique, Phys. Rev. B **90**, 195438 (2014)
425. G. Benedek, M. Bernasconi, K.-P. Bohnen, D. Campi, E.V. Chulkov, P.M. Echenique, R. Heid, I.Y. Sklyadneva, J.P. Toennies, Phys. Chem. Chem. Phys. **16**, 7159–7172 (2014)
426. K. Deguchi, Y. Takano, Y. Mizuguchi, Sci. Technol. Adv. Mater. **13**, 054303 (2012)
427. R.A- Klemm, (ed.), *Layered Superconductors*, vol. 1 (Oxford Univ. Press, 2012)
428. H. Jin, J.-H. Song, A.J. Freeman, M.G. Kanatzidis, Phys. Rev. B **83**, 41202(R) (2011)
429. F. Ortmann, S. Roche, S.O. Valenzuela, (eds.), *Topological Insulators- Fundamentals and Perspectives* (Wyley-VCH Verlag, Weinheim, 2015)
430. S. Jandl, J.L. Brebner, B.M. Powell, Phys. Rev. B **13**, 686 (1976)
431. J.J. Weiting, M. Schlüter (eds.), *Electrons and Phonons in Layered Crystal Structures* (Reidel, Dordrecht, 1979)
432. V. Ya Altshul, V. Ya Bashenov, D.I. Marakov, A.G. Petukhov, Phys. Stat. Sol. **98**, 715 (1980)
433. Y. Depeursinge, A. Baldereschi, Phys. B **105**, 324 (1981)
434. D.E. Moncton, J.D. Axe, F.J. DiSalvo, Phys. Rev. Lett. **34**, 734 (1975)
435. D.E. Moncton, J.D. Axe, F.J. DiSalvo, Phys. Rev. B **16**, 801 (1977)
436. L. Plucinski, R.L. Johnson, B.J. Kowalski, K. Kopalko, B.A. Orlowski, Z.D. Kovalyuk, G. V. Lashkarev, Phys. Rev. B **68**, 125304 (2003)
437. V. Pannella, A.L. Glebov, J.P. Toennies, C. Sebenne, C. Eckl, C. Adler, P. Pavone, U. Schröder, Phys. Rev. B **59**, 15772 (1999)
438. G. Benedek, Physica B **127**, 59 (1984)
439. L. Miglio, G. Benedek, Europhys. Lett. **3**, 619 (1987)
440. L. Miglio, L. Colombo, Phys. Rev. B **37**, 3025 (1988)
441. K. Yamaguchi, N. Nishina, Phys. B **105**, 329 (1981)
442. K. Yamaguchi, N. Nishina, J. Phys. Soc. Jpn. **53**, 4583 (1984)
443. J.A. Wilson, Phys. Rev. B **17**, 3880 (1978)
444. J.A. Wilson, F.J. DiSalvo, S. Mahajan, Adv. Phys. **24**, 117 (1975)
445. F. Levy (ed.), *Crystallography and Crystal Chemistry of Materials with Layered Structures* (Reidel Publishing Co., Dordrecht, 1976)
446. K. Motizuki (ed.), *Structural Phase Transitions in Layered Transition Metal Compounds in Physics and Chemistry of Materials with Low Dimensional Structure*. Series A, Layered Structures (D. Reidel Publishing Co., Dordrecht, Holland, 1986)

447. P. Hajiyev, C. Cong, C. Qiu, T. Yu, Sci. Rep. **3**, 2593 (2013)
448. A. Dal Corso, Surf. Sci, **637–638**, 106 (2015)
449. G. Brusdeylins, C. Heimlich, J.G. Skofronick, J.P. Toennies, R. Vollmer, G. Benedek, L. Miglio, Phys. Rev. B **41**, 5707 (1990)
450. G. Benedek, L. Miglio, G. Brusdeylins, C. Heimlich, J.G. Skofronick, J.P. Toennies, Europhys. Lett. **5**, 253 (1988)
451. J.A. Holy, M.V. Klein, L.W. McMillan, S.F. Meyer, Phys. Rev. Lett. **37**, 1145 (1976)
452. J.C. Tsang, J.E. Smith Jr., M.W. Schafer, Solid State Commun. **27**, 145 (1978)
453. G. Benedek, G. Brusdeylins, C. Heimlich, L. Miglio, J.G. Skofronick, J.P. Toennies, R. Vollmer, Phys. Rev. Lett. **60**, 1037 (1988)
454. N.V. Smith, M.M. Traum, Phys. Rev. B **11**, 2087 (1975)
455. R. Migoni, H. Bilz, D. Bäuerle, Phys. Rev. Lett. **37**, 1155 (1976)
456. R. Migoni, K.H. Rieder, K. Fischer, H. Bilz, Ferroelectrics **13**, 377 (1976)
457. H. Bilz, H. Buettner, A. Bussmann-Holder, W. Kress, U. Schroeder, Phys. Rev. Lett. **48**, 264 (1982)
458. G. Brusdeylins, C. Heimlich, J.G. Skofronick, J.P. Toennies, R. Vollmer, G. Benedek, Europhys. Lett. **9**, 563 (1989)
459. X.L. Wu, C.M. Liebor, Phys. Rev. Lett. **64**, 1150 (1990)
460. C.B. Scuby, P.M. Williams, G.S. Parry, Phil. Mag. **31**, 255 (1975)
461. G. Boato, P. Cantini, R. Colella, Phys. Rev. Lett. **24**, 1635 (1979)
462. P. Cantini, in *Dynamics of Gas-Surface Interaction*, ed. by G. Benedek, U. Valbusa (Springer, Berlin, 1982), p. 84
463. R. Bouwer, F. Jellinek, Phys. B **99**, 91 (1980)
464. R. Claessen, B. Burandt, H. Carstensenand, M. Skibowski, Phys. Rev. B **41**, 8270 (1990)
465. K.R.A. Ziebeck, B. Dorner, W.G. Stirling, R. Schöllhorn, J. Phys. **F7**, 1139 (1977)
466. G. Benedek, G. Brusdeylins, F. Hofmann, P. Ruggerone, J.P. Toennies, R. Vollmer, J.G. Skofronick, Surface Sci. **304**, 185 (1994)
467. G. Brusdeylins, J.P. Toennies, unpublished (the data, measured in 1987, have been reproduced in G. Benedek, L. Miglio, G. Seriani, in *Helium Atom Scattering from Surface*, ed. by E. Hulpke (Springer V., Heidelberg, 1992), p. 207)
468. B.M. Murphy, H. Requardt, J. Stettner, M. Müller, M. Krisch, W. Press, Phys. Rev. Lett. **95**, 256104 (2005)
469. B.M. Murphy, M. Müller, J. Stettner, H. Requardt, J. Serrano, M. Krisch, W. Press, J. Phys.: Condens. Matter **20**, 224001 (2008)
470. F. Weber, S. Rosenkranz, J.-P. Castellan, R. Osborn, R. Hott, R. Heid, K.-P. Bohnen, T. Egami, A.H. Said, D. Reznik, Phys. Rev. Lett. **107**, 107403 (2011)
471. F. Weber, R. Hott, R. Heid, K.-P. Bohnen, S. Rosenkranz, J.-P. Castellan, R. Osborn, A.H. Said, B.M. Leu, D. Reznik, Phys. Rev. B **87**, 245111 (2013)
472. M. Calandra, I.I. Mazin, F. Mauri, Phys. Rev. B **80**, 241108 R (2009)
473. G. Benedek, M. Miura, W. Kress, H. Bilz, Phys. Rev. Lett. **52**, 1907 (1984)
474. G. Benedek, M. Miura, W. Kress, H. Bilz, Surf. Sci. **148**, 107 (1984)
475. B.M. Murphy, J. Stettner, M. Traving, M. Sprung, I. Grotkopp, M. Müller, C.S. Oglesby, M. Tolan, W. Press, Phys. B **336**, 103 (2003)
476. G. Benedek, M. Bernasconi, D. Campi, J.P. Toennies, M.J. Verstraete, Surface phonons: theoretical methods and results. in *Springer Handbook of Surface Science*, ed. by M.A. Rocca, T. Rahman, L. Vattuone (Springer, 2019)
477. B. Monserrat, D. Vanderbilt, Phys. Rev. Lett. **117**, 226801 (2016)
478. R. Heid, I.Y. Sklyadneva, E.V. Chulkov, Sci. Rep. **7**, 1095 (2017)
479. A. Lock, J.P. Toennies, G. Witte, J. Electr. Spectr. Rel. Phenom. **54**(55), 309 (1990)
480. S. Miret-Artes, J.P. Toennies, G. Witte, Phys. Rev. B **54**, 5881 (1996)
481. B. Djafari-Rouhani, L. Dobryznski, G. Allan, Surf. Sci. **55**, 663 (1976)
482. G. Armand, P. Masri, Surf. Sci. **130**, 89 (1983)
483. J.E. Black, P. Bopp, Surf. Sci. **140**, 275 (1984)
484. J.E. Black, Surf. Sci. **278**, 229 (1992)

485. D. Neuhaus, F. Joo, B. Feuerbacher, Surf. Sci. Lett. **165**, L90 (1986)
486. P. Knipp, Phys. Rev. B **40**, 7993 (1989)
487. G. Witte, J. Braun, A. Lock, J.P. Toennies, Phys. Rev. B **52**, 2165 (1995)
488. H.J. Ernst, R. Folkerts, L. Schwenger, Phys. Rev. B **52**, 8461 (1995)
489. L. Niu, D.J. Gaspar, S.J. Sibener, Science **268**, 847 (1995)
490. L. Niu, D.D. Koleske, D.J. Gaspar, S.J. Sibener, J. Chem. Phys. **102**, 9077 (1995)
491. E.J. Mele, M.V. Pykhtin, Phys. Rev. Lett. **75**, 378 (1995)
492. G. Witte, J.P. Toennies, Phys. Rev. B **55**, 1395 (1997)
493. G.G. Rusina, A.V. Berch, I.Y. Sklyadneva, S.V. Eremeev, A.G. Lipnitskii, E.V. Chulkov, Fiz. Tverd. Tela **38**, 141 (1996)
494. G.G. Rusina, I.Y. Sklyadneva, E.V. Chulkov, Fiz. Tverd. Tela **38**, 1483 (1996)
495. S. Durukanoglu, A. Kara, T.S. Rahman, Phys. Rev. B **55** (1997)
496. I.Y. Sklyadneva, G.G. Rusina, E.V. Chulkov, Surf. Sci. **416**, 17 (1998)
497. F. Raouafi, C. Barreteau, M.C. Desjonquères, D. Spanjaard, Surf. Sci. **507**, 748 (2002)
498. A. Kara, S. Surukanoglu, T.S. Rahman, J. Chem. Phys. **106**, 2031 (1997)
499. A. Kara, T.S. Rahman, Surf. Sci. Rep. **56**, 159 (2005)
500. R. Claessen, R. Manzke, H. Carstensen, B. Burandt, T. Buslaps, M. Skibowski, J. Fink, Phys. Rev. B **39**, 7316 (1989)
501. D. Shechtman, I. Blech, D. Gratias, J.W. Cahn, Phys. Rev. Lett. **53**, 1951 (1984)
502. W. Steurer, T. Haibach, B. Zhang, S. Kek, R. Lück, Acta Crystallogr. B **49**, 661 (1993)
503. W. Steurer, T. Haibach, *Physical Properties of Quasicrystals*, ed. by Z.M. Stadnik (Springer, Berlin, 1999)
504. J.-B. Suck, M. Schreiber, P. Häussler (eds.), *Quasicrystals—An Introduction to Structure, Physical Properties and Applications* (Springer, Berlin, 2002)
505. P. Gille, R.-U. Barz, L.M. Zhang, in *Quasicrystals, Structure and Physical Properties* ed. by H.-R. Trebin (Wiley-VCH, Weinheim, 2003)
506. H.R. Sharma, K.J. Franke, W. Theis, P. Gille, P.h. Ebert, K.H. Rieder, Phys. Rev. B **68**, 054205 (2003)
507. W. Theis, H.R. Sharma, K.J. Franke, K.H. Rieder, Prog. Surf. Sci. **75**, 227–236 (2004)
508. A.A. Maradudin, Solid State Phys. **18**, 274 (1966); **19**, 2 (1966)
509. G. Benedek, G.F. Nardelli, J. Chem. Phys. **48**, 5242 (1968)
510. W. Steurer, A. Apfolter, M. Koch, W.E. Ernst, E. Søndergård, J.R. Manson, B. Holst, Phys. Rev. Lett. **99**, 035503 (2007)
511. W. Steurer, A. Apfolter, M. Koch, W.E. Ernst, E. Søndergård, J.R. Manson, B. Holst, Phys. Rev. Lett. **100**, 135504 (2008)
512. W. Steurer, A. Apfolter, M. Koch, W.E. Ernst, E. Søndergård, J.R. Manson, B. Holst, Phys. Rev. B **78**, 045427 (2008)
513. W. Steurer, A. Apfolter, M. Koch, W.E. Ernst, E. Søndergård, J.R. Manson, B. Holst, J. Phys.: Condens. Matter **20**, 224003 (2008)
514. O. Pilla, S. Caponi, A. Fontana, M. Montagna, L. Righetti, F. Rossi, G. Viliani, G. Ruocco, G. Monaco, F. Sette, R. Verbeni, G. Cicognani, A.J. Dianoux, Phil. Mag. B **82**, 223 (2002)
515. T. Scopigno, W. Steurer, S.N. Yannopoulos, A. Chrissanthopoulos, M. Krisch, G. Ruocco, T. Wagner, Nat. Comm. **2**, 195 (2011)
516. V.L. Gurevich, D.A. Parshin, H.R. Schober, JEPT Lett. **76**, 553 (2002)
517. A.I. Chumakov, G. Monaco, A. Monaco, W.A. Crichton, A. Bosak, R. Rüffer, A. Meyer, F. Kargl, L. Comez, D. Fioretto, H. Giefers, S. Roitsch, G. Wortmann, M. Manghnani, A. Hushur, Q. Williams, J. Balogh, K. Parliński, P. Jochym, P. Piekarz, Phys. Rev. Lett. **106**, 225501 (2011)
518. R. Zorn, Physics **4**, 44 (2011)
519. W. Steurer, A. Apfolter, M. Koch, W.E. Ernst, B. Holst, E. Søndergård, S.C. Parker, Phys. Rev. B **78**, 035402 (2008)
520. F. Sette, M.H. Krisch, C. Masciovecchio, G. Ruocco, G. Monaco, Science **280**, 1550 (1998)
521. T.S. Grigera[1], V. Martín-Mayor, G. Parisi, P. Verrocchio, Nature **422**, 289 (2003)
522. P.T. Truant, J.P. Carbotte, Phys. Rev. B **6**, 3642 (1972)

523. B.N. Brockhouse, K.R. Rao, A.D.B. Woods, Phys. Rev. Lett. **7**, 93 (1961)
524. M. El-Batanouny, S. Burdick, K.M. Martini, P. Stancioff, Phys. Rev. Lett. **58**, 2762 (1987)
525. Y.W. Wang, L.J. Fan, Langmuir **18**, 1157 (2002)
526. J.P. Toennies, in *Solvay Conference on Surface Science*, ed. by F.W. de Wette, Springer Ser. Surf. Sci. vol. 14 (Springer, Berlin, Heidelberg, 1989)
527. A. Euceda, D.M. Bylander, L. Kleinman, Phys. Rev. B **28**, 528 (1983)
528. F. Lindemann, Phys. Z. **11**, 609 (1910)
529. M. Posternak, A. Baldereschi, A.J. Freeman, E. Wimmer, Phys. Rev. Lett. **52**, 86 (1984)
530. M.O.D. Camara, A. Mauger, I. Devos, Phys. Rev. B **65**, 125206 (2002)
531. R.J. Nemanich, G. Lucovsky, S.A. Solin, Mater. Sci. Eng. **31**, 157 (1977)
532. T. Aizawa, R. Souda, Y. Ishizawa, H. Hirano, T. Yamada, K. Tanaka, Surf. Sci. **237**, 194 (1990)
533. C. Adler, P. Pevone, U. Schröder, Comp. Mat. Sci. **20**, 371 (2001)
534. A. Al Taleb, Dissertation , Universidad Autonoma De Madrid, 2017
535. J. Perdereau, J.P. Biberian, G.E. Rhead, J. Phys. F **4**, 798 (1974)
536. M.A. van Hove, R.J. Koestner, P.C. Stair, J.P. Bibérian, L.L. Kesmodel, I. Bartoš, G.A. Somorjai, Surf. Sci. **103**, 189 (1981)
537. J.C. Heyraud, J.J. Métois, J.J. Surf. Sci. **100**, 519 (1980)
538. Y. Tanishiro, H. Kanamori, K. Takayanagi, K .Yagi, G. Honjo, Surf. Sci. **111**, 395 (1981)
539. L.D. Marks, V. Heine, D.J. Smith, Phys. Rev. Lett., **52**, 656 (1984)
540. S.B. Darling, A.W. Rosenbaum, Yi Wang, S.J. Sibener, Langmuir **18**, 7462 (2002)
541. J.S. Nelson, M.S. Daw, E.C. Sowa, Phys. Rev. B **40**, 1465 (1989)
542. A. Tamtögl, P. Kraus, N. Avidor, M. Bremholm, E.M.J. Hedegaard, Bo B. Iversen, M. Bianchi, Ph. Hofmann, J. Ellis, W. Allison, G. Benedek, W. Ernst, Phys. Rev. B **95**, 195401 (2017)
543. A. Tamtögl, D. Campi, M. Bremholm, E.M.J. Hedegaard, B.B. Iversen, M. Bianchi, Ph. Hofmann, N. Marzari, G. Benedek, J. Ellis, W. Allison, arXiv 1803:11184v1 [cond.matrl. sci] (2018)

Chapter 12
Experimental Results: Vibrations of Adsorbates and Thin Films

It is true that we have really in Flatland
a Third unrecognized Dimension called 'height'.

(Edwin A. Abbott)

Abstract This chapter reviews some exemplary high resolution HAS experiments on surface phonon dispersion curves of adsorbate atoms and molecules either well dispersed or in layers on single crystal surfaces. Here also HAS provides unique information on the low energy vibrations with respect to the surface as well as within the layers. HAS experiments have also been used to follow the evolution of the phonon dynamics as the coverage is increased to produce one or more monolayers. These experiments are first illustrated for rare gases and simple molecules. The alkali metals show a number of new phenomena, including so-called organ-pipe modes which result from the confinement provided by the interface at the substrate. In the classical macroscopic realm the organ-pipe modes are known as Sezawa modes. The HAS dispersion curve measurements of thin films of the superconductor Pb on Cu(111) show a surprising plethora of modes. Recent DFPT and EAM calculations demonstrate that the dynamics throughout the entire film can only be understood by taking account of the strong electron-phonon coupling, which is discussed in detail in Sect. 8.4. Thus modes deep within the film and even modes localized at the interface, known as Stonely waves in the classical limit, make themselves noticed at the surface by a mechanism which is called the quantum sonar effect. HAS has also been used to study thin films of polyatomic organic molecules physisorbed on metals and also self-assembled chemisorbed monolayers. Interesting changes accompany the temperature induced transition from physisorption to chemisorption. The complex phonon dispersion curves of methyl radicals chemisorbed on silicon have recently been understood with the aid of DFPT calculations. The chapter closes with a brief review of HAS prospects for studying one-dimensional chains. Each section is meant as an introduction to the specific phenomena and issues characteristic to each of these different surface types, with special consideration to the questions solved and the new theoretical issues raised. This chapter is complemented by appendices which contain a comprehensive list (up to 2016) of the HAS experimental data on adsorbates on insulators and

G. Benedek and J. P. Toennies, *Atomic Scale Dynamics at Surfaces*, Springer Series in Surface Sciences 63, https://doi.org/10.1007/978-3-662-56443-1_12

semiconductors (Appendix A.5), atomic adsorbates on metals (Appendix A.6), diatomic molecules on metals (Appendix A.7) and polyatomic molecules on metals (Appendix A.8).

12.1 Introduction

Although this monograph is largely devoted to phonons at the surface of a single crystal, many of the methods both experimental and theoretical can equally well be applied to the vibrations of thin films of physisorbed and chemisorbed atoms and molecules with respect to the surface even at submonolayer coverages. The spectroscopy of these vibrations presently provide the most direct information on the potential acting between the molecule and the substrate as well as on the dynamical coupling with the substrate.

As illustrated in Fig. 12.1 for CO the vibrational modes localized inside an absorbed diatomic molecule are called *internal* or *intrinsic* modes. The remaining modes are called *external* or *extrinsic* modes and will be the main subject of this chapter. The external modes can be conveniently divided into three groups. The modes with the highest frequencies are the stretch modes and are frequently denoted as the S-mode or FT_z-mode. The next in frequency are the frustrated rotation modes designated as R-modes, which are sometimes referred to as librations. Generally the lowest frequency modes are the T-mode, or FT_x-mode. They correspond to a frustrated or hindered translation along the surface. These external modes are particularly sensitive to the potential binding the molecule to the surface, especially at low coverages. With increasing coverage the R- and the T-modes are replaced by the collective modes of the layer in which the intralayer interactions become important. The intralayer interactions lead to a dispersion of the extrinsic

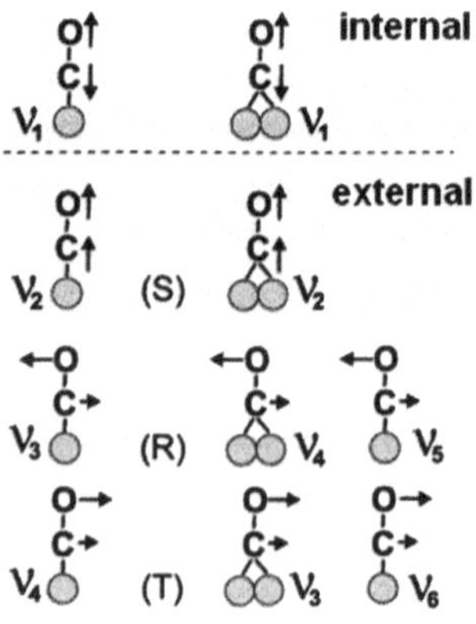

Fig. 12.1 Normal modes of a CO molecule adsorbed at an on-top (left column) and at a bridge site (right side). The frustrated rotation and the frustrated translation modes, R and T, are doubly degenerate at high-symmetry on-top or hollow sites, whereas at the bridge site their degeneracy is lifted (adapted from [1])

vibrations and, of course, in competition with the interactions with the substrate determine the layer structure. With increasing thickness these modes go over to the "half-bulk" or surface modes discussed in the preceding chapters (see e.g. Fig. 3.10).

The internal modes and the external S- and R-modes usually have energies greater than about 30 meV and have been extensively investigated by EELS [2, 3], infrared spectroscopy [3, 4] and with additional optical methods such as Raman spectroscopy sum frequency generation [4] and scanning tunnelling spectroscopy (STS) of single molecules [5–12]. The T-mode has been largely inaccessible to these methods because of its low energy and unfavourable polarisation and was observed for the first time with HAS in 1986. At that time Lahee, Toennies and Wöll [13] observed a low energy dispersionless vibrational mode at about 6 meV with HAS from randomly absorbed CO molecules covering only 5% of the Pt(111) surface. In the following years a large number of systems were studied mostly by the Göttingen group. In 1996 Hofmann and Toennies reviewed and tabulated the HAS T-mode results for more than 40 different molecule/metal surface combinations including the following molecules: CO, O_2, N_2, NO, CO_2, NH_3, CH_4, C_2H_4, C_6H_6, C_2H_6, C_6H_{14}, C_8,H_8 and $C_{10}H_{22}$ and 12 different molecule/insulator surfaces [14]. Measurements were reported both for isolated molecules and also for different ordered adsorbate submonolayer structures. In most cases, and even for monolayers with well-defined structures, the collective modes exhibit only a weak or negligible dispersion. Some notable exceptions are CO c(2×2) on Cu(001) [15] and CO_2 on NaCl(001) [16, 17] for which the force constants for the lateral coupling of the molecules were derived. These early HAS studies of the vibrations of adsorbates on metal surfaces were later reviewed by Witte [1] in 2002 and in a comprehensive review by Graham [18] in 2003. Appendix A (Tables A.5, A.6, A.7 and A.8) and a recent review [19] from 2015 provide an up-to-date list of the HAS studies of isolated adsorbates as well as of thin films. In addition to the T-mode, HAS results are now available for the S-modes and, for a few systems, also for the R-modes and, in the case of monolayers, for the vertical polarized Rayleigh waves as well as for collective longitudinal phonons. For each adsorbate/substrate combination the corresponding references to HAS data are found in the last column of Tables A.5, A.6, A.7 and A.8. For more general reviews on adsorbates on surfaces the reader is referred to, the book by Bruch, Cole and Zaremba [20] and for a review specifically on physisorption to the 2007 Review of Modern Physics article by Bruch, Diehl and Venables [21].

The discussion of these and more recent results are covered in the following sections. In the first section the studies of submonolayer coverages are illustrated by some few exemplary experiments. The following sections are devoted to the vibrations and phonons of single layer and thicker monoatomic films, while the HAS studies on organic films on metal and semiconductor surfaces are presented in the last two sections.

12.2 Vibrations of Submonolayer Adsorbates

12.2.1 CO Molecules on Metal Surfaces

Of the three external modes the low frequency, T-modes are usually the only ones which are thermally excited at ordinary temperatures. For this reason they are the only ones that play an important role in determining the entropy and thereby the thermodynamic properties of the layers [22]. Since the T-mode energy $\hbar\omega_T$ is rather low and falls in the band of the long wave bulk transverse acoustic (TA) modes, they are the only ones that can couple facilly to the T-modes. A strong resonance interaction will occur for T-modes of small parallel momentum with those bulk TA modes which have a large normal wavevector component k_z, since their amplitudes will also have a component parallel to the surface [23, 24]. Thus energy acquired by the T-modes is readily passed on to the bulk. Also because of their low frequencies the T-modes are usually the only ones which are readily excited by impinging gas phase atoms and molecules at ordinary temperatures [25]. For this reason they are usually the most important energy gateway for transferring energy from the gas phase into the adsorbed phase and ultimately into the substrate bulk [26, 27]. This also explains why helium atoms are ideal for probing the T-modes.

Figure 12.2 shows HAS measurements of the T-mode linewidth for CO on Cu (001) [18, 28], which reveal the coupling between the T-modes and the substrate. The regular increase with increasing temperature (Fig. 12.2c) is indicative of a fourth-order anharmonicity. This is expected for the T-mode when the potential energy is a symmetric function with respect to a displacement parallel to the surface (as is often the case) and the corresponding third-order anharmonicity is suppressed. From an extrapolation of the linewidth to 0 K a relaxation time of 8 psec could be determined (Fig. 12.2d) suggesting that the coupling to the substrate phonons is the dominant dissipative mechanism [28].

As for the free molecules isotope shifts provide valuable signatures for identifying the nature of the low frequency mode. This was demonstrated for the benchmark system CO at on-top sites on Cu(001). Compared to $^{12}C^{16}O$ the following isotope shifts of the T-mode were reported for $^{13}C^{16}O$ (−0.75%), $^{12}C^{18}O$ (−4.5%) and $^{13}C^{18}O$ (−5.0%) [30]. These T-mode isotope shifts found with HAS were subsequently confirmed with EELS [31]. The shifts for the same set of isotopes, measured with infrared reflection adsorption spectroscopy (IRAS) are for the R-mode, −3%, −1% and −3.9%, and for the S-mode, −1%, −2.1% and −0.7%, respectively [30]. This data combined with QHAS diffusion measurements (Sect. 13.2) was used to construct a multidimensional potential along the [110] direction describing the binding energy as a function of position above the surface and the tilt angle of the CO molecule [14, 30]. The potential for the same system calculated with density functional theory was found to be in satisfactory agreement with the empirical potential [33]. The potential surfaces of CO on both Cu(001) and

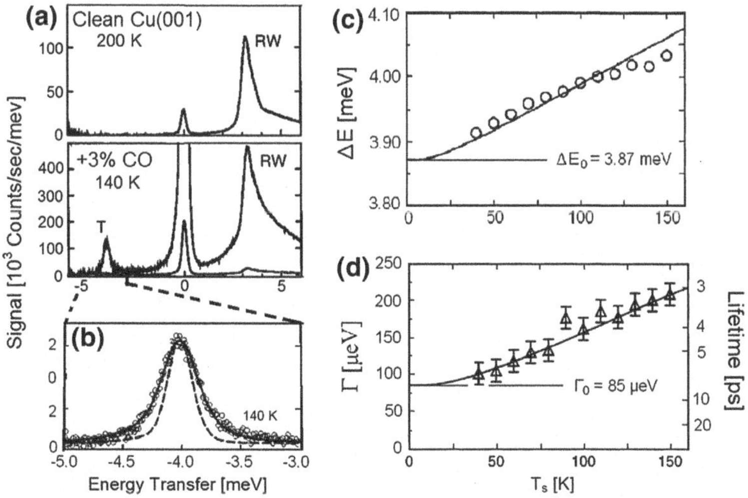

Fig. 12.2 HAS measurements for the excitation of the T-mode of 3% ML of CO molecules physisorbed on Cu(001) at an incident energy of 11.2 meV along the [100] direction of the substrate. **a** Energy transfer spectrum of the clean and absorbate covered surface showing the new peak attributed to the T mode of CO. **b** Enlarged view of the T mode with the instrumental width (dashed line curve) and a best fit of the data (solid line curve). **c** Temperature dependence of the T-mode energy transfer showing the extrapolated value for $T = 0$. **d** Temperature dependence of the peak linewidth, after subtraction of the instrumental width also showing the extrapolation to $T_s = 0$ (adapted from [28])

Cu(111) have also been studied using diffusion data from helium spin echo spectroscopy [34]. (see Sect.13.2.2).

In the initial study of HAS from low coverages of CO molecules on low index metal surfaces an interesting interference pattern was observed in the angular distributions [35, 36]. This phenomenon was attributed to the reflection of the incident He wave packet from the isolated and randomly distributed molecules each of which acts as a prominent scattering center on the otherwise very smooth low index metal surfaces. Choi, Graham, Tang and Toennies [37–40] later showed that the inelastic cross section can be determined from the measured elastic cross-section via a simple transformation. This theory, relating the elastic angular distribution to the inelastic angular distribution via a form factor, predicts not only the inelastic interference structures and their correlations with the elastic channel, but also the individual state-to-state transitions.

12.2.2 Na on Metal Surfaces

In recent years there has been very extensive activity in studying and understanding thin adsorbate layers of alkali atoms on surfaces, largely motivated by their being relatively simple model systems. The experimental and theoretical developments up to 2013 have been comprehensively reviewed by Politano et al. [41].

The early pioneering surface investigations of Langmuir and coworkers in the 1910s established that alkali atom adsorption is accompanied by a large reduction in the work function of several eV [42]. This is attributed to the positive charge on the alkali atoms resulting from the partial transfer of electrons from the adsorbate atoms to the metal surface. Together with the image charge below the surface, strong dipole moments are formed which are oriented normal to the surface. The resulting repulsion between the adsorbates has been invoked to explain the rather well defined ring structures, very similar to a Wigner lattice, which are characteristic of alkali atom films at low coverages ($\theta < 0.1$ ML) [42, 43]. Electron scattering was first used to observe the resulting ring structures in diffraction, but with HAS it has been possible to detect these structures in the angular distributions for Na coverages as low as 0.008 ML [44, 45]. Figure 12.3 shows the ring structure observed at a coverage of 0.048 and 0.125 ML Na on Cu(001). At the higher coverage the Na atoms form an ordered c($2\sqrt{2}\times4\sqrt{2}$) R45° layer [46]. The average distance between atoms from the diffraction pattern varies from 32Å at 0.008ML to about 8 Å at a 0.125 ML coverage (see inset of Fig. 12.4b).

Figure 12.4 shows HAS results for the T-mode of Na atoms at a surface temperature of $T_s = 50$ K. The TOF spectra at low coverages (Fig. 12.4a) are

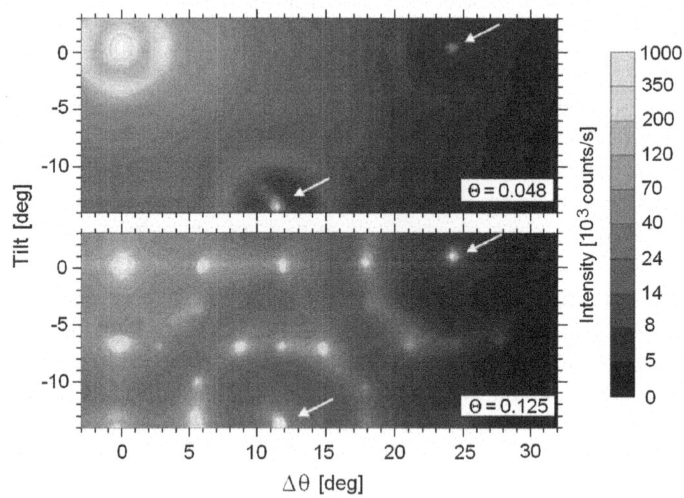

Fig. 12.3 A two-dimensional grey scale intensity plot obtained from HAS diffraction patterns of Na atoms on Cu(001) at $T = 50$ K for two different coverages $\Theta = 0.048$ (above) and 0.125 (below) as a function of scattering angle away from the specular and tilt angles [44, 45]. The rings of spots around the specular peak (upper left corner) and diffraction peaks of the substrate (arrows) correspond to ordered quasi-hexagonal adsorbate layers with two possible equivalent orientations. The substrate diffraction peak intensities grow with coverage, indicating that the Na atoms also induce a surface corrugation of the substrate. The diffraction peaks of the clean Cu(001) surface are too weak to be discernible on the gray scale of these pictures

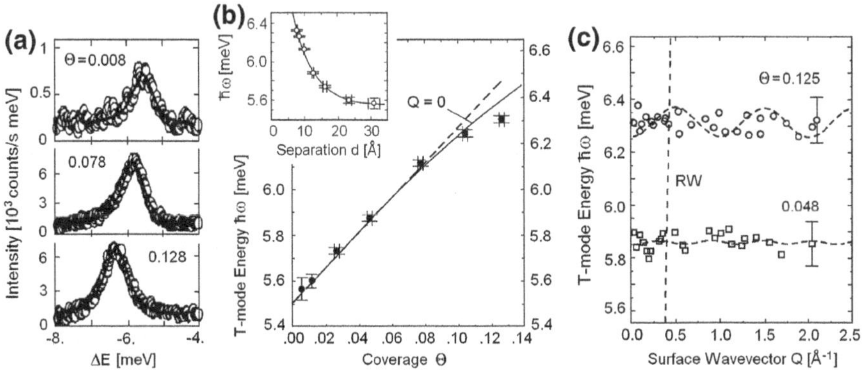

Fig. 12.4 HAS investigations of the T-mode of Na atoms on Cu(001) at a surface temperature $T_s = 50$ K for different coverages varying from 0.008 to 0.078 ML. **a** Energy transfer spectrum at 3 different coverages. **b** T-mode energy as a function of coverage from measurements over a wide range of temperatures T_s extrapolated to 0 K. The nearly linear behaviour of the T-mode energy is well reproduced by a calculation based on a model with effective long range forces originating from a charge transfer into intrinsic surface states. The inset shows the dependence of the T-mode energy on the sodium interatomic distance calculated from the coverage. **c** T-mode dispersion curves for sodium coverages of 0.048 and 0.125 ML measured along the [100] direction for an incident beam energy of 20 meV and a surface temperature of 50 K. For comparison the dashed nearly vertical line shows the dispersion of the substrate Rayleigh wave (RW) dispersion. The two broken line curves for 0.125 and 0.048 ML show the theoretical weak dispersion of the T-mode due to dipole-dipole interaction (adapted from [45])

dominated by an intense inelastic peak due to the frustrated translational (T) mode [45, 47]. With increasing coverage the frequency increases from 5.55 to 6.30 meV (Fig. 12.4a, b) [45]. This shift was found to be much larger than predicted by a dipole-dipole interaction. To rule out some other type of direct lateral interaction within the layer the dependence of the frequency shift on wavevector was also investigated. As shown in Fig. 12.4c no dispersion was found above the experimental error. Since the vibrations at the zone center correspond to a lateral translation of the entire overlayer without a change of interatomic distances the lack of any dispersion implies that the lateral bonding is collectively affected and that the intraplanar forces are not of significance at these coverages. Similar results on the dispersion at similar low coverages were also found for K on Cu(001), but in the case of Cs a noticeable dispersion was observed [47]. In Na, K and Cs the amplitude of the dispersion curves was found to increase with the coverage, which was attributed to the corrugation of the substrate [47], except for Cs, where the electrostatic forces between the adsorbates had to be invoked.

Although the dipole-dipole interaction is responsible for the structural ordering it cannot explain the coverage dependence of the T-mode which must arise from some other interaction such as a global modification of the dipole substrate coupling [45]. The frequency increase and the lack of dispersion were found [45] to be a direct consequence of Lau-Kohn long range surface mediated interactions, which are mediated by partially filled surface states [48, 49]. Upon adsorption of a Na atom its valence electron is transferred into the Cu(001) surface state which forms four pockets just below the Fermi level around the four equivalent $\overline{\mathrm{X}}$ points of the surface Brillouin zone [50]. This additional charge is trapped at the $\overline{\mathrm{X}}$-point pockets and leads to a corrugation with the same surface lattice periodicity. This in turn increases the potential curvature, and the Na atom T-mode force constant at the hollow site. This explains why the *substrate* diffraction peaks and the corrugation, which are very small for the clean surface (see Sect. 11.2), increase with coverage (Fig. 12.3, arrows). In this special case the Friedel oscillations, which are responsible for the long range Lau-Kohn interaction [48, 49] have a wavevector $2k_F \approx G$. Thus they are commensurate with and locked-into the lattice, and consequently the Lau-Kohn interaction does not contribute to the dispersion. In conclusion, although Lau-Kohn forces had been already detected from the analysis of adatom distributions [51–53], the study of the T-mode frequency as a function of coverage has provided the first evidence of their key role in the dynamics of low coverage adsorbates on metals.

12.3 Monolayer and Thicker Films

In the last few decades atomically thin films (hereafter called ultrathin (UT) films) have attracted increasing attention in connection with the nanotechnology revolution and the growing importance of sub micron-sized devices in the electronics industry [54]. Today it is straightforward to prepare thin films under ultrahigh vacuum conditions [55], and their structural characterization with atom beam diffraction proves to be an efficient tool complementary to LEED and STM [56].

Already about 1970 the future importance of finite size effects in epitaxial UT-films was anticipated by several surface lattice dynamical calculations based on force constant models for slab-shaped crystals, without [57–60] or with [61] overlayers. This development was accompanied by specific theoretical GF studies for free [62] and supported [63] films. Because of the large acoustic impedance mismatch with respect to the substrate UT-films offer a unique opportunity for single phonon studies in a (quasi) two-dimensional system. In addition to the inherent simplification in interpretation, two-dimensional systems are also expected to exhibit new phenomena suppressed by the half-space geometry. For example, the reduced dimensionality enhances the Kohn anomalies [64, 65]. Moreover, new quantum size effects [66, 67] are expected which should show up in comparisons of films with different layer thickness. For this reason it is perhaps not surprising that many of the theoretical concepts and experimental techniques developed for studies of clean surfaces have found fruitful application to the study of the lattice vibrations of UT-films.

Elastic wave propagation in thin films has been extensively studied in the continuum limit [68]. Two different types of coupling between the layer and substrate are expected depending on the ratio β of the velocities of transverse acoustic waves for the layer ($\hat{v}_t$) and substrate (v_t) materials [69],

$$\beta = \frac{\hat{v}_t}{v_t} = \left(\frac{\hat{c}_{44}\rho}{c_{44}\hat{\rho}}\right)^{1/2}, \tag{12.1}$$

where c_{44} and ρ are the shear elastic constant and the mass density of the medium and capped symbols refer to the layer material. For $\beta > \sqrt{2}$ the Rayleigh wave velocity of the covered substrate exceeds that of the clean material, and the substrate surface is said to be *stiffened.* For $\beta < 1/\sqrt{2}$ the Rayleigh wave velocity of the covered substrate is decreased with respect to the clean surface, which in this case is said to be *loaded* [69]. The latter condition has been met in most of the EELS and HAS studies. The threshold values for β provide sufficient but not necessary conditions for the stiffening or loading effects. They are only valid for isotropic materials and a RW wavelength much larger than the layer thickness. For $1/\sqrt{2} < \beta < \sqrt{2}$ special combinations of the material constants may yield a change of the free surface RW velocity as well as the appearance of a sagittal interface mode called the Stoneley wave [69].

Another related condition determining the extent of coupling between the layer vibrations and the substrate is the mismatch in the acoustic impedance Z,

$$\delta = \frac{\hat{Z}}{Z} - 1 \tag{12.2}$$

where $Z = \rho\, v_\ell$, in which v_ℓ is the substrate longitudinal velocity in a direction normal to the surface, and $\hat{Z} = \hat{\rho}\hat{v}_\ell$ refers to the layer material [70]. For large negative values of δ ($-1 < \delta < -0.8$) the acoustic waves travelling in the layer are reflected at the interface and the coupling to the substrate can be largely neglected. In this case the substrate can be approximated as being rigid [23].

In microscopic terms loading ($\beta < 1/\sqrt{2}$) and a large acoustic impedance mismatch, i.e., a weak dynamic coupling to the substrate, correspond to the situation in which the phonon frequencies of the film at sufficiently small wavevectors lie below those of the substrate. For example, in the case of UT- films of sodium atoms on a copper surface the sodium dispersion curves only intersect those of the substrate near the zone origin. At larger wavevectors with increasing frequency of the dispersion curve the film cannot interact with the substrate and the film is effectively isolated and free standing. The coupling at small wavevectors is not a serious problem since at the corresponding long wavelengths little information is expected concerning the microscopic interactions anyway.

The electronic coupling of the film to the substrate is a more complex problem [20], but at least for some metal systems it may only be of secondary importance [72]. The interaction potential of the He atoms is, however, somewhat modified by the presence of neighboring film atoms (*polarization mediated forces*) [71]. In addition there will be a weak interaction with the substrate through the film depending on the film thickness. As discussed by Gumhalter et al. [68, 73], for closed shell atom monolayers physisorbed on metals or insulators the polarization mediated forces contribute a positive (repulsive) correction to the potential [20], whereas the correction due to the He atom-substrate interaction is negative (McLachlan attractive contribution [74]). An analysis of the highly resolved phonons of Ar, Kr and Xe monolayers on Pt(111) measured by HAS, discussed in the next Section, indicates that the MacLachlan corrections are not important due to an approximate cancelation with the repulsive three-body terms [75, 76] though they can be of significance for understanding the lattice constants [20].

12.3.1 Rare-Gas Ultrathin Films

The thin films of the inert gases, especially Ne, Ar, Kr and Xe are still perhaps the most important model systems (see Tables A.5 and A.6 in the Appendix). They were also one of the first systems to be investigated by HAS. Two seminal experiments were reported by Mason and Williams in 1984 [77] and by Gibson and Sibener in 1985 [78]. The inert gases have the big advantage that the forces can be calculated by assuming the addition of pairwise interactions since the relatively weak three body effects can usually be neglected [79]. Moreover the inert gas atom-atom interactions in the gas phase have been extensively studied and are well characterized [80]. The structural studies of inert gas thin films have been surveyed in the review and book by Bruch et al. [21, 71]. In the 1984 inelastic HAS experiments energy loss spectra were observed for the inert gases and a variety of polyatomic molecules including CO_2, H_2O and C_2H_6 on Ag (110) single crystal surfaces [77]. Figure 12.5 reproduces the data of Gibson and Sibener for Kr on Ag (111), which nicely shows the evolution of the flat branch observed for one monolayer (1 ML) into a dispersed branch for 25 ML [78]. As expected from the discussion in Chap. 3 and Fig. 3.10, the latter branch is seen to be practically

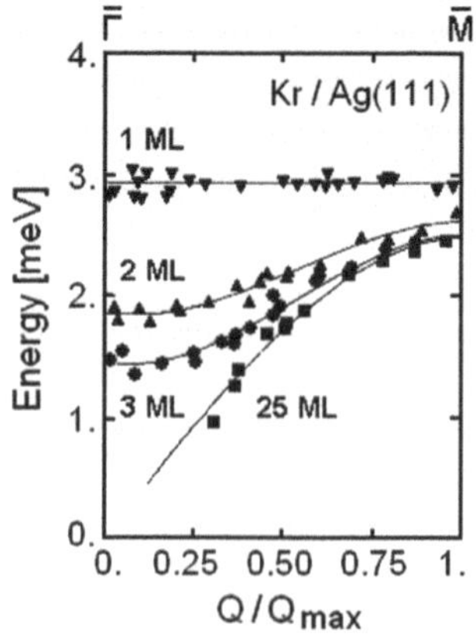

Fig. 12.5 Dispersion curves of the vertically polarized S-mode for 1, 2, 3 and 25 krypton monolayers (ML) on the Ag(111) surface in the $\overline{\Gamma M}$ direction from HAS measurements (adapted from Gibson and Sibener [78]). For thick films the S-mode branch evolves into the Rayleigh branch of the semi-infinite Kr crystal

identical to the RW branch for the semi-infinite crystal. For this reason the flat branch for one monolayer was attributed to the vertical polarized S-mode.

These and subsequent studies of physisorbed rare gas monolayers (see Table A.5 in the Appendix and [18, 81]) only provided evidence for the dispersionless S-mode. The lack of any evidence for modes of longitudinal (L) polarization caused some speculation concerning the strength of the weak lateral van der Waals interactions. New interest was aroused by the HAS experiments of Zeppenfeld et al. [82, 83] on a high order commensurate monolayer of Xe on Cu(110), which exhibited a new, nearly dispersionless mode at ≈3 meV at large wavevectors ($Q > 0.5$ Å^{-1}) which was attributed to a longitudinal mode. Figure 12.6 shows the dispersive longitudinal mode for Xe monolayers on Cu(001) and on Cu(111) which extends out to the Brillouin zone boundary of the hexagonal layer [84–86]. A lattice dynamical analysis assuming a flat rigid non-vibrating substrate indicated that the observed frequencies could only be explained by an effective lateral force constant in the Xe layers that was only 25% of that expected for the gas phase potential [86]. Subsequent full lattice dynamical calculations based on a simple force constant model by Gumhalter [68] for a ($\sqrt{3}\times\sqrt{3}$)R30° overlayer of Xe on a 40-layer Cu (111) slab and earlier work by Halls, Mills and Black [87] based on the gas phase Xe-Xe HFD-B2 potential [88–90] confirmed the extensive softening. Very similar TOF spectra and dispersion curves exhibiting very similar dispersion curves as in Fig. 12.6 have also been found for Xe monolayers on NaCl (001) [91].

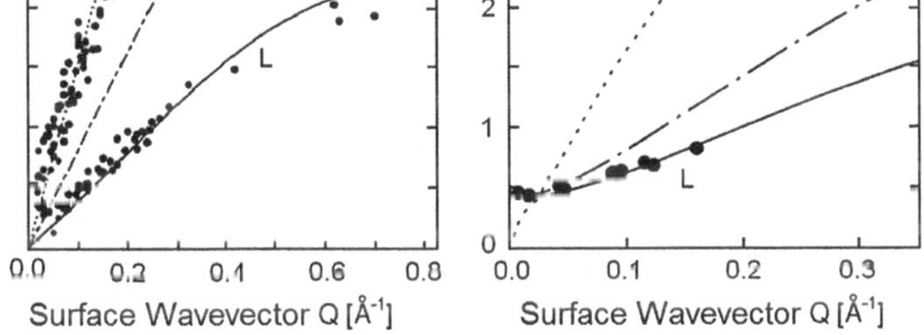

Fig. 12.6 Comparison of HAS phonon dispersion curves (filled circle) for Xe monolayers on Cu (001) and Cu(111) with theory. **a** For a high order commensurate Xe monolayer on Cu(001) along the [100] direction [84, 85, 92]. **b** For a Xe ($\sqrt{3}\times\sqrt{3}$)R30° monolayer on Cu(111) along the [11$\bar{2}$] direction [86, 92]. The dashed and full lines are theoretical best fits of vertically polarized (S) and longitudinal (L) modes of the monolayer, respectively, whereas the dashed-dotted lines are the L-branches calculated for the gas Xe-Xe potential [68]. For comparison the substrate RW branch is also shown for both surfaces (dotted lines)

As shown in Fig. 12.7 calculations based on the exponentiated Born approximation (EBA) [93–95] in the single phonon as well as in the multiphonon regimes for a Xe monolayer on Cu(111) and Cu(001) (see Sect. 13.1) reproduce the measured spectra [92, 96]. Gumhalter [68] attributed the strong softening to the wave functions of surface electronic states which extend into the region of the Xe adatoms [97].

The assignment of a longitudinal polarization to the lowest energy overlayer branch was subsequently questioned by Bruch [98], who noted that if the new mode is assigned to the SH mode, then the force constant is comparable to that for the gas phase potential and no unphysical softening would be required. This led to a new dedicated HAS study of Ar, Kr, and Xe on Pt(111) [99–103]. The diffraction patterns for Ar on Pt(111) reveal a nearly commensurate $(\sqrt{3}\times\sqrt{3})R30°$ monolayer, whereas for Kr and Xe the hexagonal close packed monolayers are rotated by 5.3° and 2.6°, respectively, with respect to the substrate. The lowest frequency dispersive mode could only be detected at low incident beam energies of about 8 meV. By assigning this lowest energy dispersive mode to the SH mode and the higher one to the longitudinal film mode, good agreement with the gas phase atom-atom force constants was achieved. Figure 12.8 illustrates these measurements for Kr/Pt(111) and shows a comparison of the measured dispersion curves with calculated dispersion curves, for the atom-atom gas phase potential. The substrate-mediated McLachlan forces lead to a very small correction in the frequencies of about 1% [73, 100, 101]. Subsequently Hansen and Bruch in wave packet calculations (Xe on Pt(111), found that the probability for exciting the SH mode increased for incident beam energies decreasing from 8 to 4 meV, which is still about 2 meV above the threshold. Presently all the modes of the inert gas

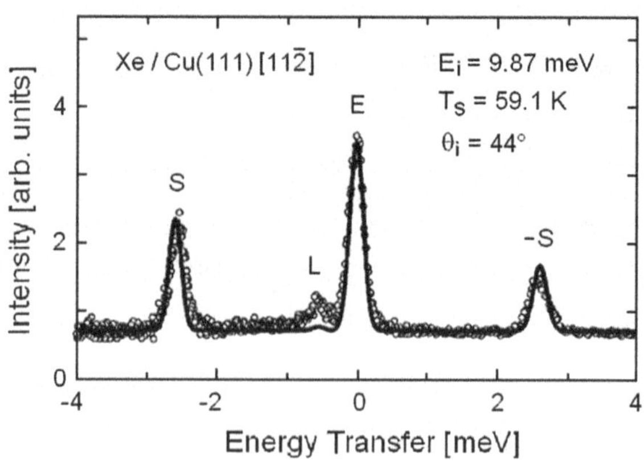

Fig. 12.7 Inelastic HAS spectra (filled circle) from a Xe $(\sqrt{3}\times\sqrt{3})R30°$ monolayer on Cu(111) along the $[11\bar{2}]$ direction measured with a low incident energy E_i = 9.87 meV for a surface temperature T_s = 59.1 K [86]. Distinct single phonon creation (S) and annihilation processes (-S) involving the S-mode and an L-mode creation process are observed. The heavy continuous line curve is from theory [92–95]

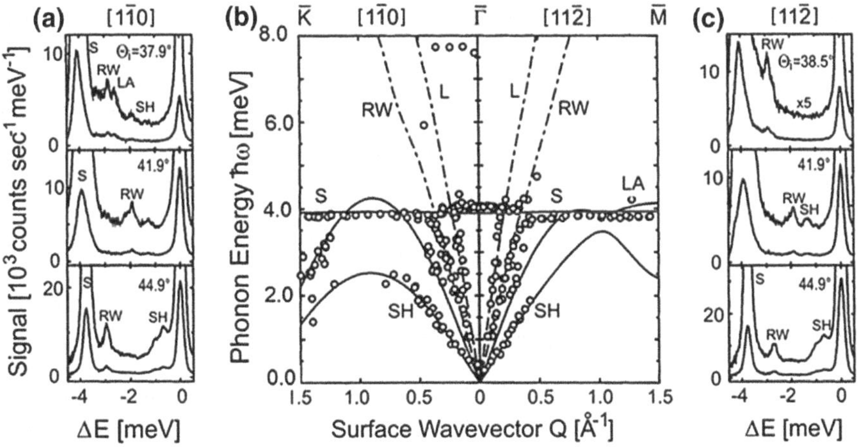

Fig. 12.8 HAS energy loss spectra and the dispersion curves for a monolayer of Kr on Pt(111) measured with a low beam energy of 8.4 meV and a surface temperature of T_s = 50 K [100]. **a** Representative energy loss spectra for the [1$\bar{1}$0] azimuth. **c** The same for the [11$\bar{2}$] azimuth. **b** The experimental dispersion curves (open circles) and best fits (solid line curves) based on gas phase potentials [100] and the substrate modes (dashed-dot line curves)

monolayers and their frequencies are in good agreement with predictions based on gas phase potentials. This being the case, the new results raised however serious questions concerning the collisional coupling mechanism responsible for the observed excitation of the "forbidden" SH modes [100–103].

In a subsequent theoretical analysis based on a time dependent wave packet formulation Bruch and Hansen [101] have shown that the excitation of the SH is made possible, with an intensity comparable to or larger than that of the L mode, by even a small misalignment of the scattering plane with respect to the symmetry directions of the surface, provided the incident energy of the He beam is sufficiently low (below 10 meV). The calculations for a Xe monolayer on Pt(111) [102] extended to much longer propagation times, so as to enable a treatment of transiently trapped helium atoms, predict a strong enhancement of the inelastic intensity from SH modes for incident energies as low as 5 meV. This analysis, however, also does not seem to be entirely conclusive. It may be argued that such small misalignments should also affect all previous measurements for clean surfaces, where surface phonon dispersion curves are well known from theory, and the symmetry forbidden SH modes have invariably not been observed. Very recent density functional calculations of the van der Waals interaction of Xe adatoms on the Cu(111) and Pt(111) surfaces by Lazić et al. [104] now appear to have solved the puzzle. They confirm the original conjecture of a strong reduction of the Xe-Xe force constant in the adsorbed layer with respect to the solid phase, as argued by Gumhalter [68] and the assignment of the lowest film branch to an L-mode. The mechanism is similar to that of the AL resonance softening due to electron-phonon interaction in metal surfaces. The surface free electron charge associated with the substrate surface states contributes an additional term in the overlayer dynamical matrix between Xe atoms. This contribution to the Xe-Xe force constant via the interposed free electron charge is similar to the last term of (5.23), where v_{ion} plays the role of the dipole potential of the highly polarizable Xe ion when displaced by its equilibrium position. The direct dipole-dipole term is already present in the dynamical matrix of solid Xe, but the presence of the interposed free charge largely softens the atom-atom interaction via the free electron susceptibility.

Although no free electrons are present in solid Xe, more recent DFPT calculations of the (111) surface of a Xe crystal, discussed in Sect. 11.1.1 [105], show that the valence electrons of heavy rare gases, despite the shell closure, play an important role in surface dynamics due to their comparatively large polarizability. This property, in combination with the electron-phonon theory of inelastic atom scattering discussed in Chap. 8, provides a more consistent and, in many respects, simpler picture of HAS from rare gas UT-films. For example, the electron-phonon mechanism permits the observation of subsurface phonons even in Xe(111), thus explaining an optical branch in the experiments which was previously attributed to two-phonon processes (Sect. 11.1.1).

12.3.2 Metal Ultrathin Films

Epitaxially grown thin films of one to several monolayers of metals on substrates of a different metal have been studied by a number of groups (see Table A.6). The early EELS work on thin films of Fe [106] and Ni [107, 108] on Cu(001) were interpreted in terms of additional stress in the first few layers [109]. Other experiments by Daum [110] on nearly incommensurate Ag layers on Ni(001) revealed a rather flat Rayleigh wave with some evidence for a gap at the $\overline{\Gamma}$-point and a relatively small 30% decrease to a minimum in the phonon frequency at a point corresponding to a reciprocal lattice vector of the substrate. A force constant fit indicated that the radial force constant within the Ag films for several layers was the same as in the bulk [111].

Figure 12.9 displays the HAS results for single monolayer films of Cs on Cu (001) studied using HAS by Witte and Toennies in 2000 [112]. The Cs surface was found to be extremely smooth and only an upper limit of 10^{-6} could be established for the ratio of the first order diffraction to the specular peak intensities. The LEED diffraction and earlier [113] measurements indicate that the Cs monolayer forms a slightly distorted hexagonal close packed overlayer with two domains rotated by 90°. The HAS energy loss distributions in Fig. 12.9a, c show two sharp peaks, labelled L and R, and a broader feature (S) at higher energy. The dispersion curves of the L- and R-modes extend over the entire Brillouin zone of the Cs layer and lie entirely below the phonon spectrum of the substrate (Fig. 12.19b) and, therefore are strongly localized in the adsorbate layer. The S-mode on the other hand is superimposed on the substrate phonons near the zone center and couples with the substrate. This explains the large broadening seen in the energy loss spectra Fig. 12.9a, c.

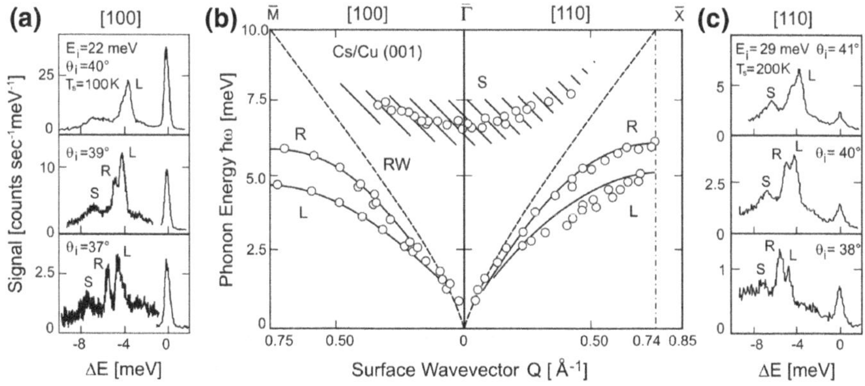

Fig. 12.9 HAS energy loss spectra and dispersion curves for a monolayer of quasi-hexagonal close packed Cs on Cu(001) [112]. **a** and **c** Some representative HAS energy loss spectra at a beam energy in the range of 22–29 meV and surface temperature of T_s = 100 K and T_s = 200 K, respectively. The solid line curves were fitted to the data using a nearest neighbor radial force constant model. The dashed line delineates the Rayleigh wave dispersion curve of the substrate

The L- and R-mode branches correspond to the adsorbate sagittal modes predicted in 1971 by Alldredge et al. for the (001) surface of a *fcc* slab loaded with heavier atoms [29]. The number of layers in this early slab calculation was not sufficient to predict the S-resonance, the existence of which, however, is implied by its effect on the dispersion of the R mode. The hybridization of the S-resonance with the substrate RW branch results in the R branch. The latter starts at $\overline{\Gamma}$ with the RW slope and ends at the overlayer zone boundary with an energy comparable to that of the S mode at $\overline{\Gamma}$. This assignment was validated with the aid of a simple Born von Karman nearest neighbour force constant model for the lattice vibrations [112]. A good fit of the experimental dispersion curves in Fig. 12.9b could be obtained with a single radial Cs-Cs force constant (2.2 N/m) which was only 25% stiffer than for the bulk [114, 115] and a radial force constant between the Cs atoms and the Cu substrate atoms of 55.0 N/m.

The close similarity of the phonon branches in the two directions suggests a very small dynamical anisotropy of the Cs overlayer. The dispersion curves in the [32] direction are smeared out because the overlapping contributions from orthogonal domains are unequal due to anisotropy. Despite this small anisotropy the slightly distorted hexagonal overlayer does not appear to lock in with the underlying substrate lattice in a commensurate phase, since no gap in the L-mode energy, which is generally expected at the $\overline{\Gamma}$-point for a commensurate overlayer, was found within the energy resolution of 0.3 meV.

The large intensity of the L-modes poses once more the problem of coupling of the He atoms with the atomic adsorbate oscillations at a metal surface. Just as in the case of the Cu surfaces the extreme smoothness of the Cs surface rules out a simple additive potential model as an explanation for the coupling to longitudinal modes, as discussed in Chap. 8. Thus these measurements and a related study of alkali atoms on graphite [46, 116] point to the importance of phonon-induced electron density deformations as a mechanism for exciting the longitudinal modes of a metal surface. In the case of a nearly free electron metal the dipolar and quadrupolar deformabilities of pseudocharges (Sect. 8.3), which are related to the dipolar (s-p hybridization) and quadrupolar (s-d hybridization) polarizabilities, respectively, are much weaker than in rare gas films due to the larger delocalization of the 6 s electrons. In this context it is interesting to point out that in a HAS study of thick Cs films of ≥ 25 ML on Cu(001) discussed in Sect. 11.3.1 the longitudinal mode is much weaker than in transition metal surfaces [117, 118]. The extreme flatness of the surface, which is a general problem for HAS studies of alkali films, also prevents the observation of diffraction peaks and therefore the determination of the surface geometry. As discussed in Sect. 11.3.1 this problem could only be circumvented by a DFPT calculation of the Cs film phonon dispersion curves for different low index surfaces. From the comparison with the HAS data the observed phonon branches could be assigned to well-defined symmetry directions.

12.3.3 Thick Metal Films: Organ-Pipe Modes and Sezawa Waves

The preparation of thick films on suitable substrates also provides new opportunities to prepare samples for surface studies which are not easily prepared by cleaving natural or custom grown crystals. For example phonon measurements on molecular beam epitaxy (MBE) grown GaSe and InSe thick films described in Sect. 11.4 produce sharper phonon dispersion curves than obtained from cleaved grown crystals [119]. Moreover, because of their greater purity, the possibility that some of the unexplained features could be due to intercalated substances of unknown origin can be ruled out. This section is devoted to thick films which provide new phenomena not observed in a semi-infinite crystal.

One of the first attempts to use thick films to access the surface phonons of the free electron alkali metals was reported by Benedek et al. for sodium in 1992 [120]. Since sodium is very soft and highly reactive it was not possible to prepare surfaces in UHV by cleaving bulk crystals. The sodium films were therefore deposited on a clean Cu(001) surface and the layer by layer growth was monitored by counting the oscillations in the HAS specular scattering intensity. Figure 12.10a shows some typical time-of-flight spectra taken under near specular final angles for slightly different incidence angles for a 10 ML sodium film of deposited on Cu(001). With increasing film thicknesses from N_L = 1–20 ML the number of peaks is found to increase and their spacing decrease. Moreover the frequencies of these modes, which correspond to near zero surface wavevector (see Fig. 12.13) correspond closely to the harmonics of an organ-pipe with a node at the interface and the open end at the surface, and therefore are designated as *organ-pipe modes*. Figure 12.11 illustrates the distribution of organ-pipe vibrational amplitudes perpendicular to the

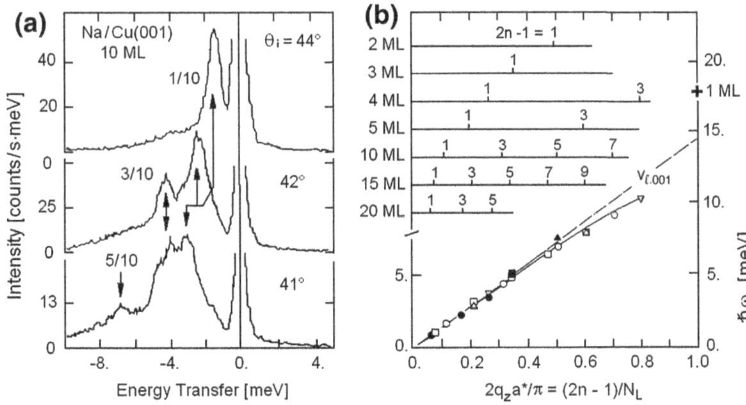

Fig. 12.10 HAS energy transfer spectra and dispersion curves for a 10 ML film of Na on Cu(001) [120]. **a** HAS energy transfer spectra measured near the specular peak ($Q \approx 0$) at three different incident angles θ_i, incident energy $E_i = 22$ meV and a surface temperature of 60 K. The peaks (arrows) are identified by values of $(2n - 1)/N_L$ where n designates the overtone and N_L the number of layers [120]. **b** HAS organ-pipe mode frequencies (right ordinate) at zero parallel wavevector as functions of the dimensionless organ-pipe wavevector q_z in the normal direction for different numbers N_L of monolayers. The open and closed symbols correspond to the numbers $2n - 1$ on horizontal lines for each thickness N_L indicated on the left ordinate at the upper part. The experimental frequencies lie along a film dispersion curve with a slope at small q_z which corresponds to the longitudinal phonon velocity v_ℓ in the [001] direction of bcc-Na of 3.02 km/s in agreement with neutron scattering data [121]

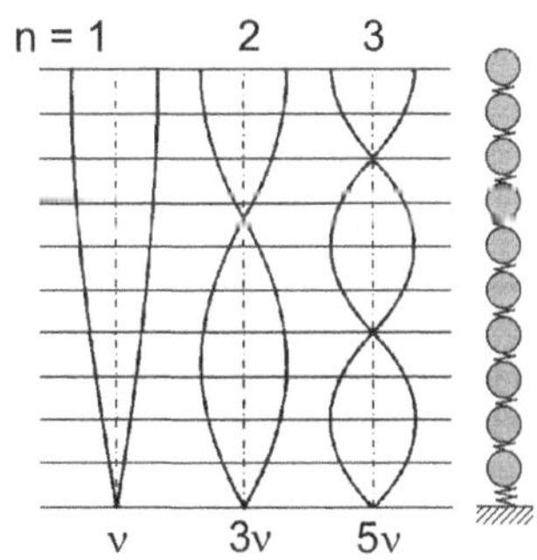

Fig. 12.11 Schematic diagram showing three different overtones of organ-pipe modes in a 10 ML thin film and a ball and spring model for a row of 10 atoms perpendicular to the plane of the layers

surface of a 10 ML film for quantum numbers $n = 1$–3. The node at the Na/Cu(001) interface is a consequence of the large negative acoustic mismatch [$\delta = -0.94$ according to (12.2)], which provides for the confinement of the film eigenmodes. The boundary conditions lead to the following expression for the frequencies:

$$\omega_n(N_L) = (2n - 1)\frac{\pi v_\ell}{2a^* N_L} = q_z v_\ell, \quad n = 1, 2, \ldots N_L, \tag{12.3}$$

where a^* is the average interlayer distance and v_ℓ is the phase velocity of longitudinal waves in the film normal to the surface with wavevector given by [120]

$$q_z = (2n - 1)\pi/2a^* N_L. \tag{12.4}$$

In Fig. 12.10b the experimental organ-pipe mode frequencies at zero parallel wavevector are plotted as functions of the reduced wavevector $2a^*q_z/\pi$ for the available values of Na film thicknesses $N_L \leq 20$. The upper part of Fig. 12.10b shows the series of observed quantum numbers $2n - 1$ labelling the organ-pipe harmonics for each thickness. The corresponding frequencies lie along a line featuring the dispersion curve of longitudinal phonons of bulk Na along the direction normal to the surface with a long wave phase velocity v_ℓ.

At large q_z the dispersion curves deviates slightly from linearity, as expected for the bulk LA branch, so that v_ℓ in (12.3) is only approximately constant. If v_ℓ is assumed to be constant then it was found that the organ-pipe eigenfrequencies divided by v_ℓ, $\omega_n(d)/\pi v_\ell = (n - \frac{1}{2})/d$, are simply a half-odd integer times the reciprocal of the film thickness $d = a^* N_L$. As such this expression holds for any thickness and is therefore valid in the continuum limit for thick films, where the same modes are known as *Sezawa waves* [122]. Like Rayleigh waves, the Sezawa waves have been first introduced to explain the so-called higher order Rayleigh waves observed in seismic diagrams [122]. Their long wavelength behaviour has since been extensively studied in thin films and SAW devices with various methods, including surface Brillouin scattering [123, 124]. In Fig. 12.12 the measured values of $\omega_n(d)/\pi v_\ell$ are plotted vs. the film thickness and compared with the curves $(n - \frac{1}{2})/d$ for Sezawa waves of order n, SWn. The agreement is excellent except at higher frequencies where the deviations from the linear dispersion are significant.

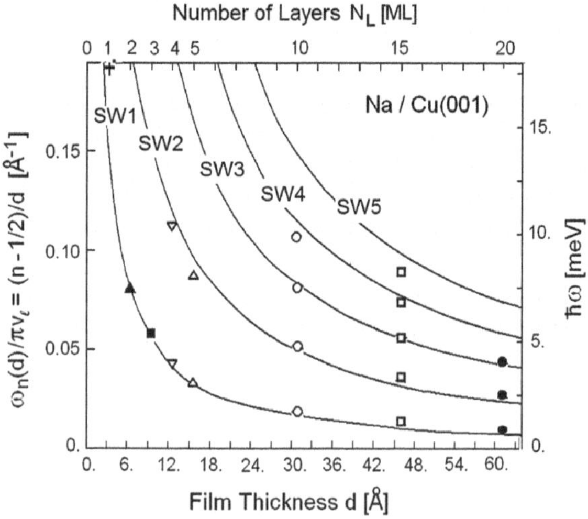

Fig. 12.12 HAS organ-pipe mode frequencies (right ordinate) for Na films on Cu (001) [120] are compared with the predicted organ-pipe frequencies of Sezawa waves of increasing order SWn (left ordinate, full line curves) as a function of the film thickness d. The same open and filled symbols as in Fig. 12.10b are used to mark the experimental frequencies

The actual value of v_ℓ can be derived from the organ-pipe frequencies, provided the average interlayer distance a^* is known. The lack of diffraction peaks in HAS from the extremely flat surface of bcc-Na films does not allow to assess whether the surface created is (001), as suggested by the substrate surface symmetry, or (110) which is a favoured growth direction. For bcc-Na(001) the average interlayer distance is $a^* = 2.15$ Å, which gives a bulk longitudinal velocity in the <001> direction of 3.02 km/s, in very good agreement with the slope derived from Woods et al. neutron scattering data (3.07 km/s) [121]. On the other hand for bcc-Na(110) it is $a^* = 3.04$ Å and $v_\ell = 4.27$ km/s, which is considerably larger than the experimental velocity of bulk Na along [110] (3.72 km/s [121]).

Figure 12.13 displays the measurements of the sodium organ-pipe frequencies as functions of the parallel wavevector Q. In these measurements the Rayleigh mode of the underlying copper substrate is also seen for thin layers up to 5 ML. For 10 ML the substrate vibrations are entirely damped but the films are now thick enough to exhibit their own Rayleigh wave. In the central part of the Brillouin zone (Fig. 12.13) the organ-pipe modes do not depend on the surface wavevector. Then at larger Q, starting near the dispersion curve of the Rayleigh wave for the Na film, the organ-pipe modes acquire some dispersion and tend to approach the film Rayleigh wave curve. This is particularly evident for the thicker films. The measured RW velocity for the film (1360 $\pm$ 40 m/s) is also in very good agreement with the RW velocity for the RW velocity of bcc-Na(001) in the [110] direction, $v_{\mathrm{RW}} = 1.38$ km/s, as derived from the experimental elastic constants [125] and the ratio $v_R/v_T = 0.56$ from Fig. 3.3 [122], with $v_T = 2.46$ km/s in the [110] direction [121]. The good agreement between the organ-pipe dispersion and that of the LA bulk modes along [001] and for the RW velocities of the (001) surface would indicate that the Na film grows on Cu(001) along the [001] direction, at least within the thickness explored in the experiments of [123]. However in [115, 120] it is argued that Na films on Cu(001) grow along a [110] direction of bcc-Na, thus creating a (110) surface. In this case, as seen above, the velocities of both organ-pipe modes and RWs would be considerably larger than those of the corresponding LA and Rayleigh waves. This can only be explained by assuming that the interface stress, which squeezes the first Na layer into a quasi-hexagonal structure, affects several layers and is responsible for the stiffening of the film vibrations with respect to those of bulk bcc-Na [120, 126].

Whether these effects are mediated by the electrons or result from the structural mismatch at the surface is not known at present. Generally, at equilibrium, electrons will be transferred from the film to the substrate or vice versa, depending on whether the work function difference $\Delta\phi$ between the film and the substrate is negative or positive. The effect is confined within the Fermi-Thomas screening length of the film, which may be, however, comparable to the film thickness of a few MLs. Thus a general increase or decrease of force constants in the film is expected to occur with respect to the bulk. For example for Na(110) on Cu(001) $\Delta\phi = -1.74$ eV [127], implying an electron depletion and increased ion-ion repulsion within the film.

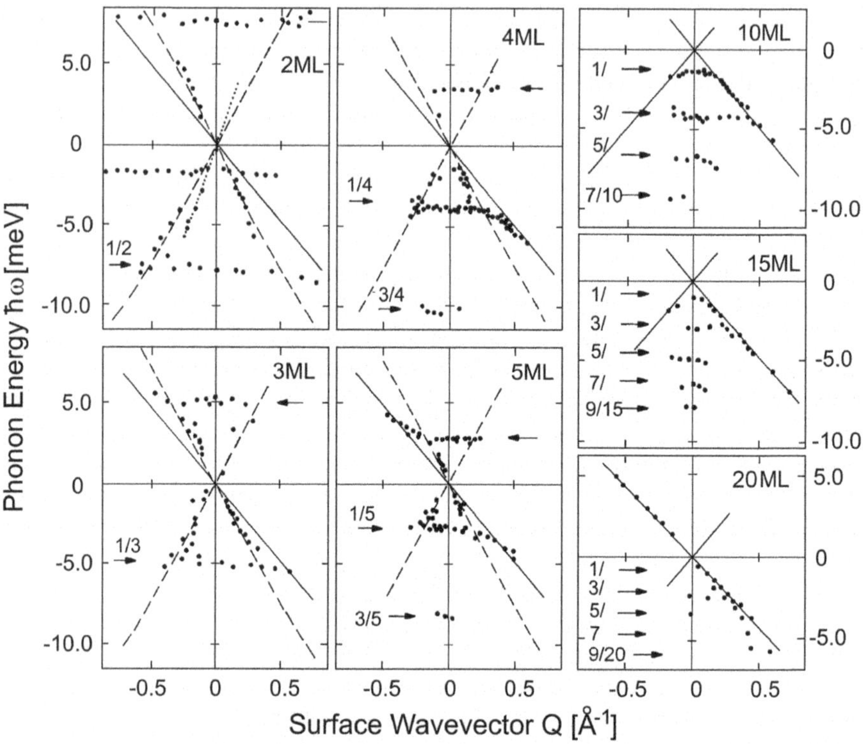

Fig. 12.13 HAS dispersion curves of *organ-pipe* modes in Na films (2–20 ML) on Cu(001) [120]. The arrows indicate the predicted positions of the organ-pipe modes. The fractions identifying the horizontal dispersion curve are values of $(2n - 1)/N_L$ where n designates the overtone and N_L the number of layers. The slopes of the Rayleigh wave branches in the film (full lines) and in the substrate (broken lines) are also shown. At large wavevectors for films with less than 5 ML a few points follow the longitudinal resonance of the substrate (dotted lines)

Organ-pipe modes have since been found for several other alkalis on metal substrates, e.g. K/Ni(001) [128] Cs/Cu(111) [128] and K/Rh(111) [129]. For the thin film systems KBr/NaCl(001) [130], Pb/Cu(111) [131–133], Ar(Kr)/Ag(111) [78] and fcc-Fe(001)/Cu(001) [134–136] similar modes were also observed at the $\overline{\Gamma}$ -point, which however had a different dependence of the phonon frequencies on layer thickness, $\omega_1(N_L) \propto N_L^{-\gamma}$. Instead of an exponent γ of unity, as predicted for the organ-pipe sequence (12.3), these systems had values of γ between 0.5 and 1.0 [126]. An alternative, perhaps more physical way to account for the deviations from (12.3), which also provides a good fit of organ-pipe modes, is to replace N_L in (12.3) with an effective thickness $N_L^* = N_L + n_0$ where n_0 is an effective penetration. This indicates less confinement of the organ-pipe modes within the film and a reduced acoustic mismatch between film and substrate. From the microscopic point of view it appears that the value of γ (or n_0) depends rather sensitively on the ratio of force constants f_i/f_0, where f_i and f_0 are the interplanar force constants at the interface and within the film, respectively [126, 128]. Equation (12.2) for the acoustic mismatch is readily expressed for cubic systems in terms of microscopic quantities as

$$\delta = \frac{A_{sub}}{A_0}\left(\frac{M_0 f_0}{M_{sub} f_{sub}}\right)^{1/2} - 1, \tag{12.5}$$

since for a cubic lattice the density $\rho = M/V_c$ and $v_\ell = a^* \sqrt{f/M}$ [137], where M is the unit cell mass, $V_c = a^* A$ is the unit cell volume and A the surface unit cell area. Here A_0 and M_0 are the surface unit cell area and unit cell mass of the film, and A_{sub}, M_{sub} and f_{sub} are the surface unit cell, unit cell mass and interlayer force constant of the substrate. A film of sodium, which is much softer and lighter than copper, has a value of $\delta \approx -1$. The interface force constant f_i does not appear explicitly in (12.5), but the reasonable approximations $f_i \cong (f_0 f_{sub})^{1/2}$ and $A_0 \cong A_{sub}$ yield $\delta + 1 \cong (f_0/f_i)(M_0/M_{sub})^{1/2}$. As discussed in [126] a necessary condition for the ideal organ-pipe sequence with $\gamma \sim 1$ (or $n_0 \sim 0$) is that $f_0/f_i \ll 1$. Thus the exponent γ, as well as the effective penetration n_0, can provide an indirect method for obtaining quantitative information above the force constants at the buried interface between the layer and the substrate. When the acoustic mismatch is not sufficiently negative ($\delta > -0.8$) the organ-pipe modes couple more strongly with the substrate and deviations from the sequence given by (12.3) are expected. On the basis of the extensive experience made with organ-pipe modes, they have been proposed as a new tool in materials science to monitor both the growth and the quality of ultrathin films [138].

12.3.4 The Quantum Sonar Effect and Stoneley Waves in Ultrathin Lead Films

Thin films of lead exhibit a number of fascinating quantum size effects, which were first discovered with helium atom scattering during the growth of multilayer films [139, 140]. Later several other manifestations of quantum size effects in lead layers on the morphology [141], roughening transition [142] and superconductivity [143–149] have been found. Since lead is a superconductor at the fairly high temperature of 7.19 K, many of these effects are related to a strong electron-phonon coupling. Thus it is not surprising that lead layers also exhibit a number of different interesting phenomena which have been observed in a series of HAS studies of Pb films on Cu(111) [131–133, 139, 150–153]. For example, as shown in Fig. 12.14a, quantum size effects lead to an overall contraction of the top bilayer spacing of about 4.7% with respect to the bulk value, and to small oscillations with layer number, which continue with increasing film thickness and reaches the bulk value beyond about 14 layers [154]. Similarly the electronic step height of an additional layer on the N_L-layer Pb film on Cu(111) (Fig. 12.14b), as measured with HAS

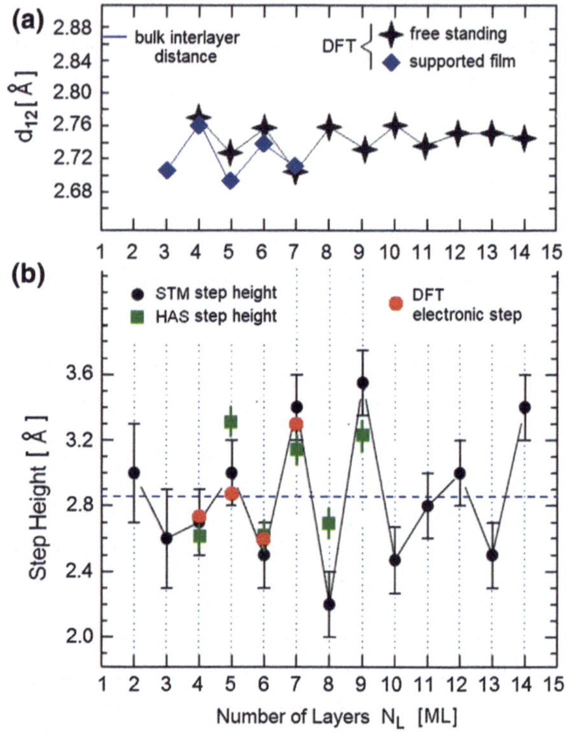

Fig. 12.14 Structural quantum size effects in UT films of Pb on Cu(111). **a** Oscillations of the surface bilayer distance as a function of the film thickness calculated with DFT for free-standing [154, 156] and supported Pb films [151]. The oscillations occur around an average value which is contracted by about 4.7% with respect the bulk interlayer distance. **b** Quantum size oscillations of the electronic step height of an additional layer on the surface of the N_L-thick film as measured with HAS [140], STM [155, 156], and calculated with DFT [151]

[140], STM [155, 156], or calculated with DFT [151], shows oscillations, which however are in anti-phase with respect to those of the bilayer spacing. This means that a larger (smaller) contraction of the bilayer spacing corresponds to more (less) charge squeezed out above the topmost plane. The important information from these studies of the quantum size effect is that the electronic step height does not correspond in general to the crystallographic one as measured, e.g., with LEED.

Figures 12.15, 12.16, 12.17, 12.18 and 12.19 illustrate a number of additional interesting phenomena observed in inelastic HAS studies of Pb films on Cu(111) which have recently been addressed by theory [131–133, 139, 150–153]. Figure 12.15c compares the dispersion curves for a 4×4 commensurate Pb 1.0 ML on Cu(111) with those of a 0.5 ML film. In both cases a shear vertical SV-mode is seen which starts out at small wavevectors near the origin with the dispersion of the Cu substrate but then deviates and assumes the frequency expected for the heavier and softer Pb. In addition another flat optical phonon branch of SV polarization is seen at 6 meV at the same energy of the zone boundary SV-mode. Both modes are attributed to the stretch vibrations of the Pb atoms against the substrate as also suggested by their avoided crossing with the substrate RW-mode at about half way to the BZ boundary. As seen both near the zone origin and also near the $\overline{K}_{Pb}$ and $\overline{M}_{Pb}$ points of the Pb(111) BZ the 1 ML film has a lower SV frequency than the 0.5

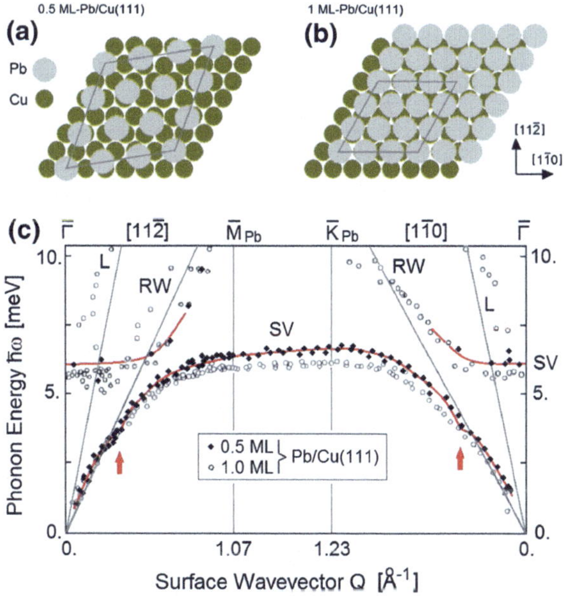

Fig. 12.15 Structures and phonon dispersion curves of 0.5 and 1 ML films of Pb on Cu(111). **a** The commensurate triangular lattice of a sub-monolayer Pb film on Cu(111) approximates a 0.5 coverage with a lattice parameter Pb-to-Cu ratio of $\sim 4\sqrt{2}/3$ and a rotation of $\sim 8.9°$. **b** The 1 ML Pb/Cu(111) commensurate 4×4 structure with the lattice parameter Pb-to-Cu ratio of 4/3. **c** HAS dispersion curves of the Rayleigh wave for the 0.5 ML (solid diamonds) and 1.0 ML (open circles) films of Pb on Cu(111) measured at a surface temperature of 140 K along the symmetry directions of the 1ML film [132]. EAM calculations [153] indicate a 2.6% compression of the 1 ML 4×4 commensurate phase which, together with additional loading of the heavier film, explains the lower frequency of the 1 ML SV mode at the zone boundary. The kinks in the 0.5 ML dispersion curves (red arrows), at 1/3 of the zone in each direction, have been associated with the film-substrate superstructure which is still perceived by HAS at 0.5 ML [132]. The straight diagonal lines starting at Γ indicate the L and RW branches of the Cu(111) substrate. The substrate RW leads to an avoided crossing of the SV branch of the Pb overlayer. Recently EAM calculations, which included the dynamics of the surface atoms, have been able to reproduce the 1 ML experimental dispersion curves (red lines) and the effect of surface temperature [153]

ML film. This is due to the compressive stress associated with the lock-in of the 1 ML film into a 4×4 structure (Fig. 12.15b, c) as opposed to the less dense 0.5 ML film with much less surface stress. The stress induced softening of the vertical vibrations is expected for a tangential force constant α_{lj} whose sign is opposite to that of the corresponding radial force constant β_{lj}, (3.22). The constant energy of the 1 ML SV branch within the boundaries $\overline{K}_{Pb}$ and $\overline{M}_{Pb}$ of the Pb(111) BZ, as shown in Fig. 12.15a, b, indicates that the Pb ML forms a regular triangular lattice which is not in registry with the underlying lattice [132, 153]. The kinks in the 0.5 ML dispersion curves (red arrows), at 1/3 of the zone in each direction (Fig. 12.15c) have been associated with the coarser distribution of Pb atoms in the extended unit cell superstructure which surprisingly is still perceived by HAS [132].

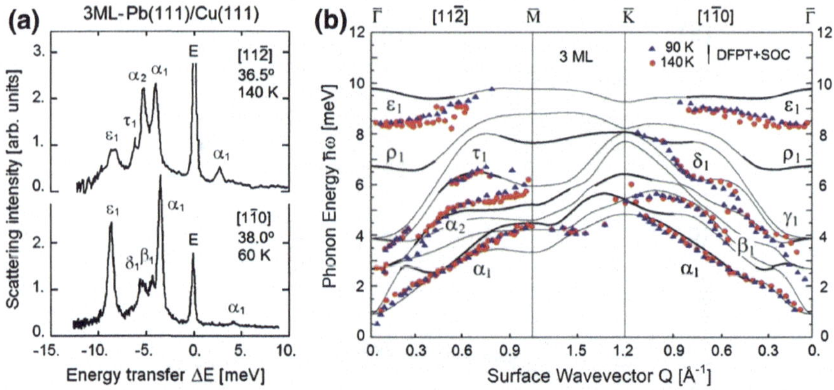

Fig. 12.16 HAS energy transfer spectra and dispersion curves for 3ML-Pb(111)/Cu(111). **a** Two exemplary HAS spectra for 3ML-Pb(111)/Cu(111) measured at similar incident angles (36.5° and 38.0°) and equal incident energy of 22.3 meV for two symmetry directions and surface temperatures. Four sharp phonon peaks are observed on the phonon creation side ($\Delta E < 0$) of the diffuse elastic peak (E). α_1 denotes the film Rayleigh wave which is also detected on the annihilation side ($\Delta E > 0$) [16, 138]. **b** The HAS surface phonon dispersion curves for a surface temperature of 90 K (blue filled triangle) [157] and 140 K (red filled circle) [132] are compared with *ab initio* DFPT calculations including spin-orbit coupling (SOC) [153]. Thick lines correspond to branches with a prevalent SV character. The greek letters label branches corresponding to the main peaks in panel (**a**)

Measurements were not possible for a 2ML film, since, as usually occurs in the growth of metal films, the 2 ML film is unstable. Only by coevaporating 20% Tl could a 2 ML film be stabilized and corresponding HAS measurements carried out [131, 157]. Nevertheless Pb films grow layer by layer with well ordered (111) films on Cu(111) from 3 ML onward at substrate temperatures below 200 K, according to the Frank-van der Merwe growth regime [132, 157]. Figure 12.16a shows two HAS TOF spectra from a 3 ML film of Pb(111) on Cu(111) measured at similar incident angles and equal incident energy for two symmetry directions and surface temperatures. The strongest inelastic peaks, denoted by α_1, are associated with the Rayleigh wave which is seen on both the creation and annihilation sides. In addition three more peaks at higher energy transfers are clearly observed. Figure 12.16b shows the dispersion curves from a large number of TOF spectra for a surface temperature of 90 K (full triangles) and 140 K (full circles). The solid line curves are from a DFPT calculation for a rigid non-vibrating substrate with the inclusion of the spin-orbit coupling (SOC) [153]. The thicker lines represent modes with prevalent SV polarization. There is in general a reasonably good agreement between the HAS data and most of the theoretical dispersion curves, except for the highest energy branch ε_1 which can be hardly associated with any of the two highest theoretical curves.

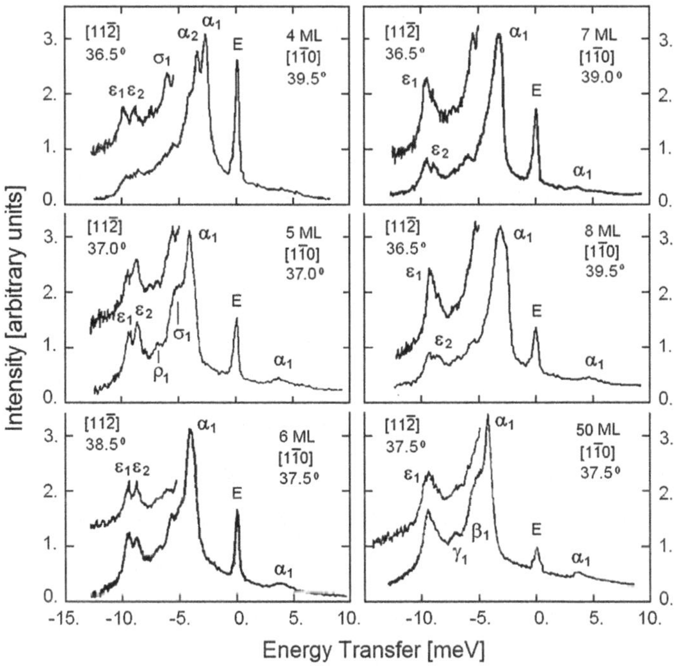

Fig. 12.17 HAS TOF spectra for 4, 5, 6, 7, 8 and 50 ML films of Pb(111) on Cu(111). For each coverage the inelastic part of the spectrum for [11$\bar{2}$], slightly shifted upwards, is shown above the spectrum for [1$\bar{1}$0] measured at about the same incident angles and an incident beam energy of 23 meV at a surface temperature of 140 K [132, 133, 152]. The greek letters label the phonon peaks as defined in [133]. ε_1 labels the highest optical mode localized at the surface bilayer, while ε_2 labels the interface mode. The latter fades away with increasing thickness and is only barely visible at 8 ML. The decrease in the elastic diffuse peak E with increasing thickness, indicates a very regular layer by layer growth of the lead films

Another important effect of the substrate is the anomalous temperature effect of some longitudinal branches, notably γ_1 and δ_1 in Fig. 12.16b, which are stiffer at the higher temperature. This is attributed to an increase in the surface stress resulting from the larger thermal expansion coefficient of lead compared to copper.

The lowest branch α_1 of Pb(111) coincides with the corresponding RW for $Q > 0.3$ Å^{-1} since from here on the penetration length is shorter than the film thickness (see Chap. 3). With decreasing Q the surface wave penetrates the substrate and gradually approaches the steeper RW curve of Cu(111). The DFPT calculation shows instead that the branch α_1 tends for $Q \to 0$ to a finite energy of about 1 meV, which is the lowest organ pipe mode, as expected for a film on a rigid non-vibrating substrate.

Figure 12.17 shows representative HAS TOF spectra for 4, 5, 6, 7, 8 and 50 ML films of Pb(111) on Cu(111) for each of the two main symmetry directions, for similar incident angles at an incident energy of 23 meV and a surface temperature

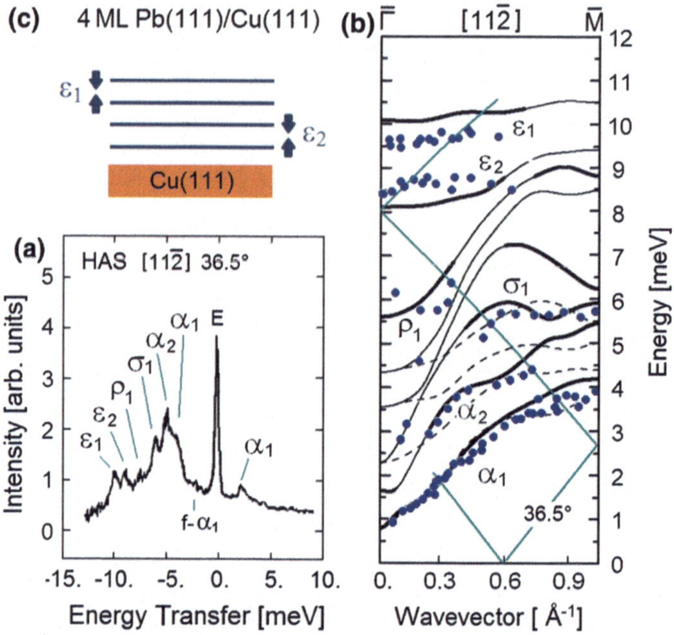

Fig. 12.18 The HAS energy transfer spectrum for a 4 ML Pb film on Cu(111) along the $[11\bar{2}]$ direction (partly shown in Fig. 12.17) with the corresponding phonon dispersion curves and the displacement patterns illustrating the interpretation of the highest energy modes. **a** The HAS spectrum at an incident angle of 36.5° and incident energy of 22 meV shows 6 peaks on the creation side and one on the annihilation side. **b** Intersections of the scan curve for the spectrum in (**a**) with the DFPT dispersion curves of quasi-SV polarization (thick lines). The scan curve has been folded into the positive energy-wavevector quadrant. The complete set of DFPT dispersion curves and HAS data (blue filled circle) are displayed in Fig. 12.19. **c** Schematic displacement patterns of the modes ε_1 and ε_2 illustrating their respective surface and interface nature

of 140 K [133, 152]. Surprisingly the spectra of the thinner films exhibit a larger number of distinct features than the 50 ML film. For example the HAS energy transfer spectrum for the 4 ML film shown in Fig. 12.18a, measured in the $[11\bar{2}]$ direction for an incident angle of 36.5° and an incident energy of 22 meV, shows distinctly all the modes accessible with the corresponding scan curve, here plotted in Fig. 12.18b after folding into the positive energy-wavevector quadrant with the DFPT dispersion curves of quasi-SV polarization (thick lines). The RW (α_1) which dominates on the creation side is the only mode observed also on the annihilation side at $\sim$2 meV. The small feature marked f-α_1 corresponds to a RW mode folded into the superstructure BZ (Fig. 12.18b), and indicates that the different periodicity of the substrate still has some effect on the surface modes of a 4 ML film.

As discussed in Sect. 8.4 (cfr. Fig. 8.13), theory explains the unexpectedly large number of dispersed modes observed in Pb films for a number of layers $N_L \geq 3$ in terms of the strong electron-phonon interaction which makes it possible for HAS to

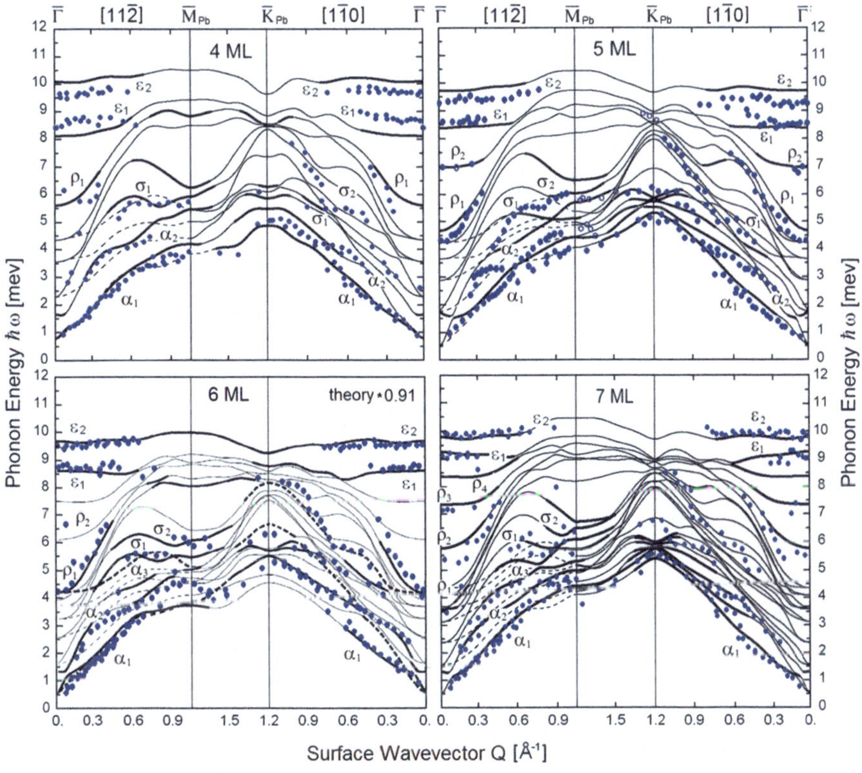

Fig. 12.19 HAS dispersion curves of the phonon modes observed in Pb (111) films of 4, 5, 6, and 7 ML on a Cu(111) at a surface temperature of 140 K [132, 157, 133] are compared with DFPT calculation without SOC and for a rigid non-vibrating substrate [151, 152]. The indexed Greek letters label surface phonon branches. The HAS data for 6 ML are for an unannealed sample and are best-fitted by DFPT after an overall softening by a factor 0.91. Thick lines: mainly SV; thin lines: mainly L; broken lines: SH

excite even deep subsurface modes [151, 152]. This mechanism explains why the number of modes observed in HAS TOF spectra increases with N_L, at least up to 8–10 ML, the limit being, besides resolution, the range of the electron-phonon interaction. This extraordinary property allows HAS spectroscopy to provide a much richer information on the dynamics of ultrathin films than other surface probes like EELS, and has been termed *quantum sonar effect* (see Sect. 8.4.2). Note that the depth explored by the quantum sonar effect is also the surface region in the semiinfinite crystal affected by oscillations in the charge density (Friedel oscillations), and therefore also by the oscillations in the interlayer distance. Above 10 layers the deepest subsurface phonons are no longer detected by HAS, and at 50 ML (Fig. 12.17) only the modes usually observed in fcc(111) metal surfaces are detected, with the correspondence (cfr. Fig. 11.28): $\alpha_1 \rightarrow S_1 \equiv RW$, $\beta_1 \rightarrow S_3$, $\gamma_1 \rightarrow S_4$, $\varepsilon_1 \rightarrow S_2$.

The highest optical modes, which for $N_L > 3$ form a doublet (ε_1, ε_2) are particularly interesting. Whereas the single optical mode ε_1 for films thicker than 3 ML has a prevalent S polarization and about the same HAS intensity, the intensity of the ε_2 mode fades away with increasing thickness. For this reason the ε_1 and ε_2 modes have been associated with vibrations localized at the surface and at the interface, their stiffening above the bulk phonon maximum being respectively due to the compression of the surface bilayer and the coupling to the more rigid substrate. This is schematically shown in Fig. 12.18c as well as by the charge density and ion displacement patterns shown in Fig. 8.6. for 5 ML Pb(111) on Cu(111). As discussed in Sect. 8.4.2, the corresponding phonon-induced CDOs, are comparable in size despite the different localization of the two modes, and therefore can explain quantitatively the observed HAS intensities. In the EAM calculation where the dynamics of the Cu substrate is included [153] the mode ε_2 although localized at the film-substrate interface actually involve also the motion of the first substrate layers. In this respect the ε_2 modes are thus the microscopic counterpart of what are known in the elastic theory as *Stoneley waves* (StW) which propagates along the solid-solid interface [158–160].

Figure 12.19 compares the full set of dispersion curves of the 4,5,6 and 7 ML Pb films on Cu(111) with the DFPT calculations for a rigid non-vibrating substrate without SOC [151, 152]. Despite the latter simplifications the agreement between theory and experiment is quite good for 4, 5 and 7 ML, while for 6 ML, for which data is only available for an unannealed sample, an excellent agreement is obtained after rescaling the DFPT results by a factor 0.91. A similar general softening of phonon energies occurring in unannealed samples was also demonstrated for 5ML Pb/Cu [133]. The HAS and DFPT dispersion curves for 50 ML, which correspond to those of the semiinfinite Pb(111), shown in Fig. 11.23, are discussed in Sect. 11.3.2.

The SV branches, marked by the thick lines in Fig. 12.19, were assigned initially to optical modes, since they tend to a finite energy and zero group velocity for $\mathbf{Q} \to 0$ (Fig. 12.20) similar to those shown in Fig. 12.12 for the sodium films. The HAS data, measured at or near $\overline{\Gamma}$ (circles) and the corresponding organ-pipe energies calculated with DFPT (crosses) [152] or EAM (squares) [158, 153] are fairly well described by (12.3) (broken lines) for an average spacing of 2.87 Å and a longitudinal acoustic velocity $v_l = 2.041$ km/s. The latter value compares well with the average longitudinal sound speed in lead (2.16 km/s), but is about 17% smaller than the longitudinal speed in the [126] direction, which indicates some softening of the UT films with respect to bulk. It should be noted however that, as anticipated in the previous Subsection, the present series of organ-pipe energies are better represented by a function of the layer number $N_L^{-\gamma}$ with $\gamma = 0.77$ (full lines). Since the localization of the modes ε_1 and ε_2 above the bulk phonon maximum are due to the local perturbations coming from the surface bilayer contraction and the interface with a stiffer material, they are not accounted for by the organ-pipe sequence (12.3).

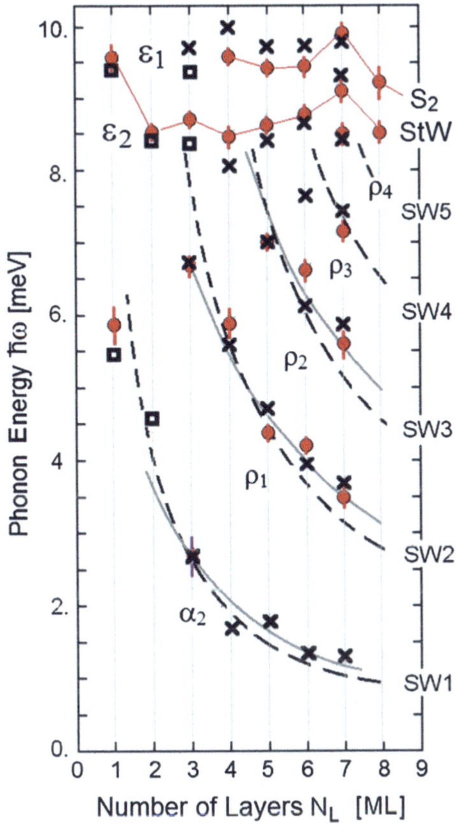

Fig. 12.20 The HAS phonon energies (yellow circles) of SV polarized modes of thin Pb films on Cu(111) at or near Q=0, plotted as a function of the film thickness N_L, follow closely the curves expected for Sezawa waves of increasing order (SWn, 1–5) [132, 133, 157]. ρ_4, ρ_3, ρ_2, ρ_1 and α_2 refer to the modes marked in Fig. 12.19. The data for $N_L = 2$ refer to the stable $Pb_{0.8}Tl_{0.2}$ film [157]. The indexed greek letters correspond to the surface phonon labels used in Figs. 12.16, 12.17, 12.18 and 12.19. The HAS data are compared with the DFPT (crosses) [152] and EAM (squares) [153] calculations shown in Figs. 12.15, 12.16 and 12.19. The broken lines are a best fit to (12.3) for an average spacing $a^* = 2.87$ Å and a longitudinal acoustic velocity $v_l = 2.041$ km/s. The full lines are best fits for a dependence $N_L^{-\gamma}$, $\gamma = 0.77$. The modes ε_1 and ε_2 with energy above the bulk phonon maximum are localized at the surface bilayer [S_2 in semiinfinite Pb(111) (Fig. 11.22)], and at the interface (Stoneley waves (StW) [158, 159]), respectively. They strongly deviate from the organ-pipe equation and are not considered in the fits

Lead ultrathin films are, so far, the only systems to exhibit such a rich display of dispersion curves. Some of the modes show small but distinct anomalies which according to an analysis based on the calculated Fermi surfaces for thin films [152, 161] appear to be Kohn anomalies related to the enhancement of electron-phonon coupling in two-dimensions. Examples of such anomalies are the sudden changes of slope of the branches α_1 and α_2, shown in Fig. 12.21a for 3, 4, and 5 ML of Pb on

Cu(111) in the $\overline{\Gamma K}$ direction [133]. The phonon wavevectors of the anomalies correspond to a nesting between Fermi contours, as shown in Fig. 12.21b for the 5ML film, either for transitions within the first Brillouin zone (as for the α_1 branch) or transitions between two different zones (*umklapp* process, as for the α_2 branch). The occurrence of Kohn anomalies is not surprising here since, as mentioned above, the electron-phonon coupling is known to be particularly strong in the superconductor lead and remains comparatively large also in ultrathin films. Figure 12.22 illustrates the dependence of the calculated mass-enhancement factor λ and the related superconducting critical temperature T_c of Pb ultrathin films on the number of layers [152]. The values of λ and T_c for the films are substantially smaller than the respective bulk values, while the experimental T_c of the films is only $\sim 10\%$ smaller than the bulk T_c in the range $4 \leq N_L \leq 8$ with a sudden drop below $N_L = 4$ [149]. It is remarkable that the same drop is predicted by theory and has been attributed to the disappearance at $N_L < 4$ of the specific phonon which contributes most to the e-ph interaction [152].

In conclusion the above HAS experiments demonstrate how HAS is able to monitor the layer by layer growth of thin films and their structure and phonon dynamics on the scale of its comparatively large coherence length. The control of the third dimension, that of the growth direction, is of great scientific and technological interest, especially when the clash of substrate and film ordering forces leads to exotic morphologies. For example growth mechanisms leading to fractal self-affine structures constitute an important chapter of statistical mechanics and have been extensively studied [164]. The mechanical properties of films, either

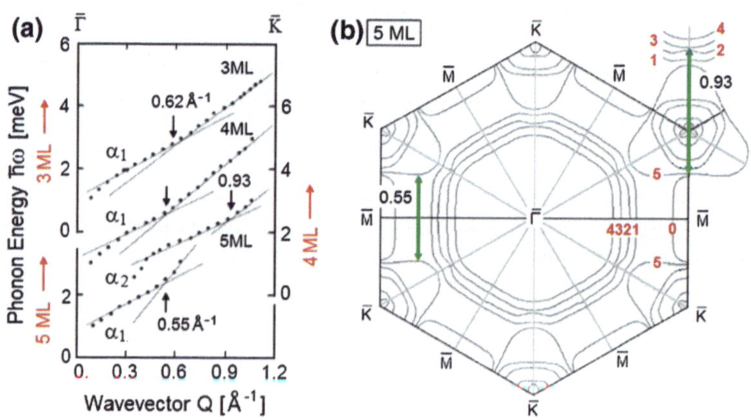

Fig. 12.21 Comparison of the experimental evidence for Kohn anomalies [133] and the corresponding nestings on the Fermi energy contours. **a** Kohn anomalies (vertical arrows) in the dispersion curves in the $\overline{\Gamma K}$ direction of the branch α_1 of 3 and 4 ML Pb films at Q = 0.62 and 0.55 $Å^{-1}$, respectively, and the branches α_1 and α_2 of 5 ML Pb film on Cu(111) at Q = 0.55 and 0.93 $Å^{-1}$, respectively. Note the different ordinate scales for the three thicknesses. **b** The latter two anomalies are tentatively assigned to transitions between nesting points on the Fermi contours for the 5 ML film (labeled by 0–5): 5↔5 for Q = 0.55 $Å^{-1}$, 5↔2 umklapp for Q = 0.93 $Å^{-1}$ [152]

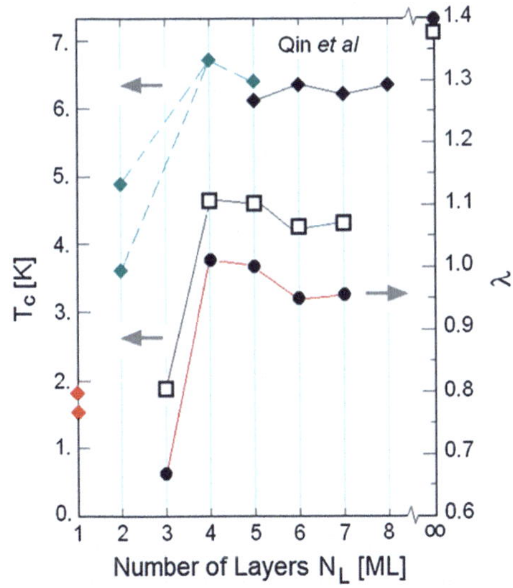

Fig. 12.22 Calculated effective e-ph coupling constant λ (filled circle, r.h. ordinate scale) and corresponding superconducting critical temperature T_c (open square, l.h. ordinate scale) for Pb ultrathin films (N_L = 3–7 ML) on a rigid non-vibrating Pb substrate layer. [152] T_c is derived from McMillan's formula [162, 163] for an effective electron-electron coupling constant $\mu^* = 0.12$. For comparison the measured T_c for ultrathin Pb films [147] and for a Pb monolayer [148] on Si(111) substrates are also shown for comparison (lozenges, l.h. ordinate scale)

amorphous or with more subtle forms of partial order like quasicrystallinity or self-affinity, and their dependence on the film morphology and chemical structure have great relevance in various fields, e.g., in tribology [165]. The HAS study of Rayleigh waves and Sezawa waves in these systems, as well as that of Stoneley waves in supported UT films and multilayers, would, for example, provide direct information on the elastic moduli at the nanoscale, well below that explored with Brillouin light scattering.

12.4 Organic Molecules

The structure, orientation and dynamics of organic molecules on solid surfaces is of fundamental relevance in many diverse areas of physics and physical chemistry, ranging from heterogeneous catalysis to tribology and polymer interface interactions and molecular electronics [166–169]. Most of the early vibrational spectroscopy experiments have been performed using either EELS or reflection absorption spectroscopies [170, 171]. In the late nineties, however, several different types of HAS studies have been carried out on a number of alkane molecules ranging from hexane (C_6H_{14}) to decane ($C_{10}H_{22}$) on a variety of metal surfaces including Cu(001), Cu(111), Pb(111), Ru(001) and on diamond C(111) [23, 172–176]. In all systems the S-mode, which is also called the vertical frustrated translation (FT_z), could be clearly identified, and in a few systems the T-mode as well. From comparisons of the line widths with various theoretical models it was concluded that the S-mode is damped predominantly by phonon processes [173, 177] unless chemical bonding is important as, for example, in the case of the highly reactive Ru(0001) surface [175].

Figure 12.23 shows the results of an early HAS investigation of normal and deuterated octane (C_8D_{18}) monolayers on Cu(111) [174]. On lowering the temperature to 162 K, which is significantly below the desorption temperature of 225 K, Fig. 12.23a reveals a surprisingly sharp drop-off of the specular intensity. The three orders of magnitude decrease of the specular intensity, which is accompanied by a rapid decrease of the diffraction peaks [174, 176], is attributed to a solid-liquid phase transition similar to that of a Kosterlitz-Thouless phase transition [178]. The HAS energy transfer spectra from deuterated d-octane (Fig. 12.23c–e) also show significant changes with the transition indicating a two-fold increase in the intensity of the S (FT_z) mode and an internal bending mode (B) of the octane molecule in the solid phase. Comparison with the clean Cu(111) surface spectrum (Fig. 12.23b) shows, moreover, that the molecular mode peaks grow at the expense of the RW and L peaks of Cu(111). In the disordered phase above T_s it was possible, using the QHAS technique, to observe diffusion (see Sect. 13.2.3) and to monitor the increase in the surface mobility of the octane molecules [174]. A similar behaviour has been observed for n-nonane [176].

These early studies illustrate the great potential of the HAS technique for even complicated organic and biological molecules. As the masses of the molecules increase the extrinsic vibrational frequencies, which for smaller molecules are normally accessible to IRS and EELS, are reduced to such an extent that they can only be measured using high resolution HAS. Moreover the gentle nature of the HAS probe is an important advantage for large easily fragmented molecules. Unlike electrons and ions, which carry an electric charge and are used as surface probes at incident energies well above the eV range, neutral atoms in the meV range cannot induce chemical changes nor alter the film morphology, and are therefore a choice probe for the structural analysis and dynamical spectroscopy of soft matter, as well as of metastable phases of reactive species. This was demonstrated by a HAS diffraction study of ice bilayers on Pt(111) [179] and a subsequent study of the phonon dispersion curves which revealed flat modes at 2.74, 3.5 and 5.85 meV attributed to H_2O monomers, dimers and islands [180]. The damage caused by LEED on a film of water on NaCl was shown by a HAS investigation of the adsorbate structure before and after exposure to the electron beam [181]. With HAS the long-standing puzzle concerning the structure of water on Ru(0001) was elucidated by a helium scattering study proving the existence of an intact (undissociated) film, in contrast with previous LEED data and first principle calculations which indicated a dissociated phase [182].

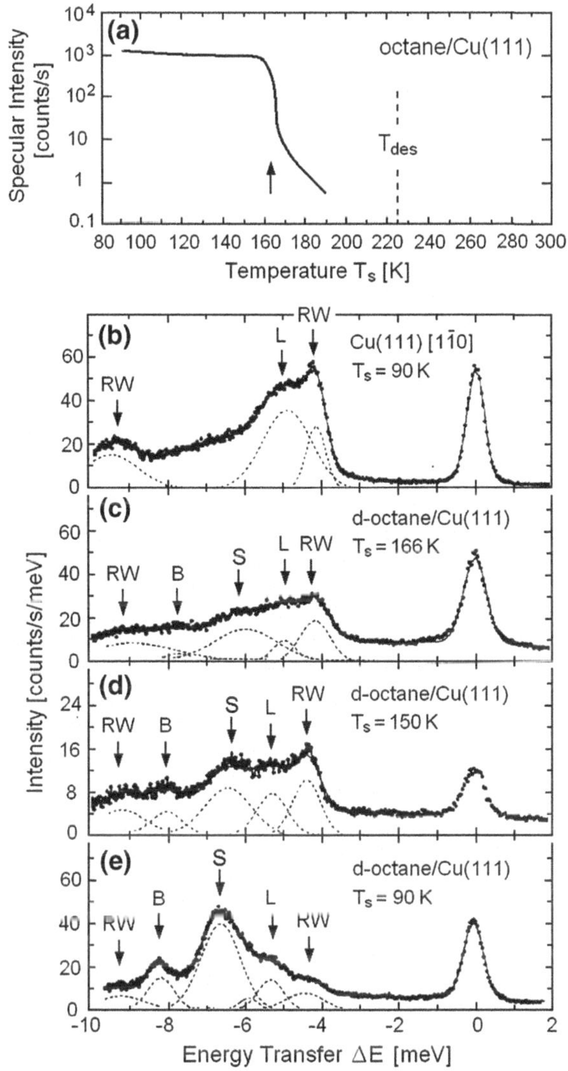

Fig. 12.23 HAS measurements of temperature dependent ordering and adsorbate vibrations of a monolayer of octane on Cu(111) (adapted from [174]). **a** The HAS specular intensity during growth of normal octane on the Cu(111) surface as a function of the surface temperature. The three orders of magnitude decrease of the specular intensity at T_s = 162 K is attributed to a transition from the ordered monolayer phase to a disordered phase which exists below the desorption temperature of T_{des} = 225 K. **b** The HAS energy transfer spectrum for the clean Cu(111) surface at E_i = 22 meV, θ_i = 40°. **c–e** The HAS energy transfer spectra for deuterated octane (d octane) on Cu(111) for decreasing surface temperatures T_s measured under identical kinematic conditions. The peaks associated with the vertical frustrated translation S (FT_z) and bending B modes of the molecule grow with decreasing temperature and are the most prominent at 90 K

12.5 Self-assembled Monolayers

An important field where the HAS technique also displays its advantages is that of organic monolayers such as the Langmuir-Blodgett (LB) or self-assembled monolayers (SAM), which form highly ordered phases [183]. Self-assembling is a method for bottom-up materials fabrication bearing the promise of many applications for optoelectronic devices and other intriguing nanotechnologies [184–187]. The organic thiols with structure of $C_nH_{2n-1}SH$ (n = 2–22) form long 20 Å chains (n = 16) which self-assemble to regular arrays of tilted chains on noble metal surfaces with the sulphur end attached to the substrate metal [188]. They are an important class of self-assembled films whose structure and evolution have been successfully studied with HAS [189–195].

The intermolecular and molecule-substrate interactions which govern self-assembling also determine the low energy spectral features of the film. Molecular dynamics simulations [196–199] suggest indeed that many relevant properties of these systems may be obtained from the study of the low frequency vibrational spectrum, which contains information on their external interactions and conformational degrees of freedom. In these systems the physisorbed phase, which allows for molecular mobility and self-assembling, may act as a metastable preliminary to the formation of a more stable and robust chemisorbed state—another feature which can be monitored by HAS. An early 1988 HAS study of a Langmuir-Blodgett film of $CH_3(CH_2)_{18}COOH$ (arachidic acid) on Ag(111) yielded a broad angular distribution as well as a broad uniform multiphonon distribution in the energy transfer spectrum with no diffuse elastic peak [200]. This was attributed to very low energy vibrations with energies of less than 0.2 meV and was confirmed by an early molecular dynamics simulations of pentadecanethiol ($S(CH_2)_{15}CH_3$) adsorbates by Hautman and Klein [196] which predicted a low energy spectrum of external vibrations in the meV range.

Subsequently the non-destructive gentleness of HAS made it possible to detect the structure of the physisorbed monolayers of another thiol on Au(111) by low energy diffraction [201, 202]. Figure 12.24a, b shows a more recent HAS diffraction study of heptanethiol (HT) on Cu(110) after deposition at 120 K (a) and after deposition at 300 K (b) [193]. The physisorbed layer in Fig. 12.24a has only a weak specular peak and a broad background compared to sharp diffraction peals indicating a well-defined structure of the chemisorbed layer. This assignment is confirmed by the resolved inelastic peaks in Fig. 12.24c. The distinct loss peak at about 6 meV in the energy transfer was found to be independent of the wavevector transfer. This dispersionless behaviour and the close agreement with the energy of the loss peak found for the nearly equal mass d-octane (C_8D_{18}, M = 132 amu) (Fig. 12.23e) gave evidence that it can be assigned to the frustrated translational mode FT_z. As seen in Fig. 12.24d, this peak was not found for the strongly bound chemisorbed layer [193].

A subsequent HAS experiment by Sibener's group [199, 203] came to a somewhat different conclusion concerning the adsorbate modes for 1-decanethiol

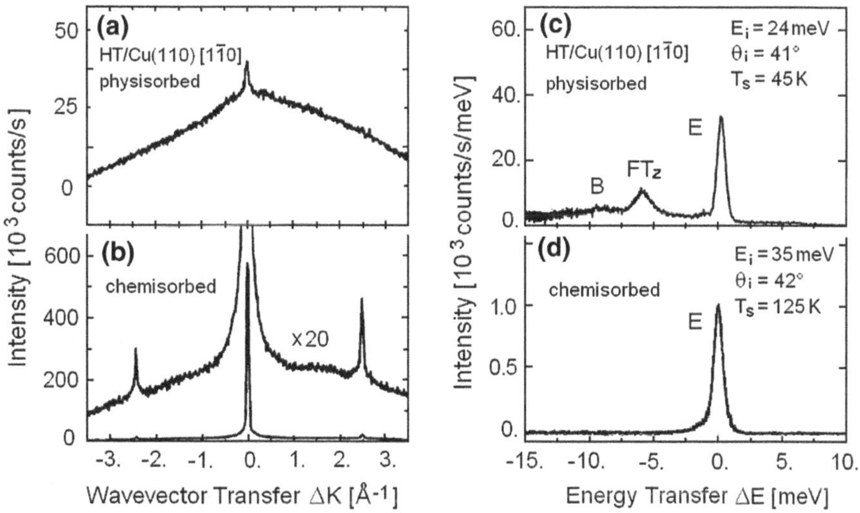

Fig. 12.24 HAS angular distributions and energy transfer spectra for a physisorbed monolayer and a chemisorbed monolayer of heptanethiol (HT) on Cu(110) measured along the [1$\bar{1}$0] direction for an incident energy E_i = 24.3 meV and a surface temperature T_s = 45 K. **a**, **b** Comparison of HAS diffraction patterns from the physisorbed and chemisorbed layers. **c** The corresponding HAS energy transfer spectra show in the physisorbed phase, besides the diffuse elastic peak (E), features corresponding to low energy vertical frustrated translation (FT_z) and bending (B) modes of the molecule. **d** In the chemisorbed phase no low energy mode is observed (adapted from [193])

($C_{10}H_{21}SH$) monolayer on the reconstructed Au(111) surface. Figure 12.25a shows the distinct HAS diffraction pattern along the [1$\bar{1}$0] direction along the chains which form a (23/2,$\sqrt{3}$) striped phase. The STM image in Fig. 12.23b shows that the molecules lie flat on the surface between the rows of the reconstructed solid structure. The HAS energy transfer spectra in (Fig. 12.25c) shows a FT_z mode at about 7 meV, nearly the same as for HT (Fig. 12.24c). Although the decanethiol molecules in this phase are known to be *chemisorbed* through the sulphur atom, a FT_z phonon and an appreciable low energy multiphonon background are observed provided the incident energy is sufficiently large. The FT_z branch is in this case also dispersionless (Fig. 12.25d), indicating a negligible intermolecular interaction. Rosenbaum et al. [199] also investigated the striped-phase self-assembled monolayers of 1-hexanethiol and 1-octanethiol on Au(111) and studied the effect of the chain length on the low energy phonons. Their observation that the longer chains have only slightly stiffer FT_z mode frequencies is in essential agreement with the results for octane and HT discussed above. These results are well accounted by a molecular dynamics simulation in which the intermolecular and molecule-substrate interactions are described by van der Waals forces and only the sulphur atom is chemically bound to the surface at the hollow site [199]. The apparent disagreement with the data for ordered alkanethiols chemisorbed monolayers on copper [193], where no low energy external mode is observed (Fig. 12.24d) was attributed by

Rosenbaum et al. [199] to the fact that the chemisorbed phase of 1-DT is lying horizontally on the gold surface (Fig. 12.25b), as clearly indicated also by an extended earlier STM investigation [204]. There is therefore a substantial difference with similar chemisorbed alkanethiols, except on copper on which the molecules tend to stand upright, possibly with only a small tilt of about 13° as indicated by NEXAFS measurements [205, 206]. These representative studies, together with the

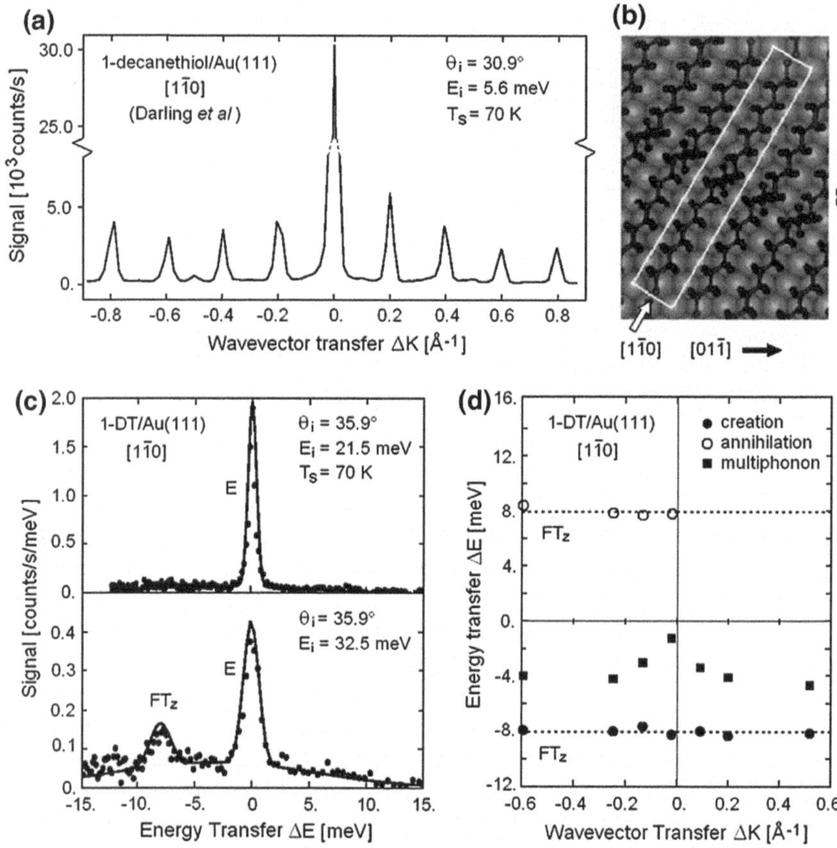

Fig. 12.25 HAS and STM investigation of the ordered self-assembled 1-decanethiol (1-DT ≡ $C_{10}H_{21}SH$) monolayer on Au(111) along a $[1\bar{1}0]$ direction at $T_s = 70$ K. **a** HAS diffraction patterns along the direction of the chains. **b** The STM image reveals that the long chain molecules lie between the rows of the reconstructed gold surface and form the $(23/2, \sqrt{3})$ striped phase inside the monolayer unit cell (white rectangle). **c** Two HAS energy transfer spectra from 1-DT/Au(111) measured at the same incident angle of 35.9° but at different incident energies of 21.9 and 32.5 meV. The FT_z phonon and the multiphonon background contribute an appreciable HAS intensity only at the higher incident energy. **d** The measured dispersion curves for the FT_z mode in creation (black filled circle) and annihilation (filled circle) and the multiphonon background (black filled square) (adapted from Darling et al. [203])

other investigations in the comprehensive list with references in Table A.8 of Appendix A demonstrate the great potential of the HAS low energy spectroscopy in the meV range. Recently the relatively new technique of spin echo ^{3}He scattering spectroscopy, with resolution in the μeV range, has yielded new insight into the dynamics of large organic and biomolecules on surfaces as discussed in Sect. 13.2.3 [207].

12.6 Chemisorbed Adlayers on Semiconductor Surfaces

The chemisorption of unsaturated hydrocarbons on semiconductor surfaces, besides providing an improved oxidative and electrochemical stability with respect to ordinary hydrogen passivation [208, 209], has stimulated a rapid evolution in the tailoring of interfacial properties with bound organic adlayers for chemical functionalization [210–214]. The low energy external degrees of freedom of the chemisorbed molecules (Fig. 12.1) and their coupling to the substrate dynamics are considered to play an important role in the structural order of the adlayer as well as in its functions, possibly via electron-phonon interaction. In this respect the combination of HAS measurements of the surface phonon dispersions across the entire surface BZ with an *ab initio* DFPT analysis can lead to a precise knowledge of bonding and dynamical properties of such decorated interfaces. A substantial number of organic molecule overlayers on semiconductor and insulator surfaces have been investigated with HAS for their structural and vibrational characterization, as summarized in Table A.8 of Appendix A.

These developments are illustrated by a recent detailed investigation of methylated Si(111) and Ge(111) surfaces with HAS and DFPT [215–219]. In Fig. 12.26 the HAS dispersion curves for CH_3/Si(111) (a) and CD_3/Si(111) (b) [216, 217] are compared with the DFPT dispersion curves and the total sagittal displacement amplitudes. For both adsorbates the amplitudes of the RW, and of the alkyl hindered rotation (HR) and two rocking librations (RL) of the chemisorbed molecules are shown on top of the surface-projected quasi-continuum of bulk phonons. The displacement patterns of these three modes are shown in Fig. 12.26c, d, e. The HR modes form a weak resonance near $\overline{\Gamma}$ at an energy of 29.0 meV for CH_3/Si(111) and 20.5 meV for CD_3/Si(111), eventually prolongating into a localized gap mode towards the zone boundaries. The substitution of H with D produces a strong isotope effect on both HR modes near $\overline{\Gamma}$ and RL modes near $\overline{K}$. The large softening of the HR energy at $\overline{\Gamma}$ on substitution of H with D, with an energy ratio of $\sim \sqrt{2}$, is indicative of a complete decoupling of the HR mode at $\overline{\Gamma}$ from the C and Si coordinates (Fig. 12.26c). The color scale indicates the extent of the contributions to the displacement pattern of the carbon atom: clearly the carbon atom is nearly at rest in the HR mode. The degenerate pair of RL modes at $\overline{\Gamma}$ shows instead a modest H/D isotope effect, from 15.8 to 14.0 meV. The corresponding effective mass ratio is 1.27, which implies some participation of the C atoms in the motion, as seen in Fig. 12.26b. However at the $\overline{\mathrm{K}}$ point the quasi-degenerate RL pair, that in CH_3/Si(111) falls in a gap at ~23 meV, exhibits a large isotope effect, with the lower mode shifted down to 16 meV in CD_3/Si(111). This also implies an energy ratio close to $\sqrt{2}$ and therefore negligible displacements of the C and Si atoms, appropriate to hindered rotations around two orthogonal axes lying on the surface. For symmetry reasons such hindered rotations are strongly coupled to the RW elliptical polarization in the sagittal plane (Fig. 12.26c), so that their dispersion curves show an important avoided crossing at about 2/3 of the BZ in the $\overline{\Gamma\mathrm{K}}$ direction. For CH_3 this is less evident because the RL branch is higher and entirely inside the bulk continuum, whereas for CD_3 the RL becomes a localized mode beyond one half of $\overline{\Gamma\mathrm{K}}$ and exchanges its character with that of the RW. The HAS data provide evidence for the strong RW-RL mixing, especially for CD_3: as shown in panel (b) The HAS data points, mostly associated with the SV component of the RW, for wavevectors approaching $\overline{\mathrm{K}}$, jump from the lower branch to the upper branch, in accord with the RW-RL character exchange. Similar information has been derived from the combined HAS and DFPT analysis of the methylated Ge (111) surface [218, 219].

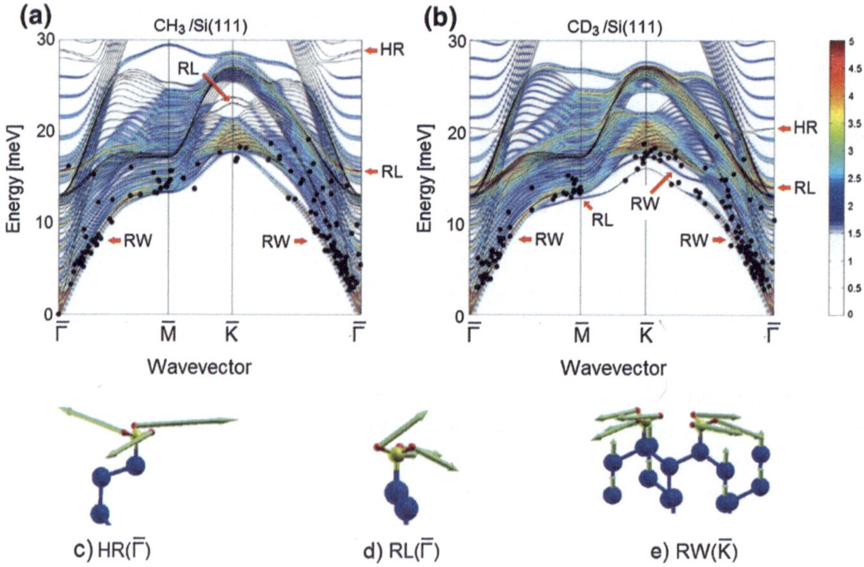

Fig. 12.26 The HAS dispersion curves are shown together with the DFPT calculated dispersion curves and the total sagittal displacement amplitudes for **a** CH_3/Si(111) and **b** CD_3/Si(111). **c–e** Displacement patterns of the hindered rotation (HR), the rocking libration (RL) motions at $\overline{\Gamma}$ and for the RW at $\overline{\mathrm{K}}$, respectively. The arrows are proportional to the real components of the mode eigenvectors [216, 217]. Both the HR energy at $\overline{\Gamma}$ and the RL energy at $\overline{\mathrm{K}}$ are strongly depressed by the isotope substitution of H with D, and the RL mode for CD_3/Si(111) shows a strong avoided crossing with the RW with exchange of character at $\overline{\mathrm{K}}$. The color scale indicates the extent of the contributions to the displacement pattern of the carbon atom

As discussed in [216], the HAS and DFPT studies of alkyl-terminated semiconductor surfaces demonstrate the importance of adlayer librations in determining the overall vibrational band structure of surface waves, and show the power of combining these techniques for examining the vibrational characteristics of a hybrid organic-semiconductor interface. In the case of silicon, alkyl termination has been a successful strategy for tailoring interfacial properties, and elucidating the dynamical coupling between adlayer and lattice vibrations which should play an important role in designing new hybrid devices operating in the THz regime.

12.7 Atomic Rows Adsorbed at Surface Steps: Quasi 1D Phonons

Stepped surfaces may act as templates for the growth of periodic arrays of atomic rows adsorbed along the steps, thus providing a convenient benchmark for several interesting phenomena induced by low dimensionality. Among these the strong enhancement in the electron-phonon interaction and its manifestations like CDW transitions, superconductivity and sharp Kohn anomalies in the quasi-1D phonon dispersion curves are the most interesting. As first found in the Pt metal atom chains in the conductor $K_2Pt(CN)_4Br_{0.30}{\cdot}xH_2O$ [220] Kohn anomalies resulting from electron-phonon coupling in one-dimensional systems are expected to be more pronounced than in two-dimensions [221]. A paradigmatic example is that of the atomic hydrogen rows adsorbed on W(110) and Mo(110) discussed in Sect. 11.3.6, although here the chain separation is still too small for being considered a real quasi 1D-system. More insight into the fundamental interactions can be expected from studies of atomic rows at the step edges of high index vicinal surfaces [222]. For example one-dimensional Kr and Xe chains have been shown to exist on the Pt (110) and along the close packed steps of the Pt(997) vicinal surface [223–225]. The electronic band structures of these chains have been investigated with XPS and UPS [226], but so far the phonons have only been studied in a preliminary investigation [227]. Obviously the interpretation of vibrational modes in one dimension will be greatly simplified. Evidence for an electron-phonon induced reconstruction to a dimerized one-dimensional chains of Tl atoms on Cu(001) from photoelectron spectroscopy was reported in 1992 [228] but the preparation of such films could not be reproduced [157]. Similarly one-dimensional chains of Cs and Pb atoms have been prepared on stepped copper (211) and (511) surfaces and characterized by LEED [229, 230].

The dissociation and chemisorption of oxygen at the step edges of metal surfaces has been the subject of several studies also as regards the phonon structure. Besides the pioneer HAS investigation by Niu et al. [231] on the effects of oxygen adsorption at the steps of Ni(977) on its vibrational structure, no other HAS studies have been made to our knowledge on the phonon dispersion curves of atomic rows adsorbed at stepped metal surfaces. Extensive HREELS studies of the vibrational features of oxygen chemisorption at the steps of Ag(001) vicinal surfaces have been reported [232–234] and theoretically analysed with DFPT *ab initio* methods [235, 236] with special attention to the interesting process of oxygen penetration into subsurface sites [237]. The chemical reactivity at steps is just one of several important dynamical aspects of gas-surface interactions [238] where a high resolution spectroscopy like HAS can help understanding the role of low energy vibrations and their contribution to electron-phonon interaction.

References

1. G. Witte, Surf. Sci. **502**, 405 (2002)
2. H. Ibach, D.L. Mills, *Electron Energy Loss Spectroscopy and Surface Vibrations* (Academic Press, New York, 1982)
3. P. Dumas, M.K. Weldon, Y.J. Chabal, G.P. Williams, Surf. Rev. Lett. **6**, 225 (1999)
4. F.M. Hoffmann, Surf. Sci. Rep. **3**, 107 (1983)
5. B.C. Stipe, M.A. Rezaei, W. Ho, Phys. Rev. Lett. **82**, 1724 (1999)
6. F. Moresco, A.G. Meyer, K.-H. Rieder, Mod. Phys. Lett. B **13**, 709 (1999)
7. Y.S.Y. Konishi, K. Kanazawa, S. Yoshida, A. Taninaka, O. Takeuchi, H. Shigekawa, Phys. Rev. B **71**, 193410 (2005)
8. R.J. Hamers, Atomic-Resolution Surface Spectroscopy with the Scanning Tunneling Microscope. Annu. Rev. Phys. Chem. **40**, 531 (1989)
9. R.M. Feenstra, J.A. Stroscio, in *Scanning Tunneling Microscopy*, ed. by W.J. Kaiser, J.A. Stroscio (Academic Press, Inc., San Diego, 1993)
10. R. Wiesendanger, *Scanning Probe Microscopy and Spectroscopy: Methods and Applications* (Cambridge University Press, Cambridge, UK, 1994)
11. R.J. Hamers, D.F. Padowitz, in *Scanning Probe Microscopy and Spectroscopy: Theory, Techniques, and Applications*, 2nd edn., ed. by D.A. Bonnell (Wiley-VCH Inc., New York, 2001)
12. H.J.W. Zandvliet, A. van Houselt, Scanning tunneling spectroscopy. Annu. Rev. Anal. Chem. **2**, 37 (2009)
13. A.M. Lahee, J.P. Toennies, Ch. Wöll, Surf. Sci. **177**, 115 (1986)
14. F. Hofmann, J.P. Toennies, Chem. Rev. **96**, 1307 (1996)
15. J. Ellis, J.P. Toennies, G. Witte, J. Chem. Phys. **102**, 5059 (1995)
16. G. Lange, D. Schmiker, J.P. Toennies, R. Vollmer, H. Weiss, J. Chem. Phys. **103**, 2308 (1995)
17. S. Picaud, P.N.M. Hoang, C. Giradet, Surf. Sci. **322**, 381 (1995)
18. A.P. Graham, Surf. Sci. Rep. **49**, 115 (2003)
19. G. Benedek, D. Campi, J.P. Toennies, *Interaction of Atoms with Surfaces and Surface Phonons,* in *Physics of Solid Surfaces,* Subvol. A, Landolt-Börnstein, Numerical Data and Functional Relationships in Science and Technology—New Ser., Subvol. 45A, ed. by P. Chiaradia, G.F. Chiarotti, Chap. 10, 572–646 (2015)
20. L.W. Brunch, M.W. Cole, E. Zaremba, *Physical Adsorption: Forces and Phenomena*, Clarendon Press, Oxford 1997, Chap. 6
21. L.W. Bruch, R.D. Diehl, J.A. Venables, Rev. Mod. Phys. **79**, 1381(2007)
22. B.N.J. Persson, Surf. Sci. Rep. **15**, 1 (1992)
23. A. Šiber, B. Gumhalter, Phys. Rev. Lett. **81**, 1742 (1998)
24. A. Šiber, B. Gumhalter, J.P. Toennies, Vacuum **54**, 315 (1999)
25. B. Gumhalter, A. Šiber, J.P. Toennies, Phys. Rev. Lett. **83**, 1375 (1999)

26. B.N.J. Persson, *Sliding Friction: Physical Principles and Applications* (Springer, Heidelberg, 1998)
27. B.N.J. Persson, Surf. Sci. Rep. **33**, 83–119 (1999)
28. A. Graham, F. Hofmann, J.P. Toennies, J. Chem. Phys. **104**, 5311 (1996)
29. G.P. Alldredge, R.E. Allen, F.W. de Wette, Phys. Rev. B **4**, 1882 (1971)
30. A.P. Graham, F. Hofmann, J.P. Toennies, G.P. William, C.J. Hirschmigl, J. Ellis, J. Chem. Phys. **108**, 7625 (1998)
31. H. Kato, J. Yoshinobu, M. Kawai, Surf. Sci. **427**, 69 (1999)
32. G.W. Farnell, in *Physical Acoustics*, vol. 6, ed. by W.P. Mason, R.N. Thurston (Academic, New York, 1970), p. 109
33. P. Fouquet, R.A. Olsen, E.J. Baerends, J. Chem. Phys. **119**, 509 (2003)
34. P.R. Kole, H. Hedgeland, A.P. Jardine, W. Allison, J. Ellis, G. Alexandrowicz, J. Phys.: Condensed Matter **24**, 104016 (2012)
35. A.M. Lahee, J.R. Manson, J.P. Toennies, Ch. Wöll, J. Chem. Phys. **86**, 7194 (1987)
36. A.M. Lahee, J.R. Manson, J.P. Toennies, Ch. Wöll, Phys. Rev. Lett. **57**, 471 (1986)
37. B.H. Choi, K.T. Tang, J.P. Toennies, J. Chem. Phys. **107**, 9487 (1997)
38. B.H. Choi, A.P. Graham, K.T. Tang, J.P. Toennies, J. Chem. Phys. **112**, 10538 (2000)
39. B.H. Choi, A.P. Graham, K.T. Tang, J.P. Toennies, J. Chem. Phys. **114**, 2883 (2001)
40. B.H. Choi, A.P. Graham, K.T. Tang, J.P. Toennies, J. Chem. Phys. **116**, 7695 (2002)
41. A. Politano, G. Chiarello, G. Benedek, E.V. Chulkov, P.M. Echenique, Surf. Sci. Rep. **68**, 305–389 (2013)
42. T. Aruga, Y. Murata, Prog. Surf. Sci. **31**, 61 (1989)
43. See articles in *Physics and Chemistry of Alkali Metal Absorption*, ed. by H.P. Bonzel, A.M. Bradshaw, G. Ertl (Elsevier, Amsterdam, 1989)
44. A.P. Graham, J.P. Toennies, Phys. Rev. B **56**, 15378 (1997-I)
45. A.P. Graham, J.P. Toennies, G. Benedek, Surf. Sci. **556**, L143 (2004)
46. J. Cui, J.D. White, R.D. Diehl, Surf. Sci. Lett. **293**, L841 (1993)
47. P. Senet, J.P. Toennies, G. Witte, Chem. Phys. Lett. **299**, 389 (1999)
48. K.H. Lau, W. Kohn, Surf. Sci. **75**, 3629 (1978)
49. P. Hyldgaard, M. Persson, J. Phys. Cond. Matter **12**, L13 (2000)
50. S.D. Kevan, Phys. Rev. B **28**, 2268 (1983); H. Bross, M. Kauzmann, Phys. Rev. B **51**, 17135 (1995)
51. E. Wahlström, I. Ekvall, H. Olin, L. Walldén, Appl. Phys. A **66**, 1107 (1998)
52. J. Repp, F. Moresco, G. Mayer, K.-H. Rieder, P. Hyldgaard, M. Persson, Phys. Rev. Lett. **85**, 2981 (2000)
53. N. Knorr, H. Brune, M. Hepple, A. Hirstein, M.A. Schneider, K. Kern, Phys. Rev. B **65**, 115420 (2002)
54. H.S. Nalwa (ed.), *Encyclopedia of Nanoscience and Nanotechnology* (American Scientific, New York, 2004)
55. J.A. Venables, *Introduction to Surface and Thin Film Processes* (Cambridge University Press, Cambridge, 2000)
56. D. Farias, K.-H. Rieder, Rep. Prog. Phys. **61**, 1575–1664 (1998)
57. R.E. Allen, G.P. Alldredge, F.W. de Wette, Phys. Rev. Lett. **23**, 1285 (1969)
58. T.S. Chen, G.P. Alldredge, F.W. de Wette, R.E. Allen, Phys. Rev. Lett. **26**, 1543 (1971)
59. R.E. Allen, G.P. Alldredge, F.W. de Wette, Phys. Rev. B **4**, 1648 (1971)
60. R.E. Allen, G.P. Alldredge, F.W. de Wette, Phys. Rev. B **4**, 1661 (1971)
61. R.E. Allen, G.P. Alldredge, F.W. de Wette, Phys. Rev. B **4**, 1682 (1971)
62. L. Wojtczak, Thin Solid Films **4**, 229 (1969)
63. O. Litzman, V. Janku, Surf. Sci. **18**, 3 (1969)
64. E. Tosatti, Festkörperprobleme XV (1975), p. 113
65. H. Böttger, *Principle of the Theory of Lattice Dynamics* (Akademie-Verlag, Berlin, 1983)
66. F.K. Schulte, Surf. Sci. **55**, 427 (1976)
67. J.P. Rogers III, P.H. Cutler, T.E. Feuchtwang, A.A. Lucas, Surf. Sci. **181**, 436 (1987)
68. B. Gumhalter, Phys. Rep. **351**, 1–159 (2001)

69. G.W. Farnell, E.L. Adler, in *Physical Acoustics Principles and Methods*, vol. IX, ed. by W. P. Mason, R.N. Thurston (Academic Press, New York, 1972), p. 35
70. B.A. Auld, *Acoustic Fields and Waves in Solids* (Wiley, New York, 1973)
71. B.M. Axilrod, E. Teller, J. Chem. Phys. **11** (1943) 299
72. S.Å. Lindgren, L. Walldén, Phys. Rev. **61**, 2894 (1988)
73. A. Šiber, B. Gumhalter, Prog. Surf. Sci. **74**, 375 (2003)
74. A.D. McLachlan, Mol. Phys. **7**, 381 (1964)
75. J.M. Hutson, C. Schwartz, J. Chem Phys. **79**, 5179 (1983)
76. S. Chung, N. Holter, M.W. Cole, Phys. Rev. B. **31**, 5179 (1985)
77. B.F. Mason, B.R. Williams, Surf. Sci. **139**, 173 (1984)
78. K.D. Gibson, S.J. Sibener, Phys. Rev. Lett. **55**, 1514 (1985)
79. B. Gumhalter, Ref. [59], Sect. 3.2
80. K.T. Tang, J.P. Toennies, J. Chem. Phys. **118**, 4976 (2003)
81. K. Kern, G. Comsa, in *Advances in Chemical Physics*, ed. by K.P. Lawley (Wiley, 1989), p. 211. This article contains a very good review on surface phase transitions in clean surface and rare gas vibrations as studied by HAS
82. P. Zeppenfeld, M. Büchel, R. David, G. Comsa, C. Ramseyer, C. Giradet, Phys. Rev. B **50**, 14667 (1994), and references therein
83. C. Ramseyer, C. Girardet, P. Zeppenfeld, J. George, M. Büchel, G. Comsa, Surf. Sci. **313**, 251 (1994)
84. A.P. Graham, M. Bertino, F. Hofmann, J.P. Toennies, Ch. Wöll, J. Chem. Phys. **106**, 6194 (1997)
85. A.P. Graham, M. Bertino, F. Hofmann, J.P. Toennies, Ch. Wöll, J. Chem. Phys. **107**, 4445 (1997)
86. J. Braun, D. Fuhrmann, A. Šiber, B. Gumhalter, Ch. Wöll, Phys. Rev. Lett. **80**, 125 (1998)
87. B. Hall, D.L. Mills, J.E. Black, Phys. Rev. **32**, 4932 (1985)
88. R.A. Aziz, M.J. Slaman, Mol. Phys. **58**, 679 (1986)
89. A.K. Dham, A.R. Allnatt, W.J. Meath, R.A. Aziz, Mol. Phys. **67**, 1291 (1989)
90. A.K. Dham, W.J. Meath, A.R. Allnatt, R.A. Aziz, M.J. Slaman, Chem. Phys. **142**, 173 (1990)
91. R. Gerlach, A.P. Graham, J.P. Toennies, H. Weiss, J. Chem. Phys. **109**, 5319 (1998)
92. A. Šiber, B. Gumhalter, J. Braun, A.P. Graham, M.F. Bertino, J.P. Toennies, D. Fuhrmann, Ch. Wöll, Phys. Rev. B **59**, 5898 (1999)
93. K. Burke, B. Gumhalter, D.C. Langreth, Phys. Rev. B **47**, 12852 (1993)
94. B. Gumhalter, K. Burke, D.C. Langreth, Surf. Rev. Lett. **1**, 133 (1994)
95. A. Bilić, B. Gumhalter, Phys. Rev. B **52**, 12307 (1995)
96. J. Braun, D. Fuhrmann, M. Bertino, A.P. Graham, J.P. Toennies, Ch. Wöll, A. Bilić, B. Gumhalter, J. Chem. Phys. **106**, 9911 and 9922 (1997)
97. M. Wolf, E. Knoesel, T. Hertel, Phys. Rev. B **54**, R5295 (1996)
98. L.W. Bruch, J. Chem. Phys. **107**, 4443 (1997)
99. L.W. Bruch, A.P. Graham, J.P. Toennies, Mol. Phys. **95**, 579 (1998)
100. L.W. Bruch, A.P. Graham, J.P. Toennies, J. Chem. Phys. **112**, 3314 (2000)
101. L.W. Bruch, F.Y. Hansen, J. Chem. Phys. **122**, 114714 (2005)
102. L.W. Bruch, F.Y. Hansen, APS March Meeting Abstr. H10.00005:Lo (2006)
103. F.Y. Hansen, L.W. Bruch, J. Chem. Phys. **127**, 204708 (2007)
104. P. Lazić, Ž. Crljen, R. Brako, B. Gumhalter, Phys. Rev. B **72**, 245407 (2005)
105. D. Campi, M. Bernasconi, G. Benedek, J.P. Toennies, J. Phys. Chem. C **119**(26), 14579 (2015)
106. W. Daum, C. Stuhlmann, H. Ibach, Phys. Rev. Lett. **60**, 2741 (1988)
107. Y. Chen, S.Y. Tang, J.S. Kim, M.H. Mohamed, L.L. Kesmodel, Phys. Rev. B **43**, 6788 (1991)
108. M.H. Mohamed, J.-S. Kim, L.L. Kesmodel, Phys. Rev. B **40**, 529 (1989)
109. H. Ibach, Surf. Sci. Rep. **29**, 193 (1997)
110. W. Daum, J. Electron Spectrosc. Relat. Phenom. **44**, 271 (1987)

111. J.E. Black, D.L. Mills, W. Daum, C. Stuhlmann, H. Ibach, Surf. Sci. **217**, 529 (1989)
112. J.G. Witte, J.P. Toennies, Phys. Rev. B **62**, R7771 (2000)
113. J. Cousty, R. Riwan, P. Soukiassan, Surf. Sci. **152**, 297 (1985)
114. J. Mizuki, C. Stassis, Phys. Rev. B **34**, 5890 (1986)
115. E. Hulpke, J. Lower, A. Reichmuth, Phys. Rev. B **53**, 13901 (1996)
116. J.D. White, J. Cui, M. Strauss, R.D. Diehl, F. Ancillotto, F. Toigo, Surf. Sci. **307–309**, 1134 (1998)
117. D. Campi, M. Bernasconi, G. Benedek, A.P. Graham, J.P. Toennies, in *15th International Conference on Vibrations at Surfaces*, ed. by T. Frederiksen (DIPC, Donostia-San Sebastián, 2015), p. 77
118. D. Campi, M. Bernasconi, G. Benedek, A.P. Graham, J.P. Toennies, Phys. Chem. Chem. Phys. **19**, 16358 (2017)
119. V. Pannella, A.L. Glebov, J.P. Toennies, C. Sebenne, C. Eckl, C. Adler, P. Pavone, U. Schröder, Phys. Rev. B **59**, 15772 (1999)
120. G. Benedek, J. Ellis, A. Reichmuth, P. Ruggerone, H. Schief, J.P. Toennies, Phys. Rev. Lett. **69**, 2951 (1992). In this paper the $\hbar\omega$ scale on the right side of Fig. 1 should be divided by a factor 2
121. A.D.B. Woods, B.N. Brockhouse, R.H. March, D.A.T. Stewart, R. Bowers, Phys. Rev. **128**, 1112 (1962)
122. K. Sezawa, *Dispersion of Elastic Waves propagating on the Surface of Stratified Bodies and on Curved Surfaces*. Bull. Earthq. Res. Inst. (Tokyo Univ.) **3**, 1 (1927); K. Sezawa, K. Kanai, *ibidem* **13**, 237 (1935), and Proc. Imp. Acad. **11**, 13 (1935)
123. V. Bortolani, A.M. Marvin, F. Nizzoli, G. Santoro, J. Phys. C **16**, 1751 (1983)
124. G.I. Stegeman, F. Nizzoli, in *Surface Excitations*, ed. by V.M. Agranovich, R. Loudon (Elsevier Science, Amsterdam, 1984), p. 195
125. G. Fritsch, F. Geipel, A. Prasetyo, J. Phys. Chem. Solids **34**, 1961 (1973)
126. N.S. Luo, P. Ruggerone, J.P. Toennies, Phys. Rev. B **54**, 5051 (1996)
127. H.L. Skriver, N.M. Rosengaard, Phys. Rev. B **46**, 7157 (1992)
128. E. Hulpke, J. Lower, A. Reichmuth, Phys. Rev. B **53**, 13901 (1996-II)
129. H. Range, Thesis (University of Göttingen, 1995 and Max-Planck-Institut für Strömungsforschung Göttingen, Bericht 3/1995)
130. S.A. Safron, G.G. Bishop, J. Duan, E.S. Gillman, J.G. Skofronick, N.S. Luo, P. Ruggerone, J. Phys. Chem. **97**, 2270 (1993)
131. J. Braun, J.P. Toennies, Surf. Sci. **368**, 226 (1996)
132. G. Zhang, Thesis (University of Göttingen 1991 and Max-Planck-Institut für Strömungsforschung, Bericht 102/1991)
133. J. Braun, P. Ruggerone, G. Zhang, J.P. Toennies, G. Benedek, Phys. Rev. B **79**, 205423 (2009)
134. W. Steinhögl, Ph.D. Thesis (University of Göttingen, 2000)
135. G. Benedek, E. Hulpke, W. Steinhögl, Phys. Rev. Lett. **87**, 027201 (2001)
136. G. Benedek, M. Bernasconi, V. Chis, E. Chulkov, P.M. Echenique, B. Hellsing, J. P. Toennies, J. Phys.: Condens. Matter **22**, 084020 (2010)
137. N.W. Ashcroft, N.D. Mermin, Solid State Physics (Brooks Cole, 1976) 1st ed.
138. G. Benedek, N.S. Luo, P. Ruggerone, A. Reichmuth, J.P. Toennies, Mater. Sci. Eng. B **23**, 123 (1994)
139. B.J. Hinch, C. Koziol, J.P. Toennies, G. Zhang, Europhys. Lett. **10**, 341 (1989)
140. J. Braun, J.P. Toennies, Surf. Sci. **387**, L858 (1997)
141. T.L. Chan, C. Wang, M. Hupalo, M. Tringides, W. Ho, Phys. Rev. Lett. **96**, 226102 (2006)
142. F. Calleja, M.G.C. Passeggi, J.J. Hinarejos, A.L.V. de Parga, R. Miranda, Phys. Rev. Lett. **97**, 186104 (2006)
143. Y. Guo, Y.-F. Zhang, X.-Y. Bao, T.-Z. Han, Z. Tang, L.-X. Zhang, W.-G. Zhu, E.G. Wang, Q. Niu, Z.Q. Qiu, J.-F. Jia, Z.-X. Zhao, Q.K. Xue, Science **306**, 1915 (2004)
144. D. Eom, S. Qin, M.Y. Chou, C.K. Shih, Phys. Rev. Lett. **96**, 027005 (2006)
145. M.M. Ozer, J.P. Thompson, H.H. Weitering, Nat. Phys. **2**, 173 (2006)

146. C. Brun, I.P. Hong, F. Patthey, I.Y. Sklyadneva, R. Heid, P.M. Echenique, K.P. Bohnen, E. V. Chulkov, W.-D. Schneider, Phys. Rev. Lett. **102**, 207001 (2009)
147. S. Qin, J. Kim, Q. Niu, C.-K. Shih, Science **324**, 1314–1317 (2009)
148. T. Zhang, P. Cheng, W.-J. Li, Y.-J. Sun, G. Wang, X.-G. Zhu, K. He, L. Wang, X. Ma, X. Chen, Y. Wang, Y. Liu, H.-Q. Lin, J.F. Jia, Q.-K. Xue, Nat. Phys. **6**, 104 (2010)
149. I.Y. Sklyadneva, R. Heid, K.-P. Bohnen, P.M. Echenique, E.V. Chulkov, Phys. Rev. B **87**, 085440 (2013)
150. N.S. Luo, P. Ruggerone, J.P. Toennies, G. Benedek, V. Celli, J. Electron Spectrosc. Relat. Phenom. **64**(65), 755 (1993)
151. I. Yu. Sklyadneva, G. Benedek, E.V. Chulkov, P.M. Echenique, R. Heid, K.-P. Bohnen, J. P. Toennies, Phys. Rev. Lett. **107**, 095502 (2011)
152. G. Benedek, M. Bernasconi, K.-P. Bohnen, D. Campi, E.V. Chulkov, P.M. Echenique, R. Heid, I.Yu. Sklyadneva, J.P. Toennies, Phys. Chem. Chem. Phys. **16**, 7159–7172 (2014)
153. G.G. Rusina, S.D. Borisova, S.V. Eremeev, I.Yu. Sklyadneva, E.V. Chulkov, G. Benedek, J. P. Toennies, J. Phys. Chem. C **120**, 22304 (2016)
154. F. Yndurain, M.P. Jigato, PRL **100**, 205501 (2008)
155. R. Otero, A.L. Vázquez de Parga, R. Miranda, Phys. Rev. B **55** 10791 (2002)
156. F. Calleja, A.L. Vázquez de Parga, E. Anglada, J.J. Hinarejos, R. Miranda, F. Yndurain, New J. Phys. **11**, 123003 (2009)
157. J. Braun, Thesis (University of Göttingen, 1997 and Max-Planck-Institut für Strömungsforschung Göttingen, Bericht 11/1997)
158. R. Stoneley, Proc. R. Soc. Lond. A **106**, 416 (1924)
159. R. Stoneley, Proc. R. Soc. London A **232**, 447 (1955)
160. I.Y. Sklyadneva, R. Heid, K.-P. Bohnen, P.M. Echenique, E.V. Chulkov, J. Phys.: Condens. Matter **24**, 104004 (2012)
161. P. Saalfrank, Surface Sci. **274**, 449 (1992)
162. W.L. McMillan, Phys. Rev. **167**, 331 (1968)
163. G. Grimvall, *The Electron-Phonon Interaction in Metals* (North-Holland, 1981)
164. A.-L. Barabasi, H.E. Stanley, *Fractal Concepts in Surface Growth* (Cambridge University Press, Cambridge, 1995)
165. B.N.J. Persson, *Sliding Friction* (Springer, Berlin, 1998)
166. A. Ulman, *An Introduction to Ultrathin Organic Films: From Langmuir-Blodgett to Self-assembly* (Academic Press, Boston, MA, 1991)
167. G. Cuniberti, G. Fargas, K. Richter (eds.), *Introducing Molecular Electronics* (Springer, Berlin, 2005)
168. J. R. Heath, Molecular electronics. Annu. Rev. Mater. Res. **39**, 1 (2009). For a more recent account on molecular electronics see the Editorial and following papers in Nat. Nanotechnol. **8**, 377 (2013)
169. C. Wöll, Physical and Chemical Aspects of Organic Electronics (Wiley-VCH, Weinheim, 2009)
170. F.P. Netzer, M.G. Ramsey, Crit. Rev. Solid State Mater. Sci. **17**(5), 397 (1992)
171. N. Sheppard, Annu. Rev. Phys. Chem. **39**, 596 (1998)
172. G. Witte, Ch. Wöll, J. Chem. Phys. **103**, 5960 (1995)
173. D. Fuhrmann, Ch. Wöll, Surf. Sci. **368**, 20 (1996)
174. D. Fuhrmann, Ch. Wöll, Surf. Sci. **377**, 544 (1997)
175. G. Witte, K. Weiss, P. Jacob, J. Braun, K.L. Koslov, Ch. Wöll, Phys. Rev. Lett. **80**, 121 (1998)
176. D. Fuhrmann, R. Gerlach, H.G. Rubahn, Ch. Wöll, Surf. Sci. **424**, 145 (1999)
177. D. Fuhrmann, Ch. Wöll, New J. Phys. **1**, 1 (1998)
178. J.M. Kosterlitz, D.J. Thouless, J. Phys. C: Solid State Physics **6**, 1181 (1973)
179. A. Glebov, A.P. Graham, A. Menzel, J.P. Toennies, J. Chem. Phys. **106**, 9382 (1997)
180. J. Braun, A. Glebov, A.P. Graham, A. Menzel, J.P. Toennies, Phys. Rev. Lett. **80**, 2638 (1998)
181. J.P. Toennies, F. Traeger, J. Vogt, H. Weiss, J. Chem. Phys. **120**, 11347 (2004)

182. M. Gallagher, A. Omer, G.R. Darling, A. Hodgson, Faraday Discuss. **141**, 231
183. F. Schreiber, Prog. Surf. Sci. **65**, 151–257 (2000)
184. M. Boncheva, G.M. Whitesides, in *Encyclopædia of Nanoscience and Nanotechnology*, ed. by J.A. Schwarz, C. Contescu, K. Putyera (Marcel Dekker, New York, 2004), p. 287
185. M. Boncheva, G.M. Whitesides, MRS Bull. **30**, 736 (2005)
186. M.V. Tirrell, A. Katz, MRS Bull. **30**, 700 (2005)
187. J. Yoon, W. Lee, E.L. Thomas, MRS Bull. **30**, 721 (2005)
188. A. Ulman, *Self-assembled Monolayers of Thiols Thin Films* (Academic Press, San Diego, CA, 1998)
189. N. Camillone III, C.E.D. Chidsey, G.-Y. Liu, T.M. Putvinski, G. Scoles, J. Chem. Phys. **94**, 8493 (1991)
190. N. Camillone III, T.Y.B. Leung, G. Scoles, Surf. Sci. **373**, 333 (1997)
191. S.M. Wetterer, D.J. Lavrich, T. Cummings, S.L. Bernasek, G. Scoles, J. Phys. Chem. B **102**, 9266 (1998)
192. K.G. Loepp, S. Vollmer, G. Witte, Ch. Wöll, Langmuir **15**, 3767 (1999)
193. S. Vollmer, P. Fouquet, G. Witte, Ch. Boas, M. Kunat, U. Burghaus, Ch. Wöll, Surf. Sci. **462**, 135 (2000)
194. S.B. Darling, A.W. Rosenbaum, Y. Wang, S.J. Sibener, Langmuir **18**, 7462 (2002)
195. P.V. Schwartz, D.J. Lavrich, G. Scoles, Langmuir **19**, 4969 (2003)
196. J. Hautman, M.L. Klein, J. Chem. Phys. **93**, 7483 (1990)
197. S.B.M. Bosio, W.L. Hase, J. Chem. Phys. **107**, 9677 (1997)
198. T. Yan, W.L. Hase, J. Phys. Chem. A **105**, 2617 (2001)
199. A.W. Rosenbaum, M.A. Freedman, S.B. Darling, I. Popova, S.J. Sibener, J. Chem. Phys. **120**, 3880 (2004)
200. V. Vogel, Ch. Wöll, Thin Solid Films **159**, 429 (1988)
201. C.E.D. Chidsey, G. Liu, P. Rowntree, G. Scoles, J. Chem. Phys. **91**, 4421 (1989)
202. N. Camillone, C.E.D. Chidsey, G. Liu, G. Scoles, J. Chem. Phys. **98**, 3503 (1993)
203. S.B. Darling, A.W. Rosenbaum, S.J. Sibener, Surf. Sci. **478**, L313 (2001)
204. G.R. Poirier, Langmuir **15**, 1167 (1999)
205. H. Rieley, G.H. Kendall, A. Chan, R.G. Jones, J. Lüdecke, D.P. Woodruff, B.C.C. Cowie, Surf. Sci. **392**, 143 (1997)
206. A. Imanishi, K. Isawa, F. Matsui, T. Tsuduki, T. Yokoyama, H. Kondoh, Y. Kitajima, T. Ohta, Surf. Sci. **407**, 282 (1998)
207. P. Fouquet, A.P. Jardine, S. Dworski, G. Alexandrowicz, W. Allison, J. Ellis, Rev. Sci. Instrum. **76**, 053109 (2005)
208. M.R. Linford, P. Fenter, P.M. Eisenberger, C.E.D. Chidsey, J. Am. Chem. Soc. **117**, 3145 (1995)
209. A. Bansal, X. Li, I. Lauermann, N.S. Lewis, S.I. Yi, W.H. Weinberg, J. Am. Chem. Soc. **118**, 7225 (1996)
210. A. Bansal, N.S. Lewis, J. Phys. Chem. B **102**, 4058 (1998)
211. T.L. Lasseter, B.H. Clare, N.L. Abbott, R.J. Hamers, J. Am. Chem. Soc. **126**, 10220 (2004)
212. W. Yang, R.J. Hamers, Appl. Phys. Lett. **85**, 3626 (2004)
213. T. Osaka, M. Matsunaga, S. Kudo, D. Niwa, Y. Shacham-Diamand, W. Jaegermann, R. Hunger, J. Electrochem. Soc. **154**, H919 (2007)
214. X. Shen, B. Sun, F. Yan, J. Zhao, F. Zhang, S. Wang, X. Zhu, S. Lee, ACS Nano **4**, 5869 (2010)
215. R.D. Brown, Q.Q. Tong, J.S. Becker, M.A. Freedman, N.A. Yufa, S.J. Sibener, Faraday Discuss. **157**, 307 (2012)
216. R.D. Brown, Z.M. Hund, D. Campi, L.E. O'Leary, N.S. Lewis, M. Bernasconi, G. Benedek, S.J. Sibener, Phys. Rev. Lett. **110**, 156102 (2013)
217. R.D. Brown, Z.M. Hund, D. Campi, L.E. O'Leary, N.S. Lewis, M. Bernasconi, G. Benedek, S.J. Sibener, J. Chem. Phys. **141**, 024702 (2014)
218. R.D. Brown, Z.M. Hund, K.J. Nihill, D. Campi, K.T. Wong, N.S. Lewis, M. Bernasconi, G. Benedek, S.J. Sibener, J. Chem. Phys. **143**, 124705 (2015)

219. Z.M. Hund, K.J. Nihill, D. Campi, K.T. Wong, N.S. Lewis, M. Bernasconi, G. Benedek, S. J. Sibener, J. Phys. Chem. C **119**, 18458 (2015)
220. R. Comes, M. Lambert, H. Launois, H.R. Zeller, Phys. Rev. B **8**, 571 (1973)
221. G. Grüner, *Density Waves in Solids* (Addison-Wesley, Reading, MA, 1994)
222. T. Jung, Y.W. Mo, F.J. Himpsel, Phys. Rev. Lett. **74**, 1641 (1995)
223. V. Marsico, M. Blanc, K. Kuhnke, K. Kern, Phys. Rev. Lett. **78**, 94 (1997)
224. W. Widdra, P. Trischberger, W. Friess, D. Menzel, S.H. Payne, H.J. Kreutzer, Phys. Rev. B **57**, 4111 (1998-I)
225. V. Pouthier, C. Ramseyer, C. Girardet, K. Kuhnke, V. Marsico, M. Blanc, R. Schuster, K. Kern, Phys. Rev. B **56**, 4211 (1997)
226. M. Weinelt, P. Trischberger, W. Widdra, K. Eberle, P. Zebisch, S. Gokhale, D. Menzel, J. Henk, R. Feder, H. Dröge, H.-P. Steinrück, Phys. Rev. B **32**, R17048 (1995)
227. B.J. Hinch, A. Lock, H.H. Madden, J.P. Toennies, G. Witte, J. Electron Spectrosc. Relat. Phenom. **51**, 673–688 (1990)
228. C. Binng, C. Norris, M.G. Barthes-Labrouse, Phys. Scr. T **45**, 283 (1992)
229. J. Braun, J.P. Toennies, G. Witte, Surf. Sci. **340**, 265 (1995)
230. F. Hofmann, U. Swenson, J.P. Toennies, Surf. Sci. **371**, 169 (1997)
231. L. Niu, D.J. Gaspar, S.J. Sibener, Science **268**, 847 (1995)
232. L. Savio, L. Vattuone, M. Rocca, Phys. Rev. Lett. **87**, 276101 (2001)
233. L. Savio, L. Vattuone, M. Rocca, J. Phys.: Condens. Matter **14**, 6065 (2002)
234. L. Vattuone, L. Savio, M. Rocca, Phys. Rev. Lett. **90**, 228302 (2003)
235. N. Bonini, A. Kokalj, A. Dal Corso, S. de Gironcoli, S. Baroni, Phys. Rev. B **69**, 195401 (2004)
236. A. Kokalj, N. Bonini, A. Dal Corso, S. de Gironcoli, S. Baroni, Surf. Sci. **599**, 1107 (2004)
237. N. Bonini, A. Dal Corso, A. Kokalj, S. de Gironcoli, S. Baroni, Surf. Sci. **587**, 50 (2005)
238. R. Diez-Muiño, H.F. Busnengo (eds.), *Dynamics of Gas-Surface Interactions* (Springer, Berlin, 2013)

Chapter 13
Experimental Results: Beyond Single Phonons

You see things; and you say "Why?" But I dream things that never were; and I say "Why not?".
(George Bernard Shaw, *Back to Methuselah*, 1921)

Abstract The previous chapters have been devoted to summarizing the vast knowledge acquired about the single phonon vibrations of atoms in or near the surface and about atoms and molecules on the surface and their dispersion curves. This new knowledge was made possible largely thanks to the development of He atom scattering spectroscopy and the progress made in theoretical analysis. Single phonon inelastic scattering processes were implicit since only in this case is it possible to extract accurate phonon dispersion curves from the data. The high resolution of HAS spectroscopy, especially in its ^{3}He-spin echo version also allows for the observation of dynamical phenomena far beyond the harmonic motion, like multiphonon processes involving low energy phonons or the diffusion of adsorbates on the surface. Moreover, the unique property of closed shell He atoms to exchange energy and momentum with the ion cores of the surface atoms via the virtual excitation of interposed electrons makes HAS also a valuable tool for investigating collective surface electron excitations in the THz domain. It therefore seems appropriate to discuss also some of these and other emerging areas of HAS research where new important insight can be expected in the future.

13.1 Multiphonon Processes

Many applications of technical interest, such as heat transfer to walls in aerodynamic flows, or the understanding of the different elementary processes in surface chemical reactions involve the energy exchange of relatively heavy collision partners such as O_2, NO and N_2 in thermal collisions with surfaces [1, 2]. In this case the inelastic interaction is frequently dominated by multiphonon interactions. The resulting smearing in the energy loss distributions makes it difficult to isolate effects arising from differences in the surface response as well as in the interaction

G. Benedek and J. P. Toennies, *Atomic Scale Dynamics at Surfaces*, Springer Series in Surface Sciences 63, https://doi.org/10.1007/978-3-662-56443-1_13

mechanisms specific to the molecules. Although the interactions of helium atoms with surfaces are not of direct technological interest, HAS provides a unique opportunity to study the transition from the regime of single phonon quantum scattering to multiphonon more classical-like scattering. Thus it is possible to isolate the influence of the phonon dispersion curves and the mechanisms responsible for the multiphonon processes. The basic theoretical and kinematical concepts of multiphonon HAS have been discussed in Sect. 7.5 and should provide a reference for the following discussion of some paradigmatic examples.

Figure 13.1a shows typical HAS energy transfer spectra for scattering from Pt (111) along the $[1\bar{1}0]$ direction, measured at a comparatively high incident energy of 69 meV and a surface temperature of 160 K. In addition to the features usually assigned to one-phonon processes at incident angles below about 38°, a large broad peak appears at energies above the maximum energy of the bulk phonon bands (Fig. 13.1b) [3]. Since this peak is not possible in a single phonon process it provides clear evidence for a multiphonon process mostly involving surface phonons according to the scheme illustrated in Fig. 7.9.

This work was subsequently extended to the transition from the single phonon quantum regime to the multiphonon classical regime in He scattering from Cu(001) in a comprehensive study by Hofmann et al. [4–6]. Figure 13.2 shows a series of

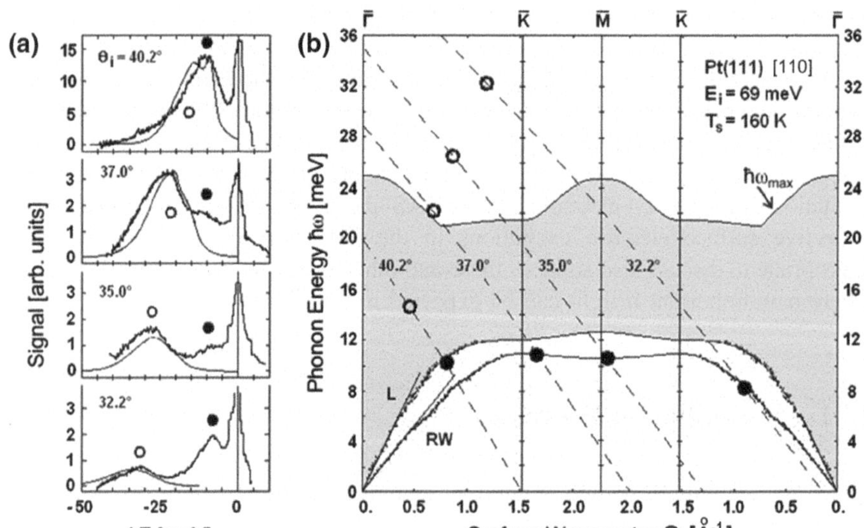

Fig. 13.1 Energy transfer and effective dispersion curves in comparatively high energy collisions of 69 meV He atoms with the Pt(111) surface at a surface temperature of 160 K. **a** Energy loss spectra for four representative incident angles. The energies of the one-phonon and multiphonon peaks are marked by full and open circles, respectively. **b** The corresponding scan curves together with the dispersion curves of the Rayleigh wave (RW), the longitudinal resonance (L) and the upper edge ($\hbar\omega_{\max}$) of the bulk phonon band (adapted from [3])

inelastic HAS spectra from the Cu(001) surface measured along [110] at the comparatively high incident energies of 82 meV (a) and 113 meV (b), for a series of increasing surface temperatures T_s [5, 6]. An incident angle $\Delta\theta = 3°$ away from the specular was chosen to avoid the much more intense specular peak. Partly resolved distinct features of the one-phonon spectrum are still visible at $T_s \sim 118$ K and 82 meV impact energy and even at $T_s \sim 119$ K and 113 meV. With increasing temperature they gradually evolve into a broad, unstructured multiphonon peak. At high temperatures and incident energies the multiphonon peak takes on a Gaussian-like shape [see Sect. 7.6.2, (7.64) and (7.65)]. The results are fitted by a many-body scattering theory which Manson has adapted from the continuum model theory for neutron scattering [7, 8]. In the continuum model the atoms are assumed to be reflected by a smooth but vibrating potential barrier rather than by a lattice of repulsive atom cores as previously postulated (Sect. 7.6). By approximating the trajectories by classical paths and assuming that the collision time is short compared to the dominant vibrational frequencies (*quick scattering limit*), Manson's theory extends from the quantum low energy limit to the classical high energy limit. As seen in Figs. 13.2 and 13.3, calculations based on the continuum model theory for parameters fitted to either the whole set of data (dashed curves) or to the low temperature single phonon data (dashed-dotted lines in Fig. 13.2) account well for the change of the peak shape with temperature. As expected, the agreement improves with increasing temperature [6]. For the comparisons in Fig. 13.2, the required parameters, β (Born Mayer repulsive range parameter), Q_c (wavevector cut off), v_R (Rayleigh wave velocity) and θ_{Ds} (Debye temperature) were fitted either to the entire set of data or to the low temperature single phonon data. In both cases the agreement is quite satisfactory.

Figure 13.3 illustrates the good agreement of the HAS spectra for five different incident angles ($T_s \sim 800$ K and $E_i = 113$ meV) with the continuum model calculations using the same parameters fitted to the entire set of data. The broad peaks at $\Delta\theta = +3°$ and $-3°$ are centered on the positive (phonon annihilation) and negative (phonon creation) energy transfer side, respectively, and are almost symmetric with respect to the elastic scattering at $\Delta E = 0$. The diffuse elastic scattering from static surface defects (see discussion of Fig. 9.5), normally found under single phonon conditions at $\Delta E = 0$, does not make a noticeable contribution at these high temperatures. The static defects are completely overshadowed by the large phonon mean-square displacements which act as dynamical defects. It is also interesting to note in Fig. 13.3 that the multiphonon peak intensity decreases rapidly with increasing angle away from the specular one. This behaviour correlates closely with the intensity dependence of the single phonon peak.

As discussed in Sect. 7.6, multiphonon atom-surface scattering evolves from the one-phonon quantum regime, valid for a Debye-Waller exponent $2W \approx 1/2$, to the semi-classical regime when $2W >> 1$. This transition has also been discussed by Manson, in the light of the various approximations usually made in the theory of atom-surface scattering [8], and by Gumhalter and Bilić in the framework of the exponentiated Born approximation (EBA) applied to the specific case of He

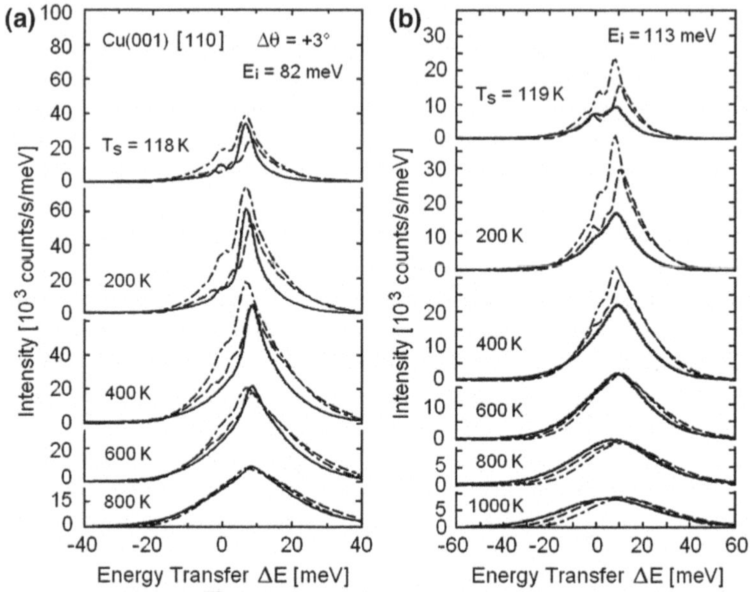

Fig. 13.2 A series of HAS energy transfer spectra from a clean Cu(001) surface along the [110] azimuth at different surface temperatures T_s for an incident angle $\Delta\theta = +3°$ away from the specular peak. **a** At an incident energy of 82 meV. **b** At an incident energy of 113 meV distinct features of the phonon spectrum, which are still visible at ~118 K gradually evolve with increasing temperatures into a broad, unstructured multiphonon peak (solid lines). The continuum model theory [7, 8] for parameters fitted to either the entire set of data (dashed curves) or to the low temperature single phonon data (dashed-dotted lines) account reasonably well for the change of the peak shape with temperature (adapted from [6])

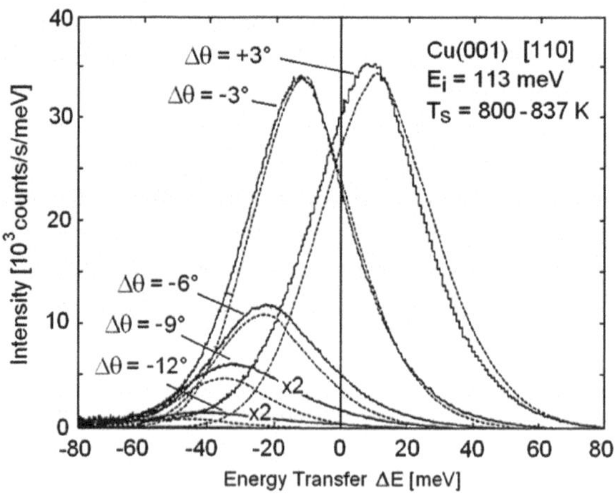

Fig. 13.3 Inelastic HAS spectra from a clean Cu(001) surface along the [110] azimuth at an incident energy of 113 meV for different incident angles away from the specular peak. The surface temperature is 800 K for $\Delta\theta = +3°$ and 837 K for the other angles. The shape and shift of the unstructured quasi-Gaussian shape of the multiphonon peaks are well reproduced by the continuum model calculation based on a fit of the entire set of data (short dashed curves) (adapted from [5])

scattering from Cu(001) [9]. Manson concluded that the multiphonon background depends very weakly on the specific form of the phonon spectrum, but carries relevant information on the atom-surface interaction potential [8].

Gumhalter has presented a strictly quantum mechanical derivation of the energy and wavevector resolved scattering spectrum [10]. A semiclassical theory has also been developed by Pollak and collaborators and applied to the multiphonon scattering of Ne, Ar and Kr from Cu(111) [11]. Gumhalter has comprehensively reviewed both the experimental and theoretical work on single and multiphonon scattering in the quantum limit as of 2001 [12].

A completely different multiphonon spectrum is found for weakly coupled adsorbate monolayers, which have flat dispersionless phonon branches, such as, e.g., the S-mode branch of rare gases on metal surfaces (Figs. 12.5, 12.6 and 12.8 b). Figure 13.4 compares HAS energy transfer spectra from a monolayer of Xe on Cu(111) at a low incident energy at which single phonon excitations occur and at about 5 times greater energy [13–16]. Similar results have also been reported and analysed for He atom scattering at 10.5 meV from Xe(111) [17]. At the high incident energy, but at a sufficiently low temperature to avoid anharmonic broadening (Fig. 13.4b), the multiphonon spectra consists of a series of well-resolved peaks located at integer multiples of the single phonon frequency on both the creation and annihilation sides. This behaviour is surprisingly similar to what is expected for the simple model of Einstein oscillators. The elastic peak is now a regular member of the series since it is mostly due to all possible multiphonon combinations which add up to a total energy transfer of zero, the contribution of surface defects being now negligible. The peak intensities follow a Poisson distribution, $P(n) = e^{-S}S^n/n!$, with respect to the phonon number n, which indicates that phonons are created or annihilated in a coherent forced harmonic oscillator state [18, 19]. A similar experimental and theoretical study has also been reported for HAS from Kr on Pt(111) [20]. A unified theory of diffractive and multiphonon atom-surface scattering from adsorbates has been developed in the quantum regime by Gumhalter and Langreth [21] and is also discussed in detail in the comprehensive review by Gumhalter [12].

Isolated adsorbates, like in the weakly coupled monolayers discussed above, also have flat phonon branches and a similar series of multiphonon peaks is expected for sufficiently high incident energies and low surface temperatures. Figure 13.5 shows the HAS spectrum of Cu(001) covered with 0.08 ML of CO molecules measured along the [100] direction at an incident angle of 27.9° [22], surface temperature of 110 K and 42 meV incident energy. Unlike the HAS data at low (11.2 meV) incident energy discussed in Sect. 12.2 (Fig. 12.2), where single phonon transitions involving the T-mode at $\Delta E = -4$ meV are observed [23], the HAS spectrum at an incident energy of 42 meV shows several sharp peaks corresponding to the zero-phonon line and to four overtones of the T-mode frequency on both the creation and annihilation sides (Fig. 13.5a). The sharp peaks are superimposed on a substantial background, which is attributed to the coupling of the T-mode oscillations of the CO molecules to the surface atoms of the substrate, as discussed in

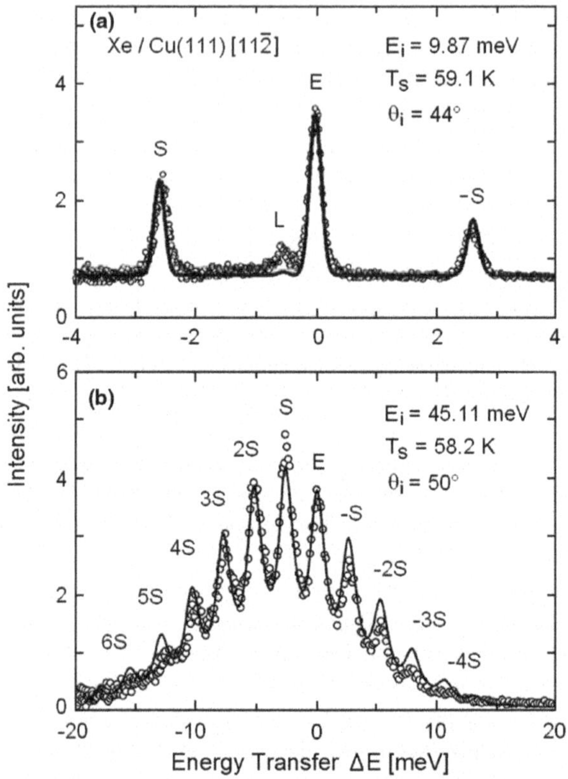

Fig. 13.4 HAS energy transfer spectra (open circle) from a Xe ($\sqrt{3}\times\sqrt{3}$)R30° monolayer on Cu (111) along the [11$\bar{2}$] direction. **a** At a low incident energy of E_i = 9.87 meV distinct single phonon creation (S) and annihilation processes (−S) involving the molecule-substrate stretching mode and a small substrate L-mode peak are observed. **b** At a high incident energy E_i = 45.11 meV and nearly the same surface temperature T_s multiphonon replicas of the S mode are observed on both creation and annihilation sides. E denotes the elastic (zero-phonon) line. An inelastic scattering theory based on the exponentiated Born approximation (EBA) (full line curve) gives a very good fit of the experimental spectra [12–16]

Sect. 12.2. Figure 13.5b shows that calculations based on (7.57–7.60), provide a very good description of the observed spectral features [22]. This calculation treats the dynamical coupling of the CO molecule with the substrate within the Green's function method (Sect. 3.5.2) in which the substrate dynamics is represented by a simple Debye model fitted to the Cu(001) surface dynamics. The model also considers the possibility that the CO molecule is tilted with respect to the surface normal [24]. A tilt of 7° gives the best fit to the experiment.

A more extensive systematic experimental investigation of the multiphonon structure in HAS spectra for small coverages of CO on Cu(001), and also for CO on Pt(111), has been carried out by Li et al. [25, 26]. The results were analysed with a temperature dependent Poisson distribution multiplied by a form factor which

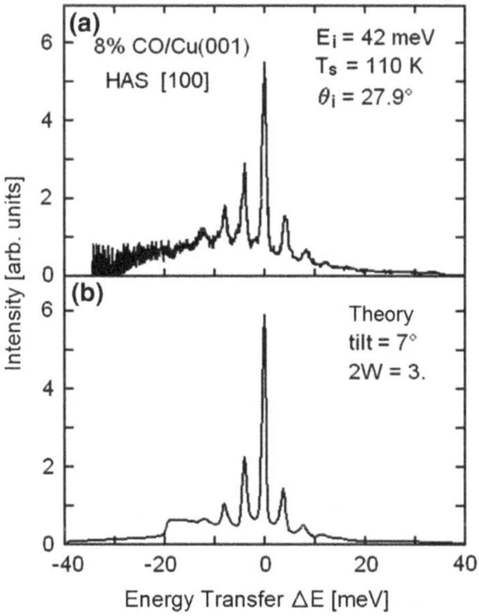

Fig. 13.5 Comparison of a HAS spectrum with theory for scattering from a 0.08 ML of CO on Cu (001) along the [100] direction at an incident angle of 27.9°, surface temperature of 110 K and incident energy of 42 meV [22]. **a** The sharp peaks in the experimental energy transfer distribution correspond to the zero-phonon line and to three overtones of the T-mode frequency on both the creation and annihilation sides. **b** The calculation of the HAS spectrum includes the CO-substrate dynamical coupling and accounts for the 7° tilt of CO molecule with respect to the surface normal [22]

describes the interaction of the He atom with the adsorbates. The theory could account for the T-mode overtones for many different incident energies, incidence angles and surface temperatures. A normal mode analysis was used to account for the interaction potentials within the molecule and with the surface in its tilted configuration. In these calculations, however, no dynamical coupling to the substrate was considered and, consistently, the comparison with the HAS spectra was made after subtraction of the associated background. However, the close correspondence between theoretical and experimental intensities seen in Fig. 13.5 indicates that, what is generically called a multiphonon background, actually originates from the coupling of the T-mode with the substrate dynamics, and its subtraction may actually result in a loss of information. Li et al. could also show why, under the experimental circumstances of the experiment, the molecule-substrate stretching (S) and the frustrated rotational (R) modes were not observed [25]. They also predicted the scattering conditions for obtaining the complete HAS spectrum including the S- and R- mode features for which there is only IRAS data and no experimental HAS data available at present [27].

13.2 Diffusion of Submonolayer Adsorbates

13.2.1 The Quasi-elastic Scattering Method

Diffusion of atoms and molecules on the surface is an important elementary step involved in the growth of thin films, equilibration of overlayers, adsorption and desorption kinetics, and catalysis. Many different methods, mostly macroscopic in which the flow of atoms and molecules are observed, have been used to quantify the diffusion process. These have been reviewed in the 2010 book by Antczak and Ehrlich [28].

The extreme sensitivity of He atoms to single adsorbates and the non-destructiveness of HAS makes it also an ideal method for studying the elementary process of diffusion at the microscopic level. This was first demonstrated by Frenken, Toennies and Wöll in 1988 [29, 30]. The method relies on the measurement of the small energetic broadening Γ (FWHM) of the quasi-elastic peak intensity as a function of the parallel wavevector change ΔK. In a classical picture the broadening is attributed to the Doppler shift in the scattering from the collective of single moving adsorbate particles on the surface. This method is generally referred to as Quasi-elastic HAS (QHAS).

Basically the QHAS method is an adaptation to the surface of the well-known method of neutron quasi-elastic scattering, which has been extensively applied to studying microscopic diffusion in the bulk, where it depends however crucially on the difference in the scattering cross section of the diffusing particle from that of the host material [31–33]. The extension of the theory for neutrons [32–34] to helium atom scattering from species diffusing on and not in a crystal was first developed by Levi and co-workers [35] and reviewed by Frenken and Hinch [36].

The differential reflection coefficient in the Born approximation is given by (7.18) and (7.29a):

$$\frac{d^2\Re(\Delta\mathbf{K},\omega)}{d\omega\, d\Omega} = \frac{k_f}{|k_{iz}|}\left|\frac{mV(\Delta\mathbf{k})}{2\pi\rho\hbar^2}\right|^2 \frac{n_d}{2\pi}\int dt\, e^{-i\omega t}\left\langle \frac{1}{N_d}\sum\nolimits_{ll'} e^{i\Delta\mathbf{K}\cdot(\mathbf{R}_l(t)-R_{l'}(0))}\right\rangle, \tag{13.1}$$

where the $\mathbf{r}_l(t) \equiv \mathbf{R}_l(t)$ are now the time-dependent positions of the N_d diffusing atoms, bound to move on the plane $z = 0$, and $n_d \equiv N_d/A$ is their surface concentration. Equation (13.1) is conveniently factorized as

$$\frac{d^2\Re(\Delta\mathbf{K},\omega)}{d\omega\, d\Omega} = n_d\, F^2\, S(\Delta\mathbf{K},\omega), \tag{13.2a}$$

where

$$F^2 = \frac{k_f}{|k_{iz}|}\left|\frac{mV(\Delta\mathbf{k})}{2\pi\rho\hbar^2}\right|^2, \tag{13.2b}$$

and

$$S(\Delta\mathbf{K},\omega)=\frac{1}{2\pi N_d}\int dt\, e^{-i\omega t}\Big\langle\sum\nolimits_{ll'} e^{i\Delta\mathbf{K}\cdot(\mathbf{R}_l(t)-R_{l'}(0))}\Big\rangle. \tag{13.2c}$$

The quantity F is the *form factor* of the diffusing atom as seen by the probe atom when its wavevector changes by $\Delta\mathbf{k}=(\Delta\mathbf{K},\Delta k_z)$. $S(\Delta\mathbf{K},\omega)$, known as the *scattering function*, is the space-time Fourier transform of the classical time-dependent *pair correlation function*

$$G(\mathbf{R},t)=\frac{1}{N_d}\Big\langle\sum\nolimits_{ll'}\delta(\mathbf{R}-\mathbf{R}_l(t)+\mathbf{R}_{l'}(0))\Big\rangle, \tag{13.3a}$$

or the time Fourier transform of the so-called *intermediate scattering function*

$$I(\Delta\mathbf{K},t)=\int G(R,t)\, e^{i(\Delta\mathbf{K}\cdot\mathbf{R}-\omega t)}\, d^2R. \tag{13.3b}$$

For quasi-elastic processes with small wavevector transfers, F in (13.2a) is approximately constant and thus the differential reflection coefficient is proportional to $S(\Delta\mathbf{K},\omega)$. The same proportionality also holds for the *transition rate* $w(\mathbf{k}_f,\mathbf{k}_i)=|V(\Delta\mathbf{k})/\hbar|^2 S(\Delta\mathbf{K},\omega)$ [cfr. (7.33 and 7.34)].

As first pointed out by van Hove [37], $G(\mathbf{R},t)$ provides the complete information on the dynamical behaviour of the ensemble of diffusing particles. At low coverages the diffusive events of different atoms are normally uncorrelated, which permits to consider only the self-correlation function $G_s(\mathbf{R},t)$

$$G_s(\mathbf{R},t)\equiv\frac{1}{N_d}\Big\langle\sum_l\delta(\mathbf{R}-\mathbf{R}_l(t)+\mathbf{R}_l(0))\Big\rangle, \tag{13.3c}$$

and its space-time Fourier transform $S_s(\Delta\mathbf{K},\omega)$. In this approximation cooperative diffusive events involving atom pairs or larger ensembles, which may occur in dense systems and quantum fluids, are neglected.

For random continuous motion on a flat surface, corresponding to Brownian motion, $G_s(\mathbf{R},t)$ has a Gaussian profile, which when substituted into (13.2a–c) leads to a Lorentzian energy profile. The half-width Γ of the Lorentzian shaped quasi-elastic energy transfer peak provides the diffusion coefficient D in the case of continuous diffusion, or the jump rates in the case of jump diffusion over a lattice with four-fold symmetry The half-widths for the two cases are respectively given by [34, 38, 39]

$$\Gamma=2\hbar D|\Delta\mathbf{K}|^2, \tag{13.4a}$$

and

$$\Gamma = 4\hbar \sum_j \frac{1}{\tau_j} \sin^2\left(\frac{\Delta\mathbf{K} \cdot \mathbf{a}_j}{2}\right), \tag{13.4b}$$

where τ_j is the inverse of the jump rate from the origin to the j-th lattice site at position $\mathbf{a}_j$. Measurements at small wavevector transfers provide information on correlations over long distances of the order $\sim 2\pi/\Delta K$ and, similarly, events with small energy transfers $\hbar\omega$ provide information over long times $t \sim h/\Gamma$. Typically broadenings down to $\Gamma \approx 10 \mu eV$ ($t < 4 \times 10^{-12}$ s) and $\Delta K > 0.03$ Å^{-1} ($\ell < 200$ Å) have been achieved in the spin echo QHAS (SE-QHAS) experiments which have the highest resolution. The time-of-flight QHAS (TOF-QHAS) and the SE-QHAS techniques and some typical results are discussed in the next two sections.

13.2.2 Time-of-Flight Quasi-elastic Helium Scattering (TOF-QHAS) Method

In the initial experiments, Frenken et al. [29, 30] used the TOF-QHAS method, to obtain the full $G(\mathbf{R}, t)$ of the highly mobile atoms in the (110) surface of lead at a temperature just below the melting point. Under these conditions earlier ion scattering experiments had shown considerable mobility and disorder at the surface [40]. Much more striking results were subsequently obtained from measurements of the diffusive motion of a low 0.028 ML coverage of Na atoms adsorbed on the Cu (100) surface shown in Fig. 13.6. In this case the easier to interpret $G_s(\mathbf{R}, t)$ could be directly investigated over a wide range of temperatures [41, 42]. Fig. 13.6a illustrates the part of the time-of-flight spectrum centered at $\Delta E = 0$ which is used to study diffusion. The overall broadening of the central peak on an expanded scale is shown as a function of temperature in Fig. 13.6b and for a given temperature as a function of $\Delta \boldsymbol{K}$ in Fig. 13.6c. To obtain the diffusion induced broadening Γ, the experimental peak shape was deconvoluted to account for the inherent peak shape of the incident beam [43]. The measurements at even the low concentrations of a few percent were made possible by the large scattering cross sections from single isolated adsorbed atoms or molecules on smooth metal surfaces [44].

Figure 13.7 shows the dependence of the broadening as a function of the wavevector transfer obtained from a series of measurements like those plotted in Fig. 13.6c. The sinusoidal dependence is close to that expected from (13.4b) and indicates that the diffusion of Na adsorbates at low coverage on Cu(001) occurs via a nearly pure single jump mechanism [41, 42]. As illustrated by the two panels in Fig. 13.7 the method is also sensitive to the direction of the diffusive jumps. From

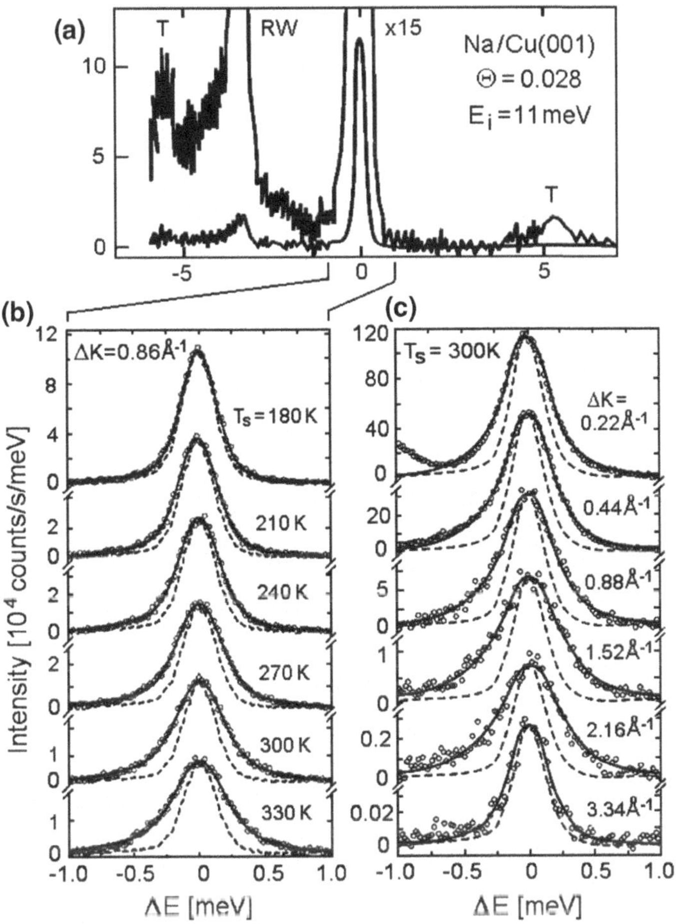

Fig. 13.6 Time-of-flight HAS spectra of the quasi-elastic peak converted to an energy transfer scale for scattering from a 2.8% monolayer coverage of Na atoms on Cu(001) measured with a 11 meV incident beam, at a surface temperature of 180 K. **a** Overview of the entire spectrum for a parallel wavevector transfer $\Delta K = 0.86$ Å^{-1} along the [100] direction. The $\times 15$ magnified spectrum shows, besides the large central peak, the weaker features of the adsorbate T-mode and the substrate RW. **b** A series of energy transfer spectra showing only the central peak for $\Delta K = 0.86$ Å^{-1} for increasing surface temperatures. The open circles show the experimental points. The solid line through the data points is a convolution of the instrument response function (dashed line) with a Lorentian-shaped peak which is fitted to the quasi-elastic broadening of the central peak. **c** The dependence of the central peak on the parallel momentum transfer at a constant surface temperature of 300 K (adapted from [41, 42])

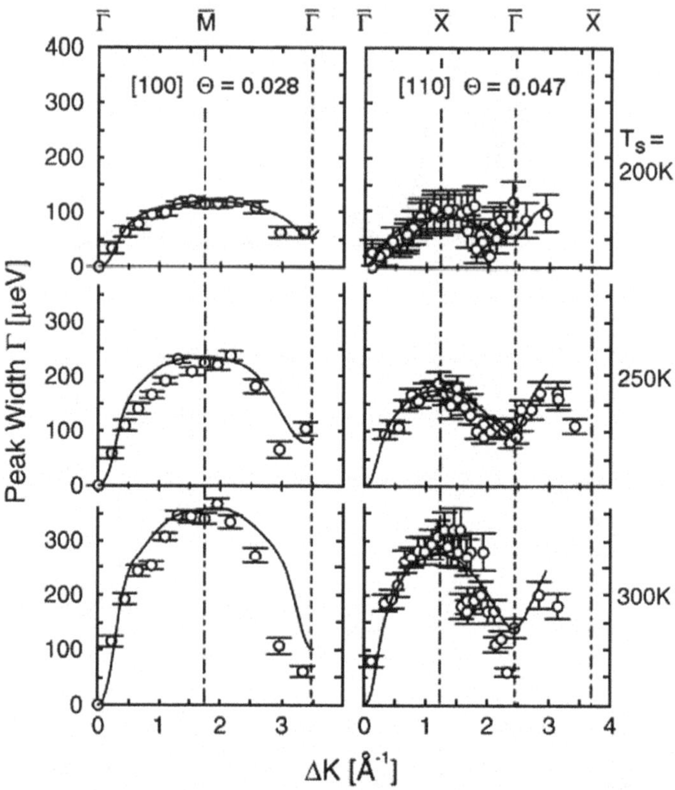

Fig. 13.7 The FWHM of the deconvoluted quasi-elastic peak as a function of ΔK along the $\overline{\Gamma M}$ and $\overline{\Gamma X}$ symmetry directions illustrates the diffusion dynamics of Na adsorbates on Cu(001) at three different surface temperatures T_s = 200 K, 250 K and 300 K. The coverage is 0.028 ML for the $\overline{\Gamma M}$ direction and 0.047 ML for the $\overline{\Gamma X}$ direction (adapted from [41, 42]). The solid lines show fits based on a Langevin molecular dynamics simulation [47]. A sinusoidal behavior with minima at $\overline{\Gamma}$ is typical for jump diffusion from one lattice site to the next

the fit of the data with (13.4b) the diffusion coefficient is estimated using the simple relationship $D = a_j^2/2\tau_j$. In this case the directionality is particularly strong because the jump rate τ_j^{-1} depends on both the jump length and the corresponding potential barrier height. The activation barrier for diffusive motion E_{diff} is usually determined from the temperature dependence of D expressed by the Arrhenius law $D = D_0 \exp(-E_{diff}/kT)$. For a 10% coverage of Na on Cu(001) the barrier was measured to be 51 ± 6 meV [45], a value which is much lower than had been reported in the past for similar systems using more conventional techniques [46].

A more detailed analysis of the Na/Cu(001) experiments (Figs. 13.6 and 13.7) relies on molecular dynamics simulations based on the Langevin equation for the two-dimensional motion of a given particle of mass m along the surface [47–49].

$$m\ddot{\boldsymbol{r}} = -\nabla V(\boldsymbol{r}) - m\eta\dot{\boldsymbol{r}} + \xi(t), \tag{13.5}$$

where the particle interacts with the substrate through a potential $V(\boldsymbol{r})$ and is subject to random impulses $\xi(t)$. The friction coefficient η is generally assumed to be independent of the position of the particle. Equation (13.5) is solved numerically for an assumed potential surface, friction parameter and random force amplitudes corresponding to the surface temperature. The velocity and positions as a function of time from a large number of trajectories are analysed to obtain, for example, the time-dependent autocorrelation functions [41, 47, 50, 51].

Figure 13.8 shows the potential surface which was found to provide a good description of the data in Fig. 13.7. The simulations reveal that the data are very sensitive to the detailed shape of the lateral potential. The reliability of the potential fitted to the diffusion data (Fig. 13.8) was subsequently confirmed by successfully simulating the diffusion barrier and also the measured temperature dependence of the T-mode energy and its width without an additional adjustment of the potential parameters [41, 50]. The simulation also provides a friction coefficient of about 1.0 ps^{-1}. Additional measurements were reported on the coverage dependence with surprising results [52] which have recently been addressed by theory [53]. Very similar results for the same system have recently been obtained with the spin echo QHAS method discussed in the next section [54].

The QHAS results available at the time of writing in 2017 are listed in Appendix C. Table C1 lists the systems studied by TOF-QHAS. In addition to the

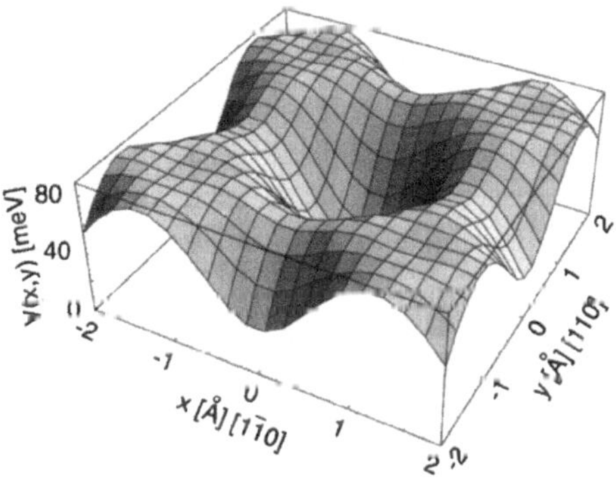

Fig. 13.8 The experimental potential energy surface $V(x, y)$ for Na atoms on Cu(001) determined via a Langevin simulation from TOF-QHAS measurements [41, 50]. The substrate Cu atoms are situated at the four corner points (x, y) = (±1.28, ±1.28) and the minima correspond to the hollow sites. The classical barrier for motion along the [110] azimuths is 75 meV, and for motion along the (diagonal) [010] azimuths is 84 meV. Even the slight minima above the copper atoms at on-top sites postulated from LEED data are reproduced

pioneering measurements of mobile Pb along on the clean surface of Pb(110) the list includes several other homogeneous metals such as Ag/Ag(110), Ni/Ni(110) and the semiconductor systems Ge/Ge(111) [55]. Similar studies have been reported for potassium atoms on a potassium surface deposited on Ni(001). The diffusion of CO has been studied on various metal surfaces including Ni(001) [56, 57], Ni(110) [56], Cu(001) [27] and on Pt(111) [58]. In addition to the atom adsorbates Na, H(D), S, and Xe atoms the molecules CO and the octane molecule (C_8H_{18}) have all been studied by TOF-QHAS.

13.2.3 ^{3}He Spin Echo Quasi-elastic Scattering Method (SE-QHAS)

With the advent of the spin echo surface scattering technique, which is an adaption of the neutron technique used to study diffusion in the bulk, a new even more powerful method has become available for investigating surface diffusion at the microscopic level. The application of this technique to inelastic scattering is discussed in Sect. 9.4.3. The spin echo technique makes it possible to directly determine the intermediate scattering function $I(\Delta\mathbf{K}, t)$ [59].

Figure 13.10 shows the spin echo apparatus, which has been constructed by the Cambridge group of Allison and Ellis [60, 61] and described in several reviews [62, 63]. The apparatus is essentially the same as in Fig. 9.12, but in addition the wave packets which play a role in measuring diffusion are explicitly shown. In these experiments as opposed to the measurement of inelastic scattering discussed in Sect. 9.4.3 the magnetic fields B in the B_1 and B_2 solenoids are set equal. Under these conditions the interaction of the polarized beam with the homogeneous field in the first solenoid (see Fig. 13.9) must be taken into account. As a result of this interaction the initial state of the atom polarized in the y-direction (perpendicular to the scattering plane) is split into two eigenstates with opposite orientations [64]. The Zeeman energy splitting in the solenoid causes the wave packets of the two eigenstates to have different velocities given by $v_\pm = v_0 \pm \gamma B/mv_0$. As a result the two wave packets arrive at the end of the solenoid of length L at times $t = t_0 \pm \tau_{sE/2}$ where τ_{SE} is called the spin echo time and is given by

$$\tau_{SE} = \frac{\gamma\hbar}{mv_0^3}\int_0^L Bdl. \tag{13.6}$$

If the He atom wave packets are coherently elastically scattered from a rigid stationary ensemble of particles the two wave packets will be oppositely affected by the reversed, otherwise identical, field as they pass through the second solenoid. Thus they will arrive at the detector at the same time ($\tau_{SE} = 0$) and recombine as if nothing had happened. Consequently the polarisation signal $P(\Delta\mathbf{K}, \tau_{SE} = 0)$

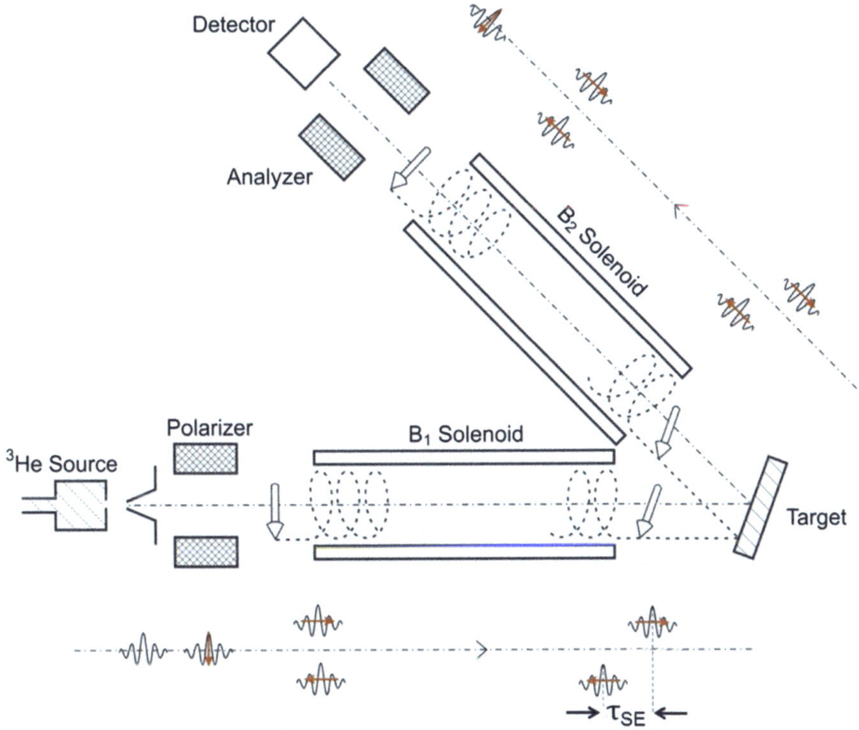

Fig. 13.9 Schematic diagram of a He[3] spin echo apparatus for diffusion measurements. The components are the same as in Fig. 9.12. In addition the relative phases of wave packets during an experiment for measuring diffusion are shown. The first B1 solenoid field splits the ^{3}He atom into two separate wave packets which arrive at the detector at different times τ_{SE}. The two components are recombined in the second B2 solenoid. If the two wave packets do not recombine as a result of particle motion during the time τ_{SE} then the polarization detected with the analyser is reduced (adapted from [65])

detected after the analyser (see Fig. 13.10a) will be unaffected. If, however, the surface atoms diffuse during the time interval τ_{SE} then the phase of the wave packet which arrives at the latter time $t = t_0 + \tau_{sE/2}$ will be affected differently than the wave packet which arrived earlier. This will lead to a reduction in the polarisation signal $P(\mathbf{\Delta K}, \tau_{SE})$ detected after the analyser (see Fig. 13.10a). By synchronously changing the magnetic field in both solenoids B_1 and B_2 the time τ_{SE} can be varied. As expected with increasing time interval between the scattering of two wave packets, the diffusing particles will move further apart and when they are recombined at the detector the polarisation signal will be reduced accordingly. It can be shown as derived, for example, in [62] that $P(\mathbf{\Delta K}, \tau_{SE}) = I(\mathbf{\Delta K}, \tau_{SE})/I(\mathbf{\Delta K}, 0)$. Thus the polarization signal provides directly the intermediate scattering function which in turn can be Fourier transformed into the scattering function $S(\mathbf{\Delta K}, \omega)$ as illustrated in Fig. 13.10b. The rapid oscillations of $P(\mathbf{\Delta K}, \tau_{SE})$ at small times of the

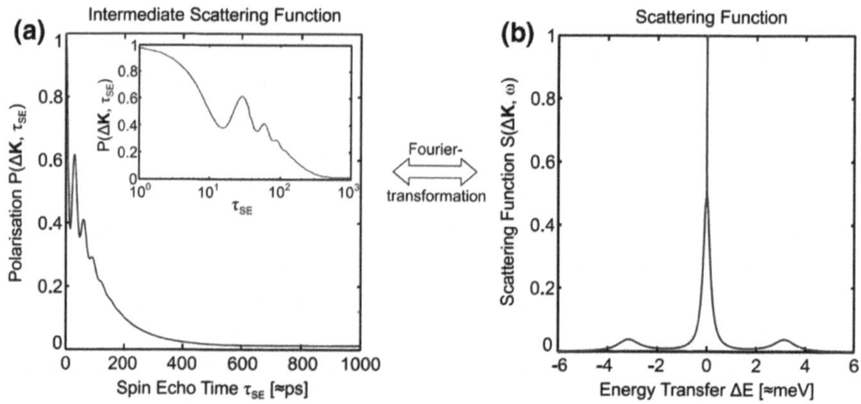

Fig. 13.10 Schematic diagram illustrating the connection between the dependence of the polarisation $P(\Delta\mathbf{K}, \tau_{SE})$ on the spin echo time and the energy domain scattering function $S(\Delta\mathbf{K},)$. **a** $P(\Delta\mathbf{K}, t)$, which is proportional to the intermediate scattering function $I(\Delta\mathbf{K}, t)$, oscillates at short times resulting from the interaction with vibrating surface atoms while the slower overall decay results from diffusion. The insert shows the short time behaviour on a logarithmic time scale. **b** The Fourier transform of $I(\Delta\mathbf{K}, t)$ yields the scattering function in which the rapid motion results in the well separated energy transfer peaks and the slow motions results in a sharp delta function-like peak related to diffusion (adapted from [62])

order of 10^{-12}s come from very fast local intra-cell motions associated with inelastic scattering from phonons, while the subsequent slow degradation of the $P(\Delta\mathbf{K}, \tau_{SE})$ polarizability contains the information on diffusion. Since each ^{3}He atom interferes with itself the spin echo interference technique has the advantage that the velocity distribution of the beams has absolutely no effect.

Table C.2 in Appendix C summarizes the SE-QHAS diffusion measurements as of 2016. The temperatures, activating energies and coverages are much the same as the parameters used in TOF-QHAS experiments. Perhaps it is surprising that only in a few cases are the data evaluated in terms of diffusion constants. This has to do with the complexity of the data obtained with SE-QHAS which is not always amenable to a simple interpretation. This is possible for simple diffusion mechanisms, such as continuous random motion or discrete jump diffusion. In these cases the temporal decay of $I(\Delta\mathbf{K}, \tau_{SE})$ is exponential and proportional to $\propto \exp[-\alpha(\Delta\mathbf{K})t]$. For this reason the data are frequently evaluated in terms of $\alpha(\Delta\mathbf{K})$.

For the simpler Na on Cu(001) system the SE-QHAS experiments are in essential agreement with the TOF-QHAS results and their interpretation. The diffusion of CO on Cu(001), which has been extensively studied by theory and macroscopic experiments in the past, has also been studied by SE-QHAS and previously by TOF-QHAS. Surprisingly in this case the two methods lead to widely differing results for the energetic barrier. The early TOF-QHAS diffusion measurements together with the extensive HAS results on the T-, R- and S-modes [27] discussed in Chap. 12 were used to derive an empirical two-dimensional energy

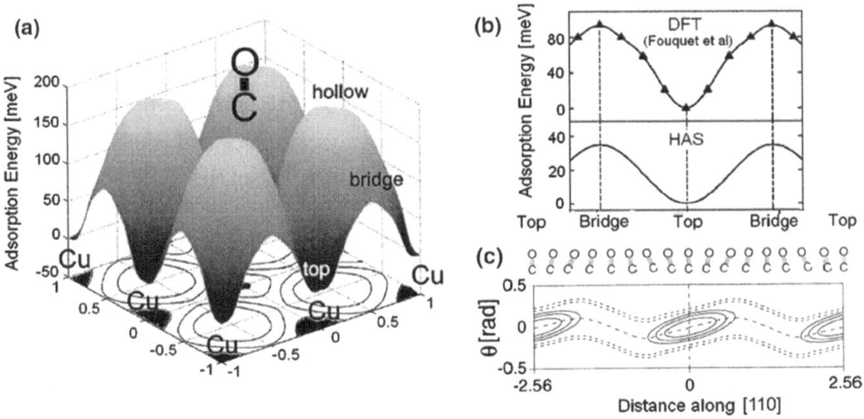

Fig. 13.11 Experimental and theoretical adiabatic lateral potential energy surface for CO molecules on Cu (001) with a c(2x2) structure [27, 42]. **a** Perspective view of the theoretical potential surface showing the preferred orientation and position of the CO molecules over the Cu atoms [69]. **b** Comparison of the potential profiles in the [110] direction as determined by a DFT calculation [69] and from the HAS analysis of T-, R- and S-modes together with TOF-QHAS data [27, 42]. **c** The experimental potential surface seen from the top: Θ is the tilt angle of the CO axis as also illustrated above the potential surface [42]

surface for the motion of a CO molecule across a Cu(001) surface along the [110] direction, where the barrier is expected to be lowest [42, 56, 57, 66]. Figure 13.11 illustrates the results and compares the empirical potential profile with a DFT calculation. From the temperature-dependent Arrenhius plot of the quasi-elastic broadening the barrier was found to be 31 ± 10 meV [27]. The data analysis indicates that the CO molecule largely retains its upright position as it moves from site to site across the surface (Fig. 13.11c). Surprisingly, the more recent spin echo SE-QHAS measurements of the temperature-dependent quasi-elastic broadening yield a much larger barrier of about 125 meV [67]. A similarly large barrier of 98 ± 5 meV from the temperature dependence of the SE-QHAS diffusion rate was also reported for CO on Cu(111) [68].

Prior to the above HAS investigations the potential energy surface of CO on Cu (001) and in particular the barrier had attracted considerable attention from theoreticians. The first comprehensive potential based on semi-empirical arguments was reported by Tully et al. 1993 [70]. In an attempt to decide between the two significantly different QHAS values and a controversial DFT calculation [66–72], Fouquet et al. carried out a new more accurate DFT calculation [69]. Their results are in qualitative agreement with the profile determined from TOF-QHAS data but their diffusion barrier energy is 115 ± 30 meV in good agreement with the experimental SE-QHAS barrier. In their article they do, however, point out that the activation energy, as determined in the experiment from an Arrenhius plot, is normally lower than their calculated adiabatic barrier [73, 74]. More recently in the latest theoretical study of the potential hypersurface Marquardt et al. in 2010, reported a DFT calculation which was fitted to a compact analytical expression [75] and compared with the potential of Tully et al. Since Jan Baerends is an author of the article of Fouquet et al. and of the new DFT article, the new calculation can be considered to be more reliable. Their result for the barrier of 58 meV, which when corrected for the large harmonic zero point energy is reduced to 33 meV, is surprisingly in remarkable agreement with the original TOF-QHAS experiment. In their article Marquardt et al. write "it was said (in Fouquet et al.) that calculated adiabatic barriers cannot be compared directly with the experimental values, because of the neglect of non-adiabatic effects. The present result indicates that such effects are probably negligible". Thus at present this special system raises important concerns about the interpretation of either the calculations or the SE-QHAS measurements.

In other systems, such as the diffusion of benzene (C_6H_6) on graphite the higher resolution of the spin echo measurements have added important details not available with TOF-QHAS. The potential shows a smooth exponential-like behaviour even at a coverage of 0.5 ML and the $\Delta\mathbf{K}$ dependence could be fitted by a simple Brownian diffusion model [76–78] with a friction coefficient $\mu = 2.2 \pm 0.2\ \mathrm{ps}^{-1}$.

If $(\Delta\mathbf{K}, \tau_{SE})$ or $\alpha(\Delta\mathbf{K}, T)$ show a more complex behaviour, then the interpretation is aided by Langevin simulations. An example is the CO on Pt(111) which was measured with coverages up to 0.3 ML [79]. Previous studies had suggested the presence of strong repulsive forces between the CO molecules. The SE-QHAS measurements backed up by extensive Langevin simulations revealed, on the contrary, uncorrelated motions which could be explained by a non-local modification of the adsorbate-substrate interaction.

The spin echo technique has also been applied to several other organic molecules each of which exhibits an unusual behaviour. For example, in the case of thiophene (C_4H_4S) on Cu(111), in addition to evidence for jump diffusion, the results provide evidence for a rotation of the molecule around the relatively strong S–Cu bond and a "flapping" of the flat molecule between states with different tilt angles [80]. Very recently, the diffusion of individual pentacene molecules on the surface of a well-ordered pentacene monolayer was interpreted to indicate the surprising ability of the diffusing molecules to switch between moving parallel and perpendicular along the rows formed by the ordered checker-board substrate [81].

Very recently the spin echo technique has been extended to operate with H_2 molecules and in initial experiments was used to constructively interfere different magnetic sublevels. In this way it was possible to control the orientation of the impinging ortho-H_2 molecules as it impacts with the surface [82].

This section is concluded by calling attention to the important impact of the QHAS method, discussed above, on the interpretation of the extensive data from the more macroscopic studies based on laser-induced desorption and refilling (LID), or linear optical diffraction (LOD), etc. [46, 83]. At high temperatures the diffusion barriers measured by QHAS generally agree with those deduced from these methods. In all cases in which the measurements are carried out at room temperature or below, the activation barriers are, however, consistently significantly smaller. For example for Na on Cu(001) the QHAS experiment yields a diffusion barrier of only $E_{diff} = 54$ meV for surface temperatures between 180 and 450 K. This barrier is much smaller than the barrier energies of about 1 eV reported earlier for alkali-metal systems [84]. The QHAS preexponential diffusion constants D_o (13.4a, b), are, on the other hand, generally much larger than previously reported, providing for some compensation for the differences in the barrier energies. These differences to previous results are attributed to the relative insensitivity of the QHAS method to the impeding effect of surface defects [85]. This was elegantly confirmed by a photoelectron emission microscopy technique with which it is possible to observe diffusion on microscopically small smooth defect free terraces without step edges or other defects [86]. The relative insensitivity of the QHAS method to defects results from its unique sensitivity to the actual microscopic motion of the particles. Since only small distances of the order of the lateral coherence length (≈100 Å) of the scattering particle or less are probed, defects, which are generally more widely spaced, cannot significantly influence the results. This explanation is also consistent with the good agreement in the high temperature measurements. At high surface temperatures the defects are much more mobile and, at the same time, the surface is continually being annealed. Thus the QHAS experiments provide for the first time a complete microscopic description of the diffusion of atoms on nearly ideal terraces on single crystal surfaces.

13.3 Anharmonicities and High Temperature Effects

The anharmonicity of surface vibrations [87–89] has long attracted much interest in connection with surface mean-square displacements (see Table 1.1) and the Debye-Waller factor discussed in Sect. 7.6 [88]. Anharmonicity is also of considerable interest in connection with the experimental linewidths of surface phonons and also for understanding the relaxation of adsorbate vibrations [89–95] which play an important role in many relevant phenomena such as adsorbate diffusion (Sect. 13.2), sliding friction [96–99] and energy accommodation processes [93, 100–102].

Anharmonicity is an inherent property of surface vibrations resulting from the lack of inversion symmetry of the vertical potential holding atoms at the surface of a crystal. As a consequence the surface atoms experience a cubic term in the potential normal to the surface and, therefore, anharmonicity is implicit for the SV motion at the surface and enhanced compared to the anharmonicity in the bulk [87]. Early theoretical studies [87] indicated for low index monoatomic surfaces that the surface anharmonicity enhancement was typically a factor two.

Figure 13.12 shows the temperature change in energy of the $\overline{\Gamma}$-point surface phonon of Cu(110) in comparison with the equivalent bulk mode. The HREELS data reported in 1991 [103] by Baddorf and Plummer show a clear downward shift in the surface vibrational frequencies due to anharmonicity. From their HREELS

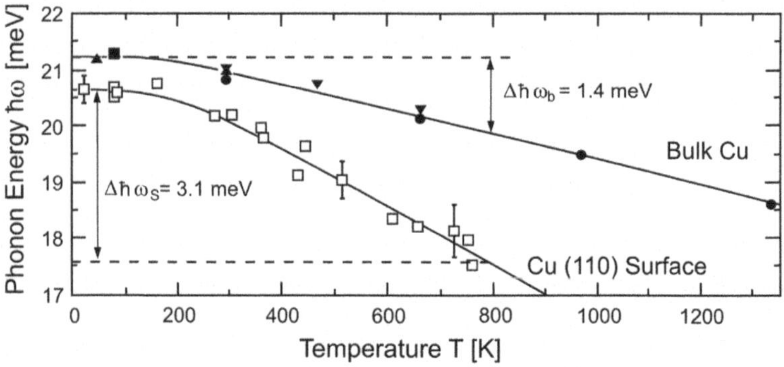

Fig. 13.12 The temperature dependence of the energy of the 20 meV surface phonon at the $\overline{\Gamma}$ point in Cu(110) measured by HREELS (open squares, [103]) is compared with the temperature dependence of the "equivalent" *bulk* longitudinal (L) phonon energy at the X = (011) point of the bulk Brillouin zone from neutron data (upward directed filled triangle [105, 106], filled squares [107], filled circles [108], downward filled triangles [109]). The difference in slope of about 2.2 is expected because of the reduced coordination at the surface [104]

measurements they claimed that the anharmonicity induced shift of an open surface such as the (110) surface of Cu can exceed that in the bulk by a factor 4–5. However, the present authors [104], through a careful comparison of the same results of Baddorf and Plummer with neutron data for bulk Cu [105–109] and with the available data for other low index metal surfaces, concluded that the surface anharmonicity enhancement is, in fact, usually of the order of two [104].

Moreover the anharmonic shift of surface phonon frequencies with temperature in Fig. 13.12 was found to be described by the following expression adapted from the theory of the vibrations of diatomic molecules [110]:

$$\omega(T) = \omega_o - x_e\omega_o(2n_o + 1) = \omega_o - x_e\omega_o \coth\frac{\hbar\omega_0}{2kT}, \tag{13.7}$$

where ω_o is the classical harmonic phonon frequency and n_o, the Bose-Einstein occupation number. The anharmonicity constant x_e is a dimensionless parameter expressed by [104]

$$x_e = \beta^2 < u^2 >_0 = \beta^2 \frac{\hbar}{2\mu\omega_0}, \tag{13.8}$$

where $<u^2>_0$ is the zero point mean-square displacement of a harmonic oscillator with effective mass μ and frequency ω_0. The inverse length β characterizes the range of the interatomic potential. For a Born-Mayer repulsive potential or a Morse potential function (Sect. 6.4) β is the prefactor of the distance in the exponent in (6.4.2). In the simplest case of a potential with the same β for bulk and surface atoms the surface enhancement of anharmonicity essentially coincides with the

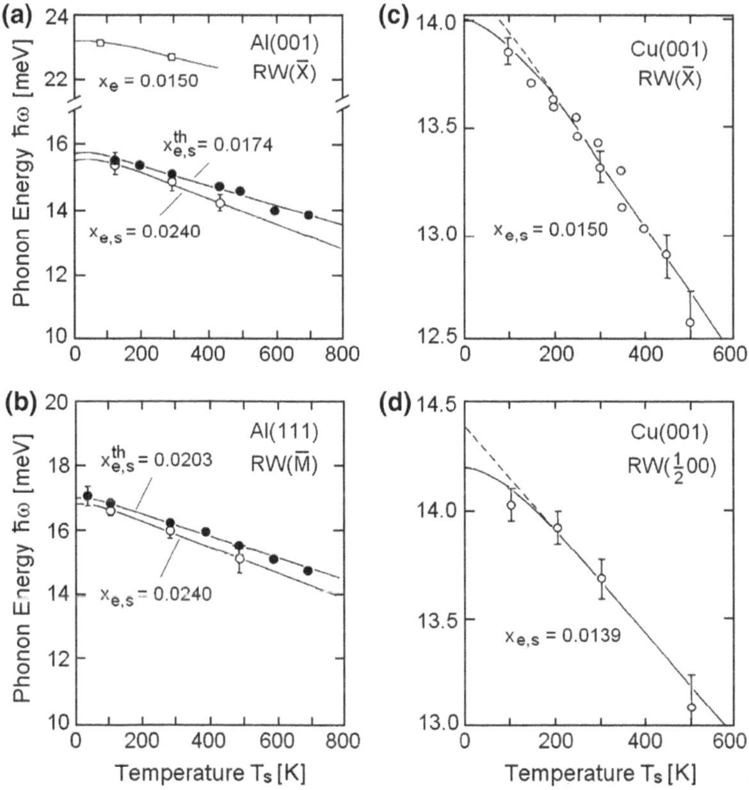

Fig. 13.13 Measurements of the Rayleigh wave (RW) frequency as a function of surface temperature at different surface wavevectors. **a**, **b** Comparison of HAS data (open circle) for Al (001) and Al(111) at the zone boundary points $\overline{X}$ and $\overline{M}$, respectively, with molecular dynamics simulations (filled circle) [104, 112]. In (a) neutron data for the bulk (open squares) [15] are also shown for comparison. **c**, **d** HAS data for Cu(001) at the $\overline{X}$ point and at one half of the [100] ($\overline{\Gamma M}$) direction as a function of the surface temperature [111]. The solid curves are fits based on (13.7) [104]. For RW data for Cu(001) at Q = 0.53 Å^{-1} see Fig. 13.19b

surface enhancement of the mean-square displacement. An analysis of the four HAS measurements in Fig. 13.13 of the RW energy as a function of temperature for Al(001), Al(111) and Cu(001) [111–112] and the HREELS data for Cu(110) in Fig. 13.12 [103] leads to values of $x_{e,s}$ at the surface of between 0.0139 and 0.0240, which are up to a factor of 2.2 greater than in the bulk. These values are in good agreement with the pioneering 1978 theoretical analysis by Ma et al. [87] and several previous and subsequent molecular dynamics simulations [88, 112–115]. This analysis provides a useful guide for assessing whether a given system shows the normal factor of two anharmonic enhancement.

The dependence on wavevector has only been studied in [114] for Al surfaces and at small wavevectors for Cu(001) using the spin echo technique. Surprisingly at

5° off the specular the drop in RW frequency was only 60 μeV from 180 to 410 K [116]. Refractory 5d metals and their compounds, which are characterized by high melting temperatures but comparatively low Debye temperatures due to their large atomic mass [117], offer an extended range of temperature where anharmonic effects can be studied in the future.

In many materials surface phase transitions of the order-order type, such as reconstructions and charge density waves occur with increasing temperature. In this case important deviations from the ordinary anharmonic effects are expected near the transition temperature (e.g., soft modes), as already discussed in connection with the phonon studies of W(001), W(110):H(1×1) and $TaSe_2(0001)$ in Sects. 11.3 and 11.4. Also order-disorder transitions such as roughening and surface melting are anticipated by important anharmonic effects, i.e., large atomic oscillations eventually leading to atomic displacements and surface self-diffusion [118, 119]. Bracco et al. have reported extensive HAS studies of the dynamics of Ag adatoms on the Ag(110) surface. Both the phonon linewidths and shifts [120–123] and self-diffusion [124] of the adatoms were investigated for temperatures up to 800 K approaching the roughening transition.

The theoretical studies of large anharmonic effects and phase transitions occurring at elevated temperatures on metal surfaces are mostly based on large scale molecular dynamics simulations and interatomic potentials including the electronic response [88, 89, 113, 114, 121, 125–128]. These studies have now become sufficiently predictive and provide a valuable complement to experiments in all situations where complex surface dynamical phenomena take place.

So far the attention has focussed on the effect of anharmonicity normal to the surface. The effect of anharmonicity parallel to the surface has only been investigated in a few cases. HAS is especially suited because of its sensitivity to only the first layer. The group of Hoinkes and Wilsch were among the first to observe how anharmonicity affects the thermal expansion of surfaces. From the analysis of selective adsorption resonances of H and D atoms on NaF(001) and LiF(001) [129, 130] they concluded that the thermal expansion surface coefficients parallel to the surface (α_s) were about two times larger than the bulk values (α_b). From the position of the HAS diffraction peak from LiF(001) Watanabe et al. reported $\alpha_s/\alpha_b = 3.6$ for temperatures up to 180°C [131]. Subsequently, Ekinci and Toennies [132] in analysing high resolution HAS diffraction data found the same expansion coefficient for the surface and bulk, $\alpha_s/\alpha_b = 1.00 \pm 0.02$. Thus LiF is, in fact, at least in this sense, after all, an ideal crystal.

Interesting unexpected high temperature effects have been found on semiconductor surfaces. A much debated case is that of the high temperature Ge(111)(1×1) semiconductor surface. Early LEED [133], ellipsometry [134] and X-ray diffraction [135] studies indicated a phase transition at $T_c = 1050$ K, with no general consensus, however, about its nature. As a metalloid germanium occupies a position between being a semiconductor at ordinary temperatures and a metal above 1211 K. First principle molecular dynamics calculations [136–138] suggested that above 1050 K the first atomic bilayer undergoes an incomplete melting,

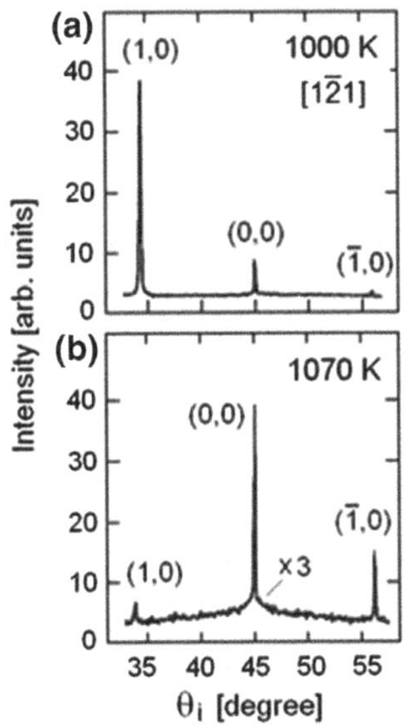

Fig. 13.14 HAS diffraction patterns along the [1$\bar{2}$1] direction from Ge(111) showing the effect of an order-order surface transition. **a** At T_s = 1000 K below the $(1\times1) \rightarrow (1\times1)$ surface phase transition at T_c = 1050 K the largest first order diffraction peak occurs at G = $(1,0)$. **b** Above T_c at T_s = 1070 K the largest first order diffraction peak occurs at G = $(\bar{1},0)$ (adapted from [144, 145])

anticipating the bulk melting at 1211 K. This liquid-like disordered state is expected to have a metallic character similar to that of liquid Ge. This was confirmed by both EELS [139, 140] and photoemission spectroscopy studies [141]. Ion scattering [142] and photoelectron diffraction [143] and high resolution EELS [139] data are also consistent with an order-disorder transition.

Figure 13.14 shows that in contrast to these observations the sharp HAS diffraction patters [144–146] below and above T_c, show an inversion of the amplitudes of the first order diffraction peaks. This observation unmistakably indicates a $(1\times1) \rightarrow (1\times1)$ order-order transition. Due to the strictly surface sensitive nature of HAS, the transition can be assigned to the topmost atomic bilayer. Moreover the reversal of the first order diffraction amplitudes can be attributed to a sliding of the first atomic layer from the normal sphalerite stacking to the wurzite stacking.

Figure 13.15 shows the results of a subsequent HAS TOF study [145] of Ge (111). Distinct RW peaks are observed both below and above the transition, and the diffuse elastic peak has about the same intensity below and above the transition. Moreover, the measured dispersion curves (Fig. 13.15b) show umklapp processes mostly occurring around the **G** vectors with the largest diffraction peaks at **G** = (1,0) below T_c and at **G** = $(\bar{1},0)$ above T_c (see Fig. 13.14) [145]. The fact that

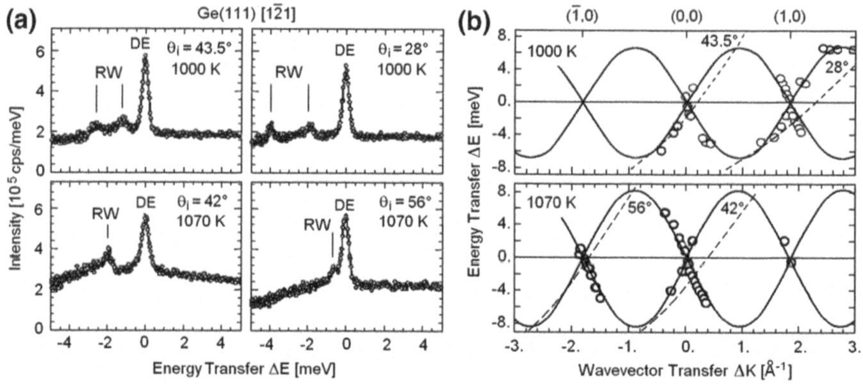

Fig. 13.15 HAS energy transfer spectra and dispersion curves from Ge(111) along the $[1\bar{2}1]$ direction at surface temperatures below (1000 K) and above (1070 K) the surface phase transition at T_c = 1050 K. **a** Energy transfer spectra at two representative incident angles θ_i and at an incident energy of 20 meV below the transition temperature (top two frames) and above the transition (bottom two frames). **b** The HAS data points (open circle), fitted by sinusoidal dispersion curves representing the RW's (full lines), for the two surface temperatures, and the scan curves (broken lines) for the spectra in panel (**a**). The sharp inelastic peaks and diffuse elastic (DE) peaks below and above the transition, suggest an order-order phase transition. The umklapp inelastic processes are preferentially observed at wavevector transfers where the corresponding diffraction peak (see Fig. 13.14) is strongest (adapted from [145])

the measured RW frequency vanishes also at $\mathbf{G} = (\bar{1}, 0)$ is further evidence against a surface liquid phase, which would instead lead to a frequency gap. Another surprising feature of the higher temperature phase is that the RW frequencies are somewhat stiffer than in the lower temperature phase, whereas anharmonicity normally implies softer frequencies at higher temperatures. Thus both the HAS diffraction patterns and inelastic spectra confirm the $(1\times1) \rightarrow (1\times1)$ order-order transition between solid phases.

A subsequent TOF-QHAS study revealed an increase in both the intensity and width of the diffuse elastic peak above T_c [146]. From this information and the QHAS analysis it was concluded that the concentration of mobile atoms increases from 25% to about 31% while the diffusion coefficient increased by about four orders of magnitude indicating a much smoother surface. This evidence and the increase of the specular peak and the average decrease of the diffraction peaks in the high temperature phase (Fig. 13.14b), indicate a smoothening of the surface which is consistent with the formation of a disordered liquid. This effect is also seen in the intensity of the diffuse elastic (DE) peaks in the two TOF spectra (left hand panels of Fig. 13.15a) which decreases by about 20% above the background across the transition, indicating less surface defects in the high temperature phase. A very similar behaviour was found for the Si(111) $(7\times7) \rightarrow (1\times1)$ transition at 1140 K [147].

The (2x1) → (1x1) phase transition of the Ge(001) surface which occurs between 900 and 1000 K has also been investigated by a careful analysis of the changes in the HAS diffraction peak intensities. In this case the data provided evidence for an order–disorder transition [148], which had been overlooked in the initial HAS study [149]. The former result has since been confirmed in a time resolved RHEED study [150].

13.4 Magnetic Effects

There is currently great interest in exploring magnetic effects at the surfaces of single crystals and in epitaxial films [151–154] because of the technical relevance to information storage systems and spintronics. The ^{4}He isotope in its ground state, currently used in HAS experiments, is devoid of a magnetic moment and is therefore unable to couple directly to surface magnetic moments. Nozzle beams of the isotope ^{3}He [155] and D_2 [156, 157] molecules, both of which do have nuclear magnetic moments, would appear to be more suitable. Their magnetic coupling to the surface is, however, expected to be very weak in view of the small probability of polarized ^{3}He atoms to undergo spin flip processes in wall collisions as indicated by the fact that they can be stored for 10^2–10^5s in metal containers [158]. Both beams have nevertheless been used in searches for magnetic effects on the anti-ferromagnetic (AF) surface of NiO(001) but without success [159].

A more promising approach involves using beams of metastable $^4He^*(2^3S_1)$ (orthohelium), which has a lifetime of about 10^6s and an electronic magnetic moment three orders of magnitude larger than the ^{3}He nuclear magnetic moment. In one of the first surface scattering experiments Hagstrum in 1954 [160] and later Conrad et al. in 1979 [161] demonstrated the extreme surface sensitivity of metastable helium to the outermost electronic orbitals by detecting the emitted electrons ejected from the adsorbates following an Auger process [162]. The efficient quenching at most metal and insulator surfaces implies that only a very small fraction of about 10^{-3}–10^{-6} of the incident beam are reflected. In 1993 El-Batanouny and colleagues [163–165] reported the observation of diffraction peaks in the angular distribution of the surviving orthohelium atoms from NiO (001). Their experiment was based on the idea that the survival probability should depend on the relative orientation of the spin of the He atom and the local electron spin at the antiferromagnetic NiO surface [166]. Other experiments were, however, not able to reproduce these interesting results [167] possibly because the survival probability is smaller than expected [168]. A more recent study of orthohelium scattering from CoO(001) by El-Batanouny et al. [169, 170], reported a half integer diffraction peak related to the antiferromagnetic structure and TOF spectra showing sharp inelastic features. They were assigned to magnetic excitations, although they fall in the region where the optical Lucas and Wallis modes are expected to occur [158, 159]. The HAS surface phonon dispersion curves of both NiO and CoO show no effect of magnetization [171, 172]. Obviously more experimental studies are called for.

A subsequent experiment in which the orthohelium atoms were initially spin polarized by a circularly polarized laser demonstrated convincingly that the survival probability on scattering from a ferromagnetic surface does, indeed, depend on the electron spin [173] and also in scattering from benzene and water adsorbed on Cu (001) [174]. The process has been dubbed Metastable De-excitation Spectroscopy (MDS). In related experiments using spin polarized orthohelium atoms (SPMDS) scattered from iron surfaces with and without adsorbates the scattered intensity was also found to depend on the spin polarization [175–177]. In a more recent apparatus the target was subjected to a strong magnetic field of 0–5 T to polarize the surface magnets. Because of the strong magnetic fields the detection had to be based on measuring the current flowing to the target resulting from the ejection of electrons [178, 179]. In this way the magnetic properties of O_2 on Cu(001) [180] and of Ho thin films [181] as well as the dangling bonds on Si(111)-(7x7) [182] could be investigated and treated by theory [183]. Finally, it should be mentioned that metastable orthohelium has found application as an etchant for creating nanostructures. These and other applications are reviewed in [184].

The surface phonons can also provide information on magnetic effects at surfaces. In the bulk the mechanical effects of magnetic interactions in solids are generally very small as compared to the effects of ordinary Coulomb and Pauli repulsive forces. In low-dimensional systems, or in systems featuring large internal stress, however, quite appreciable changes in the lattice structure and dynamics can be induced by the magnetic order. Under these conditions the steepness of the exchange potential as a function of the interatomic distance is large enough to give a measurable renormalization of the force constants and phonon frequencies.

Figure 13.16 shows the results of an HAS study of the dependence of the RW frequencies in Fe(110) on a wide range of temperatures up to temperatures slightly below the Curie point at 1044 K. Along the $[1\bar{1}0]$ azimuth the RW phonon energy decreases monotonically by about 30% for a temperature increase from 150 to 948 K (Fig. 13.16a, c). In contrast, along the [001] azimuth the RW energy decreases initially by only about 6% and then increases by about 16% (Fig. 13.16b, d). Similar azimuthal differences have been found in neutron scattering measurements of the density of phonon states [185, 186]. While a decrease of frequency with temperature can easily be attributed to normal anharmonicity, the following analysis of the multiphonon background and the anisotropy of the effect indicates that it is related to the temperature dependence of magnetization.

The analysis takes into account that the effects of magnetization can be comparable in size to the forces between atoms at the surface. Thus they can have an influence on the anharmonicity and therefore must also affect the multiphonon part of the inelastic HAS spectra. In order to account for the shape of the multiphonon background and determine the Rayleigh peak position at the higher temperatures in Fig. 13.16 the background was fitted to (7.72) in Chap. 7. Figure 13.17 shows the temperature dependence of the four parameters appearing in (7.72) as determined

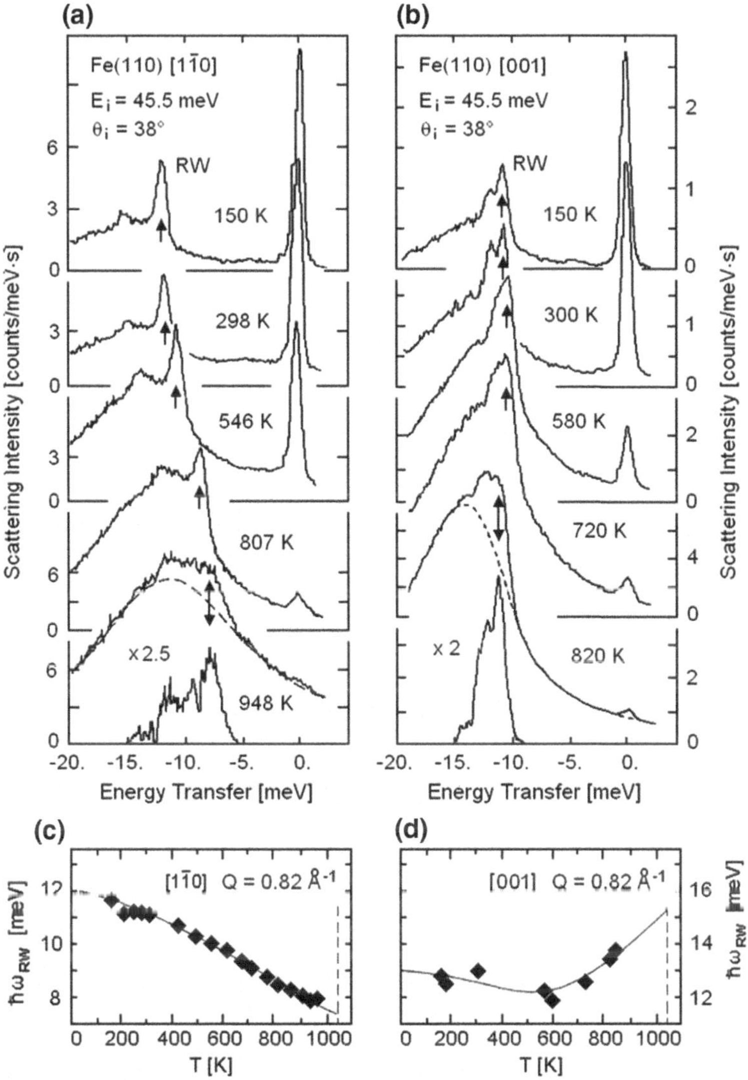

Fig. 13.16 HAS energy transfer spectra from *bcc*-Fe(110) over a broad range of surface temperatures below the Curie temperature of iron (1044 K) at a fixed incident energy of $E_i = 45.5$ meV at one-half of the Brillouin zone. **a** Along the [1$\bar{1}$0] direction the RW energy transfer (upward arrows) decreases significantly with increasing temperature. At the highest temperature 948 K the RW energy is determined by subtracting the multiphonon background (broken line). The difference magnified by a factor 2.5 gives the one-phonon spectrum at the bottom. **b** A similar series of energy transfer spectra for the [001] direction show only a slight change of RW energy with temperature. **c**, **d** The experimental RW energy at Q = 0.82 Å^{-1} is plotted as a function of temperature for the [1$\bar{1}$0] and [001] directions, respectively. The strong anisotropy in the temperature dependence is indicative of a magnetic contribution. The full lines are fits based on (7.65) (adapted from [187])

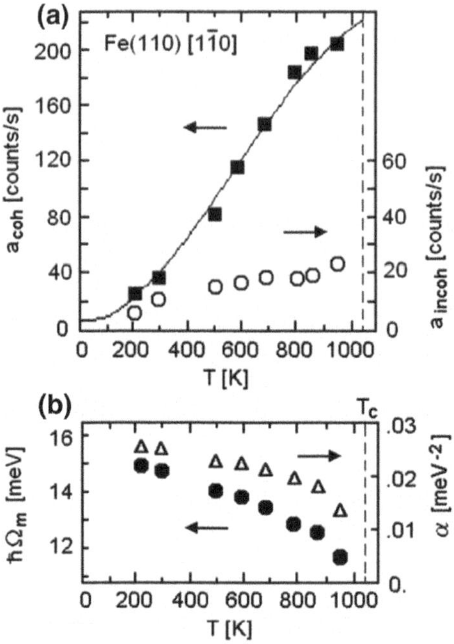

Fig. 13.17 Temperature dependence of the four parameters obtained from a fit of the multiphonon background in Fig. 13.16a to (7.72). **a** The temperature dependence of the amplitudes determining the coherent (left scale) and incoherent contribution (right scale). The solid line curve is a fit of the coherent intensity based on (7.73). **b** The maximum energy $\hbar\Omega_m$ and exponential coefficient for the coherent multiphonon background as a function of temperature

from the fit of the data in Fig. 13.16a for the $[1\bar{1}0]$ azimuth [187, 188] assuming $\alpha' = \alpha$. As seen in Fig. 13.17b with increasing temperature the average multiphonon energy $\hbar\Omega_m$ and the exponential coefficient α, which is inversely proportional to the squared half-width of the multiphonon distribution, show a decrease of about 25% thus yielding the typical broadening of the multiphonon peak. The coherent (a_{coh}) and incoherent (a_{incoh}) amplitudes in Fig. 13.17a, on the other hand show a rapid increase with temperature, towards saturation at high temperature. This is expected from the expression for the temperature dependence of a_{coh} in terms of the Debye-Waller factor as predicted by (7.73). The good fit of the experimental $a_{coh}(T)$ (Fig. 13.17a) was obtained by assuming that the surface Debye frequency appearing in the expression for the Debye-Waller factor, (7.86), is proportional to the experimental Ω_m, i.e., $\omega_{Ds}(T) = 1.7\Omega_m(T)$, which gives a $T = 0$ surface Debye temperature $\theta_{Ds} = 314$ K (see Appendix B). The approximate similarity in the size and shape of the background with the one-phonon RW signal and the decrease of the surface Debye frequency for increasing temperature can therefore be regarded as partly due to the drop of magnetization for T approaching T_c.

Such large effect lends itself to some speculation. It is noted that the magnetoelastic coupling producing visible effects of magnetization on the phonon frequencies also accounts for phonon-magnon coupling, provided the energies and momenta of the two excitations match each other [189]. Since ^{4}He atoms carry no spin they cannot excite magnons directly, but magnon excitation can substantially increase the HAS multiexcitation background with temperature via phonon-magnon coupling. Unitarity does not allow to recognize which excitations are involved in the total amplitude a_{coh}, but the rapid softening of the magnon spectrum as $T \rightarrow T_c$ may have some implications for the unusual softening of the average excitation energy $\Omega_m(T)$ [189].

A shortcoming of these early studies on Fe(110) was the large multiphonon background at T_c, which is too large to follow the surface phonon frequencies across the magnetic transition with sufficient resolution. This problem could be circumvented by investigating the RW dispersion curves of thin fcc-Fe(001) films, stabilized on a Cu(001) substrate [190], for which the Curie temperature T_c lies in the range of, or even below, room temperature [191]. Earlier studies on fcc-Fe(001) films grown on non-magnetic substrates such as Cu(001) provide evidence for a connection between the magnetic order and a structural phase transition found in this system [192]. The complex structure of this system as a function of thickness has been extensively studied by LEED and STM [193]. Moreover, it has been shown that the complex magnetic order of such films in the thickness range of a few monolayers (>3 ML), consists of alternating pairs of ferromagnetically (FM) and antiferromagnetically (AFM) coupled layers [194–197], This sequence of layers is accompanied by an oscillating interlayer magnetostriction in the range of a few percent of the topmost interlayer distance d_{12}. In similar fashion HAS interference measurements of surface step heights seen in Fig. 13.18a indicate oscillations of d_n amounting to $\pm 10\%$ with layer number N_L in the region of the transition from the ferromagnetic (FM) to the paramagnetic (PM) phases [190, 198].

Figure 13.18b shows the surface Rayleigh wave (RW) velocity as a function of the layer thickness at $T_s = 300$ K. The remarkably large changes in the RW velocity, of the order of 10% [190], are associated with the structural phase transition reported by Zharnikov et al. [192]. The observed decrease of the RW velocity at $N_L = 3$ is taken as a signature of the presence of magnetic interactions. The theoretical first principle analysis by Asada et al. [196] of all the possible stable spin configurations of fcc-Fe films predicts that the transition from the perfect FM configuration to AFM ordered configurations with a FM top bilayer takes place at coverages between 3 and 4 ML.

The contribution of ferromagnetism to the interaction between the topmost layers has been probed directly by measuring the RW frequency in a 6 ML film at a fixed wavevector of Q = 0.53 Å^{-1} for decreasing temperatures varying across the Curie temperature $T_c = 270$ K [190]. As shown in Fig. 13.19a, the RW frequency was found to increase sharply below T_c with a shape consistent with a magnetization effect. The observed increase largely exceeds the slow, uniform change due to

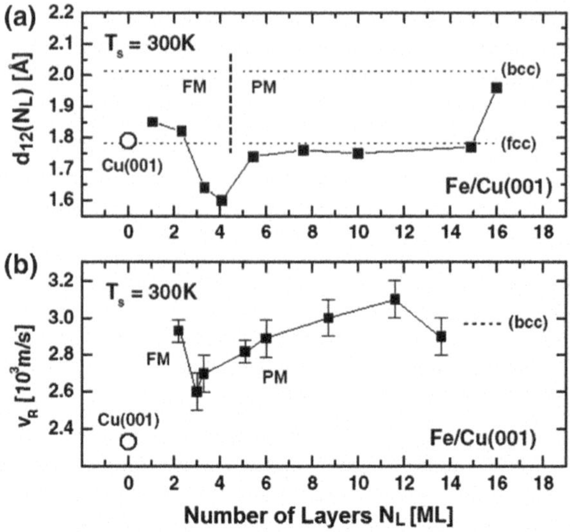

Fig. 13.18 Effect of the number of layers of fcc-Fe on the Cu(001) on the first interlayer distance and the surface Rayleigh wave (RW) velocity v_R. **a** Distance $d_{12}(N_L)$ between the first and second surface layer of the fcc-Fe films as determined by HAS interference measurements at 300 K as a function of the number N_L of layers. At 16 ML the step height jumps to the value expected for bcc-Fe, whereas for $N_L \leq 15$ it is about the same as the surface interlayer distance of Cu(001) (open circle). **b** Dependence of the Rayleigh wave phase velocity v_R on the film thickness from HAS measurements at 300 K. With increasing thickness, v_R tends to the RW phase velocity for the bcc phase of iron. For comparison the RW velocity of the clean Cu(001) surface is also shown (open circle). The dips in v_R and d_{12} at around 3–4 ML signals the transition from the ferromagnetic (FM) to the paramagnetic (PM) phase [191] (from [190])

anharmonicity, as occurs, e.g., for the corresponding RW frequency of the clean Cu (001) substrate, shown in Fig. 13.19b for comparison. The temperature dependence of the squared RW frequency on a magnetized surface is well represented by an extension of (13.7) [187]:

$$\omega(Q,T) = \omega_0(Q) - x_e\omega_0(Q)\coth\frac{\hbar\omega_0(Q)}{2kT} + b\omega_0(Q)\frac{M^2(T)}{M^2(0)}, \qquad (13.9)$$

where $\omega_0(Q)$ is the harmonic frequency for the normal (PM) phase. The coefficient $x_e = 0.0110$ is the anharmonic constant of the fcc-Fe film which fits the dependence in the normal phase. For comparison the anharmonic constant for Cu(001) is $x_{e,s} = 0.0139$ [see (13.8) and Fig. 13.13d]. The last term in (13.9) describes the effect of magnetic ordering, where $M(T)$ is the surface magnetization and b is a dimensionless constant which determines the magnitude of the magnetic shift. A good fit

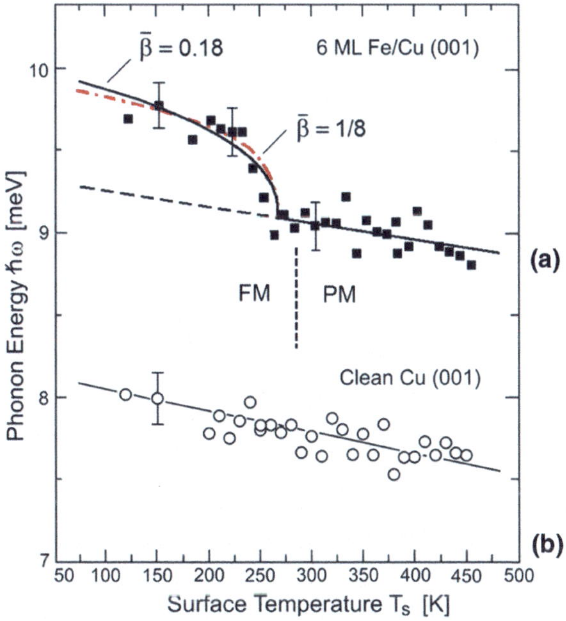

Fig. 13.19 Temperature dependence of the RW frequency measured by HAS at a fixed wavevector Q = 0.53 $Å^{-1}$ in the [110] direction for a 6 ML fcc-Fe(001) film on a Cu(001) substrate [190]. **a** The sudden change of the RW frequency at the magnetic transition from the low temperature ferromagnetic phase to the paramagnetic phase above the Curie temperature T_c = 270 K reflects the shape of the squared magnetization as indicated by the full line curve fit based on (13.9) and (13.10) for magnetization below T_c. The separation between the full line and the broken line represents the magnetic shift, which is equally well described by the experimental magnetization [199] (critical exponent $\overline{\beta}$ = 0.18, full line) or by the 2D Ising model ($\overline{\beta}$ = 1/8, dash-dotted red line). **b** The RW frequency of the clean Cu(001) substrate at the same wavevector shows the expected pure anharmonic dependence [190]

of the data in Fig. 13.19a (full line) is obtained with $\hbar\omega_0(Q)$ = 9.39 meV, b = 0.074, and the following expression for $M(T)$:

$$M(T)/M(0) = (1-x)^{\overline{\beta}\sqrt{x}}, \tag{13.10}$$

where $x \equiv T/T_c$. The analytical form of (13.10) is based on the magneto-optical data for these films reported by Li et al. [199] with the correct Bloch $T^{3/2}$-law for $T \to 0$ and the critical behavior for $T \to T_c$ given by $\overline{\beta} = 0.18$. This value of $\overline{\beta}$ is larger than the universal critical exponent $\overline{\beta} = 1/8$ for the 2D Ising model magnetization [200]. The latter value, however, gives within the experimental error an equally good fit of the magnetic frequency shift (Fig. 13.19a, dashed-dotted red line) with a slightly smaller b = 0.069. An interesting fact is that the relative magnetic shift for the *fcc*-Fe film is comparable to that for the semi infinite *bcc*-Fe

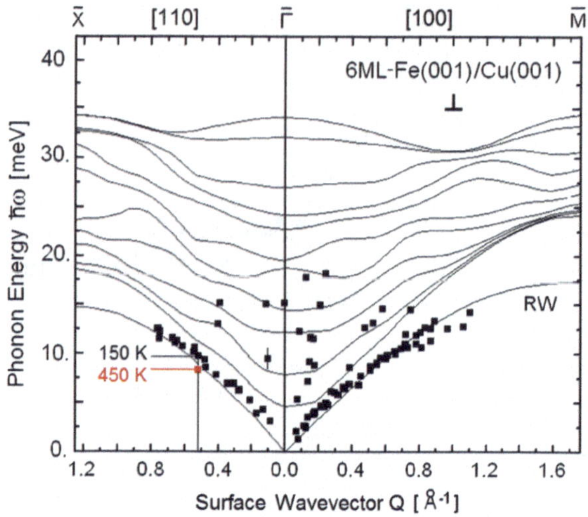

Fig. 13.20 Comparison of calculated dispersion curves of the phonons with sagittal polarization for 6 ML *fcc*-Fe(001)/Cu(001) in the magnetic phase with HAS data (black squares) measured at an incident energy of 31 meV and a surface temperature of 150 K [198]. The vertical line marks the RW phonon energies at a wavevector of 0.53 Å^{-1} shown in Fig. 13.19a. The small black and red squares are for surface temperatures of 150 and 450 K, respectively. The calculated dispersion curves are based on the nearest neighbor central force constant model [190] which includes the magnetic contribution originating from the ground state spin configuration (↑↑↓↓↑↑) [201]

(110) surface in the $[1\bar{1}0]$ direction ($b = 0.062$) obtained by fitting (13.9) to the data of Fig. 13.16c.

More detailed information on the effect of magnetic interactions was obtained from a calculation of the phonon dispersion curves of 6 ML fcc-Fe(001) on Cu(001 with a nearest neighbor force constant model which includes the magnetic contribution $\beta_{1m,ij} \equiv J^{(2)} \langle \mathbf{S}_i \cdot \mathbf{S}_j \rangle$ to the radial force constant between nearest neighbor atoms of spins $\mathbf{S}_i$ and $\mathbf{S}_j$. The magnetic force constant $J^{(2)}$ is the second order derivative of the Heisenberg exchange constant with respect to the interatomic distance [201]. The nearest neighbor central force constant model (see Sect. 3.3) is known to be satisfactory for the *fcc* substrate Cu(001) with a force constant $\beta_{1Cu} = 28.0$ N/m [202]. The same force constant model gives a good fit of the experimental RW branch of the paramagnetic (PM) *fcc*-Fe film with $\beta_{1Fe} = 38$ N/m and the interface force constant chosen as $(\beta_{1Fe} + \beta_{1Cu})/2$ with $J^{(2)} = 7$ N/m. The additional magnetic force constant $\beta_{1m,ij}$ with $J^{(2)} = 7$ N/m for the magnetic phase of the 6 ML in the ground state spin configuration (↑↑↓↓↑↑) [201] accounts for the observed magnetic shift of the RW frequency, in agreement with the first principle calculations by Sabiryanov and Jaswal [189]. The calculated 12 dispersion curves of sagittal polarization of the 6ML fcc-Fe film, not shown in [190] but reported later in [198], are plotted in Fig. 13.20. The model, besides predicting the correct RW dispersion curve, accounts fairly well for a few additional HAS data associated with the lowest organ-pipe modes near Q = 0.

The surface phonons of iron single crystals and thin iron films have also been investigated by a technique called resonant inelastic nuclear γ-ray scattering [91, 186, 203] in which x-rays are absorbed by the ^{57}Fe nucleus with subsequent K-fluorescence emission. The method provides a direct measure of the phonon density of states [204].

The above few examples of HAS experiments exploring magnetic effects on surface phonons illustrate the great potential of the method. With further improvements in resolution and sensitivity more detailed results can be expected.

13.5 HAS Spectroscopy of Low Energy Surface Electron Excitations

As discussed in Chaps. 7 and 8, the basic mechanism of HAS surface phonon spectroscopy is the exchange of energy and momentum between the He atom and the atom cores of the surface via the interposed electrons, whether localized as in semiconductors and insulators or delocalized as in conducting surfaces. More precisely virtual electron-hole pair excitations at the Fermi level behave as intermediate bosons, as illustrated in Fig. 8.3. A natural consequence is that HAS should *a fortiori* be able to detect low energy surface electronic excitations, whether single particle or collective, provided they produce sufficient charge density oscillations in the region of the scattering turning points.

13.5.1 Searching for Surface Electron-Hole Pair Excitations

Since helium atoms have no net charge and a very weak polarizability their dominant interaction with the electron gas can only be purely mechanical, via Pauli repulsion of their electron shell. The possibility that helium atoms might be able to directly excite electron-hole pairs was extensively discussed in the late seventies and lead to several theoretical investigations [205–211]. At the time of the first atom scattering TOF experiments, Gunnarson and Schönhammer [212–215] published a theory for calculating the probability for the excitation of electron-hole pair excitations by helium atoms. They assumed that the He atoms act on the substrate electrons via a repulsive pseudopotential and treated the non-adiabatic response of the surface electron gas with the lowest order of perturbation, as originally proposed by Brivio and Grimley [210, 211]. The experiment reported by Feuerbacher and Willis [216, 217] with Ne atom scattering (NeAS) from Ni(111) has been discussed in connection with the possibility that it might be due to electron-hole excitations [215]. They observed inelastic processes at zero momentum transfer which occur over a wide energy transfer range exceeding the maximum phonon frequency. As discussed below in Sect. 13.5.3, this can be regarded in hindsight as, perhaps the first evidence for acoustic surface plasmon (ASP) excitations. Their existence was theoretically predicted much later [202] and then experimentally demonstrated with HREELS [218–222].

In their early theoretical analysis Gunnarsson and Schönhammer concluded that the HAS cross section for the excitation of single electron-hole pairs is only about 10^{-5} and much smaller than that for ordinary phonon processes, the main reason being the exceedingly low electron density met by the probe atom at the turning point [215]. Similarly the attractive van der Waals interaction was shown to give a negligible contribution to electron-hole excitations despite the prominent role of the surface electron response [223]. More favourable conditions are however expected for HAS from electron-hole excitations at $\Delta K \cong 2k_F$, where in quasi-one-dimensional (quasi-1D) metallic systems the electronic susceptibility is strongly enhanced. As discussed in Sect. 11.3.3, the sharp anomalies observed by HAS (and not by EELS) in W(110):H(1×1) and Mo(110):H(1×1) [224, 225], can be interpreted as electron-hole excitation branches associated with the quasi-1D hydrogen rows.

13.5.2 Low Energy Surface Plasmons and Plasmarons in Semiconductors

A conducting surface, besides hosting localized electron-hole pair excitations, can sustain collective electron excitations, i.e., surface plasmons and the associated electromagnetic modes coupled to plasmons, called surface plasmon polaritons. The theory of surface plasmons and surface plasmon polaritons was reviewed in 2007 by Pitarke, Silkin, Chulkov and Echenique [226]. Plasmon frequencies in the meV domain occur in the accumulation layer of doped semiconductors as, for example, on the hydrogen passivated surfaces of InAs(110) [227] and ZnO [228]. As shown by Yu and Hermanson [227, 228] the presence of electronic subbands in the accumulation layers implies an equal number of intrasubband plasmon branches as well as intersubband plasmon branches. Moreover the occurrence of two-dimensional (2D) plasmon excitations within the spectral domain of optical phonon polaritons, as well as of surface optical phonons, leads to coupled plasmon-phonon states called *plasmarons* (Fig. 13.21). The energies of plasmaron states calculated by Yu and Hermanson [227, 228] agree quite well with the HREELS experiments for both InAs(110) [229] and ZnO [230]. Moreover plasmons associated with anti-phase charge density oscillations in two different sub-bands may acquire an acoustic character [*acoustic surface plasmons* (ASP)] with a linear dispersion $\hbar\omega_{\mathrm{ASP}}(Q) \propto Q$ for $Q \to 0$ (Fig. 13.21), provided that their polarization field is efficiently screened by the charge density of the lowest subband

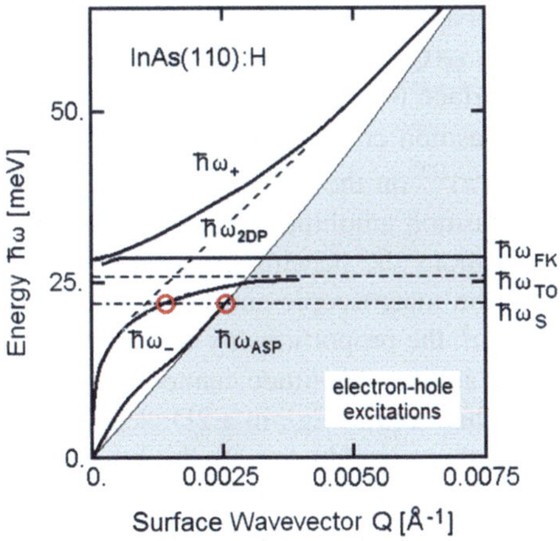

Fig. 13.21 The calculated intraband surface *plasmaron* dispersion curves at very small wavevectors associated with the accumulation layer of the hydrogen exposed InAs(110) surface. A 2D plasmon branch of energy $\hbar\omega_{\mathrm{2DP}}$, associated with the confined 2D electron gas (2DEG), has an avoided crossing with the Fuchs-Kliever (FK) surface phonon polariton branch of energy $\hbar\omega_{\mathrm{FK}}$ leading to two *plasmaron* branches $\hbar\omega_-$ and $\hbar\omega_+$, the former tending to the bulk TO phonon energy $\hbar\omega_{TO}$ at large wavevector Q. In the case of two 2DEG sub-bands a further surface plasmon branch having energy $\hbar\omega_{\mathrm{ASP}}$ proportional to Q for $Q \rightarrow 0$ (*acoustic surface plasmon* (ASP)), may appear. All plasmon branches only exist at small values of the surface wavevector. The $\hbar\omega_-$ and $\hbar\omega_{\mathrm{ASP}}$ branches, before merging into the electron-hole excitation spectrum (grey area) and losing definition due to the Landau damping, may have further intersections (encircled) with an optical surface phonon branch (e.g., $\hbar\omega_S$ = 22 meV in InAs [101, 232]). The corresponding avoided crossings yield further surface plasmaron pairs (adapted from Yu and Hermanson [228])

[227]. As discussed in the next section, ASPs were predicted [226, 231] and found by EELS [218, 220] in metal surfaces.

Given the small cross section for single electron hole (e–h) pair excitations, an intriguing question is whether HAS can excite low energy surface plasmons, as they occur, for example, in the 2D electron (hole) gas trapped at the surface accumulation layer of a doped semiconductor. The interaction of the impinging atom with a plasmon excitation can be described in the Esbjerg-Nørskov local density approximation, as expressed by (8.19–8.21), where the time-dependent inelastic scattering potential $\delta V_{pl}(t)$ is assumed to be proportional to the plasmon oscillations of the local surface charge density. By means of the theoretical derivation of Ritchie et al. [233], adapted to the case of a plane soft-wall surface, the interaction potential can be written as [234]

$$\delta V_{pl}(\mathbf{r},t) = A\beta\, e^{-\beta z} \sum_{\mathbf{Q}} \left(\frac{\hbar}{2m^{*}\omega_{pl}(\mathbf{Q})} \frac{n_a}{a_c N_c} \frac{Q}{\pi} \right)^{1/2} e^{i\mathbf{Q}\cdot\mathbf{R}} [b^{\dagger}_{pl,\mathbf{Q}}(t) + b_{pl,\mathbf{Q}}(t)], \tag{13.11}$$

where A is the Esbjerg-Nørskov constant (8.8), $1/\beta$ the softness parameter, m^* the electron effective mass, $\omega_{pl}(\mathbf{Q})$ the plasmon frequency, n_a the corresponding electron density, a_c the surface unit cell area, N_c the number of surface unit cells, $b^+_{pl,\mathbf{Q}}(t)$ and $b_{pl,\mathbf{Q}}(t)$ the plasmon creation and annihilation operators, and as usual $\mathbf{r} \equiv (\mathbf{R}, z)$. The factor $(Q/\pi)^{1/2}$ on the right hand side of (13.11) accounts for the decrease of the surface plasmon amplitude with increasing penetration length into the bulk. For surface plasmons the penetration length λ is also inversely proportional to Q, the same as the macroscopic surface modes (Chap. 3). However, for systems of finite thickness d, the proportionality breaks down when the penetration length exceeds d since the surface amplitude cannot by smaller than $1/d^{1/2}$ in order to satisfy normalization. This occurs, e.g., in a 2D electron gas such as the surface accumulation layer of doped semiconductors. In this case for plasmons with $Q < \pi/d$ the factor Q/π in (13.11) is replaced by $1/d$.

The expected inelastic scattering amplitudes from surface plasmons, (13.11) can be compared with the corresponding expression for the direct impulsive interaction with surface phonons. By combining (7.3) with (3.36) and (3.64) the inelastic scattering potential for phonon excitation becomes

$$\delta V_{ph}(\mathbf{r},t) = \sum\nolimits_{\mathbf{Q}} \frac{\partial V}{\partial z} \left(\frac{\hbar}{2N_c M \omega_{ph}(\mathbf{Q})} \right)^{1/2} e_z(0,\mathbf{Q}) e^{i\mathbf{Q}\cdot\mathbf{R}} [b^+_{ph,\mathbf{Q}}(t) + b_{ph,\mathbf{Q}}(t)], \tag{13.12}$$

where M is the surface atom mass and $e_z(0, \mathbf{Q})$ the SV component of the polarization vector at the surface layer.

For a significant comparison between the expected HAS inelastic intensities from plasmons and phonons, it is convenient to refer them to the same surface wavevector $\mathbf{Q}$. For plasmons, however, the accessible range of $\mathbf{Q}$-values is very small on the scale of the reciprocal lattice vectors, since at larger wavevectors their dispersion curves quickly merge into the continuum of electron-hole pair excitations (Fig. 13.21). In (13.11) it is assumed, for consistency with the flat soft-wall potential adopted in the plasmon case, that $\partial V/\partial z \cong -A n_0 \beta e^{-\beta z}$, where n_0 is the average density of valence electrons at the surface layer ($z = 0$). If identical kinematic conditions are assumed, then the ratio of the differential reflection coefficients (7.18) for plasmon and phonon excitations, as derived in the distorted-wave Born approximation (7.18) and (7.33), is approximately given by

$$\frac{d^2\Re_{pl}/dE_f d\Omega_f}{d^2\Re_{ph}/dE_f d\Omega_f} \cong \frac{M}{m*} \frac{\omega_{ph}(\mathbf{Q})}{\omega_{pl}(\mathbf{Q})} \frac{n_a Q}{\pi a_c n_0^2 e_z^2(0,\mathbf{Q})} \frac{2n_{pl}+1}{2n_{ph}+1}, \tag{13.13}$$

where the n_{pl} and n_{ph} are the plasmon and phonon occupation numbers, respectively.

The surface accumulation layer of hydrogen exposed InAs(110) provides a good system for estimating the above ratio. According to the calculations of Yu and Hermanson [228], $m^* = 0.022\, m_e$, $n_a \approx 10^{18}$ cm^{-3} and $d \cong 3.2 \times 10^{-6}$ cm, so that on the wavevector scale of Fig. 13.21 the factor Q in (13.13) is replaced by $\pi/d \cong 0.01$ Å^{-1}. A suitable choice for the phonon is the surface gap mode of InAs(110):H, which has an energy $\hbar\omega_{ph}(\mathbf{Q}) \cong 22$ meV at the $\overline{\Gamma}$ point and is strongly localized ($e_z^2(0, \mathbf{Q}) \approx 1$) on the As ion of the top chain (Fig. 11.16b) [232, 235]. Figure 13.21 shows that the dispersion curve of this surface gap mode $\hbar\omega_{ph}(\mathbf{Q}) \equiv \hbar\omega_S$ is perfectly flat and intersects the $\hbar\omega_{pl}(\mathbf{Q}) \equiv \hbar\omega_-$ plasmon curve at about $Q = 0.0015$ Å^{-1}. The gap-mode oscillation of the top As ion implies an oscillating electric dipole and therefore a coupling to the $\hbar\omega_-$ plasmon, as well as to the ASP, which leads to *surface plasmaron* states and avoided crossings in the intersection region (red circles in Fig. 13.21). Since the As ion vibrates alone, M is about equal to the As mass (74.9 a.u.). With $n_0 = 0.14$ Å^{-3} and $a_c = 25.8$ Å^2 for InAs(110) [235] the scattering amplitude ratio (13.13), calculated at the branch intersection, is 3.9×10^{-2}. This value is comparatively small, but well accessible to the present inelastic HAS sensitivity. A stronger confinement and a larger carrier density in the accumulation layer, as obtained with a heavier doping, would improve the amplitude ratio but also increase the plasmon energies. In this case larger incident energies would be required, with a substantial increase of the multiphonon background which might overshadow the weak plasmon features. The case of ZnO studied by Yu and Hermanson [227] represents a good compromise and probably offers better conditions for a study of surface plasmon and plasmaron excitations by means of inelastic HAS. In particular the additional plasmon branch ω_{ac}, arising from the interplay of two surface subbands and having an acoustic character (linear dispersion for Q → 0) (Fig. 13.19), would be difficult to observe with HREELS, but should be accessible to and worth investigating with Helium Atom Scattering, especially with the new ^{3}He-spin echo technique.

13.5.3 HAS from Acoustic Surface Plasmons (ASP)

The interest in observing low energy surface plasmons with HAS has been greatly stimulated by the ground breaking theoretical work of Silkin et al. [226, 231]. For atomically flat surfaces, with partly occupied electronic surface states, they called attention to a new situation which arises in comparison with a purely two-dimensional electron gas (2DEG). According to their model at the Fermi level there are two spatially-separated electron gases, one associated with a surface band at a metal surface, the other with the underlying bulk states. The interplay of the 2DEG with the 3D Fermi gas produces the acoustic branch ω_{ASP} much in the same way as two coupled 2D subbands at a semiconductor surface (Fig. 13.21). The subsequent experimental demonstration of ASPs by HREELS in Be(0001) [218] and at noble metal surfaces [219–222] raises the question about the extreme acoustic limit of ASPs, where they enter the phonon spectrum and possibly couple to the optical surface phonons near $\overline{\Gamma}$. This coupling, of great relevance in THz plasmonics, can hardly be assessed with HREELS, but is well within the domain of

TOF-HAS and SE-HAS spectroscopy. The rationale behind this expectation is that only in the spectral region where electron-hole excitations are strongly hybridized with surface phonons can the corresponding sharp non-adiabatic anomalies be observed and explained in terms of single e-h surface excitations (see Sect. 11.3.15 and Fig. 11.37). Thus a similar hybridization is expected to occur between ASP dispersion curves and the surface optical phonon branches of metal surfaces, notably, with the shear-vertical (S_2) and longitudinal Lucas (S_4) modes.

A close inspection of the HAS data from Cu(111) [188] and Ni(111) multilayers on the H-passivated diamond (111) surface [236] provides evidence for ASPs in the extreme low energy range inaccessible to HREELs. Additional evidence comes from the newly interpreted HAS data for the semiinfinite Ni(111) surfaces [237] and SE-HAS data from Cu(111) [238], with hints of an avoided crossing with the optical S_2 and S_4 surface phonons, similar to that defining the plasmarons as illustrated in Fig. 13.21. This new analysis also provides an interpretation for the oldest puzzle in inelastic atom scattering spectroscopy, the observation of elementary excitations of very small wavevector made by Feuerbacher and Willis with NeAS from Ni(111) [216, 217]. All of the old HAS and NeAS measurements were performed before the theory of ASPs was formulated, which can explain why the corresponding features were unnoticed. Another reason is the superposition, in the region explored by HAS, of the ASP dispersion curve onto the surface-projected bulk phonon continuum, which may confer to the ASP features the shape of a Fano resonance, rather than that of a sharp peak.

Figure 13.22 collects a selection of HAS energy transfer spectra measured at different incident angles on Cu(111) along the $[1\bar{1}0]$ direction [225]. A new analysis of the data reported in [188] reveals new features near where the vertical arrows indicate Q = 0. Since these small features consistently appear in all the spectra near $Q = 0$ and some exhibit variable shapes reminiscent of Fano resonances [239] they are considered as evidence for the existence of ASPs. Their energies are plotted as a function of their small momentum transfer in Fig. 13.23 (full diamonds). At very low energy a few more points (full squares) from recent dedicated high resolution SE-HAS measurements have been added [238]. Despite the comparatively large error bars of the HAS data on the very small wavevector scale, the data suggest a qualitative fit showing the avoided crossing between the theoretical S_2 and S_4 optical phonon branches [240] and the ASP linear dispersion lines. At high energies above the phonon maximum the ASP velocity $v_{ASP,\infty}$ is the same as known from HREELS data [220], while in the low energy limit the ASPs have about a three times smaller velocity $v_{ASP,0}$ due to the coupling with the phonon background. In this spectral region the electron density oscillations are coupled to the lattice atoms, due to the existence of phonons with the same parallel wavevector and energy, which increase the effective ASP dynamical mass. Again this mass-renormalization, as well as the avoided crossing gaps, depend on the strength of the electron-phonon coupling. An ASP moving across a medium, which can respond by moving its atoms with the same frequency and momentum, is similar to a photon in a

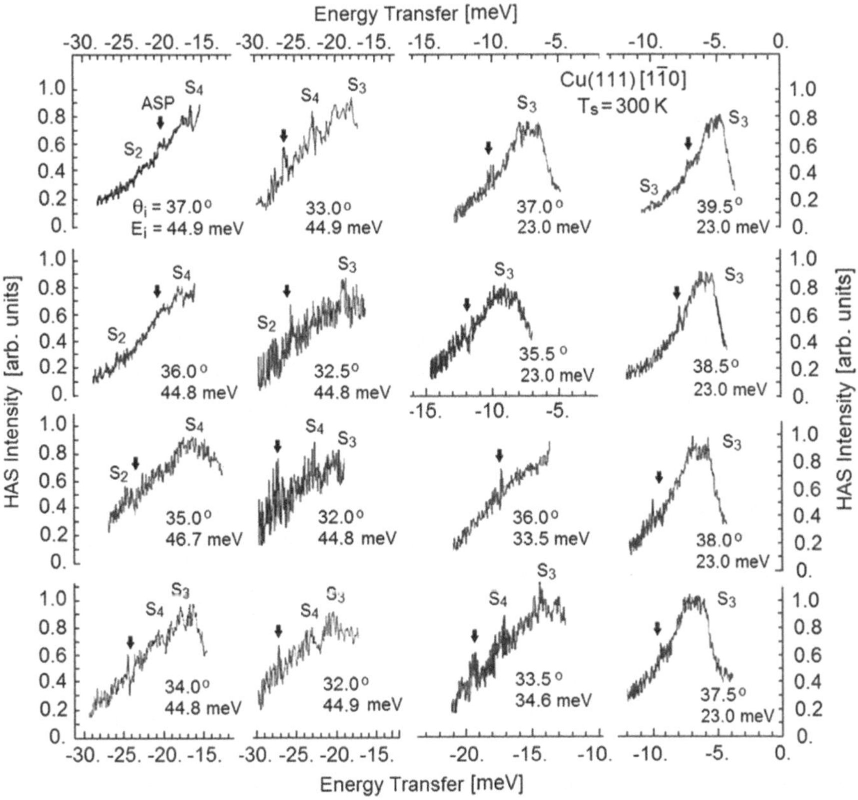

Fig. 13.22 HAS energy transfer spectra from Cu(111) in the [1$\bar{1}$0] direction for different incident angles and energies, and a surface temperature of 300 K. The spectra in the first two columns are plotted in the high energy transfer region, those in the third and fourth column in the intermediate and low energy transfer region. In addition to the broad peaks corresponding to the S_3, S_4 and S_2 phonon resonances, distinct weak features, sometimes in the form of Fano resonances, are observed at very small wavevector transfers near where the vertical arrows indicate Q = 0. They are attributed to acoustic surface plasmons (ASP) (adapted from [188])

polarizable medium. Its speed is reduced by a factor playing the role of a refractive index and proportional to the electron-phonon interaction. This strong coupling with a surface phonon leads to an avoided crossing which, seen on the energy scale has the shape of what in optics is known as anomalous dispersion.

Additional evidence for ASPs comes from a new analysis of HAS energy transfer from a 10 ML film of Ni(111) grown on the H-passivated diamond (111) surface [236]. Figure 13.24a shows HAS energy transfer spectra for two different incident angles. The resulting dispersion curves, folded on the negative energy transfer side (Fig. 13.24b) show the lowest organ-pipe (OP) mode at the zone center, evolving with increasing wavevector transfer into the RW of Ni(111) (full

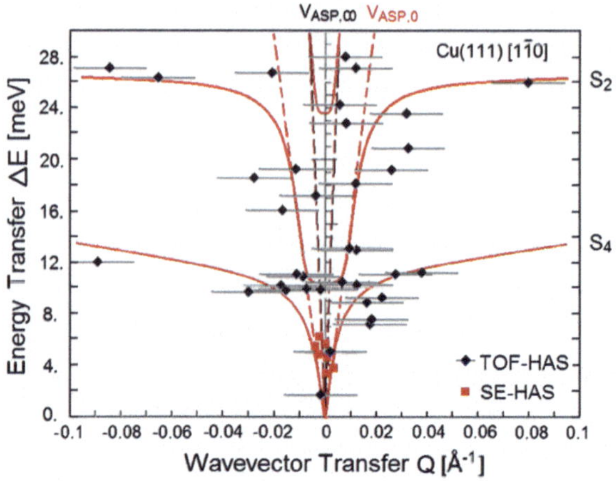

Fig. 13.23 The features marked by an arrow in the energy transfer distributions in Fig. 13.22 are plotted as a function of wavevector transfer near Q = 0. The red curves show a qualitative fit of the avoided crossing between the theoretical S_2 and S_4 optical phonon branches [240] and the ASP linear dispersion lines. The ASP velocity $v_{A\,SP,\infty}$ (dashed black line) is from HREELS data at high energies above the phonon maximum [220], The ASP velocity $v_{A\,SP,0}$ (dashed red line) is from a fit of the low energy HAS data. It is attributed to avoided crossings of the ASP with surface phonons leading to *surface plasmarons* (see Sect. 13.5.2). The apparently large error bars are due to the expanded wavevector scale and correspond to the small energy uncertainty indicated by the diamond symbols. The red squares are from SE-HAS measurements [238]

circles). The data points corresponding to the smaller features (open circles) in the inelastic spectra accumulate instead around $Q = 0$ and are therefore associated to ASPs. At the time of these measurements ASPs were not known and this branch was assigned to the RW velocity of diamond (111) surface [236]. As appears in Fig. 13.22b, the RW velocity of diamond (RW_C) [241], while much faster than in Ni(111), cannot account for the observed peaks near $Q = 0$, which now are attributed to ASPs.

The complete set of HAS and NeAS data for ASPs and surface phonons for Ni (111) [216, 217, 236, 237] is collected in Fig. 13.25 and shown together with the HREELS data for comparison. They are plotted on top of a spin-DFPT calculation of the dispersion curves for a thick Ni(111) slab [237].

The spin-DFPT calculated dispersion curves for Ni(111) in Fig. 13.25 are very similar to those for Cu(111). Thus the Cu(111) DOS projected on the SV and L components of the 1st and 2nd layer (Fig. 11.26) were used to assign the calculated dispersion curves in the following way: (i) S_1 is the RW (continuous blue line); (ii) S_2 at small wavevectors is a broad SV 2nd-layer optical resonance (broken blue line and blue area), which becomes a SV 1st-layer gap mode near the zone boundary; (iii) S_3 is a surface resonances, which starts as a longitudinal resonance in the 1st-plane (broken blue line), then turns into a strong 2nd-layer SV resonance

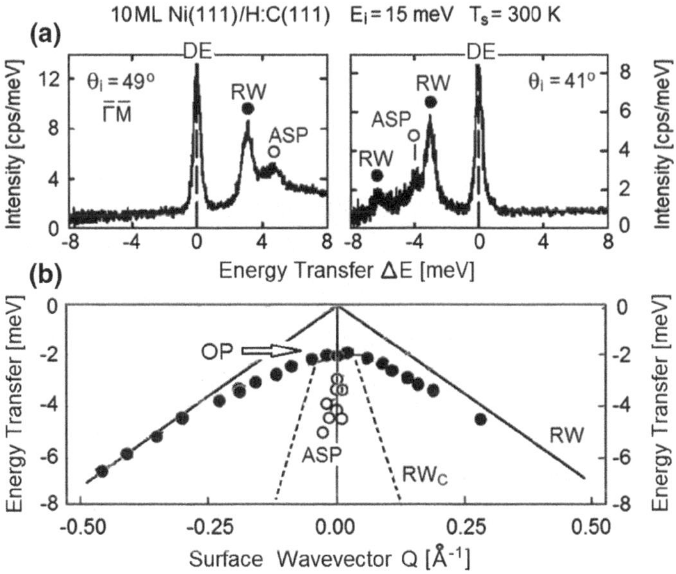

Fig. 13.24 HAS energy transfer spectra and dispersion curves from a 10 ML-Ni(111) film on a H-passivated diamond (111) surface at 300 K in the $\overline{\Gamma M}$ direction [236]. **a** Two typical energy transfer curves on both sides of the specular peak showing inelastic features on the annihilation side (θ_I = 49°) and on the creation side (θ_I = 41°) now attributed to ASPs. Previously [236] they were assigned to the Rayleigh mode (marked RW_c from [241]) of the diamond substrate. **b** The corresponding energy transfers as functions of the wavevector transfer after folding into the negative energy (creation) side. The organ-pipe (OP) mode at 2 meV evolves into the Ni(111) RW at small wavevectors. Figure 12.13 shows a similar behavior but for a Na film

near the zone boundary (blue area). (iv) S_4 is a longitudinal 2nd-layer resonance (broken blue line) known as the longitudinal Lucas mode.

The dispersion curves of the two optical 2nd-layer resonances S_4 and S_2 are qualitatively represented with their avoided crossing with the ASP dispersion curve so as to give a reasonable fit of the experimental points, as for Cu(111) in Fig. 13.23. In the present case, however, the high energy ASP velocity $v_{ASP,\infty}$ is chosen so as to correspond to the Fermi velocity of the shallow surface electronic band S_1 for majority spins as determined by Lobo-Checa et al. [243], since no direct HREELS measurements of ASPs in Ni(111) are available yet. Also in this case, as for Cu(111), the low energy limit of the ASP velocity $v_{ASP,0}$ which gives a reasonable fit of all the available data is about three times smaller than $v_{ASP,\infty}$.

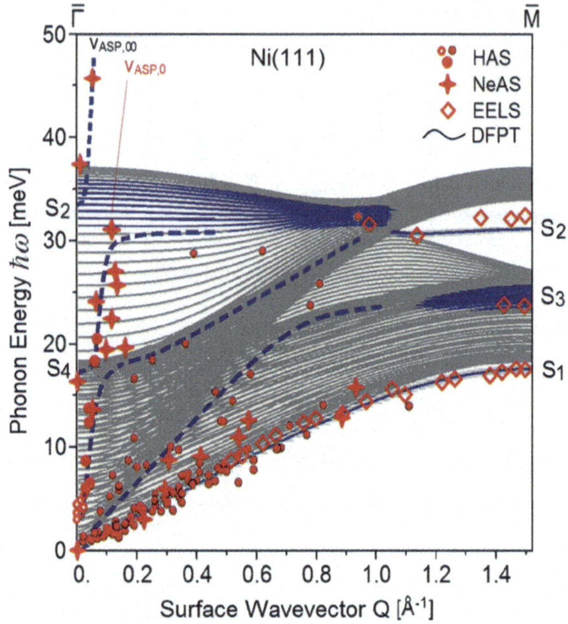

Fig. 13.25 Comparison of experimental dispersion curves of the Ni(111) surface in the $\overline{\Gamma M}$ direction from several different sources with a spin-DFPT calculation [237] for a thick Ni(111) slab. The HAS data are denoted by red circles: small closed red circles are phonons [237], the few large filled circles at small Q [237] and the small open red circles ($\hbar\omega < 5$ meV, $Q \approx 0$, from Fig. 13.24 [236]) are attributed to ASPs. The crosses are NeAs data [216, 217]. The open diamonds are from HREELS [242]. The full thick blue lines correspond to localized surface branches calculated with spin-DFPT. The dark blue areas represent strong 2nd-layer resonances of SV polarization. In addition to the well-known L acoustic resonance S_3, the broken thick blue lines indicate the 2nd-layer S_4 and S_2 resonances of dominant L and SV polarization, respectively. They show hybridization (avoided crossings) with ASPs according to a qualitative fit. The high-frequency ASP velocity $v_{ASP,\infty}$ corresponds to the Fermi velocity of a Ni(111) shallow surface electronic band as determined by Lobo-Checa et al. [243]

13.6 High Energy Atom Inelastic Scattering

HAS spectroscopy, besides exploring phonons over a broad range of energies from the ultra-slow surface dynamical processes, particularly in the SE-HAS version, up to multiphonons with energies of 100 meV, also has found interesting applications at much higher energies thanks to the development of neutral atom beams in the eV to keV energy range [244–251]. Despite the intrinsic limitations in resolution, important information can be extracted in principle about more energetic surface excitations. In the following, the authors' original analysis of these ground breaking experiments are reassessed with some new considerations about the interaction mechanisms which should help to illustrate the potential and some new possible directions of fast atom-surface spectroscopy.

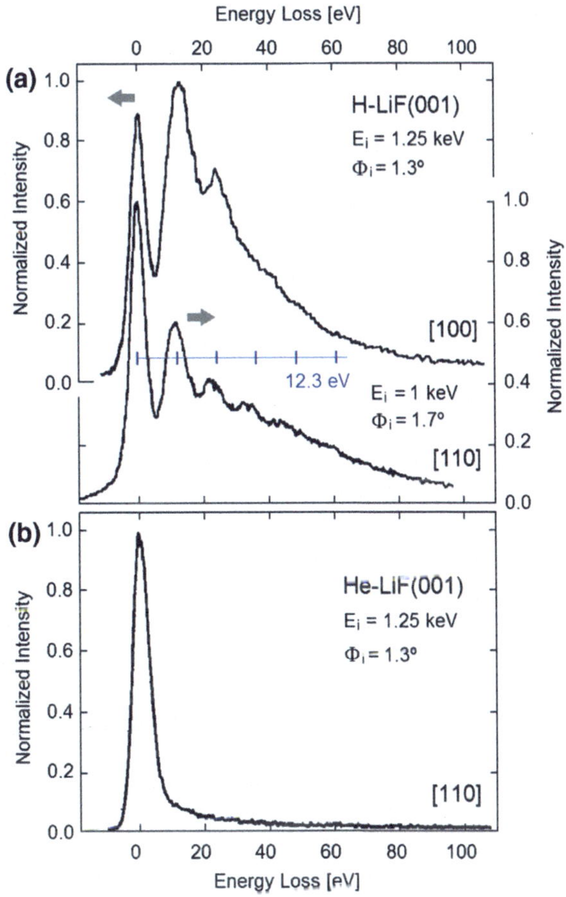

Fig. 13.26 Energy loss spectra of high energy H and He atom beams scattered from the LiF(001) surface at grazing incidence [247, 248]. **a** H atoms at $E_i = 1.25$ keV incident energy in the [100] direction and $E_i = 1$ keV in the [110] direction fly-by the surface with sufficient speed so as to create many 12.3 eV surface excitons [252]. **b** ^{4}He atoms at $E_i = 1.25$ keV in the [110] direction with comparable incident energy are too slow to create surface excitons

Figure 13.26a shows two energy loss spectra of H atoms scattered at grazing incidence from the LiF(001) surface [247, 248] in the [100] and [110] directions at slightly different incident energies of 1.25 and 1.0 keV and grazing angles $\Phi_i \equiv \pi/2 - \theta_i = 1.3°$ and 1.7°, respectively. They exhibit the typical Poisson distribution of surface excitations in Glauber coherent states [171] of the 12.3 eV surface excitons (hole-electron bound states) of LiF(001) [252]. The multiexciton spectra, starting from the 0-exciton elastic peak, are formally similar to the multi-phonon spectra described in Chap. 7 by (7.64–7.67), with the phonon DOS replaced by the surface-exciton DOS, and the DW exponent 2W replaced by the current-current matrix element of the boson propagator ((14–16) of [253]). This matrix element, here denoted by S [254], is actually the upper vertex term in the Feynman diagram of Fig. 8.3, where the two fermion lines represent the creation of the excitonic electron-hole bound state. It measures the strength of the atom-exciton interaction, more precisely the average number of excitons created in a single scattering event. After subtraction of the incoherent background (see Sect. 7.6.2) the Poisson distribution $e^{-S}S^n/n!$ for the [110] direction corresponds to $S \sim 0.6$. A similar value is expected for the [100] direction, where however the incoherent background is considerably larger. The coupling strength expressed by S and the coherent-incoherent partition a_{coh}/a_{incoh} (7.72) depend in general on the incident energy and incident angle of the atom beam. As shown in experiments by Busch et al. [249], when ^4He atoms are used instead of hydrogen at about the same incidence energy and grazing angle (Fig. 13.26b), practically no inelastic sideband is observed. Such a large suppression of the coupling to surface exciton polaritons can be understood by comparing the grazing fly-by velocities of H and ^4He atoms for the energy loss spectra of Fig. 13.26a, b, both along [110]. The 1 keV H atoms travel at $v \cong 4.4 \times 10^5$ m/s along a periodic surface made of positive and negative ions and are therefore subject to an oscillating electromagnetic (e-m) pole-induced dipole field of frequency $v/d \cong 3.1 \times 10^{15}$ Hz, where $d = 1.41$ Å is the LiF interplanar distance along [110]. The oscillating induced dipoles of the H atoms interacts strongly with the surface electrons, with an oscillation frequency which corresponds to a quantum energy of 12.8 eV, just enough to excite surface excitons almost in resonance. Note that this simple argument is based on a sinusoidal shape of the surface atom induced dipole potential and that higher **G**-vector components in this potential (see (6.38)) would imply higher harmonics of the coupling e-m field. The velocity of 1.25 keV ^4He atoms (Fig. 13.26b), is lower at 2.5×10^5 m/s and the e-m field quantum energy for the same [110] direction is reduced to 7.3 eV, which is insufficient to excite surface exciton polaritons and therefore no surface-exciton sideband is observed in the energy loss spectrum of Fig. 13.26b. It will be seen below that more energetic 12 keV ^4He atom beams are able to excite multi-exciton coherent states in LiF(001) with a comparatively high S.

As shown by Khemliche et al. [255], also the neutralization of a 600 eV proton beam impinging the LiF(001) surface at grazing incidence ($\Phi_i \equiv 3°$) yields a final H atom beam with well resolved energy loss peaks corresponding to the creation of one or more 12.3 eV surface excitons (Fig. 13.27), just as found later by Winter et al. with neutral H beams [247, 248]. The ion scattering can be accompanied by the emission of one or more surface electrons with a final energy of $\sim$2 eV, and the energy loss spectra are shifted to larger energy losses by multiples of the LiF(001) ionization energy plus 2 eV. All the 0-, 1-, and 2-electron energy loss spectra exhibit a multiexciton Poisson distribution, $P(n) = e^{-S}S^n/n!$, for $S = 1.1$, 0.9 and 0.7, respectively, with a negligible incoherent background. Note that the 0-electron, 0-exciton line occurs at an energy loss of about 3.6 eV. As explained in [255] this misfit is due to the excitation of a large number of LiF(001) surface phonon polaritons (Fuchs and Kliever (FK) modes, see Sect. 3.6) via the Coulomb forces exerted by the protons just before neutralization. In this case the coherent emission of FK modes implies $S \sim 50$. Note that the fly-by speed of 0.6 keV neutral H atoms would be largely insufficient to excite 12.3 eV surface excitons via the e-m coupling described above, whereas charged protons can excite surface polaritons through the standard dipole scattering mechanism known for EELS [256], as observed by [255].

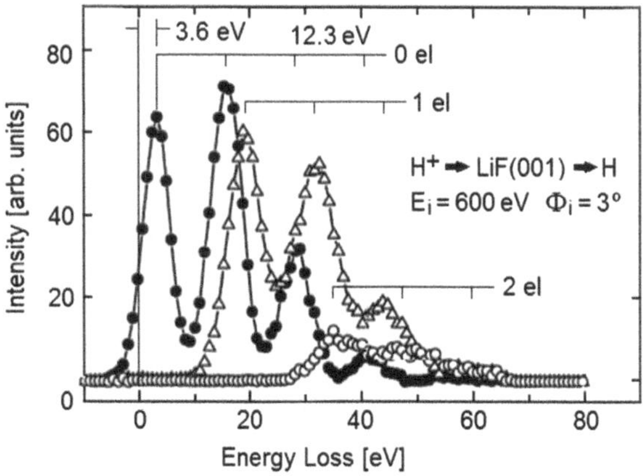

Fig. 13.27 Energy loss spectra of inelastic proton neutralization scattering from LiF(001) for an incident energy of 600 eV under a grazing angle of 3°, as measured by Khemliche et al. [255]. The spectra are correlated with the emission of one (open triangle) or two (open circle) electrons are shifted with respect to the spectrum without electron emission (filled circle) by ~16 and ~32 eV, respectively. All spectra show a Poisson distribution of the LiF(001) 12.3 eVsurface excitons [252] with S = 1.1, 0.9 and 0.7 for 0-, 1- and 2-electron emission, respectively. The decrease in coherent coupling with the increase of emitted electrons is compensated by an apparent increase in the incoherent background. The additional 3.6 eV shift of the 0-electron 0-exciton spectrum with respect to the incident energy (0 of the energy loss scale) is attributed to the excitation of an average of ~50 LiF(001) surface phonon polaritons

High energy atoms or ions at grazing incidence can also be selectively adsorbed into a surface bound state and re-emitted into the final scattering state via the exchange of surface **G**-vectors (Sect. 10.1) after oscillating one or more times within the attractive well in the normal direction. This semi-classical picture of the motion in a bound state (see Fig. 10.3), applies well to fast atoms (ions) heavier than He. Figure 13.28 shows the energy loss spectra of 1 keV Ne^+ ions and of the neutralized Ne atoms scattered from LiF(001) at $\Phi_i = 0.95°$ [00Vil]. The quasi-Gaussian shape of each ion energy loss peak in Fig. 13.28 implies a large average number of emitted bosons [cfr. (7.64)], and suggests therefore the emission of ~70 meV FK phonon polaritons of energy with S as large as ~400. Such large phonon numbers are not unusual in the optical vibronic structure of color centers with a strong electron-phonon interaction, as, e.g., for the F-centers in alkali halides at room temperature [257].

A semiclassical simulation (full line in the inset of Fig. 13.28) of the inelastic bouncing during the fast ion motion in the bound state provides a convincing explanation of the observed energy loss spectrum as due to *skipping motion*, also called *quantum ricochets*, against the potential walls in the normal direction [259]. At each bounce of the ion, surface polaritons are coherently emitted. The weaker series of energy loss peaks associated with the neutralized Ne atoms shows the same structure, apart from the shift of about −11.5 eV due to the neutralization energy gain and a slightly smaller periodicity of ~45 eV. Such similarity suggests

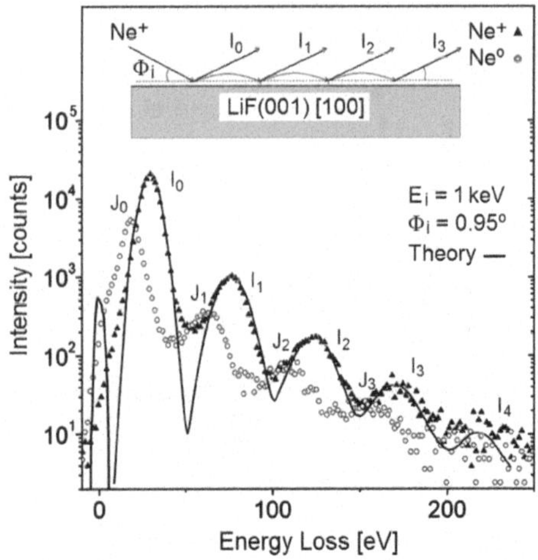

Fig. 13.28 Energy loss spectrum of Ne^+ (filled triangle) and neutralized Ne^0 atoms (open circle) after scattering of incident 1 keV Ne^+ ions from the LiF(001) surface at 0.95° grazing angle [258]. The Ne^+ peaks labeled I_0 to I_4 are regularly separated by 48 ± 1 eV, starting from the parent peak I_0 at 29 eV, while the neutralized Ne^0 series J_0 to J_3 shows a smaller separation of ~45 eV with J_0 at 17.5 eV. As illustrated in the inset and supported by theory (full line) [259], the series of peaks is due to multiple bouncing of the Ne^+ ions trapped into a surface bound state against the surface potential walls

that smaller energy losses is due to ions which are neutralized pick up the electron when leaving the surface and that the trapping into a deeper bound state with a shorter bouncing period. The neutralization in leaving the surface also explains the large difference between the inelastic scattering of protons (Fig. 13.27) and that of Ne^+, substantiated in the dominance of surface exciton polaritons in the former and surface phonon polaritons in the latter. Ne^+ ions are subject to attractive dispersion forces from surface atoms providing surface bound states, whereas protons are not, and the mechanical bouncing of the trapped ions against the potential walls favours the coupling to optical phonon polaritons.

Free electron collective excitations, considered in the previous section which compete with surface phonons in low energy HAS from metal surfaces, provide the most efficient energy loss mechanism at high energíes. For neutral He beams in the keV energy range the Pauli repulsion with the metal surface free electron gas is expected to create a multitude of surface plasmons. Figure 10.29 displays the energy loss spectrum and the analysis of grazing collisions He atoms from Ag(110) measured by Bundaleski et al. [244] for an incident energy of E_i = 2 keV and an angle of 1°. The energy loss peak (thick full line) is shifted by 8 eV with respect to the incident beam Gaussian peak (broken line) and is twice as broad. The asymmetric shape suggests a Poisson multi-boson distribution for a boson energy in the eV range. The best fit with a Poisson distribution of peaks having the incident peak width (Fig. 10.29b) yields a surface plasmon (SP) energy E_{SP} = 4.1 eV and an average plasmon number S = 2.7. This plasmon energy agrees with the surface

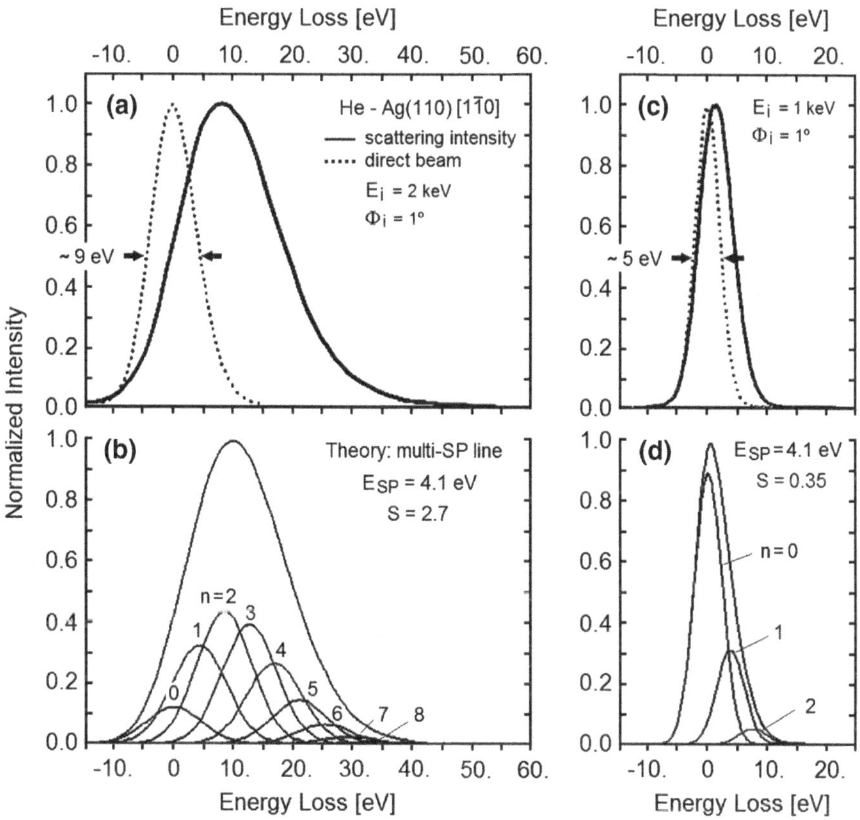

Fig. 13.29 Energy loss spectra of a high energy beam of ^{4}He from the Ag(110) surface at grazing incidence ($\Phi_i = 1°$) in the [1$\bar{1}$0] direction [244]. **a** For incident energy $E_i = 2$ keV with a FWHM of $\sim$ 9 eV. **b** Theoretical analysis of the scattered intensity in (**a**) in terms of a coherent emission (Poisson distribution with S = 2.7) of 4.1 eV surface plasmon polaritons. **c** Same as (**a**) for an incident beam energy of 1 keV with a FWHM of $\sim$5 eV. **d** Theoretical analysis as in (**b**), but for (**c**) with S = 0.35

plasmon polariton energy measured with EELS in Ag(110), which varies along the same [1$\bar{1}$0] direction from 3.86 eV near $Q = 0$ to 4.1 eV at Q = 0.4 Å^{-1} [258, 260]. At $E_i = 1$ keV (Fig. 13.29b) the much narrower energy loss peak is also well reproduced by theory with the same surface plasmon energy of 4.1 eV and $S = 0.35$ (Fig. 13.29d) [244].

The large decrease in atom-plasmon coupling when the incident energy is reduced from 2 to 1 keV (Fig. 13.29c) can be understood by considering the additional e-m interaction due to the fast He atom fly-by of the He atoms along the surface periodic coulomb potential. The He fly-by velocity for $E_i = 2$ keV is equal to 3.1×10^5 m/s and for a Ag(110) surface period in the [1$\bar{1}$0] of 2.89 Å, the atom is subject to an electromagnetic oscillation of frequency $\nu = 1.07$ PHz corresponding to photons of 4.42 eV exchanged between the surface and the oscillating

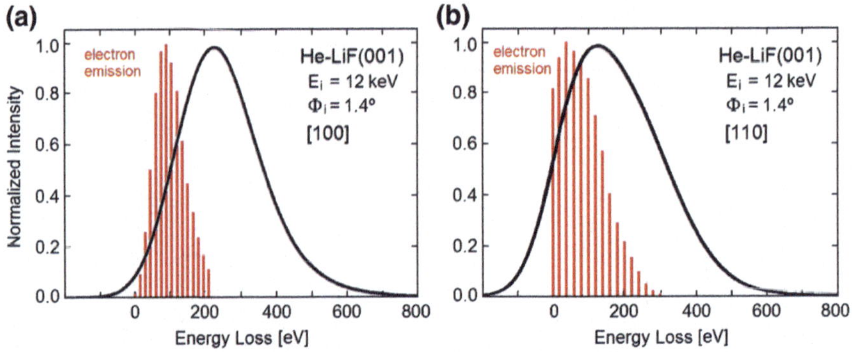

Fig. 13.30 Energy loss spectra (full line) of a high energy ^{4}He beam scattered from the LiF(001) surface with incident energy of 12 keV; and grazing angle of $\Phi_i = 1.4°$ along (**a**) the [100] direction and (**b**) along the [110] direction [248]: The energy loss, which is attributed to coherent creation of an average number of $S \cong 21$ surface exciton polaritons, is associated with resonant surface electron emission (red bars). The normalized electron emission intensities of 0, 1, 2, etc. electrons are separated by the resonance energy of 15.6 eV for the [100] and $15.6 \cdot \sqrt{2} = 22$ eV for the [110] direction

He atom dipole. This is just enough to excite the surface plasmon polaritons of Ag (110). On the contrary for $E_i = 1$ keV the electromagnetic frequency is a factor 2 smaller and only higher harmonics of the surface periodic potential can provide an effective electromagnetic interaction, besides the pure mechanical one.

Both the examples shown in Figs. 13.26b and 13.29a, b illustrate the creation of coherent surface bosons via a resonant e-m mechanism. As shown in Fig. 13.30 at much higher incident energies of 12 keV the energy loss due to the creation of coherent surface excitons by the grazing incident ^{4}He beam scattered from LiF(001) is associated with the emission of a comparatively large average number of electrons [248]. Since the emitted electrons go into a state of the free electron continuum, it may be assumed that they are extracted in resonance with the e-m field oscillations of energy 15.6 and 22 eV for the 12 keV ^{4}He atom fly-by speed along the [100] and [110] directions, respectively. Thus in Fig. 13.30 a, b the bars of the electron emission spectral distribution are plotted with a spacing of 15.6 and 22 eV for the [100] and [110] directions, respectively. The convolution of the electron emission distributions with the Poisson distributions of the 12.3 eV surface exciton polaritons permits to reproduce the total energy loss spectra in both directions with the same $S \sim 21$.

The few examples discussed above illustrate the great potential that high energy atom scattering has for the study of surface electron excitations. This kind of spectroscopy, however, does not seem to be sufficiently investigated yet, neither from the experimental nor from the theoretical point of view. In particular the study of ASPs in the intermediate (eV) energy spectral region above the phonon spectrum deserves future attention because of the potential applications in various areas of plasmonics.

13.7 The Vibrational Dynamics of Clusters

Small elemental and molecular clusters have acquired great importance among the various forms of condensed matter [261–263]. In clusters the surface forces, dominant in small clusters, favor the highest point symmetries, and compete with the bulk forces in larger aggregates and impose, under suitable thermodynamic conditions, periodic translational symmetries. This competition leads to a great variety of morphologies with the growing size of clusters, which correspond to unique physical and chemical properties.

The assessment of structural, electronic and vibrational properties of free clusters has great relevance for both fundamental and technological reasons and HAS as a sensitive surface probe is an ideal tool for their investigation. This was demonstrated by Udo Buck and coworkers in crossed beam experiments in which a He atom beam is scattered off a beam of flying clusters. In this way valuable information on the cluster size [262, 264–266] and vibrational spectrum [267] of argon [268–270], ammonia [267, 271], and water [272–274], with between 25 and 4600 particles. In the examples reproduced in Fig. 13.31a, b, c the TOF spectra of HAS from a beam of small water clusters (average molecule number $\overline{n} = 10$), measured in a 90°-crossed beam experiment with an He atom incident energy of 70.6 meV and three different scattering angles of 18.5° (a), 15° (b) and 11° (c) [273], show in addition to the diffuse elastic (DE) peak prominent inelastic peaks on the creation side with energies ranging from −7 to −21 meV. Their intensities and positions depend on the scattering angle, with the higher energies favored at larger scattering angles. The observed energies all correspond to intermolecular vibrations or surface localized modes. The larger scattering angle corresponds to a smaller impact parameter, i.e., a closer approach of the He atoms to the cluster surface. Under the assumption of a spherical cluster, the vertical component of the incident energy increases with Θ as $E_i \sin^2 \frac{1}{2}\Theta$, which favours at larger Θ the excitation of more energetic internal cluster modes. Figure 13.31d shows an inelastic scattering energy transfer distribution from large $(H_2O)_n$ ice nanocrystal clusters ($\overline{n} = 990$) [267]. Surprisingly at the scattering angle $\Theta = 11°$ inelastic features are observed which are strikingly similar to those in panel (e), observed with HAS from the surface of a 1000 Å thick amorphous ice film on Pt(111) at an incident energy of 21.7 meV [275].

A closer comparison between the $\overline{n} = 990$ cluster and the semi-infinite amorphous data indicates that the energy of ∼7 meV corresponds to the Rayleigh wave along the MK zone boundary. According to a force constant model calculation for a hexagonal H_2O (0001) slab [275] this mode gives a large contribution to the surface-projected DOS. However, as seen in [267], also for He scattering from large

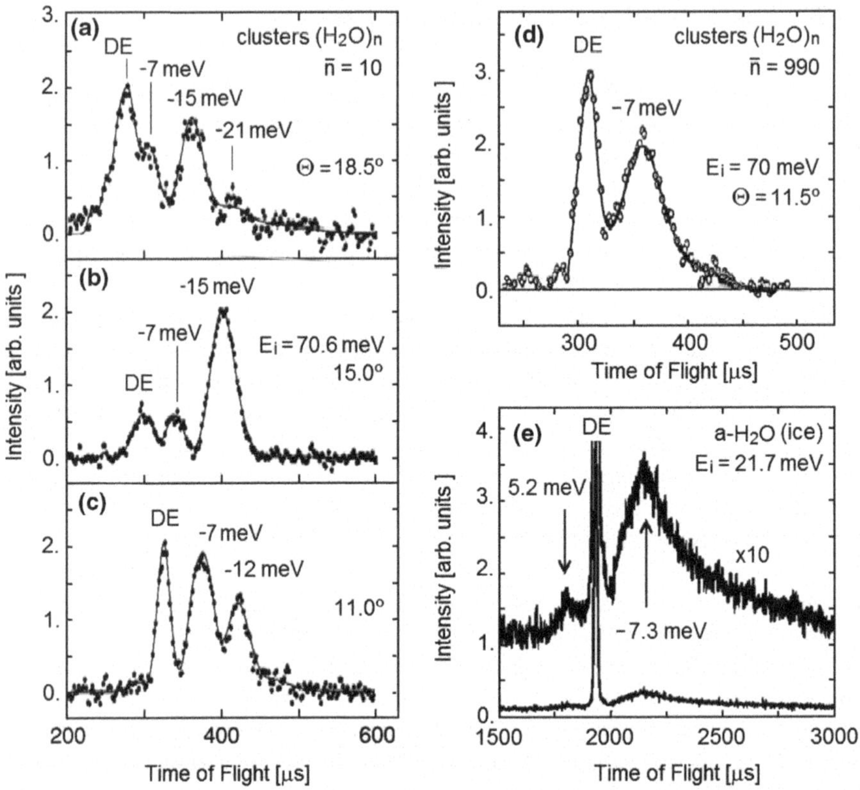

Fig. 13.31 Time-of-flight spectra of He atoms scattered from a beam of $(H_2O)_n$ clusters measured in a crossed (90°) beam experiment with an He atom incident energy of 70.6 meV. **a–c** TOF spectra from small $(H_2O)_n$ clusters (average molecule number $\bar{n} = 10$), at three different scattering angles of 18.5° (**a**), 15° (**b**) and 11° (**c**) [273]. Besides the diffuse elastic (DE) peak, prominent inelastic peaks occur on the creation side. Their intensities and positions depend on the scattering angle, with the higher energies favored at larger scattering angles. **d** TOF spectra from large ice nanocrystal clusters ($\bar{n} = 990$) [267]. At the scattering angle $\Theta = 11°$ the inelastic features are strikingly similar to those in panel (**e**), observed with HAS from the surface of amorphous ice with an incident energy of 21.7 meV [275]

free clusters like those of Fig. 13.31d an increase of the scattering angle leads to the suppression of the −7 meV peak in favor of phonons at higher energy corresponding to the phonons in the bulk of a solid.

The study of clusters and their aggregation in supracrystals on surfaces represents a promising area of future investigation for HAS and SE-HAS, as will be discussed in Sect. 14.2.4.

13.8 The Helium Microscope

Microscopy is the most direct method to explore the structure and dynamics of matter. In recent years many new methods have emerged for the microscopy of the atomic structure of surfaces such as STM and AFM in addition to the classical methods of high energy electron microscopy and field ion microscopy.

For many surfaces such as magnetic materials or fragile, weakly bonded and insulating biological materials as occur on human tissue, insects or plants, the available methods are only of limited applicability. As emphasized throughout this monograph, He atom beams provide a uniquely surface sensitive, extremely gentle non-destructive probe. Thus there has been for some time various attempts to develop a helium microscope to directly visualize the outermost structure of surfaces, which cannot be studied by the available methods.

Three different approaches have been used to develop the necessary lens needed to use He atom beams to scan surfaces with high spatial resolution: (1) Reflection mirrors as a focusing element, (2) Fresnel zone plates as an atom lens and, (3) a pin-hole. These devices are called scanning helium microscopes (SHeM). The progress in developing these different approaches and their advantages and disadvantages are analyzed below.

(1) *Reflection mirrors as a focusing element.* Doak in 1992 demonstrated the focusing of a He atom beam from a cylindrical mirror in one dimension but did not report a spot size [276]. The first successful 2D focusing was achieved by Holst and Allison in 1997 [277] who used an electrostatically concave deformed hydrogen passivated 50 μm thick Si(111) crystal to focus a beam of He atoms down to 210 ± 50 μm. With an improved apparatus an European consortium involving scientists from six countries [278] achieved a spot size of 27 × 31 μm using a stigmatically deformed ellipsoidal silicon single crystal wafer mirror. This method has the advantage of a relatively large aperture of about 10^{-1} sr, but appears to have the disadvantage that the best available mirror surfaces have a reflectivity of only about 3% [279]. Thus the signal was only about 10^3 counts/s, comparable to the signal of a Fresnel zone plate [280]. Thin films of $(\sqrt{3}\times\sqrt{3})$ R30 Pb on Si(111) have a greater reflectivity of 15%, but so far have not been used as concave mirrors [281, 282]. Recently even higher reflectivities of 25% have been obtained with graphene terminated Ru thin films on a laser polished concave fused silica surface [283]. An alternative scheme, consisting of two concave elliptical near-grazing quantum reflection mirrors, has been proposed and demonstrated in one dimension to produce spots of the order of 2 μm [284]. This method has the advantage that since the atoms undergo quantum reflection without touching the mirror surface the method has essentially 100% reflectivity, but presently with a small acceptance angle of the order of 10^{-5} rad.

(2) *Fresnel zone plates as an atom lens.* The focusing of a beam of metastable helium atoms with a 210 μm dia commercial gold zone plate (with 128 zones)

down to about 20 μm was reported in 1991 by Carnal et al. [285]. Despite the efficient nearly 100% detection of the 20 eV electronically excited atoms the spot intensity was only 0.5 counts/s. The first focusing of a neutral ground state He beam was achieved in 1999 with a custom-made SiN 270 μm dia Fresnel zone plate. The focused spot had a diameter of less than 2.0 μm with a count rate of 500 counts/s [280]. The experiment using the same lens was repeated by Koch et al. [286] with similar results and by Eder et al. [287] who reported a spot size of about 1 μm with a signal of about 100 counts/s. The disadvantage is again the extremely small acceptance solid angle of about 10^{-8} sr.

(3) *A Pin-hole*. The first realization of a He atom microscope was achieved in 2011 by Witham and Sanchez who used a simple arrangement in which the He atom beam from a 300 nm inner dia. pin-hole was scanned across the surface. In their apparatus large intensities on the target surface were obtained by locating the pinhole very close to the surface at a distance of only about 100 μ [288–291]. Surprisingly sharp images of the surface structure were achieved even though the scattered He atoms were detected over a relatively wide solid angle off to the side of the incident beam direction.

Figure 13.32 shows a schematic diagram of the pin-hole microscope and Fig. 13.33 an image of an uncoated biological sample consisting of pollen grains on an adhesive. As seen in the enlarged section of Fig. 13.33 the resolution is estimated to be about 0.35 ± 0.05 μm comparable with the spot size of about 0.3 μm which is defined by the pin-hole aperture of about the same size. The measurements are made by slowly scanning the sample across the spot. The signal is low and amounts to about 10 pixels per sec and a typical measurement takes up to half a day. Since the He atoms are not absorbed by the surface the contrast is largest at flat edges and least when the atoms get trapped in deep valleys on a rough surface.

This approach seems to be the most promising and a theoretical analysis predicts that a resolution of about 40 nm should be possible by reducing the distance between aperture and target [292]. These initial results have inspired another group in Australia [293–295] to build a similar device. Improvements can be envisaged

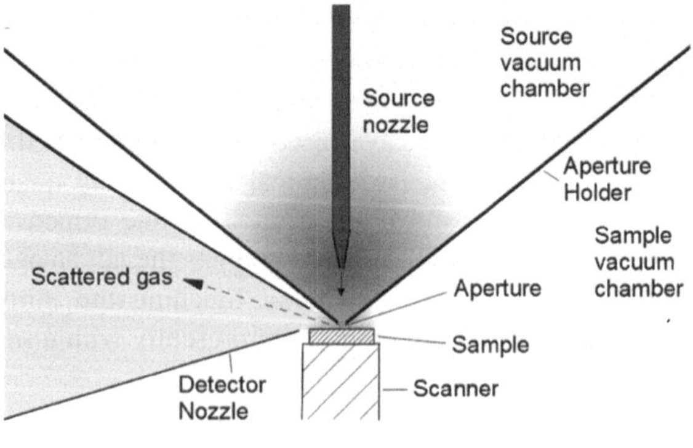

Fig. 13.32 Schematic diagram of the pin-hole helium atom microscope of Witham and Sanchez (adapted from [289]). The 1 μm dia source nozzle is about 1 mm above the 300 nm dia aperture, which depending on the sample topography is located about 100 μm from the sample. The scattered gas through a conical opening is measured in the stagnation mode by a mass spectrometer

Fig. 13.33 Two helium microscope images of *Crocusemia* pollen grains on adhesive. The resolution of better than 0.35 ± 0.05 μm was determined from the width of the change in brightness at the bright edges marked by the four white lines in the upper half of the right hand image [289]

by, for example, mounting a Fresnel zone plate to focus the scattered beam or by using new improved detection schemes [296]. In conclusion, it is presently quite reasonable to expect that ultimately in the not too distant future it will become possible with HAS to measure energy resolved scattering from the topmost outer surface of any sample non-destructively and spatially resolved.

Although not of direct relevance it is of interest to call attention to two other types of helium microscopes, using either He metastable atoms or ions. As already established in 1954 metastable He atoms on striking a surface give up their 20 eV excitation energy to eject an electron from the surface [160]. Countless studies since then have shown that the electron energy distribution depends on the chemical nature of the surface [297]. Since the atoms are deexcited by the outermost electrons at the surface the method provides a uniquely surface sensitive probe of the chemical constitution of the surface. As the metastable atoms have thermal energies the method is also non-destructive in much the same way as in the scattering of neutral helium atoms. Harada and colleagues have constructed a metastable helium atom microscope based on the same arrangement as in *low energy electron microscopy* (LEEM) [298] and as in photoelectron emission spectroscopy. The surface is illuminated with an intense beam of metastable atoms and the emitted electrons are imaged on to a microchannel plate. In the energy filtered version a resolution of 0.3 μm was reported [299]. This microscope has been given the name *metastable electron emission microscope* (MEEM) [299, 300]. Another possibility is to scan a focused beam of metastable atoms across the surface and measure the electron emission as a function of the location of the focussed spot on the surface as in the neutral ground state atom microscope described above [301].

The helium ion microscope also has unique features. As in a scanning electron microscope the ions are accelerated to a high energy (30 keV) and focused on to the surface from which either the ejected electrons, the scattered ions or the transmitted beam can be detected. Since the ions are much heavier than electrons the diffraction limit is greatly reduced and a theoretical resolution of 3 Å has been predicted [302].

In a recent publication a lateral resolution of less than 54 nm and a temporal resolution of 17 ns was reported [303]. Of course the price paid is the destruction of the surface in the wake of the massive high energy particles

This chapter at the end of the book is intended to point towards the future of Helium Atom Scattering spectroscopy by illustrating its great potential for studying not only the surface dynamics of atoms, but also diffusion, magnetic effects and moreover the collective electron dynamics in the THz domain and the important effects of electron-phonon interactions. These areas and microscopy are examples of active research fields with relevant technological implications (and expectations), where HAS can provide new valuable information hardly accessible to other spectroscopies.

References

1. See articles in *Kinetics of Interface Reactions*, ed. by M. Grunze, H.J. Kreutzer (Springer, Heidelberg, 1987)
2. J.R. Manson, *Energy Transfer to Phonons in Atom and Molecule Collisions with Surfaces*, in Handbook of Surface Science, ed., Vol. 3: Surface Dynamics, ed. by W.N. Unertl (North Holland, Amsterdam, 2008), pp. 53–93
3. V. Celli, D. Himes, P. Tran, J.P. Toennies, Ch. Wöll, G. Zhang, Phys. Rev. Lett. **66**, 3160 (1991)
4. F. Hofmann, J.P. Toennies, J.R. Manson, J. Chem. Phys. **101**, 10155 (1994)
5. F. Hofmann, J.P. Toennies, J.R. Manson, Surf. Sci. Lett. **349**, L184 (1996)
6. F. Hofmann, J.P. Toennies, J.R. Manson, J. Chem. Phys. **106**, 1234 (1997)
7. J.R. Manson, Comput. Phys. Commun. **80**, 145 (1994)
8. J.R. Manson, Phys. Rev. B **43**, 6924 (1991)
9. B. Gumhalter, A. Bilić, Surf. Sci. **370**, 47 (1997)
10. B. Gumhalter, J. Phys.: Condens. Matter **24**, 104015 (2012)
11. S. Daon, E. Pollak, J. Chem. Phys. **142**, 174102 (2015)
12. B. Gumhalter, Phys. Rep. **351**, 1–159 (2001)
13. J. Braun, D. Fuhrmann, M. Bertino, A.P. Graham, J.P. Toennies, Ch. Wöll, A. Bilić, B. Gumhalter, J. Chem. Phys. **106**, 9911 and 9922 (1997)
14. A. Šiber, B. Gumhalter, J. Braun, A.P. Graham, M.F. Bertino, J.P. Toennies, D. Fuhrmann, Ch. Wöll, Phys. Rev. B **59**, 5898 (1999)
15. J. Braun, D. Fuhrmann, J.P. Toennies, Ch. Wöll, A. Bilic, B. Gumhalter, Surf. Sci. **368**, 232 (1996)
16. A. Šiber, B. Gumhalter, Ch. Wöll, J. Phys.: Cond. Matter **14**, 5913 (2002)
17. A. Šiber, B. Gumhalter, Surf. Sci. **502–503**, 422–428 (2002)
18. R.J. Glauber, Phys. Rev. **98**, 1692 (1955); Phys. Rev. **131**, 2766 (1963)
19. N. Terzi, in *Collective Excitations in Solids*, ed. by. B. Di Bartolo (Plenum, New York, 1983), pp. 149–181; G.M. Mazzucchelli, N. Terzi, Solid State Comm. **48**, 679 (1983)
20. M. Li, J.R. Manson, A.P. Graham, Phys. Rev. B **65**, 195404 (2002)
21. B. Gumhalter, D.C. Langreth, Phys. Rev. B **60**, 2789 (1999)
22. F. Hofmann, J.P. Toennies, G. Benedek, unpublished
23. A. Graham, F. Hofmann, J.P. Toennies, J. Chem. Phys. **104**, 5311 (1996)
24. C.F. McConville, D.P. Woodruff, K.C. Prince, G. Paducci, V. Chab, M. Surman, A.M. Bradshaw, Surf. Sci. **166**, 221 (1986)
25. M. Li, J.R. Manson, A.P. Graham, Phys. Rev. B **63**, 155410 (2001)

26. J.R. Manson, A.P. Graham, M. Li, J. Phys.: Cond. Matter **14**, 6233 (2002)
27. A.P. Graham, F. Hofmann, J.P. Toennies, G.P. Williams, C.J. Hirschmugl, J. Ellis, J. Chem. Phys. **108**, 7825 (1998)
28. G. Antczak, G. Ehrlich, *Surface Diffusion* (Cambridge University Press, Cambridge, 2010)
29. J.W.M. Frenken, J.P. Toennies, Ch. Wöll, Phys. Rev. Lett. **60**, 1727 (1988)
30. J.W.M. Frenken, B.J. Hinch, J.P. Toennies, Ch. Wöll, Phys. Rev. B **41**, 938 (1990)
31. T. Springer, in *Springer Tracts in Modern Physics*, vol. 64 (Springer, Berlin, 1972), p. 2
32. T. Springer, in *Dynamics of Solids and Liquids by Neutron Scattering*, ed. by S.W. Lovesey, T. Springer (Springer, Berlin, 1977), p. 255
33. M. Bèe, *Quasielastic Neutron Scattering* (Adam Hilgar, Bristol, 1988)
34. C.T. Chudley, R.J. Elliott, Proc. Phys. Soc. **77**, 353 (1961)
35. A.C. Levi, R. Spadacini, G.E. Tommei, Surf. Sci. **121**, 504 (1982)
36. J.W.M. Frenken, B.J. Hinch, in *Helium Atom Scattering from Surfaces*, ed. by E. Hulpke, Springer Series in Surface Sciences, vol. 27 (1992), p. 287
37. L. Van Hove, Phys. Rev. **95**, 249 (1954)
38. G.H. Vineyard, Phys. Rev. **110**, 999 (1958)
39. K.S. Singwi, A. Sjölander, Phys. Rev. **119**, 863 (1960)
40. J.F. Van der Veen, J.W.M. Frenken, Surf. Sci. **251–252**, 1 (1991)
41. A.P. Graham, F. Hofmann, J.P. Toennies, L.Y. Chen, S.C. Ying, Phys. Rev. B **56**, 10567 (1997)
42. A.P. Graham, J.P. Toennies, Surf. Sci. **427**(428), 1 (1999)
43. A.P. Graham, W. Silvestri, J.P. Toennies, in *Surface Diffusion: Atomistic and Collective Processes*, ed. by M. Tringides (Plenum Press, New York, 1997), p. 565
44. B.H. Choi, A.P. Graham, K.T. Tang, J.P. Toennies, J. Chem. Phys. **112**, 10538 (2000)
45. J. Ellis, J.P. Toennies, Phys. Rev. Lett. **70**, 2118 (1993)
46. J.V. Barth, Surf. Sci. Rep. **40**, 75 (2000)
47. L.Y. Chen, S.C. Ying, Phys. Rev. Lett. **71**, 4361 (1993)
48. J. Ellis, J.P. Toennies, Surf. Sci. **317**, 99 (1994)
49. M. Diamant, S. Rahav, R. Ferrando, G. Alexandrowicz, J. Phys.: Condens. Matter **27**, 125008 (2015)
50. A.P. Graham, F. Hofmann, J.P. Toennies, L.Y. Chen, S.C. Ying, Phys. Rev. Lett. **78**, 3900 (1997)
51. T. Ala-Nissila, R. Ferrandos, S.C. Ying, Adv. Phys. **51**, 949 (2002)
52. J. Ellis, A.P. Graham, F. Hofmann, J.P. Toennies, Phys. Rev. B **63**, 195408 (2001)
53. R. Martinez-Casado, A.S. Sanz, G. Rojas Lorenzo, S. Miret-Artés, J. Chem. Phys. **132**, 054704 (2010). See also the preceding papers: R. Guantes, J.L. Vega, S. Miret-Artés, E. Pollak, J. Chem. Phys. (2003), **119**, 2780 (2003) and R. Martínez-Casado, J.L. Vega, A.S. Sanz, S. Miret-Artés, Phys. Rev. Lett. **98**, 216102 (2007)
54. G.J. Alexandrowicz, A.P. Jardine, H. Hedgeland, W. Allison, J. Ellis, Phys. Rev. Lett. **97**, 156103 (2006)
55. A.L. Glebov, J.P. Toennies, S. Vollmer, Phys. Rev. Lett. **82**, 3300 (1999)
56. F. Hofmann, W. Schöllkopf, J.P. Toennies, in *Proceedings R. A. Welch Foundation, 38th Conference on Chemical Research, Chemical Dynamics of Transient Species* (Houston TX, 1994), pp. 197–218
57. F. Hofmann, Ph.D. Dissertation (Göttingen University, Germany, 1995)
58. A.P. Graham, J.P. Toennies, Europhys. Lett. **42**, 449 (1998)
59. F. Mezei, Z. Phys. **255**, 146 (1972)
60. A.P. Jardine, J. Ellis, W. Allison, J. Phys.: Condens. Matter **14**, 6173 (2002)
61. P. Fouquet, A.P. Jardine, S. Dworski, G. Alexandrowicz, W. Allison, J. Ellis, Rev. Sci. Inst. **76**, 053109 (2005)
62. A.P. Jardine, H. Hedgeland, G. Alexandrowicz, W. Allison, J. Ellis, Prog. Surf. Sci. **84**, 323 (2009)
63. A.P. Jardine, G. Alexandrowicz, H. Hedgeland, W. Allison, J. Ellis, Phys. Chem. Chem. Phys. **11**, 3355 (2009)

64. R. Gähler, R. Golub, K. Habicht, T. Keller, J. Felber, Phys. B **229**, 1 (1996)
65. B.A.J. Lechner, *Studying Complex Surface Dynamical Systems Using Helium-3 Spin-Echo Spectroscopy*, (Springer Theses, Springer Cham, Heidelberg 2014)
66. A.P. Graham, J.P. Toennies, J. Chem. Phys. **114**, 1051 (2001)
67. G. Alexandrowicz, A.P. Jardine, P. Fouquet, S. Dworski, W. Allison, J. Ellis, Phys. Rev. Lett. **93**, 156103 (2004)
68. P.R. Kole, H. Hedgeland, A.P. Jardine, W. Allison, J. Ellis, G. Alexandrowicz, J. Phys.: Condens. Matter **24**, 104016 (2012)
69. P. Fouquet, R.A. Olsen, E.J. Baerends, J. Chem. Phys. **119**, 509 (2003)
70. J.C. Tully, M. Gomez, J. Vac. Sci. Technol. A **11**, 1914 (1993)
71. Q. Ge, D.A. King, J. Chem. Phys. **111**, 9461 (1999)
72. Q. Ge, D.A. King, J. Chem. Phys. **114**, 1053 (2001)
73. A.P. Graham, Surf. Sci. Rep. **49**, 115 (2003)
74. A.P. Jardine, J. Ellis, W. Allison, J. Chem. Phys. **120**, 8724 (2004)
75. R. Marquardt, F. Cuvelier, R.A. Olsen, E.J. Baerends, J.C. Tremblay, P. Saalfrank, J. Chem. Phys. **132**, 074108 (2010)
76. P. Fouquet, H. Hedgland, A. Jardine, G. Alexandrowicz, W. Allison, J. Ellis, Phys. B **385–386**, 269–271 (2006)
77. H. Hedgeland, P. Fouquet, A.P. Jardine, G. Alexandrowicz, W. Allison, J. Ellis, Nat. Phys. **5** (8), 561–564 (2009)
78. P. Fouquet, M.R. Johnson, H. Hedgeland, A.P. Jardine, J. Ellis, W. Allison, Carbon **47**, 2627 (2009)
79. G. Alexandrowicz, P.R. Kole, E.Y.M. Lee, H. Hedgeland, R. Ferrando, A.P. Jardine, W. Allison, J. Ellis, J. Am. Chem. Soc. **130**, 6789–6794 (2008)
80. B.A.J. Lechner, M. Sacchi, A.P. Jardine, H. Hedgeland, W. Allison, J. Ellis, S.J. Jenkins, P. C. Dastoor, B.J. Hinch, J. Phys. Chem. Lett. 1953 (2013)
81. P. Rotter, B.A.J. Lechner, A. Morherr, D.M. Chisnall, D.J. Ward, A.P. Jardine, J. Ellis, W. Allison, B. Eckhardt, G. Witte, Nat. Mater. **397** (2016)
82. O.P. Godsi, G. Corem, Y. Alkoby, J.T. Cantin, R. Krems, M. Somers, J. Meyer, G.-J. Kroes, T. Maniv, G. Alexandrowicz, Nature Comm. **8**, 15357 (2017)
83. R. Gomer, Rep. Prog. Phys. **53**, 917 (1990)
84. M.F. Bertino, F. Hofmann, W. Steinhögl, J.P. Toennies, J. Chem. Phys. **105**, 11297 (1996)
85. G.X. Cao, E. Nabighian, X.D. Zhu, Phys. Rev. Lett. **79**, 3696 (1997)
86. M. Ondrejcek, W. Stenzel, H. Conrad, V. Chab, Z. Chvoj, W. Engel, A.M. Bradshaw, Chem. Phys. Lett. **215**, 528 (1993)
87. S.K.S. Ma, F.W. de Wette, G.P. Alldredge, Surf. Sci. **78**, 598 (1978)
88. L. Yang, T.S. Rahman, M.S. Daw, Phys. Rev. B **44**, 13725 (1991)
89. A.N. Al-Rawi, A. Kara, T.S. Rahman, Phys. Rev. B **66**, 165439 (2002)
90. J. Blomquist, P. Uvdal, Phys. Chem. Chem. Phys. **12**, 14121 (2010)
91. T. Tanaka, A. Tajima, R. Moriizumi, C. Ostuma, Y. Tsunoda, M. Seto, S. Kitao, T. Mitsui, J. Phys. Soc. Japan **74**, 1762 (2009)
92. S.P. Lewis, A.M. Rappe, Phys. Rev. Lett. **77**, 5241 (1996)
93. J.L. Vega, R. Guantes, S. Miret-Artés, D.A. Micha, J. Chem. Phys. **121**, 8580 (2004)
94. T.S. Rahman, J.D. Spangler, A. Al-Rawi, Surf. Sci. **502–503**, 429–436 (2002)
95. I.S. Ulusoy, Y. Scribano, D.M. Benoit, A. Tschetschetkin, N. Maurer, B. Koslowski, P. Ziemann, Phys. Chem. Chem. Phys. **13**, 612 (2011)
96. B.N.J. Persson, Phys. Rev. B **44**, 3277 (1991)
97. B.N.J. Persson, *Sliding Friction: Physical Principles and Applications* (Springer, Heidelberg, 1998)
98. B.N.J. Persson, H. Ueba, Surf. Sci. **502–503**, 18 (2002)
99. T. Komeda, Y. Kim, M. Kawai, B.N.J. Persson, H. Ueba. Science **295**, 2055 (2002)
100. V.A. Ermoshin, A.K. Kazanski, K.S. Smirnov, D. Bougeard, J. Chem. Phys. **105**, 9371 (1996)
101. S.P. Lewis, A.M. Rappe, Phys. Rev. Lett. **77**, 5241 (1996)

102. K. Liu, S. Gao, Phys. Rev. Lett. **95**, 226102 (2005)
103. A.P. Baddorf, E.W. Plummer, Phys. Rev. Lett. **66**, 2770 (1991)
104. G. Benedek, J.P. Toennies, Phys. Rev. (Rapid Comm.) **B46**, 13643 (1992)
105. R.M. Nicklow, G. Gilat, H.G. Smith, L.J. Raubenheimer, M.K. Wilkinson, Phys. Rev. **164**, 922 (1967)
106. E.C. Svensson, B.N. Brockhouse, J.M. Rowe, Phys. Rev. **155**, 619 (1967)
107. G. Nilsson, R. Rolandson, Phys. Rev. B **7**, 2393 (1973)
108. A. Larose, B.N. Brockhouse, Can. J. Phys. **49**, 1990 (1976)
109. A.P. Miller, B.N. Brockhouse, Can. J. Phys. **49**, 704 (1971)
110. H. Ibach, D.L. Mills, *Electron Energy Loss Spectroscopy and Surface Vibrations* (Academic Press, New York, 1982)
111. A. Reichmuth, Diplom Thesis (University of Göttingen, 1991)
112. M. Gester, D. Kleinhesselink, P. Ruggerone, J.P. Toennies, Phys. Rev. B **49**, 5777 (1994)
113. L. Yang, T.S. Rahman, Phys. Rev. Lett. **67**, 2327 (1991)
114. R. Zivieri, G. Santoro, V. Bortolani, Phys. Rev. B **59**, 15959 (1999)
115. A.N. Al-Rawi, A. Kara, T.S. Rahman, Phys. Rev. B **66**, 165436 (2002)
116. G. Alexandrowicz, A.P. Jardine, J. Phys.: Condens. Matter **19**, 305001 (2007)
117. H.-J. Ernst, E. Hulpke, J.P. Toennies, Phys. Rev. B **46**, 16081 (1992)
118. C.S. Jayanthi, E. Tosatti, L. Pietronero, Phys. Rev. B **31**, 3456 (1985)
119. E. Tosatti, in *The Structure of Surfaces II*, ed. by J.F. van der Veen, et al. (Springer, Berlin, 1988), p. 535
120. G. Bracco, L. Bruschi, L. Pedemonte, R. Tatarek, Surf. Sci. **352**, 964 (1996)
121. G. Bracco, L. Bruschi, R. Tatarek, A. Franchini, V. Bortolani, G. Santoro, Europhys. Lett. **34**, 687 (1996)
122. G. Bracco, L. Bruschi, L. Pedemonte, R. Tatarek, Surf. Sci. **377**, 325 (1997)
123. L. Pedemonte, G. Bracco, Phys. Rev. B **68**, 205420 (2003)
124. L. Pedemonte, R. Tatarek, G. Bracco, Phys. Rev. B **66**, 045414 (2002)
125. A. Fasolino, E. Tosatti, Phys. Rev. B **35**, 4264 (1987)
126. C.Z. Wang, A. Fasolino, E. Tosatti, Europhys. Lett. **7**, 263 (1988)
127. C.Z. Wang, A. Fasolino, E. Tosatti, Phys. Rev. B **37**, 2116 (1988)
128. G. Piccini, J. Sauer, J. Chem. Theory Comput. **10**, 2479 (2014)
129. H. Frank, H. Hoinkes, H. Wilsch, U. Wonka, Surf. Sci. **64**, 362 (1974)
130. H.U. Finzel, H. Frank, H. Hoinkes, M. Luschka, H. Nahr, H. Wilsch, U. Wonka, Surf. Sci. **49**, 577 (1975)
131. Y. Watanabe, A. Sakai, T. Sakurai, J. Phys. Soc. Jpn. **66**(3), 649 (1974)
132. Y. Ekinci, J.P. Toennies, Surf. Sci. **563**, 127–134 (2004)
133. E.G. McRae, R.A. Malic, Phys. Rev. Lett. **58**, 1437 (1987); Phys. Rev. B **38**, 13163 (1988)
134. M. Abraham, G. Lelay, J. Hila, Phys. Rev. B **41**, 9828 (1990)
135. A. Mak, K.W. Evans-Lutterodt, K. Blum, D.Y. Noh, J.D. Brock, G.A. Held, R.J. Birgenau, Phys. Rev. Lett. **66**, 2002 (1991)
136. N. Takeuchi, A. Selloni, E. Tosatti, Phys. Rev. Lett. **72**, 2227 (1994)
137. N. Takeuchi, A. Selloni, E. Tosatti, Phys. Rev. B **49**, 10757 (1994)
138. N. Takeuchi, A. Selloni, E. Tosatti, Phys. Rev. B **55**, 15405 (1997)
139. S. Modesti, V.R. Dhanak, M. Sancrotti, A. Cantoni, B.N.J. Persson, E. Tosatti, Phys. Rev. Lett. **73**, 1951 (1994)
140. A. Di Cicco, B. Giovenali, R. Gunnella, E. Principi, S. Simonucci, Solid State Commun. **134**, 577–582 (2005)
141. A. Santoni, V.R. Dhanak, A. Goldoni, M. Sancrotti, S. Modesti, Europhys. Lett. **34**, 275 (1996)
142. A.W. Denier van der Gon, J.M. Gay, J.W.M. Frenken, J.F. van der Veen, Surf. Sci. **241**, 335 (1991)
143. T.T. Tran, S. Thevuthasan, Y.J. Kim, D.J. Friedman, A.P. Kaduwela, G.S. Herman, C.S. Fadley, Surf. Sci. **281**, 270 (1993)
144. C.A. Meli, E.F. Greene, G. Lange, J.P. Toennies, Phys. Rev. Lett. **74**, 2054 (1995)

145. A.L. Glebov, J.P. Toennies, S. Vollmer, G. Benedek, Europhys. Lett. **46**, 369 (1999)
146. A.L. Glebov, J.P. Toennies, S. Vollmer, Phys. Rev. Lett. **82**, 3300 (1999)
147. G. Lange, C.A. Meli, J.P. Toennies, Phys. Rev. B **56**, 4642 (1997)
148. D. Cvetko, L. Floreano, A. Crottini, A. Morgante, F. Tommasini, Surf. Sci. **447**, L147–L151 (2000)
149. C.A. Meli, E.F. Greene, J. Chem. Phys. **101**, 7139 (1994)
150. X. Zeng, H.E. Elsayed-Ali, Surf. Sci. **497**, 373–384 (2002)
151. H.C. Siegmann, J. Phys.: Condens. Matter **4**, 8395–8434 (1992)
152. F.J. Himpsel, J.E. Ortega, G.J. Mankey. Adv. Phys. **47**, 511–597 (1998)
153. O. Fruchart, A. Thiaville. C.R. Phys. **6**, 921–933 (2005)
154. C.A.F. Vaz, J.A.C. Bland, G. Lauhoff. Rep. Prog. Phys. **71**, 056501 (2008)
155. J.J. Hernández, S.A. Safron, J.G. Skofronick, J. Vac. Sci. Technol. A **13**, 1416 (1995)
156. W. Allison, B. Feuerbacher, Phys. Rev. Lett. **45**, 2040 (1980)
157. G. Brusdeylins, J.P. Toennies, Surf. Sci. **126**, 647 (1983)
158. R.S. Timit, J.M. Daniels, A.D. May, Can. J. Phys. **49**, 560 (1971)
159. G. Witte, Dissertation, University of Göttingen 1995 (Max Planck Institut für Strömungsforschung, Report no. 8/1995)
160. H.D. Hagstrum, Phys. Rev. **96**, 325 (1954)
161. H. Conrad, G. Ertl, J. Küppers, S.W. Wang, Phys. Rev. Lett. **42**, 1082 (1979)
162. Y. Harada, S. Masuda, H. Ozaki, Chem. Rev. **97**, 1897–1952 (1997)
163. M. Marynowski, A. Swan, W. Franzen, M. El-Batanouny, K.M. Martini, Surf. Interface Anal. **23**, 105 (1995)
164. A. Swan, M. Marynowski, W. Franzen, M. El-Batanouny, K.M. Martini, Phys. Rev. Lett. **71**, 1250 (1993)
165. A. Swan, M. Marynowski, W. Franzen, M. El-Batanouny, K.M. Martini, J. Vac. Sci. Technol. A **12**, 2219 (1994)
166. M. Marynowski, W. Franzen, M. El-Batanouny, V. Staemmler, Phys. Rev. B **60**, 6053 (1999)
167. P. Fouquet, P.K. Day, G. Witte, Surf. Sci. **400**, 140 (1998)
168. M. El-Batanouny, G. Murthy, C.R. Willis, S. Kais, V. Staemmler, Phys. Rev. B **58**, 7391 (1998)
169. M. El-Batanouny, J. Phys.: Condens. Matter **14**, 6281 (2002)
170. P. Banerjee, X. Liu, M. Farzaneh, C.R. Willis, W. Franzen, M. El-Batanouny, V. Staemmler, arXiv:cond-mat/0401130v1 (8 Jan 2004)
171. G. Witte, P. Senet, J.P. Toennies, Phys. Rev. B **58**, 13264 (1998)
172. J.P. Toennies, G. Witte, A.M. Shikin, K.H. Rieder, J. Electr. Spectr. Rel. Phenom. **64/65**, 677 (1993)
173. M. Kurahashi, T. Suzui, X. Ju, Y. Yamauchi, Phys. Rev. Lett. **91**, 267203-1 (2003)
174. M. Kurahashi, Y. Yamauchi, Surf. Sci. **590**, 21–30 (2005)
175. T. Suzuki, M. Kurahashi, X. Ju, Y. Yamauchi, Surf. Sci. **549**, 97–102 (2004)
176. M. Kurahashi, X. Sun, Y. Yamauchi, Surf. Sci. **601**, 1371–1376 (2007)
177. R. Moroni, M. Canapea, L. Mattera, Phys. Res. B **256**, 438–441 (2007)
178. M. Kurahashi, Y. Yamauchi, Rev. Sci. Instrum. **77**, 023904 (2006)
179. M. Kurahashi, S. Entani, Y. Yamauchi, Rev. Sci. Instrum. **79**, 073902 (2008)
180. M. Kurahashi, X. Sun, Y. Yamauchi, Phys. Rev. B **86**, 245421 (2012)
181. A. Pratt, M. Kurahashi, Y. Yamachi, J. Appl. Phys. **109**, 07C111 (2011)
182. M. Kurahashi, A. Pratt, Y. Yamauchi, Surf. Sci. **605**, 612–615 (2011)
183. S. Achilli, G.P. Brivio, G. Fratesi, M.I. Trioni, J. Phys. Chem. A **115**, 8498–8503 (2011)
184. B. Rohwedder, Am. J. Phys. **75**, 394 (2007)
185. S.K. Satija, R.P. Comes, G. Shirane, Phys. Rev. B **32**, 3309 (1985)
186. T. Slezak, J. Lazewski, S. Stankov, K. Parlinski et al., Phys. Rev. Lett. **99**, 066103 (2007)
187. G. Benedek, J.P. Toennies, G. Zhang, Phys. Rev. Lett. **68**, 2644 (1992)

188. G. Zhang, Dissertation (University of Göttingen 1991 and Max-Planck-Institut für Strömungsforschung, Bericht 102/1991)
189. R.F. Sabiryanov, S.S. Jaswal, Phys. Rev. Lett. **83**, 2062 (1999)
190. G. Benedek, E. Hulpke, W. Steinhögl, Phys. Rev. Lett. **87**, 027201 (2001)
191. J. Thomassen, F. May, B. Feldmann, M. Wuttig, H. Ibach, Phys. Rev. Lett. **69**, 3831 (1992)
192. M. Zharnikov, A. Dittschar, W. Kuch, C.M. Schneider, J. Kirschner, Phys. Rev. Lett. **76**, 4620 (1996)
193. L. Hammer, S. Müller, K. Heinz, Surf. Sci. **569**, 1–3 (2004)
194. T. Kraft, P.M. Marcus, M. Scheffler, Phys. Rev. B **49**, 11511 (1994)
195. R. Lorenz, J. Hafner, Thin Solid Films **281–282**, 482 (1996)
196. T. Asada, S. Blügel, Phys. Rev. Lett. **79**, 507 (1997)
197. R. Kläsges, D. Schmitz, C. Carbone, W. Eberhardt, T. Kachel, Solid State Commun. **107**, 13 (1998)
198. G. Benedek, M. Bernasconi, V. Chis, E. Chulkov, P.M. Echenique, B. Hellsing, J. P. Toennies, J. Phys.: Cond. Matter **22**, 084020 (2010)
199. D. Li, M. Freitag, J. Pearson, Z.Q. Qiu, S.D. Baker, Phys. Rev. Lett. **72**, 3112 (1994)
200. N. Memmel, Th. Detzel, Surf. Sci. **307–309**, 490 (1994)
201. D. Spišàk, J. Hafner, Phys. Rev. B **61**, 16129 (2000)
202. V.M. Silkin, A. García-Lekue, J.M. Pitarke, E.V. Chulkov, E. Zaremba, P.M. Echenique, Europhys. Lett. **66**, 260 (2004)
203. S. Stankov, R. Röhlsberger, T. Slezak, M. Sladecek, B. Sepiol, G. Vogl, A.I. Chumakov, R. Rüffer, N. Spriridis, J. Lazewski, K. Parlinski, J. Korecki, Phys. Rev. Lett. **99**, 185501 (2007)
204. W. Sturhahn, T.S. Toellner, E.E. Alp, X. Zhang, M. Ando, Y. Yoda, S. Kikuta, M. Seto, C. W. Kimball, B. Dabrowski, Phys. Rev. Lett. **74**, 3832 (1995)
205. E.G. d'Agliano, P. Kumar, W. Schaich, H. Suhl, Phys. Rev. B **11**, 2122 (1975)
206. A. Nourtier, J. Phys. (Paris) **38**, 479 (1977)
207. R. Brako, D.M. Newns, Solid State Commun. **33**, 713 (1980)
208. J.W. Gadzuk, H. Metiu, Phys. Rev. B **22**, 2603 (1980)
209. J.K. Nørskov, B.I. Lundqvist, Surf. Sci. **89**, 251 (1979)
210. G.P. Brivio, T.B. Grimley, J. Phys. C **10**, 2351 (1977)
211. G.P. Brivio, T.B. Grimley, Surf. Sci. **89**, 226 (1980)
212. K. Schönhammer, O. Gunnarsson, Z. Phys. B **38**, 127 (1980)
213. K. Schönhammer, O. Gunnarsson, Phys. Rev. B **22**, 1629 (1980)
214. O. Gunnarsson, K. Schönhammer, Phys. Rev. B **25**, 2503 (1982)
215. O. Gunnarsson, K. Schönhammer, Phys. Rev. B **25**, 2514 (1982)
216. B. Feuerbacher, R.F. Willis, Phys. Rev. Lett. **47**, 526 (1981)
217. B. Feuerbacher, in *Dynamics of Gas-Surface Interactions*, ed. by G. Benedek, U. Valbusa (Springer, Heidelberg, 1982), p. 263
218. B. Diaconescu, K. Pohl, L. Vattuone, L. Savio, P. Hofmann, V.M. Silkin, J.M. Pitarke, E.V. Chulkov, P.M. Echenique, D. Farías, M. Rocca, Nature **448**, 57 (2007)
219. S.J. Park, R.E. Palmer, Phys. Rev. Lett. **105**, 016801 (2010)
220. K. Pohl, B. Diaconescu, G. Vercelli, L. Vattuone, V.M. Silkin, E.V. Chulkov, P.M. Echenique, M. Rocca, EPL **90**, 57006 (2010)
221. L. Vattuone, G. Vercelli, M. Smerieri, L. Savio, M. Rocca, Plasmonics **7**, 323 (2012)
222. L. Vattuone, M. Smerieri, T. Langer, C. Tegenkamp, H. Pfnür, V.M. Silkin, E.V. Chulkov, P.M. Echenique, M. Rocca, Phys. Rev. Lett. **110**, 127405 (2013)
223. W.L. Schnaich, J. Harris, J. Phys. F **11**, 65 (1981)
224. E. Hulpke, J. Lüdecke, Phys. Rev. Lett. **68**, 2846 (1992)
225. E. Hulpke, J. Lüdecke, Surf. Sci. **287**(288), 83 (1993)
226. J.M. Pitarke, V.M. Silkin, E.V. Chulkov, P.M. Echenique, Rep. Prog. Phys. **70**, 1 (2007)
227. H. Yu, J.C. Hermanson, Phys. Rev. B **40**, 11851 (1989)

228. H. Yu, J.C. Hermanson, Phys. Rev. B **41**, 5991 (1990)
229. Y. Chen, J.C. Hermanson, G.J. Lapeyre, Phys. Rev. B **39**, 12682 (1989)
230. J.I. Gersten, I. Wagner, A. Rosenthal, Y. Goldstein, A. Many, R.E. Kirby, Phys. Rev. B **29**, 2458 (1984)
231. V.M. Silkin, J.M. Pitarke, E.V. Chulkov, P.M. Echenique, Phys. Rev. B **72**, 115435 (2005)
232. J. Fritsch, U. Schröder, Phys. Rep. **309**, 209 (1999)
233. R.H. Ritchie, J.C. Ashley, T.L. Ferrell, in *Electromagnetic Surface Modes*, ed. by A.B. Boardman (Wiley, Chichester, 1982), p. 119
234. G. Benedek, M. Pardo, J.P. Toennies, Theory of inelastic atom scattering from surface electron-hole and plasmon excitations, in *Highlights on Spectroscopies of Semiconductors and Nanostructures*, ed. by G. Guizzetti, A.C. Andreani, F. Marabelli, M. Patrini, Conference Proceedings, vol. 94 (SIF, Bologna, 2007), pp. 151–167
235. H. Nienhaus, Phys. Rev. B **56**, 13194 (1997)
236. J. Braun, J.P. Toennies, C. Wöll, Phys. Rev. B **60**, 11707 (1999)
237. G. Benedek, M. Bernasconi, D. Campi, P.M. Echenique, V.M. Silkin, E.V. Chulkov, D. Farías, D. Maccariello, J.P. Toennies (to be published)
238. P.R. Kole, *Dynamics and Morphology of Metal and Metal Oxide Surfaces* (Ph.D. Dissertation, University of Cambridge, 2011), and P.R. Kole, G. Benedek, ICSFS 16 Genova July 1–6, 2012—abstr. PLA 1.2
239. U. Fano, Phys. Rev. **124**, 1866 (1961)
240. V. Chis, B. Hellsing, G. Benedek, M. Bernasconi, E.V. Chulkov, J.P. Toennies, Phys. Rev. Letters **101**, 206102 (2008)
241. G. Lange, J.P. Toennies, Phys. Rev. B **53**, 9614 (1996)
242. W. Menezes, P. Knipp, G. Tisdale, S.J. Sibener, Phys. Rev. B **41**, 5648 (1990)
243. J. Lobo-Checa, T. Okuda, M. Hengsberger, L. Patthey, T. Greber, P. Blaha, J. Osterwalder, Phys. Rev. B **77**, 075415 (2008)
244. N. Bundaleski, H. Khemliche, P. Soulisse, P. Roncin, Phys. Rev. Lett. **101**, 177601 (2008)
245. A. Schüller, H. Winter, Phys. Rev. Lett. **100**, 097602 (2008)
246. H. Khemliche, P. Rousseau, P. Roncin, V.H. Etgens, F. Finocchi, Appl. Phys. Lett. **95**, 151901 (2009)
247. H. Winter, A. Schüller, Progr. Surf. Sci. **86**, 169 (2011)
248. M. Busch, J. Lienemann, J. Seifert, A. Schüller, H. Winter, Vacuum **86**, 1618 (2012)
249. J. Lienemann, D. Hülsewede, J. Seifert, K. Maass, M. Busch, H. Winter, Nucl. Instr. Meth. Phys. Res. B **315**, 30 (2013)
250. M. Debiossac, A. Zugarramurdi, P. Lunca-Popa, A. Momeni, H. Khemliche, A.G. Borisov, P. Roncin, Phys. Rev. Lett. **112**, 023203 (2014)
251. J. Seifert, J. Lienemann, A. Schuller, H. Winter, Nucl. Instrum. Meth. B **350**, 99 (2015)
252. N.-P. Wang, M. Rohlfing, P. Krüger, J. Pollmann, Phys. Rev. B **67**, 115111 (2003)
253. R.J. Glauber, Phys. Rev. **84**, 395 (1951)
254. K. Huang, A. Rhys, Proc. R. Soc. (Lond.) A **204**, 406 (1950); when referred to the multi-phonon sidebands of optical transitions S is known as Huang-Rhys exponent
255. H. Khemliche, J. Villette, P. Roncin, M. Barat, Nucl. Instrum. Methods Phys. Res. B **164**, 608 (2000)
256. A.A. Lucas, M. Sunjic M, in *Progress in Surface Science*, vol. 2, ed. by S.G. Davison (Pergamon, New York, 1971)
257. G. Benedek, E. Mulazzi, Phys. Rev. **179**, 906 (1969)
258. J. Villette, A. G. Borisov, H. Khemliche, A. Momeni, P. Roncin, Phys. Rev. Lett. **85**, (2000)
259. A.A. Lucas, M. Sunjic, G. Benedek, P.M. Echenique, New. J. Phys. **16**, 063015 (2014)
260. S. Suto, K.-D. Tsuei, E. W. Plummer, E. Burstein, Phys. Rev. Lett. **63**, 2590 (1989)
261. H. Haberland, in *Fundamental Processes of Atomic Dynamics*, ed. by J.S. Brigss, H. Kleinpoppen, H.O. Lutz (Plenum Press, New York, 1988)

262. G. Benedek, T.P. Martin, G. Pacchioni (eds.), *Elemental and Molecular Clusters* (Springer, Heidelberg, 1988)
263. G. Scoles (ed.), *The Chemical Physics of Atomic and Molecular Clusters* (North-Holland, 1991)
264. U. Buck, R. Krohne, J. Siebers, Z. Phys. D **26**, 169 (1993)
265. U. Buck, in *Physics of Electronic And Atomic Collisions: XVIII International Conference*, ed. by T. Andersen, B. Fastrup, F. Folkmann, H. Knudsen, N. Andersen, AIP Conference Proceedings, vol. 295, (1993), p. 719
266. U. Buck, R. Krohne, J. Chem. Phys. **105**, 5408 (1996)
267. U. Buck, Surf. Rev. Lett. **3**, 467 (1996)
268. U. Buck, R. Krohne, Phys. Rev. Lett. **73**, 947 (1994)
269. U. Buck, R. Krohne, P. Lohbrandt, J. Chem. Phys. **106**, 3205 (1997)
270. T. Schroder, R. Schinke, R. Krohne, U. Buck, J. Chem. Phys. **106**, 9067 (1997)
271. T.A. Beu, C. Steinbach, U. Buck, J. Chem. Phys. **117**, 3149 (2002)
272. J. Brudermann, P. Lohbrandt, U. Buck, Phys. Rev. Lett. **80**, 2821 (1998)
273. J. Brudermann, U. Buck, E. Fredj, R.B. Gerber, M.A. Ratner, J. Chem. Phys. **111**, 10069 (1999)
274. J. Brudermann, P. Lohbrandt, U. Buck, J. Chem. Phys. **112**, 11038 (2000)
275. A. Glebov, A.P. Graham, A. Menzel, J.P. Toennies, P. Senet, J. Chem. Phys. **112**, 11011 (2000)
276. R.B. Doak, in E. Hulpke (ed.) *Helium Atom Scattering from Surfaces*, (Springer, Berlin, 1992), pp. 5–24
277. B. Holst, W. Allison, Nature **390**, 244 (1997)
278. K. Fladischer, H. Reingruber, T. Reisinger et al., New J. Phys. **12**, 033018 (2010)
279. D. Barredo, F. Calleja, A.E. Weeks, P. Nieto, J.J. Hinarejos, G. Laurent, A.L. Vazquez de Parga, D. McLaren, D. Farias, W. Allison, R. Miranda, et al., Surf. Sci. **601**, 24–29 (2007)
280. R.B. Doak, R.E. Grisenti, S. Rehbein, G. Schmal, J.P. Toennies, C. Wöll, Phys. Rev. Lett. **83**, 4229 (1999)
281. D. Barredo, F. Calleja, P. Nieto, J.J. Hinarejos, G. Laurent, A.L. Vazquez de Parga, D. Farias, R. Miranda, Adv. Mater. **20**, 3492–3497 (2008)
282. D. Barredo, G. Laurent, F. Calleja, P. Nieto, J.J. Hinarejos, G. Laurent, A.L. Vazquez de Parga, D. Farias, R. Miranda, Appl. Phys. Lett. **96**, 081901 (2010)
283. G. Anemone, C. Weingarten, A. AlTaleb, C. Prieto, D. Farias, Appl. Phys. Lett. **111**, 181602 (2017)
284. H.C. Schewe, B.S. Zhao, G. Meijer, W. Schollkopf, New J. Phys. **11**, 113030 (2009)
285. O. Carnal, M. Sigle, T. Sleator, H. Takuma, J. Mlynek, Phys. Rev. Lett. **67**, 3231 (1991)
286. M. Koch, S. Rehbein, G. Schmahl et al., J. Microsc. **229**, 1–5(2008)
287. S.D. Eder, T. Reisinger, M.M. Greve, G. Bracco, B. Holst, New J. Phys. **14**, 073014 (2012)
288. P. Witham, E. Sanchez, Rev. Sci. Instr. **82**, 103705 (2011)
289. P.J. Witham, E.J. Sanchez, J. of Microscopy **248**, 223–227 (2012)
290. P. J. Witham, Master Thesis (2013), Portland State Univ
291. P. Witham, E. Sanchez, Cryst. Res. Technol. **49**, 690–698 (2014)
292. A.S. Palau, G. Bracco, B. Holst, Phys. Rev. A **94**, 063624 (2016)
293. M. Barr, A. Fahy, A. Jardine, J. Ellis, D. Ward, D.A. MacLaren, W. Allison, P.C. Dastoor, Nuc. Instr. Meth. Phys. Res. B **340**, 76–80 (2014)
294. A. Fahy, M. Barr, J. Martens, P.C. Dastoor, Rev. Sci. Instrum. **86**, 023704 (2015)
295. M. Barr, A. Fahy, J. Martens, J. Martens, A.P. Jardine, D.J. Ward, J. Ellis, W. Allison, P.C. Dastoor, Nat. Commun. **7**, 10189 (2016)
296. P. Witham, private communication
297. H. Morgner, Adv. At. Mol. Opt. Phys. **42**, 387–488 (2000)
298. E. Bauer, Rep. Prog. Phys. **57**, 895 (1994)

299. N. Ueno, H. Yasufuku, S. Kera, K.K. Okudaira, Y. Harada, Lect. Notes Phys. **588**, 131–144 (2002) (Springer, Berlin)
300. Y. Harada, S. Yamamoto, M. Aoki, S. Masuda, T. Ichinokawa, M. Kato, Y. Sakai, Nature **372**, 657 (1994)
301. R. Castillo-Garza, J. Gardner, S. Zisman, M.G. Raizen, ACS Nano, 4378 (2013)
302. B.W. Ward, J.A. Notte, N.P. Economou, J. Vac. Sci. Technol. B **24**, 2871 (2006)
303. N. Klingner, R. Heller, G. Glawacek, J. Von Borany, J. Notte, J. Huang, S. Facsko, Ultramicroscopy **162**, 91–97 (2016)

Chapter 14
Future

Prediction is very difficult, especially if it's about the future.

(Niels Bohr)

Abstract New technological developments in molecular beam methods provide the basis for designing future HAS apparatus with greatly improved energy resolution and intensity. Some of these technological advances are reviewed and the gains to be expected are discussed. The greater intensity can be expected to lead to a great improvement in the inelastic scattering spatial resolution from the present resolution of the order of several square mm. In addition to the many areas of HAS research covered in the preceding chapters there are many more opportunities for HAS-based investigations, which although accessible with present technology, would greatly profit from the new technological developments. These new areas include the surfaces of liquids, organic films and surfaces coated with biomolecules as well as THz plasmonic systems. The present state of these research areas are briefly surveyed and possible HAS experiments are suggested. Finally the expected new challenges for the theory will be briefly outlined.

14.1 Future Helium Atom Scattering Apparatus

14.1.1 *The Ultimate Machine*

The majority of the experimental results described in Chaps. 11–13, were obtained with a time-of-flight helium atom scattering (TOF-HAS) apparatus similar to the Göttingen apparatus described in Sect. 9.4.1, using technology available at about 1980. The absolute energetic resolution, which at the time was considered to be the best possible, amounted to about 0.2–0.3 meV (Sect. 9.8). This resolution was achieved by expanding He gas at very high pressures P_0 of several hundred bar through a tiny $d = 5$ µm dia thin walled orifice into vacuum. In this way speed ratios of $S \approx 200$ could be obtained for the first time [1]. The corresponding beams

G. Benedek and J. P. Toennies, *Atomic Scale Dynamics at Surfaces*, Springer Series in Surface Sciences 63, https://doi.org/10.1007/978-3-662-56443-1_14

had relative FWHM velocity and energetic widths of $\Delta v/v = 7 \times 10^{-3}$ and $\Delta E/E = 1.4 \times 10^{-2}$ [2, 3].

For many of the future applications of TOF-HAS an improved energy resolution as already achieved with spin echo HAS (SE-HAS), but without the complications and limitations of SE-HAS (see Sect. 9.4.3), would be highly desirable. The other problem is the low detection efficiency of only 10^{-4}–10^{-6}, which is inherently limited by the electron impact ionizer currently used in the mass spectrometer detector [4, 5]. Another related problem with the present technology is the lack of spatial resolution so that it is not possible to analyze nanostructures or other surface objects of small size. At present a relatively wide incident beam with an area at the target surface of about 10 mm^2 is required to achieve a signal intensity in the Rayleigh peak of about $I_{RW} = 10^2$–10^3 counts/s (see, e.g. Fig. 9.5) and measure a dispersion curve in a reasonable period of several hours before the surface is contaminated. In addition the inability of this type of detector to distinguish between the atoms in the beam and the rest gas in the detector complicates the construction of the apparatus by, for example, the need for many pumping stages to reduce the residual gas pressure. These two main limitations of (1) energy resolution, and (2) detector sensitivity of the TOF-HAS method can now be overcome with the currently available technology as briefly described in the following.

(1) *Energy resolution*: Much larger speed ratios of 1500 were reported in 1988 by Giese and Gentry in a pulsed beam apparatus with a beam flux $\phi \propto$ 12.0 Torr cm^2, which is a factor 40 greater than in the Göttingen apparatus [6]. The energetic width of such a beam at the same beam velocity as in the Göttingen apparatus is 40 μeV which is comparable to the 70 μeV, presently achieved with SE-HAS in phonon spectroscopy (Fig. 9.13). A smaller but still significant improvement in energy resolution corresponding to $S \approx 500$ was, in fact, already achieved with a continuous beam in an unpublished experiment on a Göttingen apparatus [1]. A surface scattering apparatus with the Giese and Gentry source has so far not been constructed and would certainly be more complicated and expensive compared to the Göttingen apparatus. In such a pulsed beam apparatus the improvement in beam flux would be partly compensated by the reduced duty cycle and the need to increase the time-of-flight resolution. This latter problem can, however, be resolved by using a new more sensitive detection scheme discussed next.
(2) *Detector sensitivity*: In an attempt to improve the detector sensitivity several groups have developed an electron impact ionizer based on confining the electrons by a magnetic field along the direction of the incoming beam of atoms [7–10]. In this way the number of ionizing collisions is increased by the greater effective path of the spiraling electrons [11]. An optimized ionizer of this type has been reported to have a sensitivity of 10^{-3} [8, 9], which is an increase of several orders of magnitude compared to the conventional ionizer. Ionizers of this type are, however, not suited for direct use in TOF experiments because of the long response time of about 100 ms [9] and the inherent relatively large length of the ionization region. To achieve a good time-of-flight resolution one possible solution would be to use an additional chopper placed immediately in

front of the detector rotating synchronously in phase with the first chopper behind the source. This chopper would then define the time-of-flight of the inelastically scattered atoms, so that the resolution would no longer depend on the greater response time and length of the magnetic field ionizer. Since this double chopper scheme is a standard in neutron scattering [12] it should be straight forward to apply to HAS experiments.

A more promising method to achieve much higher sensitivities of about 10^{-1} is based on single photon ionization using either XUV photons provided by a free electron laser, e.g., available at FLASH in Hamburg (<110 fs at 18.5 nm, 100 Hz) or using intense pulsed electric fields commercially available with fs pulses from a Ti: Sa laser (40 fs $\approx 2 \times 10^{15}$ Wcm^{-2} at 780 nm, 10 kHz) [13, 14]. A special coaxial arrangement has been envisaged which would allow the laser to ionize a large fraction of each TOF pulse [15]. Photon ionization has the additional advantage over electron ionization that the nascent ions are not deflected so that they travel along the initial direction of the atoms where they can be nearly 100% detected, i.e., by an open electron multiplier. The residual gas background problem is virtually completed eliminated since only a minute fraction of the background ions travel in the forward direction. Both laser techniques have recently been used with great success for detecting and measuring the size of the helium dimer and trimer [13, 16].

There are many other less spectacular possibilities to increase the resolution and sensitivity starting from a Göttingen-type arrangement. For example, the sensitivity can be increased by the greater flux in each TOF pulse by using a recently developed new type of pulsed nozzle beam source [17]. There are also presently a number of successful experiments which illustrate how a more compact and less expensive apparatus could be constructed [18–20].

Once these improved techniques have been introduced both the increased resolution and beam flux will open up new areas for the application of the HAS techniques. In addition to improving the resolution the increase in intensity will lead to a considerable increase in the spatial and temporal resolution. Thus at some point in the future it should be possible to analyze nanostructures and even small biological samples for their surface dynamics. Or, if the spatial resolution is of no concern, it might be possible to monitor the time evolution of surface processes such as an order-order phase transformation. Some additional suggestions for future study by the new and present day HAS techniques are outlined in the following sections.

14.2 Other Future Applications of Helium Atom Scattering

14.2.1 Liquid Surfaces

Compared to solid surfaces very much less is known about the structure and the phonon excitations of liquid surfaces [21]. The elementary vibrational waves on a liquid surface are known as *ripplons*, which are also frequently referred to as capillary waves. Most of the sparse literature on ripplons is devoted to the surface of superfluid He II. The remarkable helium atom scattering experiments from the surface of liquid helium by David Edwards and colleagues [22, 23] and the theoretical work by Echenique and Pendry [24] provided the first evidence for the excitation of ripplons and rotons at the surface of a quantum liquid [25]. The only available dispersion curve of a ripplon comes from inelastic neutron scattering off of thin films of liquid helium on the surface of graphite [26].

The study of the surface dynamics of ordinary liquids has only been accomplished with light scattering and has been reviewed in [27, 28]. The surfaces of gallium [29, 30] and indium [31] have been investigated by light and x-ray scattering, respectively. Both metals have the advantage that they are liquid at room temperatures. All the above experiments are restricted to low ripplon energies by the very long wavelengths of light and x-rays.

One reason for the scarcity of data comes from the difficulties of preparing a clean non-vibrating surface. Molecular beam scattering experiments have recently been used to investigate the ways in which gas molecules are captured by and react at the surface of industrially important liquids [32]. In these experiments thin films of liquids with not too high vapor pressures were prepared by an arrangement first introduced in about 1975 by Delahay et al. [33]. In this method a rotating disc is continually passed through the liquid and as the disc rises above the liquid level the liquid adhering to the disc can be used as a target. More recently the rotating disc enabled measurements of the angular and energy distributions of Ne, Ar and Xe scattered from the liquid metal surfaces of Ga, In and Bi, which have a sufficiently low vapor pressure at room temperature for molecular beam experiments [34, 35]. Experiments with He atoms have not been reported except on nematic liquid crystal films [36]. There have also been a number of theoretical investigations stimulated by the experiments on liquid Ga [37].

To facilitate studies of high vapor pressure liquids micron-sized liquid jets have also been introduced [38–41]. The small size of the cylindrical jets and the rapidly decreasing radial vapor density distribution of the evaporating atoms make even aqueous solutions and water surfaces accessible to vacuum sensitive atomic and molecular beam scattering [32, 40]. Thus the experimental problems which one would expect to arise from the small size and curved surface of such a target appear to be surmountable and it is indeed conceivable that high resolution scattering of He atoms from micron-sized liquid jets is possible. Evidence for this comes from successful *inelastic* scattering experiments of He atom beams from even much smaller gas phase Ar or H_2O clusters [42] (see Sect. 13.7).

Helium atom scattering from liquid metal surfaces would be very rewarding because of the unique interactions of the He atoms with the electrons at the surfaces of these metals. We are only aware of a number of theoretical publications in the 1970s in which the electron distributions at liquid surfaces including also metal surfaces were studied [43, 44]. Also there have been some early theoretical studies of phonons and plasmons at liquid metal surfaces [45, 46].

14.2.2 Organic Electronics

In Chap. 12 several examples of past HAS studies of organic molecules and monolayer films, mostly in the form of the alkanes and a few self-assembled monolayers, were presented. However, none of these studies touched upon recent modern developments in organic electronics which have in recent years found wide spread applications [47]. These include organic light emitting devices (OLEDS), organic field effect transistors (OFETS), organic solar cells [48] and organic spin valves [49]. Of particular interest for organic electronics are the multilayers of the acene molecules (polyclyclic aromatic molecules with between 2 (naphthalene) and 7 (heptacene) benzene rings) [50]. This field has also been called *two-dimensional crystal engineering* since the weak non-covalent interactions and ligand-metal substrate interactions lead to structures with regular arrays consisting of a wide variety of different arrangements of organic molecules on single crystal metal surfaces [51, 52]. For many of the future technological applications it is important to produce well-defined and ordered films by molecular beam deposition [53, 54]. Ideal candidates for future HAS studies are the porphins and the structurally related phthalocyanines and their many metal complexes [55]. Many have been studied by STM, LEED and reflection absorption infrared spectroscopy (RAIRS). In a few cases structural studies with HREELS have been successful. In many of these films the chirality of the surface layer can be controlled and in some cases the chiral surfaces can catalyze the chiral outcome of surface reactions [56].

A phase transition turning a chiral surface into its specular image with respect to a vertical plane can in principle be detected by HAS as a time-reversal symmetry break, i.e., as a $\mathbf{G} \leftrightarrow -\mathbf{G}$ asymmetry in the diffraction pattern along the direction normal to that plane. A clear example, which could be readily extended to the case of ordered chiral adsorbates, is that of the high temperature order-order phase transition of Ge(111), where the largest HAS diffraction peak in the $[1\bar{2}1]$ direction moves abruptly from $\mathbf{G}$ to $-\mathbf{G}$ across the critical temperature T_c [57]. Similarly the HAS phonon dispersion curves corresponding to *umklapp* scattering processes switch from $\mathbf{G}$ to $-\mathbf{G}$ [58].

Considerable effort has been devoted to understanding the electronic structure of molecular films of the large planar PTCDA molecules ($C_{24}H_8O_6$) and the related NTCDA which have a number of unique electrical and photoelectrical properties [59]. Comparative studies using LEED, STM and X-ray diffraction of these molecules on Cu(111), Ag(111) and Au(111) reveal different ordered structures for each of these systems [51, 60–62]. Ultraviolet photoelectron spectroscopy shows a clear correlation between the electronic structures and the interface bonding [63]. The charge transfer of PTCDA to the metal substrate increases as the bond distance decreases in the sequence Au–Ag–Cu [64]. For the latter two systems charge transfer led to new occupied molecular states of PTCDA. On Ag(111) a hybrid state was found at the Fermi level so that the monolayer has metallic character [63, 65]. Thus this system would be extremely interesting for a HAS study because of the sensitivity of the He atoms to the electronic distortions and their effect on the phonons and the binding to the surface. Also the quantum sonar effect might be useful for understanding the charge reorganization occurring between copper-II-phthalocynine absorbed on PTCDA on Ag(111) [66].

As far as we are aware the only HAS study in this area was a SE-HAS experiment devoted to cyclooctatetracene C_8H_8 (a benzene like molecule but with 8 carbon atoms) which is one of the first discovered conducting polymers. The ordered monolayer structure, several inelastic features, and the diffusional behavior on Cu(111) could all be observed [67]. The largest molecule studied so far with TOF-HAS is hexatriacontane ($C_{38}H_{74}$) a very large alkane which is also a polymer [68]. Some evidence for diffraction peaks and a sharp FT mode at 5 meV as observed for the smaller alkanes was found (see Sect. 12.5). The only other information on the vibrations of organic films comes largely from EELS and RAIRS. The former has the disadvantage that the delicate structures can easily be perturbed and disordered. RAIRS is limited to high energy vibrations in the range of 600–3600 cm^{-1}. Thus it will be very informative to use the non-destructive HAS method to access the low energy vibrations, especially those between the various constituent groups within the large molecules and with the substrate. With sufficiently large coherent patches of the molecules it may even be possible to measure the dispersion curves of long wavelength vibrations. Future HAS experiments will profit from the extensive literature on these complex structures from the standard techniques such as LEED, STM and RAIRS [51, 52, 69]. We therefore anticipate that HAS experiments in combination with these methods and refined DFT calculations will bring new insight into the intramolecular, intermolecular and molecule-surface interactions.

14.2.3 Biological Films

Currently there is great interest in investigating the low frequency vibrational modes of large biomolecules, such as the peptides, enzymes, immunoglobulines, and oligonucleotides in the tetraherz range [70, 71]. The major interest is of course directed at understanding the dynamics of biomolecules in vitro [72] which can be investigated using neutron scattering [73]. As in the case of organic films there is also interest in the well-ordered structures of biomolecules on surfaces partly because of medical applications [74]. Many of the films of large organic molecules mentioned above, for example, of the porphins and the structurally related phthalocyanines, as well as their many metal complexes, are also of biological interest as centers of biological activity in many areas of biology ranging from blood cells to photosynthesis in plants. Also well-defined surfaces provide ideal platforms for facilitating and monitoring the self-organization of biomolecular structures with novel applications [52, 75].

The preparation of films of the small biological molecules such as the amino acids is quite similar to that of the organic films discussed above. But the preparation of large biological molecules such as DNA molecules presents special challenges. Experiments on small amino acid molecules on metal surfaces, such as Cu(110), have been extensively reviewed by Barlow and Raval [51], mentioned in the previous section, and by Costa, Pradier, Tielens and Savio [76]. In the latter review also studies of amino acids on oxide and peptides on metal surfaces are discussed. As with the organic films their monolayer structures are well known from STM, LEED investigations. In view of this early work the amino acids appear to be ideal objects for HAS investigations. We are aware of only two HAS studies of amino acid films. One was devoted to clarifying the self-assembled structure of L-methionine $\{(CH_3SCH_2CH_2CH(NH_3)^+(COO)^-\}$ on Ag(111) [77]. In the other explorative investigation of L-alanine ($C_2O_2NH_{11}$) on Cu(110) [68] sharp diffraction peaks were resolved and found to be compatible with earlier LEED studies. The time-of-flight energy transfer distribution revealed two maxima at 5.3 and 7.5 meV. These were assigned to two different COO^--metal stretch vibrations.

Finally we mention one particularly intriguing motivation for the study of films of amino acid molecules on mineral surfaces in connection with the search for the mechanism for the polymerization of amino acids to form long hetero-peptides or hetero-nucleotides in prebiotic times [78]. Bernal's 1951 [79] conjecture that the polymerization occurred on surfaces that were likely present on the early earth such as silicates, oxides and sulfides is still being discussed.

Until recently it has not been possible to easily prepare larger biomolecules such as the peptides or proteins on surfaces. Recently the group of Kern and Rauschenbach [80] have demonstrated that it is possible to use electrospray ionization (ESI) for the selection and deposition of specific very large molecules on metal surfaces. With this technique they have prepared two-dimensional self-assembled honeycomb networks of oligopeptides on Au(111) [81] and have been able to prepare folded and unfolded very large $\approx$1000 Dalton polypeptides [82] and $\approx$12,000 Dalton proteins [83] on Cu single crystal surfaces. With the advent of this technique even these extremely large molecules are now accessible for the investigation of their low frequency vibrations and diffusion by HAS methods.

14.2.4 Clusters and Supracrystals

Supracrystals and nanocrystals are another large class of objects which can in principle be studied by HAS [84, 85]. They are currently of considerable interest as constituents of novel microelectronic devices, as catalysts, special surface coatings and in medicine. Their properties are different from those of the bulk and by changing their sizes and by growing supracrystals in any crystalline structure it is possible to tailor their properties for special applications. The recent developments in the self-assembling of well-ordered 2D and 3D arrays of clusters and nanocrystals of uniform size (*supracrystals*) [86–90] is providing a new class of interesting materials with yet unexplored possibilities in GHz to THz nanotechnologies. Coated with a dodecanethiol layer nanocrystals with sizes of 5–6 nm containing 500–1000 atoms with a size standard deviation of 8–10% [91] can be prepared with a high degree of crystallinity to form supracrystals with different morphologies [87]. The mechanical properties, e.g., the Young modulus, vary over a wide range from MPa to GPa depending on the organic molecules bridging nearest neighbor nanoparticles. The supracrystal dynamics is characterized by acoustic vibrations [92] and hindered rotations of the nanoparticles considered as rigid entities, which typically fall in the high GHz spectrum and is well separated from the spectrum of internal nanoparticle vibrations, well inside the THz domain [93]. They are also expected to have interesting plasmon modes and exhibit interesting plasmon-phonon couplings [94]. Uncoated metal clusters can be produced in gas expansions, which however exhibit broad lognormal size distributions [95, 96].

The availability of regular nanoparticle arrays over mesoscopic scales suggests that HAS, especially in its spin echo version, is an ideal tool to investigate the structural and dynamical properties of 2D supracrystals and 3D supracrystal surfaces. An early test of this possibility is represented by the 1997 HAS study of the structural and vibrational properties of the (111) surface of fcc-C_{60} (fullerite) [97, 98]. In this study the dispersion curves of the librational modes at $\sim$1.5 meV and their overtones were clearly observed by HAS. Also the Rayleigh waves were measured and found to be in agreement with BCM calculations. In comparison the nanocrystals assembled in supracrystals are usually two to three orders of magnitude heavier than C_{60} which would lower the energy of librations well below the meV domain, a spectral region which would be however easily accessible to SE-HAS.

14.2.5 He Atom as a Mechanical Probe of Electronic Systems

The electron interacts with an electromagnetic field via its charge and, to a minor extent, its magnetic moment, whereas its interaction with a closed shell atom like He essentially occurs via its being a fermion and the exclusion principle and consequent short range repulsion. In this respect He atoms act as nanometric fingers which probe locally the electron gas, and are therefore complementary to electromagnetic fields, as in essence pointed out by Fermi in 1934 [99]. As discussed in Chap. 8 and Sect. 13.5, a low energy He atoms flying by a free electron surface act as a scattering center for the electrons (Fig. 8.3), thus inducing electron-hole excitations either virtual, then recombining into a phonon or a low energy plasmon, or real as is the case of deep non-adiabatic Kohn anomalies (Figs. 11.37 and 11.38).

Electronic transitions at the Fermi level with wavevector changes concentrated around specific values, as is the case of inter-pocket transitions in semimetals, can in principle be detected by HAS time-of-flight measurements and even in angular distributions as apparent diffraction processes for the corresponding parallel wavevector transfers. Since the pocket carrier populations may have a comparatively strong temperature dependence, HAS diffraction spectra at different temperatures may provide evidence of interpocket electronic transitions. For a purely elastic process the electronic energies of the initial and final pocket states are the same, and a stationary superposition of degenerate symmetry-equivalent pocket states is actually a charge density wave producing the observed diffractive scattering of He atoms. In case of strong spin-orbit interaction the HAS-induced interpocket electron scattering may be accompanied by a spin polarization change and a consequent asymmetry (time-reversal symmetry break) in the corresponding diffraction intensities with respect to the specular peak. These expected effects suggest HAS as a valuable tool for a deeper insight into the interesting surface electronic properties of semimetal materials such as thermoelectrics and topological insulators.

Among the surface electronic transitions induced by He atom scattering, those across a superconducting gap should deserve more attention. This is another inelastic process in the energy range of HAS spectroscopy which can in principle be detected from the temperature dependence of TOF spectra. Preliminary HAS studies on the (001) surface of the orthorhombic high-T_c superconductors $YBa_2Cu_3O_{7-x}$ for $x < 0.3$ and $Bi_2Sr_2Ca_1Cu_2O_{8-x}$ [100] show a rich vibrational structure, but no analysis is available on the temperature dependence across the transition temperature. The study of low energy surface excitations in low-dimensional superconductors, and specifically of their mode selected electron-phonon interaction, is another domain where HAS can bring new knowledge, hard to obtain with other means.

14.3 Future Challenges for Theory

Since the early days of HAS spectroscopy it was clear that the complexity of high resolution TOF spectra contains subtle information on surface dynamics which in most cases cannot be disentangled without considerable theoretical effort. A classical example is that of the early measurements on noble metal surfaces which revealed several unexpected features [101]. They then stimulated the development of new theoretical concepts, like, e.g., the multipole expansion (ME) method [102, 103], and ultimately led, by means of DFPT [104], to recognize that inelastic HAS intensities from metal surfaces provide a direct measure of the electron-phonon interaction for each observed phonon (see Chap. 8 [103, 105, 106]).

Present DFPT calculations, as implemented in some of the most widely used simulation codes like Quantum Espresso [107], ABINIT [108], VASP [109], and CASTEP [110], have at hand a choice of efficient functionals with ultrasoft fully relativistic pseudopotentials [111, 112] and can deal with atomic systems characterized by a large spin-orbit interaction [113] and, with reasonable accuracy, dispersion forces [114]. As seen in Chaps. 8 and 11 and in [103], the DFPT calculations of the electron-phonon effects as detected by HAS in all classes of metal and semimetal surfaces, and also in highly polarizable insulators [115] are quite accurate and reliable. All these calculations rely however on the adiabatic approximation. The analysis and treatment of deep non-adiabatic anomalies like, e.g., those of H-covered W(110) and Mo(110) surfaces, are beyond the present DFPT methods. Only models like the ME method, where pseudocharges can be endowed with *ad hoc* effective masses, are presently available.

First principle theories also allow for a precise calculation of the energy and momentum dependent electron susceptibility, so that the dispersion curves of collective excitations such as surface plasmons can be derived within the random-phase approximation. However in the THz spectral region where acoustic surface plasmons with a linear dispersion [116–118], or 2D plasmons with a square-root dispersion, cross the phonon spectrum, the coupling effects cannot be fully treated with present density functional methods. Also in this case models remain at the moment the only resource.

The new insight gained into the role of electrons in mediating the energy transfer of atomic collisions discussed in this monograph has important implications for the microscopic theory of reactions with and on metal surfaces as occur in catalysis [119–121]. For some time it has been realized that electron-hole pair excitation is an important process in energy transfer to surfaces [120, 121]. The role of phonons in energy transfer to surfaces has only very recently been accounted for explicitly with detailed account of the different modes, their dispersion and density of states [122]. Only recently has the strong coupling of radicals to electron-hole pairs been experimentally observed [123]. The first principles theory of the microscopic theory of these phenomena and others related to understanding surface reactions present many challenges for the future.

14.4 Conclusion

As we have tried to demonstrate in the previous chapters the He atom scattering probe has a number of unique features. Originally it was thought to be the surface analog of neutrons with very similar particle-wave features, but ideal for the investigation of the phonon dispersion curves at surfaces. Its neutral character and non-destructive gentleness made He atom scattering an ideal tool for studying the structure and dynamics of sensitive surface films. In the course of these studies it was also realized that as opposed to other scattering probes such as electrons the He atoms were remarkably sensitive to even the smallest concentration of defects at surfaces, to step edges and minute structural details. As a result they are the ideal probe for the microscopic analysis of diffusion on surfaces. The phonon studies of metal surfaces revealed an anomalous mode which in many metals was not seen by EELS. This led ultimately to the realization that the He atoms collided largely with the loosely bound electrons and could in systems with a large electron phonon coupling excite indirectly second layer or even phonons further removed from the surface. In some systems it appears that helium atom scattering intensities are a direct measure of phonon mode electron-phonon coupling constants. The most recent studies have revealed that the He atoms can also couple to the plasmonic excitations at the surface. This remarkable sensitivity not shared by any other surface probe will certainly in the future continue to reveal many new unexpected fundamental surface phenomena some of which have been outlined in this chapter.

Finally it is hoped that the wealth of new information which has been and can be obtained on the dynamics at surface from He atom scattering, whether with TOF-HAS, or in the SE-HAS version, or with high energy atom beams, will stimulate new experimental work, with the development of novel apparatus and techniques, and new insight from a further progress in first-principle theoretical methods.

References

1. G. Brusdeylins, H.D. Meijer, J.P. Toennies, K. Winkelmann, in *Rarefied Gas Dynamics*, Proc. X Symp., in: Progress in Astronautics and Aeronautics, vol. 51, part II, J.L. Potter, ed (AIAA, New York, 1977) p. 1047
2. J.P. Toennies, K. Winkelmann, J. Chem. Phys. **66**, 3965 (1977)
3. L.W. Bruch, W. Schöllkopf, J.P. Toennies, J. Chem. Phys. **117**, 1544 (2002)
4. B. Samelin, Diplomarbeit, Georg-August-Universität Göttingen (1993)
5. K. Kuhnke, K. Kern, R. David, G. Comsa, Rev. Sci. Instrum. **65**, 3458 (1994)
6. J. Wang, V.A. Shamamian, B.R. Thomas, J.M. Wilkinson, J. Riley, Phys. Rev. Lett. **60**, 696 (1988)
7. M. DeKieviet, D. Dubbers, M. Klein, U. Pielesm, C. Schmidt, Rev. Sci. Instr. **71**, 2015 (2000)
8. A.V. Kalinin, L.Yu. Rusins, J.P. Toennies, Instr. Exp. Techn. **49**, 709 (2006)
9. D.M. Chisnall, Dissertation, Homerton College, University of Cambridge (2012)
10. A.R. Alderwick, A.P. Jardine, H. Hedgeland, D.A. MacLaren, W. Allison, J. Ellis, Rev. Sci. Instrum. **79**, 123301 (2008)
11. J.H. Malmberg, C.F. Driscoll, Phys. Rev. Lett. **44**, 654 (1980)
12. M. Arai, in *Neutron Scattering Fundamentals*, ed. by F. Fernandez-Alonso, D.L. Price (Elsevier, Amsterdam, 2013)
13. M. Kunitski, S. Zeller, J. Voigtsberger, A. Kalinin, L.P.H. Schmidt, M. Schöffler, A. Czasch, W. Schöllkopf, R. Grisenti, T. Jahnke, D. Blume, R. Dörner, Science **348**, 551 (2015)
14. W. Schöllkopf, private communication
15. O. Kornilov, private communication
16. J. Voigtsberger, S. Zeller, J. Becht, N. Neumann, F. Sturm, H.-K. Kim, M. Waitz, F. Trinter, M. Kunitski, A. Kalinin, J. Wu, W. Schöllkopf, D. Bressanini, A. Szasch, J.B. Williams, K. Ullmann-Pfleger, L.P.H. Schmidt, M.S. Schöffler, R.E. Grisenti, T. Jahnke, R. Dörner, Nat. Commun **6765**, 1–6 (2014)
17. U. Even, Hindawi Pub. Corp. vol. 2014, Art. ID 636042
18. R.B. Doak, J. Phys.: Condens. Matter **16**, 2863 (2004)
19. D.P. DePonte, S.D. Kevan, F.S. Patton, Rev. Sci. Instrum. **77**, 055107 (2006)
20. M. Barr, K.M. O'Donnell, A. Fahy, W. Allison, P.C. Dastoor, Meas. Sci. Technol. **23**, 105901 (2012)
21. J. Penfold, Rep. Prog. Phys. **64**, 777–814 (2001)
22. D.O. Edwards, P. Fatouros, G.G. Ihas, P. Mrozinski, S.Y. Shen, F.M. Gasparini, C.P. Tam, Phys. Rev. Lett. **34**, 1153 (1975)
23. D.O. Edwards, J.R. Eckardt, F.M. Gasparini, Phys. Rev. A. **9**, 2070 (1974)

24. P.M. Echenique, J.B. Pendry, Phys. Rev. Lett. **37**, 561 (1976), and J. Phys. C: Solid State Phys. 9, 3183 (1976)
25. D.O. Edwards, W.F. Saam, Prog. Low-Temp. Phys. **7 A**, 293 (1978)
26. H.J. Lauter, H. Godfrin, V.L.P. Frank, P. Leiderer, Phys. Rev. Lett. **68**, 2484 (1992)
27. T.K. Barik, P.R. Chaudhhuri, A. Roy, S. Kar, Meas. Sci. Technol. **17**, 1553 (2006)
28. Y. Minami, K. Sakai, Rev. Sci. Instrum. **80**, 014902 (2009)
29. V. Kolevzon, G. Gerbeth, J. Phys. D Appl. Phys. **29**, 2071–2081 (1996)
30. V. Kolevzon, J. Phys.: Condens. Matter **11**, 8785–8792 (1999)
31. B. Wehinger, M. Krisch, H. Reichert, New J. Phys. **13**, 023021 (2011)
32. J.A. Faust, G.M. Nathanson, Chem. Soc. Rev. **45**, 3609 (2012)
33. L. Nemec, L. Chia, P. Delahay, J. Phys. Chem. **79**, 2935–2940 (1975)
34. W.R. Ronk, D.V. Kowalski, M. Manning, G.M. Nathanson, J. Chem. Phys. **104**, 4842 (1996)
35. M. Manning, J.A. Morgan, D.J. Castro, G.M. Nathanson, J. Chem. Phys. **119**, 12593 (2003)
36. E.R. Waclawik, M.C. Goh, D.J. Donaldson, J. Chem. Phys. **110**, 8098 (1999)
37. W.W. Hayes, J.R. Manson, Phys. Rev. B **89**, 045406 (2014)
38. M. Faubel, S. Schlemmer, J.P. Toennies, Z. Phys. D: At., Mol. Clusters **10**, 269 (1988)
39. K.R. Wilson, B.S. Rude, J. Smith, C. Cappa, D.T. Co, R.D. Schaller, M. Larsson, T. Catalano, R.J. Saykally, Rev. Sci. Instrum. **75**, 725 (2004)
40. B. Winter, M. Faubel, Chem. Rev. **106**, 1176 (2006)
41. D.P. dePonte, U. Weierstall, K. Schmidt, J. Warner, D. Starodub, J.H.C. Spence and R.B. Doak, J. Phys. D: Appl. Phys. **41**, 195505 (2008)
42. U. Buck, R. Krohne, P. Lohbrandt, J. Chem. Phys. **106**, 3205–3215 (1997)
43. J.W. Allen, S.A. Rice, J. Chem. Phys. **67**, 5105 (1977)
44. J.W. Allen, S.A. Rice, J. Chem. Phys **68**, 5053 (1978)
45. N.H. March, M.P. Tosi, Phil. Mag. **28**, 91–102 (1973)
46. N.H. March, M.P. Tosi, Proc. R. Soc. Lond. Ser. A Math. Phys. Sci. **330**, 373 (1972)
47. Ch. Wöll (ed.), in *Organic Electronics: Structural and Electronic Properties of OFETs*, Wiley-VCH Verlag GmbH & Co. KGaA, Weinheim (2009)
48. S. Günes, H. Neugebauer, N.S. Sariciftci, Chem. Rev. **107**, 1134 (2007)
49. V.A. Dediu, L.E. Hueso, I. Bergenti, C. Taliani, Nat. Mat. **8**, 707 (2009)
50. J.E. Anthony, Angew. Chem. Int. Ed. **47**, 452–483 (2008)
51. S.M. Barlow, R. Raval, Surf. Sci. Rep. **50**, 201–341 (2003)
52. J.A.A.W. Elemans, S. Lei, S. DeFeyter, Angew. Chem. Int. Ed. **48**, 7298–7332 (2009)
53. G. Witte, C. Wöll, J. Mat. Res. **19**, 1889–1916 (2004)
54. G. Witte, Ch. Wöll, in *Organic Electronics: Structural and Electronic Properties of OFETs*, Ch. Wöll (ed.) Wiley-VCH Verlag GmbG & Co. KGaA, Weinheim (2009) p. 207
55. J.M. Gottfried, Surf. Sci. Rep. **70**, 259–379 (2015)
56. K.-H. Ernst, Phys. Status Solidi B **249**, 2057–2088 (2012)
57. C.A. Meli, E.F. Greene, G. Lange, J.P. Toennies, Phys. Rev. Lett. **74**, 2054 (1995)
58. A.L. Glebov, J.P. Toennies, S. Vollmer, G. Benedek, Europhys. Lett. **46**(3), 369–375 (1999)
59. P. Gregory, J. Prophyrins Phthalocyanines, 432 (2000)
60. F.S. Tautz, Progr. Surf. Sci. **82**, 479–520 (2007)
61. L. Kilian, A. Hauschild, R. Remirov, S. Soubatch, A. Schöll, A. Bendounan, F. Reinert, T.-L. Lee, F.S. Tautz, M. Sokolowski, E. Umbach, Phys. Rev. Lett. **100**, 136103 (2008)
62. C.R. Braats, G. Öhl, P. Jakob, J. Chem. Phys. **136**, 134706 (2012)
63. S. Duhm, A. Gerlach, I. Salzmann, B. Bröler, R.L. Johnson, F. Schreiber, N. Koch, Science Direct. Org. Electron. **9**, 111–118 (2008)
64. L. Romaner, D. Nabok, P. Puschnig, E. Zojer, C. Ambrosch-Draxl, New J. Phys. **11**, 053010 (2009)
65. Y. Zou, L. Kilian, A. Schöll, Th. Schmidt, R. Fink, E. Umbach, Surf. Sci **600**, 1240 (2006)
66. B. Stadtmüller, M. Gruenewald, J. Peuker, R. Forker, T. Fritz, C. Kumpf, J. Phys. Chem. **118**, 28592–28602 (2014)
67. J.A. Lau, Masterarbeit, Georg-August-Universität Göttingen (2016)

68. D. Fuhrmann, Bericht 14/1998, Max-Planck-Institut für Strömungsforchung, Göttingen (1998)
69. Y. Wang, K. Wu, J. Kröger, R. Berndt, AIP Adv. **2**, 041402 (2012)
70. A. Paciaroni, A. Orecchini, M. Haertlein, M. Moulin, N. Conti, A. DeFrancesco, C. Petrillo, F. Sacchetti, J. Phys. Chem. B **116**, 3861–3865 (2012)
71. G. Acbas, K.A. Niessen, E.H. Snell, A.G. Markelz, Nat. Commun. **5**, 3076 (2014)
72. B.S. Kalanoor, M. Ronen, Z. Oren, D. Gerber, Y.R. Tischler, ACS Omega **2**, 1232 (2017)
73. M. Diehl, W. Doster, W. Petry, H. Schober, Biophys. J. **73**, 2726–2732 (1997)
74. B. Kasemo, Surf. Sci. **500**, 656–677 (2002)
75. J.V. Barth, Annu. Rev. Phys. Chem. **58**, 375–407 (2007)
76. D. Costa, C.-M. Pradier, F. Tielens, L. Savio, Surf. Sci. Rep. **70**, 449–553 (2015)
77. A. Schiffrin, A. Riemann, W. Auwärter, Y. Pennec, A. Weber-Bargioni, D. Devetko, A. Cossaro, A. Morgante, J.V. Barth, PNAS **13**, 5279–5284 (2007)
78. J.-F. Lambert, Orig. Life Evol. Biosph. **38**, 211–242 (2008)
79. J.D. Bernal, *The Physical Basis of Life* (Routledge and Kegan Paul, London, 1951)
80. S. Rauschenbach, M. Ternes, L. Harnau, K. Kern, Annu. Rev. Annal. Chem. **9**, 16.1–16.26 (2016)
81. S. Abb, L. Harnau, R. Gutzler, S. Rauschenbach, K. Kermn, Nat. Commun. **10335**, 1–7 (2015)
82. S. Rauschenbach, G. Rinke, R. Gutzler, S. Abb, A. Albarhash, D. Le, T.S. Rahman, M. Dürr, L. Harnau, K. Kern, ACS Nano **11**, 2420–2427 (2017)
83. G. Rinke, S. Rauschenbach, L. Harnau, A. Albarghash, M. Pauly, K. Kern, Nano Lett. **14**, 5609–5615 (2014)
84. G.V. Hartland, Chem. Rev. **111**, 3858 (2011)
85. A.E. Saunders, B.A. Korgel, in *Handbook of Nanophysics: Nanoparticles and Quantum Dots*, ed. K.D. Sattler (CRC Press, 2010) p. 20/1
86. N. Goubet, M.P. Pileni, J. Phys. Chem. Lett. **2**, 1024–1031 (2011)
87. M.P. Pileni, J. Phys.: Condens. Matter **23**, 503102 (2011)
88. M.P. Pileni, Account. Chem. Res. **45**, 1965–1972 (2012)
89. M.P. Pileni, J. Col. Interf. Sci. **388**, 1–8 (2012)
90. M.P. Pileni, EPL **109**, 58001 (2015)
91. A. Courty, I. Lisiecki, M.P. Pileni, J. Chem. Phys. **116**, 8074 (2002)
92. V. Mankad, P.K. Jha, T.R. Ravindran, J. Appl. Phys. **113**, 074303 (2013)
93. L. Saviot, D.B. Murray, Phys. Rev. B **79**, 214101 (2009)
94. J.J. Wie, P. Yang, H. Portales, P.-A. Albouy, M.P. Pileni, J. Phys. Chem. C, 13732–13738 (2016)
95. P. Piseri, H.V. Tafreshi, P. Milani, Curr. Op. Solid State and Mat. Sci. **8**, 195–202 (2004)
96. J.B. Chen, J.F. Zhou, A. Häfele, C.R. Yin, W. Kronmüller, M. Han, H. Haberland, Eur. Phys. J. D **34**, 251 (2005)
97. A. Glebov, V. Senz, J.P. Toennies, Ph. Lambin, G. Gensterblum, P. Senet, A.A. Lucas, Phys. Rev. B **56**, 9874 (1997)
98. A. Glebov, V. Senz, J.P. Toennies, G. Gensterblum, J. Appl. Phys. **82**, 2329 (1997)
99. E. Fermi, Nuovo Cimento **4**, 157 (1934)
100. U. Paltzer, D. Schmicker, F.W. de Wette, U. Schröder, J.P. Toennies, Phys. Rev B **54**, 11989 (1996)
101. U. Harten, J.P. Toennies, Ch. Wöll, Faraday Discuss. Chem. Soc. **80**, 137 (1985)
102. C.S. Jayanthi, H. Bilz, W. Kress, G. Benedek, Phys. Rev. Lett. **59**, 795 (1987)
103. G. Benedek, D. Campi, J.P. Toennies, Interaction of atoms with surfaces and surface phonons, in *Physics of Solid Surfaces*, Subvol A, Landolt-Börnstein, Numerical Data and Functional Relationships in Science and Technology – New Ser., Subvol. 45A, ed. by P. Chiaradia, G.F. Chiarotti, Chap. 10, 572–646 (2015) ISSN 1615-1925; ISBN 978-3-662-47735-9
104. S. Baroni, S. De Gironcoli, A. Dal Corso, Rev. Mod. Phys. **73**, 515 (2001)

105. V. Chis, B. Hellsing, G. Benedek, M. Bernasconi, E.V. Chulkov, J.P. Toennies, Phys. Rev. Letters, **101**, 206102 (2008); *ibidem* **103**, 069902 (2009) (E)
106. I.Yu. Sklyadneva, G. Benedek, E.V. Chulkov, P.M. Echenique, R. Heid, K.-P. Bohnen, J. P. Toennies, Phys. Rev. Lett. **107**, 095502 (2011)
107. P. Giannozzi, S. Baroni, N. Bonini, M. Calandra, R. Car, C. Cavazzoni, D. Ceresoli, G.L. Chiarotti, M. Cococcioni, I. Dabo, A. Dal Corso, S. De Gironcoli, S. Fabris, G. Fratesi, R. Gebauer, U. Gerstmann, C. Gougoussis, A. Kokalj, M. Lazzeri, L. Martin-Samos, N. Marzari, F. Mauri, R. Mazzarello, S. Paolini, A. Pasquarello, L. Paulatto, C. Sbraccia, S. Scandolo, G. Sclauzero, A.P. Seitsonen, A. Smogunov, P. Umari, R.M. Wentzcovitch, J. Phys.:Condens. Matter, **2009**, 21, 395502
108. X. Gonze et al., Comput. Phys. Commun. **180**, 2852 (2009)
109. G. Kresse, J. Furtmüller, Phys. Rev. B **54**, 11169 (1996)
110. Clark, S.J., Segall, M.D., Pickard, C.J., Probert, P.J., Refson, K., Payne, M.C.: Z. Kristall. **220**(5–6), 567 (2005)
111. D. Vanderbilt, Phys. Rev. B **32**, 8412 (1985)
112. J.P. Perdew, K. Burke, M. Ernzerhof, Phys. Rev. Lett. **77**, 3865 (1996)
113. R. Heid, K.P. Bohnen, I.Y. Sklyadneva, E.V. Chulkov, Phys. Rev. B **81**, 174527 (2010)
114. S. Grimme, J. Comp. Chem. **27**, 1787 (2006)
115. D. Campi, M. Bernasconi, G. Benedek, J.P. Toennies, J. Phys. Chem. C **119**(26), 14579 (2015)
116. J.M. Pitarke, V.M. Silkin, E.V. Chulkov, P.M. Echenique, Rep. Prog. Phys. **70**, 1 (2007)
117. V.M. Silkin, J.M. Pitarke, E.V. Chulkov, P.M. Echenique, Phys. Rev. B **72**, 115435 (2005)
118. B. Diaconescu, K. Pohl, L. Vattuone, L. Savio, Ph Hofmann, V.M. Silkin, J.M. Pitarke, E.V. Chulkov, P.M. Echenique, D. Farías, M. Rocca, Nature **448**, 57 (2007)
119. G.J. Kroes, Phys. Chem. Chem. Phys. **14**, 14966 (2012)
120. D.J. Auerbach, A.M. Wodtke, in *Dynamics of Gas-Surface Interactions*, ed. by R. Díez Muiño, H.F. Busnengo, Springer Ser. Surf. Sci. 50 (Springer, Berlin, 2013) p. 267
121. R. Díez Muiño, H.F. Busnengo (eds.), *Dynamics of Gas-Surface Interactions* (Springer, Berlin, 2013)
122. K. Shakouri, J. Behler, J. Meyer, G.-J. Kroes, J. Phys. Chem. Lett. **8**, 2131 (2017)
123. O. Buenermann, H. Jiang, Y. Dorenkamp, A. Kandratsenka, S.M. Janke, D.J. Auerbach, A. M. Wodtke, Science **350**, 1346 (2015)

Correction to: Theory of Atom Scattering from Surface Phonons: Basic Concepts and Temperature Effects

Correction to:
Chapter 7 in: G. Benedek and J. P. Toennies,
***Atomic Scale Dynamics at Surfaces*,**
Springer Series in Surface Sciences 63,
https://doi.org/10.1007/978-3-662-56443-1_7

Equations (7.33a) and (7.82) of Chapter 7 were incorrect. These had been corrected.

The updated version of this chapter can be found at
https://doi.org/10.1007/978-3-662-56443-1_7

G. Benedek and J. P. Toennies, *Atomic Scale Dynamics at Surfaces*, Springer Series in Surface Sciences 63, https://doi.org/10.1007/978-3-662-56443-1_15

Appendix A
Tables of HAS Surface Phonon Experiments and Related Theories

Introduction to the Appendices A.1–A.8

As a complement to the discussion of the surface phonon dispersion curves of clean surfaces in Chap. 11, the Appendices A.1–A.4 provide a complete list as of 2017 of all the surfaces for which HAS data are available in the literature or in reports. The surfaces are listed in alphabetic order according to the chemical formula, then according to the surface faces and structure. For each system the method used in the available theoretical calculations are also listed with references. The different types of solids are collected in different tables as follows: insulator surfaces (A.1), semiconductor surfaces (A.2), layered crystals (A.3), metal, metal alloy and elemental semi-metal surfaces (A.4), according to the order of their appearance in Chap. 11. The HAS measurements of the surface phonons of diamond and graphite surfaces are listed in Table A.2 with the semiconductor surfaces. The available SE-HAS and NeAS measurements are included in the tables and marked by * and **, respectively. Measurements of SH branches of NaCl(001) with in-plane ortho-H_2 rotation-flip scattering spectroscopy are also included in Appendix A.1.

The Appendices A.5–A.8 provide an analogous list as of 2017 of the available HAS data on the dynamical characterization of atomic and molecular overlayers, ordered adsorbates and ultrathin films on different substrates. The data and references are listed in alphabetic order according to, first, the chemical formula of the substrate, the surface face, and then according to the adsorbate atom or molecule and the film thickness of the substrate. The data for different types of substrates are in different tables as follows: adsorbates on insulator surfaces (A.5), adsorbates on metal and semimetal surfaces (A.6), diatomic adsorbates on metal surfaces (A.7), and polyatomic adsorbates on metal surfaces (A.8) according to their appearance in

G. Benedek and J. P. Toennies, *Atomic Scale Dynamics at Surfaces*, Springer Series in Surface Sciences 63, https://doi.org/10.1007/978-3-662-56443-1

Chap. 12. The data on the low-energy dynamics of graphene are listed in Appendix A.8 together with the large polyaromatic adsorbates.

Acronyms Used in Appendices A.1–A.8

Acronyms for inelastic atom-scattering experimental methods for measuring the surface phonon dispersion curves, the surface phonon density and other surface dynamical processes

HAS	Helium atom scattering (also IHAS: inelastic HAS) spectroscopy
SE-HAS	3He spin echo scattering spectroscopy (references with *)
NeAS	Neon atom scattering (references with **)

Acronyms for theoretical methods for the calculation of the surface phonon dispersion curves

3D	Three-dimensional (also referred to film growth regime)
BCM	Bond charge model
BMB	Brownian molecular dynamics
BSGF	Breathing-shell-model Green's function calculation
BSM	Breathing shell model
CE	Continuum elastic theory
CFM	Continued fraction method
CPM	Car-Parrinello method
DFPT	Density functional perturbation theory
DLC	Dispersive linear chain
DR	Dielectric response theory
EA	Embedded atom method
EAMD	Embedded atom molecular dynamics
EBOA	Extended band orbital approximation
EM	Effective medium theory
FC	Force constant model
FP	First principle calculation
FS	Finnis-Sinclair potentials
GF	Green's function method
GGA	Generalized gradient approximation
GM	Glue model
HF	Helmann-Feynman forces
KM	Keating model
LCT	Local coupling transfer method
LDA	Local density approximation
MD	Molecular dynamics
ME	Multipole expansion method
PC	Pseudocharge model
PDF	Partial density formalism

RI	Rigid ion model
SAM	Self-assembled monolayer
SCMBP	Sutton-Chen many body potential
SETE	Semi-empirical total-energy approach:
SM	Shell model
SMM	Scattering matrix method
SO	Spin-orbit (interaction)
TB	Tight-binding
TBSM	Tight-binding second-moment semi-empirical total energy method
vdW	Van der Waals (interaction, forces, potential)
VFF	Valence force field

Appendix A.1

Table A.1 Surface phonon dispersion curves of insulator surfaces from HAS measurements and theory

Surface	Structure	HAS experiment	Theory	
		References	Method	References
AgBr(001)	fcc	[84]	–	–
Ar(*lmn*)	Clusters	[76]	–	–
BaF_2(111)	fcc	[77]	–	–
CaF_2(111)	fcc	[70]	RI	[16]
			SM	[62]
CoO(001)	fcc	[68, 81]	FC	[81]
FeO(001)	fcc	[83]		
H_2O(0001)	hexagonal	[78, 87]	FC FP–MD	[87] [106]
$KAl_3Si_3O_{10}(OH)_2$ (001)	Monoclinic muscovite mica	[69]		
KBr(001)	fcc	[44, 50]	BSGF	[50, 22, 56]
			SM	[32, 38]
KBr(001)	fcc, film on NaCl(001)	[57, 63]	FC	[63]
KBr(001)	fcc, film on RbCl(001)	[64, 65]	SM	[65]
KCl(001)	fcc	[20, 18, 25, 101]	BSGF	[22, 56, 40]
			SM DFPT	[32, 38] [109]
KCN(001)	fcc	[67]		
$KMnF_3$(001)	Cubic	[60]	SM	[41, 42, 53]
$KTaO_3$(001)	Cubic	[110]	FP-MD	[110]

(continued)

Table A.1 (continued)

Surface	Structure	HAS experiment	Theory	
		References	Method	References
LiF(001)	fcc	[5**, 15, 19, 25, 36, 47]	BSGF	[12, 22, 24, 56]
			SM	[6, 13, 27, 38, 39, 61]
MgO(001)	fcc	[26, 28, 46, 58, 90]	NIS	[1]
			RI	[9, 17]
			BSM	[14, 94]
			SM	[11, 13, 21, 31, 32, 38, 61]
NaCl(001)	fcc	[23, 54, 51, 59, 72, 74,75, 73, 98 (ortho H_2)]	BSGF	[10, 12, 22, 23, 24, 56, 54, 59]
			BSM	[73, 74, 72]
			RI	[3, 4]
			RI-SMM	[2, 7]
			SM	[7, 13, 61]
NaF(001)	fcc	[26, 30, 34]	BSGF	[56, 29, 40, 22]
			SM	[13, 33, 61]
NaI(001)	fcc	[55]	BSGF	[56, 40, 12]
			SM	[13]
NiO(001)	fcc	[45, 66, 82]	FC	[82]
			MD-RI	[89]
RbBr(001)	fcc	[45, 52]	SM	[43, 38, 61]
RbCl(001)	fcc	[45, 100]	SM	[7, 61]
RbI(001)	fcc	[52]	SM	[32, 38, 39, 61]
α-SiO_2(0001)	α-quartz	[102]	SM	[102, 111]
SiO_2	Amorphous	[99, 103, 104, 105]	–	–
TiO_2(110)	rutile	[95]	–	–
Xe(111)	fcc 160 ML/Pt (111)	[93, 92]	DFPT	[107]

(**: NeAS data; diamond and graphite surfaces in Table A.2; graphene in Table A.8)

References

1. K.H. Rieder, E.M. Hörl, Phys. Rev. Lett. **20**, 200 (1968)
2. B.N.N. Achar, G.R. Barsch, Phys. Rev. **188**, 1361 (1969)
3. S.Y. Tong, A.A. Maradudin, Phys. Rev. **181**, 1318 (1969).
4. W.E. Jones, W.E. Fuchs, Phys. Rev. B **4**, 3581 (1971).
5. B. R. Williams, J. Chem. Phys. **55**, 1315 (1971)**

6. T.S. Chen, G.P. Alldredge, F.W. de Wette, R.E. Allen, Phys. Rev. B **6**, 627 (1972)
7. T.S. Chen, G.P. Alldredge, F.W. de Wette, Solid State Commun. **10**, 941 (1972)
8. T.S. Chen, G.P. Alldredge, F.W. de Wette, Phys. Lett. A **40**, 401 (1972)
9. L. Genzel, T.P Martin, Phys. Status Solidi (b) **51**, 101 (1972)
10. G. Benedek, Phys. Status Solidi B **58**, 661 (1973)
11. T.S. Chen, G.P. Alldredge, F.W. de Wette, Phys. Lett. A **46**, 91 (1973)
12. G. Benedek, Surf. Sci. **61**, 603 (1976)
13. T.S. Chen, F.W. de Wette, G.P. Alldredge, Phys. Rev. B **15**, 1167 (1977)
14. R.N. Barnett, R. Bass, Phys. Rev. B **19**, 4259 (1979)
15. G. Brusdeylins, R.B. Doak, J.P. Toennies, Phys. Rev. Lett. **44**, 1417 (1980)
16. G. Laksmi, F.W. de Wette, R. Srinivaran, Surf. Sci. **94**, 232 (1980)
17. G. Laksmi, F.W. de Wette, Phys. Rev. B **22**, 5009 (1980)
18. G. Benedek, U. Valbusa, (eds.), in *Dynamics of Gas-Surface Interaction*. Springer Series in Chemical Physics, vol. 21 (Springer, Berlin, Heidelberg, 1981)
19. G. Brusdeylins, R.B. Doak, J.P. Toennies, Phys. Rev. Lett. **46**, 437 (1981)
20. R.B. Doak, Max-Planck-Institut für Stömungsforschung, Göttingen, Report 14, 1981.
21. G. Lakshmi, F.W. de Wette, Phys. Rev. B **23**, 4208 (1981)
22. G. Benedek, G.P. Brivio, L. Miglio, V.R. Velasco, Phys. Rev. B **26**, 497 (1982)
23. G. Benedek, G. Brusdeylins, R.B. Doak, J.G. Skofronick, J.P. Toennies, Phys. Rev. B **28**, 2104 (1983)
24. G. Benedek, L. Miglio, in *Ab-Initio Calculation of Phonon Spectra*, ed. by J.T. Devreese, V. E. van Doren, P.E. Van Camp (Plenum, New York, 1983)
25. G. Brusdeylins, R.B. Doak, J.P. Toennies, Phys. Rev. B **27**, 3662 (1983)
26. G. Brusdeylins, R.B. Doak, J.G. Skofronick, J.P. Toennies, Surf. Sci. **128**, 191 (1983)
27. E.R. Cowley, J.A. Barker, Phys. Rev. B **28**, 3124 (1983)
28. P. Cantini, E. Cevasco, Surf. Sci. **148**, 37 (1984)
29. G. Platero, V.R. Velasco, F. Garcia-Moliner, G. Benedek, L. Miglio, Surf. Sci. **143**, 244 (1984)
30. G. Brusdeylins, R. Rechsteiner, J.G. Skofronick, J.P. Toennies, G. Benedek, L. Miglio, Phys. Rev. Lett. **54**, 466 (1985)
31. P.A. Masym, Surf. Sci. **149**, 157 (1985)
32. F.W. de Wette, W. Kress, U. Schröder, Phys. Rev. B **32**, 4143 (1985)
33. F.W. de Wette, W. Kress, U. Schröder, Phys. Rev. B **33**, 2835 (1986)
34. G. Benedek, L. Miglio, G. Brusdeylins, J.G. Skofronick, J.P. Toennies, Phys. Rev. B **35**, 6593 (1987)
35. G. Benedek, L. Miglio, J.G. Skofronick, G. Brusdeylins, C. Heimlich, J.P. Toennies, J. Vac. Sci. Technol. A **5**, 1093 (1987)
36. G. Bracco, M. D'Avanzo, C. Salvo, R. Tatarek, F. Tommasini, Surf. Sci. **189/190**, 684 (1987)
37. G. Bracco, R. Tatarek, F. Tommasini, U. Linke, M. Persson, Phys. Rev. B **36** (1987)
38. W. Kress, F.W. de Wette, A.D. Kulkarni, U. Schröder, Phys. Rev. B **35**,, 5783 (1987)
39. W. Kress, F.W. de Wette, A.D. Kulkarni, U. Schröder, Phys. Rev. B **35**,, 2467 (1987)
40. L. Miglio, G. Benedek, in *Structure and Dynamics of Surfaces II*, ed. by W. Schommers, P. von Blanckenhagen (Springer, Heidelberg, 1987), p. 35
41. R. Reiger, J. Prade, U. Schröder, F.W. de Wette, W. Kress, J. Electr. Spectr. Rel. Phenom. **44**, 403 (1987)
42. T.L. Reinecke, S.C. Ying, Phys. Rev. Lett. **58**, 242 (1987)
43. F.W. de Wette, A.D. Kulkarni, U. Schröder, W. Kress, Phys. Rev. B **35**, 2467 (1987)
44. G. Chern, J.G. Skofronick, W.P. Brug, S.A. Safron, Phys. Rev. B **39**, 12828 (1989)
45. G. Chern, W.P. Brug, S.A. Safron, J.G. Skofronick, J. Vac. Sci. Technol. A **7**, 2091 (1989)
46. J. Cui, D.R. Jung, D.R. Frankl, Phys. Rev. B **42**, 9701 (1990)
47. R.B. Doak, Superlattices Microstr. **7**, 201 (1990)

48. R.B. Doak, Y.J. Chabal, G.S. Iligashi, P. Dumas, J. Electr. Spectr. Rel. Phen. **54/55**, 291 (1990)
49. R.B. Doak, J. Electr. Spectr. Rel. Phen. **54/55**, 281 (1990)
50. S.A. Safron, G. Chern, W.P. Brug, J.G. Skofronick, G. Benedek, Phys. Rev. B **41**, 10146 (1990)
51. S.A. Safron, W.P. Brug, G. Chern, J. Duan, J.G. Skofronick, J. Vac. Sci. Technol. A **8**, 2627 (1990)
52. S.A. Safron, W.P. Brug, G.G. Bishop, J. Duan, G. Chern, J.G. Skofronick, J. El. Spectr. Rel. Phenom. **54/55**, 343 (1990)
53. U. Schröder, J. Prade, F.W. de Wette, A.D. Kulkarni, W. Kress, Superlattices Microstruct. **7**, 247 (1990)
54. J.G. Skofronick, G.G. Bishop, W.P. Brug, G. Chern, J. Duan, S.A. Safron, Superlattices Microstruct. **7**, 239 (1990)
55. J.G. Skofronick, W.P. Brug, G. Chern, J. Duan, S.A. Safron,G. Benedek, J. Vac. Sci. Technol. A **8**, 2632 (1990)
56. G. Benedek, L. Miglio, in *Surface Phonons,* ed. by de F.W. Wette, W. Kress (Springer, Berlin, Heidelberg, 1991), p. 37
57. J. Duan,G.G. Bishop, E.S. Gillman, G. Chern, S.A. Safron, J.G. Skofronick, Surf. Sci. **272**, 220 (1991)
58. D.R. Jung, J. Coi, D.R. Frankl, J. Vac. Sci. Technol. A **9**, 1589 (1991)
59. S.A. Safron, W.P. Brug, G. Chern, J. Duan, J.G. Skofronick, Vac. Sci. Technol. A **9**, 1657 (1991)
60. J.P. Toennies, R. Vollmer, Phys. Rev. B **44**, 9833 (1991)
61. F.W. de Wette, in *Surface Phonons,* ed. by W. Kress, F.W. de Wette (Springer, Heidelberg, 1991), p. 67
62. A. Jockisch, U. Schröder, F.W. de Wette, W. Kress, J. Phys.: Condens. Matter **5**, 5401 (1993)
63. S.A. Safron, G.G. Bishop, J. Duan, E.S. Gillman, J.G. Skofronick, N.S. Luo, P. Ruggerone, J. Chem. Phys. **97**, 2270 (1993)
64. S.A. Safron, J. Duan, G.G. Bishop, E.S. Gillmann, J.G. Skofronick, J. Phys. Chem. **97**, 1749 (1993)
65. J.G. Skofronick, G.G. Bishop, J. Duan, E.S. Gillman, S.A. Safron, D. Bonart, C. Schröder, J. Electr. Spectr. Rel. Phenom. **64/65**, 747 (1993)
66. J.P. Toennies, G. Witte, A.M. Shikin, K.H. Rieder, J. Electr. Spectr. Rel. Phenom. **64/65**, 677 (1993)
67. E.S. Gillman, J.J. Hernandez, G.G. Bishop, J. Baker, S.A. Safron, J.G. Skofronick, J.R. Manson, S. Weera, Bull. Am. Phys. Soc. **39**, 254 (1994)
68. J.J. Hernandez, G.G. Bishop, E.S. Gillman, J. Baker, J.G. Skofronick, S.A. Safron, Bull. Am. Phys. Soc. **39**, 466 (1994)
69. G. Brusdeylins, D. Schmicker, Surf. Sci. **331–333**, 237 (1995)
70. G. König, A. Lehmann, K.H. Rieder, Surf. Sci. **331–333**, 1430 (1995)
71. S. Weera, J.R. Manson, J. Baker, E.S. Gillman, J.J. Hernandez, G.G. Bishop, S.A. Safron, J. G. Skofronick, Phys. Rev. B **52**, 14185 (1995)
72. G. Benedek, R. Gerlach, A. Glebov, G. Lange, W. Silvestri, J.G. Skofronick, J.P. Toennies, Vuoto **25**, 26 (1996)
73. G. Benedek, A. Glebov, W. Silvestri, J.G. Skofronick, J. Vac. Sci. Technol. A **14**, 1522 (1996)
74. A. Glebov, W. Silvestri, J.P. Toennies, G. Benedek, J.G. Skofronick, Phys. Rev. B **54**, 17866 (1996)
75. A. Glebov, J.P. Toennies, H. Weiss, Surf. Sci. **351**, 200 (1996)
76. U. Buck, R. Krohne, P. Lohbrandt, J. Chem. Phys. **106**, 3205 (1997)
77. A. Lehmann, G. Fahsold, G. Konig, K.-H. Rieder, Surf. Sci. **369**, 289 (1997)

78. J. Braun, A. Glebov, A.P. Graham, A. Menzel, J.P. Toennies, Phys. Rev. Lett. **80**, 2638 (1998)
79. J. Braun, D. Fuhrmann, A. Šiber, B. Gumhalter,C. Wöll, Phys. Rev. Lett. **80**, 125 (1998)
80. J. Braun, J. Weckesser, J. Ahner, D. Mocuta, J.T. Yates, C. Wöll, J. Chem. Phys. **108**, 5161 (1998)
81. G. Witte, P. Senet, J.P. Toennies, Phys. Rev. B **58**, 13264 (1998)
82. G. Witte, K. Weiss, P. Jakob, J. Braun, K.L. Kostov, C. Wöll, Phys. Rev. Lett. **80**, 121 (1998)
83. G. Fahsold, G. Konig, W. Theis, A. Lehmann, K.-H. Rieder, Appl. Surf. Sci. **137**, 224 (1999)
84. A. Glebov, U. Panella, J.P. Toennies, Phys. Rev. B **60**, 2046 (1999)
85. A.L. Glebov, J.P. Toennies, S. Vollmer, G. Benedek, Europhys. Lett. **46**, 369 (1999)
86. A. Glebov, A.P. Graham, A. Menzel, Surf. Sci. **427**, 22 (1999)
87. A. Glebov, A.P. Graham, A. Menzel, J.P. Toennies, P. Senet, J. Chem. Phys. **112**, 11011 (2000).
88. A.L. Glebov, V. Panella, J.P. Toennies, F. Traeger, H. Weiss, S. Picaud, P.N.M. Hoang, C. Girardet, Phys. Rev. B **61**, 14028 (2000)
89. T.E. Karakasidis, D.G. Papageorgiou, G.A. Evangelakis, Appl. Surf. Sci. **162**, 233 (2000)
90. G. Benedek, G. Brusdeylins,V. Senz, J.G. Skofronick, J.P. Toennies, F. Traeger, R. Vollmer, Phys. Rev. B **64**, 125421 (2001)
91. G. Benedek, E. Hulpke, W. Steinhögl, Phys. Rev. Lett. **87**, 027201 (2001)
92. B. Gumhalter, Phys. Rep. **351**, 1–159 (2001)
93. A. Šiber, B. Gumhalter, A.P. Graham, J.P. Toennies, Phys. Rev. B **63**, 115411 (2001)
94. B. Wassermann, K.-H. Rieder, Phys. Rev. Lett. **88**, 045501 (2002)
95. E.A. Akhadov, S.A. Safron, J.G. Skofronick, D.H. Van Winkle, F.A. Flaherty, R. Fatema, Phys. Rev. B **68**, 035409 (2003)
96. E.A. Akhadov, T.W. Trelenberg, S.A. Safron, J.G. Skofronick, D.H. Van Winkle, F.A. Flaherty, W. Theis, Phys. Rev. B **67**, 113406 (2003)
97. J.G. Skofronick, J.P. Toennies, F. Traeger, H. Weiss, Phys. Rev. B **67**, 035413 (2003)
98. G. Benedek, F. Traeger, J.P. Toennies, Phys. Rev. Lett. **94**, 086103 (2005)
99. W. Steurer, A. Apfolter,M. Koch, W.E. Ernst, E. Søndergård, J.R. Manson, B. Holst, Phys. Rev. Lett. **99**, 035503 (2007)
100. G.G., Bishop, J. Duan, E.S. Gillman, S.A. Safron, J.G. Skofronick, J. Vac. Sci. Technol. **11**, 2008 (1993)
101. R. Fatema, D.H. VanWinkle, J.G. Skofronick, S.A. Safron, F.A. Flaherty, Phys. Rev. B **77**, 024305 (2008)
102. W. Steurer, A. Apfolter, M. Koch, W.E. Ernst, B. Holst, E. Søndergård, S.C. Parker, Phys. Rev. B **78**, 035402 (2008)
103. W. Steurer, A. Apfolter, M. Koch, W.E. Ernst, E. Søndergård, J.R. Manson, B. Holst, Phys. Rev. Lett. **100**, 135504 (2008)
104. W. Steurer, A. Apfolter, M. Koch, W.E. Ernst, E. Søndergård, J.R. Manson, B. Holst, Phys. Rev. B **78**, 045427 (2008)
105. W. Steurer, A. Apfolter, M. Koch, W.E. Ernst, E. Søndergård, J.R. Manson, B. Holst, J. Phys.: Condens. Matter **20**, 224003 (2008)
106. T.A. Pham, P. Huang, E. Schwegler, G. Galli, J. Phys. Chem. A **116**, 9255 (2012)
107. D. Campi, M. Bernasconi, G. Benedek, J.P. Toennies, J. Phys. Chem. C **119**, 14579 (2015)
108. F.A. Flaherty, T.W. Trelenberg, J.A. Li, R. Fatema, J.G. Skofronick, D.H. Van Winkle, S.A. Safron, L.A. Boatner, Phys. Rev. B **92**, 035414 (2015)
109. H.M. Tütüncü, G.P. Srivastava, Surf. Sci. **637–638**, 101 (2015)
110. J.A. Li, E.A. Akhadov, J. Baker, L.A. Boatner, D. Bonart, J. Fritsch, S.A. Safron, U. Schröder, J.G. Skofronick, T.W. Trelenberg, Phys. Rev. **68**, 045402 (2003)
111. C. Büchner, S.D. Eder, T. Nesse, D. Kuhness, P. Schlexer, G. Pacchioni, J.R. Manson, M. Heyde, B. Holst, H.-J. Freund, Phys. Rev. Lett. **120**, 226101 (2018)

Appendix A.2

Table A.2 Surface phonon dispersion curves of semiconductor surfaces from HAS measurements and theory. Graphite is included. For graphene data see Table A.8

SURFACE	Structure	HAS experiment	Theory	
		References	Method	References
C(111):H(1x1)	fcc diamond	[54]	SETE	[66, 47, 48, 73]
			FP	[37]
C(111):D(1x1)	fcc diamond	[73]	SETE	[73]
C(0001)	hex graphite	[4, 5, 9, 35, 87]	FC	[2, 6]
			BCM	[38, 40]
			FP-MD	[46]
C_{60}(111)	fcc	[62]	FC	[62, 88]
GaAs(110)	fcc	[14, 13, 30]	BCM	[30, 58, 68, 72]
			FC	[27, 1]
			DFPT	[39, 60, 61, 77]
			CPM	[44]
			FP-HF	[49, 55]
			TB	[17, 22, 33]
GaP(110)	fcc	[83]	BCM	[81]
			TB	[17]
			FP-HF	[49, 55]
			DFPT	[59, 61, 77]
Ge(001) p(2x1)	fcc	[82, 67]	DFPT	[67, 74]
			BCM	[75]
			FC-MD	[36]
Ge(001) p(2x2)	fcc	[67]	DFPT-LCT	[67]
			DFPT	[74]
Ge(001) c(4x2)	fcc	[67]	DFPT-LCT	[67]
			DFPT	[74]
Ge(111) (1x1)	fcc (high T)	[79]	BCM	[79]
Ge(111) p(2x1)	fcc	[50]	BCM	[34]
Ge(111) (2x8)	fcc (room T)	[85]	FP-MD	[90]
InAs(110)	fcc	[83]	BCM	[68, 69]
			TB	[17]
			FP-HF	[49]
			DFPT	[52, 59, 60, 77]
InSb(110)	fcc	[43]	DFPT	[43]
			BCM	[76]

(continued)

Table A.2 (continued)

SURFACE	Structure	HAS experiment	Theory	
		References	Method	References
Si(111) (1x1)	fcc diamond (high temperature)	[65]	FC-GF	[7]
			FC	[24]
			EBOA	[23]
			FP	[84]
Si(111):H(1x1)	fcc diamond	[16, 28, 29]	BCM	[16]
			CPM	[42]
			DFPT	[53, 64]
			SETE	[47, 48]
Si(111) (2x1)	fcc diamond	[10]	BCM	[25]
			CPM	[26, 31, 32]
			EBOA	[23]
			TB	[8, 11, 12, 15]
			FP	[84]
			PDF	[41]
			DFPT	[70, 89]
Si(111) (7x7)	fcc diamond	[71]	FC	[71, 72]
			CPM, FP-TB	[45]
			CPM	[51, 56, 57]
α-SiO_2(0001)	α-quartz	[86]	SM	[86]

References

1. V. Bortolani, F. Nizzoli, G. Santoro, Phys. Rev. Lett. **41**, 39 (1978)
2. F. de Rouffignac, G.P. Alldredge, F.W. de Wette, Phys. Rev. B **23**, 4208 (1981)
3. A.D. Boardman (ed.), *Electromagnetic Surface Modes* (Wiley, New York, 1982)
4. P. Cantini, R. Tatarek, Surf. Sci. **114**, 471 (1982)
5. P. Cantini, in *Dynamics of Gas Surface Interactions,* ed. by G. Benedek, U. Valbusa (Springer, Heidelberg, 1982) p. 273
6. G.P. Alldredge, E. de Rouffignac, B. Firey, F.W. de Wette, Phys. Rev. B **29**, 3712 (1984)
7. W. Goldhammer, W. Ludwig, W. Zierau, C. Falter, Surf. Sci. **141**, 138 (1984)
8. O.L. Alerhand, D.C. Allan, E.J. Mele, Phys. Rev. Lett. **55**, 2700 (1985)
9. G. Benedek, G. Brusdeylins, C. Heimlich, J.P. Toennies, U. Valbusa, Surf. Sci. **178**, 545 (1986)
10. U. Harten, J.P. Toennies, Ch. Wöll, Phys. Rev. Lett. **57**, 2947 (1986)
11. O.L. Alerhand, E.J. Mele, Phys. Rev. B **35**, 5533 (1987)
12. O.L. Alerhand, E.J. Mele, Phys. Rev. Lett. **59**, 657 (1987)
13. R.B. Doak, D.B. Nguyen, J. Electr. Spectr. Rel. Phenom. **44**, 205 (1987)
14. U. Harten, J.P. Toennies, Europhys. Lett. **4**, 833 (1987)
15. O.L. Alerhand, E.J. Mele, Phys. Rev. B **37**, 2536 (1988)
16. U. Harten, J.P. Toennies, C. Wöll, L. Miglio, P. Ruggerone, L. Colombo, G. Benedek, Phys. Rev. B **38**, 3305 (1988)
17. Y.R. Wang, C.B. Duke, Surf. Sci. **205**, L755 (1988)

18. X.Q. Wang, G.L. Chiarotti, F. Ercolessi, E. Tosatti, Phys. Rev. B **38**, 8131 (1988)
19. X.W. Wang, C.T. Chan, K.-H. Ho, W. Weber, Rev. Lett **60**, 2066 (1988)
20. C.Z. Wang, E. Tosatti, A. Fasolino, Phys. Rev. Lett. **60**, 2661 (1988)
21. C.Z. Wang, A. Fasolino, E. Tosatti, Phys. Rev. B **37**, 2116 (1988)
22. C.B. Duke, Y.R. Wang, J. Vac. Sci. Technol. B **7**, 1 (1989)
23. Z. Liu, K. Wang, Acta Phys. Sin. **38**, 891 (1989)
24. W. Ludwig, Jpn. J. Appl. Phys. Suppl. 2, Pt. 2 (1974) 879; Festkörperprobleme **29**, 107 (1989)
25. L. Miglio, P. Santini, P. Ruggerone, G. Benedek, Phys. Rev. Lett. **62**, 3070 (1989)
26. F. Ancillotto, W. Andreoni, A. Selloni, R. Car, M. Parrinello, Phys. Rev. Lett. **65**, 3148 (1990)
27. P.K. Das, R.E. Allen, in *20th International Conference on Physics of Semiconductors*, vol. 2 ed. by E.M. Anastassakis, J.D. Ioannopoulos, (World Scientific, Singapore, 1990), p. 1973
28. R.B. Doak, Y.J. Chabal, G.S. Iligashi, P. Dumas, J. Electr. Spectr. Rel. Phen. **54/55**, 291 (1990)
29. R.B. Doak, J. Electr. Spectr. Rel. Phen. **54/55**, 281 (1990)
30. P. Santini, L. Miglio, G. Benedek, U. Harten, P. Ruggerone, J.P. Toennies, Phys. Rev. B **42**, 11942 (1990)
31. F. Ancillotto, W. Andreoni, A. Selloni, R. Car, M. Parrinello, Phys. Scr. (T) **35**, 21 (1991)
32. F. Ancillotto, A. Selloni, W. Andreoni, S. Baroni, R. Car, M. Parrinello, Phys. Rev. B **43**, 8930 (1991)
33. T.J. Godin, J.P. la Femina, C.B. Duke, J. Vac. Sci. Technol. B **9**, 2282 (1991)
34. P. Santini, L. Miglio, G. Benedek, P. Ruggerone, Surf. Sci. **241**, 346 (1991)
35. J.G. Skofronick, J.P. Toennies, in *Surface Properties of Layered Structures,* ed. by G. Benedek (Kluver, Amsterdam, 1992), p. 151
36. P.C. Weakliem, E.A. Carter, J. Chem. Phys. **96**, 3240 (1992)
37. X. Zhu, S.G. Louie, Phys. Rev. B **45**, 3940 (1992)
38. G. Benedek, G. Onida, Phys. Rev. B **47**, 16471 (1993)
39. J. Fritsch, P. Pavone, U. Schröder, Phys. Rev. Lett. **71**, 4194 (1993)
40. G. Benedek, F. Hoffmann, G. Onida, P. Ruggerone, L. Miglio, Surf. Sci. Rep. **20**, 3 (1994)
41. H. Büscher, Ph. D. Thesis (University of Münster, 1994)
42. H. Gai, G.A. Voth, J. Chem. Phys. **101**, 1734 (1994)
43. M. Buongiorno Nardelli, D. Cvetko, V. de Renzi, L. Floreano, A. Morgante, M. Peloi, F. Tommasini, Phys. Rev. B **52**, 16720 (1995)
44. R. di Felice, A.I. Shkrebtii, F. Finocchi, C.M. Bertoni, G. Onida, J. Electr. Spectr. Rel. Phenom. **64/65**, 697 (1995)
45. J. Kim, F.S. Khan, M. Yeh, J.W. Wilkins, Phys. Rev. B **52**, 14709 (1995)
46. G. Kresse, J. Furthmüller, J. Hafner, Europhys. Lett. **32**, 729 (1995)
47. B. Sandfort, A. Mazur, J. Pollmann, Phys. Rev. B **51**, 7150 (1995)
48. B. Sandfort, A. Mazur, J. Pollmann, Phys. Rev. B **51**, 7168 (1995)
49. W.G. Schmidt, F. Bechstedt, G.P. Srivastava, Phys. Rev. B **52**, 2 (1995)
50. S. Schmidt, Diploma-Thesis (University of Göttingen, 1995, unpublished)
51. I.Štich, K. Terakura, B.E. Larson, Phys. Rev. Lett. **74**, 4491 (1995)
52. C. Eckl, J. Fritsch, P. Pavone, U. Schroeder, in *Proceedings of 23rd International Conference on the Physics of Semiconductors Berlin 1996*, ed. by M. Scheffler, R. Zimmermann (World Sci., Singapore, 1996), p. 229
53. R. Honke, P. Pavone, U. Schröder, Surf. Sci. **367**, 75 (1996)
54. G. Lange, J.P. Toennies, Phys. Rev. B **53**, 9614 (1996)
55. W.G. Schmidt, F. Bechstedt, G.P. Srivastava, Surf. Sci. **352**, 83 (1996)
56. I Štich, J. Kohanoff, K. Terakura, Phys. Rev. B **54**, 2642 (1996)
57. I. Štich, Surf. Sci. **368**, 152 (1996)
58. H.M. Tütüncu, G.P. Srivastava, Surf. Sci. **377**, 304 (1996)
59. C. Eckl, J. Fritsch, P. Pavone, U. Schröder, Surf. Sci. **394**, 47 (1997)
60. C. Eckl, R. Honke, J. Fritsch, P. Pavone, U. Schröder, Z. Phys. B **104**, 715 (1997)
61. J. Fritsch, C. Eckl, P. Pavone, U. Schröder, Festkörperprobleme/Adv. Solid State Phys. **36**, 135 (1997)

62. A. Glebov, V. Senz, J.P. Toennies, P. Lambin, G. Gensterblum, P. Senet, A.A. Lucas, Phys. Rev. B **56**, 9874 (1997)
63. A. Glebov, R.E. Miller, J.P. Toennies, J. Chem. Phys. **106**, 6499 (1997)
64. R. Honke, A.P. Mayer, P. Pavone, U. Schröder, Surf. Sci. **377**, 317 (1997)
65. G. Lange, C.A. Meli, J.P. Toennies, E.F. Green, Phys. Rev. B **56**, 204 (1997)
66. A. Mazur, B. Sandfort, V. Gräschus, J. Pollmann, Festkörperprobleme/Adv. Solid State Phys. **36**, 181 (1997)
67. W. Stigler, P. Pavone, U. Schröder, J. Fritsch, G. Brusdeylins, Y. Wach, J.P. Toennies, Phys. Rev. Lett. **79**, 1090 (1997)
68. H.M. Tütüncu, G.P. Srivastava, J. Phys. Chem. Sol. **58**, 685 (1997)
69. H.M. Tütüncu, G.P. Srivastava, Surf. Sci. **377**, 304 (1997)
70. M. Zitzlsperger, R. Honke, P. Pavone, U. Schröder, Surf. Sci. **377**, 108 (1997)
71. G. Lange, J.P. Toennies, P. Ruggerone, G. Benedek, Europhys. Lett. **41**, 647 (1998)
72. G. Benedek, in *Ultrafast Dynamics of Quantum Systems: Physical Processes and Spectroscopic Techniques,* ed. by B. Di Bartolo, G. Gambarota (Plenum Press, New York, 1998), pp. 295–331
73. A. Glebov, J.P. Toennies, S. Vollmer, S.A. Safron, J.G. Skofronick, V. Gräschus, A. Mazur, J. Pollmann, Phys. Rev. B **57**, 10082 (1998)
74. W. Stigler, P. Pavone, J. Fritsch, Phys. Rev. **58**, 13686 (1998)
75. H.M. Tütüncu, S.J. Jenkins, G.P. Srivastava, Phys. Rev. B **57**, 4649 (1998)
76. H.M. Tütüncu, M. Cakmak, G.P. Srivastava, Appl. Surf. Sci. **123**, 146 (1998)
77. J. Fritsch, U. Schröder, Phys. Rep. **309**, 209 (1999)
78. A. Glebov, U. Panella, J.P. Toennies, Phys. Rev. B **60**, 2046 (1999)
79. A.L. Glebov, J.P. Toennies, S. Vollmer, G. Benedek, Europhys. Lett. **46**, 369 (1999)
80. A. Glebov, A.P. Graham, A. Menzel, Surf. Sci. **427**, 22 (1999)
81. H.M. Tütüncu, G.P. Srivastava, Physica B **263**, 400 (1999)
82. T. Wach, J.P. Toennies, unpublished data, reproduced in ed. by J. Fritsch, U. Schröder, Phys. Rep. **309**, 209 (1999)
83. H. Tröger, W. Theis, G. König, K.-H. Rieder, Surf. Sci. **490**, 251 (2001)
84. W. Wolf, J. Stitch, A. Mavromaras, P. Saxe, E. Wimmer, Materials Sci. Poland **23**, 2 (2005)
85. J. Lobo, D. Farias, M. Garcia, E. Hulpke, J.P. Toennies, Phys. Rev. B **74**, 035303 (2006)
86. W. Steurer, A. Apfolter, M. Koch, W.E. Ernst, B. Holst, E. Søndergård, S.C. Parker, Phys. Rev. B **78**, 035402 (2008)
87. H. Shichibe, Y. Satake, K. Watanabe, A. Kinjyo, A. Kunihara, Y. Yamada, M. Sasaki, W. W. Hayes, J.R. Manson, Phys. Rev. B **91**, 155403 (2015)
88. G. Onida, G. Benedek, "Dynamics of fcc-C60 Fullerite" in *Fullerenes: Status and Perspectives*, ed. by G. Ruani, C. Taliani, R. Zamboni (World Scientific, Singapore 1992) p. 181
89. C.H. Patterson, S. Banerjee, J.F. McGilp, Phys. Rev. B **84**, 155314 (2011)
90. N. Takeuchi, A. Selloni, E. Tosatti, Phys. Rev. B **51**, 10844 (1995)

Appendix A.3

Table A.3 Surface phonon dispersion curves of layered compound surfaces from HAS measurements and theory (*: SEHAS measurements. Graphite is included in Appendix A.2; and graphene in Appendix A.8)

SURFACE	Structure	HAS experiment	Theory	
		References	Method	References
$Bi_2Se_3(111)$	Rhombohedral	[1, 2]	ME (PC) DFPT	[1] [3, 4]
$Bi_2Sr_2CaCu_2O_8$ (001) (Bi-2212) (BiO plane)	Orthorhombic	[5–7]	SM	[7–9]
$Bi_2Te_3(111)$	Rhombohedral	[10, 11*, 12*]	ME (PC) DFPT	[10] [3, 4, 11–14]
GaSe(0001)	Hex	[5, 15, 16]	DLC DFPT	[5, 15, 17–19] [16, 20]
InSe(0001)	Hex	[16]	DFPT	[16, 20]
$KAl_2Si_3AlO_{10}(OH)_2$ (001)	Monoclinic (muscovite mica)	[21]	–	–
$LiCu_2O_2(001)$	Orthorhombic	[22, 23]	SM	[23]
$NbSe_2(0001)$	Hex 2H polytype	[24]	DFPT	[25–28]
$Sr_2CuO_2Cl_2$ (001)	bc-tetragonal (SrCl plane)	[29]	SM	[29]
$TaS_2(0001)$	Hex $1T_a$ polytype	[5, 30, 31]	FC	[5, 32]
			PC	[31]
			ME (PC)	[19, 31]
$TaSe_2(0001)$	Hex 2H polytype	[32–36]	DLC	[19, 32, 33, 35, 36]
$YBa_2Cu_3O_{7-x}(001)$	Orthorhombic ($x < 0.3$) YBCO(123)	[6]	SM	[37]

References

1. Zhu, X., Santos, L., Sankar, R., Chikara, S., Howard, C., Chou, F.C., Charmon, C., El-Batanouny, M.: Phys. Rev. Lett. **107** (2011) 186102
2. Zhu, X.T., Santos, L., Howard, C., Sankar, R., Chou, F.C., Chamon, C., El-Batanouny, M.: Phys. Rev. Lett. **108** (2012) 185501
3. W. Cheng and S.-F. Ren, Phys. Rev. B **83**, 094301 (2011)
4. D. Campi. PhD Thesis (University of Milano-Bicocca, 2016)
5. Skofronick, J.G., Toennies, J.P.: in: Benedek, G.(ed.): *Surface Properties of Layered Structures*, Amsterdam: Kluver, 1992, p. 151
6. Schmicker, D.: Thesis, University of Göttingen, 1992

7. Paltzer, U., Schmicker, D., de Wette, F.W., Schröder, U., Toennies, J.P.: Phys. Rev. B **54** (1996) 11989
8. Paltzer, U., de Wette, F.W., Schröder, U., Kress, W.: Bull. Am. Phys. Soc. **39** (1994) 569
9. Paltzer, U., de Wette, F.W., Schroder, U., Rampf, E.: Physica C **301** (1998) 55
10. Howard, C., El-Batanouny, M., Sankar, R., Chou, F.C.: Phys. Rev. B **88** (2013) 035402
11. A. Tamtögl, P. Kraus, N. Avidor, M. Bremholm, E. M. J. Hedegaard, Bo B. Iversen, M. Bianchi, Ph. Hofmann, J. Ellis, W. Allison, G. Benedek, and W. Ernst, , Phys. Rev. B **95**, 195401 (2017)
12. A. Tamtögl, D. Campi, M. Bremholm, E. M. J. Hedegaard, B. B. Iversen, M. Bianchi, Ph. Hofmann, N. Marzari, G. Benedek, J. Ellis, and W. Allison, arXiv 1803:11184v1 [con.matrl. sci] (2018)
13. Huang, G. Q., EPL **100**, 17001 (2012)
14. M. Wiesner, A. Trzaskowska, B. Mroz, S. Charpentier, S. Wang, Y. Song, F. Lombardi, P. Lucignano, D. Campi, M. Bernasconi, G. Benedek, F. Guinea, and A. Tagliacozzo, Sci. Rep. **7**, 16449 (2017)
15. Brusdeylins, G., Rechsteiner, R., Skofronick, J.G., Toennies, J.P., Benedek, G., Miglio L.: Phys. Rev. B **34** (1986) 902
16. Panella, V., Glebov, A.L., Toennies, J.P., Sebenne, C., Eckl, C., Adler, C., Pavone, P., Schröder, U.: Phys. Rev. B **59** (1999) 15772
17. Miglio, L., Benedek, G.: Europhys. Lett **3** (1987) 619
18. Miglio, L., Colombo, L.: Phys. Rev. B **37** (1988) 3025
19. Benedek, G., Hoffmann, F., Onida, G., Ruggerone, P., Miglio, L.: Surf. Sci. Reports **20** (1994) 3
20. Adler, C., Honke, R., Pavone, P., Schröder, U.: Phys. Rev. B **57** (1998) 3726
21. Brusdeylins, G., Schmicker, D.: Surf. Sci. **331** (1995) 23
22. Y. Yao, X. Zhu, H. C. Hsu, F. C. Chou, and M. El-Batanouny, Surf. Sci. **604** (2010) 692
23. Zhu, X., Yao, Y., Hsu, H.C., Chou, F.C., El-Batanouny, M.: Surf. Sci. **605** (2011) 376
24. Brusdeylins and G., Toennies, J.P (unpublished) in: Benedek, G., Miglio, L., Seriani, G.: in: Hulpke, E. (ed.): *Helium Atom Scattering from Surfaces*, Heidelberg: Springer-Verlag, 1992, p. 207
25. M. Calandra, I. I. Mazin, and F. Mauri, Phys. Rev. B **80**, 241108 R (2009)
26. F. Weber, S. Rosenkranz, J.-P. Castellan, R. Osborn, R. Hott, R. Heid, K.-P. Bohnen, T. Egami, A. H. Said, and D. Reznik, Phys. Rev. Lett. **107**, 107403 (2011)
27. F. Weber, R. Hott, R. Heid, K.-P. Bohnen, S. Rosenkranz, J.-P. Castellan, R. Osborn, A. H. Said, B. M. Leu, and D. Reznik, Phys. Rev. B **87**, 245111 (2013)
28. R. He, J. van Baren, J.-A. Yan, X. Xi, Z. Ye, G. Ye, I.-Hsi Lu, S. M. Leong, C. H. Lui, *2D Mater.* 3 031008 (2016)
29. Farzaneh, M., Liu, X.-F., El-Batanouny, M., Chou, F.C.: Phys. Rev. B 72 (2005) 085409. G.: Europhys. Lett. **9** (1989) 563
30. Brusdeylins, G., Hofmann, F., Ruggerone, P., Toennies, J.P., Vollmer, R., Benedek, G., Skofronick, J.G.: in: Hunklinger, S., Ludwig, W., Weiss, G. (eds.): *Phonons 89*, Singapore: World Scientific, 1990, p. 892
31. Benedek, G., Brusdeylins, G., Hofmann, F., Ruggerone, P., Toennies, J.P., Vollmer, R., Skofronick, J.G.: Surf. Sci. **304** (1994) 185
32. Brusdeylins, G., Helmlich, C., Skofronick, J.G., Toennies, J.P., Vollmer, R., Benedek, G., Miglio, L.: Phys. Rev. B **41** (1990) 5707
33. Benedek, G., Brusdeylins, G., Heimlich, C., Miglio, L., Skofronick, J.G., Toennies, J.P., Vollmer, R.: Phys. Rev. Lett. **60** (1988) 1037
34. Brusdeylins, G., Heimlich, C., Skofronick, J.G., Toennies, J.P., Vollmer, R., Benedek, G., Miglio, L.: Phys. Rev. B **41** (1990) 5707
35. Benedek, G., Miglio, L., Skofronick, J.G., Brusdeylins, G., Heimlich, C., Toennies, J.P.: J. Vac. Sci. Technol. A **5** (1987) 1093
36. Benedek, G., Miglio, L., Brusdeylins, G., Heimlich, C., Skofronick, J.G., Toennies, J.P.: Europhys. Lett. **5** (1988) 253
37. W. Kress, U. Schröder, J. Prade, A. D. Kulkarni, F. W. de Wette, Physica C, 153–155, 221 (1988)
38. Bruch, L.W., Glebov, A., Toennies, J.P., Weiss, H.: J. Chem. Phys. **103** (1995) 5109

Appendix A.4

Table A.4 Surface phonon dispersion curves of metal, metal alloy and elemental semi-metal surfaces from HAS measurements and theory

SURFACE	Structure	HAS Experiment	Theory	
		References	Method	References
Ag(001)	fcc	[20, 125]	ME(PC)	[125]
			FC	[1, 2, 21, 79]
			FP-HF	[73, 84, 85, 128]
			FP-GF	[109]
			MD-EA	[91]
Ag(110)	fcc	[29, 64, 141, 142, 153]	FC	[1, 142, 82]
			GF	[109]
			EA	[129, 168]
			MD-EA	[117]
Ag(111)	fcc	[14, 22]	FC	[18, 21, 124]
			FC-GF	[9]
			EA	[60, 48]
			EA-MD	[174]
			ME (PC)	[34]
			FP-HF	[73, 110, 128, 158, 165]
			FP-GF	[109]
			DFPT	[172]
Al(001)	fcc	[87, 126]	FP-DR	[87]
			DFPT	[189, 197]
			EA	[123, 131]
			FC	[10]
			FC-MD	[173]
			FP-DR	[56, 71, 111]
			FP-HF	[73]
Al(110)	fcc	[39]	FP-DR	[71, 111]
			EA	[123, 130, 168]
			FP-HF	[23, 44]
			FP-DR	[42, 88, 212]
Al(111)	fcc	[45, 71 87, 126, 111]	FP-DR	[71, 87, 126, 111]
			EA	[123, 131]
			EM	[86]
			FC	[3, 7, 10, 45,]
			FC-MD	[173]
			FP-HF	[73, 136, 158]
Al(221)	fcc	[77]	FC	[77]
			EA	[148, 168]

(continued)

Table A.4 (continued)

SURFACE	Structure	HAS Experiment	Theory	
		References	Method	References
Au(110) (2x1)	fcc	[58]	FC (SCMBP)	[171]
			FP-HF	[33, 73, 58, 65]
			GM-MD	[49]
Au(111)	fcc	[13, 22, 38, 18, 204]	EA	[48]
			FC	[3, 7, 18, 38]
			FC-MD	[61, 81, 90]
			ME (PC)	[34]
			GM	[68, 69]
Ba(0001)	1 to 3ML on Cu(001), hex	[178]	–	–
Bi(111)	rhombohedral	[198, 206]	DFPT	[206, 202, 213, 214]
Cr(110)	bcc	[114, 127]	FC	[3, 7]
Cs(110)	bcc 5 ML on Cu(111)	[144]	–	–
Cu(001)	fcc	[1, 107, 108, 115 188]	ME (PC)	[107]
			FC	[3, 7, 10, 21, 108]
			FS-MD	[115]
			TBSM	[183]
			EA	[46, 47, 78, 60, 59, 80]
			EA-MD	[91]
			EM	[86]
			FP-HF	[73, 85, 128]
			FP-GF	[109]
			DFPT	[180]
Cu(110)	fcc	[16, 53, 115]	FS-MD	[115]
			EA	[46, 129]
			EM	[86]
			EA-MD	[92]
			FC	[28]
			FP-HF	[119, 120]
			FP-GF	[109]
Cu(111)	fcc	[22]	ME (PC)	[34, 103]
			DFPT	[184, 193, 197]
			EA	[19, 48, 46, 60, 78]
			EM	[70]
			FC	[3, 7, 10, 21, 43, 181]
			FP-HF	[73, 110, 128]
			FP-GF	[109]
			FS-MD	[115]

(continued)

Table A.4 (continued)

SURFACE	Structure	HAS Experiment	Theory	
		References	Method	References
Cu(211), (511)	fcc	[139]	EA	[155, 148, 168, 187]
			FC	[139]
			TBSM	[183]
$Cu_3Au(001)$	Cubic	[210, 97]	FC	[210, 98]
α-Ga(010)	Base-c orthorhomb.	[185]		
Fe(001)	fcc films on Cu(001)	[169, 179, 197]	FC FP-DFT	[169, 179, 197 191]
Fe(110)	bcc	[95, 116]	PC FP-DFT	[116] [191, 190, 192]
K(001)	bcc	[163]	FC	[3, 7]
K(110)/Rh(111)	bcc supported films	[135]	–	–
K(110)/Ni(001)	bcc supported fims	[144]	–	–
Li(110)	bcc supported films	[176]	EA	[150, 177]
Mo(001)	bcc	[57, 75, 76]	FC	[3, 4,7, 66, 83]
			FP-TB	[50]
			FP-MD	[51]
Mo(110)	bcc	[112, 127]	FC	[127]
Mo(110): H(1x1)	bcc	[112, 113]	FP-HF	[145, 146]
$Mo_{1-x}Re_x(110)$: H(1x1)	bcc	[182]	FP-HF	[145]
Na(110)	bcc supported films	[94, 144]	FC EA FP	[147] [150, 215] [62, 63]
Nb(001)	bcc	[76]	PC	[100]
Ni(001)	fcc	[27]	FC	[3, 7, 11, 17, 27, 216, 217, 218]
			EA	[159]
			EA-MD	[122]
			GF	[24]
			FP-GF	[109]
			FS-MD	[115, 72]
Ni(110)	fcc	[151]	FC	[17]
			GF-CFM	[67]
			EA-MD	[122]
			FP-GF	[109]
			FS-MD	[115]
Ni(111)		[5,** 8**, 208*, 209]	FC	[12]

(continued)

Table A.4 (continued)

SURFACE	Structure	HAS Experiment	Theory	
		References	Method	References
Ni(977)	fcc	[133, 134, 164]	FC	[161]
			CE	[132]
			EA-GF	[156, 187]
Pb(110)	fcc	[41, 32]	–	–
Pb(111)	fcc supported films up to 50 ML	[195]	FC	[195]
			DFPT	[194, 199, 201, 205]
Pd(110)	fcc	[157]	FC	[157]
			EA	[129, 149, 159]
			FP	[93]
Pd(111)	fcc	[74]	FC	[3, 7]
			EA	[149, 159]
			DFPT	[196]
Pd(111):H(1x1)	fcc	[74]	–	–
Pt(001) (1x1)	fcc	[88]	FC	[3, 7, 21]
Pt(001)-hex-rot	fcc	[88]	–	
Pt(111)	fcc	[15, 25, 26, 104]	FC	[3, 7, 54, 104, 186]
Pt(111):H(1x1)	fcc	[54]	FC	[54]
Rb	bcc supported films	[162]		
Rh(110)	fcc	[105]	FP-HF	[170]
Rh(111)	fcc	[121]	FC	[137, 3, 7, 167]
			FP-HF	[140]
Rh(111):H(1x1)	fcc	[138]	FC	[166]
Ru(0001)	hex	[154]	–	–
Ru(0001)+H (1x1)	hex	[154]	–	–
Tl(0001)	hex (supported thin films)	[152]	DFPT	[200]
W(001)	bcc	[30, 55, 96, 76, 99]	FC	[2, 211, 3, 7, 31, 35, 36, 37, 89, 118]
			B-MD	[175]
			FP-TB	[40]
			FP MD	[52]
W(110)	bcc	[101, 102]	DFPT	[143]
W(110):H(1x1)	bcc	[101, 102, 112, 113]	DFPT FP-HF	[143] [145, 146]

(*: SE-HAS measurements; **: NaAS measurements. Graphite is included in Appendix A.2, and graphene in Appendix A.8)

References

1. G.P. Alldredge, R.E. Allen, F.W. de Wette, J. Acoust. Soc. Am. **49**, 1453 (1971)
2. D. Castiel, L.Dobrzynski, D. Spanjaard, Surf. Sci. **59**, 252 (1976)
3. J.A. Black, D.A. Campbell, R.F. Wallis, Surf. Sci. **105**, 629 (1981)
4. A. Fasolino, G. Santoro, E. Tosatti, J. Physique **42**, C6-84 (1981)
5. B. Feuerbacher, R.F. Willis, Phys. Rev. Lett. **47**, 526 (1981)**
6. B.F. Mason, B.R. Williams, Phys. Rev. Lett. **46**, 1138 (1981)
7. J.A. Black, D.A. Campbell, R.F. Wallis, Surf. Sci. **115**, 161 (1982)
8. B. Feuerbacher, in *Dynamics of Gas-Surface Interactions,* ed. by G. Benedek (Springer, Valbusa Heidelberg, 1982), p. 263**
9. G. Armand, P. Masri, Surf. Sci. **130**, 89 (1983)
10. J.A. Black, F.C. Shanes, R.F. Wallis, Surf. Sci. **133**, 199 (1983)
11. V. Bortolani, A. Franchini, F. Nizzoli, J. Electr. Spectr. Rel. Phenom. **29**, 219 (1983)
12. V. Bortolani, A. Franchini, F. Nizzoli, Surf. Sci. **128**, 249 (1983)
13. M. Cates, D.R. Miller, J. Electr. Spectr. Rel. Phenom. **30**, 157 (1983)
14. R.B. Doak, U. Harten, J.P. Toennies, Phys. Rev. Lett. **51**, 578 (1983)
15. U. Harten, J.P. Toennies, C. Wöll, G. Zhang, Phys. Rev. Lett. **55**, 2308 (1983)
16. B.F. Mason, K. McGreer, B.R. Williams, Surf. Sci. **130**, 282 (1983)
17. J.E. Black, P. Bopp, Surf. Sci. **140**, 275 (1984)
18. V. Bortolani, A. Franchini, F. Nizzoli, G. Santoro, Phys. Rev. Lett. **52**, 429 (1984)
19. M.S. Daw, M.I. Baskes, Phys. Rev. B **29**, 6443 (1984)
20. W.R. Lambert, P.I. Trevor, R.B. Doak, M.J. Cardillo, J. Vacuum Sci. Technol. A **2**, 1066 (1984)
21. V. Bortolani, A. Franchini, G. Santoro, in *Electronic, Dynamic and Quantum Structural Properties of Condensed Matter,* ed. by J.T. Devreese, P.E. van Camp (Plenum, New York, 1985), p. 401
22. U. Harten, J.P. Toennies, Ch. Wöll, Faraday Disc. Chem. Soc. **80**, 137 (1895)
23. K.M. Ho, K.P. Bohnen, Phys. Rev. Lett. **56**, 934 (1986)
24. H. Ibach, S. Lehwald, T.S. Rahman, Surf. Sci. **171**, 632 (1986)
25. K. Kern, R. David, R.L. Palmer, G. Comsa, T.S. Rahman, Phys. Rev. B **33**, 4334 (1986)
26. D. Neuhaus, F. Joo, B. Feuerbacher, Surf. Sci. Lett. **165**, 90 (1986)
27. R. Berndt, J.P. Toennies, C. Wöll, J. Electr. Spectr. Rel. Phenom. **4**, 183 (1987)
28. J.E. Black, A. Franchini, V. Bortolani, G. Santoro, R.F. Wallis, Phys. Rev. B **36**, 2996 (1987)
29. G. Bracco, R. Tatarek, F. Tommasini, U. Linke, M. Persson, Phys. Rev. B **36**, 2928 (1987)
30. H.-J. Ernst, E. Hulpke, J.P. Toennies, Phys. Rev. Lett. **58**, 1941 (1987)
31. A. Fasolino, E. Tosatti, Phys. Rev. B 35, 4264 (1987)
32. J.W.H. Frenken, J.P. Toennies, C. Wöll, in *Proceedings of ICSOS II*, ed. by J.F. Van der Veen, M.A. van Hove (Springer, Amsterdam, 1987, Heidelberg, 1987)
33. K.M. Ho, K.P. Bohnen, Europhys. Lett. **4**, 345 (1987)
34. C.S. Jayanthi, H. Bilz, W. Kress, G. Benedek, Phys. Rev. Lett. **59**, 795 (1987)
35. A.D. Kulkarni, F.W. de Wette, J. Prade, Phys. Rev. B **36**, 6190 (1987)
36. A.D. Kulkarni, F.W. de Wette, J. Prade, J. Elect. Spectr. Rel. Phenom. **44**, 397 (1987)
37. T.L. Reinecke, Ying, S.C.: Phys. Rev. Lett. 58 (1987) 242.
38. G. Santoro, A. Franchini, V. Bortolani, U. Harten, J.P. Toennies, C. Wöll, Surf. Sci. 183 (1987) 180.
39. J.P. Toennies, Ch. Wöll, Phys. Rev. B **36**, 4475 (1987)
40. X.W. Wang, W. Weber, Phys. Rev. Lett. **58**, 1452 (1987)
41. C. Wöll, Thesis, University of Göttingen, 1987, and Max-Planck-Institut für Strömungsforschung Göttingen, Bericht 18/1987
42. A.G. Eguiluz, A.A. Maradudin, R.F. Wallis, Phys. Rev. Lett. **60**, 309 (1988)

43. B.M. Hall, D.L. Mills, M.H. Mohamed, L.L. Kesmodel, Phys. Rev. B **38**, 5856 (1988)
44. K.M. Ho, K.P. Bohnen, Phys. Rev. B **38**, 12897 (1988)
45. A. Lock, J.P. Toennies, C. Wöll, V. Bortolani, A. Franchini, G. Santoro, Phys. Rev. B **37**, 7087 (1988)
46. N. Luo, W. Xu, S. Shen, Solid State Commun. **67**, 837 (1988)
47. J.S. Nelson, E.C. Sowa, M.S. Daw, Phys. Rev. Lett. **61**, 1977 (1988)
48. P.R. Underhill, Surf. Sci. **200**, L441 (1988)
49. X.Q. Wang, G.L. Chiarotti, F. Ercolessi, E. Tosatti, Phys. Rev. B **38**, 8131 (1988)
50. X.W. Wang, C.T. Chan, K.-H. Ho, W. Weber, Phys. Rev. Lett. **60**, 2066 (1988)
51. C.Z. Wang, E. Tosatti, A. Fasolino, Phys. Rev. Lett. **60**, 2661 (1988)
52. C.Z. Wang, A. Fasolino, E. Tosatti, Phys. Rev. B **37**, 2116 (1988)
53. P. Zeppenfeld, K. Kern, R. David, K. Kuhnke, G. Comsa, Phys. Rev. B **38**, 123 (1988)
54. V. Bortolani, A. Franchini, G. Santoro, J.P. Toennies, C. Wöll, G. Zhang, Phys. Rev. B **40**, 3524 (1989)
55. H.-J. Ernst, E. Hulpke, J.P. Toennies, Europhys. Lett. **10**, 747 (1989)
56. J.A. Gaspar, A.G. Eguiluz, Phys. Rev. B **40**, 11976 (1989)
57. E. Hulpke, D.-M. Smilgies, Phys. Rev. B **40**, 1338 (1989-I)
58. A.M. Lahee, J.P. Toennies, C. Wöll, K.P. Bohnen, K.M. Ho, Europhys. Lett. **10**, 261 (1989)
59. N. Luo, W. Xu, Acta Phys. Sin. **38**, 900 (1989)
60. S. Nelson, M.S. Daw, E.C. Sowa, Phys. Rev. B **40**, 1465 (1989)
61. R. Ravelo, M. El-Batanouny, Phys. Rev. B **40**, 9574 (1989)
62. T. Rodach, K.P. Bohnen, Phys. Rev. B **35**, 3446 (1985)
63. T. Rodach, K.P. Bohnen, K.M. Ho, Surf. Sci. **209** 481 (1989)
64. R. Tatarek, G. Bracco, F. Tommasini, A. Franchini, V. Bortolani, G. Santoro, R.F. Wallis, Surf. Sci. **211/212**, 314 (1989)
65. B. Voigtländer, S. Lehwald, H. Ibach, K.P. Bohnen, K.M. Ho, Phys. Rev. B **40**, 8068 (1989)
66. J. Zi, K. Zhang, X. Xie, Acta Phys. Sin. **38**, 1475 (1989)
67. J.E. Black, in *Dynamical Properties of Solids*, ed. by G.K. Horton, A.A. Maradudin (Elsevier, Amsterdam, 1990), Chap. 5
68. V. Bortolani, F. Ercolessi, E. Tosatti, A. Franchini, G. Santoro, Europhys. Lett **12**, 149 (1990)
69. V. Bortolani, A. Franchini, G. Santoro, F. Ercolessi, E. Tosatti, Vacuum **41**, 428 (1990)
70. P.D. Ditlevsen, J.K. Nørskov, J. El. Spectr. Rel. Phenom. **54/55**, 237 (1990)
71. A.G. Eguiluz, J.A. Gaspar, M. Gester, A. Lock, J.P. Toennies, Superlattices Microstruct. **7**, 223 (1990)
72. J.L. Erskine, E.-J. Jeong, J. Vac. Sci. Technol. A **8**, 2649 (1990)
73. K.M. Ho, K.P. Bohnen, J. Electr. Spectr. Rel. Phenom. **54/55**, 229 (1990)
74. C.H. Hsu, M. El-Batanouny, K.M. Martini, J. Electr. Spectr. Rel. Phenom. **54/55**, 353 (1990)
75. E. Hulpke, D.-M. Smilgies, Phys. Rev. B **42**, 9203 (1990)
76. E. Hulpke, J. Electr. Spectr. Rel. Phenom. **54/55**, 299 (1990)
77. A. Lock, J.P. Toennies, G. Witte, J. Electr. Spectr. Rel. Phenom. **54/55**, 309 (1990)
78. N. Luo, W. Xu, S. Shen, Phys. Stat. Solidi (b) **158**, 493 (1990)
79. P. Moretto, M. Rocca, U. Valbusa, J. Black, Phys. Rev. B **41**, 12905 (1990)
80. J.S. Nelson, M.S. Daw, E.C.S. Sowa, Superlattices Microstr. **7**, 259 (1990)
81. R. Ravelo, M. El-Batanouny, J. Electr. Spectr. Rel. Phenom. **54/55**, 255 (1990)
82. Y.E. Yates, A.D. Kulkarni, F.W. de Wette, J.L. Erskine, J. Electr. Spectr. Rel. Phenom. **54/55**, 395 (1990)
83. J. Zi, K. Zhang, X. Xie, Chin J. Infrared Res. **9**, 35 (1990)
84. Y. Chen, S.Y. Tong, M. Rocca, P. Moretto, U. Valbusa, K.P. Bohnen, K.M. Ho, Surf. Sci. Lett. **250**, 389 (1991)
85. Y. Chen, S.Y. Tong, J.S. Kim, L.L. Kesmödel, T. Rodach, K.P. Bohnen, K.M. Ho, Phys. Rev. B **44**, 11394 (1991)
86. P.D. Ditlevsen, J.K. Nørskov, Surf. Sci. **254**, 261 (1991)
87. J.A. Gaspar, A.G. Eguiluz, M. Gester, A. Lock, J.P. Toennies, Phys. Rev. Lett. **66**, 337 (1991)

88. K.-E. Kuhnke, Bericht des Forschungszentrums Jülich; 2490, June 1991
89. T.L. Reinecke, S.C. Ying, Phys. Rev. B **43**, 12234 (1991)
90. X.Q. Wang, Phys. Rev. Lett. **67**, 1294 (1991)
91. L. Yang, T.S. Rahman, M.S. Daw, Phys. Rev. B **44**, 13725 (1991)
92. L. Yang, T.S. Rahman, Phys. Rev. Lett. **67**, 2327 (1991)
93. W. Zhong, Y.S. Li, D. Tomanek, Phys. Rev. B. **44**, 13053 (1991)
94. G. Benedek, J. Ellis, A. Reichmuth, P. Ruggerone, H. Schief, J.P. Toennies, Phys. Rev. Lett. **69**, 2951 (1992)
95. G. Benedek, J.P. Toennies, G. Zhang, Phys. Rev. Lett. **68**, 2644 (1992)
96. H.-J. Ernst, E. Hulpke, J.P. Toennies, Phys. Rev. B **46**, 16081 (1992)
97. B. Gans, S.K. King, P.A. Knipp, D.D. Koleske, S.J. Sibener, J. Electr. Spectr. Rel.Phenom. **54/55**, 333 (1990)
98. B. Gans, P.A. Knipp, D.D. Koleske, S.J. Sibener, Surf. Sci. **264**, 81 (1992)
99. W.K. Han, S.C. Ying, Phys. Rev. B **46**, 1849 (1992)
100. E. Hulpke, M. Hüppauf, D.-M. Smilgies, A.D. Kulkarni, F.W. de Wette, Phys. Rev. B **45**, 1820 (1992)
101. E. Hulpke, J. Lüdecke, Surf. Sci. **272**, 289 (1992)
102. E. Hulpke, J. Lüdecke, Phys. Rev. Lett. **68**, 2846 (1992)
103. C. Kaden, P. Ruggerone, J.P. Toennies, G. Zhang, G. Benedek, Phys. Rev. B **46**, 13509 (1992); Nuovo Cim. D **14**, 627 (1992)
104. D. Neuhaus, in *Helium Scattering from Surfaces,* ed. by E. Hulpke (Springer, Heidelberg, 1992), p. 265
105. A. Bellman, V.R. Dhanalz, A. Morgante, M. Polli, F. Tommasini, D. Cvetko, K.C. Prince, in *Inelastic Energy Transferin Interactions with Surfaces and Adsorbates,* ed. by B. Gumhalter, A.C. Levi, F. Flores (World Scientific, Singapore, 1993), p. 49
106. G. Benedek, G. Onida, Phys. Rev. B **47**, 16471 (1993)
107. G. Benedek, J. Ellis, N.S. Luo, A. Reichmuth, P. Ruggerone, J.P. Toennies, Phys.Rev. B **48**, 491 (1993)
108. V. Bortolani, D. Cvetko, F. Floreano, A. Franchini, A. Lansi, A. Morgante, M. Peloi, G. Santoro, F. Tommasini, T. Zambelli, A.F. Bellman, J. Electr. Spectr. Rel. Phenom. **64/65**, 671 (1993)
109. H. Büscher, W. Klein-Hessling, W. Ludwig, Ann. Phys. **2**, 159 (1993)
110. Y. Chen, S.Y. Tong, K.P. Bohnen, T. Rodach, K.M. Ho, Phys. Rev. Lett. **70**, 603 (1993)
111. A. Franchini, V. Bortolani, G. Santoro, V. Celli, A.G. Eguiluz, J.A. Gaspar, M. Gester, A. Lock, J.P. Toennies, Phys. Rev. B **47**, 4691 (1993)
112. E. Hulpke, J. Lüdecke, Surf. Sci. **287/288**, 837 (1993)
113. E. Hulpke, J. Lüdecke, J. Electr. Spectr. Rel. Phenom. **64/65**, 641 (1993)
114. H. Köster, Diploma Thesis (University of Göttingen, 1993) unpublished
115. D.D. Koleske, S.J. Sibener, Surf. Sci. **290**, 179 (1993)
116. D. Paulmann, Diploma Thesis (University of Göttingen, 1993), unpublished
117. T.S. Rahman, Z.J. Tian, J. Electr. Spectr. Rel. Phenom. **64/65**, 651 (1993)
118. T.L. Reinecke, S.C. Ying, Phys. Rev. B **48**, 5679 (1993)
119. T. Rodach, K.P. Bohnen, K.M. Ho, Surf. Sci. **286**, 66 (1993)
120. T. Rodach, K.P. Bohnen, K.M. Ho, Surf. Sci. **296**, 123 (1993)
121. G. Witte, Thesis, University of Göttingen, 1995, and Max-Planck-Institut für Strömungsforschung Göttingen, Bericht 8/1995
122. Y. Beaudet, L.J. Lewis, M. Persson, Phys. Rev. B **50**, 12084 (1994)
123. A.V. Bertsch, S.V. Eremeev, A.G. Lipnitskii, I.Y. Sklyadneva, A.V. Chulkov, Fiz Tverd. Tela **36**, 2935 (1994); Engl. transl.: Phys. Solid State **36**, 1562 (1994)
124. V. Bortolani, G. Santoro, U. Harten, J.P. Toennies, Surf. Sci. **148**, 82 (1994)
125. N. Bunjes, K.S. Lao, P. Ruggerone, J.P. Toennies, G. Witte, Phys. Rev. B **50**, 8897 (1994-II)
126. M. Gester, D. Kleinhesselink, P. Ruggerone, J.P. Toennies, Phys. Rev. B **49**, 5777 (1994-II)

127. J. Lüdecke, Dissertation, University of Göttingen (1994), Max-Planck-Institut für Strömungsforschung, Report 15/1994
128. S.Y. Tong, Y. Chen, K.P. Bohnen, T. Rodach, K.M. Ho, Surf. Rev. Lett. **1**, 97 (1994)
129. A.V. Bertsch, E.V. Chulkov, S.V. Eremeev, A.G. Lipnitskii, G.G. Rusina, I.Y. Sklyadneva, Vacuum **46**, 625 (1995)
130. A.V. Bertsch, E.V. Chulkov, S.V. Eremeev, A.G. Lipnitskii, G.G. Rusina, I.Y. Sklyadneva, Surf. Sci. **331–333**, 1414 (1995)
131. E.V. Chulkov, I.Y. Sklyadneva, Surf. Sci. **331–333**, 1414 (1995)
132. E.J. Mele, M.V. Pykthin, Phys. Rev. Lett. **75**, 387 (1995)
133. L. Niu, D.J. Gaspar, S.J. Sibener, Science **268**, 847 (1995)
134. Niu, L., Koleske, D.D., Gaspar, D.J., Sibener, S.J.: J. Chem. Phys. **102**, 9077 (1995)
135. H. Range, Dissertation, University of Göttingen, 1995, Max-Planck-Institut für Strömungsforschung, Report 3/1995
136. J. Schöchlin, K.P. Bohnen, K.M. Ho, Surf. Sci. **324**, 113 (1995)
137. J.P. Toennies, G. Witte, C. Wöll, Surf. Sci. **323**, 228 (1995)
138. G.W. Witte, Dissertation, University of Göttingen, 1995, Max-Planck-Institut für Strömungsforschung, Report 8/1995
139. G. Witte, J. Braun, A. Lock, J.P. Toennies, Phys. Rev. B **52**, 2165 (1995)
140. K.-P. Bohnen, A. Eichler, J. Hafner, Surf. Sci. **368**, 222 (1996)
141. G. Bracco, L. Bruschi, L. Pedemonte, R. Tatarek, Surf. Sci. **352**, 964 (1996)
142. G. Bracco, L. Bruschi, R. Tatarek, A. Franchini, V. Bortolani, G. Santoro, Europhys. Lett. **34**, 687 (1996)
143. C. Bungaro, S. de Gironcoli, S. Baroni, Phys. Rev. Lett. **77**, 2491 (1996)
144. E. Hulpke, J. Lower, A. Reichmuth, Phys. Rev. B **53**, 13901 (1996)
145. B. Kohler, P. Ruggerone, M. Scheffler, E. Tosatti, Z. Phys. Chemie **197**, 193 (1996)
146. B. Kohler, P. Ruggerone, M. Scheffler, Surf. Sci. **368**, 213 (1996)
147. N.S. Luo, P. Ruggerone, J.P. Toennies, Phys. Rev. B **54**, 5051 (1996)
148. G.G. Rusina, A.V. Berch, I.Y. Sklyadneva, S.V. Eremeev, A.G. Lipnitskii, E.V. Chulkov, Fiz. Tverd. Tela **38**, 141 (1996)
149. G.G. Rusina, I.Y. Sklyadneva, E.V. Chulkov, Fiz. Tverd. Tela **38**, 1483 (1996)
150. I.Y. Sklyadneva, E.V. Chulkov, A. V. Bertsch, Surf. Sci. **352**, 25 (1996)
151. M.F. Bertino, G. Witte, Surf. Sci. **385**, L984 (1997)
152. J. Braun, Dissertation, University of Göttingen, 1997, Max-Planck-Institut für Strömungsforschung, Report 11/1997
153. G. Bracco, L. Bruschi, L. Pedemonte, R. Tatarek, Surf. Sci. **377**, 325 (1997)
154. J. Braun, K.L. Kostow, G. Witte, J.G. Skofronick, S.A. Safron, L. Surnev, C. Wöll, Surf. Sci. **372**, 132 (1997)
155. S. Durukanoglu, A. Kara, T.S. Rahman, Phys. Rev. B **55**, 13894 (1997)
156. A. Kara, S. Surukanoglu, T.S. Rahman, J. Chem. Phys. **106**, 2031 (1997)
157. A. Lahee, J.P. Toennies, Ch. Wöll, Surf. Sci. **191**, 529 (1987)
158. S. Narashimhan, M. Scheffler, Z. Phys. Chemie **202**, 253 (1997)
159. I.Y. Sklyadneva, G.G. Rusina, E.V. Chulkov, Phys. Low. Dimens. Struct. **4–5**, 87 (1997)
160. I.Y. Sklyadneva, G.G. Rusina, E.V. Chulkov, Surf. Sci. **377**, 313 (1997)
161. G. Witte, J.P. Toennies, Phys. Rev. B **55**, 1395 (1997)
162. B. Flach, E. Hulpke, W. Steinhögl, Surf. Sci. **412**, 12 (1998)
163. D. Fuhrmann, E. vHulpke, W. Steinhögl, Phys. Rev. B **57**, 4798 (1998)
164. D.J. Gaspar, A.T. Hanbicki, S.J. Sibener, J. Chem. Phys. **109**, 6947 (1998)
165. S. Narashimhan, Surf. Sci. **417**, L1166 (1998)
166. G. Santoro, A. Franchini, V. Bortolani, Nuovo Cim. D **20**, 1055 (1998)
167. G. Santoro, A. Franchini, V. Bortolani, Phys. Rev. Lett. **80**, 2378 (1998)
168. I.Y. Sklyadneva, G.G., Rusina, E.V. Chulkov, Surf. Sci. **416**, 17–36 (1998)
169. W. Steinhögl, Thesis (University of Göttingen 1998, Max-Planck-Institut für Stromungsforschung Göttingen Bericht 10/1998)

170. J.J. Xie, M. Scheffler, Phys. Rev. B **57**, 4768 (1998)
171. I. Vilfan, F. Lancon, E. Adam, Surf. Sci. **440**, 279 (1999)
172. J. Xie, S. de Gironcoli, S. Baroni, M. Scheffler, Phys. Rev. B **59**, 970 (1999)
173. R. Zivieri, G. Santoro, V. Bortolani, Phys. Rev. B **59**, 15959 (1999)
174. A.N. Al-Rawi, A. Kara, T.S. Rahman, Surf. Sci. **446**, 17 (2000)
175. M.R. Baldan, E. Granato, S.C. Ying, Phys. Rev. B **62**, 2146 (2000)
176. B. Flach, Dissertation (University of Göttingen, 1999, Max-Planck-Institut für Strömungsforschung, Report 1/2000)
177. D.M. Riffe, G.K. Wertheim, Phys. Rev. B **61**, 2302 (2000)
178. S. Bartholmei, P. Fouquet, G. Witte, Surf. Sci. **473**, 227 (2001)
179. G. Benedek, E. Hulpke, W. Steinhögl, Phys. Rev. Lett. **87**, 027201 (2001)
180. A. dal Corso, Phys. Rev. B **64**, 235118 (2001)
181. G. Santoro, A. Franchini, V. Bortolani, D.L. Mills, R.F. Wallis, Surf. Sci. **478**, 99 (2001)
182. M. Okada, B. Flach, E. Hulpke, W. Steinhögel, E.W. Plummer, Surf. Sci. **498**, L78 (2002)
183. F. Raouafi, C. Barreteau, M.C. Desjonquères, D. Spanjaard, Surf. Sci. **507**, 748 (2002)
184. R. Heid, K.-P. Bohnen, Physics Reports **387**, 151 (2003)
185. N. Pertaya, Dissertation (Freie Universität Berlin 2004) www.diss.fu-berlin.de
186. S. Hong, T.S. Rahman, R. Heid, K.P. Bohnen, Surf. Sci. **587**, 41 (2005)
187. A. Kara, T.S. Rahman, Surf. Sci. Rep. **56**, 159 (2005)
188. G. Alexandrowicz, A.P. Jardine, J. Phys.: Condens. Matter **19**, 305501 (2007)
189. V. Chis, B. Hellsing, G. Benedek, M. Bernasconi, J.P. Toennies, J. Phys.: Condens.Matter **19**, 305011 (2007)
190. T. Slezak, J. Lazewski, S. Stankov, K. Parlinski, R. Reitinger, M. Rennhofer, R. Rüffer, B. Sepiol, M. Slezak, N. Spiridis, M. Zajac, A.I. Chumakov, J. Korecki, Phys. Rev. Lett. **99**, 066103 (2007)
191. J. Lazewski, J. Korecki, K. Parlinski, Phys. Rev. B **75**, 054303 (2007)
192. S. Stankov, R. Rohlsberger, T. Slezak, M.Sladecek, B. Sepiol, G. Vogl, A.I. Chumakov, R. Rüffer, N. Spiridis, J. Lazewski, K. Parlinski, J. Korecki, Phys. Rev. Lett. **99**, 185501 (2007)
193. V. Chis, B. Hellsing, G. Benedek, M. Bernasconi, E.V. Chulkov, J.P. Toennies, Phys. Rev. Lett. **101**, 206102 (2008); Erratum: Phys. Rev. Lett. **103**, 069902 (2009)
194. F. Yndurain, M.P. Jigato, Phys. Rev. Lett. **100**, 205501 (2008)
195. J. Braun, P. Ruggerone, G. Zhang, J.P. Toennies, G. Benedek, Phys. Rev. B **79**, 205423 (2009)
196. I.Y. Sklyadneva, R. Heid, K.-P. Bohnen, E.V. Chulkov, New J. Phys. **11**, 103038 (2009)
197. G. Benedek, M. Bernasconi, V. Chis, E. Chulkov, P.M. Echenique, B. Hellsing, J. P. Toennies, J. Phys.: Condens. Matter **22**, 084020 (2010)
198. A. Tamtögl, M. Mayrhofer-Reinhartshuber, N. Balak, W.E. Ernst, K.H. Rieder, J. Phys.: Condens. Matter **22**, 304019 (2010)
199. I.Y. Sklyadneva, G. Benedek, E.V. Chulkov, P.M. Echenique, R. Heid, K.-P. Bohnen, J. P. Toennies, Phys. Rev. Lett. **107**, 095502 (2011)
200. I.Y. Sklyadneva, R. Heid, K.-P. Bohnen, P.M. Echenique, G. Benedek, E.V. Chulkov, J. Phys. Chem. A **115**, 7352 (2011)
201. I.Y. Sklyadneva, R. Heid, K.-P. Bohnen, P.M. Echenique, E.V. Chulkov, J. Phys.: Condens. Matter **24**, 104004 (2012)
202. V. Chis, G. Benedek, P.M. Echenique, E.V. Chulkov, Phys. Rev. B **87**, 075412 (2013)
203. A. Graham, J.P. Toennies, to be published
204. E.M. McIntosh, P.R. Kole, M. El-Batanouny, D.M. Chisnall, J. Ellis, W. Allison, Phys. Rev. Lett. **110**, 086103 (2013)
205. I.Y. Sklyadneva, R. Heid, K.-P. Bohnen, P.M. Echenique, E.V. Chulkov, Phys. Rev.B **87**, 85440 (2013)
206. A. Tamtögl, P. Kraus, M. Mayrhofer-Reinhartshuber, W.E. Ernst, D. Campi, M. Bernasconi, G. Benedek, Phys. Rev. B **87**, 035410 (2013); Erratum: Phys. Rev. B **87**, 159906 (2013)
207. D. Campi, M. Bernasconi, G. Benedek, A.P. Graham, J.P. Toennies, Phys. Chem. Chem. Phys. **19**, 16358 (2017)

208. A. Tamtögl, E. Bahn, J. Zhu, P. Fouquet, J. Ellis, W. Allison, J. Phys. Chem. C **119**, 25983 (2015)*
209. M. Al-Taleb, D. Farìas, J. Phys.: Condens. Matter **28**, 103005 (2016)
210. B. Gans, S.K. King, P.A. Knipp, D.D. Koleske and S.J. Sibener, J. Electr. Spectr. Rel. Phenom. 54/55, 333 (1990)
211. A. Fasolino, G. Santoro, E. Tosatti, Phys. Rev. Lett. 44, 1684 (1980)
212. A.G. Eguiluz, A.A. Maradudin and R.F. Wallis, in Solvay Conference on Surface Physics, F. W. de Wette Ed. (Springer, Heidelberg 1988) p. 291
213. G.Q. Huang, and J.Yang, J. Phys. Cond. Matt. 25, 175004 (2013)
214. M.A. Ortigoza, I.Y. Sklyadneva, R. Heid, E.V. Chulkov, T.S. Rahman, K.-P. Bohnen, P.M. Echenique, Phys. Rev. B 90 195438 (2014)
215. R B Wilson and D M Riffe J. Phys.: Condens. Matter 24, 335401 (2012)
216. Mila F., Szeftel J., J. Phys. C: Solid State Phys. 21, 177 (1988)
217. J Szeftel, A Khater, F Mila, S d'Addato and N Auby, J. Phys. C: Solid State Phys. 21, 211 (1988)
218. Szeftel J., Mila F., J. Phys. C: Solid State Phys. 21, L1131 (1988)

Appendix A.5

Table A.5 Vibrations of adsorbates on insulator and semiconductor surfaces from HAS. Hydrogen-passivated surfaces are in Table A.1

Substrate	Surface	Adsorbate	Thickness [ML]	References
BaF_2	(111)	H_2O	≥ 1	[37]
C (diamond)	(111)	H	1	[19]
C (diamond)	(111)	D	1	[23]
Ge	(001)	NaCl	≤ 1, 2, 3, 10	[5, 28]
Ge	(111)	CH_3, CD_3	1	[44, 46]
Ge	(111)	Sn	1	[31, 35, 50]
Ge	(111)	Pb $\sqrt{3}\times\sqrt{3}$	4/3, 3–8	[47, 48, 52]
Ice Ih	(0001)	CHF_3	≤ 1	[26]
Ice Ih	(0001)	Xe	≤ 1	[32]
KCl	(001)	C_2H_2 (acetylene)	≤ 1	[27]
KCl	(001)	C_2H_2 (acetylene)	15	[30]
KBr	(001)	KCN	≤ 1	[14, 25]
LiF	(001)	n-H_2, pH_2	≤ 1	[36]
Mica	(001)	C_{60}	1.7	[4]
MgO	(001)	CO	≤ 1	[11]
MgO	(001)	CO_2	≤ 1	[7]
MgO	(001)	n-H_2, p-H_2, HD, D_2	<1	[33, 36]
MgO	(001)	$^{14}N_2$, $^{15}N_2$	<1	[29]
NaCl	(001)	n-H_2, p-H_2	≤ 1	[34, 36, 40, 43]
NaCl	(001)	CH_4 (methane)	≤ 1	[13]
NaCl	(001)	C_2H_2 (acetylene)	≤ 1	[20, 24]

(continued)

Table A.5 (continued)

Substrate	Surface	Adsorbate	Thickness [ML]	References
NaCl	(001)	CO	≤ 1	[12, 7]
NaCl	(001)	CO_2	≤ 1	[13]
NaCl	(001)	n-H_2, p-H_2, HD, D_2	≤ 1	[34, 36]
NaCl	(001)	H_2O	≤ 1	[10]
NaCl	(001)	KBr	2, 3, 4, 7	[6, 8]
NaCl	(001)	OCS	≤ 1	[18, 21, 15]
NaCl	(001)	Xe	≤ 1	[22]
RbCl	(001)	KBr	2,3	[9, 16, 17]
Si	(111)	CH_3, CD_3	≤ 1	[39, 41, 42, 44]
Si	(111)	H	1	[1, 2, 3]
Si	(111)	Pb $\sqrt{3}\times\sqrt{3}$	4/3, 3–5	[49, 48, 51]
SiOx/Si(100)	(100)	Polymethyl-methacrylate	Thick films	[38]
SiOx/Si(100)	(100)	PMMA	Thick films	[38]
SiOx/Si(100)	(100)	Polystyrene,	Thick films	[38]
SiOx/Si(100)	(100)	Polybutadiene	Thick films	[38]

References

1. U. Harten, J.P. Toennies, Ch. Wöll, L. Miglio, P. Ruggerone, L. Colombo, G. Benedek, Phys. Rev. B **38**, 3305 (1988)
2. R.B. Doak, Y.J. Chabal, G.S. Iligashi, P. Dumas, J. Electr. Spectr. Rel. Phen. **54/55**, 291 (1990)
3. R.B. Doak, J. Electr. Spectr. Rel. Phen. **54/55**, 281 (1990)
4. D. Schmicker, S. Schmidt, J.G. Skofronick, J.P. Toennies, R. Vollmer, Phys. Rev. B **44**, 10995 (1991)
5. G. Brusdeylins, N.S. Luo, P. Ruggerone, D. Schmicker, J.P. Toennies, R. Vollmer, T. Wach, Surf. Sci. **272**, 358 (1992)
6. J. Duan, G.G. Bishop, E.S. Gillman, G. Chern, S.A. Safron, J.G. Skofronick, Surf. Sci. **272**, 220 (1992)
7. J. Heidberg, E. Kampshoff, R. Künemuth, O. Schönekäs, G. Lange, D. Schmicker, J. P. Toennies, R. Vollmer, H. Weiss, J. Electron. Spectrosc. Relat. Phenom. **64/65**, 803 (1993)
8. S.A. Safron, G.G. Bishop, J. Duan, E.S. Gillman, J.G. Skofronick, N.S. Luo, P. Ruggerone, J. Chem. Phys. **97**, 2270 (1993)
9. S.A. Safron, J. Duan, G.G. Bishop, E.S. Gillmann, J.G. Skofronick, J. Phys. Chem. **97**, 1749 (1993)
10. L.W. Bruch, A. Glebov, J.P. Toennies, H. Weiss, J. Chem. Phys. **103**, 5109 (1995)
11. R. Gerlach, A. Glebov, G. Lange, J.P. Toennies, H. Weiss, Surf. Sci. **331–333**, 1490 (1995)
12. G. Lange, D. Schmicker, J.P. Toennies, R. Vollmer, H. Weiss, J. Chem. Phys. **103**, 2308 (1995)
13. H. Weiss, Habilitation Thesis (University of Hannover, 1995)
14. J. Baker, J.J. Hernández, J. Li, J.G. Skofronick, S.A. Safron, E.S. Gillman, Surf. Sci. **357/358**, 639 (1996)
15. J. Dohrmann, A. Glebov, J.P. Toennies, H. Weiss, Surf. Sci. **368**, 118 (1996)
16. E.S. Gillman, J. Baker, J.J. Hernández, G.G. Bishop, J.A. Li, S.A. Safron, J.G. Skofronick, D. Bonart, U. Schröder, Physica B **219+220**, 428 (1996)

17. E.S. Gillman, J. Baker, J.J. Hernández, G.G. Bishop, J.A. Li, S.A. Safron, J.G. Skofronick, D. Bonart, U. Schröder, Phys. Rev. B **33**, 13891 (1996-II)
18. A. Glebov, J.P. Toennies, H. Weiss, Surf. Sci. **351**, 200 (1996)
19. G. Lange, J.P. Toennies, Phys. Rev. B **53**, 9614 (1996)
20. A. Glebov, R.E. Miller, J.P. Toennies, J. Chem. Phys. **106**, 6499 (1997)
21. S. Picaud, C. Girardet, A. Glebov, J.P. Toennies, J. Dohrmann, H. Weiss, J. Chem. Phys. **106**, 5271 (1997)
22. R. Gerlach, A.P. Graham, J.P. Toennies, H. Weiss, J. Chem. Phys. **109**, 5319 (1998)
23. A. Glebov, J.P. Toennies, S. Vollmer, S.A. Safron, J.G. Skofronick, V. Gräschus, A. Mazur, J. Pollmann, Phys. Rev. B **57**, 10082 (1998)
24. S. Picaud, P.N.M. Hoang, C. Girardet, A. Glebov, R.E. Miller, J.P. Toennies, Phys. Rev. B **57**, 10090 (1998)
25. J. Baker, J.A. Li, J.G. Skofronick, S.A. Safron, Phys. Rev. B **60**, 2011 (1999-I)
26. A.P. Graham, A. Menzel, J.P. Toennies, J. Chem. Phys. **111**, 1169 (1999)
27. A.L. Glebov, V. Panella, J.P. Toennies, F. Traeger, H. Weiss, S. Picaud, P.N.M. Hoang, C. Girardet, Phys. Rev. B **61**, 14028 (2000)
28. N.S. Luo, P. Ruggerone, J.P. Toennies, Phys. Rev. B **54**, 5051, (1996)
29. F. Träger, Dissertation (University of Göttingen, 2000)
30. J.P. Toennies, F. Traeger, H. Weiss, S. Picaud, P.N.M. Hoang, Phys. Rev. B **65**, 165427 (2002)
31. D. Farías, W. Kaminski, J. Lobo, J. Ortega, E. Hulpke, R. Pérez, F. Flores, E.G. Michel, Phys. Rev. Lett. **91**, 016103 (2003)
32. A.P. Graham, Surf. Sci. Reports **49**, 115 (2003)
33. J.G. Skofronick, J.P. Toennies, F. Traeger, H. Weiss, Phys. Rev. B **67**, 035413 (2003)
34. F. Traeger, J.P. Toennies, J. Phys. Chem. B **108**, 14710 (2004)
35. J. Lobo, D. Farías, E. Hulpke, E.G. Michel, Phys. Rev. B **71**, 205402 (2005)
36. J.P. Toennies, F. Traeger, J. Phys.: Condens. Matter **19**, 305009 (2007)
37. J. Vogt, J. Chem. Phys. **126**, 244710 (2007)
38. M.A. Freedman, J.S. Becker, A.W. Rosenbaum, S.J. Sibener, J. Chem. Phys. **129**, 044906 (2008)
39. J.S. Becker, R.D. Brown, E. Johansson, N.S. Lewis, S.J. Sibener, J. Chem. Phys. **133**, 104705 (2010)
40. L.W. Bruch, F.Y. Hansen, F. Traeger, J. Chem. Phys. **134**, 194308 (2011)
41. R.D. Brown, Q. Tong, J.S. Becker, M.A. Freedman, N.A. Yufa, S.J. Sibener, Faraday Discuss. **157**, 307 (2012).
42. R.D. Brown, Z.M. Hund, D. Campi, L.E. O'Leary, N.S. Lewis, M. Bernasconi, G. Benedek, S.J. Sibener, Phys. Rev. Lett. **110**, 156102 (2013)
43. F. Y. Hansen, L. W. Bruch, B, J. Dammann, Chem. Phys. **138**, 107405 (2013)
44. R.D. Brown, Z.M. Hund, D. Campi, L.E. O'Leary, N.S. Lewis, M. Bernasconi, G. Benedek, S.J. Sibener, J. Chem. Phys. **141**, 024702 (2014)
45. Z.M. Hund, K.J. Nihill, D. Campi, K.T. Wong, N.S. Lewis, M. Bernasconi, G. Benedek, and S.J. Sibener, J. Phys. Chem. C **119**, 18458–18466 (2015).
46. R.D. Brown, Z.M. Hund, K.J. Nihill, D. Campi, K.T, Wong, N.S. Lewis, M. Bernasconi, G. Benedek, S.J. Sibener, J. Chem. Phys. **143**, 124705 (2015)
47. S. Schmidt, Thesis (University of Göttingen, 1992) (unpublished)
48. G. Benedek , I. Yu. Slyadneva , E.V. Chulkov , P.M. Echenique ,R. Heid , K.-P. Bohnen , D. Schmicker , S. Schmidt , J.P. Toennies , Surface Science (2018) https://doi.org/10.1016/j.susc.2018.02.009
49. S. Schmicker, Thesis (University of Göttingen, 1992) (unpublished)
50. D. Farías, W. Kamínski, J. Lobo, and E.G. Michel, Appl. Surf. Sci. **237**, 86, (2004)
51. S. Sakong, P. Kratzer, S. Wall, A. Kalus, and M. Horn-von Hoegen, Phys. Rev. B **88**, 115419 (2013)
52. L. Floreano, D. Cvetko, F. Bruno, G. Bavdek, A. Cossaro, R. Gotter, A. Verdini, and A. Morgante, Progr. Surface Sci. **72**, 135 (2003)

Appendix A.6

Table A.6 Vibrations of adsorbates on metal and semimetal surfaces from HAS. Graphene is included in Sect. A.8; hydrogen-passivated surfaces in Sect. A.4

Substrate	Surface	Adsorbate	Thickness [ML]	References
Ag	(110)	Kr	≤1	[5]
Ag	(110)	Xe	≤1	[4, 5]
Ag	(111)	Ar	≤1, 2, 3, 25	[6, 7, 14]
Ag	(111)	Kr	≤1, 2, 3, 25	[6, 7, 8, 14]
Ag	(111)	Xe	≤1, 2, 3, 25	[6, 7, 14]
Au	(111)	Kr	≤1, 2	[11]
Au	(111)	Xe	≤1, 2–7	[11]
Au	(111)	Co	≤1, 2–3.5	[57]
C graphite	(0001)	Cs	≤1, 2, 3	[36, 29, 82]
C graphite	(0001)	K	≤1	[36, 29, 82]
C graphite	(0001)	Rb	1.2	[36, 29, 82]
C graphite	(0001)	Xe	≤1	[17]
Cu	(001)	Ba	≤1, 2, 3	[61]
Cu	(001)	Co	≤1, 8	[18]
Cu	(001)	Cs	≤1, 2, 3.4	[60, 82]
Cu	(001)	Cs	≤1	[55, 74, 82]
Cu	(001)	Fe	2, 3	[18]
Cu	(001)	Fe	≤1, 2–14	[62, 78]
Cu	(001)	Fe	3, 8, 13	[86]
Cu	(001)	Fe	0–5	[68]
Cu	(001)	Li	0–2	[77**, 80]
Cu	(001)	K	≤1, 2–5	[38, 82]
Cu	(001)	K	≤1	[55, 76, 82]
Cu	(001)	Kr	≤1	[3]
Cu	(001)	N	≤1	[79]
Cu	(001)	Na	≤1	[49, 50, 51, 87, 82]
Cu	(001)	Na	2–5, 10, 15, 20	[24, 30, 33, 34, 55, 72, 82]
Cu	(001)	Ni	≤1, 3	[18]
Cu	(001)	Ni	1–7, 3D	[70]
Cu	(001)	Tl	≤1	[44]
Cu	(001)	Xe	≤1	[1, 2]
Cu	(001)	Xe	≤1	[48, 56]
Cu	(110)	Kr	≤1	[4]
Cu	(110)	Xe	2	[4]

(continued)

Table A.6 (continued)

Substrate	Surface	Adsorbate	Thickness [ML]	References
Cu	(110)	Xe	$\le$1	[37, 52]
Cu	(110)	Xe	$\le$1	[67]
Cu	(111)	Cs	2–5	[43, 82]
Cu	(111)	Pb	$\le$1	[39]
Cu	(111)	Pb	3–6.5	[44, 75, 81]
Cu	(111)	Pb	$\le$1, 3–10, 300	[23, 75, 81, 85]
Cu	(111)	Xe	$\le$1	[53, 56]
Mo	(110)	H	1	[31, 32]
$Mo_{1-x}Re_x$	(110)	H	1	[66]
Ni	(001)	K	2–5	[43, 35, 82]
Ni	(001)	K	10	[35, 82]
Ni	(001)	K	74	[86, 82]
Ni	(001)	Pt	3D	[71]
Pd	(111)	H	1	[19, 83]
Pd	(311)	H	1	[83]
Pt	(111)	Ar	$\le$1, 2, 3	[16, 20, 21, 22]
Pt	(111)	Ar	$\le$1	[58, 13]
Pt	(111)	H	1	[15]
Pt	(111)	Na	$\le$1	[63, 82]
Pt	(111)	Kr	$\le$1, 2, 3, 25	[16, 13, 22, 12]
Pt	(111)	Kr	$\le$1	[58, 65, 13]
Pt	(111)	Xe	$\le$1, 2, 3, 25	[22, 9, 10, 73]
Pt	(111)	Xe	$\le$1	[58, 54, 13]
Pt	(111)	Xe	160	[64]
Rh	(111)	H	1	[38]
Ru	(0001)	H	1	[17]
Sb	(111)	H	$\le$1	[84]
W	(110)	Li	$\le$1, 2–9	[59, 82]
W	(110)	H	1	[27, 28, 31, 32]

References

1. B.F. Mason, B.R. Williams, Phys. Rev. Lett. **46**, 1138 (1981).
2. B.F. Mason, B.R. Williams, Surf. Sci. **111**, 609 (1981)
3. B.F. Mason, R. Caudano, B.R. Williams, J. Chem. Phys. **77**, 562 (1982)
4. B.F. Mason, B.R. Williams, Surf. Sci. **130**, 295 (1983)
5. B.F. Mason, B.R. Williams, Surf. Sci. **139**, 173 (1984)
6. K.D. Gibson, S.J. Sibener, B.M. Hall, D.L. Mills, J.E. Black, J. Chem. Phys. **83**, 4256 (1985)
7. K.D. Gibson, S.J. Sibener, Faraday Discuss. Chem. Soc. **80**, 203 (1985).
8. K.D. Gibson, S.J. Sibener, Phys. Rev. Lett. **55**, 1514 (1985)

9. K. Kern, R. David, R.L. Palmer, G. Comsa, Phys. Rev. Lett. **56**, 2823 (1986).
10. K. Kern, R. David, R.L. Palmer, G. Comsa, Surf. Sci. **175**, L669 (1986).
11. B.F. Mason, B.R. Williams, Surf. Sci. **177**, L925 (1986)
12. K. Kern, P. Zeppenfeld, R. David, G. Comsa, Phys. Rev. B **35**, 886 (1987).
13. K. Kern, P. Zeppenfeld, R. David, G. Comsa, J. Electron. Spectrosc. Relat. Phenom. **44**, 215 (1987).
14. K.D. Gibson, S.J. Sibener, J. Chem. Phys. **88**, 7862 (1988)
15. V. Bortolani, A. Franchini, G. Santoro, J.P. Toennies, Ch. Wöll, G. Zhang, Phys. Rev. B **40**, 3524 (1989)
16. B. Hall, D.L. Mills, P. Zeppenfeld, U. Becher, K. Kern, G. Comsa, Phys. Rev. B **40**, 6326 (1989-II)
17. J.P. Toennies, R. Vollmer, Phys. Rev. B **40**, 3495 (1989)
18. J. Ellis, E.M. McCash, W. Allison, J. Electron. Spectrosc. Relat. Phenom. **54–55**, 325 (1990)
19. C.H. Hsu, M. El-Batanouny and K.M. Martini, J. Electr. Spectr. Rel. Phenom. **54/55**, 353 (1990)
20. K. Kern, U. Becher, P. Zeppenfeld, G. Comsa, B. Hall, D.L. Mills, Chem. Phys. Lett. **167**, 362 (1990).
21. P. Zeppenfeld, U. Becher, K. Kern, R. David, G. Comsa, Phys. Rev. B **41**, 8549 (1990 II)
22. P. Zeppenfeld, U. Becher, K. Kern, G. Comsa, J. Electron. Spectrosc. Relat. Phenom. **54**, 265 (1990)
23. G. Zhang, Dissertation, University of Göttingen, 1990, Max-Planck-Institut für Strömungsforschung, Report 102/1991.
24. G. Benedek, J. Ellis, A. Reichmuth, P. Ruggerone, H. Schief, J.P. Toennies Phys. Rev. Lett. **69**, 2951 (1992)
25. G. Benedek, J.P., G. Zhang, Phys. Rev. Lett. **68**, 2644 (1992)
26. G. Brusdeylins, N.S. Luo, P.D. Ruggerone, J.P. Toennies, R. Vollmer, T. Wach, Surf. Sci. **272**, 358 (1992)
27. E. Hulpke, J. Lüdecke, Surf. Sci. **272**, 289 (1992)
28. E. Hulpke, J. Lüdecke, Phys. Rev. Lett. **68**, 2846 (1992)
29. J. Cui, J.D. White, R.D. Diehl, Surf. Sci. Lett. **293**, L841 (1993)
30. Ellis, J., J.P. Toennies, Phys. Rev. Lett. **70**, 2118 (1993)
31. Hulpke, E., J. Lüdecke, Surf. Sci. **287/288**, (1993) 837
32. E. Hulpke, J. Lüdecke, J. Electr. Spectr. Rel. Phenom. **64/65**, 641 (1993)
33. G. Benedek, J. Ellis, P. Ruggerone, J.P. Toennies: Vuoto **23**, 30 (1994)
34. G. Benedek, N.S. Luo, P. Ruggerone, A. Reichmuth, J.P. Toennies, Mater. Sci. Eng. B **23**, (1994) 123
35. A. Reichmuth, Dissertation, University of Cambridge, 1994
36. J.D. White, J. Cui, M. Strauss, R.D. Diehl, F. Ancilotto, F. Toigo, Surf. Sci. **307**, (1994) 1134
37. P. Zeppenfeld, M. Büchel, R. David, G. Comsa, C. Ramseyer, C. Girardet, Phys. Rev. B **50**, 14667 (1994)
38. G.W. Witte: Dissertation (University of Göttingen, 1995, Max-Planck-Institut für Strömungsforschung, Report 8/1995)
39. J. Braun, J.P. Toennies, Surf. Sci. **368**, 226 (1996)
40. G. Bracco, L. Bruschi, L. Pedemonte, R. Tatarek, Surf. Sci. **352**, 964 (1996)
41. G. Bracco, L. Bruschi, R. Tatarek, A. Franchini, V. Bortolani, G. Santoro, Europhys. Lett. **34**, 687 (1996)
42. J. Braun, A.P. Graham, F. Hofmann, W. Silvestri, J.P. Toennies, G. Witte, J. Chem. Phys. **105**, 3258 (1996)
43. E. Hulpke, J. Lower, A. Reichmuth, Phys. Rev. B **53**, 13901 (1996)

44. J. Braun, Dissertation (University of Göttingen, 1997), Max-Planck-Institut für Strömungsforschung, Report 11/1997
45. J. Braun, K.L. Kostov, G. Witte, C. Wöll, J Chem. Phys. **106**, 8262 (1997)
46. G. Bracco, L. Bruschi, L. Pedemonte, R. Tatarek, Surf. Sci. **377**, 325 (1997)
47. J. Braun, K.L. Kostow, G. Witte, J.G. Skofronick, S.A. Safron, L. Surnev, C. Wöll, Surf. Sci. **372**, 132 (1997)
48. A.P. Graham, M.F. Bertino, J.P. Toennies, C. Wöll, J. Chem. Phys. **106**, 6194 (1997)
49. A.P. Graham, F. Hofmann, J.P. Toennies, L.Y. Chen, S.C. Ying, Phys. Rev. B **56**, 10567 (1997)
50. A.P. Graham, F. Hofmann, J.P. Toennies, L.Y. Chen, S.C. Ying, Phys. Rev. Lett. **78**, 3900 (1997)
51. A.P. Graham, J.P. Toennies, Phys. Rev. B **56**, 15378 (1997)
52. C. Ramseyer, V. Pouthier, C. Girardet, P. Zeppenfeld, M. Büchel, V. Diercks, G. Comsa, Phys. Rev. B **55**, 13203 (1997)
53. J. Braun, A. Glebov, A.P. Graham, A. Menzel, J.P. Toennies, Phys. Rev. Lett. **80**, 2638 (1998)
54. L.W. Bruch, A.P. Graham, J.P. Toennies, Mol. Phys. **95**, 579 (1998)
55. P. Senet, J.P. Toennies, G.Witte, Chem. Phys. Lett. **299**, 389 (1999)
56. A. Šiber, B. Gumhalter, J. Braun, A.P. Graham, M.F. Bertino, J.P. Toennies, D. Fuhrmann, C. Wöll, Phys. Rev. B **59**, 5898 (1999-II)
57. C. Tolkes, R. David, G. Comsa, P. Zeppenfeld, J. Electron. Spectrosc. **105**, 37 (1999)
58. L.W. Bruch, A.P. Graham, J.P. Toennies, J. Chem. Phys. **112**, 3314 (2000)
59. B. Flach, Dissertation (University of Göttingen, 1999), Max-Planck-Institut für Strömungsforschung, Report 1/2000
60. G. Witte, J.P. Toennies, Phys. Rev. B **62**, R7771 (2000)
61. S. Bartholmei, P. Fouquet, G. Witte, Surf. Sci. **473**, 227 (2001)
62. G. Benedek, E. Hulpke, W. Steinhögl, Phys. Rev. Lett. **87**, 027201 (2001)
63. A.P Graham, J. Chem. Phys. **115**, 524 (2001)
64. A. Šiber, B. Gumhalter, A.P. Graham, J.P. Toennies, Phys. Rev. B **63**, 115411 (2001)
65. M. Li, J.R. Manson, A.P. Graham, Phys. Rev. B **65**, 195404 (2002)
66. M. Okada, B. Flach, E. Hulpke, W. Steinhögel, E.W. Plummer, Surf. Sci. **498**, L78 (2002)
67. C. Boas, M. Kunat, U. Burghaus, B. Gumhalter, C. Wöll, Phys. Rev. B **68**, 075403 (2003)
68. P.C. Dastoor, W. Allison Phys. Rev. B **67**, 245403 (2003)
69. A.P. Graham, Surf. Sci. Reports **49**, 371 (2003)
70. C. Huang, D.A. MacLaren, W. Allison, J. Phys. Conf. Ser. **19**, 182 (2005)
71. D.A. MacLaren, W. Allison, J. Phys. Cond. Matter **17**, 7433 (2005)
72. G. Alexandrowicz, A.P. Jardine, H. Hedgeland, W. Allison, J. Ellis, Phys. Rev. Lett. **97**, 156103 (2006)
73. F.Y. Hansen, L.W. Bruch, J. Chem. Phys. **127**, 204708 (2007)
74. A.P. Jardine, G. Alexandrowicz, H. Hedgeland, R.D. Diehl, W. Allison, J. Ellis, J. Phys.: Cond. Matter **19**, 305010 (2007)
75. J. Braun, P. Ruggerone, G. Zhang, J. P. Toennies, G. Benedek, Phys. Rev. B **79**, 205423 (2009)
76. H. Hedgeland, P.R. Kole, H.R. Davies, A.P. Jardine, G. Alexandrowicz, W. Allison, J. Ellis, G. Fratesi, G.P. Brivio, Phys. Rev. B **80**, 125426 (2009)
77. A C Levi, C. Huang, W. Allison D.A MacLaren, J. Phys. Cond. Matter **21,** 225009 (2009)**
78. G. Benedek, M. Bernasconi, V. Chis, E. Chulkov, P.M. Echenique, B. Hellsing, J. Peter Toennies, J. Phys.: Cond. Matter **22**, 084020 (2010)
79. E.Z. Ciftlikli, L.V. Goncharova, B.J. Hinch, M.A. Ortigoza, S. Hong, T.S. Rahman, Phys. Rev. B **81,** 115465 (2010)

80. C. Huang, D.A MacLaren, R.T Bacon, W Allison, J. Phys. Cond. Matter **23**, 355001 (2011)
81. I.Y. Sklyadneva, G. Benedek, E.V. Chulkov, P.M. Echenique, R. Heid, K.-P. Bohnen, J. P. Toennies, Phys. Rev. Lett. **107**, 095502 (2011)
82. A. Politano, G. Chiarello, G. Benedek, E.V. Chulkov, P. M. Echenique, Surf, Sci.Rep., **68**, 305–389 (2013)
83. I.G. Shuttelworth, Chem. Phys. Lett. **592**, 14 (2014)
84. P. Kraus, C. Gösweiner, A. Tamtögl, F. Apolloner, W.E. Ernst, EPL **114**, 56001 (2016)
85. G.G. Rusina, S.D. Borisova, S.V. Eremeev, I.Y. Sklyadneva, E.V. Chulkov, G. Benedek, J. P. Toennies, J. Phys. Chem. C **120**, 22304 (2016)
86. W. Steinhögl, Thesis (University of Göttingen 1998, Max-Planck-Institut für Strömungsforschung Göttingen Bericht 10/1998)
87. A. P. Graham, J. P. Toennies and G. Benedek , Surf. Sci. Lett. 556, L143 (2004)

Appendix A.7

Table A.7 Vibrations of diatomic adsorbates on metal surfaces from HAS

Substrate	Surface	Adsorbate	Thickness [ML]	References
Al	(111)	CO	<1	[14, 24]
Cu	(001)	CO	≤ 1	[1, 8, 10, 13, 14, 20, 24, 25*]
Cu	(001)	NO	≤ 1	[14, 24]
Cu	(001)	CN	≤ 1	[26]
Cu	(110)	CO	≤ 1	[17]
Cu	(111)	CO	<1	[14]
Cu	(211)	CO	≤ 1	[12, 14]
Cu	(511)	CO	≤ 1	[12]
Fe	(110)	CO	≤ 1	[7, 14, 24]
Ge	(001)	NaCl	2, 3, 10	[6, 21]
Ni	(001)	CO	≤ 1	[3, 14]
Ni	(001)	N_2	<1	[14, 24]
Ni	(110)	CO	≤ 1	[15]
Ni	(001)	N_2	<1	[14, 24]
Ni	(111)	CO	≤ 1	[5]
Pt	(111)	CO	≤ 1	[2, 4, 18, 19]
Pt	(111)	NO	≤ 1	[22]
Pt	(111)	N_2	1	[24]
Rh	(111)	CO	≤ 1	[23]
Rh	(111)	CO + O_2	–	[11, 14]
Rh	(111)	NO	≤ 1	[11, 14]
Rh	(111)	CO + NO	≤ 1	[23]
Ru	(0001)	CO	≤ 1	[16]
Ru	(0001)	CO + H	≤ 1	[16]
Ru	(0001)	CO + O	≤ 1	[16]
W	(110)	CO	≤ 1	[9, 14, 24]

*: SE-HAS measurements

References

1. B.F. Mason, R. Caudano, B.R. Williams, J. Chem. Phys. **77**, 562 (1982)
2. A.M. Lahee, J.P. Toennies, C. Wöll, Surf. Sci. **177**, 371 (1986)
3. R. Berndt, J.P. Toennies, C. Wöll, J. Electr. Spectr. Rel. Phenom. **4**, 183 (1987)
4. G. Hähner, J.P. Toennies, C. Wöll, Appl. Phys. A: Solids Surf. A **51**, 208 (1990)
5. J.S. Ha, S. Sibener, Surf. Sci., **256**, 281 (1991)
6. G. Brusdeylins, N.S. Luo, P. Ruggerone, D. Schmicker, J.P. Toennies, R. Vollmer, T. Wach, Surf. Sci. **272**, 358 (1992)
7. J.P. Toennies, C. Wöll, G. Zhang, J. Chem. Phys. **96**, 4023 (1992)
8. M. Bertino, J. Ellis, F. Hofmann, J.P. Toennies, J.R. Manson, Phys. Rev. Lett. **73**, 605 (1994)
9. J. Lüdecke, Dissertation (University of Göttingen 1994), Max-Planck-Institut für Strömungsforschung, Report 15/1994
10. J. Ellis, G. Witte, J.P. Toennies, J. Chem. Phys. **102**, 5059 (1995)
11. H. Range, Dissertation (University of Göttingen 1995), Max-Planck-Institut für Strömungsforschung, Report 3/1995
12. J. Braun, A.P. Graham, F. Hofmann, W. Silvestri, J.P. Toennies, G. Witte, J. Chem. Phys. **105**, 3258 (1996)
13. A.P. Graham, M.F. Bertino, F. Hofmann, J.P. Toennies, J. Chem. Soc., Faraday Trans. **92**, 4749 (1996)
14. F. Hofmann, J.P. Toennies, Chem. Rev. **96**, 1307 (1996)
15. M.F. Bertino, G. Witte, Surf. Sci. **385**, L984 (1997)
16. J. Braun, K.L. Kostov, G. Witte, C. Wöll, J. Chem. Phys. **106**, 8262 (1997)
17. J. Braun, J. Weckesser, J. Ahner, D.Mocuta, J.T. Yates, C. Wöll, J. Chem. Phys. **108**, 5161 (1998)
18. A.P. Graham, J. Chem. Phys. **109**, 9583 (1998)
19. A.P. Graham, J.P. Toennies, Europhys. Lett. **42**, 449 (1998)
20. A.P. Graham, F. Hofmann, J.P. Toennies, G.P. Williams, C.J. Hirschmugl, J. Ellis, J. Chem. Phys. **108**, 7825 (1998)
21. N.S. Luo, P. Ruggerone, J.P. Toennies, Phys. Rev. B **54**, 5051 (1996)
22. P. Fouquet, A.P. Graham, G. Witte, J. Chem. Phys. **112**, 7600 (2000)
23. G. Witte, J. Chem. Phys. **115**, 2757 (2001)
24. A.P. Graham, Surf. Sci. Rep. **49**, 115 (2003)
25. G. Alexandrowicz, A.P. Jardine, P. Fouquet, S. Dworski, W. Allison, J. Ellis, Phys. Rev. Lett. **93**, 156103 (2004)*
26. E.Z. Ciftlikli, Rutgers, Thesis (The State University of New Jersey, 2012), URL http://dx.doi.org/doi:10.7282/T34F1PJN

Appendix A.8

Table A.8 Vibrations of polyatomic adsorbates including graphene on metal and semimetal surfaces from HAS

Substrate	Surface	Polyatomic adsorbate	Thickness [ML]	References
Ag	(110)	C_2H_6 (ethane)	$\leq$1, 2–4	[2, 3]
Au	(111) 23 × √3	C_6H_{13} SH (1-hexanethiol)	1 SAM	[24]
Au	(111) 23 × √3	n-hexylthiolate	1 SAM	[21**]
Au	(111) 23 × √3	C_8H_{17} SH (1-octanethiol)	1 SAM	[24]
Au	(111) 23 × √3	$C_{10}H_{21}$ SH (1-decanethiol)	1 SAM	[24, 19, 23**]
Au	(111)	$[C_2C_1Im][Tf_2N]$	<1	[30*]
C (graphite)	(0001)	C_6H_6 (benzene)	<1	[27]
Cu	(001)	CH_4 (methane)	$\leq$1	[12]
Cu	(001)	C_2H_4 (ethane)	$\leq$1	[10]
Cu	(001)	C_2H_6 (ethane)	$\leq$1	[12]
Cu	(001)	C_6H_6 (benzene)	$\leq$1	[6]
Cu	(001)	C_6H_{12} (cyclohexane)	$\leq$1	[8]
Cu	(001)	C_6H_{14} (n-hexane)	$\leq$1	[6]
Cu	(001)	C_8H_{18} (n-octane)	$\leq$1	[8]
Cu	(001)	$C_{10}H_{22}$ (n-decane)	$\leq$1	[6]
Cu	(001)	C_2H_5 SH (ethanethiol)	1 SAM	[18]
Cu	(001)	C_7H_{15} SH (heptanethiol)	1 SAM	[18]
Cu	(001)	CH_3CN (acetonitrile)	$\leq$1	[25]
Cu	(001)	C_2N_2 (cyanogen)	$\leq$1	[26]
Cu	(001)	PF_3	6.25	[14]
Cu	(001)	TCNQ (tetracyanoquinodimethane)	1	[29]
Cu	(110)	CH_4 (methane)	$\leq$1	[1]
Cu	(110)	C_2H_6 (ethane)	<1, 2	[1]
Cu	(110)	C_2H_5SH (ethanethiol)	1 SAM	[18]
Cu	(110)	C_7H_{15} SH (heptanethiol)	1 SAM	[18]
Cu	(111)	C_8H_{18} (d-octane, n-octane)	$\leq$1	[9,11, 15]
Cu	(111)	C_9H_{20} (n-nonane)	$\leq$1	[9,11, 15]
Cu	(111)	$C_{10}H_{22}$ (n-decane)	$\leq$1	[9,11]
Cu	(111)	graphene	1	[31, 36]

(continued)

Table A.8 (continued)

Substrate	Surface	Polyatomic adsorbate	Thickness [ML]	References
Fe	(110)	NH_3	≤ 1, 2	[4]
Ni	(110)	CO_2	≤ 1	[7]
Ni	(111)	graphene	1	[35**, 36, 37]
Pb	(111)	C_8H_{18} (d-octane, n-octane)	≤ 1	[9]
Pt	(111)	H_2O	≤ 1	[16]
Pt	(111)	H_2O	250	[17]
Pt	(111)	CH_3COOH (acetic acid)	≤ 1	[20]
Pt	(111)	C_2H_5COOH (propionic acid)	≤ 1	[20]
Pt	(111)	$C_{24}H_{50}$ (tetracosane)	≤ 1	[22]
Pt	(111)	$C_{30}H_{62}$ (squalane)	≤ 1	[22]
Pt	(111)	graphene	1	[34]
Rh	(111)	CO_2	≤ 1	[7]
Rh	(111)	C_6H_6 (benzene)	≤ 1	[5]
Rh	(111)	C_6H_6/CO	≤ 1	[6]
Rh	(111)	graphene	1	[28]
Ru	(0001)	C_8H_{18} (n-octane)	≤ 1	[13]
Ru	(0001)	graphene	1	[32, 34, 33, 36]
Ru + H (1x1)	(0001)	C_8H_{18} (n-octane)	≤ 1	[13]

*: SE-HAS measurements; ** NeAS measurements

References

1. B.F. Mason, B.R. Williams, Surf. Sci. **130**, 295 (1983)
2. B.F. Mason, B.R. Williams, Surf. Sci. **130**, L329 (1983)
3. B.F. Mason, B.R. Williams, Surf. Sci. **139**, 173 (1984)
4. J.P. Toennies, C. Wöll, G. Zhang, J. Chem. Phys. **96**, 4023 (1992)
5. G. Witte, H. Range, J.P. Toennies, C. Wöll, Phys. Rev. Lett. **71**, 1063 (1993)
6. G. Witte, Thesis (University of Göttingen 1995 and Max-Planck-Institut für Strömungsforschung Göttingen, Bericht 8/1995)
7. H. Range, Dissertation (University of Göttingen 1995) Max-Planck-Institut für Strömungsforschung Göttingen, Report 3/1995
8. G. Witte, C. Wöll, J. Chem. Phys. **103**, 5860 (1995)
9. D. Fuhrmann, C. Wöll, Surf. Sci. **368**, 20 (1996)
10. A.P. Graham, M.F. Bertino, F. Hofmann, J.P. Toennies, J. Chem. Soc., Faraday Trans. **92**, 4749 (1996)
11. D. Fuhrmann, C. Wöll, Surf. Sci. **377**, 544 (1997)

12. A.P. Graham, M.F. Bertino, F. Hofmann, W. Silvestri, J.P. Toennies, J. Chem. Phys. **106**, 2502 (1997)
13. G. Witte, K. Weiss, P. Jakob, J. Braun, K.L. Kostov, C. Wöll, Phys. Rev. Lett. **80**, 121 (1998)
14. J. Braun, G.G. Bishop, A.V. Ermakov, L.V. Goncharova, B.J. Hinch, J. Chem. Phys. **110**, 5337 (1999)
15. D. Fuhrmann, R. Gerlach, H.G. Rubahn, C. Wöll, Surf. Sci. **424**, 145 (1999)
16. A. Glebov, U. Panella, J.P. Toennies, Phys. Rev. B **60**, 2046 (1999)
17. A. Glebov, A.P. Graham, A. Menzel, J.P. Toennies, P. Senet, J. Chem. Phys. **112**, 11011 (2000)
18. S. Vollmer, P. Fouquet, G. Witte, C. Boas, M. Kunat, U. Burghaus, C. Wöll, Surf. Sci. **462**, 135 (2000)
19. S.B. Darling, A.W. Rosenbaum, S.J. Sibener, Surf. Sci. **478**, L313 (2001)
20. A.P. Graham, J. Chem. Phys. **115**, 524 (2001)
21. T. Yan, N. Isa, K.D. Gibson, S.J. Sibener, W.L. Has, J. Phys. Chem. A **107**, 10600 (2003)**
22. D. Fuhrmann, A.P. Graham, J. Chem. Phys. **120**, 2439 (2004)
23. N. Isa, K.D. Gibson, T. Yan, W. Hase, S.J. Sibener, J. Chem. Phys. **120**, 2417 (2004)**
24. A.W. Rosenbaum, M.A. Freedman, S.B. Darling, I. Popova, S.J. Sibener, J. Chem. Phys. **120**, 3880 (2004)
25. I.G. Shuttleworth, Surf. Rev. Lett. **14**, 387 (2007)
26. I.G. Shuttleworth, Surf. Rev. Lett. **14**, 863 (2007)
27. H. Hedgeland, P. Fouquet, A.P. Jardine, G. Alexandrowicz, W. Allison, J. Ellis, Nature Phys. **5**, 561 (2009)
28. K.D. Gibson, S.J. Sibener, J. Phys. Chem. C **118**, 29077 (2014)
29. K. Fladischer, A. Politano, W.E. Ernst, D. Farìas, R. Miranda, Surf. Sci. **620**, 65 (2014)
30. E.M. McIntosh, J. Ellis, A.P. Jardine, P. Licence, R.G. Jones, W. Allison, Chem. Sci. **5,** 667 (2014)*
31. A. Al-Taleb, H.K. Yu, G. Anemone, D. Farías, A.M. Wodtke, Carbon **95**,731 (2015)
32. D. Maccariello, D. Campi, A. Al Taleb, G. Benedek, D. Farías, M. Bernasconi, R. Miranda, Carbon **93**, 1 (2015)
33. A. Politano, Surf. Sci. **634**, 44 (2015)
34. H. Shichibe, Y. Satake, K. Watanabe, A. Kinjyo, A. Kunihara, Y. Yamada, M. Sasaki, W.W. Hayes, J.R. Manson, Phys. Rev. B **91**, 155403 (2015)
35. A. Tamtögl, E. Bahn, J. Zhu, P. Fouquet, J. Ellis, W. Allison, J. Phys. Chem. C **119**, 25983 (2015)*
36. M. Al-Taleb, D. Farìas, J. Phys.: Condens. Matter **28**, 103005 (2016)
37. A. Al-Taleb, G. Anemone, D. Farías, R. Miranda, Carbon **99**, 4162016 (2016)

Appendix B
Surface Debye Temperatures from HAS data with some values from other methods (hydrogen atom scattering (HyAS), D_2 scattering (D_2S), low energy electron diffraction (LEED, low energy ion diffraction (LEIS), medium-energy ion diffraction (MEIS))

The data are compared with the bulk melting temperature T_m and the bulk Debye temperature θ_B. Additional data on surface Debye temperatures from photoemission, ion-scattering and x-ray scattering experiments can be found in C. Waldfried, D.N. McIlroy, J. Zhang, P.A. Dowben, G.A. Karich and E.W. Plummer, Surf. Sci. 363, (1996) 296–302 and in W. Martienssen, H. Warlimont (Eds.) Springer Handbook of Condensed Matter and Materials Data (Springer 1985), Tables 5.2–21, 22 [65].

In connection with free-electron surfaces, such as on metals, it is noted that the surface Debye temperatures as derived from HAS or LEED data have similar values although they have a slightly different physical meaning. As discussed in Chap. 8 the interaction of the He atoms with surface phonons is mediated by the electron-phonon interaction. In LEED experiments, however, the electrons probe directly the vibrations of the surface atom cores. Since they are nearly equal their surface Debye temperatures can both be used for estimating the electron-phonon mass-enhancement factor λ. This follows from (8.26):

$$\lambda \approx \frac{-d\ln I}{dT}\frac{\phi}{6Zk_B}\frac{k_F^2}{k_{iz}^2} \tag{B.1}$$

where Z and k_F are the number of surface free electrons per atom and their Fermi wavevector, respectively and ϕ is the surface work function. (B.1) provides an approximate but important connection between the surface temperature dependent attenuation of the specular intensity and the mass-enhancement factor λ. At high temperatures the beam attenuation in (B.1) is directly related to the Debye Waller factor 2W for surface vibrations by $-d\ln I/dT \cong 2W/T$. Next it is recalled that the

G. Benedek and J. P. Toennies, *Atomic Scale Dynamics at Surfaces*, Springer Series in Surface Sciences 63, https://doi.org/10.1007/978-3-662-56443-1

Debye Waller factor is related to the surface Debye frequency ω_{Ds} and the surface Debye temperature by (7.86). Finally (B.1) and (7.86) lead to the following expression:

$$\lambda \cong \frac{2k_F^2\phi}{ZM\omega_{Ds}^2}, \quad T > 0.7\theta_{Ds}, \quad \omega_{Ds} \equiv k_B\theta_{Ds}/\hbar, \tag{B.2}$$

where M is the surface atom mass.

Surface Debye temperatures from photoemission, ion-scattering and x-ray scattering experiments can be found in C. Waldfried, D.N. McIlroy, J. Zhang, P.A. Dowben, G.A. Karich and E.W. Plummer, Surf. Sci. **363**, 296-302 (1996). The data are compared with the bulk melting temperature T_m and the bulk Debye temperature θ_B

Surfaces	T_m [K]	θ_D [K]	θ_{Ds} [K]	Method	References for θ_{Ds}
Insulators					
AgBr(001)	705	168	95	HAS	[1]
BaF_2(111)	1641	282	125 ± 6	LEED	[63]
CaF_2(111)	1691	510	196 ± 19	LEED	[63]
CoO(001)	2208	…	320 ± 12	HAS	[2]
Ice Ih(0001)	273	218	175	HAS	[3]
KCl(001)	1049	235	(130–145) ± 10	LEED	[4]
			165 ± 10	HAS	[5]
KCN(001)	907	…	123	HAS	[6]
LiF(001)	1115	730	415	HAS	[7]
			520	HAS	[8]
			513	HyAS[(a)]	[8]
			610	HyAS[(a)]	[8]
$KAl_3Si_3O_{10}(OH)_2$(001) mica	1300[(b)]	…	328	HAS	[9]
NaCl(001)	1074	320	250	HAS	[10]
			240 ± 20	HAS	[11]
NaF(001)	1261	492	370	HyAS[(a)]	[12]
			416	HAS	[13]
			411–458	HAS	[14]
			425± 20	HAS	[15]
NiO(001)	2263	…	384 ± 15	HAS	[2]
Metals					
Ag(001)	962	225	104	LEED	[16, 65]
Ag(110)	962	225	195	HAS	[69]

(continued)

(continued)

Surfaces	T_m [K]	θ_D [K]	θ_{Ds} [K]	Method	References for θ_{Ds}
			152	LEED	[16, 65]
Ag(111)	962	225	253 ± 10	HAS	[17]
			155	LEED	[18, 37]
			150	LEED	[19–21]
Al(001)	933	426	300	LEED	[23]
		428	200	LEED	[65]
		426	356	LEED	[24]
Al(110)	933	428	140	LEED	[65]
Al(111)	933	790	440	LEED	[22]
		428	200	LEED	[65]
Be(0001)	1548	1050	525	LEED	[58]
Cr(001)	1857	440	175	LEED	[46]
Cr(110)	1857	440	333	LEED	[49]
10 ML Cs(110)/Cu (111)	301	43	48	HAS	[25]
Cu(001)	1083	350	270	HAS	[9]
			267	HAS	[30]
			230	HAS	[31]
			229	HAS	[32]
		343	230	LEED	[26, 65]
			249	LEED	[27, 28]
Cu(110)	1083	350	236	HAS	[31, 32]
		343	260	LEIS	[65]
Cu(111)	1083	350	237	HAS	[31, 32]
			171–212	LEED	[33]
		343	245	LEED	[65]
Ir(001)	2410	420	236	LEED	[44]
		285	216	LEED	[46]
Ir(110)	2410	420	150	LEED	[65]
Ir(111)	2410	420	170	LEED	[65]
Mo(001)	2617	380	213	HAS	[70]
			239	LEED	[46]
			230	LEED	[50]
Nb(001)	2468	281	193	HAS	[71]
			106	LEED	[51]
Nb(110)	2468	281	106	LEED	[51]

(continued)

(continued)

Surfaces	T_m [K]	θ_D [K]	θ_{Ds} [K]	Method	References for θ_{Ds}
Ni(001)	1453	450	344	LEED	[64]
			415 ± 20	LEED	[34]
Ni(110)	1453	450	374 ± 12	HAS	[35]
			387	D_2MS (c)	[35]
			300	LEED	[36]
			348	LEED	[64]
			395	MEIS	[65]
Ni(111)	1453	450	357	LEED	[64]
Pb(111)	328	90	56	HAS	[68]
			55 ± 10	LEED	[37, 39]
			60	LEED	[38, 39]
Pb(110)	328	90	37	LEED	[38, 39]
Pd(001)	1552	273	140 ± 10	LEED	[37, 39]
Pd(110)	1552	273	140 ± 10	LEED	[37, 39]
Pd(111)	1552	273	140 ± 10	LEED	[37, 39]
Pt(001)	1772	234	118 ± 10	LEED	[40]
Pt(110)	1772	234	107 ± 10	LEED	[40]
Pt(111)	1772	234	232 ± 3	HAS	[41]
			250	HAS	[42]
			231	HAS	[43]
			111 ± 10	LEED	[40]
Rh(001)	1966	350	260	LEED	[45]
Rh(111)	1966	350	255	LEED	[45]
V(001)	1890	380	250	LEED	[52]
W(001) 1x1	3410	400	150	HAS	[68]
			183	LEED	[53]
			220	LEED	[54, 55]
			210	LEED	[65]
W(110)	3410	400	190–235	LEED	[65]
			245	LEED	[56, 57]
Semiconductors and semimetals					
Bi(111)	271	120	84 ± 8, 75 ± 8	HAS	[47, 67]
			~48	LEED	[38]
			71(+7/−5)	LEED	[66]
Bi(100)	271	120	~40	LEED	[38]
Bi_2Te_3(111)	853	145	81 ± 6	HAS	[72]
C(0001) graphite		800	690 ± 10	LEED	[60]

(continued)

(continued)

Surfaces	T_m [K]	θ_D [K]	θ_{Ds} [K]	Method	References for θ_{Ds}
Ge(111)	1211	354	138 T<1050K;	HAS	[61]
			107 T>1050K		
			191 T<1000K;	LEED	[62]
			105 T>1060K		
Sb(111)	904	211	150 ± 6	HAS	[48]
Si(111) 7×7	1687	543–645	380–441	HAS	[59]

(a)HyAS stands for hydrogen atom scattering
(b)Decomposition temperature
(c)D_2MS stands for D_2 molecule scattering

References

1. A. Glebov, V. Panella, J.P. Toennies, Phys. Rev. B **60**, 2046 (1999)
2. G. Witte, P. Senet, J.P. Toennies, Phys. Rev. B **58**, 13264 (1998-I)
3. J. Braun, A. Glebov, A.P. Graham, A. Menzel, J.P. Toennies, Phys. Rev. Lett. **80**, 2638 (1998)
4. 4.S.A. Knyazev, G.K. Zyryanov, Soviet Physics, Solid State **22**, 909 (1980)
5. F. Traeger, Dissertation, University of Göttingen (2000)
6. S.M. Weera, J.R. Manson, G.G. Bishop, E.S. Gillman, J. Baker, J.J. Hérnandez, S.A. Safron, J.G. Skofronick, Phys. Rev. B **52**, 14185 (1995)
7. H. Hoinkes, H. Nahr, H. Wilsch, Surf. Sci. **33**, 516 (1972)
8. G. Armand, J. Lapujoulade, Y. Lejay, Surf. Sci. **63**, 143 (1977)
9. G.G. Bishop, E.S. Gillman, Jeff Baker, J.J. Hérnandez, S.A. Safron, J.G. Skofronick, S.M. Weera, J.R. Manson, Phys. Rev. B **52**, 13229 (1995).
10. G. Brusdeylins, D. Schmicker, Surf. Sci. **331–333**, 237–242 (1995)
11. A. Glebov, J.P. Toennies, H. Weiss, Surf. Sci. **351**, 200 (1996)
12. L. Jannotta, G. Scoles, U. Valbusa, Surf. Sci. **161**, 411 (1985)
13. Wilsch, H.-U. Finzel, H. Frank, H. Hoinkes, H. Nahr, in *Proceedings of the Second International Conference on Solid Surfaces*, Kyoto, 1974, Japan, J. Appl. Phys. Suppl. **2**, Pt. 2, 567 (1974)
14. B. Wood, B.F. Mason, B.R. Williams, J. Chem. Phys. **61**, 1435 (1974); H. Frank, H. Hoinkes, H. Wilsch, Surf. Sci. **63**, 121 (1977)
15. V. Krishnaswamy, G. Derry, D. Wesner, T.J. O'Gormann, D.R. Frankl, Surf. Sci. **77**, 493 (1978)
16. J.M. Morabito, Jr., R.F. Steiger, G.A. Somorjai, Phys. Rev. **179**, 638 (1969)
17. J.M. Horne, C. Yerkes, D.R. Miller, Surf. Sci. **93**, 47 (1980)
18. E.R. Jones, J.T. McKinney, M.B. Webb, Phys. Rev. **151**, 476 (1966)
19. D.W. Jepsen, P.M. Marcus, F. Jona, Surf. Sci. **41**, 223 (1974)
20. C. Corotte, P. Ducros, A. Mascall, Compt. Rend. (Paris) B **267**, 544 (1968)
21. G. Rovida, M. Torrini, E. Zanazzi, Il Nuovo Cimento **4B**, 97 (1971)
22. J.R. Noonan, H.L. Davis, J. Vac. Sci. Technol. A **8**, 2671 (1990)

23. G.E. Laramore, C.R. Duke, Phys. Rev. B **2**, 4783 (1970)
24. M. Van Hove, S.Y. Tong, and N. Stoner, Surf. Sci. 54 (1976) 259.
25. E. Hulpke, J. Lower, A. Reichmuth, Phys. Rev. B **53**, 13901 (1996)
26. R.J. Reid, Phys. Status Solidi (a) **2**, K109 (1970)
27. R.J. Reid, Surf. Sci. **29**, 623 (1972)
28. D.P. Woodruff, M.P. Seah, Phys. Status Solidi (a) **1**, 429 (1970)
29. F. Hofmann, J.R. Manson, J.P. Toennies, J. Chem. Phys. **106**, 1234 (1997)
30. F. Hofmann, J.P. Toennies, J.R. Manson, J. Chem. Phys. **101**, 10155 (1994)
31. J. Lapujoulade, J. Perreau, A. Kara, Surf. Sci. **129**, 59 (1983)
32. G. Armand, J. Lapujoulade, Y. Lejay, In Proceedings 7th Intern. Vacuum Congr. and 3rd Intern. Conf. on Solid Surfaces, Vienna, 1977, Ed. R. Dobrozemsky et al., p. 1361 (1977)
33. D.P. Woodruff, M.P. Seah, Phys. Lett. **30A**, 263 (1969)
34. S. Andersson, B. Kasemo, Solid State Comm. **8**, 1885 (1970)
35. M.F. Bertino, F. Hofmann, J.P. Toennies, J. Chem. Phys. **106**, 4327 (1997)
36. A. Liebsch, J. Harris, Surf. Sci. **111**, L721 (1981)
37. R.M. Goodman, H.H. Farrell, G.A. Somorjai, J. Chem. Phys. **48**, 1046 (1968)
38. R.M. Goodman, G.A. Somorjai, J. Chem. Phys. **52**, 6325 (1970)
39. R.M. Goodman, Ph.D. Thesis (University of California, Berkeley, 1969)
40. H.B. Lyon, G.A. Somorjai, J. Chem. Phys. **44**, 3707 (1966)
41. V. Bortolani, V. Celli, A. Franchini, J. Idiodi, G. Santoro, K. Kern, B. Poelsma, G. Comsa, Surf. Sci. **208**, 1 (1989)
42. J.R. Manson, Phys. Rev. B **43**, 6924 (1991)
43. V. Celli, D. Himes, P. Tran, J.P. Toennies, Ch. Wöll, G. Zhang, Phys. Rev. Lett. **66**, 3160 (1991)
44. M.A. Van Hove, R.J. Koestner, P.C. Stair, J.P. Biberian, L.L. Kesmodel, I. Bartoš, G.A. Somorjai, Surface Sci. **103**, 218 (1981)
45. D.G. Castner, G.A. Somorjai, J.E. Black, D. Castiel, R.F. Wallis, Phys. Rev. B **24**, 1616 (1981)
46. D. Tabor, J.M. Wilson, T.J. Bastow, Surf. Sci. **26**, 471 (1971)
47. P. Kraus, A. Tamtögl, M. Mayrhofer-Reinhartshuber, F. Apolloner, Ch. Gösweiner, S. Miret-Artés, W.E. Ernst, J. Phys. Chem C **119**, 17236 (2015)
48. A.Tamtögl, M. Mayrhofer-Reinhartshuber, P. Kraus, W. E. Ernst, Surf. Sci. **617**, 225 (2013)
49. G.A. Somorjai, Chemistry in Two Dimensions: Surfaces (Cornell University Press, Ithaca, 1981), p. 170
50. L.J. Clarke, Surf. Sci. **91**, 131 (1980)
51. D. Tabor, J. Wilson, Surf. Sci. **20**, 203 (1970)
52. D.J. Chueng, R.F. Wallis, C. Megerle, G.A. Somorjai, Phys. Rev. B **12**, 5599 (1975)
53. P.J. Estrup, in *The Structure and Chemistry of Solid Surfaces*, ed. by G.A. Somorjai (John Wiley, New York, 1969), p. 19-1
54. L.J. Clarke, L. Morales de la Garza, Surf. Sci. **99**, 419 (1980)
55. J.K. Debe, K.A. King, F.S. Marsch, Surf. Sci. **68**, 437 (1977)
56. J. Aldag, R.M. Stern, Phys. Rev. Lett. **14**, 857 (1965)
57. R.M. Stern, Trans. Am. Cryst. Ass. **4**, 14 (1968)
58. H.L. Davis, J.B. Hannon, K.B. Ray, E.W. Plummer, Phys. Rev. Lett. **68**, 2631 (1992)
59. G. Lange Dissertation (University of Göttingen 1996), Max-Planck-Institut für Strömungsforschung, Bericht 12/1996
60. G. Albinet, J.P. Biberian, M. Bienfait, Phys. Rev. B **3**, 2015 (1971)
61. C.A. Meli,,E.F. Greene, G. Lange, J.P.Toennies, Phys. Rev. Lett. **74**, 2054 (1995)
62. E.G. McRae, R. Malic, Phys. Rev. Lett. **58**, 1437 (1987)
63. J. Vogt, J. Henning, H. Weiss, Surf. Sci. **578**, 57 (2005)
64. A Grudniewski, S Morz J. Phys. C: Solid State Phys. **18**, 3387 (1985)
65. W. Martienssen, H. Warlimont (Eds.) Springer Handbook of Condensed Matter and Materials Data (Springer 1985) Tables 5.2–21,22

66. Mönig, H.; Sun, J.; Koroteev, Y.M.; Bihlmayer, G.; Wells, J.; Chulkov, E. V.; Pohl, K.; Hofmann, P., Phys. Rev. B: Condens. Matter Mater. Phys. **72**, 085410 (2005)
67. Mayrhofer-Reinhartshuber, M.; Tamtögl, A.; Kraus, P.; Rieder, K.H.; Ernst, W.E. , J. Phys.: Condens. Matter **24**, 104008 (2012)
68. J.R. Manson, G. Benedek, Salvador Miret-Artès, J. Phys. Chem. Lett. **7**, 1016–1021 (2016); E: 7, 1691 (2016)
69. C. Capozzi, S.M. Francis, D. Roscoe, N.V. Richardson, S. Holloway, J. Vac. Sci.Technol. A **5**, 1049 (1987)
70. D.-M. Smilgies, Ph. D. Thesis, Univ. of Göttingen 1991 (Max-Planck Inst. Bericht 6/1991, ISSN 0436-1199)
71. M.W. Hüppauf, Thesis, Univ. of Göttingen 1989 (unpublished)
72. A. Tamtögl, P. Kraus, N. Avidor, M Bremholm, E.M.J. Hedegaard, Bo B. Iversen, M. Bianchi, Ph. Hofmann, J. Ellis, W. Allison, G. Benedek, W. Ernst, Phys. Rev. B **95**, 195401 (2017)

Appendix C
Diffusion Coefficients from Time-of-Flight Quasi-elastic ^{4}He Atom Scattering (TOF-QHAS) and from Spin Echo Quasi-elastic ^{3}He Atom Scattering (SE-QHAS) Measurements

The Tables C.1 and C.2 list the systems in alphanumeric order of the adsorbates for which time-of-flight quasi-elastic ^{4}He atom scattering (TOF-QHAS) and spin echo quasi-elastic ^{3}He Atom scattering (SE-QHAS), respectively, are available as of the end of 2017 with references. The table complements the discussion in Sect. 13.2 of Chap. 13. The table lists the range of the surface temperatures T_s, and the corresponding values of the diffusion coefficient D, the activation barrier E_b and the coverage θ. The activation barrier E_b is determined from the Arrhenius behavior.

Table C.1 Time-of-flight quasi-elastic ^{4}He atom scattering (TOF-QHAS)

System	T_s (K)	D (cm^2s^{-1})	E_b (meV)	Coverage θ (ML)	References
Ag/Ag(110)	300–800	$0.9 - 2.5 \times 10^{-5}$	290 ± 20	Clean	[33]
CO/Ni (110) < 001>	240–330	$1.3 - 2.1 \times 10^{-5}$	35 ± 4	0.15	[34]
CO/Ni (110) < 1$\bar{1}$0>	200–360	$0.6 - 2.9 \times 10^{-5}$	57 ± 4	0.15	[34]
CO/Ni(001)	120–330	$0.4 - 2.4 \times 10^{-5}$	27 ± 2	0.15	[35]
CO/Cu(001)	100–150	–	32	0.03	[11, 36]
CO/Pt(111)	400	3.2×10^{-6}	130 ± 20	0.05	[37]
Octane C_8H_{18} /Cu(111)	200	1.2×10^{-5}	–	1.0	[5]
D/Pt(111)	140–250	–	76 ± 7	0.1	[11]
Ge/Ge(111)	1050–1160	1.2×10^{-4}		Clean	[10]
H/Pt(111)	140–250	$3.0–30 \times 10^{-6}$	68 ± 5	0.1	[11]
K/K/Ni(001)	185–320	$\sim 2 \times 10^{-4}$	63 ± 15	0.2	[7]
Na/Cu(001)	175–400	–	51 ± 6	0.1	[38]
Na/Cu(001)	150–390	$0.1–5.8 \times 10^{-4}$	52.5 ± 0.9	0.028	[8, 39, 40]

(continued)

G. Benedek and J. P. Toennies, *Atomic Scale Dynamics at Surfaces*, Springer Series in Surface Sciences 63, https://doi.org/10.1007/978-3-662-56443-1

Table C.1 (continued)

System	T_s (K)	D (cm^2s^{-1})	E_b (meV)	Coverage θ (ML)	References
Na/Pt(111)	100–350		21.5 ± 1	0.04	[13]
Ni/Ni(110) < 1 $\bar{1}$ 0>	900–1200	1–5 × 10^{-5}	424–536 ± 40	Clean	[8]
Pb/Pb(110)	446–521	0.1–1 × 10^{-4}	1000	Clean	[1-3,41]
S/Cu(111)	820	2.9 × 10^{-5}	–	<0.16	[4]
Xe/Pt(111)	–	–	–	–	[9, 11]

Table C.2 Spin echo quasi-elastic ^{3}He atom scattering (SE-QHAS)

System	T_s (K)	D (cm^2s^{-1})	E_b (meV)	Coverage θ (ML)	References
CH_3CH_2S/Cu(111) (ethanethiolate)	150–325	2.2 ± 0.2 × 10^{-4}	86 ± 5	0.03	[24]
C_4H_4S/Cu(111) (thiophene)	105–165	–	59 ± 2 62 ± 4	0.015–0.022	[25, 26]
C_4H_5N/Cu(111) (pyrrole)	125–170	–	50 ± 3	0.033	[25, 27, 30]
C_5H_5/Cu(111) (cyclopentadienyl)	114–400	–	40 ± 3	0.03	[23, 25, 29, 30]
C_6H_6/graphite	140	5.39 × 10^{-5}	17 ± 12		[15, 19, 20, 22]
Cs/Cu(001)	80, 130	–	130	0.014–0.044	[16, 21]
CO/Cu(001)	190	–	125	0.1	[22, 42]
Coronene/Au(111) $C_{24}H_{12}$	95–437	–	30 ± 5		[14, 22]
CO/Pt(111)	340	–	–	0.065–0.3	[17, 22]
H/Ru(0001)	75–250	–	–	–	[28]
Na/Cu(001)	155	–	–	0.02–0.08	[21, 22, 43]
Na/Cu(115)	115–155	–	53 ± 5	0.09	[31]
Pentacene/pentacene 1 ML/Cu(111)	300	–	–	0.23, 0.44	[32]
Propane/Pt(111)	115	–	57 ± 4	0.02	[18, 22]

References

1. J.W.M. Frenken, J.P. Toennies, C. Woll, Self-diffusion at a melting surface observed by He scattering. Phys. Rev. Lett. **60**(17), 1727–1730 (1988)
2. J.W.M. Frenken, B.J. Hinch, J.P. Toennies, He scattering study of diffusion at a melting surface. Surf. Sci. **211**(1–3), 21–30 (1989)

3. J.W.M. Frenken, Surface melting of Pb(110)—a compilation of experimental results. J. Vac. Sci. Technol. Vac. Surf. Films **7**(3), 2147–2151 (1989)
4. B.J. Hinch, J.W.M. Frenken, G. Zhang, J.P. Toennies, Sulfur adatom diffusion on the Cu (111) surface. Surf. Sci. **259**(3), 288–300 (1991)
5. D. Fuhrmann, C. Wöll, The melting of a two-dimensional lattice of flexible rod-shaped molecules: structure and dynamics studied by He atom scattering. Surf. Sci. **377**(1–3), 544–550 (1997)
6. A.P. Graham, F. Hofmann, J.P. Toennies, L.Y. Chen, S.C. Ying, Experimental and theoretical investigation of the microscopic vibrational and diffusional dynamics of sodium atoms on a Cu(001) surface. Phys. Rev. B **56**, 10567–10578 (1997)
7. D. Fuhrmann, E. Hulpke, J. Chem. Phys. **106**, 3407 (1997)
8. A.P. Graham, W. Silvestri, J.P. Toennies, *Elementary processes of surface diffusion studied by quasielastic helium atom scattering*. In Tringides, H.C. (editor) Surface Diffusion: Atomistic and Collective Processes, vol. 360 of NATO ASI Series B, Physics, Plenum Press (1997): p. 565 - 580
9. J. Ellis, A.P. Graham, J.P. Toennies, Phys. Rev. Lett. **82**, 5072 (1999)
10. A.L. Glebov, J.P. Toennies, S. Vollmer, Phys. Rev. Lett. **82**, 3300 (1999)
11. A.P. Graham, J.P. Toennies, Determination of the lateral potential energy surface of single adsorbed atoms and molecules on single crystal surfaces using helium atom scattering. Surf. Sci. 1999. **427–428**, 1–10
12. J. Ellis, A.P. Graham, F. Hofmann, J.P. Toennies, Coverage dependence of the microscopic diffusion of Na atoms on the Cu(001) surface: a combined helium atom scattering experiment and molecular dynamics study. Phys. Rev. B **63**(19) (2001)
13. A.P. Graham, J.P. Toennies, Macroscopic diffusion and low-frequency vibrations of sodium on Pt(111) investigated with helium atom scattering. J. Phys. Chem. B **105**(18), 4003–4009 (2001)
14. M. DeKieviet, D. Dubbers, S. Hafner, F. Lang, in *Atomic and Molecular Beams: The State of the Art*, ed. by R. vol. 161 (Springer, Campargue, 2001)
15. P. Fouquet, H. Hedgeland, A. Jardine, G. Alexandrowicz, W. Allison, J. Ellis, Measurements of molecule diffusion on surfaces using neutron and helium spin echo. Physica B-Condens. Matter **385**, 269–271 (2006)
16. G. Alexandrowicz, A.P. Jardine, Helium spin-echo spectroscopy: studying surface dynamics with ultra-high-energy resolution. J. Physics-Condens. Matter **19**(30), 305001 (2007)
17. G. Alexandrowicz, P.R. Kole, E.Y.M. Lee, H. Hedgeland, R. Ferrando, A.P. Jardine, W. Allison, J. Ellis, Observation of uncorrelated Microscopic motion in a Strongly Interacting Adsorbate System. J. Am. Chem. Soc., 2008, **130**, 6789–6794
18. A.P. Jardine, et al., Probing molecule-surface interactions through ultra-fast adsorbate dynamics: propane/Pt(111). New J. Phys. **10**, 125026 (2008)
19. P. Fouquet, M.R. Johnson, H. Hedgeland, A.P. Jardine, J. Ellis, W. Allision, Molecular dynamics simulations of the diffusion of benzene sub-monolayer films on graphite basal plane surfaces. Carbon **47**(11), 2627–2639 (2009)
20. H. Hedgeland, P. Fouquet, A.P. Jardine, G. Alexandrowicz, W. Allison, J. Ellis, Measurement of single-molecule frictional dissipation in a prototypical nanoscale system. Nat. Phys. **5**(8), 561–564 (2009)
21. A.P. Jardine, G. Alexandrowicz, H. Hedgeland, W. Allison, J. Ellis, Studying the microscopic nature of diffusion with helium-3 spin-echo. Phys. Chem. Chem. Phys. **11** (18), 3355–3374 (2009)
22. A.P. Jardine, H. Hedgeland, H. Alexandrowicz, W. Allison, J. Ellis, Helium-3 spin-echo: Principles and application to dynamics at surfaces. Prog. Surf. Sci. **84**(11–12), 323–379 (2009)
23. H. Hedgeland, B.A.J. Lechne, F.E. Tuddenham, A.P. Jardine, W. Allison, J. Ellis, M. Sacchi, S.J. Jenkins, B.J. Hinch, Weak intermolecular interactions in an ionically bound molecular adsorbate: cyclopentadienyl/Cu(111). Phys. Rev. Lett. **106**(18), 186101 (2011)

24. S. Paterson, W. Allison, H. Hedgeland, J. Ellis, A.P. Jardine, Rotation and translational motion prior to self-assembly: dynamics of ethanethiolate on Cu(111). Phys. Rev Lett. **106**, 256101 (2011)
25. B.A.J., Lechner, A.S. deWijn, H. Hedgeland, A.P. Jardine, B.J. Hinch, W. Allison, J. Ellis, Atomic scale friction of molecular adsorbates during diffusion. J. Chem. Phys. **138**(19), 194710 (2013)
26. B.A.J. Lechner, M. Sacchi, A.P. Jardine, H. Hedgeland, W. Allison, J. Ellis, S.J. Jenkins, P. C. Dastoor, B.J. Hinch, Jumping, rotating, and flapping: the atomic-scale motion of thiophene on Cu(111). J. Phys. Chem. Lett. **4**(11), 1953–1958 (2013)
27. B.A.J. Lechner, H. Hedgeland, J. Ellis, W. Allison, M. Sacchi, S.J. Jenkins, B.J. Hinch, Quantum Influences in the Diffusive Motion of Pyrrole on Cu(111). Angew. Chem. Int. Ed. **52**, 5085–5088 (2013)
28. E.M. McIntosh, K.T. Wikfeldt, J. Ellis, A. Michaelides, W. Allison, Quantum effects in the diffusion of hydrogen on Ru(0001). J. Phys. Chem. Lett. **4**(9), 1565–1569 (2013)
29. B.A.J. Lechner, P.R. Kole, H. Hedgeland, A.P. Jardine, W. Allison, B.J. Hinch, J. Ellis, Ultra-high precision determination of site energy differences using a Bayesian method. Phys. Rev. B **89**(12), 121405 (R) (2014)
30. B.A.J. Lechner, *Studying Complex Surface Dynamical Systems Using Helium-3 Spin-Echo Spectroscopy*, Springer Theses, 2014
31. O. Godsi, G. Corem, T. Kravchuk, C. Bertram, K. Morgenstern, H. Hedgeland, A.P. Jardine, W. Allison, J. Ellis, G. Alexandrowicz, How atomic steps modify diffusion and inter-adsorbate forces: empirical evidence from hopping dynamics in Na/Cu(115). J. Phys. Chem. Lett. **6**(20), 4165–4170 (2015)
32. P. Rotter, B.A.J. Lechner, A. Morherr, D.M. Chisnall, D.J. Ward, A.P. Jardine, J. Ellis, W. Allison, B. Eckhardt, G. Witte, Coupling between diffusion and orientation of pentacene molecules on an organic surface. Nat. Mater. **15**(4), 397-+ (2016)
33. L. Pedemonte, R. Tatarek, G. Bracco, Phys. Rev. B **66**, 045414 (2002)
34. M.F. Bernito, F. Hofmann, W. Steinhögl, J.P. Toennies, J. Chem. Phys. **105**, 11297 (1996)
35. F. Hofmann, W. Schöllkopf, J.P. Toennies, in *Proc. R. A. Welch Foundation, 38th Conf. on Chemical Dynamics of Transient Species* 197–218 (Houston TX, 1994)
36. A.P. Graham, F. Hofmann, J.P. Toennies, G.P. Williams, C.J. Hirschmugl, J. Ellis, J. Chem. Phys. **108** 7825 (1998)
37. A.P. Graham, J.P. Toennies, Europhys. Lett. **42** 449 (1998)
38. J. Ellis, J.P. Toennies, Phys. Rev. Lett. **70** 2118 (1993)
39. A.P. Graham, F. Hofmann, J.P. Toennies, L.Y. Chen, S.C. Ying, Phys. Rev. B **56** 10567 (1997)
40. A.P. Graham, F. Hofmann, J.P. Toennies, L.Y. Chen, S.C. Ying, Phys. Rev. Lett. **78** 3900 (1997)
41. J.W.M. Freken, B.J. Hinch, J.P. Toennies, Ch. Wöll, Phys. Rev. B **41** 938 (1990)
42. G. Alexandrowicz, A.P. Jardine, P. Fouquet, S. Dworski, W. Allison, J. Ellis, Phys. Rev. Lett. **93** 156103 (2004)
43. G.J. Alexandrowicz, A.P. Jardine, H. Hedgeland, W. Allison, J. Ellis, Phys. Rev. Lett. **97** 156103 (2006)

Appendix D: Energy Conversion Factors

	Frequency ν	Angular frequency $\omega = 2\pi\nu$	Wavenumber ν/c	Energy[(a)] $h\nu = \hbar\omega$		Temperature $\hbar\omega/k_B$
	THz	10^{13} rad s^{-1}	cm^{-1}	meV	10^{-21} J = zJ	°K
THz	1	0.6283	33.36	4.136	0.6626	47.99
10^{13} rad s^{-1}	1.592	1	53.09	6.582	1.055	76.39
cm^{-1}	0.02998	0.01884	1	0.1239	0.01985	1.439
meV	0.2418	0.1519	8.066	1	1.6022	11.603
10^{-21} J = zJ	1.509	0.9479	50.38	0.6241	1	72.41
°K	0.02084	0.01309	0.6949	0.08618	0.01381	1

[(a)]Other energy units: 1 a.u. = 2 Ry = 27.21 eV = 2.195×10^5 cm^{-1}; 1 kcal/mol = 43.4 meV

Helium Atom Beam Energies and Wavevectors

1. Kinetic energy of a strongly expanded He atom beam from a source at temperature $T_0 \geq 50\ K^*$:$E = 5/2\ k_B T_0$. In practical units $E[meV]$ = 0.216 T_0 $[K]$
2. Conversion wavevector k ↔ energy E. $E[meV]$=0.5230 $k^2[Å^{-2}]$; k[Å] = 1.383 $\sqrt{E[meV]}$

*L.W. Bruch, W. Schöllkopf and J.P. Toennies , J. Chem. Phys. 117, 1544 (2002)

G. Benedek and J. P. Toennies, *Atomic Scale Dynamics at Surfaces*, Springer Series in Surface Sciences 63, https://doi.org/10.1007/978-3-662-56443-1

Subject Index

G. Benedek and J. P. Toennies, *Atomic Scale Dynamics at Surfaces*, Springer Series in Surface Sciences 63, https://doi.org/10.1007/978-3-662-56443-1

B

E